CRC Handbook of Incineration of Hazardous Wastes

Editor

William S. Rickman

Project Manager
San Diego, California

CRC Press
Taylor & Francis Group
Boca Raton London New York

CRC Press is an imprint of the
Taylor & Francis Group, an **informa** business

First published 1991 by CRC Press
Taylor & Francis Group
6000 Broken Sound Parkway NW, Suite 300
Boca Raton, FL 33487-2742

Reissued 2018 by CRC Press

© 1991 by Taylor & Francis
CRC Press is an imprint of Taylor & Francis Group, an Informa business

No claim to original U.S. Government works

A Library of Congress record exists under LC control number: 90002564

Publisher's Note
The publisher has gone to great lengths to ensure the quality of this reprint but points out that some imperfections in the original copies may be apparent.

Disclaimer
The publisher has made every effort to trace copyright holders and welcomes correspondence from those they have been unable to contact.

ISBN 13: 978-1-138-55006-3 (hbk)
ISBN 13: 978-1-138-55958-5 (pbk)
ISBN 13: 978-0-203-71254-2 (ebk)

Visit the Taylor & Francis Web site at http://www.taylorandfrancis.com and the CRC Press Web site at http://www.crcpress.com

DEDICATION

This book is dedicated to my children, Will and Melissa, and to my parents, Bill and Jane, whose ongoing encouragement, love, and fellowship were constant sources of inspiration and renewal to me.

PREFACE

Hazardous-waste incineration technologies have been developed to meet the needs of a market that has been created by the proliferation of hazardous waste in modern society. These hazardous wastes are continuously produced as by-products of many industries.

A well-documented overview of hazardous waste incineration is presented first, followed by an analysis of strategic market trends that will continue to shape the industry for years to come. A review of regulatory requirements is presented along with a section on the permitting and operation of incinerator facilities. Siting issues and public acceptance is a key section for any new facility. The wide spectrum of actual incinerator technologies is then presented, ranging from conventional units to innovative techniques still in the laboratory. Combustion calculations allow the reader to size an incinerator for a specific application, and a trial burn section details the methodologies used for assessing incinerator performance.

A number of typical case histories of incinerators are presented. Appendices provide a convenient reference to physical properties, combustion parameters and detailed equipment performance nomographs.

This handbook provides a reference for the potential user of an incinerator as well as a source of design data for incinerator vendors, consultants, and regulators.

THE EDITOR

William S. Rickman is a project manager and environmental consultant in San Diego, California. He received his Bachelor of Science degree in Chemical Engineering (Magna Cum Laude) from Tennessee Technological University in 1971, where he was named Outstanding Chemical Engineering Graduate and was a member of the honorary engineering and mathematics societies as well as a member of the American Institute of Chemical Engineers. Mr. Rickman completed the Executive Program for Scientists and Engineers at the University of California — San Diego in 1985.

Mr. Rickman has worked for 20 years in developing and commercializing hazardous waste treatment technologies with an emphasis on incineration. He has presented 23 papers on hazardous waste incineration at national and international conferences and holds three patents in that field.

CONTRIBUTORS

Geoffrey G. Back, M. S.
Project Manager
ICF Kaiser Engineers
Fairfax, Virginia

Gary S. Dietrich
Senior Vice President
ICF Technology Incorporated
Fairfax, Virginia

Dennis R. Engler, P.E.
Senior Engineer
San Diego, California

Paul Gorman, Ph.D.
Midwest Research Institute
Kansas City, Missouri

Roger Hathaway, Ph.D.
ABB Environmental
Chapel Hill, North Carolina

George L. Huffman
U.S. Environmental Protection Agency
Cincinnati, Ohio

Stephen C. James
U.S. Environmental Protection Agency
Cincinnati, Ohio

G. Scott Koken
Senior Engineer
Fluor Daniel, Inc.
Irvine, California

C. C. Lee, Ph.D.
U.S. Environmental Protection Agency
Cincinnati, Ohio

Wendy S. Lessig
Senior Engineer
Ogden Environmental Services
San Diego, California

Gregory Ondich
U.S. Environmental Protection Agency
Cincinnati, Ohio

E. Timothy Oppelt
Director
Hazardous Waste Engineering Research
  Laboratory
U.S. Environmental Protection Agency
Cincinnati, Ohio

Richard W. Raushenbush, J.D.
Attorney
Latham and Watkins, Attorneys-at-Law
San Diego, California

William S. Rickman
Project Manager
San Diego, California

Martha A. Rozelle, Ph.D.
Director
Public Involvement Services
Dames and Moore
Phoenix, Arizona

Joseph L. Tessitore, P.E.
Vice President
Cross-Tessitore Associates
Orlando, Florida

Andrew Trenholm, Ph.D.
Midwest Research Institute
Kansas City, Missouri

Dennis Wallace
Midwest Research Institute
Cary, North Carolina

Mark White
Ogden Environmental Services
San Diego, California

TABLE OF CONTENTS

Chapter 1
Introduction 1
William S. Rickman

Chapter 2
Overview 3
E. Timothy Oppelt

Chapter 3
Market Trends 59
Geoffrey G. Back and Gary S. Dietrich

Chapter 4
Regulatory Requirements and the Permitting Process 67
G. Scott Koken and Richard W. Raushenbush

Chapter 5
Siting Issues and Public Acceptance 183
Martha A. Rozelle

Chapter 6
Established Technologies 213
Joseph L. Tessitore and Wendy S. Lessig

Chapter 7
Innovative Thermal Destruction Technologies 255
C. C. Lee, George L. Huffman, Gregory Ondich, and Stephen C. James

Chapter 8
Combustion Calculations 293
Joseph L. Tessitore and Dennis Engler

Chapter 9
Trial Burn 309
P. Gorman, R. Hathaway, D. Wallace, and A. Trenholm

Appendix A
Case Histories 351
    1.   Site Remediation 351
    2.   In-House Wastes 397

Appendix B
Data Base 417
    1.   Physical Properties of Materials 418
    2.   Combustion Parameters 505
    3.   Equipment Performance 547

Index 579

Chapter 1

INTRODUCTION

William S. Rickman

Hazardous waste incineration technologies have been developed to meet the needs of a market which has been created by the proliferation of hazardous waste in modern society. These hazardous wastes are continuously produced as by-products of many industries.

New regulations and economic incentives have driven the generators of wastes as well as the owners of the hazardous waste landfills toward treatment of wastes prior to their disposal in a safe and responsible manner (Figure 1). Hazardous wastes are treated with four major goals: to protect the public by *detoxification* and *reduction in waste volumes* and by the recovery of both *energy* and *valuable chemicals*.

The U.S. Environmental Protection Agency has formulated a hierarchy for coping with wastes (Figure 2). This hierarchy suggests that it is most desirable not to generate as many wastes as are currently generated. It is least desirable to dispose of the wastes generated in a landfill with no treatment at all. In between these two most and least desirable actions are the four goals of waste treatment prior to final disposal.

Advanced treatment technology is required to accomplish these four goals of detoxification, volume reduction, recovery of energy, and recovery of chemicals. There are a number of treatment technologies such as biological treatment, solidification, ion exchange, and solvent extraction which meet some of these goals. However, incineration is capable of achieving all four of these goals.

In the U.S. there is over 1 ton/year of hazardous waste generated for every man, woman, and child. This amounts to over 200 million ton/year of hazardous waste generated in the U.S. Of this amount roughly 30%, or over 60 million ton/year, is capable of being treated via incineration in order to detoxify or reduce the volume of material. Today, however, according to various estimates, only between 2 to 8% of the waste is being incinerated.

The waste incinerator market is composed of several major segments including site cleanups, continuous process plant wastes, and commercial waste facilities.

Site cleanups are batch jobs with no requirement for heat recovery since they are primarily the treatment of soil. Site cleanups are suited to a standard sizing because they are not an ongoing continuous stream of material but are more typically a 1- to 3-year cleanup of a contaminated site. There have been many such sites identified by the Environmental Protection Agency, with over 1000 on their National Priority List.

Continuous process plant-type incinerators are characterized as being permanently sited, usually with heat recovery. There is a matrix of demands on a continuous process unit including throughput and feed composition. For these reasons, the design normally requires some customizing for each application. Commercial waste incinerators are large central facilities accepting a broad spectrum of waste types. There are fewer than ten such units nationwide.

Vendors offer a wide range of incinerator products and services. Their minimum scope is as an equipment supplier, providing such items as combustors, heat recovery systems, control and instrumentation, and training and start-up services. Vendors may also offer to supply all of the above equipment plus fans, feeders, filters, structural work, civil work, and the erection of the unit on the site. This would be a total system supply. Finally, vendors may offer a turnkey service for processing a client's waste at the site where it is generated, including system installation and operation.

In summary, this handbook provides a reference point for the potential user of an incinerator as well as a valuable source of design data for incinerator vendors, consultants, and regulators.

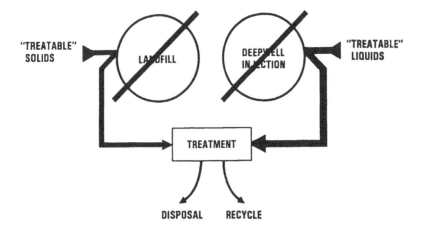

DEMANDS IMPROVED TREATMENT TECHNOLOGY

FIGURE 1. U.S. policy forcing treatment of hazardous wastes.

**WASTE TREATMENT HIERARCHY ACCEPTS
DISPOSAL OF TREATED WASTES**

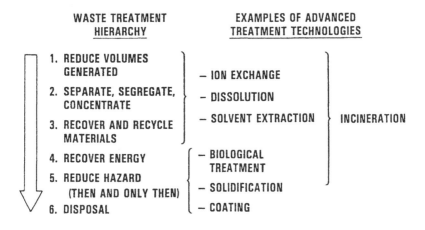

FIGURE 2. Waste treatment hierarchy accepts disposal of treated wastes.

Chapter 2

OVERVIEW*

E. Timothy Oppelt

TABLE OF CONTENTS

I.   Introduction 4

II.   Background 5
        A.   Historical Perspective 5
        B.   Regulations 6

III.   Current Incineration Practice 8
        A.   Incineration Practice 8
        B.   Incineration Technology 12
                1.   Waste Preparation and Feeding 12
                2.   Combustion Chambers 14
                3.   Air Pollution Control 19
                4.   Residue and Ash Handling 20
        C.   Other Hazardous Waste Thermal Destruction Systems 21

IV.   Measuring Process Performance 22
        A.   Performance Measurement 23
        B.   Process Monitoring 27

V.   Emissions from Hazardous Waste Incineration 28
        A.   RCRA-Regulated Performance and Emissions 29
        B.   Metal Emissions 32
        C.   Combustion By-Product Emissions 34
        D.   Dioxin and Furan Emissions 35
        E.   Ash and Air Pollution Control Residue Quality 37

VI.   Predicting and Assuring Incinerator Performance 38
        A.   Surrogates 40
        B.   Performance Indicators 40
        C.   Predicting Performance 41

VII.   Environmental and Public Health Implications 42
        A.   Risks from Single-Event Emissions 42
        B.   Methods for Assessing Risks from Recurring Emissions 43
        C.   Overall Risks from Long-Term Air Pollution Emissions from
                Hazardous Waste Incinerators 44

VIII.   Conclusions 47

IX.   Remaining Issues and Research Needs 48
        A.   Destruction Effectiveness on Untested and Unique Wastes 49
        B.   Control of Heavy Metal Emissions 49
        C.   Emissions of Combustion By-Products 49

* Reprinted with permission from *The Journal of the Air and Waste Management Association*, Pittsburgh, PA.

D.   Real-Time Performance Assurance 50
E.   Role of Innovative Technology...50

References...51

I. INTRODUCTION

Over the last 10 years, concern over improper disposal practices of the past has manifested itself in the passage of a series of federal- and state-level hazardous waste cleanup and control statutes of unprecedented scope. The impact of these various statutes will be a significant modification of waste management practices. The more traditional and lowest cost methods of direct landfilling, storage in surface impoundments, and deep-well injection will be replaced, in large measure, by waste minimization at the source of generation, waste reuse, physical/chemical/biological treatment, incineration, and chemical stabilization and solidification methods. Of all the "terminal" treatment technologies, properly designed incineration systems are capable of the highest overall degree of destruction and control for the broadest range of hazardous waste streams. Substantial design and operational experience exists, and a wide variety of commercial systems are available. Consequently, significant growth is anticipated in the use of incineration and other thermal destruction methods. The objective of this review is to examine the current state of knowledge regarding hazardous waste incineration in an effort to put these technological and environmental issues into perspective.

Hazardous waste management was the environmental issue of the 1980s and continues to be a major continuing environmental theme of the 1990s. Discovery of the numerous environmental catastrophes resulting from the improper disposal practices of the past have elevated public awareness and concern. Over the last 10 years, this concern has manifested itself in the passage of a series of federal- and state-level hazardous waste cleanup and control statutes of unprecedented scope and impact. At the federal level these laws include the Resource Conservation and Recovery Act of 1976 (RCRA) and its "cradle to grave" provisions for controlling the storage, transport, treatment, and disposal of hazardous waste. In 1979, the polychlorinated biphenyl (PCB) regulations promulgated under Section 6(e) of the Toxic Substances Control Act (TSCA), prohibited the further manufacture of PCBs after July 2, 1979, established limits on PCB use in commerce, and established regulations for proper disposal. Cleanup of the uncontrolled waste sites created by poor disposal practices of the past was provided for in the Comprehensive Environmental Response, Compensation, and Liability Act of 1980 (CERCLA) which established a national fund (Superfund) to assist in remedial actions. The 1986 Superfund Amendments and Reauthorization Act (SARA) not only reauthorized the Superfund program but greatly expanded the provisions and funding of the initial act.

The most significant of all of these statutes were the 1984 amendments and reauthorization of RCRA. Termed the Hazardous and Solid Waste Act of 1984 (HSWA), these amendments established a strict time line for restricting untreated hazardous waste from land disposal. By 1990, most wastes were restricted and pretreatment standards were established based on the treatment levels achievable by Best Demonstrated Available Technology (BDAT).[1]

The impact of these various statutes will be a significant modification of waste management practices. The more traditional and lowest cost methods of direct landfilling, storage in surface impoundments, and deep-well injection will be replaced, in large measure, by waste minimization at the source of generation, waste reuse, physical/chemical/biological treatment, incineration, and chemical stabilization and solidification methods.

Of all the "terminal" treatment technologies, properly designed incineration systems are capable of the highest overall degree of destruction and control for the broadest range of hazardous waste streams. Substantial design and operational experience exists and a wide variety of commercial systems are available. Consequently, significant growth is anticipated in the use of incineration and other thermal destruction methods.[2]

While thermal destruction offers many advantages over existing hazardous waste disposal practices and may help to meet the anticipated need for increased waste management capacity, public opposition to the permitting of new thermal destruction operations has been strong in recent years.[3] The environmental awareness and activism which spawned the major hazardous waste laws of the 1980s have, in many respects, switched to skepticism over the safety and effectiveness of the technological solutions which the laws were designed to implement. Citizen distrust of the waste management facility owners and operators remains. The ability of government agencies to enforce compliance is questioned. Reports of trace quantities of chlorinated dioxins, chlorinated furans, and other combustion by-products in the stack emissions of municipal solid waste and PCB incinerators have raised questions in the minds of some concerning whether the RCRA incinerator standards are sufficient to protect public health and the environment. Yet, waste generators, faced with the specter of complying with the HSWA land disposal restrictions or faced with the prospect of future multimillion dollar environmental damage settlements over contaminated groundwater, are looking to ultimate destruction techniques such as incineration as their only viable alternative.

The objective of this review is to examine the current state of knowledge regarding hazardous waste incineration in an effort to put these technological and environmental issues into perspective. In doing so, it will be important to review:

- Current and emerging regulations and standards for hazardous waste incinerators
- Current incineration technology and practice
- Capabilities and limitations of methods for measuring process performance
- Destruction efficiency and emissions characterization of current technology
- Methods for predicting and assuring incinerator performance
- Environmental and public health implications of hazardous waste incineration
- Remaining issues and research needs

While the focus of this chapter is on hazardous waste incineration, it is important to understand that many of the same issues relate to municipal waste incineration and to the use of hazardous waste as a fuel in industrial boilers and furnaces. Where possible and appropriate, the performance and emissions of these systems will be compared and contrasted with hazardous waste incinerators.

II. BACKGROUND

A. HISTORICAL PERSPECTIVE

Purification by fire is an ancient concept. Its applications are noted in the earliest chapters of recorded history. The Hebrew word for hell, *Gehenna*, was actually derived from the ancient phrase *ge-ben Hinnom* or the valley of the son of Hinnom, an area outside of Jerusalem which housed the smoldering town dump and was the site of propitiatory sacrifices to Moloch II.[4] Today, waste fires on the ground or in pits are still used by nomadic tribes.

In the Middle Ages, an early innovation to waste fires was the "fire wagon", the first mobile incinerator.[5] It was a simple rectangular wooden wagon protected by a clay lining. The horse-drawn wagon traveled the streets allowing residents to throw their refuse into the moving bonfire.

Incineration as we know it today began slightly more than 100 years ago when the first municipal waste "destructor" was installed in Nottingham, England.[5] Incineration use in the U.S. grew rapidly also, from the first installation on Governor's Island in New York to more than 200 units in 1921. Most of these were poorly operated batch feed units, some with steam recovery. Until the 1950s, incinerators and their attendant smoke and odors were considered a necessary evil, and their operations were generally undertaken in the cheapest possible manner. However, as billowing smoke stacks became less of a symbol of prosperity and air pollution regulations began to emerge, incineration systems improved dramatically.[6] These improvements included continuous feed, improved combustion control, the use of multiple combustion chambers, designs for energy recovery, and the application of air pollution control systems.[7]

Incineration has been employed for disposal of industrial chemical wastes (hazardous waste) for slightly more than 20 years. Initial units borrowed on municipal waste technology, but poor performance and adaptability of these early grate-type units led to the subsequent use of rotary kilns. Many of the earliest rotary kiln facilities were in West Germany. One of the first U.S. kiln units was at the Dow Chemical Company facility in Midland, MI.

B. REGULATIONS

The first U.S. federal standards for control of incineration emissions were applied to municipal waste combustors under the New Source Performance Standards (NSPS) provisions of the Clean Air Act of 1970. The NSPS established a time-averaged (2 h) particulate emission limit of 0.08 grains per dry standard cubic foot (gr/dscf), corrected to 12% CO_2, for all incineration units constructed after August 1971 and having charging rates greater than 50 tons per day (tpd). Opacity limits were not promulgated at the federal level, but many states now have their own opacity limits and, in some cases, more stringent particulate control requirements.

Hazardous waste incinerators were not regulated until the passage of the Resource Conservation and Recovery Act of 1976 (RCRA). Technical standards for incinerators were proposed in December 1978 under Section 3004 of RCRA.[8] These standards provided both performance and operating requirements. The performance standards included requirements for acceptable levels of combustion efficiency, destruction efficiency, halogen removal efficiency, and an emission limit for particulate matter. Operational standards required semicontinuous monitoring of process variables (e.g., CO) and specific minimum temperature and combustion gas residence time levels. During the allowable comment period of the proposed rules, the Environmental Protection Agency (EPA) received extensive comment on the scope of the standards and the adequacy of the combined EPA and industrial data base used to set the standards.

Based on the public comment, EPA subsequently proceeded down a three-phase regulatory path:

- Phase I (May 19, 1980). Interim status standards were proposed outlining operating procedures to be followed by existing incinerator facilities.[9]
- Phase II (January 23, 1981). Performance standards were proposed for new incineration facilities, requiring specific levels of organic hazardous constituent destruction and removal, exhaust gas HCl removal, and maximum particulate emission concentration.[10]
- Phase III (June 24, 1982). Interim final standards were published for both new and existing incinerators, incorporating and modifying somewhat the provisions of the Phase I and Phase II rules.[11]

The provisions of the final incinerator standards, which are of great importance to this paper, are the performance standards which are now listed in the Code of Federal Regulations (CFR) under 40 CFR 264.343. These standards require than a facility to receive a RCRA permit must attain the following performance levels:

1.   99.99% destruction and removal efficiency (DRE) for each principal organic hazardous constituent (POHC) in the waste feed:

$$DRE = [(W_{out})/W_{in}] \times 100$$

where W_{in} = mass feed rate of the POHC in the waste stream fed to the incinerator and W_{out} = mass emission rate of the POHC in the stack prior to release to the atmosphere.

2.   At least 99% removal of hydrogen chloride from the exhaust gas if hydrogen chloride stack emissions are greater than 1.8 kg/h.

3.   Particulate matter emissions no greater than 180 mg per standard cubic meter corrected to 7% oxygen in the stack gas. The measured particulate matter concentration is multiplied by the following correction factor to obtain the corrected particulate matter emissions:

$$Correction\ factor = 14/(21 - Y)$$

where Y = measured oxygen concentration in the stack gas on a dry basis (expressed as a percentage).

The concept and selection of POHCs is an important part of the incineration regulations. POHCs, which are to be sampled during "trial burns" to assess attainment of the standards, are to be selected from the RCRA Appendix VIII constituents present in the wastes.[10] Appendix VIII is a list of approximately 400 organic and inorganic hazardous chemicals first published in Part 261 of the May 19, 1980 *Federal Register*.[9] The list is updated semiannually in 40 FR 261.

POHC selection guidance suggests that Appendix VIII constituents which are in the highest concentration in the waste feed and which are the most difficult to incinerate are the most likely and appropriate to be selected as POHCs.[12] This selection approach, particularly the concept of hazardous compound incinerability, has been the subject of considerable scientific debate since the guidance was first proposed in 1981. These issues will be examined in greater detail later.

It is important to note that EPA chose not to apply the incineration standards to the practice of disposing of hazardous waste as a fuel in industrial boilers and furnaces.[9] This exemption was based on a lack of sufficient information about the practice and on the fact that energy recovery constituted a beneficial use of wastes. Considerable data have been assembled since the exemption was granted in 1980. EPA began to control the practice first in 1985 with issuance of RCRA regulations on the use of waste oil for energy recovery.[13] This rule provides a basis for distinguishing between a used oil and a hazardous waste for energy recovery purposes and provides a used-oil specification that limits the types of boilers that can burn used oils that fail the specification. EPA is developing regulations which will cover the disposal of hazardous waste in industrial boilers and other industrial process furnaces. These rules are currently scheduled for proposal in 1987.[14]

EPA has also promulgated regulations for the incineration of specific wastes. Incineration of PCBs is controlled under TSCA rules promulgated in May 1979.[15] These rules require that whenever disposal of PCBs is undertaken, they must be incinerated, unless the PCB concentration is less than 50 parts per million (ppm). If the concentration is between 50 and 500 ppm, the rule provides for certain exceptions that allow alternatives to the incineration requirements, such as use as fuel in high efficiency boilers. Where the concentration exceeds 500 ppm, PCBs must be disposed of in incinerators which achieve a 99.9% combustion efficiency (CE) and meet a number of specific incinerator operating conditions (combustion temperature, residence time, stack oxygen concentration).

The incineration of certain wastes containing certain chlorinated dibenzo-*p*-dioxins, chlorinated dibenzofurans, and chlorinated phenols is regulated under RCRA rules promulgated January 14, 1985. The so-called "dioxin rule" limits the incineration of these specific wastes (EPA waste codes F020-F028) to incinerators which have been "certified" as being capable of achieving 99.9999% DRE for chlorinated dioxins or similar compounds.[16]

Current municipal waste incineration standards under the Clean Air Act provide only limits on particulate emissions, as previously stated. The 1984 Hazardous and Solid Waste Act Amendments (Section 102), however, require EPA to prepare a report to Congress on the extent of risks due to dioxin emissions from municipal waste incinerators and on appropriate methods for reducing these emissions. EPA also plans to expand their report to include data on cancer risks and controls associated with additional pollutants emitted by these incinerators. EPA is committed by an agreement with the Natural Resources Defense Council to issue an announcement by May 1987 on what actions EPA plans to take regarding risks from municipal waste incineration.

III. CURRENT INCINERATION PRACTICE

A. INCINERATION PRACTICE

Incineration is an engineered process that employs thermal decomposition via thermal oxidation at high temperature (usually 900°C or greater) to destroy the organic fraction of the waste and reduce volume. Generally, combustible waste or wastes with significant organic content are considered most appropriate for incineration. However, technically speaking, any waste with a hazardous organic fraction, no matter how small, is at least a functional candidate for incineration. For instance, significant amounts of contaminated water are currently incinerated in the U.S.[17] Contaminated soils are also being incinerated with increasing frequency. EPA, for example, has employed a mobile incinerator to decontaminate 40 ton of Missouri soil which had been contaminated with 4 lb of chlorinated dioxin compounds.[18] Many other designs for mobile incineration facilities have emerged and are also being applied in the field for decontamination of soil and debris.[19]

Since the promulgation of the RCRA interim status incinerator standards in 1980, a number of surveys and studies have been conducted to assess the quantity and types of hazardous waste generated in the U.S. as well as the quantities and types of wastes being managed by various treatment storage and disposal facilities.[20-26] These studies often reveal significant differences in what would seem to be relatively straightforward statistics. While frustrating to those in government and industry who are evaluating waste management alternatives and economic impacts, the deficiencies in the data base are not surprising. They have resulted from many factors: changes and uncertainties in regulatory definitions of hazardous waste terms, differences in methods and assumptions employed in the various surveys, and incomplete or inaccurate responses by facility owners and operators. Continuing changes in waste generation and the number and permit status of facilities which have occurred in response to regulatory changes and economic factors have also made it difficult to project accurately waste management practice from one point in time to another.

In spite of these deficiencies and limitations, it is possible to construct a reasonable picture of hazardous waste generation and incineration practice from the aggregate of the studies. Total annual hazardous waste generation in the U.S. appears to be approximately

265 million metric tons (MMT). This number was first projected by EPA in the so-called Westat mail survey[24] and later confirmed in separate studies by the Congressional Budget Office (CBO),[25] and the Congressional Office of Technology Assessment (OTA).[26] Only a small fraction of this waste (<1%) was believed to have been incinerated. EPA estimated that 1.7 MMT was disposed in incinerators in 1981,[14] and CBO projected this amount at 2.7 MMT in 1983.[25]

Precise information on the exact types of wastes actually going to incineration facilities is not available. Many facilities operate on an intermittent basis and handle mixtures of wastes which are difficult to describe in terms of EPA standard waste codes. A 1983 EPA study examined data on 413 waste streams going to 204 incineration facilities in the U.S.[17] The major waste streams incinerated were spent nonhalogenated solvents (EPA waste code F003) and corrosive and reactive wastes contaminated with organics (EPA waste codes D002 and D003). Together, these accounted for 44% of the waste incinerated. Other prominent wastes included hydrocyanic acid (P063), acrylonitrile bottoms (K011), and nonlisted ignitable wastes (D001).

While only a small fraction of available hazardous waste is currently managed by incineration, many believe that implementation of the HSWA land disposal restriction regulations and generator concern for long-term liability will result in increased utilization of incineration for ultimate disposal. EPA has estimated that nearly five times more hazardous waste could have been thermally destroyed in incinerators and industrial furnaces in 1981 than actually was.[24] Numerous other studies have indicated that the actual use and demand for incineration technologies to manage hazardous waste will increase significantly.[2,25,27-29]

The CBO study, however, offers the best perspective of potential hazardous waste incineration practice.[25] These data, which are based on industrial output models, are the only available source of comprehensive waste generation estimates which are aggregated on the basis of waste type. This allows more precise estimation of incinerable waste quantities. CBO also examined the potential impact of waste reduction and recycling activities on waste available for incineration. The results of these analyses (Table 1) indicated that even after recycling and reduction, as much as 47 MMT/year could have been available for incineration in 1983. This estimate, however, did not include potentially incinerable wastes from uncontrolled hazardous wastes sites.

It is clear that considerable potential exists for expansion of incineration practice. This assumes, however, sufficient RCRA-permitted capacity can be made available. This is, of course, a significant issue and one which has been given attention in a number of studies.[2,20-23] While capacity appears to be adequate in the near term (Table 2), waste quantities received for incineration appear to be increasing at a faster rate than capacity is being added.[20] Beyond this, future increases in demand will be primarily for organic solids and sludges (e.g., wastes restricted from land disposal or resulting from uncontrolled site cleanup). Liquid capacity will likely remain sufficient for a longer time, especially if the capacity represented by potential disposal in cement kilns and industrial boilers is included.[2]

One of the major barriers to increased incineration capacity is public opposition to the permitting and siting of new facilities, especially the off-site commercial facilities which would be necessary to handle much of the solids and sludges which will increasingly require suitable disposal. Public opposition has been strong in recent years. The normal time required for permitting new incineration facilities is 3 years. This time, as well as the expense of obtaining a permit, may be greatly increased if public opposition exists.

This has created considerable uncertainty for waste generators, equipment manufacturers, and commercial waste disposers. Since 1981, for instance, almost 100 incinerators have withdrawn from the RCRA system because they either ceased operation or decided no longer to handle hazardous wastes.[22] Of the 57 companies identified as marketing hazardous waste incinerators in 1981, 23 have gone out of business, left the hazardous waste incinerator business, or put considerably less emphasis on this activity.[22]

TABLE 1
Quantities of Incinerable Wastes Generated in the U.S., 1983.[27]

Type of waste	Quantity generated (MMT)_a	Current percent recycled/recovered_b	Quantity after recycled/recovered (MMT)_b
Liquids			
Waste oils	14.25	11	12.68
Halogenated solvents	3.48	70	1.04
Nonhalogenated solvents	12.13	70	3.64
Other organic liquids	3.44	2	3.37
Pesticides/herbicides	0.026	55	0.012
PCSs	0.001	0	0.001
Total liquids	33.33	38	20.74
Sludge and solids			
Halogenated sludges	0.72	0	0.72
Nonhalogenated sludges	2.24	0	2.24
Dye and paint sludges	4.24	0	4.24
Oily sludges	3.73	5	3.54
Halogenated solids	9.78	0	9.78
Nonhalogenated solids	4.58	0	4.58
Resins, latex, monomer	4.02	65	1.41
Total sludges/solids	29.31	10	26.51
Total incinerable wastes	62.64	25	47.25
Total hazardous wastes	265.60	6	249.28

a   MMT = millions of metric tons.
b   These waste recycling and recovery practice estimates were derived by the Congressional Budget Office from information obtained directly through surveys of industrial waste generators and the waste recovery industry.[25]

The amount of public opposition to proposed permits for land-based incinerator facilities varies by location and type of waste. On-site facilities that directly serve a single waste generator have greater public acceptance than off-site, commercial incinerators that serve multiple generators in a large market area. Off-site facilities are often not perceived as providing sufficient economic benefits to the local community to offset the risks associated with the importation of wastes from other areas. On-site facilities are more clearly perceived as being linked to businesses that are important to the local economy and are generally not perceived as being importers of hazardous waste. Opposition has tended to focus primarily on new off-site facilities, including incinerator ships, and on new applications to burn PCBs, which critics view as particularly hazardous.

In an effort to assess the dilemma of perceived benefits versus public concerns, EPA conducted an assessment in 1985 to determine if there was a need for a change in the approach toward regulating thermal destruction.[3] The major concern reported by citizens included concern for the following:

- Hazardous material spills in storage, treatment, and handling
- Environmental and health impacts of land-based and ocean facilities
- Poor site selection processes
- Distrust of incinerator owners and operators
- Inability of government agencies to enforce compliance

The study concluded that public opposition to both land and ocean incineration may decline somewhat if regulators address more fully some citizen concerns regarding national

TABLE 2

Estimation of Available Hazardous Waste Incinerator Capacity by Incinerator Design[22]

Incinerator design	Number of units		Reported average design capacity[a] (MM Btu/h)	Reported utilitzation[b] (%)	Projected available capacity[c] (MM Btu/h)	Percent with air pollution control equipment
	Reported	Projected[d]				
Rotary kiln	42	45	61.37	77	635	90
Liquid injection	95	101	28.26	55	1284	42
Fume	25	26	33.14	94	52	40
Hearth	32	34	22.75	62	294	38
Fluidized bed and other	14	15	19.29	—	95[e]	38
Total or average values	208	221	32.37	67	2360	50

[a] 154 incinerators reporting.
[b] 71 incinerators reporting.
[c] Calculated by multiplying projected number of units × average design capacity × (100-utilization)/100.
[d] Includes units planned and under construction.
[e] For this projection, average value of 67% utilitzation is used.

regulatory strategy, local community impact, equity of facility siting, public decision-making processes, and, especially, enforcement plans and capacity. It was also concluded that there is a need to communicate better how health and environmental concerns and priorities are reflected in regulations and standards. Better communication of regulatory policy, strategy, and other activity related to decisions on proposed permits for individual incinerator facilities or vessels is certainly desirable since improved communication with the public can enhance the credibility of regulatory and enforcement agencies.

B. INCINERATION TECHNOLOGY

Different incineration technologies have been developed for handling the various types and physical forms of hazardous waste. A recent study identified 221 hazardous waste incinerators operating under the RCRA system in the U.S.[21] Some of the results of this study are displayed in Table 2. The four most common incinerator designs (in order of use) are liquid injection (sometimes combined with fume incineration), rotary kiln, fixed hearth, and fluidized bed incinerators. These units are located at 189 separated facilities. Only 18 facilities are commercial off-site operations,[31] the balance of incineration practice being located at the site of waste generation.

The process of selection and design of hazardous waste incineration systems can be very complex. Fortunately, considerable industrial manufacturing experience exists, and many useful design guides have been published.[7,32-34] Thus, while a detailed examination of design principles is beyond the scope of this book, a generalized review of the most prominent features of incineration systems and important design factors will be helpful in understanding their operation and emissions performance.

The major subsystems which may be incorporated into a hazardous waste incineration system are

1. Waste preparation and feeding
2. Combustion chamber(s)
3. Air pollution control
4. Residue and ash handling

The normal orientation of these subsystems is shown in Figure 1 along with typical process component options. The selection of the appropriate system combination of these components is primarily a function of the physical and chemical properties of the waste stream or streams to be incinerated.

1. Waste Preparation and Feeding

The physical form of the waste determines the appropriate feed method.[32] Liquids are blended, then pumped into the combustion chambers through nozzles or via specially designed atomizing burners. Wastes containing suspended particles may need to be screened to avoid clogging of small nozzle or atomizer openings. While sustained combustion is possible with waste heat content as low as 4000 Btu/lb, liquid wastes are typically blended to a net heat content of 8000 Btu/lb or greater. Blending is also used to control the chlorine content of the waste fed to the incinerator. Wastes with chlorine content of 70% and higher have been incinerated;[35] however, most operators limit chlorine content to 30% or less. Blending to these levels provides the best combustion control and limits the potential for formation of hazardous free chlorine gas in combustion gases.

Sludges are typically fed using progressive cavity pumps and water-cooled lances. Bulk solid wastes may require shredding for control of particle size. They may be fed to the combustion chamber via rams, gravity feed, airlock feeders, vibratory or screw feeders, or belt feeder. Containerized waste is typically gravity or ram fed.

13

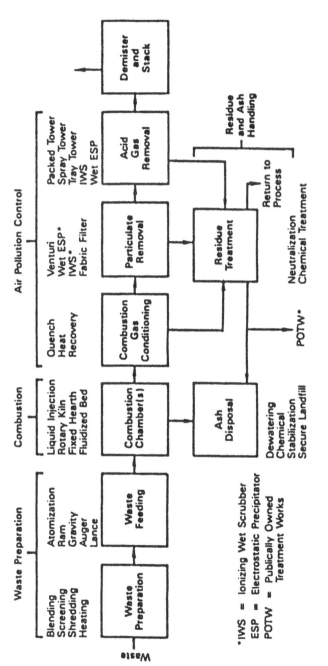

FIGURE 1. General orientation of incinerator subsystems and typical process component options.

TABLE 3
Applicability of Major Incinerator Types to Wastes of Various Physical Form[32]

	Liquid injection	Rotary kiln	Fixed hearth	Fluidized bed
Solids				
Granular, homogeneous		X	X	X
Irregular, bulky (pallets, etc.)		X	X	
Low melting point (tars, etc.)	X	X	X	X
Organic compounds with fusible ash constituents		X		
Unprepared, large, bulky material		X		
Gases				
Organic vapor laden	X	X	X	X
Liquids				
High organic strength aqueous wastes	X	X		X
Organic liquids	X	X		X
Solids/liquids				
Waste contains halogenated aromatic compounds 2200°F minimum	X	X		
Aqueous organic sludge		X		X

2. Combustion Chambers

The physical form of the waste and its ash content determine the types of combustion chamber selected. Table 3 provides general selection considerations for the four major combustion chamber (incinerator) designs as a function of wastes of different forms.[32] Most incineration systems derive their name from the type of combustion chamber employed.

Liquid injection incinerators or combustion chambers are applicable almost exclusively for pumpable liquid waste. These units (Figure 2) are usually simple, refractory-lined cylinders (either horizontally or vertically aligned) equipped with one or more waste burners. Liquid wastes are injected through the burner(s), atomized to fine droplets, and burned in suspension. Burners as well as separate waste injection nozzles may be oriented for axial, radial, or tangential firing. Improved utilization of combustion space and higher heat release rates, however, can be achieved with the utilization of swirl or vortex burners or designs involving tangential entry.[36]

Good atomization is critical to achieving high destruction efficiency in liquid combustors. Nozzles have been developed to produce mists with mean particle diameters as low as 1 m,[37] compared to typical oil burners which yield droplets in the 10- to 50-m range.[38] Atomization may be attained by low pressure air or steam (1 to 10 psig), high pressure air or steam (25 to 100 psig), or mechanical (hydraulic) means using specially designed orifices (25 to 450 psig).

Vertically aligned liquid injection incinerators are preferred when wastes are high in inorganic salts and fusible ash content, while horizontal units may be used with low ash waste. The typical capacity of liquid injection incinerators is roughly 28×10^6 Btu/h heat release. Units, however, range as high as 70 to 100×10^6 Btu/h.

Rotary kiln incinerators (Figure 3) are more versatile incinerators in the sense that they are applicable to the destruction of solid wastes, slurries, and containerized waste as well as liquids. Because of this, these units are most frequently incorporated into commercial off-site incineration facility design. The rotary kiln is a cylindrical refractory-lined shell that is mounted on a slight incline. Rotation of the shell provides for transportation of waste through the kiln as well as for enhancement of waste mixing. The residence time of waste solids in the kiln is generally 1 to 1.5 h. This is controlled by the kiln rotation speed (1 to 5 rpm), the waste feed rate, and, in some instances, the inclusion of internal dams to retard the rate of waste movement through the kiln. The feed rate is generally adjusted to limit also the amount of waste being processed in the kiln to at most 20% of the kiln volume.

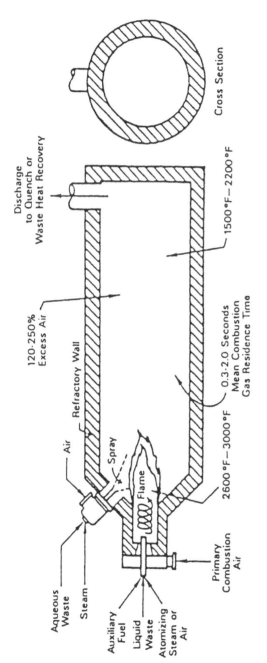

FIGURE 2. Typical liquid injection combustion chamber.

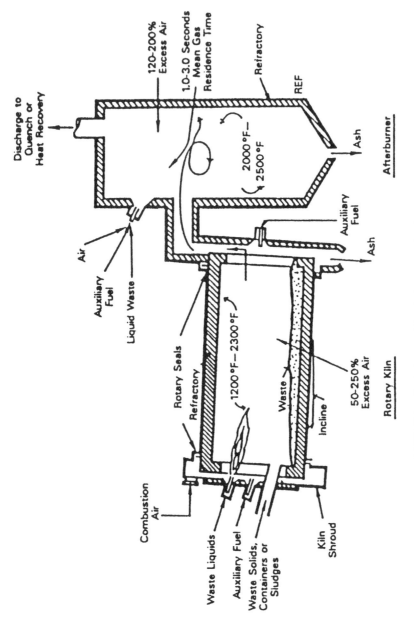

FIGURE 3.   Typical rotary kiln/afterburner combustion chamber.

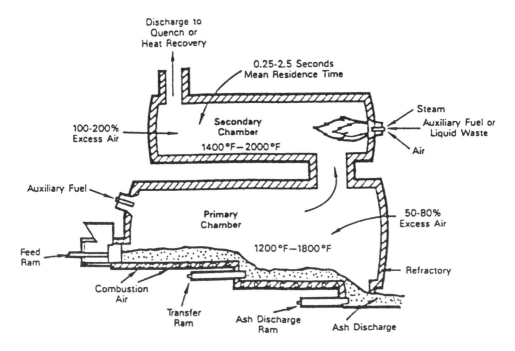

FIGURE 4. Typical fixed hearth combustion chamber.

The primary function of the kiln is to convert solid wastes to gases, which occurs through a series of volatilization, destructive distillation, and partial combustion reactions. However, an afterburner is necessary to complete the gas phase combustion reactions. The afterburner is connected directly to the discharge end of the kiln, whereby gases existing in the kiln turn from a horizontal flow path upward to the afterburner chamber. The afterburner itself may be horizontally or vertically aligned and essentially functions much on the same principles as a liquid injection incinerator. In fact, many facilities also fire liquid hazardous waste through separate waste burners in the afterburner. Both the afterburner and kiln are usually equipped with an auxiliary fuel firing system to bring the units up to and maintain the desired operating temperatures. Rotary kilns have been designed with a heat release capacity as high as 90×10^6 Btu/h in the U.S. On average, however, units are typically 60×10^6 Btu/h.

Fixed hearth incinerators, also called controlled air, starved air, or pyrolytic incinerators, are the third major technology in use for hazardous waste incineration today. These units employ a two-stage combustion process, much like rotary kilns (Figure 4). Waste is ram fed into the first stage, or primary chamber, and burned at roughly 50 to 80% of stoichiometric air requirements. This starved air condition causes most of the volatile fraction to be destroyed pyrolytically, with the required endothermic heat provided by the oxidation of the fixed carbon fraction. The resultant smoke and pyrolytic products, consisting primarily of volatile hydrocarbons and carbon monoxide, along with products of combustion, pass to the second stage, or secondary chamber. Here, additional air is injected to complete the combustion, which can occur either spontaneously or through the addition of supplementary fuels. The primary chamber combustion reactions and turbulent velocities are maintained at low levels by the starved-air conditions to minimize particulate entertainment and carryover. With the addition of secondary air, total excess air for fixed hearth incinerators is in the 100 to 200% range.

Fixed hearth units tend to be of smaller capacity than liquid injection or rotary kiln incinerators because of physical limitations in ram-feeding and transporting large amounts

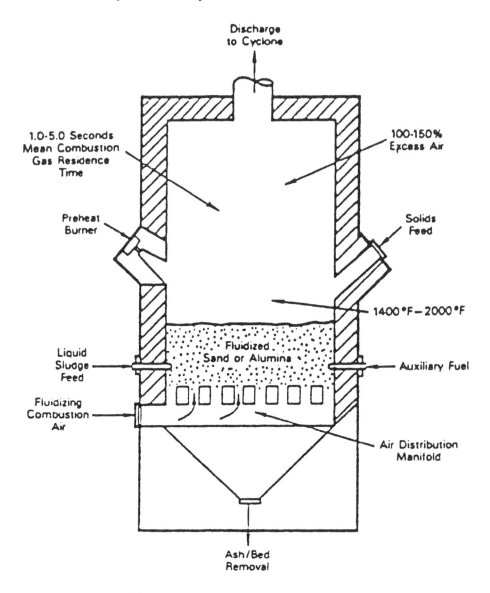

FIGURE 5.   Typical fluidized bed combustion chamber.

of waste material through the combustion chamber. These lower relative capital costs and potentially reduced particulate control requirements make them more attractive than rotary kilns for smaller on-site installations.

Fluidized beds have long served the chemical processing industry as a unit operation. This type of combustion system has only recently begun to see application in hazardous waste incineration. Fluidized bed incinerators may be either circulating or bubbling bed designs.[39] Both types consist of a single refractory-lined combustion vessel partially filled with particles of sand, alumina, sodium carbonate, or other materials. Combustion air is supplied through a distributor plate at the base of the combustor (Figure 5) at a rate sufficient to fluidize (bubbling bed) or entertain the bed material (circulating bed). In the circulating bed design, air velocities are higher and the solids are blown overhead, separated in a cyclone, and returned to the combustion chamber. Operating temperatures are normally maintained in the 1400 to 1600°F range, and excess air requirements range from 100 to 150%.

Fluidized bed incinerators are primarily used for sludges or shredded solid materials. To allow for good distribution of waste materials within the bed and removal of solid residues from the bed, all solids generally require prescreening or crushing to a size less than 2 in. in diameter. Fluidized bed incinerators offer high gas to solids ratios, high heat transfer efficiencies, high turbulence in both gas and solid phases, uniform temperatures throughout the bed, and the potential for *in situ* acid gas neutralization by lime or carbonate addition. However, fluid beds also have the potential for solids agglomeration in the bed if salts are present in waste feeds and may have a low residence time for fine particulates.

Regardless of the incinerator type selected, the chemical and thermodynamic properties of the waste determine the sizing of the combustion chamber and its operating conditions (temperature, excess air, flow rates) and determine the nature of air pollution control and ash/residue handling systems. Elemental composition and moisture content data are necessary to determine stoichiometric combustion air requirements and to predict combustion gas flow and composition. These parameters are important in determining combustion temperature and residence time conditions, the efficiency of waste/fuel/air mixing, and in the type and size of air pollution control equipment. Typical operating temperatures, gas (and solid) residence times, and excess air rates for each of the four major incinerator types are indicated in Figure 5. It is important to understand, however, that significant deviation from these values has been observed in actual field practice without detrimental effect on waste destruction and removal efficiency.[40]

3. Air Pollution Control

Following incineration of hazardous wastes, combustion gases may need to be further treated in an air pollution control system. The presence of chlorine or other halogens in the waste will generally signal a need for a scrubbing or absorption step for combustion gases to removal HCl and other halo-acids. Ash in the waste is not destroyed in the combustion process. Depending on its composition, ash will either exist as bottom ash (at the discharge end of a kiln or hearth, for example) or as particulate matter suspended in the combustion gas stream (fly ash). Particulate emissions from most hazardous waste combustion systems generally have particle diameters less than 1 μm and require high efficiency collection devices to meet the RCRA emission standards. In addition, gas cleaning systems provide some limited additional buffer against accidental release of incompletely destroyed waste products. Such systems, however, are not a substitute for good combustion and operating practices.

The most common air pollution control equipment employed in hazardous waste facilities is summarized in Table 4.[21] Most often, several of these devices are employed in series. The most common system used is a quench (gas cooling and conditioning), followed by high-energy Venturi scrubber (particulate removal), a packed tower absorber (acid gas removal), and a demister (visible vapor plume elimination). It is interesting to note, however, that more than half of the incinerators employ no air pollution control system at all (Table 2). This could be because these facilities are handling low ash, low halogen content liquid waste streams for which such control may not be necessary.

Venturi scrubbers involve the injection of a scrubbing liquid (usually water or a water/caustic solution) into the exhaust gas stream as it passes through a high velocity constriction, or throat. The liquid is atomized into fine droplets which entrain fine particles and a portion of the absorbable gases in the gas stream. The major advantage of Venturi scrubbers is their reliability and relative simplicity of operation. On the other hand, maintaining the significant pressure drop across the Venturi throat (60 to 120 in. of water column) required for hazardous waste combustion particular matter control represents a significant percentage of the total cost of operation of incineration facilities employing Venturi scrubbing.

Acid gas removal is generally accomplished in packed bed or plate tower scrubbers. Packed bed scrubbers are generally vessels filled with randomly oriented packing material

TABLE 4
Distribution of Air Pollution Control Devices
among Hazardous Waste Incinerators

APCD[a] type	Number	Precent
Quench	21	23.3
Venturi scrubber	32	35.6
Wet scrubber	7	7.8
Wet ESP	5	5.5
Ionizing wet scrubber	5	5.5
Other nonspecified scrubber	12	13.3
Packed tower absorber	18	20.0
Spray tower absorber	2	2.2
Tray tower absorber	1	1.1
Other absorbers	2	2.2
Non/unknown	31	34.4
Total incinerator systems surveyed	90	

Note: Total number types of air pollution control devices are greater than systems surveyed because many incinerators report more than one device.

[a] APCD = air pollution control device.

such as polyethylene saddles or rings. The scrubbing liquid is fed to the top of the vessel, with the gas flowing in either concurrent, countercurrent, or cross-flow modes. As the liquid flows through the bed, it wets the packing material and, thus, provides the interfacial surface area for mass transfer with the gas phase which is required for effective acid gas absorption.

Like packed bed scrubbers, plate scrubbers also rely on absorption for the removal of contaminants. The basic design is a vertical cylindrical column with a number of plates or trays inside. The scrubbing liquid is introduced at the top plate and flows successively across each plate as it moves downward to the liquid outlet at the tower bottom. Gas comes in at the bottom of the tower and passes through openings in each plate before leaving through the top. Gas absorption is promoted by the breaking up of the gas phase into small bubbles which pass through the volume of liquid in each plate.

Packed bed or plate tower scrubbers are commonly used at liquid injection incinerator facilities, where absorption of soluble gaseous pollutants (HCl, SO_x, NO_x) is most important and particulate control is less critical. However, at rotary kiln or fixed hearth facilities, or liquid injection facilities where high ash content wastes are incinerated, Venturi scrubbers are often used in series with packed bed or plate tower scrubbers.

Many designs in recent years have begun to incorporate waste heat boilers as a substitute for gas quenching and as a means of energy recovery.[41,42] Wet electrostatic precipitators (ESP), ionizing wet scrubbers (IWS), and fabric filters are also being incorporated into new systems[43] largely due to their high removal efficiencies for small particles and lower pressure drop.

4. Residue and Ash Handling

The inorganic components of hazardous wastes are not destroyed by incineration. These materials exit the incineration system either as bottom ash from the combustion chamber, as contaminants in scrubber waters and other air pollution control residues, and in small amounts in air emissions from the stack. Under RCRA, residues generated from the incineration of hazardous waste should be managed carefully.

Ash is commonly either air-cooled or quenched with water after discharge from the combustion chamber. From this point ash is frequently accumulated on-site in storage lagoons or in drums prior to disposal in a permitted hazardous waste land disposal facility. Dewatering or chemical fixation/stabilization may also be applied prior to disposal.

Air pollution control residues are generated from the combustion gas quenching, particulate removal, and acid gas absorption steps in an incineration system. These residues are typically aqueous streams containing entrained particulate matter, absorbed acid gases (usually as HCl), and small amounts of organic contaminants. These streams are often collected in sumps or recirculation tanks where the acids are neutralized with caustic and returned to the process. Eventually, a portion of all of these waters must be discharged for treatment and disposal (generally, when the total dissolved solids level exceeds 3%). Many facilities discharge neutralized waters to settling lagoons or a chemical precipitation step to allow for suspended contaminants to be concentrated and ultimately send to land disposal. Depending on the nature of the dissolved contaminants and their concentration after treatment, water may be either returned to the process or discharged to sewers.

C. OTHER HAZARDOUS WASTE THERMAL DESTRUCTION SYSTEMS

Other types of systems are also being employed to destroy hazardous waste thermally. These include ocean incineration vessels, mobile incinerators, and high temperature industrial furnaces.

Ocean incineration involves the thermal destruction of liquid wastes at sea in specially designed tanker vessels outfitted with high-temperature incinerators. The principle of operation of these units is identical to that of land-based liquid injection incinerators with the exception that current ocean incinerators are not equipped with air pollution control systems. Acid gas produced from incinerating chlorinated wastes is discharged to the air without treatment to be neutralized by contact with seawater, which has a naturally high buffering capacity.

Ocean incineration has been routinely used in Europe since 1969.[44] A total of six different vessels have conducted hundreds of waste burns in the North Sea. Although several test burns and research studies involving herbicide orange, mixed organochlorine waste, and PCBs have occurred under U.S. sponsorship, ocean incineration has never been used on a routine commercial basis in the U.S.[45] This has been largely due to vocal public concern over potential environmental effects that could result from incinerator emissions and spills of hazardous materials during loading and transport of wastes.

The importance of ocean incineration, however, is not that it is currently a major factor in the U.S. hazardous waste management but, rather, that it has served as a focal point for public and scientific debate over the state of knowledge of the character and potential effects of emissions from incineration processes in general. These issues have been explored in several major studies on ocean and land-based incineration conducted over the past 2 years.[30,46,47] While these studies have found incineration to be the effective technology currently available for organic hazardous waste destruction, many of the important technical and policy issues examined will significantly impact the direction of incineration practice and research over the next decade.

A number of companies are marketing mobile or transportable incineration systems. Most of these are scaled down, trailer-mounted versions of conventional rotary kiln or fluidized bed incinerators. The thermal capacities of most mobile systems range from 10 to 20 million Btu/h.

The first mobile rotary kiln incinerator was designed and tested by EPA as a potential solution to on-site cleanup at uncontrolled waste sites.[19,48] Other rotary kiln systems have since been developed and employed at waste sites.[19,49-51] Mobile fluidized bed systems are also being marketed.[52]

Overall, the performance of these mobile systems has been shown to be comparable to equivalent stationary facilities. Current experience suggests, however, that waste incineration is more expensive in mobile units, on a unit cost basis, than it is in stationary units. The principal advantage of mobile systems appears to be that they are more sociopolitically acceptable than removal and transportation of cleanup residues to commercial facilities. In the instance of soil decontamination, on-site incineration may also be more cost-effective than transportation of large amounts of contaminated material to central incineration facilities.

More substantial than either ocean or mobile incineration practice is the use of hazardous wastes as fuels in industrial boilers and furnaces. In 1981, these operations disposed of more than twice the amount of waste that was disposed of via incinerators.[2] Processes that have burned or do burn hazardous waste materials as fuels include industrial boilers, cement kilns, iron-making furnaces, and light-weight aggregate and asphalt plants. The principal attractions to this approach include exemption (currently) from RCRA incineration standards, fuel and waste transportation cost savings, and waste disposal cost savings since hazardous waste used as a fuel can be sold.

The most recent source of information on waste fuel use in industrial processes was compiled for EPA in 1984.[53] The study synopsizes results of a national questionnaire of waste fuel and waste oil use in 1983. The study revealed that there were 1300 facilities using hazardous waste-derived fuels (HWDF), accounting for a total of 230 million gal/year. The chemical industry (Standard Industrial Classification 28) accounted for 67% of this while operating only 12.4% of the facilities using HWDF. Other industries employing significant quantities of hazardous waste as fuel included SICs 26 (paper), 29 (petroleum), 32 (stone, clay, glass, concrete), and 33 (primary metals). The majority (69%) of the waste was burned in large quantities by a few facilities representing only a small fraction (1.6%) of the 1300 facilities. These included medium- to large-size industrial boilers, cement and aggregate kilns, and iron-making furnaces.

While waste code-specific data on HWDF are not readily available, recent data indicate that of the HWDF burned in 1983, 30% was organic solvents, and 45% was other hazardous organics.[53] Most of this waste was generated on-site, and 74% of the balance arrived directly from an off-site generator rather than through an intermediary.

IV. MEASURING PROCESS PERFORMANCE

Proper and accurate measurement of the emissions from incineration systems is a critical issue. Great demands have been placed upon sampling and analysis technology by the RCRA incinerator regulations. Fortunately, significant progress has been made in adopting measurement methods to the rigors of specific compound identification and the level of detection and accuracy which are often necessary to assess compliance with the RCRA incineration standards. These methods will rarely be a limitation in assessing incinerator performance if proper attention is given to quality assurance and quality control, if adequate advanced planning is conducted, and if experienced personnel are involved in sampling and analysis activities.

Performance measurement may have any of the following three purposes:

- To establish compliance with performance standards (e.g., trial burns)
- To monitor process performance and direct process control (e.g., continuous monitoring)
- To conduct performance measurements for research and equipment development purposes

Methods employed in assessing regulatory compliance are generally official methods which have been standardized and published in the *Federal Register* or in EPA guidance documents. Routine performance monitoring for process control often involves the use of continuous monitors for emissions and facility-specific engineering parameters (e.g., temperature, pH, kiln rotation). Research and equipment assessment investigations may involve any of the above techniques in combination with standard and often nonstandard sampling and analysis techniques designed for rapid screening of performance or for ultralow detection of specific materials.

A. PERFORMANCE MEASUREMENT

Figure 6 illustrates sampling points which may be involved in assessing incinerator performance. In the case of trial burn activities, the main focus of sampling activities is on collection of waste feed and stack emission samples. Ash and air pollution control system residues are also sampled and analyzed. Sampling of input and output streams around individual system components (e.g., scrubbers) may also be conducted in research testing or equipment evaluation studies.

The main focus of analytical activities is on POHCs. Stack gas analysis also includes determination of HCl and particulate emissions and may be extended to a determination of other organic compound emissions as well as metals of concern. In the case of particulate emissions, the size distribution of stack particles may also be of interest. The size of emitted particulate affects its transportation and fate in the atmosphere and influences the likelihood of inhalation, an important factor in health effects assessment. Few hazardous waste incinerator tests have actually collected particle size data, primarily due to time and funding limitations.

EPA has provided guidance on the types and methods of sampling and analysis to be used in trial burns designed to measure facility compliance with the RCRA incineration standards.[54,55] Additional guidance is being prepared. Similar guidance has also been provided for PCB incinerators.[56] Table 5 outlines sampling and analysis methods typically involved in RCRA trial burns. For any trial burn, at any one set of operating conditions and waste feed conditions, three replicate runs (i.e., as identical as possible) are usually recommended to obtain a representative assessment of incinerator performance.[55]

The sampling method numbers in Table 5 refer to methods identified in a manual combustion sampling and analysis methods compiled by EPA.[57] This manual expands on and augments the information in EPA SW-846, "Test Methods for Evaluating Solid Waste: Physical/Chemical Methods"[59] and "Samplers and Sampling Procedures for Hazardous Waste Streams."[60] Together, these references are the best sources from which to identify methods to be used in incinerator performance evaluations.

Analytical methods for specific hazardous compounds are often of greatest interest. Analytical methods for Appendix VIII compounds[10] in these references are generally based on high-resolution fused-silica-capillary column gas chromatography (GC) in combination with mass spectrometry (MS) for specific compound detection. High-performance liquid chromatography (HPLC) is recommended for determination of compounds that are inappropriate for detection by GC/MS. Application of analytical methods has been evaluated for 240 of the approximately 400 Appendix VIII Compounds.[61,62] The methods showed acceptable precision in determination of most of the compounds. Detection limits in synthetic samples were on the order of 1 to 10 ng per injection, but detection in actual waste samples will be dependent on the nature of interferences in the waste matrix.

While all emissions from hazardous waste incinerators are important, the greatest interest is most often placed on stack emissions. The accuracy and reliability of stack sampling results are central to the entire issue of incinerator performance and environmental safety. Existing methods have been the subject of substantial research, debate, and, in some cases, criticism.[63]

Stack emissions are sampled to determine stack gas flow rate, HCl, particulate concentration, and the concentration of organic compounds of interest. Determination of stack gas

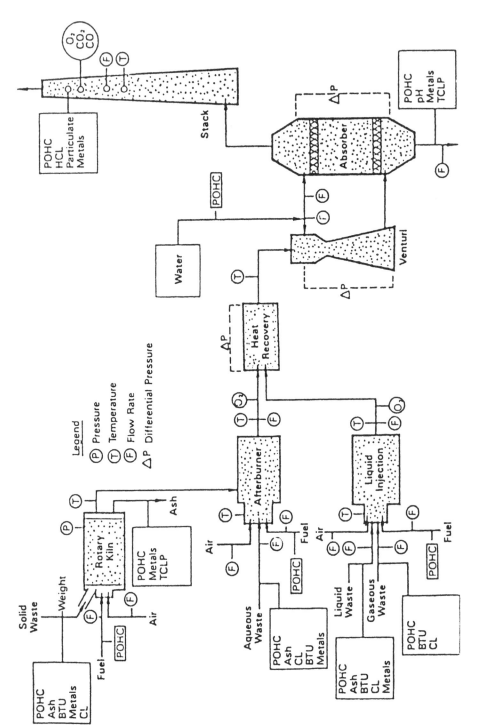

FIGURE 6.   Potential sampling points for assessing incineration performance.

TABLE 5
Sampling Methods and Analysis Parameters

Sample	Sampling frequency for each run	Sampling method[a]	Analysis parameter[b]
Liquid waste feed	Grab sample every 15 min	SOO4	V&SV-POHCs, Cl, ash ult. anal., viscosity, HHV
Solid waste feed	Grab sample from each drum	SOO6, SOO7	V&SV-POHCs, Cl, ash, HHV
Chamber ash	Grab 1 sample after all 3 runs are completed	SOO6	V&SV-POHCs, TCLP[c]
Stack gas	Composite	MM5 (3 h)	SV-POHCs, particulate H_2O, HCl
	Three pairs of traps 40 min/pair	VOST (2 h)	V-POHCs
	Composite in Tedlar gas bag	SO11	V-POHCs[d]
	Composite in Mylar gas bag	M3 (1—2 h)	CO_2 by Orsat
	Continuous (3 h)	Continuous monitor	CO (by plants monitor)

[a] VOST denotes volatile organic sampling train; MM5 denotes EPA Modified Method 5; M3 denotes EPA Method 3; SXXX denotes methods found in "Sampling and Analysis Methods for Hazardous Waste Combustion."[55]

[b] V-POHCs denotes volatile principal organic hazardous constituents (POHCs); SV-POHCs denotes semivolatile POHCs; HHV denotes heating value.

[c] TCLP — toxicity characteristic leaching procedure.[58]

[d] Gas bag samples may be analyzed for V-POHCs only if VOST samples are saturated and not quantifiable.

flow rate and particulate emissions is performed using the conventional stack sampling method commonly referred to as Method 5 (M5). This method encompasses EPA Methods 1 through 5 and is defined in detail in 40 CFR Part 60, Appendix A. HCl emissions are sampled by modifying the Method 5 train to include a caustic impinger. A specialized sampling and analytical method has also been developed to further speciate and quantify hydrogen halide and halogen emissions.[64]

The technology of incinerator stack sampling for trace organic compounds is sophisticated. While the basic technology is well developed, many pitfalls await those who attempt the job without sufficient knowledge or experience. Sampling of stack effluent for organics in order to determine DRE may require from one to three separate methods (or more), depending on the number of compounds to be quantified and the characteristics, and on the detection limits that are required to prove a DRE of 99.99% or establish levels of incomplete combustion by-products. Special attention must be given to sampling rate and duration in planning for emission tests to insure that a sufficient amount of sample is collected to meet detection limit objectives and to allow for all necessary analyses to be completed.[55]

The three methods for hazardous waste incinerator sampling are

1.   Modified Method 5 (MM5)
2.   Volatile Organic Sampling Train (VOST)
3.   Gas bags

The Modified Method 5 (MM5) train is used to capture semivolatile (boiling point 100 to 300%C) and nonvolatile (boiling point >300°C) organic compounds. The MM5 is merely a simple modification of the M5 train involving insertion of a sorbent module (XAD-2 resin) between the filter and the first impinger.[65] It is recommended that a separate M5 train for particulate determination be used in tandem with the MM5 train since drying of the filter for particulate determination may invalidate analysis of organic compounds on the filter.[55] Like the M5, the MM5 involves isokinetic traversing of the stack with a sampling probe. Water-cooled sample probes are necessary for sampling of hot combustion gases in regions ahead of quenching. Where it is desirable to collect larger amounts of sample for more extensive analysis or lower detection limits, the much larger Source Assessment Sampling System (SASS) may be used instead of the MM5.[55] SASS involves single point (pseudo-isokinetic) sampling at a rate of 110 to 140 l/min (4 to 5 cfm) compared to the 14 to 28 l/min (0.5 to 1 cfm) rate of the MM5. The same sorbent resin (XAD-2) is also used. Because of its more convenient sample size and potability and its multipoint isokinetic sampling, the MM5 train is generally preferred over the SASS train.

For volatile organic compounds (boiling point 30 to 130°C), the Volatile Organic Sampling Train (VOST) is used. The VOST was developed by EPA in 1981 to enable detection of stack concentrations of volatile organic compounds as low as 0.1 ng/l.[66] This detection limit was deemed necessary to be able to demonstrate greater than 99.99% DRE for volatile organic compounds at concentrations as low as 100 ppm in the waste feed. The VOST system involves drawing a single stack gas sample through two sorbent tubes in series. The first tube contains Tenax resin and the second, Tenax and activated charcoal. Up to six pairs of sorbent tubes operating at 1 l/min for 20 min each may be needed to achieve the lowest detection levels.[67] For higher stack gas concentrations, however, the VOST may be operated at lower flow rates with fewer pairs of tubes (longer sampling times for each pair).

Various types of gas sampling bags may also be used to sample for volatile organic compounds. These are generally appropriate only for higher organic concentrations. The accuracy of sampling with this method is a function of the sampling and storage characteristics of the bags.[68] The use of extensive quality assurance and quality control procedures is required with both plastic bags and the VOST to avoid sample contamination in the field

TABLE 6
Summary of Continuous Emission Monitors

Pollutant	Monitor type	Expected concentration range	Available range[a]
O_2	Paramagnetic	5—14%	0—25%
	Electrocatalytic (e.g., zirconium oxide		
CO_2	NDIF[b]	2—12%	0—21%
CO	NDIR	0—100 ppm	0—5,000 ppm
NO_x	Chemiluminescent	0—4,000 ppm	0—10,000 ppm
SO_2	Flame photometry	0—4,000 ppm	0—5,000 ppm
	Pulsed fluorescence		
	NDUV[c]		
SO_3	Colorimetric	0—100 ppm	0—50 ppm
Organic compounds	Gas chromatography (FID)[d]	0—50 ppm	0—100 ppm
	Gas chromatography (ECD)[e]		
	Gas chromatography (PID)[f]		
	IR absorption		
	UV absorption		
	GC/MS		

[a] For available instruments only. Higher ranges are possible through dilution.
[b] Nondispersion infrared.
[c] Nondispersion ultaviolet.
[d] Flame ionization detector.
[e] Electron capture detector.
[f] Photo-ionization detector.

and in transit.[69] This problem was not fully appreciated in some of the early field tests employing the VOST.

Both the MM5 and VOST sampling methods have been subjected to laboratory and field validation studies for selected compounds.[70-72] These studies have demonstrated that excellent results are possible with these methods. It is important to note, however, that modifications of these methods may be required for certain POHC compounds which become chemically or physically altered in the sampling systems. High water-soluble compounds (e.g., acetonitrile) and water-reactive compounds (e.g., phthalic anhydride), for instance, present special challenges to current sampling methods.

B. PROCESS MONITORING

Measurement of a wide variety of incinerator operating parameters may be necessary to maintain thermal destruction conditions which are equivalent to those observed during a successful trial burn. These measures are used as indicators of the performance of the incineration system and serve as input to automatic and manual process control strategies. There are nearly two dozen potential measurements, including such parameters as combustion temperature, waste feed rate, oxygen and carbon monoxide (CO) concentration in the stack, gas flow rate at strategic points, and scrubber solution pH. These parameters and their use are described in detail in a number of resource documents.[32,54,55,73]

Continous emission monitors (CEM) are often used or required in measuring combustion gas components such as carbon monoxide (CO), oxygen (O_2), nitrogen oxides (NO_x), and total unburned hydrocarbons (TUHC). If properly interpreted, combustion gas components may be indicators of the completeness of the thermal destruction reaction. These methods typically require extraction of gas samples from the gas stream of interest and measurement with a remote instrument. Some parameters such as CO and O_2 may be measured *in situ* (in the stack). Table 6 summarized monitor types and available concentration measurement ranges for a number of CEMs.[74]

RCRA incinerator operating requirements stipulate that the permit specifies an operating limit for CO concentration in the stack gas. While continuous CO monitoring is required, specific CEM requirements are not specified. However, permit writers frequently cite Reference Method 10 from the EPA New Source Performance Standard (NSPS) as a guide (40 CFR 60, Appendix A). CO concentration in stack gas is an indicator of combustion efficiency. However, this parameter is also being examined as a possible real-time indicator of DRE for hazardous organic compounds.

TUHC emissions are also being considered as a possible DRE indicator. Three types of analyzers are available: flame ionization detector (FID), photoionization detector (PID), and the automated total gaseous nonmethane organics analyzer (TNMO). FID issued most typically. All three methods have limitations. The theoretical minimum detectability of all three techniques is significantly below 10 g/dscf (detection in the 10 to 700 g/dscf range would be required for DRE correlation purposes). However, due to electrical noise, sampling limitations and component degradation, the practical field detectability limit is from 40 to 200 g/dscf, if care is exercised.[74] The individual methods are also limited by the fact they do not respond to all classes or organic compounds of potential concern.

V. EMISSIONS FROM HAZARDOUS WASTE INCINERATION

Ideally, the primary products from combustion are carbon dioxide, water vapor, and inert ash. In reality, what appears outwardly to be a straightforward, simple process is actually an extremely complex one involving thousands of physical and chemical reactions, reaction kinetics, catalysis, combustion aerodynamics, and heat transfer. This complexity is further aggravated by the complex and fluctuating nature of the waste feed to the process. While combustion and incineration devices are designed to optimize changes for completion of these reactions, they never completely attain the ideal. Rather, small quantities of a multitude of other products may be formed, depending on the chemical composition of the waste and the combustion conditions encountered. These products along with potentially unreacted components of the waste become the emissions from the incinerator.

Hydrogen chloride (HC1) and small amounts of chlorine (CL_2), for example, are formed from the incineration of chlorinated hydrocarbons. Hydrogen fluoride (HF) is formed from the incineration of organic fluorides, and both hydrogen bromide (HBr) and bromine (BR_2) are formed from the incineration of organic bromides. Sulfur oxides (SO_x), mostly as sulfur dioxide (SO_2), but also including 1 to 50% sulfur trioxide (SO_3), are formed from the sulfur present in the waste material and auxiliary fuel. Highly corrosive phosphorus pentoxide (P_5O_5) is formed from the incineration of organophosphorus compounds. In addition, oxides of nitrogen (NO_x) may be formed by fixation of nitrogen from nitrogen compounds present in the waste material or in the combustion air. Suspended particulate emissions are also produced and include particles of mineral oxides and salts from the mineral constituents in the waste material. A wide range of organic compounds may also be formed from the incomplete thermal destruction of organic compounds in the waste and auxiliary fuel.

Until recently, there were only limited data available on waste destruction performance and pollutant emissions from hazardous waste thermal destruction devices. Studies by EPA and others in the early to late 1970s employed a variety of evolving trace organic pollutant sampling and analysis techniques and were often targeted only toward measuring macro-destruction and combustion efficiencies[75-77] rather than the performance standards now required under the Resource Conservation and Recovery Act of 1976. Since 1981, however, EPA has conducted a substantial program of performance testing at thermal destruction facilities. The testing was designed to estimate the environmental impact of these operations and to provide information on the ability of these facilities to control emissions to the degree required by the 1982 incinerator performance standards. These test facilities, test procedures, and performance results have been summarized[78] for the facilities tested (incinerators, industrial boilers, and industrial process kilns). Complete test reports have been published for the incinerators,[79] industrial boiler,[80,81] and cement/aggregate kilns[82,83] tested. These data as well as trial burn results from 14 additional RCRA incinerators have been summarized recently in an EPA report, "Permit Writer's Guide to Test Burn Data — Hazardous Waste Incineration."[84]

The following sections summarize these data in five areas:

- RCRA-regulated performance and emissions (DRE, particulate matter, and HCl)
- Metal emissions
- Combustion by product emissions
- Dioxin and furan emissions
- Ash and air pollution control residue quality

A. RCRA-REGULATED PERFORMANCE AND EMISSIONS

Tables 7, 8, and 9 summarize waste destruction efficiency, HCl, and particulate emissions results for the incinerators, industrial boilers, and cement kilns tested. The tables also summarize certain process-operating parameters as well as emissions of CO and O_2 and in some instances $NO_x SO_x$.

These data reveal that well-operated incinerators, industrial boilers, and process kilns are capable of achieving a 99.99% DRE, the RCRA performance standard. All of the incinerators tested by EPA achieved this level of performance for candidate POHC compounds in concentrations greater than 1200 ppm in the waste feed.[85] Candidate POHC compounds between 200 and 1200 ppm frequently were not destroyed to a 99.99% DRE, and no compounds below 200 ppm in the waste feed met the RCRA DRE limit. In fact, regression analysis of the pooled data suggested that statistically significant correlations (correlation coefficients were 0.76 and 0.84) existed between compound penetration (1-DRE) and compound feed concentration, showing that DRE increased with waste feed concentrations.[85]

This phenomenon, which has been observed in tests of other thermal destruction devices, was not anticipated. A number of possible explanations have been advanced.[2,79] The most frequently stated theory postulates that at the very low stack emission concentrations (<1 ng/l) necessary to demonstrate greater than 99.99% DRE for a sub-1200 ppm compound, sufficient amounts of that compound may actually be formed as an incomplete combustion or recommendation by-product from other compounds in the wastes to effect a reduction of the DRE below 99.99%. Others argue that limitations of current stack sampling and analysis techniques for such low levels of trace organic compounds are responsible.

EPA is conducting research to assess this concentration phenomenon. From a regulatory standpoint, however, this is not currently perceived as an issue. Few, if any, of the low concentration compounds in the wastes identified in the EPA test program would have actually been selected as "principal" organic hazardous constituents in trial burns if existing EPA guidance on POHC selection were employed. It is also important to note that even though DRE declines with lower initial compound concentrations in the waste, the amount absolute of compound emitted also declines. In fact, the DRE vs. concentration correlation noted above actually predicts that the net emissions resulting from a reduced DRE for a 100-ppm compound will actually be slightly less than those for a 99.99% DRE for the compound at 1000 ppm in the waste.

Table 8 indicates that industrial boilers, particularly the larger water tube units, typically attain 99.99% DRE. Cement kilns, lime kilns, and light-weight aggregate kilns with good combustion control and waste atomization all met or exceeded the 99.99% DRE (Table 9).

All incinerators and industrial process kilns tested met or approached the RCRA HCl

TABLE 7
Incinerator Performance and Stack Emissions Data

Data Reported as Averages for Each Facility

Facility type	O_2 (%)	CO (ppm)	TUHC (ppm)	DRE (%)	Particulate (mg/m³)	HCl control (%)
Commercial rotary kiln/liquid incinerator	10.5	6.2	1.0	99.999	152	99.4
Commercial fixed hearth, two-stage incinerator	11.4	6.9	1.0	99.994	400	98.3
On-site two-stage liquid incinerator	8.1	9.4	6.0	99.994	143	99.7
Commercial fixed hearth, two-stage incinerator	11.0	327.7	18.7	99.997	60	a
On-site liquid injection incinerator	13.2	11.9	1.0	99.999	186	a
On-site rotary kiln incinerator	9.7	554.0	61.7	99.999	23	99.9
Commercial two-stage fixed hearth incinerator	13.4	26.8	1.8	99.996	168	98.3
On-site rotary kiln	b	794.5	NAᶜ	99.998	184	99.7
On-site liquid injection incinerator	9.7	66.3	7.8	99.994	95	a
On-site rotary kiln incinerator	10.7	5.8	NA	99.996	404	99.9
On-site rotary kiln incinerator	14.1	323.0	NA	99.996	NA	99.8
On-site liquid injection incinerator	12.4	31.9	1.9	99.999	163	98.6
On-site liquid injection incinerator	9.3	1.0	NA	99.996	40	a
On-site fluidized bed incinerator	3.6	67.4	NA	99.996	259	a
On-site fixed hearth incinerator	12.9	NDᵈ	NA	99.999	93	a
On-site liquid injection incinerator	4.5	358.0	NA	99.995	99	a
On-site liquid injection incinerator	3.6	28.4	NA	99.998	12	99.3
Commercial rotary kiln incinerator	9.4	8.0	0.5	99.999	172	99.9
On-site liquid injection incinerator	3.1	779.3	NA	99.999	88	99.6
On-site liquid furnace incinerator	6.4	56.3	NA	99.999	4	99.9
On-site fixed hearth incinerator	13.5	5.0	NA	99.999	150	98.4

ᵃ   HCl emissions 4 lb/h.
ᵇ   Reported only as a range (3.1—16.7%).
ᶜ   NA — not available.
ᵈ   Not detected.

removal standard of 99%. Industrial boilers typically have no existing controls for HCl, but none exceeded the 1.8 kg/h emission standard because wastes with low net chlorine content were employed.

Achieving the RCRA particulate emission standard of 0.08 gr/dscf was a problem for a number of the incinerators tested by EPA. Four of the eight units tested failed to meet the RCRA standard. Two of those facilities were marginally above the emission limit and could likely meet the standard with minor operating adjustments. There remaining two facilities appeared to need significant design and/or operational changes.[86] In some cases, failure of the particulate emission standard may be attributed to dissolved neutralization salts in mist carryover from caustic scrubbers.

TABLE 8
Summary of Boiler Performance

	(%)	(%)	Release time(s)	Avg. vol. heat rel. rate (kW/m³)	DRE[a]	W/F[b] (%)	NO_x (ppm)[c]	CO (ppm)[c]
Watertube stoker	100	6—16	1.2	509	99.98	40	163—210	900—1200
Packaged firetube	25	4—6	0.8	739	99.991	0.1—0.5	40—65	47—88
Field erected watertube	26	10	2	78	99.999	37	61—96	18—21
Converted stoker	78	4—6	1.1	339	99.998	18—48	193—250	75—127
Packaged watertube	36—73	6—7	1.1—0.5	960	99.995	19—56	164—492	83—138
Converted watertube	53	7—11	2	107	99.98	8.7—10.1	243—328	109—139
Modified firetube	44	8	0.4	807	99.998	100	67—74	146—170
Tangentially fired watertube	100	6	2	180	99.991	2.4—4.3	393—466	142—201
Packaged watertube	65	2	1.8	343	99.998	8.2	64—78	410—1125[d]
								46—750
Packaged firetube	50—100	3—8	0.7—0.3	1240	99.999	100	85—203	20—135
Packaged watertube	82	4	1.8	269	99.999	49	154—278	102—119

a Mass weighted average for all POHCs in the waste >100 ppm.
b W/F = waste heat input as a percent of total heat input.
c Range of average values across individual sites and runs including baseline.
d Higher values are for high nitrogen content waste firing.

TABLE 9
Summary of Industrial Kiln Performance and Stack Emissions Data

Data Reported as Averages for Each Facility

Facility type	Test[a]	DRE (%)	DRE (kg/MG)[b]	Particulate (kg/h)	Particulate (ppm)	HC (ppm)	HC W/F (%)[c]
Wet process cement kiln	W	99.200	0.27	0.36	68	450	25
(nonatomized waste)	B	—	0.26	0.09	136	279	—
Wet process cement kiln	W	99.996	0.27	2.1	478	265	15
(atomized waste)	B	—	0.26	0.6	371	636	—
Dry process cement kiln	W	99.998	—	11.5	814	19	45
(nonatomized waste)	B	—	—	1.3	620	7	—
Dry process cement kiln	W	99.992	—	0.47	486	27	15
(atomized waste)	B	—	—	0.25	680	27	—
Lime kiln	W	99.997	0.11	0.20	446	596	30
(atomized waste)	B	—	0.10	0.09	386	553	—
Shale aggregate kiln (atomized waste)	W	>99.99	0.33	2.1	—	—	100
Clay aggregate kiln (atomized waste)	W	99.998	0.58	0.023	162	1130	59
Clay products kiln	W	>99.99	0.002	0.84	—	—	100

[a] W = waste testing; B = baseline (fossil fuel only).
[b] Particulate emissions are expressed as kg particulate per metric ton (MG) of product produced (e.g., cement, lime).
[c] W/F = waste fuel heat input expressed as a percent of total heat input.

No significant changes in particulate emissions were observed for industrial boilers and certain of the industrial process kilns when they fired waste fuels compared to emissions for fossil fuels only.[87,88] Some increased emissions were observed in kilns employing electrostatic precipitators for particulate control. These increases were attributed to changes in the electrical resistivity of the particles due to the presence of increased chloride levels. Adjustments in ESP operation should correct this in most cases.

B. METAL EMISSIONS

Metals such as arsenic, barium, beryllium, chromium, cadmium, lead, mercury, nickel, and zinc are of possible concern in waste incineration because of their presence in many hazardous wastes and because of possible adverse health effects from human exposure to emissions. Incineration will change the form of metal fractions in waste streams, but will not destroy the metals. As a result, metals are expected to emerge from the combustion zone essentially in the same total quantity as the input. The principal environmental concern, therefore, centers around where and in what physical or chemical form the metals end up in the combustion system, i.e., bottom ash, in APCD residues, or sack emissions.

Most interest has traditionally focused on stack emissions of metals. Increasing attention, however, is now being given to the quality of residuals from incineration of metal-bearing wastes since disposal of these materials may be subject to tough restrictions on land disposal under HSWA.

Metals present in the feed to combustion devices are typically emitted in combustion gases as particles rather than vapors. However, some of the more volatile elements (e.g., mercury and selenium) or their chemical compounds may be released to the atmosphere partially in the vapor state. The processes involved in the formation of particles are very complex and are only partially understood at present. Most of the current state of knowledge on metal behavior in combustion has come from research on coal combustion.[89-92]

In general, data on metal emissions and partitioning for hazardous waste incineration

TABLE 10
Average Stack Emissions of Metals from Five Hazardous Waste Incinerators

		Emission rate (g/kJ) and concentration on particulate (g/g)[a]				
		Plant B				
Metals	Plant A	Uncontrolled	Controlled	Plant C	Plant D	Plant E
Sb	0.32 (8,380)[a]	0.26 (300)	b	b	—	—
As	—	b	b	b	—	c
Be	0.052 (840)	0.19 (250)	0.11 (930)	b	0.056 (1.490)	0.050 (470)
Be	—	(5)[b]	b	b	—	c
Cd	0.055 (890)	0.11 (140)	0.019 (150)	b	0.012 (1,120)	0.36 (4,000)
Cr	0.14 (2,300)	0.73 (590)	0.19 (1,500)	2.5 (47,500)	c	0.094 (1,100)
Pb	5.4 (85,500)	2.3 (3,100)	0.64 (5,300)	b	0.24 (25,600)	9.0 (98,000)
Ng	d	b	b	b	—	—
Ni	0.024 (400)	0.50 (650)	0.087 (740)	2.7 (49,000)	0.052 (4,570)	c
Se	—	7.0 (9,200)	0.45 (3,700)	b	0.29 (34,6000)	—
Ag	0.0008 (11.4)	b	b	0.33 (620)	0.0076 (9170)	—
Ti	0.0089 (140)	b	b	b	—	—

a Numbers in parentheses represent values for g of metal/g of emitted particulate.
b All values below detection limit.
c Some values below detection limits, average not calculated.
d Emissions are reported as gram of particulate emitted per hour divided by the design heat release rate of the incinerator in kJ/h.

Adapted from Reference 17.

are limited and often incomplete. Organic emissions have been the focus of most historical emissions assessments of these facilities. Data on air pollution control device effectiveness for metals are even more scarce.

In 1982, Gorman and Ananth collected data on metal emissions from the Cincinnati MSD incinerator.[93] Metals data were also collected in three earlier tests sponsored by EPA in 1976.[94-96] The best source of metal emissions data, however, is the series of incinerator tests more recently conducted by EPA in support of the Regulatory Impact Analysis for the 1982 RCRA incinerator standards.[79] Average emission rates for these latter tests are shown in Table 10. Actual values varied over a considerable range.

Wallace et al. recently reviewed all of the available metal emissions data for hazardous waste incinerators and compared them to emissions for other conventional combustion sources.[97] The five most frequently detected metals are Ba, Cd, Cr, Pb, and Ni. Hg was not found in any of the emissions, but this was because only particulate metal and not vapor was sampled in the studies. The relatively volatile metals (Sb, Cd, Pb) generally show enrichment in fine particulate emissions (i.e., higher concentration relative to their concentration on larger particles). This enrichment phenomenon is an important consideration for health assessment studies, since the fine particulates are also more likely to be inhaled.

Metal emissions from the hazardous waste incinerators tested were equivalent to those reported for municipal solid waste incinerators.[99] The emissions appear to be from 2 to 20 times higher than those from sewage sludge incinerators.[100] It is important to note, however, that it is difficult to extend the findings from these few tests to hazardous waste incinerators in general. This is because metal emissions are a function of the amount of metal input to the incinerator (which is highly variable day to day and facility to facility) as well as the efficiency of the incinerator in controlling metal emissions. Some of the incinerators tested, for instance, employed no air pollution control equipment.

Even where air pollution control equipment is used, it is difficult to draw conclusions on air pollution control system efficiency for metals because of the uncertainty of the sampling methods usually employed for control device input rates, the lack of particle size information, and the relatively low quantities of metals available in the collection samples. EPA is in the process of assembling a report on all of the available air pollution control device efficiency data for metals.[101]

Metals emissions may also be of concern for high-temperature industrial processes employing hazardous waste as a combustion[102,103] in small boilers. Lead is the primary element of concern here since 50 to 60% of the combustion chamber input generally exists in the stack. Stack concentrations of lead of 5000 to 72,000 g/m^3 have been observed in tests of small boilers of sizes ranging from 0.50 to 15 $\times$ 10^6 Btu.[103] EPA has promulgated rules controlling the metals level in used oils.[13] In addition, restrictions on waste metal content are being considered in proposed rules for industrial boilers and furnaces to be published in 1987.[14]

C. COMBUSTION BY-PRODUCT EMISSIONS

The current RCRA incineration standards regulate destruction and removal only for the major hazardous compounds in the waste. However, even under good combustion conditions, incomplete combustion byproducts may be emitted. One of the concerns expressed by some scientists and environmentalists regarding hazardous waste thermal destruction is the possible impact on human health and the environment of emissions of potentially hazardous products of incomplete combustion (often referred to as PICs). While many of the incinerator field tests conducted to date have attempted to quantify by-product emissions, these data have been criticized as being incomplete and insufficient for the purpose of a full risk assessment.[46] Testing has focused largely on identification of Appendix VIII organic compounds only. Comparison of total hydrocarbon emissions with the total quantity of specific organic compounds identified in the emissions has revealed that only a relatively small percentage of the total hydrocarbon emissions may have been identified.[104]

Incomplete combustion by-products from hazardous waste incineration have been recognized for some time. Early pilot-scale studies of the thermal destruction of the pesticide, Kepone, found emissions of hexachlorobenzene and several other "daughter products" which had been predicted from previous laboratory-scale studies.[105] Similar thermal decomposition studies followed for PCBs[106] and dozens of other compounds.[107-110]

While the RCRA incinerator standards do not currently regulate incomplete combustion by-products, the earlier proposals did recognize and discuss this issue. The January 1981 Phase I rule proposed that emissions of incomplete combustion by-products be limited to 0.01% of the POHC input to hazardous waste incinerators.[10] Although no final action was taken on that aspect of the rule, researchers, regulators, and environmentalists have pursued the question of PICs with great vigor since then, including various attempts to compare actual field performance results to the proposed standard.[111-113]

One of the basic problems in assessing the results of laboratory and, particularly, field studies of PIC emissions is the fact there is no standardized definition of what a PIC is. While a POHC is defined in the RCRA regulations, a PIC is not, in a rigorous sense. Thus, there is often confusion even among scientists working in the area. Strictly speaking, PICs are organic compounds which are present in the emissions from the incineration process, which were not present or detectable in the fuel or air fed to the incinerator. In EPA's test program, compounds were considered to be PICs if they were regulated organic compounds (i.e., listed in Appendix VIII of CFR 40 Part 261) which were detected in stack emissions, but not present in the waste feed at concentrations greater than 100 ppm.[111]

Compounds in the emission stream which are identified as PICs may actually result from any one of the following phenomena:

1.   Compounds resulting from the incomplete destruction of the POHCs, i.e., fragments of the original POHCs.

2. New compounds 'created' in the combustion zone and downstream as the result of partial destruction followed by radical-molecule reactions with other compounds or compound fragments present. These compounds may also result from the incomplete combustion of non-Appendix VIII compounds in the waste. This aspect may be especially significant where fossil fuel is used in incineration and where waste is fired into conventional industrial furnaces as only a percentage of the heat input.

3. An Appendix VIII compound originally present in the feed stream before incineration but not specifically identified as a POHC.

4. Compounds from other sources (e.g., ambient air pollutants in combustion air). In some field tests, compounds identified in the stack emissions as PICs were actually found to have come from contaminants (trihalomethanes) in the potable water used for scrubber water make-up.[111]

Given the complexity of sources of potential PIC compounds, it is not surprising that a consensus PIC definition has been difficult to achieve. Consequently, for the purpose of this review, it seems more productive to examine the issue of combustion by-products separately from any type of specific definition by ignoring the source or cause of the emission of particular compounds and considering all organic compound emissions (including POHCs) as combustion by-products (CBs). A recent EPA study has examined CBs in this fashion.[112] The study examined field test data from 23 EPA-sponsored emissions tests at thermal destruction facilities. Included were eight incinerators, nine industrial boilers, and six industrial kilns. Organic emissions from hazardous waste facilities were compared to emissions where facilities were burning fossil fuel only. The organic emissions were also compared to organic emissions from municipal solid waste incinerators and coal-fired utility boilers.

The EPA studies of thermal destruction systems identified 55 Appendix VIII compounds (28 volatile and 27 semivolatile) in stack emissions. These compounds were emitted at normalized rates that span over five orders of magnitude, 0.09 to 13,000 ng of emissions per kilojoule (ng/kJ) of combustor heat input (1 ng/kJ = 2.34×10^{-6} lb/million Btu).

The greatest number compounds were emitted in the 10 to 100 ng/kJ range. Only 9 of the 23 facilities emitted identifiable hazardous compounds at rates exceeding 100 ng/kJ.

The volatile compounds tended to be detected more often and in significantly higher concentrations than in the semivolatile compounds. The compounds that occurred most frequently and in the highest concentrations, nine volatile and six semivolatile, are listed in Table 11. Emission rates for incinerators, boilers, and kilns are shown in Table 12 for 12 of these compounds for which sufficient data are available for comparison. The data show that values from test run to test run varied considerably. Thus, these data do not allow prediction of levels for all three combustion devices. Many of the volatile compounds showed higher levels for boilers, and semivolatile compounds tended to be higher for incinerators.

Data were also available from several baseline (no waste firing) tests on boilers and kilns which allowed comparison of emissions from hazardous waste combustion with combustion of other fuels. While there was a wide range in values from test to test, the data suggested there is little inherent difference between waste and fuel and combustion emissions.[112]

Sufficient data for five semivolatile compounds were available to compare their emissions when burning hazardous waste vs. their emissions from municipal incinerators and coal-fired power plants. Table 13 presents this comparison. The four phthalate compounds in the table all show very similar emission rates from all three sources. Naphthalene emissions were lower for power plants than the other two sources. Again, the data suggest that for these compounds there is little inherent difference between combustion sources.

D. DIOXIN AND FURAN EMISSIONS

Without doubt, the greatest amount of scientific and public attention has been given to

TABLE 11
**Most Frequent Thermal Destruction
Process Stack Emissions**

Volatile compounds	Semivolate compounds
Benzene	Naphthalene
Toluene	Phenol
Carbon tetrachloride	Bis(2-ethylhexy)phthalate
Chloroform	Diethyphthalate
Methylene chloride	Butylbenzylphthalate
Trichloroethylene	Dibutylphthalate
Tetrachloroethylene	
1,1,1-Trichloroethane	
Chlorobenzene	

TABLE 12
**Emission Rates of Specific Compounds from Incinerators, Boilers, and Kilns
$(ng/kJ)^a$**

	Incinerators		Boilers		Kilns	
	Mean	Range	Mean	Range	Mean	Range
Benzene	87	2—980	30	0—300	580	290—1000
Toluene	1.6	1.5—4.1	280	0—1200		No data
Carbon tetrachloride	0.8	0.3—1.5	1.8	0—7.2		No data
Chloroform	3.8	0.5—8.4	120	0—1700		No data
Methylene chloride	2.2	0—9.6	180	0—5800		No data
Trichloroethylene	5.2	2.3—9.1	1.2	0—13	1.3	0.7—2.8
Tetrachloroethylene	0.3	0—1.3	63	0—780		No data
1,1,1-Trichloroethane	0.3	0—1.3	7.5	0—66	2.4	(One value)
Chlorobenzene	1.2	0—6.0	63	0—1000	152	33—270
Naphthalene	44	0.7—150	0.6	0.3—2.1		No data
Phenol	7.8	0—16	0.3	0—0.8	0.02	0—0.05
Diethylphthalate	3.7	2.8—4.8	0.4	0.04—1.6		No data

[a]   Expressed as ng of emission per kJ of combustor heat input (1 ng/kJ = 2.34 × 10^{-6} lb/MM Btu).

TABLE 13
**Semivolatile Compound Emission Rates from Hazardous Waste Combustion,
Municipal Incinerators, and Coal Power Plant**

	Hazardous waste		Municipal waste		Coal power plant	
	Mean	Range	Mean	Range	Mean	Range
Naphthalene	17	0.3—150	71	0.4—400	0.5	0.06—1.8
Bis(2-ethylhexyl)phthalate	4.6	0—21	4.6	0.4—12	7.6	0.2—24
Diethylphthalate	1.2	0.04—4.8	0.5	0.0.9	2.8	0.4—5.7
Butylbenzyphthalate	3.7	0.7—23	No data		0.5	0.3—1.0
Dibutylphthalate	0.3	0—1.1	3.9	1.5—7.6	3.0	0.09—8.7

[a]   Expressed as ng of emission per kJ of combustor heat input (1 ng/kJ = 2.34 × 10^{-6} lb/MM Btu).

one class of incinerator combustion by-products, the dioxins and furans. The terms dioxin and furan refer to families of 75 related chemical compounds known as polychlorinated dibenzo-*p*-dioxins (PCDDs) and 135 related chemical compounds known as polychlorinated dibenzofurans (PCDFs), respectively. These compounds are not intentionally made for any purpose; they are unavoidable by-products created in the manufacture of other chemicals

TABLE 14
Dioxin/Furan Emissions Hazardous Waste Thermal Destruction Facilities (ng/m³)

Facility type	Sample/(Waste)[a]	2,3,7,8 TCDD	PCDD	PCDF	Ref.
Commercial rotary kiln/liquid injection combustion incinerator	FG/FA[b] (HW)	ND[c]	ND	ND—1.7	80
Fixed hearth incinerator	FG/FA (HW)	ND	16	56	80
Liquid injection incinerator	FG/FA (HW)	ND	ND	ND	80
Horizontal liquid injection incinerator	FG/FA (HW)	ND	ND	7.3	80
Incinerator ship	FG/FA (PCB)	ND	ND	0.3—3	126
4 lime/cement kilns	FG (HW)	ND	ND	ND	127—129
Fixed hearth incinerator	FG/FA (HW)	ND	ND	ND	130
Rotary kiln/liquid injection	FG (PCB)	ND	ND—48	0.6—95	131
Industrial boiler	FG/FA (PCP)	ND	75—76	ND	125
Industrial boiler	FG/FA (HW)	ND	0.64—0.8	ND	125
Industrial boiler	FG/FA (HW)	ND	ND	ND	125
Industrial boiler	FG/FA (HW)	ND	ND	ND	125
Industrial boiler	FG/FA (HW)	ND	1.1	ND	125

[a] Information in parentheses describes waste feed; HW = hazardous waste; PCB = polychlorinated biphenyl pentachlorophenol waste.
[b] FG = flue gases analyzed; FA = flue gas particulate analyzed.
[c] ND = not detected.

such as some pesticides, or as a result of incomplete combustion of mixtures containing certain chlorinated organic compounds. Since the first published report of PCDD and PCDF emissions from a municipal incinerator by Olie et al.,[114] a large number of studies have been carried out to examine this phenomenon, including work by Buser and Rappe,[115] Eiceman et al.,[116] Karasek,[117] Bumb et al.,[118] Cavallaro et al.,[118] and Lustenhouwer et al.[120]

Most of the interest has been placed on municipal waste incineration. A number of excellent summaries of field emission data have been prepared.[121-123] EPA has reviewed available PCDD/PCDF emissions data for a broad range of combustion sources including fossil fuel and wood combustion and a wide range of industrial furnances,[124] and has reported the results of recent emissions testing at 11 additional facilities.[125]

Dioxin/furan emissions data are somewhat less available for hazardous waste incineration facilities. EPA tests have examined dioxin/furan emissions at five incinerators,[79] six industrial boilers,[126] and three calcining kilns employing hazardous waste as a fuel.[127-129] Data are also available from test burns at PCB incinerators.[130,131] Dioxin/furan emissions from these data sources are summarized in Table 14.

Of the 17 facilities only five emitted detectable levels of PCDD or PCDF. None of the facilities tested was emitting detectable levels of the most hazardous isomer, 2,3,7,8-TCDD. The highest PCDD levels reported were for an industrial boiler using a creosote/PCP sludge as a fuel,[126] where PCDDs were present in the waste feed. In most cases, no PCDD or PCDF was detected in hazardous waste incineration emissions for the facilities tested. By comparison, PCDDs and PCDFs have been found in emissions from 22 municipal waste incinerators for which complete stack emissions data are available.[123] Average emissions of PCDD and PCDF (3300 ng/m³, respectively), were nearly three orders of magnitude greater than the highest values reported for hazardous waste incineration units.

E. ASH AND AIR POLLUTION CONTROL RESIDUE QUALITY

Facilities which incinerate hazardous wastes containing significant ash or halogen content will generate combustion chamber bottom ash and various types of residues collected by subsequent air pollution control equipment. Under RCRA, these ashes and residues are generally classified as hazardous waste also. Thus, facility operators must assess the characteristics of these materials to determine the proper method of disposal. The principal contaminants of interest are heavy metals and any undestroyed organic material.

Most, but not all, operating hazardous waste incinerators which generate combustion chamber ash, quench the ash (usually in water) before discharge. Since air pollution control equipment using wet collection methods predominates incinerator practice, most additional ash and haloacid (e.g., HCl) is also collected in aqueous effluent from a scrubber, absorber, or wet ESP.

Only limited characterization data are available for combustion chamber ash and air pollution control residues. Combustion chamber ash and scrubber wastes were analyzed for several of the incinerators tested by EPA as part of the incineration Regulatory Impact Analysis (RIA) program.[79] Recently, ten additional incinerators were sampled to characterize ash and residues.[132]

In the RIA study,[79] incinerator ash and scrubber waters were analyzed for organic constituents. Only two facilities had ash concentrations of organic compounds at levels greater than 35 g/g. When organic compounds were detected, they tended to be toluene, phenol, or naphthalene at concentration less than 10 g/g. The same compounds were also detected in scrubber waters, usually at concentrations below 20 g/l.

The results of the recent ten-incinerator test program generally confirmed the RIA results.[132] While more organic compounds were detected across all of the facilities (19 volatile and 24 semivolatile compounds), levels in ash were typically at or well below 30 g/g. One facility had a toluene concentration of 120 g/g and phenol concentration of 400 g/g. These levels were believed to have resulted from the facility's use of chemical manufacturing plant wastewater for ash quenching.

More compounds were detected in scrubber waters across the ten facilities than in the RIA study (nine volatiles and five semivolatiles) and in higher concentrations. Semivolatiles ranged from 0 to 100 g/l while volatile compounds were much higher (0 to 32 mg/l).

Combustion chamber ash and scrubber waters were also analyzed for metals in both the RIA study and the ten-incinerator study. Detected concentrations varied widely and were a function of the amount of metal in the input waste stream at each facility and how the residues were processed. However, only 3 of 104 measurements of these metals in leachates from residues and ashes exceeded allowable toxicity characteristic levels using the EPA's extraction procedure test (EP). If a leachate from a waste material exceeds these levels (CFR 40 Pat 261.24), the waste will be designated as a hazardous waste.

Overall, the data from both test programs have suggested that very small amounts of residual organic compounds remain in incinerator ash and incinerator air pollution control residues. Thus, the destruction and removal efficiencies reported for incinerators are almost entirely a result of destruction, rather than removal, of organic compounds. Levels of metals in ashes and air pollution control residues varied widely but generally appear not to exhibit the RCRA toxicity characteristic. On the other hand, it should be recognized that available data represent short-term samples from less than 10% of the total hazardous waste incinerator population in the U.S. Use of these data to project residue and ash quality for specific wastes and incinerator combinations is not possible. Metal and organic concentrations are highly waste and facility specific. They will likely be influenced strongly by waste characteristics and by operating conditions (e.g., scrubber water recycle rate, solids residue time in the combustion chamber, and contaminants in scrubber water and quench make-up water.)

VI. PREDICTING AND ASSURING INCINERATOR PERFORMANCE

Existing data indicate that well-operated hazardous waste incinerators and other thermal destruction facilities are capable of achieving high levels of organic hazardous material

TABLE 15
Parameters Typically Employed to Trigger Fail-Safe
Corrective Action for Incinerators

| | Basis for corrective action | | |
	Excess emissions	Worker safety	Equipment protection
Parameter			
High CO in stack gas	X		
Low chamber temperature	X		
High combustion gas flow	X		
Low pH of scrubber water			X
Low scrubber water flow	X		X
Low scrubber P	X		
Low sump levels	X		X
High chamber pressure		X	
High chamber temperature		X	X
Excessive fan vibration			X
Low burner air pressure	X		
Low burner fuel pressure	X		
Burner flame loss	X	X	X

destruction which equal or exceed current RCRA performance standards. Putting aside arguments over the adequacy of the DRE standard for protecting public health and the environment, virtually any well-designed thermal destruction unit should be capable of demonstrating high DRE if sufficient operating temperature, oxygen, and feed control are provided. While convincing trial burn performance data can be presented, uncertainty and distrust may exist regarding the reliability of thermal destruction systems in day-to-day operation after a permit is approved and when regulators are not present. Little is known quantitatively about the impact of normal process upsets or failure modes upon emissions. This is often a concern of the public in hearings on permit actions.

Currently, permit conditions are primarily based on process operating conditions which are documented during the conduct of a successful trial burn; i.e., one which demonstrated that the facility achieved or exceeded the RCRA performance standards. These operating conditions are then also used to established fail-safe controls for the facility, which designate corrective action to be taken in the event process operation deviates from the demonstrated set points. Corrective actions could include changing auxiliary fuel addition rates, shutting off waste feed, increasing combustion airflow, etc., to control emissions, or other actions to protect worker safety and process equipment.[32] Table 15 shows typical shutdown parameters which may be used to trigger fail-safe controls.

Operating conditions (e.g., combustion temperature, O_2, and CO in stack emissions, etc.) must be used as surrogates for continued high-destruction performance after the trial burn since there is currently no real-time method to determine DRE for specific POHCs. DRE can only be determined with certainty via expensive ($50,000 to $150,000), often multiday testing procedures. Analysis results may take weeks or months to complete. While EPA believes the current permit approach is reasonable and protective of public health and the environment, many argue that the availability of a real-time monitoring technique to detect process upsets and alert operators to take corrective action automatically would significantly increase public acceptance of thermal destruction technology.

Two general classes of performance estimation techniques exist. The first of these involves the use of compounds which are either identified in the waste or added to it to serve as ''surrogates'' for the destruction of other important compounds in the waste. The second approach involves the use of indicator emissions such as CO or unburned hydrocarbons to mirror waste destruction efficiency. Both concepts are used to some extent in incinerator permitting currently.

A. SURROGATES

The surrogates concept involves identifying an easily detected organic compound which is more difficult to destroy thermally than any of the other hazardous compounds in a waste mixture. It is then assumed that if destruction efficiency for this compound is known for a given facility, then all other compounds in the waste will be destroyed at least to that degree. This concept, therefore, involves developing an incinerability ranking of compounds.

The RCRA permit guidance for selecting POHCs in wastes actually employs this approach. However, the identification of compound incinerability has proved difficult, and possible unreliable EPA has suggested the use of compound heat of combustion (Hc) as a ranking of compound incinerability.[12] This ranking method has received considerable criticism, and alternative scales, which have also been criticized, have been proposed. These ranking approaches have been recently reviewed and compared by Dellinger et al.[133] They include autoignition temperature,[134] theoretical flame mode, kinetics,[135] experimental flame failure modes,[136] ignition delay time,[137] and gas phase (nonflame) thermal stability.[138] The rankings of compounds by each of these indices were compared to their observed incinerability in actual waste incineration tests in ten pilot- and field-scale units.[133] Each index failed to predict field results except for the nonflame thermal stability method. This method, based on experimentally determined thermal stability for mixtures of compounds under low oxygen concentration conditions, showed a statistically significant correlation for the compounds evaluated.

While the low oxygen thermal stability concept appears promising, data are available for only 28 compounds. Correlation for other important Appendix VIII compounds over a range of compound concentrations will be necessary before the method can be used reliably for POHC designation or as a basis for establishing continuous monitoring systems for specific surrogate compounds.

As a result of the uncertainty over incinerability rankings, the use of ''additives'' is being considered for overcoming the limitations of the single POHC compound approach. This concept involves the addition of a single, well-characterized compound or small group or ''soup'' of compounds to a waste stream, with subsequent continuous monitoring of the emissions of the compound(s) to serve as a measure of destruction performance. Compounds such as various freons[139,140] and sulfur hexafluoride (SF_6) have been proposed.[141-143] Conceptually, these types of materials would be ideal additives since they rarely occur in hazardous wastes, can be detected in emissions using on-line instruments, and are not likely to be formed as combustion by-products.

Combustion by-product formation has caused difficulty in interpretation of incinerability data for mixtures of conventional POHC candidate compounds.[141,144,145] While laboratory-scale studies have shown some promise, attempts in correlating field incinerator performance with additive behavior results have been inconclusive to date. Additional testing is needed.

B. PERFORMANCE INDICATORS

Carbon monoxide (CO) and total unburned hydrocarbons (TUHC) are emitted from all combustion systems in varying amounts. Because CO is the final combustion intermediate prior to the formation of CO_2 in the combustion process, it has been used in the determination of combustion efficiency. Unburned hydrocarbon emission values do not include all incompletely combusted hydrocarbons. Rather this is an instrumentation-derived value resulting from the passage of gaseous emissions through a hydrogen flame ionization detector (HFID), which is commonly used with gas chromatographs. THE HFID responds to the number of carbon-hydrogen and carbon-carbon bonds in residuals in the combustion gas. Because it does not respond to oxidized products such as O_2, CO, CO_2, and H_2O, it has been used as an indicator of residual fuel emissions.

Because CO is an indicator of the degree of completion of combustion and TUHC may be reflective of the amount of incompletely combusted material in the exhaust gas, these measures have been considered as possible indicators of incinerator performance. Continuous monitoring of CO is required by the RCRA incinerator standards for this reason. TUHC measurements, however, are not required.

The use of CO and TUHC in hazardous waste incineration has been studied by several groups[136,141,146,149] and criticized by others.[145,148] Watergland obtained pilot-scale data which indicated correlations of the fractional penetration of POHCs (1% DRE/100) with CO and THC.[146] Kramlich et al.[136] and LaFond et al.[147] found that increases in CO preceded increases in the penetration of POHCs in a laboratory-scale turbulent flame reactor as parameters such as air/fuel ratio, atomization, and degree of thermal quenching were varied. At the same time, TUHC tended to increase as POHC penetration increased. In a test of a pilot-scale circulating-fluidized bed combustor, Chang et al.[141] indicated penetration of combustion by-products without a corresponding increase in CO. The converse was not true, i.e., increases in CO were observed on some occasions without a corresponding increase in combustion by-product penetration. POHC destruction efficiency was high throughout this series of tests and did not appear to correlate well with either TUHC or CO. Daniels et al., although critical of the use of CO as a surrogate for POHC DRE or as an indicator of incinerator performance, presented data obtained from a full-scale rotary kiln, which in five out of six cases indicated increased POHC penetration with increased CO concentration.[148]

Analysis of the pooled data from the EPA incinerator test program revealed that there was no absolute level of mean combustion temperature, mean gas phase residence time, or carbon monoxide emission concentration which correlated with achieving 99.99% DRE.[79] Residence times ranged from 0.1 to 6.5 s in the facilities tested. Temperatures ranged from 648 to 1450°C. Carbon monoxide (CO) levels were as high as 600 ppm, but at most plants ranged from 5 to 15 ppm. It was concluded that the relationships between DRE and these parameters are, in all likelihood, facility specific and that waste characteristics, waste atomization, and combustion chamber mixing likely play equally important roles in achieving high DRE. Timing, funding, and facility constraints, however, did not allow for collection of sufficient performance data under varying conditions at each site tested to allow for such relationships to be quantified. In particular, few of the test conditions produced DRE significantly below 99.99%.

Dellinger and Hall[145] have suggested that one reason for difficulties in correlating CO with DRE is that the assumed rapid oxidation of hydrocarbons to CO may not be correct for complex hazardous wastes containing large halogenated and heteroatom molecules. For these wastes, formation of stable intermediate organic reaction products may delay the production of CO. This delay would tend to distribute or move the CO production maximum relative to fuel (waste) destruction efficiency and tend to negate the usefulness of CO measurement in the region of 99.99% DRE. Hall et al. have conducted laboratory studies of CO formation vs. compound destruction for several complex mixtures and found no correlation.[149]

C. PREDICTING PERFORMANCE

Based on current knowledge it would appear that no single performance indicator or surrogate is sufficient as a predictor of organic compound destruction in incinerators. While low oxygen thermal stability data show promise as a predictor of compound incinerability, the data base is still not sufficient to extend this concept to POHC selection or to the development of standard POHC soups for trial burns or compliance monitoring. Data on additives are also insufficient to project a DRE correlation. CO may be useful in setting an upper bound condition on compound penetration (1-DRE), but there is no demonstrated "correlation" between CO emissions and DRE. Elevated TUHC emissions are indicative of an increase in incomplete combustion by product emissions, but not necessarily a decline in DRE or even an increase in hazardous combustion by-product emission, in part, because the HFID is less sensitive to halocarbon compounds.

One of the additional limitations placed on attempts to correlate surrogates and indicators with DE is the lack of a significant data base on incinerator operation under failure conditions. A failure condition can be defined as a normal or accidental operational deviation which results in failure of the facility to achieve a 99.99% DRE. Most of the field incinerator data used to make TUHC and CO correlation have been taken under steady-state operating conditions. The impact of failure modes such as nozzle clogging and kiln overcharging on CO and TUHC emissions has not been adequately quantified, largely because of limitations on test time and funding and, in particular, permit constraints which prohibit off-design operation of facilities. EPA is, therefore, conducting failure-mode testing at its bench- and pilot-scale research facilities in Jefferson, AR, Cincinnati, OH, and Research Triangle Park, NC. Preliminary results from these tests have been reported recently.[150,151]

EPA has also conducted non-steady-state operational assessments at three boilers employing hazardous waste as a fuel.[152] The impact of typical non-steady-state operating conditions (e.g., start-up soot blowing, load change) on DRE, combustion by-product emissions, CO, and TUHC was studied. While elevated CO emissions were observed at two of the sites under off-design operation, attempts to correlate DRE with CO, NO_x, and O_2 emissions were unsuccessful, largely because 99.99% DRE was achieved under both good and off-design operation. The testing, however, acknowledges some of the difficulties in conducting off-design studies. In some cases, the duration of the process transient to be studied may be shorter than the sampling time required to collect a sufficient sample to assess DRE. The large volume and high surface area (boiler tubes) in boilers tended to delay emissions of organics from one off-design test to the next, making it difficult to separate cause and effect. This so-called ''hysteresis effect'' may also cause difficulty to some degree in interpreting the results of studies of transient operation in incinerators.

Most testing under non-steady-state is needed, particularly for incinerators. While attempts to correlate performance with single indicators and surrogates have been largely unsuccessful to date, taken in some appropriate combination they may prove useful as real-time indicators of the onset of process failure.

VII. ENVIRONMENTAL AND PUBLIC HEALTH IMPLICATIONS

Regardless of the apparent capabilities of hazardous waste incinerators to meet or exceed the RCRA performance standards, the ultimate public test involves demonstration that there is no unacceptable increase in public health risk from the emissions to the environment. While any of the emissions from an incinerator may potentially be of environmental interest, most attention has been directed toward air pollution emissions. This is because they appear to represent the most important source of off-site human exposure and because there is no opportunity for secondary containment or treatment of emissions once they leave the stack. Ash and scrubber residues, however, are lower in volume and can be contained, examined, and, if necessary, treated prior to discharge or disposal. In addition to chronic exposure to recurring emissions, there are also environmental and public health impacts which could result from potential single-event or catastrophic emissions at incineration facilities.

A. RISKS FROM SINGLE-EVENT EMISSIONS

As with any industrial facility, there are risks from potential accidents at incineration facilities, such as fires, explosions, spills of raw waste, and similar single-point events. These events are probabilistic in nature, and their evaluation in a risk assessment is handled differently from continuous pollutant emissions from stacks. For instance, the U.S. Department of Transportation maintains statistics on the frequency of releases of cargo from vehicular accidents involving trucks. For tank trucks of all types, for instance, this is estimated to be 0.35 releases per million miles traveled.[47] Similar values may be identified for accidents involving storage facilities and transfer operations.

Little specific information on these types of accidents is available for hazardous waste incineration facilities. Ingiwersen et al. evaluated the potential off-site impacts of five hypothetical accidents at a planned hazardous waste incineration facility.[153] It was estimated that no long-term adverse effects could be expected from chronic exposure for the nearest residents (0.5 mi) and that any effects due to acute exposures to the HC1 emissions from the accidents were expected to be short-term and reversible. Actual accidents at an operating European facility have been documented.[154] Seven accidents occurred over an 11-year period. One employee was injured, and no off-site effects were reported for any of the incidents, which generally involved storage and handling operations. EPA has also recently examined transportation and spill-related risk for ocean-based incineration and found that these risks were greater than the risk from incineration.[44] In the absence of accident data specific for incineration facilities, statistics from related industrial practice are probably adequate in assessing these risks.

B. METHODS FOR ASSESSING RISKS FROM RECURRING EMISSIONS

The major concern of this discussion is the risk associated with recurring air pollution emissions from incinerators. The assessment of risk to human health rather than environmental damage is generally believed to be of greatest interest. Four general steps are involved in assessing the impact on public health from stack emissions from an incinerator.[155]

- Identify the health effects of constituents of concern as a function of concentration level.
- Predict the concentrations of these constituents to which the public may be exposed.
- Estimate the health impact of these concentration exposures.
- Conduct an uncertainty analysis.

Identification of the constituents of concern in stack emissions and the health effects of these constituents is, of course, a function of the waste streams and incineration facility of interest. In general, any of the constituents on Appendix VIII of the RCRA standards are of possible interest. However, other organic compounds frequently found in combustion emissions (certain polynuclear aromatics and polycyclic aromatic compounds) may be of concern also. The major health effects of concern are for low-level chronic exposure to these materials. These effects are generally carcinogenicity, mutagenicity, teratogenicity, or target organ toxicity (e.g., sterility, behavioral effects).

Predicting the potential levels of human exposure to pollutants requires information on the frequency, intensity, duration, and continuity of exposure.[155] Exposure assessment generally requires the use of mathematical models which stimulate the transport and dispersion of emissions from the stack to the exposed population. Of the air dispersion models available, EPA has most often used the Industrial Source Complex Long Term Model (ISC-LT) for predicting annual average concentration for hazardous waste incinerator facility studies.[156,157] The Oak Ridge National Laboratory has linked the ISC-LT model with computerized meteorology and population data bases and programs to form the Inhalation Exposure Methodology (IEM).[158,159] The IEM employs U.S. population data from the 1980 Census and local meteorological data along with the ISC-LT to estimate air pollutant concentrations and human inhalation exposures in the vicinity of hazardous waste incinerators located anywhere in the U.S. The IEM has been used extensively by EPA in assessing regulatory alternatives for hazardous waste incinerators.[160,161]

The exposure information generated by models such as the IEM may then be employed to estimate human health risk. The individuals at highest risk of developing adverse health effects are of most interest. The risk to this "maximally exposed population" is estimated from the modeled exposure at the point of highest annual average pollutant ground-level concentration outside the facility. For each exposed individual, cancer risk is expressed as the cumulative risk over a 70-year (lifetime) period of continuous exposure.

A variety of estimators are available to quantify the health risks of substances. Carcinogen potency factors have been developed by EPA based on "assumed no safe" level.[162] For noncarcinogenic effects, "no observable adverse effect" levels (NOAELs) have been used to derive reference dose (RfD) levels.[163]

There is considerable uncertainty involved in conducting risk assessment. Numerous assumptions must be made regarding pollutant emission levels, pollutant effects, dispersion factors, etc. Only a fraction of the needed tests of the effects of chronic, low-level exposures to environmental pollutants have been done. There is also considerable uncertainty in extrapolating effects from high doses which cause effects in animals to low doses in humans. Linearity assumptions are typically used in making such extrapolations. Some investigators have questioned the wisdom of such assumptions, however.[164]

Beyond this, very little is known about how, or even if, this information can be used to estimate the effects of complex mixtures of the substances usually present in incinerator stack emissions. For these reasons and other limitations, most assessments adopt assumptions and risk-estimate values which produce an estimate of a worst-case effect. In order to promote consistency in risk assessments, EPA has recently published in the *Federal Register* a six-part guidance on risk and exposure assessment methodologies.[165] This guidance is an excellent resource to those conducting or evaluating risk-assessment studies.

C. OVERALL RISKS FROM LONG-TERM AIR POLLUTION EMISSIONS FROM HAZARDOUS WASTE INCINERATORS

Risk assessment and risk management have been used increasingly by industry and government over the past 10 years in evaluating control technology and regulatory options for managing hazardous waste.[166] The initial 1978 RCRA incineration standards, for instance, were almost entirely design and performance oriented. In the 1981 proposal, however, EPA incorporated risk assessment into what was called the best engineering judgment (BEJ) approach to regulating and permitting incinerators.[10] The operating and performance standards for incinerators were to apply to facilities unless a site-specific risk assessment indicated that a higher degree of control was necessary. This risk assessment proposal, however, was not included in the final rule in 1982, largely because of concern from the regulated community over the uncertainty of risk assessment approaches. Rather, risk assessment and cost-benefit analysis became a more intergral part of the development of hazardous waste control technology standards through the conduct of Regulatory Impact Analyses (RIA) of all proposed standards as required by Executive Order 12291.

A number of risk assessments have been conducted for specific hazardous waste incinerators and for incineration on a national basis. Using the results of emissions data from nine full-scale incinerator tests,[79] EPA conducted a risk assessment as part of its Incinerator RIA in 1982. The objective was to examine the economic impact of the regulations on the regulated community and to estimate the health and environmental effects of the regulations.[167] The risks due to principal organic hazardous constituents (POHC), combustion by products (PICs), and metal emissions were developed (Table 16).

While these results show that the human health risks from most incinerator emissions are low, risks from metal emissions show the greatest potential for exceeding a 10^{-6} cancer risk. The risks from metal emissions ranged up to two to six orders of magnitude higher than values for POHCs and PICs and dominated the total risk values. Risks from residual POHC and PIC from the incinerators tested were low and were essentially equivalent.

TABLE 16
Total Excess Lifetime Cancer
Risk to the Maximum Exposed
Individual for Incinerator
Emissions[167]

POHCs	10^{-7}—10^{-10}
PICs	10^{-7}—10^{-11}
Metals	10^{-5}—10^{-8}
Total	10^{-5}—10^{-8}

Taylor et al. reported the results of a risk assessment for metal emissions using the same test data,[168] but employing somewhat different assumptions. Using the IEM methodology, carcinogenic and noncarcinogenic risks were examined. Interestingly, these results showed even lower cancer risks than the EPA study. Individual lifetime cancer risks for the maximum exposed population ranged from a low of 4.48×10^{-6} for chromium. Noncarcinogenic risks were also small. All values were well below the respective ADI (acceptable daily intake) values. Lead intake was highest, estimated at 2% of the ADI.

Kelly reported similar conclusions for a risk assessment of stack emissions from a hazardous waste incinerator in Biebesheim, West Germany.[169] Maximum ground level air concentrations for 24 metals (and for PCB) were estimated using the IEM. All levels (including PCB) were less than 2% of the corresponding continuous exposure limit (CEL) value.

Holton et al. examined the significance of various exposure pathways for air pollution emissions from three sizes of land-based incinerators located at three hypothetical sites in the U.S.[170-172] For certain organic chemicals, the food chain pathway may be an important contributor to total human exposure. However, the study concluded that the human health risk from emissions was small for all of the chemicals studied irrespective of the exposure pathway.

Fugitive emissions from auxiliary facilities at incinerators (e.g., storage tanks) were also estimated to be an important contributor to total pollutant emissions.[171,172] Few studies have quantified fugitive emission levels at incinerators. The studies which have been done, however, have not shown that ambient levels are a cause for concern.[79,173,174]

The risks associated with incineration of hazardous wastes at sea have been recently compared to risks from land-based incineration.[47] While risks of marine and terrestrial ecological damage were estimated, the direct human health risks from stack emissions are of greatest interest in this discussion (Table 17). The incremental cancer risk to the most exposed individual was determined for POHCs, PICs, and metals for two scenarios: a PCB waste and an ethylene dichloride (EDC) waste. Not surprisingly, the human health risk of stack emissions from ocean incineration were less than those of land-based systems, largely due to distance from population. The land-based incinerator risk values were similar to those estimated in the EPA incinerator RIA.

POHC and PIC releases showed low risk, generally one to five orders of magnitude less than those for metals. Risks from metals accounted for from 90% to almost all of the identified risk from either system, and exceeded the 10^{-6} risk level for only the land-based scenario. The study notes, however, that the assumptions used in the assessment overstate the likely levels of carcinogenic metals in the hypothetical wastes used in the assessment and, therefore, likely overestimate emissions and risk level.

In another study, Holton[175] compared the potential differences in human exposure to emissions from identical PCB incinerators in ocean and land-based contests. Land-based incineration showed higher inhalation exposure. The only human exposure pathway considered for at-sea incineration was ingestion of contaminated fish and shellfish. Accidental

TABLE 17
Incremental Cancer Risk to the Most Exposed
Individual by Type of Stack Release[47]

Systems	PCB waste	EDC waste
Ocean-bases		
POHCs	1.45×10^{-10}	5.51×10^{-10}
PICs	1.68×10^{-12}	3.36×10^{-9}
Metals	6.37×10^{-7}	1.06×10^{-6}
Total stack	6.37×10^{-7}	1.06×10^{-6}
Land-based (two sites)		
POHCs	5.13×10^{-6}	1.43×10^{-7}
PICs	1.79×10^{-6}	2.59×10^{-8}
Metals	2.65×10^{-5}	3.12×10^{-5}
Total stack	2.74×10^{-5}	3.14×10^{-5}

spills were not assessed. Inhalation exposure for land units was two orders of magnitude higher, terrestrial food chain ingestion exposure was a factor of two higher, and drinking water ingestion exposure was estimated to be about the same as that for consuming fish and shellfish. No estimate of absolute risk was attached to any of the estimates.

All of these risk assessment studies point to a conclusion that stack emissions from incineration of hazardous waste pose little risk to human health. However, as previously stated, the emissions data base on which many of the assessments were based has been criticized by the EPA Science Advisory Board (SAB) as being insufficient.[46] SAB has recommended that a more complete assessment of the quantity and physical and chemical character of incineration emissions be done to provide a basis for a more complete risk assessment than has been possible to date. The SAB points to the fact that only a portion of the organic mass emissions has been identified in past studies. Many believe that most of this unidentified mass in nonchlorinated C_1-C_5 hydrocarbons, which are of little concern from a risk standpoint. However, test data are only just beginning to emerge to confirm or disprove this belief. EPA has recently completed a full-scale incinerator emissions test where as much of the mass emissions as possible will be specifically identified. The testing included steady-state and typical upset conditions for a large rotary kiln incinerator. Results should be available by the summer of 1987.[176]

It is not clear, however, that even this level of emission information will really answer the question of how much absolute risk is associated with incinerator emissions. From the standpoint of the lay public, it may be more useful and productive to compare these emissions with other types of combustion emissions whose risks we have accepted in daily life. Lewtas[177] and others have done interesting work on the comparative cancer potency of complex mixtures of pollutants (e.g., power plant emissions, automobile exhaust, cigarette smoke). Using short-term bioassays of organics attracted from actual emissions, the relative cancer potency emissions have been estimated.

Comparative mutagenic emissions rates (expressed as revertants per mile or joule) have been determined from testing of mobile sources and stationary sources. Experimental work to date suggests that the mutagenic emission rates of wood stoves, for instance, are as much as four orders of magnitude greater than those for conventional coal-fire utility power plants.[177] It is also apparent from the data base that variations in organic emission rate affect the net carcinogenic emission rate more than variations in the carcinogenic or mutagenic potency of the emissions. Similar sampling and testing is needed for incineration emissions so that their potency can be compared to everyday sources such as wood stoves, oil furnaces, and utility power plants. It should be noted, however, that use of comparative mutagenic emission rates alone does not account for variation in potential human health impact that occurs due to differences in exposure level to emission from sources of different types or as a result of different routes of exposure.

VIII. CONCLUSIONS

The body of knowledge concerning hazardous waste incineration has been expanding rapidly since 1980. This review has examined some of the most significant aspects of this information. A number of conclusions may be drawn on the status of incineration technology, current pratice, monitoring methods, emmissions and performance, and public health risks. Beyond these, a number of remaining issues and research needs can also be identified.

Based on this review, the following conclusions may be drawn:

1. Incineration is a demonstrated, commercially available technology for hazardous waste disposal. Considerable design experience exists, and design and operating guidelines are available on the engineering aspects of the systems.

2. A variety of process technologies exist for the range of hazardous wastes appropriate for thermal destruction. The most common incinerator designs incoroporate one of four major combustion chamber designs: liquid injection, rotary kiln, fixed hearth, or fluidized bed. The most common air pollution control system involves combustion gas quenching followed by a Venturi scrubber (for particulate removal), a packed tower absorber (for acid gas removal), and a mist eliminator. However, more than half of the existing incinerators employ no air pollution control equipment at all.

3. Uncertainty exists regarding the exact scope of current hazardous waste incineration practice in the U.S. The best information is available for 1983, when between 1.0 and 2.7 MMT of hazardous waste is believed to have been incinerated in 208 incineration units across the U.S. As much as 47 MMT of incinerable waste was generated in 1983.

4. Implementation of HSWA and SARA as well as industrial concerns for limiting long-term environmental instability will encourage an increasing amount of hazardous waste to be directed in incineration facilities. While current capacity appears adequate, a near-term shortfall in commercial incineration capacity may develop, particularly for facilities which can handle hazardous waste sludges and solids.

5. The technology of stack sampling for trace organic compounds is relatively sophisticated. Considerable experience and attention to quality assurance and quality control are needed. Documented sampling and analysis methods are available for most of the parameters interest in incineration performance assessment. Methods have been validated for a number of compounds. With proper planning of test activites, detection limits are not a limiting factor in assessing incinerator performance.

6. Continuous emission monitors are available with adequate operating ranges for many of the combustion emissions of interest (CO, CO_2, O_2, TUHC, NO_x). However, continuous monitors for specific organic compounds are not available. No real-time monitor exists for measuring destruction and removal efficiency.

7. Incinerators and most industrial processes employing hazardous waste as a fuel can attain the RCRA destruction and removal efficiency requirment and the HCl emission limit.

8. Certain incinerators have had difficulty achieving the RCRA particulate matter emission limit of 180 mg/m^3. Data suggest that improved air pollution control technology or operating practices will enable these facilities to be upgraded to meet the standard, however.

9. Insufficient data are available on the fate of heavy metals in incineration systems and the efficiency of typical hazardous waste incinerator air pollution control equipment to control emissions of specific metals and their salts.

10. Insufficient data are available on the impact of typical upset or off-design operating conditions on incinerator and industrial furnace emissions.

11. Considerable uncertainty exists over the definition and significance of incomplete combustion by-products. An insufficient data base exists on the full spectrum of potentially hazardous compounds which may be in incinerator emissions. Comparative emissions data for a limited number of compounds suggest that incinerator emissions are similar in character and emission rate to emissions from fossil fuel combustions.

12. Based on current data, chlorinated dioxin and furan emissions are not significant for hazardous waste incinerators. The most hazardous dioxin isomer (2,3,7,8-TCDD) has not been detected in emissions at 17 facilities tested for these compounds. Levels of all PCDD and PCDF emissions from hazardous waste incinerators are approximately three orders of magnitude less than those reported for municipal waste incinerators.

13. Limited data on incinerator ash and air polution control residues suggest that organic compound levels are low and that destruction is the primary reason for high destruction and removal efficiencies, not removal. Metal concentration in ash and residues vary widely, depending on metal input rate to the incinerator and process operation (e.g., scrubber water recycle and make-up rates).

14. General process control systems and strategies exist to control incinerator performance. However, none of the available real-time monitoring performance indicators appear to correlate with actual organic compound DRE. No correlation between indicator emissions of CO or TUHC and DRE has been demonstrated for field-scale incinerator operations, although CO may be useful as an estimator of a lower bound of acceptable DRE performance. It may be that combinations of several potential real-time indicators (CO, TUHC, surrogate compound destruction) may be needed to predict and assure incinerator DRE performance more accurately on a continuous basis.

15. Available approaches for estimating compound incinerability have not been correlated with field experience. The best approach appears to involve the use of experimentally derived nonflame thermal stability data. However, an insufficient data base is currently available to extend this method for use in revised guidance on POHC selection.

16. There appears to be little increased human health risk from hazardous waste incinerator emissions, based on assessments done to date. Metal emissions appear to be most significant in the risk values which have been derived. However, a complete assessment of all of the potentially hazardous materials in incinerator emissions has not been completed. This information is needed to enable a comprehensive risk assessment of incinerator emissions.

17. In spite of the demonstrated destruction capabilities of hazardous waste incinerators and the apparent low incremental risk of emissions, there is considerable public opposition to the siting and permitting of these facilities. Permits require 3 years to finalize, on average. Uncertainty over permitting and public acceptance will likely result in a near-term shortfall in needed capacity, particularly for commercial facilities which could incinerate solids and sludges.

IX. REMAINING ISSUES AND RESEARCH NEEDS

While thermal destruction represents the most effective and widely applicable control technology available today for organic hazardous waste, a number of issues remain concerning its use in the long term. These include:

- Destruction effectiveness on untested and unique wastes
- Detection of process failure
- Control of heavy metal emissions

- Emissions of combustion by-products
- Real-time performance assurance
- The role on innovative technology

A. DESTRUCTION EFFECTIVENESS ON UNTESTED AND UNIQUE WASTES

All of the performance data which have been used in the development and assessment of thermal destruction regulations and standards to date have been collected for waste/thermal technology combinations typical of current pratice. However, the character of wastes which may be subjected to incineration in the near future will begin to change, perhaps dramatically. These changes will be influenced by EPA action to restrict many wastes from land disposal under the Hazardous and Solid Waste Act Amendments of 1984 and by increased emphasis on remedial action at Superfund sites. Incineration will emerge as a feasible technological alternative for destruction of many of these wastes and site cleanup residues. However, EPA and industry will have considerably less experience in handling these wastes. Wastes will tend to have higher solid and water content, be more complex in their physical and chemical composition, have lower heating value, and/or potentially contain higher levels of hazardous metals and high hazard organics compared to wastes which are typically incinerated today.

Consequently, while incineration is capable of achieving high levels of destruction for the wastes of today, future practice may place new performance on current technology. For instance 99.9999 % DRE is now required for wastes bearing chlorinated dioxin and furan compounds. Many of these wastes appear at Superfund sites. EPA is also studying the need for heavy metals control regulations for incinerators. That effort may suggest the need for improved air pollution control systems or waste pretreatment.

Thus, performance testing of incinerators and other thermal destruction devices must continue in order to assure destruction and removal effectiveness for these untested wastes, to assess process limitations and waste pretreatment requirements, to determine the safety of process residues, and to improve our ability to predict incinerator performance on new waste materials.

B. CONTROL OF HEAVY METAL EMISSIONS

While the human health risk of incinerator emissions appears to be small, metal emissions have been the dominant component of the risk levels identified thus far. Metal emissions are controlled only indirectly by current standards through the RCRA incinerator particulate emission limit. The particulate standard, however, has proved difficult for a number of operating facilities to achieve. In addition, the metal content of wastes which may be subjected to incineration in the future may be higher as a result of the implementation of the HSWA land disposal restrictions and increased ultimate cleanup actions at Superfund sites. Consequently, while metal emissions may not pose a risk now, they may in future practice.

Insufficient data exist on the physical and chemical character of particulate matter generated by hazardous waste incineration systems. Few tests have examined the particle size distribution of emissions or the specific metal removal capability of the various air pollution control systems available. Likewise, insufficient data exist on the fate or partitioning of these materials in incineration systems. This information is needed to examine the potential impact of metals in wastes on net environmental emissions and to evaluate various regulatory strategies which may be necessary to control increased emissions. These strategies could include metal input limits for waste or specific metal emission limits.

C. EMISSIONS OF COMBUSTION BY-PRODUCTS

Current information suggests that organic combustion by-product emissions identified by incineration of hazardous waste do not represent a significant risk to public health. Some, however, have questioned the completeness of emissions data and, therefore, the adequacy of risk assessments performed using these data. This issue has emerged as a concern in numerous public meetings on incinerator permits and facility siting.

There is little doubt that none of the emissions testing efforts conducted to date has identified all compounds in incinerator stack emissions. The same is true, however, for virtually any other source of air or water pollution. Because hazardous waste facilities are perceived as being more hazardous than many other types of pollution control or industrial facilities, more attention is given to their emissions. Thus, while it is unlikely that any major, highly hazardous components of emissions have been overlooked, the data are not available to prove this to all who may be concerned. On the other hand, the task of finding all potentially hazardous compounds is an open-ended one, ultimately limited by expense.

Another issue concerning combustion by-products emissions is that few tests have examined the level, and the chemical character of emissions for periods of time facilities may be operating under upset conditions (transients or failure modes). More experimental work is clearly needed here. Some testing has shown that there is little change in POHC DRE over significant operating ranges or under "apparent" failure conditions. At the same time, emissions of unburned hydrocarbons have increased. Research is necessary to determine if these "failure-mode" emissions pose a hazard.

One approach to resolving both the question of data completeness and failure mode impacts is to examine the relative potency of emissions using short-term bioassays and to use bioassay-directed chemical analysis as means of more cost-effectively identifying the chemical compounds (perhaps previously unidentified) which are primarily causing the potency. While short-term bioassays have their own set of constraints and limitations, they have proved useful in comparing the cancer potency of mixtures of compounds from other combustion sources. Testing of a reasonable range of hazardous waste types under "good" and off-design conditions would give an indication of the range of potency of emissions as a function of operational conditions and in comparison with conventional combustion sources whose risks and character are more familiar to the general public.

D. REAL-TIME PERFORMANCE ASSURANCE

Once the public health significance of incinerator emissions is verified, methods must be available to assure that effective operation is maintained. A variety of surrogates and indicators of incinerator performance are being evaluated. None is fully satisfactory, and little evaluation has been done under true failure conditions.

It may not be possible to find a set of easily monitored parameters which "correlate" with incinerator performance. However, it may be possible to identify parameters which may be sufficient to identify the one set of process failure. Research is necessary to examine the suitability of existing real-time monitoring systems and approaches to predict process failure reliably. Availability of such techniques may have a significant impact on public acceptance of these facilities and form a technical basis for more effective compliance monitoring by regulatory agencies.

E. ROLE OF INNOVATIVE TECHNOLOGY

A wide range of innovative hazardous waste technology has emerged since the passage of RCRA.[178] A number of these technologies are thermal destruction processes. The potential destruction capabilities and cost-effectiveness of these processes has been well publicized, although many of the techniques must be considered to be only in the developmental stage.

Many argue, particularly in public hearings, that decisions on permitting conventional incineration facilities should be postponed in favor of adopting more innovative approaches, whose inventors often claim higher destruction efficiency at lower cost than conventional systems. For specific waste streams (e.g., contaminated soils, PCBs), a number of innovative systems have demonstrated DREs equivalent to those of conventional systems. Some systems appear to offer advantages for handling specific (although sometimes limited) waste streams. Considerable uncertainty exists as to the true cost-effectiveness of some systems, since practical field experience is often not yet available to aid in identifying operating limitations.

Many of these emerging systems will find a role in future hazardous waste management strategies. Policy makers, public officials, and industrial decisions makers should be careful, however, in delaying action on currently available, demonstrated, thermal destruction systems until the need, benefit, and operability of such innovative systems are clearly established.

REFERENCES

Background
1. **White, D. C.**, TEA program for treatment alternatives for hazardous wastes, *J. Air Pollut. Control Assoc.*, 34, 369, 1985.
2. **Oppelt, E. T.**, Hazardous waste destruction, *Environ. Sci. Technol.*, 20, 312, 1986.
3. U.S. EPA, Assessment of Incineration as a Treatment Method for Liquid Organic Hazardous Wastes—Background Report V; Public Concerns Regarding Land-Based and Ocean-Based Incineration, 1985-526-778-/30376, U.S. Government Printing Office, Washington, D.C., March 1985.
4. *The Bible* (Old Testament), Kings 23:10.
5. **Hering, R. and Greelly, S. A.**, *Collection and Disposal of Municipal Refuse*, McGraw-Hill, New York, 1921.
6. **Corey, R. C.**, *Principles and Practices of Incineration*, John Wiley & Sons, New York, 1969.
7. **Cross, F. L. and Hesketh, H. E.**, *Controlled Air Incineration*, Technomic, Lacaster, PA, 1985.
8. U.S. EPA, Hazardous Waste Proposed Guidelines and Regulations and Proposal on Identification and Listing, *Fed. Regist.*, 43(243, Part IV), 58946, 1978.
9. U.S. EPA, Hazardous Waste Management System: Identification of Listing of Hazardous Waste, *Fed. Regist.*, 45(98, Part III), 33084, 1980.
10. U.S. EPA, Incinerator Standards for Owners and Operators of Hazardous Waste Management Facilities; Interim Final Rule and Proposed Rule, *Fed. Regist.*, 46(15, Part IV), 7666, 1981.
11. U.S. EPA, Standards Applicable to Owners and Operators of Hazardous Waste Treatment Facilities, *Fed. Regist.*, 47(122, Part V), 27516, 1982.
12. U.S. EPA, Guidance Manual for Hazardous Waste Incinerator Permits, EPA-SW-966, NTIS No. PB84-100577, National Technical Information Service, Washington, D.C., 1983.
13. U.S. EPA, Hazardous Waste Management System; Burning of Waste Fuel and Used Oil Fuel in Boilers and Industrial Furnaces, *Fed. Regist.*, 50(230), 49164, 1985.
14. **Holloway, R.**, U.S. EPA, Office of Solid Waste, personal communication, January 1987.
15. U.S. EPA, Polychlorinated Biphenyls (PCBs) Manufacturing, Processing, Destruction in Commerce and Use Prohibitions, *Fed. Regist.*, 44, 31514, 1979.

Current Incineration Practice
16. U.S. EPA, Dioxin Rule, *Fed. Regist.*, January 14, 1985.
17. **Vogel, G. et al.**, Composition of Hazardous Waste Streams Currently Incinerated, Mitre Corp., U.S. EPA, April 1983.
18. **Tejada, S.**, The blue goose flies!, *EPA J.*, October 1985.
19. Camp, Dresser and McKee, Inc., Superfund Treatment Technologies: A Vendor Inventory, U.S. EPA 540/2-80/004, July 1986.
20. U.S. EPA, Hazardous Waste Generation and Commercial Hazardous Waste Management Capacity, SW-894, December 1980.
21. **Vogel, G. et al.**, Incineration and cement kiln capactiy for hazardous waste treatment, in *Proc. 12th Annu. Res. Symp.*, Incineration and Treatment of Hazardous Wastes, Cincinnati, OH, April 1986.
22. U.S. EPA, A Profile of Existing Hazardous Waste Incineration Facilities and Manufacturers in the United States, NTIS PB 84-157072, National Technical Information Service, Washington, D.C., February 1984.
23. U.S. EPA, Assessment of Incineration as a Treatment Method for Liquid Organic Hazardous Wastes — Background Report. III. Assessment of the Commercial Hazardous Waste Incineration Market, 1985-526-778/30376, U.S. Government Printing Office, Washington, D.C., March 1985.

24. U.S. EPA, National Survey of Hazardous Waste Generator and Treatment, Storage and Disposal Facilities Regulated Under RCRA in 1981, 055000-0023908, U.S. Government Printing Office, Washington, D.C., 1984.

25. U.S. Congress, Congressional Budget Office, Hazardous Waste Management: Recent Changes and Policy Atlernatives, U.S. Government Printing Office, Washington, D.C., 1985.

26. U.S. Congress, Office of Technology Assessment, Technologies and Management Strategies for Hazardous Waste Control, OTA-196, U.S. Government Printing Office, Washington, D.C., March 1983.

27. Environmental Resources Management, Inc., New Jersey Hazardous Waste Facilities Plan, New Jersey Waste Facilities Siting Commission, Trenton, NJ, March 1985, 170.

28. ICF, Inc., Survey of Selected Firms in the Commercial Hazardous Waste Management Industry: 1984 Update, U.S. EPA, September 1985.

29. Office of Appropriate Technology, Alternatives to the Land Disposal of Hazardous Wastes, An Assessment for California, Toxic Waste Assessment Group, Governor's Office of Appropriate Technology, Sacramento, CA, 1981.

30. U.S. Congress, Office Of Technology Assessment, Ocean Incineration: Its Role in Managing Hazardous Waste, OTA-0-313, U.S. Government Printing Office, Washington, D.C., August 1986.

31. U.S. EPA, 1986 National Screening Survey of Hazardous Waste Treatment, Storage, Disposal and Recycling Facilities, December 1986.

32. U.S. EPA, Engineering Handbook on Hazardous Waste Incineration, SW-889, NTIS PB 81-248163, National Technical Information Service, Washington, D.C., September 1981.

33. **Niessen, W. R.,** *Combustion and Incineration Processes,* Marcel Dekker, New York, 1978.

34. **Brunner, C. A.,** *Incineration Systems and Selection and Design,* Van Nostrand Reinhold, New York, 1984.

35. **Ackerman, D. G., McGaughey, J. F., and Wagoner, D. E.,** At-Sea Incineration of PCB-Containing Wastes on Board the M/T Vulcanus, 600/7-83-024, U.S. EPA, April 1983.

36. **Kiang, Y. H.,** Total hazardous waste disposal through combustion, *Ind. Heat.,* 14, 9, 1977.

37. **Kiang, Y. H., and Metry, A. A.,** *Hazardous Waste Processing Technology,* Ann Arbor Science, Ann Arbor, MI, 1982.

38. **Reed, R. D.,** *Furnace Operations,* 3rd Ed., Gulf Publishing, Houston, 1981.

39. **Rickman, W. S., Holder, N. D., and Young, D. T.,** Circulating bed incineration of hazardous wastes, *Chem. Eng. Prog.,* 81, 34, 1986.

40. **Trenholm, A.,** Summary of testing program at hazardous waste incinerators, in Proc. 11th Annu. Res. Symp. *Incineration and Treatment of Hazardous Waste,* EPA/600/-9-83/028, September 1985.

41. **Santoleri, J. J.,** Energy recovery — a by-product of hazardous waste incineration systems, in Proc. 15th Mid-Atlantic Industrial Waste Conf. Toxic and Hazardous Waste, 1983.

42. **Novak, R. G., Troxier, W. L., and Dehnke, T. H.,** Recovering energy from hazardous waste incineration, *Chem. Eng. Prog.,* 91, 146, 1984.

43. **Moller, J. J., and Christiansen, O. B.,** Dry scrubbing of hazardous waste incinerator flue gas by spray dryer adsorption, in Proc. 77th annual APCA Meeting, 1984.

44. A.D. Little, Inc., Overview of Ocean Incineration, U.S. Congress, Office of Technology Assessment, May 1986.

45. **Ackerman, D. G., Jr. and Venezia, R. A.,** Research of at-sea incineration in the United States, *Wastes Ocean,* 5, 1982.

46. U.S. EPA, Science Advisory Board, Report on the Incineration of Liquid Hazardous Wastes, April 1985.

47. U.S. EPA, Assessment of Incineration as a Treatment Method of Liquid Organic Hazardous Waste Background Report IV, Comparison of Risks from Land-Based and Ocean-Based Incinerators, March 1985.

48. **Yezzi, J. J., et al.,** Results of the initial trail burn of the EPA-ORD mobile incineration system, in Proc. ASME 1984 National Waste Processing Conference, Orlando, FL, June 1984.

49. **Dietrich, G. N. and Martini, G. F.,** Today's technology offers on-site incineration of hazardous waste, *Hazardous Mater. Waste Manage.,* January-February, 1986.

50. **Hatch, N. N. and Hayes, E.,** State-of-the-art remedial action technology used for the Sydney mine waste disposal site cleanup, in Proc. 5th Natl. Conf. Management of Uncontrolled Hazardous Waste Sites, Washington, D.C., November 1984.

51. **Noland, J. and Sisk, W. E.,** Incineration of explosives contaminated soils, in Proc. 5th Natl. Conf. Management of Uncontrolled Hazardous Waste Sites, Washington, D.C., November 1984.

52. **Vrable, D. L. and Engler, D. R.,** Transportable circulating bed combustor for the incineration of hazardous waste, in Proc. 6th Natl. Conf. in Management of Uncontrolled Hazardous Waste Sites, Washington, D.C., November 1985.

53. Westat, Inc., Used or Waste Oil and Waste-Derived Fuel Material Burner Questionnaire for 1983, U.S. EPA, 1984.

Measuring Process Performance

54. U.S. EPA, Guidance Manual for Hazardous Waste Incinerator Permits, March 1983.
55. **Gorman, P. et al.**, Practical Guide — Trail Burns for Hazardous Waste Incineration, U.S. EPA, NTIS PB 86-190246/AS, National Technical Information Service, Washington, D.C., November 1985.
56. **Ackerman, D. G., et al.**, Guidelines for the Disposal of PCB's and PCB Items by Thermal Destruction, U.S. EPA, May 1980.
57. **Harris, J. C., Larsen, D. J., Rechsteiner, C. E., and Thrun, K. E.**, Sampling and Analysis Methods for Hazardous Waste Combustion, EPA-600/8-84-002, PB 84-155846, U.S. EPA, Februray 1984.
58. U.S. EPA, CRF 40, Part 268, Appendix I — Toxicity Characteristic Leaching Procedures, *Fed. Regist.*, 51(216), 40643, 1986.
59. U.S. EPA, Test Methods for Evaluating Solid Waste, Physical/Chemical Methods, 2nd Ed., SW-846, July 1982.
60. **deVera, E. R., Simmons, B. P., Stephens, R. D., and Storm, D. L.**, Samplers and Sampling Procedures for Hazardous Waste Streams, EPA-60/2-80-018, PB 80-135353, U.S. EPA, January 1980.
61. **James, R. H., Adams, R. E., Finkel, J. M., Miller, H. C., and Johnson, L. D.**, Evaluation of analytical methods for the determination of POHC in combustion products, *J. Air Pollut. Control Assoc.*, 35, 959, 1985.
62. **Adams, R. E., James, R. H., Farr, L. B., Thomason, M. M., Miller, H. C., and Johnson, L. D.**, Analytical methods for determination of selected principal organic hazardous constituents in combustion products, *Environ. Sci. Technol.*, 20, 711, 1986.
63. **Bond, D. H.**, At-sea incineration of hazardous waste, *Environ. Sci. Technol.*, 18, 5, 1984.
64. **Stern, D. A., Myatt, B. M., Lachowshi, J. F., and McGregor, K. T.**, Specification of halogen and hydrogen halide compounds in gaseous emissions, in Proc. 9th Annu. Res. Symp. Incineration and Treatment of Hazardous Wastes, EPA-600/9-84-015, PB 84-234525, U.S. EPA, July 1984.
65. **Schlickenrieder, L. M., Adams, J. W., Thrun, K. E.**, Modified Method 5 Train and Source Assessment Sampling System Operators Manuel, EPA-600/8-85-003, PB 85-169878, U.S. EPA, February 1985.
66. **Jungclaus, G. A. et al.**, Development of a volatile organic sampling train (VOST), in Proc. 9th Annu. Res. Symp. on Land Disposal, Incineration, and Treatment of Hazardous Waste, EPA-600/9-84-015, PB 84-234525, July 1984.
67. **Hansen, E. M.**, Protocol for the Collection and Analysis of Volatile POHC's Using VOST, EPA-600/8-84-007, PB 84-170042, U.S. EPA, March 1984.
68. **Johnson, L. D. and Merrill, R. G.**, Stack sampling for organic emissions, *Toxicol. Environ. Chem.*, 6, 109, 1983.
69. **Johnson, L. D.**, Detecting waste combustor emissions, *Environ. Sci. Technol.*, 20, 223, 1986.
70. **Prohoska, J., Logan, T. J., Fuerst, R. G., and Midgett, M. R.**, Validation of the Volatile Organic Sampling Train (VOST) Protocol. Vol. 2. Field Validation Phase, EPA-600/4-86/014a, PB 86-145547, U.S. EPA, January 1986.
71. **Prohoska, J., Logan, T. J., Fuerst, R. G., and Midgett, M. R.**, Validation of the Volatile Oragnic Sampling Train (VOST) Protocol. Vol. 2. Field Validation Phase, U.S. EPA, EPA-600-4-86/014b, PB 86-145554, January 1986.
72. **Bursey, J., Hartman, M., Homolya, J., McAllister, R., McGaughey, J., and Wagoner, D.**, Laboratory and Field Evaluation of the Semi-Vost Method. Vol. 1, EPA-600/4-85-075a, PB 86-123551/AS; Vol. 2, EPA-600/4-85-075b, PB 86-123569/AS. U.S. EPA, September 1985.
73. U.S. EPA, Guidance on Trial Burn Reporting and Setting Permit Conditions, draft report, November 1986.
74. **Podlenski, J., Peduto, E., et al.**, Feasibility Study for Adapting Present Combustion Source Continuous Monitoring Systems to Hazardous Waste Incinerators, EPA-600/8-84-011a, U.S. EPA, March 1984.

Emissions from Hazardous Waste Incineration

75. U.S. EPA, Final Standards for Hazardous Waste Incineration-Background Document, January 1981.
76. U.S. EPA, Determination of Incineration Operating Conditions Necessary for Safe Disposal of Pesticides, July 1975.
77. U.S. EPA, Destroying Chemical Wastes in Commercial-Scale Incinerators, 1977.
78. **Oppelt, E. T.**, Perfomrance Assessment of Incinerators and High Temperature Industrial Processes Disposing Hazardous Waste in the U.S., in *Hazardous and Industrial Solid Waste Testing and Disposal*, ASTM Standard Publ. 933, American Society for Testing and Materials, Philadelphia, 1986.
79. U.S. EPA, Performance Evaluation of Full-Scale Hazardous Waste Incineration, Vol. 1—5, PB 85-129500, National Technical Information Service, November 1984.
80. U.S. EPA, Engineering Assessment Report — Hazardous Waste Co-Firing in Industrial Boilers, Vol. 1 and 2, June 1984.
81. U.S. EPA, A Technical Overview of the Concept of Disposing of Hazardous Wastes in Industrial Boilers, June 1984.

82. U.S. EPA, Draft Summary Report on Hazardous Waste Combustion in Calcining Kilns, March 1985.
83. U.S. EPA, Evaluation of Hazardous Waste Incineration in Aggregate Kilns, February 1985.
84. U.S. EPA, Permit Writers Guide to Test Burn Data — Hazardous Waste Incineration, EPA-625/6-86-012, September 1986.
85. **Trenholm, A., Gorman, P.G., Smith, B., and Oberacker, D. A.,** Emission Test Results for a Hazardous Waste Incineration RIA, EPA-600/9-84/015, U.S. EPA, 1984.
86. **Gorman, P. G., Trenhom, A., Oberacker, D. A., and Smith, B.,** Particulate and Hydrogen Chloride Emissions from Hazardous Waste Incinerator, EPA-600/9-84/015, U.S. EPA, 1984.
87. **Mournighan, R. E., Peters, J. A., and Branscome, M. R.,** Effects of Burning Hazardous Wastes in Cement Kilns on Conventional Pollutant Emissions, in Proc. 78th APCA Annu. Meet., Vol 5, 1985, 85-86.1.
88. **Olexsey, R. A.,** Air emission from industrail boilers burning hazardous waste material, in Proc. 77th APCA Annu. Meet., Vol. 1, 1984, 84-183.
89. **Edwards, L. et al.,** Trace Metals and Stationary Conventional Combustion Processes, Vol. 1. Technical Report, EPA-600/7-80-155a, U.S. EPA, 1980.
90. **Kaakinen, J. et al.,** Trace elements behavior in coal-fired power plants, *Environ. Sci. Technol.,* 9, 862, 1975.
91. **Klein, D. et al.,** Pathways of thirty-seven trace elements through coal-fired power plant, *Environ. Sci. Technol.,* 9, 973, 1975.
92. **Smith, R. D.,** The trace element chemistry of coal during combustion and the emissions from coal-fired plants, *Prog. Energy Combustion Sci.,* 6, 53, 1980.
93. **Gorman, P. G. and Ananth, K. P.,** Trail Burn Potocol Verification at a Hazardous Waste Incinerator, EPA-600/2-84/048, U.S. EPA, February 1984.
94. **Clausen, J. F., Honson, R. J., and Zee, C. A.,** Destroying Chemical Waste in Commercial-Scale Incinerators. Facility Report No. 1 — The Marquardt Company, PB 265-541, U.S. EPA, April 1977.
95. **Ackerman, D. G., Clausen, J. F., Johnson, R. J., and Zee, C. A.,** Destroying Chemical Waste in Commercial-Scale Incinerators. Facility Report No. 3 — Systems Technology, U.S. EPA, April 1977.
96. **Ackerman, D. G., Clausen, J. F., et al.,** Destoying Chemical Wastes in Commercial Incinerators. Facilities Report No. 6 — Rollins Environmental Services, Inc., U.S. EPA, 1977.
97. **Wallace, D. D., Trenholm, A., and Lane, D. D.,** Assessment of Metal Emissions from Hazardous Waste Incinerators, in Proc. 78th APCA Annu. Meet., Detroit, June 1985.
98. **Davidson, R. L., Natusch, D., and Wallace, J.,** Trace elements in fly ash, dependence on particle size, *Environ. Sci. Technol.,* 8, 1974.
99. **Greenberg, R., Zoller, W., and Gordon, G.,** Composition and size distributions of particles released in refuse incineration, *Environ. Sci. Technol.,* 12, 566, 1978.
100. **Bennett, R. and Knapp, K.,** Characterization of particulate emissions from municipal wastewater sludge incinerators, *Environ. Sci. Technol.,* 16, 831, 1982.
101. **Turgeon, M.,** Office of Solid Waste, U.S. EPA, personal cummunication, March 1987.
102. **Hall, R. E., Cooke, M. W., and Barbour, R. L.,** Comparison of air pollutant emissions from vaporizing and air atomizing waste oil heaters, *J. Air Pollut. Control Assoc.,* 33, 683, 1983.
103. **Fennely, P., McCabe, M., and Hall, J.,** Environmental characterzation of water oil combustion in small boilers, *Hazardous Waste,* 1, 489, 1984.
104. **Oppelt, E. T.,** Remaining issues over hazardous waste thermal destruction, presented at the 79th APCA Annu. Meet., Minneapolis, June 24, 1986.
105. **Duval, D. S., and Ruby, W. A.,** Laboratory Evaluation of High Temperature Destruction of Kepone and Related Pesticides, EPA-600/2-76-299, U.S. EPA, December 1976.
106. **Duval, D. S., and Ruby, W. A.,** Laboratory Evaluation of High Temperature Destruction of Polychlorinated Biphenyls and Related Compounds, EPA 600/2-77-228, U.S. EPA, December 1977.
107. **Dellinger, B. et al.,** Determination of the Thermal Decomposition Properties of 20 Selected Hazardous Organic Compounds, U.S. EPA, EPA-600/2-84-138, August 1984.
108. **Senken, S. et al.,** Combustion/Incineration Characteristics of Chlorinated Hydrocarbons, AIChE Annu. Meet., New Orleans, November 8 to 12, 1981.
109. **Dellinger, B. et al.,** PIC Formation Under Pyrolytic and Starved Air Conditions, U.S. EPA, October 1985.
110. **Kramlich, J. C. et al.,** Laboratory-Scale Flame Mode Hazardous Waste Thermal Destruction Research, in Proc. 9th Annu. Res. Symp. Land Disposal. Incineration and Treatment of Hazardous Waste, EPA-600/9-84/015, July 1984.
111. **Trenholm, A. and Hathaway, R.,** Products of Incomplete Combustion from Hazardous Waste Incinerators, in Proc. 10th Annu. Hazardous Waste Research Symp. Incineration and Treatment of Hazardous Waste, EPA-600/9-84-022, U.S. EPA, December 1984.
112. Midwest Reseach Insitute, Products of Incomplete Combustion from Hazardous Waste Combustion, final report, U.S. EPA, June 1986.

113. **Oleexsey, R. A., et al.,** Emission and Control of By-Product from Hazardous Waste Combustion Processes, in Proc. 11th Annu. Hazardous Waste Research Symp. Incineration and Hazardous Waste, EPA-600/9-85-028, U.S. EPA, September 1985.

114. **Olie, K., Vermeulen, P. L., and Hutzinger, O.,** Chlorodibenzodioxins and chlorodibenzofurans are trace components of ash and flue gas of some municipal incinerators in the Netherlands, *Chemosphere,* 6, 455, 1977.

115. **Buser, H. R., and Rappe, C.,** Identification of substitution patterns in polychlorinated dibenzo-*p*-dioxins (PCDD's) by mass spectrometry, *Chemosphere,* 7, 199, 1978.

116. **Eiceman, G. A., Clement, R. E., and Karasek, F. W.,** Analysis of fly ash from municipal incinerators for trace organic compounds, *Anal. Chem.,* 51, 2343, 1979.

117. **Karasek, F. W.,** Dioxins from garbage: previously unknown source of toxic compounds is being uncovered using advanced analytical instrumentaion, *Can. Res.,* September 1980.

118. **Bumb, R. R. et al.,** Trace chemistries of fire: a source of chlorinated dioxins, *Science,* 210, 1980.

119. **Cavallaro, A., Bandi, G., Invernizzi, G., Luciani, L., Mongini, E., and Gorni, A.,** Sampling, occurence an evaluation of PCDD's and PCDF's from incinerated solid urban waste, *Chemosphere,* 9, 611, 1980.

120. **Lustenhouwer, J. W., Olie, K., and Hutzinger, O.,** Chlorinated dibenzo-*p*-dioxins and related compounds in incinerator effluents: a review of measurements and mechanisms of formation, *Chemosphere,* 9, 501, 1980.

121. **A. D. Little, Inc.,** Study on the State-of-the Art of Dioxin from Combustion Sources, American Society of Mechanical Engineers, New York, 1981.

122. **Camp, Dresser and McKee, Inc.,** Dioxin Emissions from Resource Recovery Facilities and Summary of Health Effects, U.S. EPA, November 1986.

123. Ontario Ministry of the Environment, Scientific Criteria Document for Standard Development. No. 4-84. Polychlorinated Dibenzo-*p*-Furans, September 1985.

124. Radian Corporation, National Dioxin Study, Tier 4-Combustion Sources — Final Literature Review, U.S. EPA, September 1985.

125. **Miles, A. et a.,** Draft Engineering Analysis Report — National Dioxin Study, Tier 4: Combustion Sources, U.S. EPA, November 1986.

126. **Castaldini, C.,** Dioxin Emissions from Industrial Boilers Burning Hazardous Materials, U.S. EPA, April 1985.

127. **Branscome, M. et al.,** Evaluation of Waste Combustion in a Wet Process Cement Kiln at General Portland, Inc., Paulding, Ohio, U.S. EPA, February 1985.

128. **Branscome, M. et al.,** Evaluation of Waste Combustion in a Dry Process Cement Kiln at Lone Star Industries, Oglesby, Illinois, U.S. EPA, December 1984.

129. **Day, D. R., Cox, L. A., and Mournighan, R. E.,** Evaluation of Hazardous Waste Incineration in a Lime Kiln: Rockwell Lime Company, EPA-600/82-84-132, U.S. EPA, November 1984.

130. **Stretz, L. A. et al.,** Controlled Air Incineration of PCP-Treated Wood, U.S. EPA, September 1982.

131. U.S. EPA, PCB Disposal by Thermal Destruction, PB 82-241860, National Technical Information Service, Washington, D.C., June 1981.

132. **Van Buren, D., Poe, G., and Castaldini, C.,** Characterization of Hazardous Waste Incineration Residuals, U.S. EPA, January 1987.

Predicting and Assuring Incinerator Performance

133. **Dellinger, B. et al.,** Examination of Fundamental Incinerability Indices for Hazardous Waste Destruction, in Proc. 11th Annu. EPA Res. Symp. Land Disposal, Remedial Action, Incineration and Treatment of Hazardous Waste, EPA/600/9-85/028, September 1985.

134. **Cudahy, J. J. and Troxier, W. L.,** Autoignition temperature as an indicator of thermal oxiation stability, *J. Hazardous Mater.,* 8, 1983.

135. **Tsang, W. and Shaub, W.,** Chemical processes in the incineration of hazardous materials, in *Detoxification of Hazardous Waste,* Ann Arbor Science, Ann Arbor, MI, 1982, 41.

136. **Kramlich, J. C., Heap, M. P., Seeker, W. R., and Samuelsen, G. S.,** Flame Mode Destruction of Hazardous Waste Compounds, in Proc. Int. Symp. on Combustion, 1985.

137. **Miller, D. L. et al.,** Incinerability characteristics of selected chlorinated hydrocarbons, in Proc. 9th Annu. Res. Symp. on Solid and Hazardous Waste Disposal, EPA 600/9-84/015, September 1983.

138. **Dellinger B. et al.,** Determination of the thermal decomposition properties of 20 selected hazardous organic compounds, *Hazardous Waste,* 1, 137, 1984.

139. *Chemical and Engineering News,* American Chemcial Society, Washington, D.C., January 30, 1984, p. 24.

140. **Graham, J. L., Hall, D. L,. and Dellinger, B.,** Laboratory investigation of thermal degradation of a mixture of hazardous organic compounds, *Environ. Sci. Technol.,* 20, 703, 1986.

141. **Chang, D. P. Y. et al.,** Evaluation of a pilot-scale circulating bed combustor as a potential hazardous waste incinerator, presented at the 78th Annu. Meet. Air Pollution Control Association, Detroit, June 1985.

142. **Berger, M. L., and Proctor, C., II,** Sulfur hexafluoride as a surrogate for verification of destruction removal efficiency, presented at the Conf. Chemical/Physical Processes of Combustion, 1985.

143. **Berger, M. L. and Proctor, C. L.,** A tracer gas ratio technique to determine boiler suitability for hazardous waste incineration, presented at the Conf. Chemical/Physical Processes of Combustion, 1985.

144. **Wolbach, D. C.,** Parametric Experimentation with pilot-scale boiler burning hazardous compounds, in Proc. 11th Annu. Res. Symp. Land Disposal, Remedial Action, Incineration and Treatment of Hazardous Waste, EPA 600/9-84/015, May 1985.

145. **Dellinger, B. and Hall, D.,** The viability of using surrogate compounds for monitoring the effectiveness of incineration systems, *J. Air Pollut. Control Assoc.,* 36, 179, 1986.

146. **Waterland, L. R.,** Pilot-scale investigation of surrogate means of determining POHC destruction, presented at the 77th Annu. Meet. Air Pollution Control Association, San Francisco, June 1984.

147. **Lafond, R. K. et al.,** Evaluation of continuous performance monitoring techniques for hazardous waste incinerators, *JAPCA,* 35, 658, 1985.

148. **Daniels, S. L. et al.,** Experience in continuous monitoring of a large-roatry kiln incinerator for CO, CO_2, and O_2, presented at the 78th Annu. Meet. Air Pollution Control Association, Detroit, June 1985.

149. **Hall, D. et al.,** Thermal decomposition of a twelve-component organic mixture, *Hazardous Waste Hazardous Mater.,* 1985.

150. **Staley, L. J.,** Carbon monoxide and DRE: how well do they correlate, in Proc. 11th Annu. Res. Symp. Incineration and Treatment of Hazardous Waste, EPA/600/9-85/028, September 1985.

151. **Linak, W. P. et al.,** On the occurence of transient puffs in a rotary kiln incinerator simulator, *J. Air Pollut. Control Assoc.,* 37, 54, 1987.

152. **Lips, H. and Castaldini, C.,** Engineering Assessment Report: Hazardous Waste Cofiring in Industrial Boilers Under Non-steady State Operating Conditions, U.S. EPA, August 1986.

Environmental and Public Implications

153. **Ingiwersen, J. C. et al.,** Analysis of the off-site impacts of hypothetical accidents at a hazardous waste incineration facility, in Proc. 3rd Int. Symp. Operating European Hazardous Waste Management Facilities, Odense, Denmark, September 1986.

154. **Toffer-Clausen, J.** Safety programs, in Proc. 3rd Int. Symp. on Operating European Hazardous Waste Management Facilities, Odense, Denmark, September 1986.

155. **Kelly, K. E.,** Methodologies for assessing the health risks of hazardous waste incinerator stack emissions to surrounding populations, *Hazardous Waste,* 1, 507, 1984.

156. U.S. EPA, Health Risk Assessment Methodologies for RCRA Regulatory Analysis, Draft Rep., November 1983.

157. **Bowers, J. F. et al.,** Industrial Source Complex (ISC) Dispersion Model User's Guide, EPA-450/4-79-030, U.S. EPA, December 1979.

158. **O'Donnell, F. R. and Holton, G. A.,** Automated Methodology for Assessing Inhalation Exposure to Hazardous Waste Incinerator Emissions, EPA-600/9-84/015, U.S. EPA, 1984.

159. **O'Donnel, F. R. et al.,** Users Guide for the Automated Inhalation Exposure Methodology (IEM), EPA-600/2-83-29, U.S. EPA, June 1983.

160. **Holton, G. A., Travis, C. C., Etnier, E. L., O'Donnell, F. R.,** Inhalation pathway risk assessment of hazardous waste incineration facilities, in Proc. 77th Annu. Meet. American Pollution Control Association, Vol. 6, 1984, 84-102.3.

161. Peer Consultants, Supporting Documentation for the RCRA Incinerator Regulations 40 CFR 264, Subpart 0 — Incinerators, U.S. EPA, October 1984.

162. U.S. EPA, Health Assessment Document for Polychlorinated Dibenzo-*p*-Dioxins, EPA/600/8-84.914F, September 1985, 11.

163. U.S. EPA, Intergrated Risk Information System (IRIS) — Appendix A. Reference Dose (RfD): Description and Use in Health Risk Assessments, 1986, A 1-15.

164. **Goldstein, B. D.,** Toxic substances in the atmospheric environment, *J. Air Pollut. Control Assoc.,* 33, 454, 1983.

165. *Fed. Regist.,* 51, 185, 33992-34504, 1986.

166. **Ruckelhaus, W. D.,** Science risk and public policy, *Science,* 221, 1026, 1983.

167. U.S. EPA, Draft Regulatory Analysis for Proposed RCRA Regulations on Burning of Hazardous Wastes, January 1984.

168. **Taylor, C. C. et al.,** Health risk assessment for metals emissions from hazardous waste incinerators, presented at 79th Annu. Meet. APCA, Minneapolis, June 1986.

169. **Kelly, K. E.,** Comparison of metal emissions data from hazardous waste incineration facilities, presented at 78th Annu. Meet. APCA, Detroit, June 1985.

170. **Holton, G. A. et al.,** Inhalation pathway risk assessment of hazardous waste incineration facilities, in Proc. 77th Annu. Meet. APCA, San Francisco, June 1984.

171. **Holton, G. A., Travis, C. C., Etnier, E. L., O'Donnel, F. R., Hetrick, D. M., and Dixon, E.**, Mutliple-Pathways Screening-Level Assessment of a Hazardous Waste Incineration Facility, ORNL/TM-8652, DE85000165, Oak Ridge National Laboratory, Oak Ridge, TN.

172. **Travis, C. C., Ethier, E. L., Holton, G. A., O'Donnell, F. R., Hetrick, D. M., Dixon, E., and Harrington, E. S.**, Inhalation Pathway Risk Assessment of Hazardous Waste Incineration Facilities, ORNL/TM-9096, Oak Ridge National Laboratory, Oak Ridge, TN, 1984.

173. **Anatas, M. Y.**, Control Technology Assessment of Hazardous Waste Disposal Operations in Chemicals Manufacturing: In-depth Survey Report of E.I. duPont de Nemours and Company, Chambers Works Incinerator, Deepwater, New Jersey, NTIS PB 84-242577, U.S. EPA, 1984.

174. **Anatas, M.**, Control Technology Assessment of Hazardous Waste Disposal Operations in Chemicals Manufacturing: At 3M Company Chemolite Incinerator, Cottage Grove, MN, In-depth Survey Report, NTIS PB 84-241959, U.S. EPA, 1984.

175. **Holton, G. A.**, Human exposure assessment on on-land versus at-sea incineration of PCB· presented at the 78th Annu. Meet. APCA, Detroit, June 1985.

176. **Oberacker, D.**, Hazardous Waste Engineering Research Laboratory, U.S. EPA, personal communication, February 1987.

177. **Lewtas, J.**, Comparative potency of complex mixtures: use of short-term genetic bioassays in cancer risk assessment, in *Short-Term Bioassays in the analysis of Complex Environmental Mixture IV*, Plenum Press, New York, 1985.

178. **Freeman, H.**, *Innovative Thermal Process for Treating Hazardous Wastes*, Technomic, Lancaster, PA, 1986.

Chapter 3

MARKET TRENDS

Geoffrey G. Back and Gary S. Dietrich

TABLE OF CONTENTS

I.   Introduction 60

II.   Character of the Commercial Incineration Industry 60

III.   Regulatory Effects on the Incineration Market 62

IV.   Ability to Respond to Market Growth 63

V.   Reality of Incineration Capacity Shortfalls 64

VI.   Conclusions 65

References 66

I. INTRODUCTION

Predicting future trends and market response to major market-shaping legislation and regulations has proved difficult even for those in the business of commercial hazardous waste management. Part of the problem, even given the resources committed to market research, is that the regulatory system does not generate sufficient data to answer strategic planning questions.

Another part of the problem is that the dust of regulatory change has not settled as yet and still obscures the horizon. The rules and requirements under which the commercial operators and their clients must operate seem to be ever-changing and fraught with considerable uncertainties concerning future regulatory decisions that can chill investments in facilities and new technologies. Prime examples of the latter are how hard EPA will push for the permanent cleanups and the use of on-site treatment technologies vs. off-site options to clean up Superfund sites, and whether more land disposal restriction variances will be issued and current variances extended. Anyone who can reliably answer these questions could provide some key variables in the strategic planning equations of commercial hazardous waste management or site remediation services firms.

Even with better data and few changes on the regulatory front, however, the key to accurate predictions of trends and market response is ultimately the ability to predict behavior, especially the behavior of waste generators. For example, one must predict how quickly and strongly generators will react to regulatory, economic, and liability pressures toward minimizing their waste volumes. So far, the available evidence suggests that there has been considerable movement by generators to reduce their waste volumes. One must also predict the behavior of regulators toward enforcing existing regulations and predict how and when regulators, politicians, interest groups, and the public will resolve the current impasse in siting new hazardous waste management facilities. In the latter case, the consensus among commercial waste management firms is it will take a crisis situation to force any compromise.

Against this backdrop then, the authors offer their analysis, opinions, and predictions of near- and long-term trends for commercial hazardous waste incineration in the following sections. Some of the observations have been drawn from surveys of selected firms in the commercial hazardous waste management industry that ICF Technology Inc. has conducted for the U.S. Environmental Protection Agency (EPA),[1-3] as well as from a variety of strategic planning and market study assignments the authors have completed for private clients.

II. CHARACTER OF THE COMMERCIAL INCINERATION INDUSTRY

The commercial fixed-base incineration industry was developed initially by entrepreneurs with small waste companies or nonwaste companies risking investment in an activity that had no clearly defined market. Over the past 5 years or so, the industry has matured significantly. Many of the original players have gone out of business or have been acquired, and several have emerged as well established, medium-sized hazardous waste management companies. It seems evident that the commercial, fixed-base, incineration industry will soon be dominated, if it is not already dominated, by several large and medium-sized hazardous waste management companies. Future entrants will probably find it difficult to compete with these established firms since the initial cost and lead time to permit, build, and shake down a new incinerator is becoming very expensive. In addition, there is evidence that the real or perceived shortfall in commercial incineration capacity may be replaced by a situation where there may be excess capacity when the incinerators currently being proposed are placed into operation (see Section V).

Until recently, only a few generators of organic hazardous waste have chosen to own and operate on-site incinerators. This situation also appears to be changing. A fair number of major petrochemical companies have proposed and/or are in the process of permitting on-site incinerators to serve their company-wide needs. In a few cases, these companies have hinted at the possibility that these dedicated incinerators might accept wastes on a commercial basis.[4] Several state hazardous waste facility siting commissions have encouraged this trend as a solution for their capacity shortfalls and difficulties in siting new facilities.

Most observers believe, however, that few generators will open up their facilities for commercial business because permitting hurdles and/or public opposition would prove fatal to the idea. Generators are more likely to want to run their own show, observers say, and it makes no sense for a generator to assume the liability that made them discontinue using commercial facilities in the first place. Also, when the economics are examined closely, the commercial operators say that most midsized waste generators find that the economics favor using commercial service firms. Even so, other observers note that the profitability of the commercial waste management business could still be a sufficient lure for generators to enter the market. If this happens, most expect this trend to have the greatest effect on the market for incineration services.

Another significant change in the character of the commercial incinerator market has been the emergence of burning hazardous waste as fuels by cement kilns, light aggregate kilns, and industrial boilers and furnaces. Their entry into the thermal treatment market has resulted in a significant competitive challenge to the traditional hazardous waste incinerators in two very interrelated ways. First, these waste fuel burners have contributed significant capacity for the burning of liquid hazardous wastes at very competitive prices. States eager to expand in-state capacity for hazardous waste management have regarded this contribution quite favorably, although concerns are voiced about the environmental risks of this option.[5] In ICF's 1986—1987 survey of commercial hazardous waste management firms, respondents noted that waste fuel burner capacity, so far, is nowhere near being used up. These firms also believe that if these waste fuel burners are left largely unregulated, and given that many new traditional incinerator units are coming on-line in the next 3 years, there will be significant excess commercial thermal treatment capacity by 1991.

The second market effect linked, at least partially, to the emergence of the waste fuel burners has been a change in the character of the wastes sent to traditional hazardous waste incinerators. Able to charge lower prices, the waste fuel burners have effectively consumed much of the volume of high-Btu liquid hazardous wastes. Accordingly, commercial incinerators have seen a greater percentage of their waste receipts comprised of the more difficult-to-handle, lower-Btu solid and sludge wastes.*

The loss of this waste fuel for supporting the combustion of the lower-Btu wastes has meant that commercial incinerators must use more virgin fuel and special waste-handling equipment to handle the more difficult solid and sludge wastes. This raises their costs of operation. It has also been a significant factor in their decision to pursue expanding incinerator capacity for solids and sludges. Some industry observers also believe that this market trend was partially responsible for Chemical Waste Management's decision not to pursue ocean incineration in the U.S. since this was basically a liquid incineration technology.

The origins of the commercial incineration industry are being revisited today with the emergence of the mobile/transportable incineration industry. This industry also has been developed primarily by entrepreneurs and today is still being actively developed by risk investors and small firms. Respondents to ICF's industry 1986—1987 survey commented that "everybody and their brother are building mobile and transportable incinerator units today". However, this industry is showing signs of maturing much as the fixed-base industry has matured. With the potential market taking on greater definition because of EPA policy

---

* Other market trends also have contributed to commercial waste management firms receiving more solid and sludge wastes, for example, more on-site pretreatment, and recovery of wastes by generators.

(and certain state policies) to consider seriously on-site incineration at Superfund sites and with several successful projects completed, the well-established commercial hazardous waste management and site cleanup firms are investing in mobile/transportable incinerators and may soon dominate the industry, making it difficult for new entrants to this market. It is also not a market without financial risks because its growth potential is tied to regulatory developments. What some observers view as the U.S. EPA's healthy hesitancy to adopt permanent cleanup technologies and the effect that has had on lowering demand and slowing cleanups has been blamed for the bankruptcy of one transportable incinerator unit builder recently, although others dispute this analysis.[6]

III. REGULATORY EFFECTS ON THE INCINERATION MARKET

The commercial hazardous waste management industry, especially firms providing incineration services, owes much of its past success and profitability to federal and state regulatory requirements. This phenomenon was illustrated most dramatically by the development of the polychlorinated biphenyl (PCB) rules under the Toxic Substances Control Act (TSCA). The requirement that virtually all PCB materials must be incinerated created, almost overnight, a substantial and very profitable market for incineration that did not exist previously and most likely would not have developed but for those rules. The rules also eliminated a considerable cost and competitive disadvantage for incineration as compared to land disposal.

Regulatory requirements will continue to shape the future of the commercial incineration industry. The PCB market phenomenon is currently being repeated, albeit less dramatically, with the serial promulgation of land disposal restrictions under the Resource Conservation and Recovery Act (RCRA). Many of these restrictions will require, in part or in whole, that specified hazardous wastes be treated by incineration in lieu of or before they are land disposed. To date, these restrictions have created, or at least strengthened, the incineration market for the listed hazardous waste solvents. Most observers expect that the land disposal restrictions will continue to foster the incineration of other organic-based hazardous wastes. In a recent paper, two Chemical Waste Management, Inc. incineration marketing managers have predicted that roughly 25% of all wastes disposed of in commercial landfills in 1986 will ultimately be restricted, and that 60% of this quantity will be incinerated — an estimated volume of 525,000 to 600,000 ton/year.[7] The authors generally concur with their analysis, but suggest it will be important to watch carefully whether EPA decides to continue to extend the current variances from the land disposal restrictions.

Another example of regulations shaping markets was the recent EPA efforts to clarify the rules discussed earlier covering burning of hazardous wastes as fuels in industrial boilers and furnaces. Initially, the absence of clear-cut rules governing burning waste fuels in industrial furnaces and boilers had been a damper on the growth of this practice. Now, as regulations governing this practice are contemplated and viewed as confirming the acceptability of this waste management option, caution has given way to a flurry of activity by industrial furnace and boiler operators to seek interim status or permits for their facilities. Similarly, the recent articulation of new EPA policy to consider deliberately the feasibility of on-site incineration at Superfund sites strengthened substantially the market for mobile/transportable incineration.

New regulations can, of course, also create markets for alternative treatment technologies that can compete with incineration. The PCB regulations, for example, encouraged the development and use of dechlorination technologies for the reclamation of PCB-contaminated mineral oils in operating transformers, thus reducing the market share for PCB incineration. The land disposal restrictions on waste solvents appear to have (1) spurred on-site solvent recovery by generators; (2) encouraged generators to substitute for or reduce outright the use of solvents, particularly halogenated solvents; and (3) increased the burning of hazardous wastes as fuel (helped by the decline in halogenated solvent use).

IV. ABILITY TO RESPOND TO MARKET GROWTH

Permitting delays, reversals in state program authorizations, regulatory constraints and uncertainties, and public opposition to siting or expanding facilities top commercial firms' list of factors that constrain their ability to respond to market growth. Other factors mentioned in ICF's 1986—1987 commercial industry survey included: difficulties in obtaining financing; complications in developing new technologies in the face of changing regulations; the Comprehensive Environmental Response, Compensation, and Liability Act (CERCLA) program's unclear signals on how often on-site treatment technologies will be selected as part of the site remedy; regulators' unwillingness to act in less than black and white situations; and the difficulty of predicting or modeling market reactions to the land disposal restrictions. Permitting delays of 3 to 6 years or more, however, were mentioned by many respondents in ICF's industry survey as the single greatest factor limiting their ability to serve growing or new markets. While some were willing to acknowledge the technical complexity of the permitting process as a factor in the delays, most blamed inadequate state permit program resources, lack of state and federal agency coordination, differing state and federal interpretations of the same issue, or high staff turnover as the sources of the problem.

Obtaining a permit for a new fixed-base incineration facility can take from 1 to 2 years from initial application to issuance of a draft permit. Aother 1 to 2 years can be spent in administrative appeals or litigation if either the applicant or others are dissatisfied with the terms of the permit. Typically, at least 1 year is required to build an incinerator once the final permit is issued, and another 6 to 12 months is needed to shake down the operation of the unit, conduct the trial burn, and obtain permitting agency approval of the trial burn results.* In total, $2^1/_2$ to 6 years can elapse before final approval is obtained for a new, or an expansion of an existing, fixed-base incinerator. For an investor in a new facility, this is a long time to sustain the expense of financing a project without generating a revenue stream to offset these costs (unless other revenue-generating services are established), and often, the financial strain cannot be supported. Consequently, the development of new incineration facilities today is being attempted, almost exclusively and with considerable caution, by companies with the financial capital and backing needed to survive a lengthy and expensive permitting process.

The permitting process for mobile/transportable incinerator units provides for slightly less financial strain as the permitting agencies are allowing fabrication of the units before a permit is issued, saving some time and, therefore, some money. In addition, the permitting process itself and the adjudication and/or litigation of permit terms for these units has, to date, consumed less time — from 1 to 2 years — than has been the case for the larger fixed-base units, particularly when the mobile/transportable unit is to be used for a waste site cleanup.

It was clear from ICF's 1986—1987 industry survey, however, that there have been important improvements on the permitting front for commercial firms. Several firms expect final permits sometime in 1988. Respondents were also asked for their opinions on whether the new rules for permit modifications and transportable treatment units* represented an improvement in their ability to respond to market change and growth. In general, most firms expressed a wait-and-see attitude, although several were more positive about the permit

---

* Commercial operation of the incinerator unit can begin after completion of the trial burn while the operator awaits regulatory approval, unless there is some indication that the trial burn did not meet the performance standards established in the regulation.

* EPA proposed new rules that are intended to make the process of obtaining permit modifications more efficient. EPA has also proposed a rule that would provide for state-wide permitting of transportable treatment units.

modification rule or felt the process had to bet better because it couldn't get any worse. Their hesitation seems to reflect a concern that implementation of the rules will not live up to their promise on paper. In particular, several firms believe that the authorized states will not be as flexible as the language of the rule would have them be. These firms believe, for example, that the type of facility modifications needed to respond to market change and growth will remain labeled as major modifications under most state rules, and, therefore, subject to the full-blown permitting process.

The ground swell of public opposition to hazardous waste management facilities, in fact, to almost any kind of locally undesirable land use, of course, has had considerable influence toward lengthening the permitting process. Public opposition creates considerable pressures on permitting agencies to be more cautious and to take more time in their review of permit applications and often necessitates longer public comment and public hearing periods. Public opposition also can cause delays in obtaining zoning approvals, building permits, or air/water discharge permits.

The firms in the ICF 1986—1987 survey were asked for their advice on the possible roles of the federal government toward resolving the present impasse on siting new facilities. All of the firms believed that the situation will get much worse before there is any hope it will get better and that it will take a crisis to find viable solutions to the problem. Some firms did offer suggestions including (1) more aggressive public education and outreach programs, (2) strictly enforcing the adequate capacity demonstration requirement under SARA,* (3) encouraging states to control the siting process on the basis of technical criteria only, (4) encouraging generators to open up dedicated TSDFs to commercial use, and (5) withdrawing RCRA program authorization in states that enact waste import restrictions. Many firms expressed disappointment over what they view as the very limited success of state facility siting commissions.

V. REALITY OF INCINERATION CAPACITY SHORTFALLS

There was a significant turnaround in opinion among participants in ICF's 1986—1987 industry survey[3] on the question of whether there is a shortfall in commercial hazardous waste incineration capacity. Respondents 2 years prior were almost unanimous in their views of a "chronic" and "severe" shortfall in commercial incineration capacity, especially for waste solids and sludges.[2] The majority opinion now seems to be that incineration capacity shortfalls are not real or, if they do exist, will be short-lived as more permits are issued between now and 1991. Several firms even suggested a surplus in incineration capacity now exists or will exist by 1991 or 1992 and will continue through the end of the century if the permit applications already in or soon to enter the pipeline are finalized fairly quickly. Several commercial incineration firms also reported that their backlogs are down appreciably and that they are having trouble filling up increased capacity already in place.

Much of the turnaround in opinion by survey respondents was attributed to three trends. The first has been the continuing decline in PCB liquids incineration. The decline in this business, they report, has released incineration capacity for RCRA hazardous wastes.

The second trend, as discussed above, has been the entry of cement kilns, light aggregate kilns, and industrial boilers into the thermal treatment services market. These waste fuel burners filled in with a great deal of thermal capacity, and they have moved to widen the window of wastes they can accept to be able to handle more waste solids. Even several of the firms who believe a shortfall in incineration capacity still exists admit that the deficit is closing rapidly.

---

* This provision of the Superfund Amendments and Reauthorization Act requires states to demonstrate adequate capacity for a 20-year period to manage hazardous wastes generated within their borders or face a cutoff in federal funds to clean up Superfund sites in the state.

The third trend mentioned by respondents was the number of commercial incineration facilities soon to receive their final permit and the large number of new units or expansions planned to be operational by 1991. Respondents to ICF's 1986—1987 industry survey[3] reported 16 new fixed incinerator unit projects, most to be added to existing facilities, most ranging in size from 100 to 150 million Btu/h and expected to increase incineration capacity anywhere from 75 to 150% depending on the facility. Several other replacement kiln and existing kiln expansion projects were also reported. While many of these projects at existing facilities must await final regulatory approval, no firm seemed to believe that getting a permit would be impossible. Significantly, about 10% of the fixed-based incinerator unit projects were reported by firms not in the commercial hazardous waste incineration business presently.

Both sides of the issue, however, agreed on the key uncertainties in their predictions: (1) the speed and successes of the permitting process over the next 3 years, (2) the volume of site remediation wastes sent off-site for treatment and disposal and what is selected as best demonstrated available treatment for these wastes, (3) whether EPA will continue to string out some of the variances from the land disposal restrictions, (4) how stringently cement and light aggregate kilns and industrial boilers are regulated, and (5) how aggressively generators pursue waste minimization. In particular, respondents were uncertain whether EPA will extend the current variance from the land disposal restrictions for CERCLA and RCRA soils and debris or whether incineration will be the only best demonstrated available technology (BDAT) for the wastes. If incineration is the only BDAT option or one of several options for these wastes, some firms predict a considerable shortfall in incineration capacity. Other firms believed that a move by EPA to string out the effective dates of the land disposal restrictions would have a chilling effect on investment in new incineration capacity. The market effects of expiring land ban variances will, however, likely vary by region of the country with the West Coast and the Northeast likely to experience capacity shortfalls sooner than the Midwest, Gulf Coast, and Southwest, where much of the current commercial capacity is concentrated.

Several firms stressed that they have detected clear evidence of waste minimization by generators in the form of even larger volumes of solid and sludge wastes and as an overall decline in waste volumes shipped off-site for handling. They predict this trend will continue and may negate or soften the impact of increases in the demand for incineration services when land ban extensions expire and the pace of waste site cleanups accelerates.

VI. CONCLUSIONS

The fact that commercial hazardous waste incineration does and will continue to play an important role in the hazardous waste management system is not disputed. Less certain is how quickly demand will grow for incineration services and whether this growth rate will outstrip the pace at which commercial firms are able to bring new or expanded incineration capacity on-line. Regulatory and market uncertainties are seen as limiting factors. Considering the substantial investment and risk of resources needed for an incinerator, commercial firms are hesitant to make any commitment on the basis of a possible "ghost" market. Combine regulatory uncertainty with an inability to predict market responses, and these firms are even more cautious. Despite considerable research and effort, several firms admit that they have trouble determining what will be an effective response to major market-shaping regulations.

Although there has clearly been a lag in the market response, it now appears that commercial firms have responded to the market need. Significant incineration capacity increases are expected soon to be in place and operational. Most firms, if only on faith, have decided that regulators will enact and enforce requirements that should increase demand for incineration services and have invested in the necessary equipment and services. The authors believe, supported by results from ICF's 1986—1987 commercial industry survey,[3] that nationwide commercial hazardous waste incineration capacity by 1990 will be more than adequate to meet current demand (although there still may be capacity shortfalls on a regional basis for some time to come). Accordingly, the potential investor in any new hazardous waste incineration project not yet in the permitting pipeline would be well advised to reassess his analysis of need.

EDITOR'S ADDENDUM

By 1990, the market had evolved essentially as projected herein. Land bans were being increasingly enforced for certain wastes that had formerly been landfilled. Toxicity Characteristic Leaching Protocols were reclassifying additional wastes as hazardous. A few new central incinerator facilities were being brought on line, but regional capacity shortfalls continued in the West Coast and the Northeast. Federally mandated Capacity Assurance Plans indicated possible excess capacity in regions such as the Gulf Coast and the Midwest. The reader is directed to trade journals for the most current information on market trends.

REFERENCES

1. ICF Incorporated, Survey of Selected Firms in the Commercial Hazardous Waste Industry: 1984 Update, Final report, U.S. Environmental Protection Agency, Office of Policy Analysis, September 1985.
2. ICF Incorporated, 1985 Survey of Selected Firms in the Commercial Hazardous Waste Industry, Final Rep., U.S. Environmental Protection Agency, Office of Policy Analysis, November 1986.
3. ICF Incorporated, 1986—1987 Survey of Selected Firms in the Commercial Hazardous Waste Management Industry, U.S. Environmental Protection Agency, Office of Policy Analysis, in preparation.
4. Three Firms Indicate Interest in Building New Jersey Incinerator, *HazTech News*, p. 196, December 31, 1987.
5. New York Panel Backs Cement Kilns for Burning Hazardous Wastes, *HazTech News*, p. 26, February 25, 1988.
6. Inside EPA, EPA's Reluctance to Use New Technologies Slows Development, Say Critics, p. 13, February 12, 1988.
7. **Schofield, W. R. and Rafferty, P.,** Incineration: technology; marketplace, in Proc. 3rd Annu. Hazardous Materials Management Conference/West, p. 426, 1987.

Chapter 4

REGULATORY REQUIREMENTS AND THE PERMITTING PROCESS

G. Scott Koken and Richard W. Raushenbush

TABLE OF CONTENTS

I.   Introduction 71
    A.   Brief History of Environmental Law and Regulation 71
        1.   Definition of Law and Regulation 71
        2.   Environmental Problems and Laws 71
            a.   Pre-Silent Spring and National Environmental
                  Policy Act (NEPA) Environmental Degradation
                  and Laws and Regulations 71
            b.   Post-NEPA Environmental Degradation and
                  Laws and Regulations 72
            c.   Statutory Scheme Includes Federal, State,
                  County, and City Law 74
            d.   Incineration: Part of the Problem/Part of the
                  Solution 75
                i.   Self-Regulation 75
                ii.   Forced Regulation 75
                iii.   Input Controls 75
                iv.   Output Controls 76
    B.   Hazardous Waste Incineration Statutory and Regulatory General
        Overview 77
        1.   Federal Laws Impacting Incineration Include CAA,
            RCRA, TSCA, CERCLA, CWA 77
            a.   The Clean Air Act 77
            b.   The Resource Conservation and Recovery Act 78
            c.   The Toxic Substances Control Act 80
            d.   Comprehensive Environmental Response,
                Compensation, and Liability Act 80
            e.   The Clean Water Act 81
         2.   State Laws Impacting Incineration 82
        3.   Local Laws and Regulations Impacting Incineration 83
            a.   Local Air Pollution Control Districts or Agencies 83
            b.   Master Plans, Zoning, and Land Use 83
            c.   Other 83
II.   The Major Laws and Regulations Affecting Incineration of Hazardous
    Wastes 83
    A.   Clean Air Act 83
        1.   Statutory and Regulatory Overview 83
        2.   Regulations 85
    B.   Resource Conservation and Recovery Act 85
        1.   Statutory Overview 85
            a.   Introduction 85

b.   Performance Standards 87
c.   Permitting Requirements 88
d.   State Authorization 90
e.   Enforcement Mechanisms 90
2.   Application of RCRA to Hazardous Waste Incineration
Facilities 91
a.   "Hazardous Waste" Under RCRA 91
b.   Definition of "Facility" 91
c.   Definitions of "Treatment", "Storage", and
"Disposal" 91
d.   Definition of "Incinerator" 92
e.   Definition of "Container" 93
f.   Hazardous Waste Incinerator Facility as
"Generator" 93
g.   Exemptions from RCRA Regulations 93
3.   TSD Facility Standards of General Applicability 93
a.   General Facility Standards 94
i.   ID Number 94
ii.   Required Notices 94
iii.   Waste Analysis 94
iv.   Facility Security 96
v.   Inspection Requirements 96
vi.   Personnel Training 97
vii.   General Requirements for Ignitable,
Reactive or Incompatible Waste 98
viii.   Location Standards 98
b.   Preparedness and Prevention 99
i.   Design and Operation of Facilities 99
ii.   Required Equipment 99
iii.   Testing and Maintenance of Equipment 99
iv.   Access to Communications Equipment 99
v.   Required Aisle Space 100
vi.   Arrangements with Local Authorities 100
c.   Contingency Plan and Emergency Procedures 100
i.   Purpose and Implementation of
Contingency Plan 100
ii.   Content of Contingency Plan 100
iii.   Copies of Contingency Plan 101
iv.   Amendment of Contingency Plan 101
v.   Emergency Coordinator 101
vi.   Emergency Procedures 101
d.   Manifest System, Recordkeeping, and Reporting 102
i.   Use of Manifest System 102
ii.   Manifest Discrepancies 103
iii.   Operating Record 103
iv.   Availability, Retention, and Disposition
of Records 103
v.   Biannual Report 104
vi.   Unmanifested Waste Report 104
vii.   Additional Reports 104
e.   Releases from Solid Waste Management Units 105
f.   Closure and Postclosure 105

i.   Closure Performance Standard 105
ii.   Closure Plan 105
iii.   Closure Schedule 107
iv.   Disposition of Contaminated Equipment,
        Structures, and Soils...........................107
v.   Closure Certificate 107
g.   Financial Requirements...................................108
i.   Closure Cost Estimate 108
ii.   Financial Assurance for Closure 109
        (a).   Closure Trust Fund.......................109
        (b).   Surety Bond Guaranteeing
                Payment into a Closure Trust
                Fund......................................111
        (c).   Surety Bond Guaranteeing
                Performance of Closure 112
        (d).   Closure Letter of Credit 113
        (e).   Closure Insurance 114
        (f).   Financial Test and Corporate
                Guarantee for Closure 115
iii.   Third-Party Liability Coverage
        Requirements 117
        (a).   Liability Insurance.......................117
        (b).   Financial Test or Guarantee for
                Liability Coverage........................117
        (c).   Letter of Credit for Liability
                Coverage 118
        (d).   Surety Bond for Liability
                Coverage 118
        (e).   Trust Fund for Liability Coverage.........118
        (f).   Combination of Mechanisms 119
iv.   Bankruptcy.......................................119
v.   Use of State-Required Mechanisms 120
vi.   State Assumption of Responsibility 120
h.   RCRA Standards for Incinerators.........................121
i.   Applicability 121
ii.   Waste Analysis 121
iii.   Performance Standards...........................122
iv.   Hazardous Waste Incineration Permits 123
v.   Operating Requirements 123
vi.   Monitoring and Inspections 124
vii.   Closure 124

III.   The Permit Application Process, Requirements, and Conditions 124
A.   General Permit Application Process For All Environmental
        Permits 124
1.   Applicable Regulations 124
2.   Overview of the Permitting Process.................124
3.   Submittal of the Application 125
4.   Agency Review 125
5.   Issuance and Review of the Draft Permit 128
6.   Public Review 129
7.   U.S. EPA Discretionary Review Power........................130

8. Permit Issuance 130
9. Administrative Appeals 130
10. Panel F Hearing Procedures 131
    a. Subpart F Hearing Procedures 131
    b. Subpart E Evidentiary Hearings (Adjudicatory) 132
B. Specific Law/Regulation Additional Permitting Process
    Requirements 134
    1. The RCRA Application Process 134
        a. Administrative and Nontechnical General
            Requirements 134
            i. Preliminary Notification of Hazardous
                Waste Activity 134
            ii. Contents of Part A 134
            iii. Administrative and Nontechnical Contents
                of Part B 136
                (a). General Description of the Facility 136
                (b). Chemical and Physical Analyses
                     of the Hazardous Waste(s) to Be
                     Handled at the Facility 136
                (c). Waste Analysis Plan 138
                (d). Security Procedures 139
                (e). General Inspection Requirements 140
                (f). Personnel Training 141
                (g). Preparedness and Prevention 148
                (h). Contingency Plan and Emergency
                     Procedures 148
                (i). Environmental/Safety Design/
                   Construction/Operating Procedures 150
                (j). Precautions for Ignitable,
                   Reactive, or Incompatible Waste 150
                (k). Vehicular Traffic 151
                (l). Facility Location Information 151
                (m). Closure/Postclosure Plan 152
                (n). Financial Requirements and
                   Closure Cost Estimate 154
        b. Subpart O Technical Requirements 155
            i. Permit Conditions 155
                 (a). Trial Burn Plan 155
                (b). POHCs and PICs 156
             ii. Performance Standards 157
                 (a). Destruction and Removal
                   Efficiency 157
                (b). Volatile and Semivolatile Organics 157
                (c). Hydrochloric Acid 157
                (d). Particulates 163
            iii. Permit and Trial Burn 163
    2. The TSCA Application Process 171
        a. Liquid PCBs 172
        b. Nonliquid PCBs 173
        c. Maintenance of Data and Records 173
            i. Continuous and Short-Interval Data 173

References 174

I. INTRODUCTION

A. BRIEF HISTORY OF ENVIRONMENTAL LAW AND REGULATION

1. Definition of Law and Regulation

Law can be defined as a binding custom or practice of a community, a rule of conduct or action prescribed or formally recognized as binding or enforced by a controlling authority.[1] Law originated as a method of solving problems and disputes between two or more people and a method of maintaining community (not necessarily complete community) harmony. In the U.S., the customs or practices of the community that, in part, bind us together as a nation (or a city or state) are laws primarily concerned with the rights of people and their rights of property. Our law is divided into two categories: "statutory" or "common". Statutory law is given as a rule to the people by the consensus of the lawgivers, the legislators of the federal, state, and local governments. Common law arises from situations not related to statutory law but handed down by the court based on custom and precedent.

Regulation is a principle, rule, or law designed to control or govern behavior. It can be a governmental order having the force of law. In the U.S., at present, Congress has enacted laws designed to protect the environment. These laws sometimes contain rules or regulations, but, in general, give authority to government agencies to formulate and enact regulations that conform to the intent of the law and that have the same force as the law.

2. Environmental Problems and Laws

a. Pre-Silent Spring and National Environmental Policy Act (NEPA) Environmental Degradation and Laws and Regulations

The first federal statutory "environmental laws" were in response to navigation interests. In 1886, a federal statute (24 Stat. 329) prohibited refuse dumping in New York Harbor. The impetus of this statute was to protect marine merchants' interests in conducting navigable commerce in the harbor. The Rivers and Harbors Act of 1899 expanded the prohibition by prohibiting the discharge of waste materials into any navigable waterway without a permit granted under the Section 13 of the Act. This section became known as the Refuse Act (33 U.S.C. and 407). While this legislation further protected marine commerce in all navigable waters, limited dumping was allowed by the permit system to protect both marine and shore-based economic interests that would be affected by a complete ban on this method of waste disposal. Almost concurrently with the trash disposal problem hindering commerce in navigable waterways, local governments were beginning to wrestle with the effects of smoke in Midwest and Eastern cities. Chicago and Cincinnati in 1881 and Pittsburgh in 1895 passed their first air pollution laws. Boston's City Board of Health passed a smoke control law in 1901 to protect streets, public places, and private residences from nuisance smoke and dust arising from private or commercial activities.

The above early environmental laws are examples of the protection of property rights (economic use of the nation's waterways) and the protection of personal rights (freedom from undue constraint or personal harm). Environmental laws through the first half of this century remained mainly local in nature (mainly enacted by municipal and then county governments which were close to the problem). Several occurrences changed the direction of environmental legislation and regulation and the focus of national interests. These included the following:

1. Closing of the country's geographic frontier near the turn of the century creating a "closed" national (as well as natural) environment for the populace
2. The Denora, PA pollution incident of 1948 which killed 20 people and left almost half of the townspeople severely ill
3. Federal monetary assistance to states and local entities for air pollution control/studies starting in 1955

4.   Pollution problems becoming household topics due to the publishing of a book in 1962
      entitled *Silent Spring,* written by Rachael Carson
5.   The enactment of the National Environmental Policy Act (NEPA) of 1969
6.   The creation of the U.S. EPA in 1970
7.   The discovery of the hazardous waste site in Love Canal, NY in 1977

The closing of the country's geographic frontier marked the beginning of the end to the
nation's "endless" dilution and buffering capacities for the wastes of people's increasingly
consumptive lives and exponential technological progress. The effects of the unrestrained
consumption and waste generation began to become apparent with the Denora, PA air
pollution catastrophe in 1948 and the already surfacing problems of smog in Los Angeles.
These and other occurrences caused the public to pressure the lawmakers. However, it wasn't
until 1955 that Congress passed the first national air pollution control law. The issue of state
and local authority rights interfered with passage of several bills in the early 1950s. The
1955 law was passed to lend financial assistance to the states and local entities to formulate
programs and conduct research. Pollution control was left up to the state and local governments.

In 1962, the book *Silent Spring* further awakened the public to the fragility of the natural
environment. The effects of man's activities on the land, water, and air ecosystems were
plainly outlined in terms most of the populace could understand and relate to. Air pollution
legislation was passed in 1963, 1965, 1966, 1967, and 1969. Congress enacted the National
Environmental Policy Act of 1969 (NEPA). NEPA established for the first time a national
environmental policy and direction. All federal actions were required to have preconstruction
reviews that incorporated an environmental impact statement (EIS). An EIS includes as one
element a risk assessment and is supposed to examine the environmental effects of a proposed
project. Although limited to federal projects, it has become a model for many state and
local laws.

b. Post-NEPA Environmental Degradation and Laws and Regulations

Many of the ideas of NEPA have been approximated in post-NEPA federal regulations
dealing with other than federal projects. Since this act, all present and probably future federal
legislation and regulation has or will establish federal standards and enforcement procedures,
and states, although free to make their regulations and procedures more stringent, must
model their regulations based on the federal standards.

Immediately following NEPA was the creation of the U.S. Environmental Protection
Agency (EPA) in 1970 which serves to carry out the mandates of the federal legislation by
promulgating (issuing) regulations enforceable by law. Also following NEPA were the
enactments of the Clean Air Act of 1970 and the 1972 Amendments to the Water Pollution
Control Act. Both these laws incorporated the NEPA idea of preconstruction reviews entitled
"new source review". In 1972, Congress also passed the Federal Insecticide, Fungicide
and Rodenticide Act (FIFRA) to control pesticides. The Kepone incident in 1975 may have
been a further impetus in amendments to this Act in 1975 and 1978. Table 1 outlines the
history of air pollution legislation.

Although most of the current federal and state air, water, and pesticide legislation can
at least be partly traced to the public awareness due to *Silent Spring,* in 1977 another tragedy,
Love Canal, startled the public into action against hazardous waste, which was soon to be
classified as one of the nation's higher priorities and basic problems. The Congress acted
to strengthen substantially the Resource, Conservation and Recovery Act of 1976 (RCRA)
and the amendments of 1980 with the Hazardous and Solid Waste Amendments to the Act
of 1984 (HSWA). Between 1950 and 1975 it has been estimated that 6 billion ton of hazardous
wastes have been deposited on or in the land at an estimated 26,000 to 50,000 different

TABLE 1
Federal Air Pollution Legislation Summary

Legislation and statutory designation (a)	Description
Air Pollution Control Act PL 84-159, 6/14/55, 69 Stat. 3221	Temporary authority and $5 million annually for 5 years for federal program of research in air pollution and technical assistance to state and local governments
Air Pollution Control Act Extension, PL 86-365, 9/22/59, 73 Stat. 646	Extended the 1955 Act for four more years
Motor Vehicle Exhaust Study Act, PL 86-493, 6/8/60, 74 Stat. 162	
Air Pollution Control Act PL 87-761, 10/9/62, 76 Stat. 760	
Clean Air Act of 1963 PL 88-206, 12/17/63, 77 Stat. 392	The Act granted permanent authority for federal air pollution control activities and authorized expenditures of $95 million over 3 years. Major provisions included: 1. Federal grants to states and local control agencies to establish and improve control programs. 2. Federal action to abate interstate air pollution through a system of hearings, conferences, and court actions 3. Provision for an expanded federal research and development program with particular emphasis on motor vehicle pollution and sulfur oxide emissions from
Vehicle Air Pollution Control Act, PL 89-272, 10/20/65, 79 Stat. 992	1. Provided for the promulgation of national standards relating to motor vehicle pollution and initially applied to the 1968 model year 2. Provided for cooperation with Canada and Mexico to abate international air pollution
Clean Air Act Amendments of 1966. PL 89-675, p	Authorized 3-year, $186 million expansion of air pollution program, including funds to operate local control agencies
Air Quality Act of 1967, PL 90-148, 11/21/67, 81 State 485	Amended the 1966 Act, enacted a national policy of air quality enhancement, provided a procedure for designation of air quality control regions, set standards of cooperation between federal and state governments, added provisions for registering fuel additives
Clean Air Act Amendments of 1969, PL 9-137	Extended authorization for research on low-emission fuels and motor vehicles
1969 Presidential Study (Nixon) on Reorganization of Federal Pollution Control Programs	
1970 Presidential Advisory Panel Recommendation to establish special pollution control agency	
Nixon Reorganization Plan #3 sent to Congress (December 2, 1970)	
EPA established December 9, 1970	
Clean Air Act Amendments of 1970 (CAA), with technical amendments in the Comprehensive Health Manpower Training Act of 1971-PL-604. 12/31/70	1. Provided for the establishment of national ambient air quality standards (NAAQS) and their achievement by July 1, 1975, through the implementation plans of air quality control regions (AQCRs) and states 2. Provided for 90% reduction of automotive hydrocarbon and carbon monoxide emissions from 1970 levels by the 1975 model year and 70% reduction of nitrogen oxide emissions from 1971

TABLE 1 (continued)
Federal Air Pollution Legislation Summary

Legislation and statutory designation (a)	Description
	levels by the 1976 model year (with 1 year extensions if necessary) 3. Provided for studies of aircraft emissions and noise pollution
Clean Air Act Extension. PL 93-15. 4/5/78 87 Stat. 11	
Energy Supply and Environmental Coordination Act of 1974 (ESECA), PL93-319, 6/24/74. 98 Stat. 246	

Clean Air Act Amendments of 1977 (CAAA), with technical amendments in the Safe Drinking Water Act of 1977 (PL-95-190), PL-95-95, 8/7/77, 91 Stat. 685	1. Established prevention of significant deterioration (PSD) regulations
	2. Allowed the EPA to use emissions offsets, banking, and bubble policies to allow companies to buy, sell, trade, and "bank" permission to pollute up to certain levels in specified areas
	3. Allowed for extensions for compliance with emissions limitations for power plants ordered to convert to coal
	4. Required EPA to:
	  a. Set national maximum emission standards, which are termed new source performance standards (NSPS), on an industry-by-industry basis for new plants or major expansions of existing plants
	  b. Identify and set national emission standards for stationary sources emitting hazardous air pollutants. These standards are termed the National Emission Standards for Hazardous Air Pollutants (NESHAPS). As of November, 1988 only eight hazardous substances have had standards established. They include asbestos, arsenic, beryllium, mercury, vinyl chloride, benzene, sulfuric acid, and radioactive isotopes.

sites. Under the Comprehensive Environmental Response and Liability Act of 1980 and the Amendments to this Act in 1986 (CERCLA, commonly called Superfund), which was legislated, in part, to respond to abandoned disposal sites, the EPA has, as of 1987, named 951 sites to a priority cleanup list. The top four states include New Jersey with 94, Michigan with 66, New York with 64, and California with 53. Many of the sites have contaminated or threaten contamination of the groundwater. The final priority list may include from 2,000 to 10,000 sites and cost up to $100 billion over 50 years.

c. *Statutory Scheme Includes Federal, State, County, and City Law*

The U.S. has now in effect a network of evolving national, state, county, and city legislation, ordinances, and regulations that address environmental pollution and degradation. This statutory scheme is provided a framework by the national legislation and regulations. States in most cases are given the opportunity to implement their jurisdiction through enacting at least equivalent legislation and promulgation of regulations. In turn, many counties and cities are receiving authority from their states to address their own unique problems within state minimum requirements.

d. Incineration: Part of the Problem — Part of the Solution

The interaction between society and technology during the environmental activism period of 1962 through the present is important to isolate as an issue for the incineration of hazardous wastes. Society must first decide, hopefully, on a sound basis, the direction in which to proceed to solve a problem. Action, in this country, occurs by society establishing laws either through the Congress and/or state and local governments or through the court system by common law. The direction society chooses is often influenced by the actions of individuals and interested groups (John Muir, Rachael Carson, the Sierra Club, Audubon Society, industry lobbyists, people in government, etc.). Depending on the degree of public and society involvement, the direction society chooses may or may not involve the whole society consensus. Once a direction is chosen, society must decide the method of maintaining the direction and momentum and of solving a problem.

All possible methods to solve current ongoing or future pollution problems involve behavioral modifications and some degree of technology use. The differences in the methods lies in whose behavior is being modified and to what degree the behavior modification is lessened by the role of technology as a tool. The behavioral changes to date have mainly involved changes in industrial practices, e.g., industry cannot freely dump waste into rivers or the air without treatment or limitations, a fact which has led to cost increases for products society desires and pays for. These cost increases have not led to major changes in the general population's behavior in how or what they buy (e.g., the automobile).

Behavioral modifications fall into one of two general categories: self-regulation or forced regulation. Both self and forced regulation are accomplished by using one or both of the concepts of input and output control.

i. Self-Regulation

The act of regulating one's self, whether pertaining to an individual citizen or a giant corporation, is guided by self-interest which can vary from self-preservation to economic health or combinations of these and/or other self-advancement concerns. Actions under self-regulation for individuals might include using mass transit instead of their automobiles to reduce air pollution and minimize resource destruction necessary to create and use the automobile. It might mean using alternative materials instead of Styrofoam cups and practicing garbage separation to recover recyclable materials. For corporations, it might mean redesigning processes, implementing recycling, and utilizing waste exchanges to reduce raw material use and waste, thereby saving money. Self-regulation usually requires a well-informed entity regarding the external and personal effects the entity's actions have on their surroundings and the effect the changed environment has or will have on themselves if certain behaviors continue. These entities usually have an understanding of history, can understand trends, and have an interest in long-term survival and improvement.

ii. Forced Regulation

Forced regulation is accomplished from outside the polluting entity, whether the entity is an individual or industry, by using incentives, penalties, taxes, and/or other economic or regulatory inducements to reduce pollution. Forced regulation is necessary for entities that fail to realize or recognize the detrimental effects of their actions on their surrounding environment affecting themselves and others.

iii. Input Controls

Input pollution control uses methods to restrict or minimize potential pollutants from entering the environment by using raw materials more efficiently, decreasing the output of waste material into the environment. The specific methods may involve no, low, or high technology.

General input controls which include major behavioral modifications and varying degrees of technology include:

1. Control population growth
2. Reduce unnecessary waste of resources (metals, paper, etc., build products so they last longer)
3. Reduce energy usage by forgoing some activities and/or use energy more efficiently
4. Use different energy sources or additional energy sources to reduce use of those causing pollution*

Input controls for specific problems which actually reduce the severity of behavioral modification necessary and that use varying degrees of technology are exemplified below for reducing SO_2 pollution:**

1. Burn low-sulfur coal in power plants
2. Remove sulfur from coal
3. Convert coal to a gas or liquid fuel
4. Remove sulfur during combustion for power by fluidized bed combustion and limestone injection (see fluidized bed and circulating bed technology)

iv. Output Controls

Output control attempts to solve pollution problems after pollution has been produced. Controls that absolutely minimize pollution are extremely costly and would cause major behavior modifications due to the economic effect on the public. However, to date, most controls have not caused severe economic impacts and, hence, no major behavior modification. Pollution reduction has been achieved mainly through technological removal or destruction. There are four major methods of output control which are

1. Removing pollutants to harmless levels or converting them to harmless or less harmful substances (physical, chemical, biological treatment)
2. Destroying harmful wastes by proper incineration to a degree that is considered environmentally safe
3. Burying the wastes in the ground
4. Recycling or reusing waste material by recategorizing the waste as a raw material

Output control may again involve no, low, or high technology, but usually requires some level of technology.

For situations in which we are cleaning up past environmental pollution occurrences, behavior modification plays an insignificant part, although self-regulating entities will use the knowledge of past mistakes in selecting regulation methods. A particular technology chosen becomes the method of solving the problem.

The inevitable conclusion to the environmental problems facing society is the balanced use of methods to minimize pollution. Major behavioral changes will be necessary from private and corporate citizens. The most likely changes for private citizens will be

1. A reduction of unnecessary waste of resources
2. A reduction in energy usage

---

* Note: Most of these choices reduce energy use, the likelihood of premature resource depletion, and pollution.

** Note: Energy use is not reduced, thereby not decreasing the chance of a resource shortage; but pollution is reduced.

3. That costs of materials and services (consumer items, transportation, water, energy) will be higher, reflecting their true cost of production and disposal

The most likely changes for industry will be

1. Product and raw material substitutions, process changes, and optimization
2. Recycling or reusing waste material, recategorizing the waste as a raw material
3. Reduction of energy usage, cogeneration, increased energy efficiency
4. Cleaner production of energy, use of different energy sources
5. Increased use of destruction methods such as incineration and physical/chemical/biological conversion of wastes

Incineration has in the past been associated with ugly, smelly smoke of the uncontrolled emissions of the steel mills and power plants prior to the 1970s. It is a history not easily overcome. The effort to convince the populace that technology improvements have created clean-burning incinerators continues. One of the main remaining concerns of incineration opponents is with the proper operation of incinerators and emission constituent detection to verify proper operation. Another concern is that not enough emphasis is being put into pollution input controls such as waste minimization and recycling. Waste minimization and recycling should be primary in both the private and industrial sectors. However, even major behavioral modifications and maximum use of waste minimization and recycling will not eliminate the need for some waste destruction capacity in the future. Waste destruction capacity is also needed for the huge backlog of waste generated and improperly disposed of for which waste minimization and recycling will not help. With the proper staffing, training, operating controls, and monitoring, incineration will contribute to the cleaning up of America.

B. HAZARDOUS WASTE INCINERATION STATUTORY AND REGULATORY GENERAL OVERVIEW

1. Federal Laws Impacting Incineration Include CAA, RCRA, TSCA, CERCLA, CWA

The federal statutes impacting hazardous waste incineration include the Clean Air Act (CAA), the Resource Conservation and Recovery Act (RCRA), the Toxic Substances Control Act (TSCA), the Comprehensive Environmental Response, Compensation, and Liability Act (CERCLA), and the Clean Water Act (CWA).

a. The Clean Air Act

The CAA is discussed first, not just because of its historical position, but because it often causes regulatory and permitting problems when overlooked in the almost overwhelming shadow of RCRA. The CAA provides a framework for, and delineates goals related to, the attainment of national clean air, and authorizes the EPA to implement procedures through their regulatory powers to attain the goals.

The CAA Amendments of 1970 and 1977 established National Ambient Air Quality Standards (NAAQS), Air Quality Control Regions (AQCRs), the concept of the Prevention of Significant Deterioration (PSD), national maximum emission standards termed New Source Performance Standards (NSPS), and National Emission Standards for Hazardous Air Pollutants (NESHAPS). These concepts and associated regulations are important to understand due to their effect on incineration economics acting through the decision process as it relates to siting, plant expansion, process modifications, and probable future requirements for pollution control.

TABLE 2
RCRA Subtitle C Hazardous Waste Management

Section 3001	Identification and listing of hazardous wastes
Section 3002	Standards applicable to generators of hazardous waste
Section 3003	Standards applicable to transporters of hazardous waste
Section 3004	Standards applicable to owners and operators of hazardous waste treatment, storage, and disposal facilities
Section 3005	Permits for treatment, storage, or disposal of hazardous waste
Section 3006	Authorized state hazardous waste programs
Section 3007	Inspections
Section 3008	Federal enforcement
Section 3009	Retention of state authority
Section 3010	Effective date
Section 3011	Authorization of assistance to states
Section 3012	Restrictions on recycled oil
Section 3013	Monitoring, analysis, and testing

b. The Resource Conservation and Recovery Act

Through Subtitle C of RCRA Congress crafted a two-step response to the nation's hazardous waste problem. First, RCRA establishes a comprehensive federal program of hazardous waste management administered by the EPA. Second, Congress encouraged states to adopt their own hazardous waste management programs that meet at least the minimum requirements of the federal program, and seek authorization to administer such programs under RCRA in lieu of the federal program.

Through RCRA, Congress intended to ensure "that hazardous waste management practices are conducted in a manner which protects human health and the environment". The regulatory requirements of Subtitle C are outlined in 13 sections. These sections are listed in Table 2.

There are eight "driver" sections. Section 3001 defines what material is a hazardous waste. Section 3010 requires agency notification of hazardous waste activity. Section 3005 establishes the requirement for permits for treatment, storage, and/or disposal (TSD) of hazardous waste. Sections 3006, 3009, 3011 establish authorization for states meeting equivalency criteria to oversee the RCRA program. Section 3008 establishes inspection and enforcement authority for the regulations. Section 3013 requires monitoring, analysis, and testing.

Sections 3002 and 3004 establish the framework of standards that generators, owners, and operators of TSD facilities (TSDFs) must meet and is the substance of the law impacting the economics (siting, design, construction and operation, and safety and health requirements).

The regulations promulgated by EPA based on the sections of the act are contained in Title 40 of the Code of Federal Regulations (40 CFR). Parts of Title 40 concerned with RCRA are listed in Table 3. Those parts listed in Table 3 that are at least in part concerned with incineration are listed in Table 4. Incineration is specifically addressed by 40 CFR 264 Subpart O. Table 5 lists Subpart O topics. There is a tendency for personnel concerned with incinerators to concentrate their knowledge base in Subpart O. In reality, incinerators rarely stand alone and are usually part of a TSDF.

Incinerator designers, project engineers, construction engineers, and operations/maintenance supervisory personnel should be thoroughly familiar with the broader regulations regarding incineration and TSDFs in Table 4.

State authorization for the RCRA program is transferred in phases from regional EPA offices to each state in the region. The requirements are detailed in 40 CFR Part 271. The state must first give public notice of its intent to apply for authority. The state must then supply a proposed program for interim authorization to the EPA. Interim authorization

TABLE 3
40 CFR Parts Concerning RCRA

Reference	Description
40 CFR Part 124	Procedures for decision making
40 CFR Part 260	Hazardous waste management system
40 CFR Part 261	Identification and listing of hazardous waste
40 CFR Part 262	Standards applicable to generators of hazardous waste
40 CFR Part 263	Standards applicable to transporters of hazardous waste
40 CFR Part 264	Standards for owners and operators of hazardous waste treatment, storage, and disposal facilities
40 CFR Part 265	Interim status standards for owners and operators of hazardous waste treatment, storage, and disposal facilities
40 CFR Part 267	Interim standards for owners and operators of new hazardous waste land disposal facilities
40 CFR Part 270	EPA administered permit programs: the hazardous waste permit program
40 CFR Part 271	Requirements for authorization of state hazardous waste programs

TABLE 4
RCRA Incinerator Regulations

Reference	Description
Part 124	General permit application and decision-making process
Part 260	Regulatory overview and definitions
Part 261	Identifies hazardous waste by the characteristics of ignitability, corrosivity, reactivity, EP toxicity, and by waste numbers grouped into chemical class and waste source
Part 262	Generator standards including pollution control waste streams from incinerators
Part 264	Performance standards and facility requirements for permitted facilities including incineration
Part 265	Performance standards and facility requirements for interim status facilities including incineration
Part 270	Federal permit process for hazardous waste facilities; should be used in conjunction with Part 124, includes permit application information requirements and applicable general permit conditions, incinerator data required are outlined
Part 271	State program authorization procedures

TABLE 5
Incinerators — 40 CFR Subpart O

Reference	Description
Section 264.340	Applicability
Section 264.341	Waste analysis
Section 264.342	Principal organic hazardous constituents (POHCs)
Section 264.344	Permits
Section 264.347	Monitoring and inspection
Section 264.351	Closure

consists of two phases, and the state may apply for one or both phases at the same time. Phase I allows states to administer the program with respect to:

1. Identification and listing of hazardous waste (40 CFR Part 261)
2. Generators (Part 262)
3. Transporters (Part 263)

Phase II allows the states to administer the program with respect to hazardous waste treatment, storage, and disposal facilities.

Due to continuing amendment of the Act and promulgation of regulations, phase II can be implemented in components.

While state programs must meet the minimum requirements of the federal RCRA, they are free to be both more stringent in and have more breadth to their law and regulations. This fact plus the fact that the state program is implemented in phases, with the EPA still having authority over unapproved phases, allows dual review of permit applications and compliance and enforcement until the full program is approved. Also, as amendments to the federal Act are enacted, the EPA retains authority over that part of the program until the state program is amended.

c. The Toxic Substances Control Act

TSCA is generally concerned with the regulation of chemical substances and mixtures in commerce to protect human health and the environment. Disposal of such substances, as a waste, is usually covered by RCRA, with the exception of asbestos and polychlorinated biphenyls (PCBs). Probably for historical and political reasons the disposal of these two substances is regulated by TSCA. For incineration, PCB disposal is of the only interest. PCB incineration and facility requirements are similar to RCRA requirements with a few differences. The main difference is the brevity of the incinerator and facility regulations. Due to the brevity of the regulations, TSCA regulators have greater flexibility overall than RCRA regulators, although some incinerator requirements are more restrictive. This flexibility can be at different times both a help and a hindrance in the permitting and operation of a facility. Normally, TSCA regulators will follow RCRA regulations when TSCA regulations do not specify requirements for certain situations. However, the regulators have the latitude to be more restrictive, if in their opinion, the restriction is justified for protecting human health or the environment. In other cases, stumbling blocks to permitting and operation of a facility that are sometimes found in RCRA can occasionally be circumvented in TSCA when any adverse impacts can be shown to be nonexistent or negligible.

TSCA addresses PCB disposal in Section 6(e) of the Act and merely states that the EPA will promulgate rules to prescribe methods for disposal of PCBs and require PCBs to be marked with clear and adequate warnings and instructions with respect to their processing, distribution in commerce, use, disposal, or with respect to any combination of such activities.

The promulgated regulations for storage and disposal of PCBs are contained in 40 CFR Part 761 Subpart D. Disposal by incineration is regulated by 40 CFR Part 761.60. These regulations are brief enough that the reader is directed to Section III of this chapter where the permitting process is discussed in detail.

d. Comprehensive Environmental Response, Compensation, and Liability Act

CERCLA was enacted due to the need for a national response to real or potentially harmful releases of chemicals to the environment or ones that would affect public safety. The national response provided by CERCLA was to do the following:

1. Establish quantities of chemicals that, if released in equal or greater quantities, must be reported.
2. Establish federal authority to act in accordance with the National Contingency Plan, and the National Priority List established by the plan, to remove, or take any other action necessary, to mitigate releases that pose a threat to the environment or human safety and health. The Act establishes the procedure of conducting remedial investigation and feasibility studies (RI/FS) prior to the remediation of a release.
3. Establish the concept of liability for remediation costs incurred by government directed response action. It also establishes penalties for failure to report releases of reportable quantities of chemicals.

4. Establish the Agency for Toxic Substances and Disease Registry, with one of its functions, in cooperation with the EPA, to prepare a list, in order of priority, of hazardous substances which are commonly found at facilities or sites on the National Priority List and that pose the most significant threat to human health.
5. Establish public participation by requiring public notice and a description of the proposed remediation plan. It stipulated the requirement for providing proper opportunity for public comment.
6. Provide the framework for promulgating cleanup standards.
7. Provide methods of funding the program by taxing chemical corporations and creating the Hazardous Substances Superfund — hence, the common name of this Act as "Superfund".

One of the methods used in the remediation of Superfund sites is incineration, either on-site or off-site. The Act, or the regulations, does not mention particular technologies. Incineration is recognized as a technology that creates a permanent solution, one of the considerations in selecting a particular remediation method. The regulations also affect incineration indirectly in other ways. The regulations are listed in 40 CFR Subpart J (40 CFR Parts 300 to 399). Pertinent regulations that affect incineration include Part 300.68, Part 300.70, and Appendix B (National Priority List By Rank).

CERCLA remediation efforts do not require federal, state, and local permits for Superfund financed remediation. Remedial actions that involve off-site storage treatment, and/or disposal of hazardous substances or other pollutants must only be accomplished at facilities that are operating under appropriate federal or state permits or authorization and other legal requirements. In addition, the remedial program manager (RPM) must carry out his responsibilities in a remedial action as delineated in 300.33(b). As such and under 300.68(d) and (e) and (e)(2), consideration of environmental risk must be demonstrated in any remedial action. These considerations and the stipulation for public participation allow the RPM and lead agency to request that the remediation method proposal provide similar information about the remediation process to what is normally submitted with a RCRA or TSCA permit application. As an example, for incineration, a trial burn plan and trial burn might be two of the requirements stipulated to demonstrate that the remediation effort will comply with environmental laws and regulations and the protection of the environment. Requirements that will apply to a superfund remediation will be spelled out in bidding documents supplied by the lead agency/RMP. It is assumed that the requirements will parallel CAA and RCRA regulations, and the reader is directed to the CAA and RCRA discussions in Section II and III of this chapter.

e. The Clean Water Act

The CWA was enacted with the original intent of eliminating wastewater discharges and, hence, water pollution. As a first step toward this goal, a wastewater permit system was established by the Act and regulations. This system was titled the National Pollution Discharge Elimination System (NPDES). Each and every wastewater discharge into the waters of the U.S. must have a permit that limits the types and amounts or concentration of pollutants discharged. As in RCRA, states may assume the program responsibility once approved by the EPA. Incinerators or facilities that discharge wastewater must have a permit to do so (even under CERCLA). Applicable wastestreams for incineration would include wet scrubber discharges. In many states, the state counterpart NPDES permit program (usually with the state initials instead of the N) includes wastewater and oil discharges to the land of the state. Bed ash and fly ash slurry systems may be included in some instances. Application requirements include a wastestream characterization of the types and amounts of pollutants upstream and downstream of any treatment, annual and average wastestream volumes, a description of any wastewater treatment systems (BACT), and the expected annual and average discharge volume and contaminant concentration/loadings. The emphasis in the act was originally on receiving body water quality and then shifted to technologically based criteria. The emphasis is partially swinging back to a water quality basis. In some cases toxicity tests (fish or other aquatic organism bioassays) may be required in addition to or in lieu of technology-based discharge limits. Pertinent sections of this act include:

1. Title III: Standards and Enforcement
    *   Section 301: Effluent Limitations
    *   Section 302: Water Quality Related Effluent Limitations
    *   Section 303: Water Quality Standards
    *   Section 307: Toxic and Pretreatment Standards
    *   Section 308: Inspections, Monitoring and Entry
    *   Section 309: Federal Enforcement
    *   Section 311: Oil and Hazardous Substance Liability
2. Title IV: Permits and Licenses
    *   Section 402: NPDES System
3. Title V: General Provisions
    *   Section 507: Definition of Point Source
    *   Section 510: State Authority

Pertinent sections of the regulations include:

*   40 CFR 112: Regulations on Oil Pollution Prevention
*   40 CFR 122: NPDES Permit Regulations
*   33 CFR 153—157: Coast Guard Regulations on Oil Spills
*   40 CFR 116: Regulations on Designations of Hazardous Substances
*   40 CFR 117: Regulations on Determination of Reportable Quantities of Hazardous Substances
*   40 CFR 124: Regulations on the Permitting Process
*   40 CFR 125: Regulations on Criteria and Standards for NPDES
*   40 CFR 403: Regulations on Pretreatment Standards
*   40 CFR 136: Regulations for Test Procedures for the Analysis of Pollutants

2. State Laws Impacting Incineration

The federal laws CAA, RCRA, and CWA have sections authorizing state programs (implemented by state legislation) to assume authority for the respective programs if they fulfill the requirements stipulated by the regulations. There is no state counterpart for TSCA. It is strictly a federal program. While there is no CERCLA state authorization, the Federal Government encourages state participation, especially as lead agency in remediation efforts. Some states have enacted state superfund laws of their own to clean up sites within their states that have high state priority but are not necessarily on the National Priority List. Some states have enacted additional laws that require environmental impact reviews of projects similar to the federal NEPA law, but not necessarily limited to government actions or government holdings. Hazardous waste facility siting laws are becoming common, and some of these laws may require environmental impact statements and risk management assessments including hazard analyses. Due to Congress mandating that states show enough capacity for various forms of waste treatment and disposal to handle their own hazardous waste generation, some states have legislated that government entities within the state must originate plans to provide enough waste minimization, recycling, and treatment and disposal facilities to handle their own hazardous waste. State and local laws must be crafted in order to avoid preemption by the federal laws, especially RCRA. State and local laws designed exclusively to placate NIMBY activists cannot be used to thwart the intent of Congress to provide an adequate number of safe hazardous-waste treatment facilities.

3. Local Laws and Regulations Impacting Incineration

Local laws and regulations that may impact incineration include the following:

a. Local Air Pollution Control Districts or Agencies

Some states through their CAA legislation have authorized their air quality control regions (AQCRs) to subdivide into local districts, which may by in the form of individual countywide districts that have rules, that while generally conforming to the policies of the AQCR and State and Federal legislation/regulations, may be tailored to the particular mix of pollution problems in the area in relation to local air quality and the SIP (state implementation plan). These local rules must be thoroughly investigated and understood as they relate to a particular incinerator and facility process(es).

b. Master Plans, Zoning, and Land Use

Siting of new incinerators should, if possible, be consistent with a local government's master development plan (the master plan could have poison NIMBY elements). The master plan is usually implemented by zoning or land use ordinances. Zoning and the master plan are not always in agreement with each other. While master plans are usually updated periodically, zoning changes to reflect the updated master plan may change more slowly. Individual facilities may have to apply to have zoning changed if the use is consistent with the master plan. Local governments may also have permitting requirements for commercial and industrial entities such as conditional use permits.

c. Other

Other local laws and regulations are local health codes, building codes, and fire codes all of which may require permits. Historical, cultural, and archaeological guidelines may impact siting and operation. Hazard regulations, although relegated mostly to new laws at state levels, may be increasingly important locally. Agricultural lands may also be zoned and protected from development. Coastal zone regulations are important to investigate in states that are bordered by the ocean. In states bordering other countries it is important to investigate any international environmental and development agreements.

II. THE MAJOR LAWS AND REGULATIONS AFFECTING INCINERATION OF HAZARDOUS WASTES

A. CLEAN AIR ACT

1. Statutory and Regulatory Overview

Section 107 of the Act requires each state to subdivide their geographic area into regions called air quality control regions (AQCRs). These regions are responsible for promulgating area specific regulations peculiar to their own circumstances and pollution sources and problems. However, they are restrained by the fact they are charged with maintaining or improving the air quality of the area. The various AQCRs are listed in 40 CFR Part 81. The EPA periodically publishes results on which AQCRs are within compliance with national regulations that form the framework for the regulatory efforts of the area.

Section 108 of the Act establishes the overall framework for regulation by establishing "criteria pollutants" that the EPA believes causes or contributes to air pollution and that can be anticipated to endanger the public health or welfare.

Section 109 empowers and directs the EPA to establish national ambient air quality standards (NAAQS) for each criteria pollutant. Each criteria pollutant has two standards set. The primary ambient air quality standard is designed to protect public health. The secondary ambient air quality standard is designed to protect the public welfare. These standards for the criteria pollutants are listed in 40 CFR Part 50. These are minimum standards. States are free to set more stringent standards.

Section 110 requires each state to adopt and submit to the EPA for approval a state implementation plan (SIP) which would, if adhered to, cause each AQCR to meet the primary standard. As part of this plan or as a separate plan, states must submit a plan which would cause the AQCR to meet the secondary standard. The plan(s) are to contain emissions limitations, schedules, and timetables for compliance. One of these provisions is the pre-construction review of direct sources and establishes regulation by permit of the modification, construction, and operation of stationary sources. Emissions limitations set to maintain or meet the primary standards vary widely throughout the country depending on the AQCR performance relative to the primary and secondary air standards, although the national minimum limit is to some extent effective in preventing one area of the country from becoming more of an air pollution problem than another.

Section 111 requires the EPA to promulgate emission standards for new stationary sources or for modifications of existing sources. These standards are termed new source performance standards (NSPS), and their purpose is to establish minimum standards for emissions from new sources. Again, a state may impose stricter standards. These regulations are codified as 40 CFR Part 60 with incinerators specifically as Part 60.50. The provisions of the subpart are applicable to incinerators with a charging rate of 45 t (50 ton) and built after August 17, 1971. The only national standards for incinerators is a maximum discharge of 0.08 gr per dry standard cubic foot (dscm) particulates corrected to 12% CO_2 and the requirement to record the daily charge rates and hours of operation.

Section 112 sets the national emission standards for hazardous air pollutants (NESHAPs). A hazardous air pollutant is defined by the act as one that has no applicable ambient air quality standard (not a criteria pollutant) and not a NSPS, but in the judgment of the EPA causes or contributes to air pollution which may be reasonably anticipated to result in an increase in mortality or serious illness.

The purpose of the NESHAPs emission standards is to limit the emissions of these pollutants from certain existing as well as new source categories. Currently, the following pollutants have NESHAPs limitations: asbestos, beryllium, mercury, vinyl chloride, arsenic from glass-manufacturing plants, primary copper smelters, metallic arsenic production plants, and radionuclides from federal or federally licensed facilities. The EPA is considering benzene, coke oven emissions, lead, fluorides, cadmium, and POM for incorporation into this list. The EPA has authority to delegate this program to qualifying states.

Section 160 discusses the prevention of significant deterioration of air quality (PSD) in an attainment AQCR. PSD review is triggered by major emitting facilities which are defined as sources that emit or have the potential to emit 100 ton/year or more of any pollutant regulated under the Act. PSD allows for only a small increment above the baseline con-centrations of SO_2 and particulates with the increment size depending on the area classifi-cation. The EPA is under pressure to expand the list of pollutants included. The effect of siting a facility in one area on an adjacent area is also considered. Air modeling will have to be done in many siting decisions. A detailed preconstruction review is often required under PSD depending on the throughput and corresponding emissions of the incinerator. The following steps are included in the process of siting a facility:

1. Determine the applicability of PSD and, if applicable, the PSD class.
2. Perform meteorological monitoring and determination of baseline characteristics of the pollutant.

3. Determine emission characteristics of the source.
4. Model emissions to determine if the applicable increments in the area are exceeded.
5. Show through modeling the effects on adjacent PSD areas.
6. Demonstrate the application of "best available control technology".
7. Submit application and receive back approval before construction.

Section 129 discusses nonattainment areas. Congress and the EPA devised the concept of "emissions offsets". All new source construction or modification in a nonattainment area is not allowed without an emission offset or reduction so that the overall emissions do not increase. States may require a net decrease in emissions of various amounts. New source construction or modification must utilize pollution control equipment that accomplishes the "lowest achievable emission rate" (LAER). LAER is defined as "the most stringent emission limitation which is contained in the implementation plan of any state for such class or category of sources or the most stringent emission limitation which is achieved in practice by such class or category of source, whichever is more stringent". All other facilities owned by the applicant in the state must be in compliance with pollution control laws. This offset policy has given rise to the concept of "banking" emissions. A company reducing emissions below that required can register the portion of the emissions reduction below the required level with the state and receive emissions reduction credit (ERC) for expansion, or once registered can even sell the rights to the emissions credit to another company wishing to locate or expand in the nonattainment area.

The concepts involved in PSD and nonattainment areas overlap. An AQCR may be a nonattainment area for one pollutant and a PSD for another.

44 FR 71780 Alternative Emissions Reductions Options discusses what is commonly called the "EPA Bubble Policy". The bubble policy concept allows the individual polluter to consider the emissions of a pollutant from an entire facility instead of just from each individual point source in decisions on how to reduce the facilities emissions. The advantage of this concept is that it may be more cost effective to reduce the emissions of one point source in excess of another so that the sum total of reductions is equivalent to reducing both point sources equally. In order to utilize this concept, the approved state SIP must incorporate its provisions, and the pollutant source must bear the responsibility of demonstrating the advantages of utilizing the concept as well as the costs of study and implementation. The pollutants in the exchange must be equivalent (CO for CO), the air quality under the bubble must be the same or better than under the point source regulations, which may require air modeling for demonstration, the emissions must be quantifiable and enforceable, cost savings apparent, and the demonstration should have credible contingency factors in the analysis in its favor. The bubble concept, if allowed by state SIPs, is a viable economic tool to avoid PSD review of plant modifications. The goal of the PSD area concept is to keep areas clean that are already clean, and the bubble concept does not increase pollution.

2. Regulations

The Clean Air Act regulations consist of 40 CFR Subchapter C air programs, and the 40 CFR references are listed in Table 6. Regulations pertinent to hazardous waste incineration are included (or perhaps a better word would be "hidden") in 40 CFR 50, 51.165, and 51.166, Appendix S of Part 51, and Parts 52 subpart A General Provisions, 53, 58, 60, 61, 62, and 81.

B. RESOURCE CONSERVATION AND RECOVERY ACT
1. Statutory Overview
a. Introduction

Through Subtitle C of the Resource Conservation and Recovery Act of 1976, as amended (RCRA),[2] Congress crafted a two-step response to the nation's hazardous waste crisis. First,

TABLE 6
40 CFR Subchapter C

Part 50	National primary and secondary ambient air quality standards
Part 51	Requirements for preparation, adoption, and submittal of implementation plans
Part 52	Approval and promulgation of implementation plans
Part 53	Ambient air monitoring reference and equivalent methods
Part 54	Prior notice of citizen suits
Part 55	Reserved
Part 56	Regional consistency
Part 57	Primary nonferrous smelter orders
Part 58	Ambient air quality surveillance
Part 59	Reserved
Part 60	Standards of performance of new stationary sources
Part 61	National emissions standards for hazardous air pollutants
Part 62	Approval and promulgation of state plans for designated facilities and pollutants
Part 63—64	Reserved
Part 65	Delayed compliance orders
Part 66	Assessment and collection of noncompliance penalties by EPA
Part 67	EPA approval of state noncompliance penalty program
Part 68	Reserved
Part 69	Special exemptions from requirements of the Clean Air Act
Part 70—78	Reserved
Part 79	Registration of fuels and fuel additives
Part 80	Regulation of fuels and fuel additives
Part 81	Designation of area for air quality planning purposes
Part 82	Protection of stratospheric ozone
Part 83—84	Reserved
Part 85	Control of air pollution from motor vehicles and motor vehicle engines
Part 86	Control of air pollution from new motor vehicles and new motor vehicle engines: certification and test procedures
Part 87	Control of air pollution from aircraft and aircraft engines
Part 88—89	Reserved

RCRA establishes a comprehensive federal program of hazardous waste management administered by the EPA. Second, Congress encouraged states to adopt their own hazardous waste management programs that meet the minimum requirements of the federal program and to seek authorization to administer such programs under RCRA in lieu of the federal program.

Through RCRA, Congress intended to ensure "that hazardous waste management practices are conducted in a manner which protects human health and the environment."[3]

Subtitle C of RCRA is the primary federal statute regulating management of hazardous wastes.[4] The regulatory requirements of RCRA Subtitle C are keyed to RCRA § 3001, 42 U.S.C.A. § 6921, which requires the EPA to promulgate regulations identifying the characteristics of hazardous wastes (and listing particular hazardous wastes) that are subject to Subtitle C regulations. Sections 3002 and 3003 of RCRA, 42 U.S.C.A. §§ 6922 and 6923, require the EPA to promulgate regulations establishing standards applicable to generators and transporters, respectively, of hazardous waste identified or listed under RCRA Subtitle C,[5] as may be necessary to protect human health and the environment. Importantly, RCRA § 3002, 42 U.S.C.A. § 6922(a)(5), requires establishment of a manifest system to track hazardous wastes from generation to disposal.

Section 3004 of RCRA, 42 U.S.C.A. § 6924, requires the EPA to "promulgate regulations establishing such performance standards, applicable to owners and operators of facilities for the treatment, storage, or disposal of hazardous waste identified or listed under [RCRA Subtitle C], as may be necessary to protect human health and the environment."[6] RCRA regulation of such treatment, storage, and disposal (TSD) facilities is applicable to HWIFs.

Specifically, Congress directed that performance standards for TSD facilities shall include requirements respecting:

1. Maintenance of records reflecting the disposition of all hazardous wastes treated, stored, or disposed of at the facility
2. Satisfactory reporting, monitoring, and inspection and compliance with the manifest system established by RCRA § 3002, 42 U.S.C.A. § 6922(5), to track the movement of hazardous wastes
3. Treatment, storage, or disposal of all hazardous waste received by the TSD facility pursuant to operating methods, techniques and practices approved of by the EPA
4. Location, design, and construction of TSD facilities
5. Contingency plans for effective action to minimize unanticipated damage from any hazardous waste treatment, storage, or disposal
6. Maintenance and operation of such facilities, including such additional qualifications as to ownership, continuity of operation, personnel training, and financial responsibility (including financial responsibility for corrective action) as may be deemed necessary or desirable by the EPA
7. Compliance with RCRA permitting requirements for TSD facilities[7]

Congress directed that, in establishing TSD facility standards, the EPA distinguish, where appropriate, between requirements appropriate for new facilities and requirements appropriate for facilities in existence on the date of promulgation of such regulations.[8]

Section 3005 of RCRA, 42 U.S.C.A. § 6925, requires the EPA to promulgate regulations "requiring each person owning or operating an existing facility or planning to construct a new facility for the treatment, storage, or disposal of hazardous waste identified or listed under [RCRA Subtitle C] to have a permit issued pursuant to this section" (a "RCRA permit").[9] No *new* TSD facility may be built or may treat, store, or dispose of such hazardous waste except in accordance with a RCRA permit.[10] Section 3005 provides a procedure by which *existing* TSD facilities newly covered by RCRA may attain "interim status" and continue to operate during EPA consideration of their RCRA permit applications.[11] Each RCRA permit must contain such terms and conditions as the EPA (or an authorized State) determines are necessary to protect human health and the environment.[12]

In enacting RCRA, Congress hoped to establish a federal-state partnership to manage hazardous wastes successfully. Thus, in section 3006 of RCRA, 42 U.S.C.A. § 6926, Congress provided that, under certain conditions, states may be delegated the authority to administer a state hazardous waste management program under RCRA in lieu of the federal program. Because a state program must be equivalent to and consistent with the federal program to be "authorized" under this section,[13] authorized state programs are very similar to the federal program. An increasing number of states have obtained such "RCRA authorization."[14] The EPA retains ultimate oversight authority over each such State program.[15]

Sections 3007 and 3008 of RCRA, 42 U.S.C.A. §§ 6927 and 6928, provide for the EPA to enforce the standards promulgated under RCRA §§ 3004 and 3005 by directing the EPA to inspect TSD facilities and authorizing the EPA to issue orders assessing civil penalties, requiring compliance, or both, or to seek civil or criminal penalties in federal court for violations of RCRA regulations or a RCRA permit.

b. Performance Standards

In accordance with RCRA § 3004, 42 U.S.C.A. § 6924, the EPA has promulgated and continues to promulgate regulations that establish performance standards for the treatment, storage, and disposal of hazardous waste.[16] The EPA has issued one set of regulations generally applicable to TSD facilities that operate pursuant to a RCRA permit (permitted facilities)[17] and another set for TSD facilities operating under interim status (interim status facilities). The standards governing permitted facilities and interim status facilities are set forth in Title 40 of the Code of Federal Regulation (40 CFR), Parts 264 and 265, respectively. Because the purpose of both sets of standards is to protect human health and the environment, most of the requirements imposed on permitted facilities and interim status facilities are the same. Where the standards differ, the differences will be identified and discussed.

Many of the standards are generally applicable to all TSD facilities. In brief, these regulations relate to:

1.   General facility standards
2.   Preparedness for and prevention of releases of hazardous waste
3.   A contingency plan and emergency procedures
4.   The manifest system, record keeping and reporting
5.   Responses to releases of hazardous waste
6.   Closure and postclosure
7.   Financial requirements[18]

These general requirements are incorporated into RCRA permits issued for TSD facilities.[19]

The EPA has also promulgated specific standards for each of ten different types of hazardous waste facility units and a separate standard for certain miscellaneous units. These standards cover the use and management of the following:

1.   Containers
2.   Tanks and tank systems
3.   Surface impoundments
4.   Waste piles
5.   Land treatment
6.   Landfills
7.   Incinerators[20]
8.   Thermal treatment[21]
9.   Chemical, physical, and biological treatment[22]
10. Underground injection wells[23]
11. Miscellaneous units[24]

The specific standards applicable to incinerators are set forth in 40 CFR Parts 264 and 265, Subparts O, and are discussed in Section III.B.1.b.[25]

Section 3004 also requires the EPA to "promulgate such regulations for the monitoring and control of air emissions at hazardous waste [TSD] facilities . . . as may be necessary to protect human health and the environment." 42 U.S.C.A. § 6924(n). Some of the regulations specifically applicable to incinerators are based on this section. Further, the EPA issued proposed regulations governing emissions of volatile organics in February 1987 and intends to publish the final regulations in August 1989. The proposed regulations address leaks in equipment used with hazardous wastes containing greater than 10% total organics by weight. When finalized, the regulations will be set forth in 40 CFR Part 269.

c. Permitting Requirements

As noted above, RCRA § 3005, 42 U.S.C.A. § 6925, requires the EPA to promulgate regulations requiring every owner or operator of an existing or planned TSD facility to obtain a RCRA permit. Those regulations are set forth in 40 CFR Part 270. The treatment, storage, or disposal of any hazardous waste and, with one exception,[26] the construction of any new TSD facility is prohibited except in accordance with such a permit.[27]

There are three ways to satisfy the permit requirement. First, the owner or operator of a TSD facility may obtain an individual RCRA permit from either the EPA, an "authorized" state, or both. In some states, both the EPA and an authorized state are involved in the permitting process because the state program has not yet been fully authorized by the EPA or because the state program has not yet incorporated the amendments made to RCRA in 1984.

Each application for a RCRA permit shall contain such information as may be required under EPA regulations.[28] The EPA or authorized state shall issue a RCRA permit to the owner or operator of a TSD facility or a determination by the EPA or authorized State of compliance by the TSD facility with the requirements of RCRA § 3005 and § 3004.[29] In the case of each RCRA permit application submitted by an HWIF before November 8, 1984, the EPA or authorized State shall issue a final permit or a final denial no later than November 8, 1989.[30]

RCRA permits must be for a fixed term and cannot exceed 10 years in the case of any incineration facility. Under certain circumstances, the EPA may review and modify a permit during its term.[31] Review of any application for a permit renewal must consider improvements in the state of control and measurement technology as well as changes in applicable regulations.[32] On determination by the EPA or authorized State that a permitted facility has not complied with the requirements of RCRA § 3004 or § 3005, the EPA or authorized State shall revoke that facility's RCRA permit.[33]

One special type of RCRA permit is worthy of note. In the 1984 amendments to RCRA, Congress provided that the EPA "may issue a research, development, and demonstration permit for any hazardous waste treatment facility which proposes to utilize an innovative and experimental hazardous waste treatment technology or process for which permit standards for such experimental activity have not been promulgated under this subchapter."[34] Although the EPA may modify or waive many general permit requirements (except those governing financial responsibility and permitting public participation), a research, development, and demonstration (RD&D) permit must include such requirements as the EPA deems necessary to protect human health and the environment.[35]

RCRA RD&D permits may provide for the construction of an RD&D facility and for operation of the facility for not longer than 1 year.[36] An RD&D permit may be renewed up to three times, with each renewal period being not more than 1 year.[37] The RD&D permit will provide for the receipt and treatment by the facility of only those types and quantities of hazardous waste that the EPA deems necessary for purposes of determining the efficacy and performance capabilities of the technology or process, and the effects of such technology or process on human health and the environment.[38]

The second way that a TSD facility may satisfy the permit requirement is by obtaining "interim status". Congress recognized that the permit application and issuance process would take time and that existing TSD facilities could not be shut down during that process. Thus, an *existing* TSD facility may satisfy the permit requirement by qualifying for "interim status" under RCRA § 3005(e), 42 U.S.C.A. 6925(e).

To obtain "interim status", the owner or operator of a TSD facility that either was in existence on November 19, 1980 or is in existence on the effective date of statutory or regulatory changes under RCRA that render the facility subject to a RCRA permit requirement must (1) file with the EPA (or with authorized states) a notification describing the activities subject to regulation and (2) apply for a permit.[39] An owner or operator that has complied with such requirements shall be treated as having been issued a RCRA permit until such time as final administrative disposition of its application is made, unless the EPA or other plaintiff proves that final administrative disposition of such application has not been made because of the failure of the applicant to furnish information reasonably required or requested to process the application.

Interim status will have little application to most HWIFs at this time. Existing facilities that were required to obtain a permit under the initial RCRA regulations published on May 19, 1980 were required to submit Part A of their permit applications by November 19, 1980. For each application submitted prior to November 8, 1984, the EPA is required to issue a final permit or final denial by November 8, 1989.[40] Thus, all facilities handling hazardous wastes initially identified or listed under 40 CFR Part 261 (and not excluded elsewhere) already must have applied for a RCRA permit (and attained interim status) and will soon learn the outcome of their applications. Because the Part 261 regulations are very broad in identifying "hazardous wastes", few HWIFs will find themselves incinerating wastes newly covered by amended RCRA regulations expanding the definition of hazardous wastes.

The third way for a TSD facility to satisfy the permit requirement is to qualify for a "permit-by-rule". The EPA has provided that TSD facilities that are subject to certain other environmental permitting programs, and that comply with certain standards promulgated under RCRA § 3004, shall be deemed to have a RCRA permit.[41]

d. State Authorization

In RCRA § 3006, 42 U.S.C.A. § 6926, Congress provided that any State which seeks to administer and enforce a hazardous waste program pursuant to RCRA may develop and, after notice and opportunity for public hearing, submit to the EPA an application, in such form as the EPA may require, for authorization of such program. The EPA shall authorize such program unless the EPA finds that (1) such state program is not equivalent to the federal program under RCRA, (2) such program is not consistent with the federal program or state programs applicable in other states, or (3) such program does not provide adequate enforcement of compliance with the requirements of RCRA. If the state program is authorized, such state is authorized to carry out such program in lieu of the federal program under RCRA in such state. An "authorized state" shall issue and enforce permits for the storage, treatment, or disposal of hazardous waste.[42]

Until a state receives either interim or final authorization for its hazardous waste management program, the EPA will directly apply the federal program in that state. States that do not qualify for either interim or final authorization of a state program reflecting the 1984 RCRA amendments may enter a cooperative agreement with the EPA to assist in the administration of those amendments.

e. Enforcement Mechanisms

Section 3008 of RCRA, 42 U.S.C.A. § 6928, provides that whenever the EPA determines that any person has violated or is in violation of any RCRA requirement, the EPA may (1) issue an order assessing a civil penalty for any past or current violation, requiring compliance immediately or within a specified time period, or both, or (2) commence a civil action in the U.S. District Court in the district in which the violation occurred for appropriate relief, including a temporary or permanent injunction. If such violation occurs in a state with an authorized state hazardous waste program, the EPA shall give notice to the state prior to issuing an order or commencing a civil action under this section. Section 3008 also provides for criminal penalties.

RCRA § 7002, 42 U.S.C.A. § 6972, also provides that any person may commence a civil action on his behalf: (1)(A) against any person who is alleged to be in violation of any permit, standard, regulation, condition, requirement, prohibition, or order which has become effective pursuant to RCRA; or (B) against any person who has contributed or who is contributing to the past or present handling, storage, treatment, transportation, or disposal of any solid or hazardous waste which may present an imminent and substantial endangerment to health or the environment; or (2) against the EPA where there is alleged a failure of the EPA to perform any act or duty under RCRA which is not discretionary with the EPA.

RCRA § 7003, 42 U.S.C.A. § 6973, provides that the EPA may bring suit on behalf of the U.S. against any person who has contributed or who is contributing to the handling, storage, treatment, transportation, or disposal of hazardous waste that may present an imminent and substantial endangerment to health or the environment. The EPA may seek (1) to restrain such person from such handling, storage, treatment, transportation, or disposal, (2) to order such person to take such action as may be necessary, or (3) both.

2. Application of RCRA to Hazardous Waste Incineration Facilities
a. "Hazardous Waste" Under RCRA

The requirements of RCRA § 3004 and § 3005 are applicable to "facilities" engaged in the "treatment, storage, or disposal" of "hazardous waste identified or listed" under Subtitle C of RCRA.[43] In determining whether an incineration facility falls under RCRA regulation, the key inquiry is whether or not it is handling hazardous waste as defined by EPA regulations promulgated in 40 CFR Part 261.

The regulations set forth in Part 261 identify those *wastes* that are subject to regulation as *hazardous waste* under RCRA regulations governing performance standards and permitting. Facilities handling hazardous wastes are subject to such regulations and a requirement that they notify the EPA or an authorized State of their activities.[44] Nonetheless, materials not identified or listed as a hazardous waste under Part 261 may still constitute a hazardous waste subject to EPA authority to inspect and monitor facilities handling "hazardous waste", and authority to abate imminent hazards, if the material meets the broad statutory definition of hazardous waste set forth in RCRA § 1004(5), 42 U.S.C.A. § 6903(5).

Although a detailed discussion of what materials constitute hazardous waste under Part 261 is beyond the scope of this book,[45] a brief explanation is that a solid waste is a hazardous waste if it (1) is not specifically excluded from regulation as hazardous waste under 40 CFR § 261.4(b) *and* (2) either exhibits any of the characteristics of hazardous waste identified in 40 CFR Part 261, Subpart C *or* is listed in 40 CFR Part 261, Subpart D.

b. Definition of "Facility"

A "facility" means "all contiguous land, and structures, other appurtenances, and improvements on the land, used for treating, storing, or disposing of hazardous waste."[46] A facility may consist of several treatment, storage, or disposal operational units. This definition is presently limited to nonmobile units. Thus, a new "facility" is created, and a new RCRA permit required, at each location that a portable unit is operated.

c. Definitions of "Treatment", "Storage", and "Disposal"

Under RCRA, the term "treatment", when used in connection with hazardous waste, means:

> any method, technique, or process, including neutralization, designed to change the physical, chemical, or biological character or composition of any hazardous waste so as to neutralize such waste, or so as to recover energy or material resources from the waste, or so as to render such waste nonhazardous; safer to transport, store, or dispose of; or amenable for recovery, amenable for storage, or reduced in volume. Such term includes any activity or processing designed to change the physical form or chemical composition of hazardous waste so as to render it nonhazardous.[47]

This definition clearly includes incineration of hazardous waste.

The term "storage", when used in connection with hazardous waste, "means the containment of hazardous waste, either on a temporary basis or for a period of years, in such a manner as not to constitute disposal of such hazardous waste."[48] An HWIF will engage in "storage" of hazardous waste when collecting waste for incineration.

Under RCRA, the term "disposal" means:

the discharge, deposit, injection, dumping, spilling, leaking, or placing of any solid waste or hazardous waste into or on any land or water so that such solid waste or hazardous waste or any constituent thereof may enter the environment or be emitted into the air or discharged into any waters, including ground waters.[49]

The EPA, however, has defined "disposal facility" to mean "a facility or part of a facility at which hazardous waste is *intentionally* placed into or on any land or water, and at which waste will remain after closure".[50] An HWIF will not ordinarily fall within the definition of "disposal facility".[51]

d. Definition of "Incinerator"

Finally, the EPA has promulgated regulations identifying and regulating specific types of TSD facilities. Under RCRA, "incinerator" means "any enclosed device using controlled flame combustion that neither meets the criteria for classification as a boiler nor is listed as an industrial furnace".[52] Although owners and operators of boilers or industrial furnaces may choose to be regulated under EPA regulations specifically applicable to incinerators,[53] such facilities will otherwise be subject to regulations governing facilities burning hazardous waste for energy recovery.[54] Thus, it is important to determine whether a facility is an incinerator or either a boiler or industrial furnace. It is important to note that some states do not allow the energy recovery exemption.

Under the RCRA regulations of the EPA, "boiler" means "an enclosed device using controlled flame combustion and having the following characteristics":

1. The unit must have physical provisions for recovering and exporting thermal energy in the form of steam, heated fluids, or heated gases;
2. The unit's combustion chamber and primary energy recovery section(s) must be of integral design. To be of integral design, the combustion chamber and the primary energy recovery section(s) (such as waterwalls and superheaters) must be physically formed into one manufactured or assembled unit. A unit in which the combustion chamber and primary energy recovery section(s) are joined only by ducts or connections carrying flue gas is not integrally designed; however, secondary energy recovery equipment (such as economizers or air preheaters) need not be physically formed into the same unit as the combustion chamber and the primary energy recovery section. The following units are not precluded from being boilers solely because they are not of integral design: process heaters (units that transfer energy directly to a process stream) and fluidized bed combustion units;
3. While in operation, the unit must maintain a thermal energy recovery efficiency of at least 60%, calculated in terms of the recovered energy compared with the thermal value of the fuel;
4. The unit must export and utilize at least 75% of the recovered energy, calculated on an annual basis. In this calculation, no credit shall be given for recovered heat used internally in the same unit. (Examples of internal use are the preheating of fuel or combustion air, and the driving of induced or forced draft fans or feedwater pumps);
5. The unit is one which the regional administrator has determined, on a case-by-case basis, to be a boiler, after considering the standards in 40 CFR § 260.32.[55]

EPA regulations define "industrial furnace" to mean any of the following enclosed devices that are integral components of manufacturing processes and that use controlled flame devices to accomplish recovery of materials or energy:

1. Cement kilns
2. Lime kilns

3.   Aggregate kilns
4.   Phosphate kilns
5.   Coke ovens
6.   Blast furnaces
7.   Smelting, melting, and refining furnaces (including pyrometallurgical devices such as cupolas, reverberator furnaces, sintering machine, roasters, and foundry furnaces)
8.   Titanium dioxide chloride process oxidation reactors
9.   Methane reforming furnaces
10. Pulping liquor recovery furnaces
11. Combustion devices used in the recovery of sulfur values from spent sulfuric acid
12. Other devices that the administrator may, after notice and comment, add to this list on the basis of one or more of the following factors:

   a. The design and use of the device primarily to accomplish recovery of material products
   b. The use of the device to burn or reduce raw materials to make a material product
   c. The use of the device to burn or reduce secondary materials as effective substitutes for raw materials, in processes using raw materials as principle feedstocks
   d. The use of the device to burn or reduce secondary materials as ingredients in an industrial process to make a material product
   e. The use of the device in common industrial practice to produce a material product
   f. Other factors, as appropriate[56]

e. Definition of "Container"

Because an HWIF may employ containers to store hazardous waste prior to incineration, and use of such containers is regulated under RCRA, an owner or operator of an HWIF must recognize that, under EPA regulations, "container" means any portable device in which a material is stored, transported, treated, disposed of, or otherwise handled.[57]

f. Hazardous Waste Incinerator Facility as "Generator"

An owner or operator who initiates a shipment of hazardous waste from a treatment, storage, or disposal facility must comply with the generator standards established in 40 CFR Part 262.[58] In exporting incinerator ash, and/or scrubber effluent, an HWIF will likely be subject to such standards unless such streams are delisted.

g. Exemptions from RCRA Regulations

A number of TSD facilities otherwise subject to RCRA regulation are specifically exempted by statute or regulation.[59] Most of these exemptions are not relevant to an HWIF. A potentially applicable "exemption" is that RCRA regulations do not apply to TSD facilities located in a state with a "authorized" state hazardous waste management program. Those states with programs authorized or partially authorized under RCRA are identified at 40 CFR Part 272. However, these states will have regulations at least as stringent as the federal RCRA.

3. TSD Facility Standards of General Applicability

As noted above, there are two sets of performance standards generally applicable to owners and operators of TSD facilities, one applicable to permitted facilities and the other applicable to interim status facilities. The following section will walk through these TSD facility standards of general applicability and, where appropriate, identify differences between standards applicable to RCRA permitted facilities and those applicable to interim status facilities.

a. General Facility Standards

i. ID Number

Every owner or operator of a TSD facility must apply to the EPA for an EPA identification number.[60] A generator of hazardous waste may not offer its hazardous waste to a TSD facility that has not received an EPA identification number.[61] A copy of the EPA notification form is shown as Figure 1.

ii. Required Notices

A TSD facility owner or operator that has arranged to receive hazardous waste from a foreign source must notify the EPA regional administrator[62] in writing at least 4 weeks before the facility expects to receive the waste. Notice of subsequent shipments of the same waste from the same foreign source is not required.[63]

The owner or operator of a permitted facility, but not an interim status facility, that receives hazardous waste from an off-site source (except where the owner or operator is also the generator) must inform the generator, in writing, that the owner or operator has the appropriate permit(s) for and will accept the hazardous waste that the generator is shipping. The TSD facility owner or operator must retain a copy of this written notice as part of its operating record.[64]

Before transferring ownership or operation of a TSD facility during its operating life,[65] or of a disposal facility during the postclosure care period, the existing owner or operator must notify the new owner or operator in writing of the requirements of the applicable TSD facility performance standards in either Part 264 or Part 265, whichever is applicable, and the permit requirements under Part 270.[66] An existing owner or operator's failure to so notify the new owner or operator may subject the existing owner or operator to both criminal and civil liability, but in no way will relieve the new owner or operator of its obligation to comply with all applicable requirements.[67]

iii. Waste Analysis

Before an owner or operator of a TSD facility treats, stores, or disposes of any hazardous waste, it must obtain a detailed chemical and physical analysis of a representative sample of the waste. This analysis must contain all the information that must be known to treat, store, or dispose of the waste in accordance with either Part 264 or Part 265, as applicable, and Part 268, *or* in accordance with the conditions of a RCRA permit, if any.[68] The analysis may utilize data developed under Part 261 and other existing data on the hazardous waste or hazardous waste generated from similar processes.[69]

The analysis must be repeated as often as is necessary to ensure that it is accurate and up to date. The waste analysis must be repeated when the owner or operator is notified or has reason to believe that the process or operation generating the hazardous waste has changed, and when the TSD facility's inspection of hazardous waste received at the facility indicates that the waste does not match the waste described on the accompanying manifest or shipping paper.[70]

The owner or operator of an off-site TSD facility may arrange for the generator of the hazardous waste to supply part or all of the information necessary for the waste analysis. Nonetheless, the owner or operator remains responsible for obtaining the information required to comply with the regulations.[71] If the generator does not supply the information, and the TSD facility chooses to accept a hazardous waste, the TSD facility must obtain the information required for the waste analysis. Even if a generator performs the necessary waste analysis, the owner or operator must inspect and, if necessary, analyze each hazardous waste received at the facility to determine whether it matches the identity of the waste specified on the accompanying manifest or shipping paper.[72]

Each owner or operator must develop and follow a written waste analysis plan that

Form Approved OME No 2050-0028 Expires 9-30-86
GSA No 0246-EPA-07

Please print or type with ELITE type (12 characters per inch) in the unshaded areas only

United States Environmental Protection Agency Washington, DC 20460	Please refer to the *Instructions for Filing Notification* before completing this form. The information requested here is required by law *(Section 3010 of the Resource Conservation and Recovery Act).*

♻EPA Notification of Hazardous Waste Activity

For Official Use Only

Comments

C

C

Installation's EPA ID Number	Approved	Date Received *(yr. mo. day)*	
C	T/A C		
F	1		

I. Name of Installation

II. Installation Mailing Address

Street or P.O. Box

C
3

City or Town	State	ZIP Code
C 4		

III. Location of Installation

Street or Route Number

C
5

City or Town	State	ZIP Code
C 6		

IV. Installation Contact

Name and Title *(last, first, and job title)*	Phone Number *(area code and number)*
C 2	

V. Ownership

A. Name of Installation's Legal Owner	B. Type of Ownership *(enter code)*
C R	

VI. Type of Regulated Waste Activity *(Mark 'X' in the appropriate boxes. Refer to instructions.)*

A. Hazardous Waste Activity	B. Used Oil Fuel Activities
☐ 1a. Generator ☐ 1b. Less than 1,000 kg/mo.	☐ 6. Off-Specification Used Oil Fuel *(enter 'X' and mark appropriate boxes below)*
☐ 2. Transporter	☐ a. Generator Marketing to Burner
☐ 3. Treater/Storer/Disposer	☐ b. Other Marketer
☐ 4. Underground Injection	☐ c. Burner
☐ 5. Market or Burn Hazardous Waste Fuel *(enter 'X' and mark appropriate boxes below)*	☐ 7. Specification Used Oil Fuel Marketer *(or On site Burner)* Who First Claims the Oil Meets the Specification
☐ a. Generator Marketing to Burner	
☐ b. Other Marketer	
☐ c. Burner	

VII. Waste Fuel Burning: Type of Combustion Device *(enter 'X' in all appropriate boxes to indicate type of combustion device(s) in which hazardous waste fuel or off-specification used oil fuel is burned. See instructions for definitions of combustion devices.)*

☐ A. Utility Boiler   ☐ B. Industrial Boiler   ☐ C. Industrial Furnace

VIII. Mode of Transportation *(transporters only — enter 'X' in the appropriate box(es)*

☐ A. Air   ☐ B. Rail   ☐ C. Highway   ☐ D. Water   ☐ E. Other *(specify)*

IX. First or Subsequent Notification

Mark 'X' in the appropriate box to indicate whether this is your installation's first notification of hazardous waste activity or a subsequent notification. If this is not your first notification, enter your installation's EPA ID Number in the space provided below.

	C. Installation's EPA ID Number
☐ A. First Notification ☐ B. Subsequent Notification *(complete item C)*	

EPA Form 8700-12 (Rev. 11-85) Previous edition is obsolete.   Continue on reverse

FIGURE 1.   EPA notification of hazardous waste activity.

describes the procedures utilized to ensure that the TSD facility is complying with the waste analysis requirements.[73] The waste analysis plan must include:

1. The parameters for which each hazardous waste will be analyzed and the rationale for the selection of those parameters, including a discussion of how analysis for such parameters will provide sufficient information to allow the TSD facility to properly store, treat, or dispose of the waste
2. The test methods that will be used to test for the selected parameters
3. The sampling method that will be used to obtain a representative sample of the waste to be analyzed[74]
4. The frequency with which the initial analysis of the waste will be reviewed or repeated to ensure that the analysis is accurate and up to date
5. The waste analyses that off-site hazardous waste generators have agreed to supply
6. Where applicable, the methods that will be used to meet the specific waste analysis requirements for particular types of TSD facilities, including those governing HWIFs
7. Particular requirements applicable to surface impoundments
8. For off-site TSD facilities, the procedures that will be used to inspect and, if necessary, analyze each movement of hazardous waste received at the facility to ensure that it matches the identity of the waste designated on the accompanying manifest or shipping paper[75]

The waste analysis plan must be kept at the facility,[76] and a copy must be submitted with Part B of the TSD facility's permit application.[77]

iv. Facility Security

The owner or operator of a TSD facility must prevent the unknowing entry, and minimize the possibility of the unauthorized entry, of persons or livestock into the "active portion" of the facility.[78] A TSD facility may be exempted from the security requirements if it can demonstrate to the EPA Regional Administrator that (1) physical contact with the waste, structures, or equipment within that portion of the facility will not injure persons or livestock; *and* (2) disturbance of the waste or equipment by such persons or livestock will not cause a violation of EPA requirements under Part 264 or Part 265, whichever is applicable.[79]

A nonexempt TSD facility must control entry into the "active portions" of the facility. The owner or operator may control entry in either of two ways. A TSD facility may utilize a 24-h surveillance system, such as television monitoring or surveillance by guards or facility personnel, that continuously monitors and controls entry into such portions of the facility. Second, a TSD facility may prevent unknowing or unauthorized entry by use of an artificial or natural barrier, such as a fence that completely surrounds the active portions of the facility, and a means to control entry through gates or other entrances at all times.[80] These security requirements are satisfied if the facility within which the active portion is located itself has a surveillance system, or a barrier and a means to control entry, that complies with such requirements.[81]

Each nonexempt TSD facility must post written warnings, stating "Danger — Unauthorized Personnel Keep Out", at each entrance to an active portion of the facility and elsewhere in sufficient numbers to be seen from any approach to the active portion. The warnings must be written in English and in any other language predominant in the area surrounding the facility, and must be legible from a distance of at least 25 ft.[82]

v. Inspection Requirements

A TSD facility owner or operator must inspect its facility for malfunctions and deterioration, operator errors, and discharges which may be causing or lead to either a release of hazardous waste constituents into the environment or a threat to human health. These inspections must be conducted often enough to identify problems in time to correct them before they result in harm to human health or the environment.[83] Although an owner or operator is required to undertake timely remedial action when an inspection reveals an environmental or human health hazard,[84] the goal of the inspection program is preventative.

As part of the general inspection program mandated above, a TSD facility must develop and follow a written schedule for inspecting monitoring equipment, safety and emergency equipment, security devices, and operating and structural equipment that are important to prevent, detect, or respond to environmental or human health hazards. The inspection schedule must be submitted with Part B of the facility's permit application.[85]

The inspection schedule must be kept at the facility, identify the types of problems that are to be looked for during the inspection, and designate the frequency of inspection for the various items on the schedule.[86] The frequency of inspection for individual items should be based on the rate of possible deterioration of the equipment and the probability of an environmental or human health hazard if the problem goes undetected between inspections. The regulations require that areas subject to spills, such as loading and unloading areas, be inspected daily when in use. Further, for an HWIF, the inspection schedule must include the terms and frequencies specifically required for incineration facilities by 40 CFR § 264.347 or § 265.347, as applicable.[87]

If an inspection reveals any deterioration or malfunction of equipment or structures, the owner or operator must remedy the problem on a schedule that ensures protection of human health and the environment. Where an environmental or human health hazard is imminent or has already occurred, remedial action must be taken immediately.[88]

A TSD facility owner or operator must record inspections in an inspection log or summary. Those records must be retained for at least 3 years after the date of inspection. The records must include the date and time of the inspection, the name of the inspector, a notation of the observations made, and the date and nature of any repairs or other remedial actions.[89]

vi. Personnel Training

TSD facility personnel must successfully complete a program of classroom instruction or on-the-job training that teaches them to perform their duties so as to ensure facility compliance with the requirements of Part 264 or Part 265, whichever is applicable, within 6 months after beginning employment or taking a new position at the facility. Until employees have completed this training, they may not work in an unsupervised position. All personnel must complete an annual review of the training program.[90]

The training program must be directed by a person trained in hazardous waste management procedures and must include instruction in such procedures relevant to the employee's position.[91] At a minimum, the training program must ensure that personnel are familiar with implementation of the facility's "contingency plan",[92] emergency procedures, emergency equipment, and emergency systems, including, where applicable:

1. Procedures for using, inspecting, repairing and replacing the facility's emergency and monitoring equipment
2. Key parameters for automatic waste feed cutoff systems
3. Communications or alarm systems
4. Response to fires or explosions
5. Response to groundwater contamination incidents
6. Shutdown of operations[93]

A TSD facility owner or operator must maintain personnel documents and records at the facility. Records must include the name and the job title of each employee working with hazardous waste, a written job description for each job title, a written description of the type and amount of both introductory and continuing training to be given to an employee under each job title, and documentation of the training or job experience given to or completed by facility personnel.[94] Training records on current personnel must be kept until closure of the facility. Training records on former employees must be kept for at least 3 years after the employee last worked at the facility.[95]

vii. General Requirements for Ignitable, Reactive, or Incompatible Waste
    A TSD facility owner or operator must take precautions to prevent accidental ignition or reaction of ignitable or reactive waste.[96] Such waste must be separated from other wastes and shielded from sources of ignition or reaction including, but not limited to: open flames, smoking, cutting and welding, hot surfaces, frictional heat, sparks, spontaneous ignition, and radiant heat. While such wastes are being handled, smoking and open-flame activities must be confined to specially designated locations. "No Smoking" signs must be conspicuously placed wherever there is a hazard from ignitable or reactive waste.[97]
    Where specifically required by other sections or Part 264 or Part 265, respectively, TSD facilities that treat, store, or dispose of ignitable or reactive waste, mix incompatible waste, or mix incompatible wastes and other materials, must take precautions to prevent reactions that:

1.   Generate extreme heat or pressure, fire or explosions, or violent reactions
2.   Produce uncontrolled toxic mists, fumes, dust, or gases in sufficient quantities to threaten human health and the environment
3.   Produce uncontrolled flammable fumes or gases in sufficient quantities to pose a risk of fire or explosions
4.   Damage the structural integrity of the device or facility
5.   Through other like means threaten human health or the environment[98]

    The owner or operator of a permitted TSD facility, but not an interim status TSD facility, must document compliance with the foregoing requirements. The documentation may be based on references to published scientific or engineering literature, data from trial tests, waste analyses, or the results of the treatment of similar wastes by similar treatment processes under similar operating conditions.[99]

viii. Location Standards
    The EPA has placed certain limits on where a *new* TSD facility may be located. To minimize the danger of a release caused by earthquake, the active portions of a new TSD facility may not be located within 61 m (200 ft) of a "fault which has had displacement in Holocene time".[100] Existing TSD facilities are not subject to this requirement.
    The EPA also has sought to protect against the danger of floods. A TSD facility located in a "100-year floodplain" must be designed, constructed, operated, and maintained to prevent "washout" of any hazardous waste by a "100-year flood", unless the owner or operator can satisfy the EPA regional administrator that one of two circumstances exists.[101]
    An owner or operator may demonstrate that existing procedures will cause all hazardous waste at the facility to be removed safely, before floodwaters can reach the facility, to an elevation where the waste will not be vulnerable to floodwaters.[102] The location to which such waste is moved must be a RCRA-permitted facility, a facility authorized to manage hazardous waste by an authorized state, or an interim status facility.[103] The second exception is applicable only to existing surface impoundments, waste piles, land treatment units, landfills, and miscellaneous units.[104]

Other federal laws that may affect the location and permitting of *new* TSD facilities are set forth in 40 CFR § 270.3; they include the Wild and Scenic Rivers Act, the National Historic Preservation Act of 1966, the Endangered Species Act, the Coastal Zone Management Act, and the Fish and Wildlife Coordination Act.

Both existing and new TSD facilities are forbidden to place any noncontainerized or bulk liquid hazardous waste in any salt dome formation, salt bed formation, underground mine or cave, except through the Department of Energy Waste Isolation Pilot Project in New Mexico.[105]

b. Preparedness and Prevention

i. Design and Operation of Facilities

Not surprisingly, the EPA requires that permitted TSD facilities must be designed, constructed, maintained, and operated to minimize the possibility of a fire, explosion, or any unplanned sudden or nonsudden release of hazardous waste or hazardous waste constituents to air, soil, or surface water that could threaten human health or the environment.[106] Interim status TSD facilities must be maintained and operated so as to minimize the possibility of such dangers.[107]

ii. Required Equipment

All TSD facilities must be equipped with certain emergency equipment, including:

1. An internal communications or alarm system capable of providing immediate emergency instruction (voice or signal) to facility personnel
2. A device, such as a telephone (immediately available at the scene of operations) or a hand-held two-way radio, capable of summoning emergency assistance from local police departments, fire departments, or state or local emergency response teams
3. Portable fire extinguishers, fire control equipment (including special extinguishing equipment, such as that using foam, inert gas, or dry chemicals), spill control equipment, and decontamination equipment
4. Water at an adequate volume and pressure to supply waterhose streams, or foam producing equipment, or automatic sprinklers, or waterspray systems.[108]

A TSD facility may avoid providing such equipment if it demonstrates to the EPA Regional Administrator that none of the hazards posed by waste handled at the facility *could* require any particular piece of the above-listed equipment.[109] An owner or operator seeking to make that demonstration must do so with Part B of its permit application.[110]

iii. Testing and Maintenance of Equipment

All TSD facility communications or alarm systems, fire protection equipment, spill control equipment, and decontamination equipment, where required as above, must be tested and maintained as necessary to ensure its proper operation in time of emergency.[111]

iv. Access to Communications Equipment

Whenever hazardous waste is being handled at a TSD facility, all personnel involved in the operation must have immediate access to an internal alarm or emergency communication device, either directly or through visual or voice contact with another employee, unless the EPA regional administrator has ruled that such a device is not required.[112]

If only one employee is on the premises during operation of a TSD facility, that employee must have immediate access to a device, such as a telephone or hand-held two-way radio, capable of summoning external emergency assistance, unless the EPA regional administrator has ruled that such a device is not required.[113]

v. Required Aisle Space

A TSD facility must maintain aisle space sufficient to allow the unobstructed movement of personnel, fire protection equipment, spill control equipment, and decontamination equipment to any area of facility operation in an emergency, unless the TSD facility can demonstrate to the EPA regional administrator that such aisle space is not needed for any of those purposes.[114] An owner or operator that seeks to make such a demonstration must do so with Part B of its permit application.[115]

vi. Arrangements with Local Authorities

The EPA requires a TSD facility owner or operator to attempt to make arrangements with local authorities for potential emergencies. The owner or operator must attempt to make the following arrangements, as appropriate for the type of waste handled at its facility and the potential need for the services of the following organizations:

1.   Arrangements to familiarize police, fire departments, and emergency response teams with the layout of the facility, properties of hazardous waste handled at the facility and associated hazards, places where facility personnel would normally be working, entrances to and roads inside the facility, and possible evacuation routes
2.   Where more than one police and fire department might respond to an emergency, agreements designating primary emergency authority to a specific police and a specific fire department, and agreements with any others to provide support to the primary emergency authority
3.   Agreements with state emergency response teams, emergency response contractors, and equipment suppliers
4.   Arrangements to familiarize local hospitals with the properties of hazardous waste handled at the facility and the types of injuries or illnesses which could result from fires, explosions, or releases at the facility[116]

If state or local authorities refuse to enter into such arrangements, the owner or operator must document that refusal in its operating record.

c. Contingency Plan and Emergency Procedures
i. Purpose and Implementation of Contingency Plan

The EPA requires each owner or operator of a TSD facility to maintain a "contingency plan" for the facility designed to minimize hazards to human health or the environment from fires, explosions, or any unplanned release of hazardous waste or constituents thereof to the air, soil, or surface water.[117] The contingency plan must be implemented immediately whenever there is a fire, explosion, or release of hazardous waste or constituents thereof that could threaten human health or the environment.[118]

ii. Content of Contingency Plan

The contingency plan must describe the emergency procedures that facility personnel will follow to minimize human health and environmental hazards resulting from emergency events and to implement the procedures required by 40 CFR § 264.56 or § 265.56, whichever is applicable.[119] The contingency plan must describe the arrangements made with local police departments, fire departments, hospitals, contractors, and state and local emergency response teams to coordinate emergency services, as described above.[120]

The contingency plan must contain an up-to-date list of the names, addresses, and phone numbers of all persons qualified to act as the emergency coordinator and must identify the primary emergency coordinator and alternates in the order in which they would assume responsibility.[121] The contingency plan must also include a list of all required emergency equipment at the facility, its location, its physical description, and a brief explanation of its capabilities.[122] Finally, the contingency plan must include an evacuation plan for facility personnel where there is a possibility that evacuation could be necessary.[123]

iii. Copies of Contingency Plan

A copy of the contingency plan, and all revisions to it, must be maintained at the TSD facility and submitted to the local and state authorities that may be called on to provide emergency services.[124]

The contingency plan must be submitted to the EPA regional administrator with Part B of the facility's permit application pursuant to 40 CFR Part 270 and, after modification or approval, will become a condition of any permit issued.[125]

iv. Amendment of Contingency Plan

The contingency plan must be reviewed and, if necessary, immediately amended whenever (1) the facility's permit is revised; (2) the plan fails in an emergency; (3) the facility changes in its design, construction, operation, maintenance, or other circumstances so as to materially increase the potential for emergency events or to change the response necessary to deal with an emergency event; (4) the list of emergency coordinators changes; (5) or the list of emergency equipment changes.[126]

v. Emergency Coordinator

Each TSD facility must have, at all times, at least one emergency coordinator at the facility or able to reach the facility within a short period of time. The emergency coordinator must be thoroughly familiar with the facility's contingency plan, the facility's operations, the location and characteristics of hazardous waste at the facility, and the facility's layout. Finally, the emergency coordinator must have the authority to carry out the contingency plan.[127]

vi. Emergency Procedures

Whenever there is an imminent or actual emergency situation, the emergency coordinator must (1) immediately activate internal alarms or communication systems to notify all facility personnel and (2) notify appropriate state and local agencies if their help is needed.[128]

Whenever there is an actual release, fire, or explosion, the emergency coordinator must immediately ascertain the character, the exact source, amount, and areal extent of any released materials.[129] At the same time, the emergency coordinator must assess whether the release, fire, or explosion may result in possible hazards to human health or the environment, either directly or indirectly.[130]

If the emergency coordinator determines that the facility has had a release, fire, or explosion that could threaten human health or the environment outside the facility, the coordinator must immediately report certain information to the government official designated as the on-scene coordinator for that geographic area under 40 CFR Part 1510 or the National Response Center at (800) 424-8802, including:

1. Name and telephone number of the reporter
2. Name and address of facility
3. Time and type of incident
4. Name and quantity of material(s) involved, to the extent known
5. The extent of injuries, if any
6. The possible hazards to human health or the environment outside the facility[131]

Further, if the emergency coordinator's assessment indicates that evacuation of local areas may be advisable, he must immediately notify the appropriate local authorities and remain available to assist the appropriate officials in deciding whether to evacuate such local areas.[132]

During an emergency, the emergency coordinator must take all reasonable measures to ensure that fires, explosions, and releases do not occur, recur, or spread to other hazardous waste at the facility. These measures shall include, where necessary, stopping the operations of the facility, collecting and containing released waste, and isolating containers of waste.[133] If the operations of the facility are stopped in response to an emergency event, the emergency coordinator must monitor the facility for leaks, pressure buildup, gas generation, or ruptures in valves, pipes, or other equipment, wherever appropriate.[134]

Immediately after an emergency, the coordinator must provide for the treatment, storage, or disposal of recovered waste, contaminated soil or surface water, and other material resulting from the emergency.[135] The coordinator is responsible for ensuring that no waste that may be incompatible with the released materials is treated, stored, or disposed of in affected areas of the facility until cleanup procedures are completed. The coordinator also must ensure that all emergency equipment listed in the contingency plan is cleaned and fit for its intended use before the operations of the facility in such areas are resumed.[136] Further, the owner or operator must notify the EPA regional administrator, and appropriate State and local authorities, that the above requirements have been met before operations are resumed in the affected areas of the facility.[137]

The owner or operator must note in the facility's operating record the time, date, and details of any incident that requires implementation of the facility's contingency plan. Within 15 d of any such incident, the owner or operator must submit a written report to the EPA regional administrator that includes:

1. Name, address, and telephone number of the owner or operator
2. Name, address and telephone number of the facility
3. Date, time, and type of incident
4. Name and quantity of material(s) involved
5. The extent of injuries, if any
6. An assessment of actual or potential hazards to human health or the environment, where applicable
7. Estimated quantity and disposition of recovered material that resulted from the incident[138]

d. Manifest System, Recordkeeping and Reporting
i. Use of Manifest System

If a TSD facility receives hazardous waste accompanied by a manifest from an off-site source, the owner or operator must:

1. Sign and date each copy of the manifest to certify that the hazardous waste covered by the manifest was received
2. Note any significant discrepancies in the manifest on each copy of the manifest
3. Immediately give the transporter at least one copy of the signed manifest
4. Within 30 d after delivery, send a copy of the manifest to the generator
5. Retain at the facility a copy of the manifest for at least 3 years from the date of delivery[139]

If a facility receives, from a rail or water transporter, hazardous waste that is accompanied by a shipping paper containing all the information required on a manifest (excluding the EPA identification numbers, generator's certification, and signature), the owner or operator must follow the procedures delineated above, utilizing the shipping paper if a manifest has not been received.[140]

The EPA does not require the owner or operator to perform a waste analysis before signing the manifest or shipping paper and giving it the transporter even if the facility's waste analysis plan requires analysis of that shipment. Nonetheless, the generator and, if necessary, the EPA regional administrator must be notified of any discrepancy discovered during later analysis.[141]

If a TSD facility ships hazardous waste from the facility, the owner or operator must comply with the requirements of 40 CFR Part 262.[142]

ii. Manifest Discrepancies

The EPA defines "manifest discrepancies" as differences between the quantity or type of hazardous waste designated on the manifest or shipping paper and the quantity or type of hazardous waste actually received by the facility. "Significant discrepancies in quantity" are variations greater than 10% in weight for bulk waste and any variation in piece count for batch waste. "Significant discrepancies in type" are obvious differences that can be discovered by inspection or waste analysis, such as waste solvent substituted for waste acid, or toxic constituents not recorded on the manifest or shipping paper.[143]

On discovering a significant discrepancy, the owner or operator must attempt to reconcile the discrepancy with the waste generator or transporter. If the discrepancy is not resolved within 15 d after receiving the waste, the owner or operator must report the discrepancy to the EPA regional administrator. The report must describe attempts to reconcile the discrepancy and include a copy of the manifest or shipping paper at issue.[144]

iii. Operating Record

The EPA requires each owner or operator of a TSD facility to keep a written operating record at the facility.[145] The following information must be recorded as it becomes available and must be maintained in the operating record until closure of the facility:

1. A description and the quantity of each hazardous waste received and the methods and dates of its treatment, storage, or disposal
2. The location of each hazardous waste within the facility and the quantity at each location, including cross-references to specific manifest document numbers, if the waste was accompanied by a manifest
3. Records and results of waste analyses performed pursuant to the waste analysis plan and specific provisions, including those applicable to HWIFs at 40 CFR § 264.341 or 265.341, whichever is applicable
4. Summary reports and details of all incidents that require implementation of the contingency plan
5. Records and results of inspections (except that such data need be kept only 3 years)
6. Monitoring testing or analytical data, and corrective action where required
7. For off-site facilities, the required notices to generators
8. Required closure cost estimates
9. An annual certification by a permitted facility that a program is in place to reduce the volume and toxicity of hazardous waste that the permittee generates to the extent economically practicable
10. For an off-site treatment facility, a copy of the notice required from a generator under 40 CFR § 268.7(a)(1)[146]

iv. Availability, Retention, and Disposition of Records

Any EPA officer, employee, or representative, who is duly designated by the EPA administrator, may request that the owner or operator of a TSD facility furnish all records required to be maintained under 40 CFR Part 264 or Part 265, whichever is applicable.

Further, such records must be made available for inspection at all reasonable times.[147] The retention period for such records is automatically extended during any unresolved enforcement action relating to the facility or as requested by the EPA administrator.[148]

v. Biannual Report

A TSD facility owner or operator must prepare and submit a copy of a biannual report, utilizing EPA form 8700-13B, to the EPA regional administrator by March 1 of each even-numbered year. The biannual report must cover facility activities during the previous calendar year and include:

1. The EPA identification number, name, and address of the facility
2. The calendar year covered by the report
3. For off-site TSD facilities, the EPA identification number of each hazardous waste generator from which the facility received hazardous waste during the year (for imported shipments of foreign hazardous waste, the report must provide the name and address of the foreign generator)
4. A description and the quantity of each hazardous waste received during the year, listed by EPA identification number of each generator
5. Method of treatment, storage, or disposal of each hazardous waste
6. The most recent closure cost estimate under 40 CFR § 264.142
7. A certification signed by the owner or operator of the TSD facility or his authorized representative.[149]

vi. Unmanifested Waste Report

If a facility has accepted for treatment, storage, or disposal any hazardous waste from an off-site source that was not accompanied by a manifest or shipping paper, and if the waste is not excluded from the manifest requirement by 40 CFR § 261.5, then the owner or operator must submit a single copy of a report on EPA form 8700-13B to the EPA regional administrator within 15 d after receiving the waste. The report must be designated ''Unmanifested Waste Report'' and include the following information:

1. The EPA identification number, name, and address of the facility
2. The date the facility received the waste
3. The EPA identification number, name, and address of the generator and the transporter, if available
4. A description and the quantity of each unmanifested hazardous waste received
5. The method of treatment, storage, or disposal of each hazardous waste
6. A certification signed by the owner or operator of the facility or his authorized representative
7. A brief explanation of why the waste was unmanifested, if known[150]

The EPA has noted that small quantities of hazardous waste are excluded from regulation under 40 CFR § 261.5 and, therefore, do not require a manifest. The EPA suggests that, where a facility received unmanifested hazardous wastes, the owner or operator obtain from each generator a certification that the waste qualifies for exclusion. Otherwise, the EPA suggests that the owner or operator file an unmanifested waste report for each such hazardous waste movement.[151]

vii. Additional Reports

In addition to the biannual reports and unmanifested waste reports required as above, a TSD owner or operator must also report to the EPA regional administrator releases, fires, and explosions as set forth in 40 CFR § 264.56(j) or § 265.56(j), which ever is applicable.[152] Owners or operators are required to report facility closures as specified in 40 CFR § 264.115 and § 265.115, whichever is applicable.[153]

A permitted TSD facility is required to make additional reports as mandated by other specific provisions.[154] An interim status TSD facility is required to report groundwater contamination monitoring data as set forth in 40 CFR § 265.93 and § 265.94.[155]

e. Releases from Solid Waste Management Units

An owner or operator of an HWIF seeking a permit must institute corrective action as necessary to protect human health and the environment for all releases of hazardous waste or constituents from any solid waste management unit at the facility, regardless of when waste was placed in such unit.[156] If corrective action is required, the permit will specify the corrective action to be taken (and a schedule for compliance if it cannot be completed before issuance of the permit) and contain requirements for financial assurances that the action will be completed.[157]

If necessary, the owner or operator must take corrective actions beyond the facility's property boundary unless the owner or operator cannot obtain permission to do so. If off-site access is denied, the owner or operator will be required to take whatever on-site actions may remedy such releases.[158]

f. Closure and Postclosure

The EPA has promulgated specific regulations governing closure of TSD facilities. Additional regulations govern postclosure care of hazardous waste disposal facilities, but such regulations ordinarily will not be applicable to an HWIF. The EPA defines "disposal facility" to mean a facility or part of a facility where hazardous waste is intentionally placed into or on any land or water and where waste will remain after closure.[159] If an HWIF is a part of a larger facility including disposal facilities, the additional regulations may be applicable to the owner or operator of the entire facility.[160]

i. Closure Performance Standard

An owner or operator must close a TSD facility in a way that:

1. Minimizes the need for further maintenance
2. Controls, minimizes, or eliminates, to the extent necessary to protect human health and the environment, postclosure escape of hazardous waste, hazardous constituents, leachate, contaminated runoff, or hazardous waste decomposition products to the ground, surface waters, and atmosphere
3. Complies with the closure requirements set forth in various RCRA regulations, including those specifically applicable to HWIFs[161]

ii. Closure Plan

The owner or operator of a hazardous waste management facility, including a TSD facility, must have a written closure plan.[162] Until final closure has been certified to be complete, a copy of the closure plan and revisions thereto must be furnished to the EPA regional administrator on request, including request by mail.[163]

The closure plan must be submitted with the facilities permit application,[164] and approved by the EPA regional administrator as part of the permit issuance procedures under 40 CFR Part 124. The approved closure plan will become a condition of any RCRA permit.[165] Approval of the closure plan must insure that the approved plan is consistent with the applicable RCRA regulations.[166]

A TSD facility without an approved plan must provide the most current closure plan to designated EPA officers, employees, or representative on the day of any site inspection.[167]

A closure plan must identify all measures necessary to complete partial and/or final closure of the facility at any point during its active life. At a minimum, the closure plan must include:

1. A description of how each hazardous waste management unit at the facility will be closed in accordance with the closure performance standard noted above
2. A description of how the facility will be closed in accordance with the closure performance standard noted above
3. An estimate of the maximum inventory of hazardous waste that will ever be on-site during the active life of the facility, and a detailed description of the methods to be used during partial closures and final closure, including methods for removing, transporting, treating, storing, or disposing of all hazardous waste likely to be on-site
4. A detailed description of the measures needed to remove or decontaminate all hazardous waste residues and contaminated containment system components, equipment, structures, and soils. A closure plan must describe procedures for cleaning equipment, removing contaminated soils, sampling and testing surrounding soils, and criteria for determining the extent of decontamination required to satisfy the closure performance standard
5. A detailed description of other activities needed to insure compliance with the closure performance standards, including groundwater monitoring, leachate collection, and surface water controls
6. A schedule for closure of each hazardous waste management unit and for final closure of the facility, including an estimate of the time required to close each unit
7. An estimate of the expected year of final closure for facilities that use trust funds to meet financial assurance requirements and are expected to close prior to expiration of their permit or, in the case of interim status facilities, those facilities whose remaining operating life is less than 20 years[168]

An owner or operator may seek to amend the closure plan at any time before notification of partial or final closure of the facility. The owner or operator of a permanent facility must submit a written notification of or request for a permit modification to authorize a proposed change to an approved closure plan. Similarly, the owner or operator of an interim status facility with an approved closure plan must submit a written notification of or request to the EPA regional administrator to authorize a proposed change to an approved closure plan. In either event, the written notification or request must include a copy of the amended closure plan for approval by the EPA regional administrator.[169]

An owner or operator *must* amend the closure plan whenever (1) proposed changes in operating plans or facility design may effect the closure plan, (2) there is a change in the expected year of closure, or (3) unexpected events during partial or final closure activities require modification of the approved closure plan.[170]

The owner or operator must seek to amend the closure plan at least 60 d prior to any proposed change in facility design or operation, or no later than 60 d after an unexpected event has occurred that requires modification of the closure plan. If the unexpected event occurs during the partial or final closure, the owner or operator must seek to amend the closure plan no later than 30 d after the unexpected event.[171] The regional administrator will approve, disapprove, or modify the proposed amended closure plan in accordance with the procedures in 40 CFR Parts 124 and 270. The approved amended closure plan will become a condition of the facility's RCRA permit.[172]

An owner or operator must notify the regional administrator in writing at least 45 d before the date on which it expects to begin final closure of a facility containing only incinerator units to be closed.[173] An owner or operator must close the facility no later than either (1) 30 d after the date on which the hazardous waste management unit receives its known final volume of hazardous waste or (2) if there is a reasonable possibility that the unit will receive additional hazardous wastes, no later than 1 year after the date on which the unit last received hazardous waste. The regional administrator may approve an extension to this 1-year limit if the owner or operator can demonstrate that the unit or facility has the capacity to receive additional hazardous waste and all steps necessary to prevent threats to human health and the environment have been taken.[174]

An owner or operator of a permitted facility need not notify the regional administrator of the anticipated closure of the facility if the facility's RCRA permit is terminated or if the facility is ordered by judicial decree or a final administrative order to stop receiving hazardous wastes.[175] Despite the notification requirements, an owner or operator may remove hazardous wastes, and decontaminate and dismantle equipment, in accordance with an approved closure plan at any time before or after notification of partial or final closure.[176]

iii. Closure Schedule

Within 90 d after receiving the final volume of hazardous waste at a hazardous waste management unit or facility, the owner or operator must treat, remove from the unit or facility, or dispose of on-site, all hazardous waste in accordance with the approved closure plan.[177] Within 180 d after receiving the final volume of hazardous waste, the owner or operator must complete partial and final closure activities in accordance with the approved closure plan.[178]

The regional administrator may extend the 90- and 180-d deadlines set under two circumstances. First, an extension may be granted if the activities required to close the facility properly will, of necessity, take longer than the 90 or 180 d otherwise permitted. Second, an extension may be granted if the owner or operator demonstrates:

1.   The unit or facility has the capacity to receive additional hazardous wastes.
2.   There is a reasonable likelihood that the present or a new owner or operator will recommence operation of the unit or facility within 1 year.
3.   Closure of the unit or facility would be incompatible with continued operation of the site.

Under either circumstance, the owner or operator must demonstrate that it has taken and will continue to take all steps necessary to prevent threats to human health and the environment, including compliance with all applicable permit requirements at a permitted facility.[179]

Any extension of the 90- or 180-d deadlines must be sought at least 30 d prior to the expiration of the relevant deadline.[180]

iv. Disposition of Contaminated Equipment, Structures, and Soils

During the partial and final closure periods, all contaminated equipment, structures, and soils must be properly disposed of or decontaminated. An owner or operator removing hazardous waste or hazardous constituents during closure will likely become a generator of hazardous waste and must comply with all applicable requirements of 40 CFR Part 262.[181]

v. Closure Certificate

Within 60 d of completion of final closure, an owner or operator must submit to the regional administrator, by registered mail, a certification that the hazardous waste management unit or facility, as applicable, has been closed in accordance with the approved closure plan. The certification must be signed by the owner or operator and by an independent registered professional engineer. Documentation supporting the independent registered professional engineer's certification must be furnished to the regional administrator on request until the regional administrator releases the owner or operator from the financial assurance requirements for closure.[182]

g. Financial Requirements

As required by RCRA Section 3004, 42 U.S.C.A. § 6924(a), the EPA has promulgated regulations establishing financial responsibility requirements for owners and operators of TSD facilities, including financial responsibility for corrective actions. RCRA provides that financial responsibility may be demonstrated through any one or a combination of insurance, guarantee, surety bond, letter of credit, or qualification as a self-insurer.[183] RCRA authorizes the EPA to specify policy or other contractual terms, conditions, or defenses governing evidence of financial responsibility.[184]

The EPA has promulgated regulations governing two aspects of financial responsibility. Regulations requiring "financial assurance for closure" are designed to ensure that the facility will be properly closed in accordance with its closure plan and, if required, that the owner or operator will provide proper postclosure care. Regulations establishing "liability coverage" requirements are intended to ensure financial responsibility for bodily injury and property damage to third parties caused by operation of the facility.

The EPA has promulgated financial responsibility regulations applicable to owners and operators of all TSD facilities, including both permitted and interim status facilities, except states and the federal government.[185] Additional financial responsibility requirements are applicable only to owners and operators of disposal facilities and certain other types of TSD facilities not including HWIFs.[186]

i. Closure Cost Estimate

An owner or operator must have a detailed written estimate, in current dollars, of the cost of closing the facility in accordance with its closure plan and specific provisions imposing additional closure requirements.[187] A provision applicable only to an HWIF requires that, at closure, the owner or operator remove all hazardous waste and hazardous waste residues (including, but not limited to, ash, scrubber waters, and scrubber sludges) from the incinerator site.[188]

The closure cost estimate must equal the cost of final closure at the point in the facility's active life when the extent and manner of its operation would make closure the most expensive under its closure plan.[189] The estimate must be based on the costs of hiring a third party to close the facility, and the estimate may utilize costs for on-site disposal only if the owner or operator can demonstrate that on-site disposal capacity will exist at all times over the life of the facility.[190] The closure cost estimate may not incorporate any salvage value of hazardous wastes, facility structures or equipment, land or other assets of the facility.[191] Further, the estimate may not utilize a zero cost for hazardous waste that might have economic value.[192]

During the active life of the facility, the closure cost estimate must be adjusted for inflation each year. Owners or operators utilizing a financial instrument to comply with the financial responsibility requirements must adjust the closure cost estimate for inflation within 60 d prior to the anniversary date of the establishment of such financial instrument. Owners and operators utilizing self-insurance to demonstrate financial responsibility must adjust the closure cost estimate for inflation within 30 d after the close of that corporation's fiscal year and before the required submission of updated information to the Regional Administrator.[193]

The adjustment for inflation may be made by recalculating the maximum cost of closure in current dollars, or by multiplying the most current closure cost estimate by an inflation factor to reach an "adjusted closure cost estimate". The inflation factor is derived by dividing the most recent implicit price deflator for gross national product published by the U.S.

Department of Commerce in its *Survey of Current Business* by the deflator for the previous year. Subsequent adjustments are made by multiplying the latest adjusted closure cost estimate by the latest inflation factor.[194]

The owner or operator must also revise the closure cost estimate no later than 30 d after the regional administrator has approved a request to modify the facility's closure plan if the modification increases the cost of closure. The revised closure plan estimate must be adjusted for inflation as set forth above.[195]

During the operating life of the facility, the owner or operator must keep at the facility both the latest closure cost estimate and the latest inflation-adjusted closure cost estimate.[196]

ii. Financial Assurance for Closure

The owner or operator of a TSD facility must demonstrate financial assurance for closure of the facility. The owner or operator of a permitted facility may meet this requirement through the use of one or a combination of a trust fund, a surety bond guaranteeing payment into a trust fund, a surety bond guaranteeing performance of closure, a letter of credit paying into a standby trust fund, insurance, or compliance with a financial test demonstrating that the owner or operator has the ability to self-insure or has obtained a corporate guarantee from a parent corporation able to self-insure.[197] The EPA believes that closure trust funds may be used to demonstrate financial responsibility for closure even though they are not specifically mentioned in RCRA.[198] An interim status facility may demonstrate financial assurance for closure through use of the same mechanisms, except use of a surety bond guaranteeing performance of closure.[199]

An owner or operator may satisfy the financial responsibility requirements governing closure by establishing more than one financial mechanism per facility. The combination may consist only of trust funds, surety bonds guaranteeing payment into a closure trust fund, letters of credit, and insurance; not surety bonds guaranteeing performance of closure or self-insurance measures. The combination of mechanisms must provide financial assurance for an amount at least equal to the current closure cost estimate. A single trust fund or standby trust fund may be used for all mechanisms. The regional administrator may use any or all of the mechanisms to provide for closure of the facility.[200]

An owner or operator of more than one facility may utilize a single financial mechanism to demonstrate compliance with the financial assurance for closure requirements for more than one facility.[201]

The owner or operator is released from the requirements of the financial assurance for closure provisions when the facility has been properly closed. Within 60 d after receiving certification from the owner or operator and an independent registered professional engineer that final closure has been completed in accordance with the approved closure plan, the regional administrator will notify the owner or operator that he is released from such requirements unless the regional administrator has reason to believe that final closure has not been in accordance with the approved closure plan. If the regional administrator so believes, he shall provide the owner or operator with a detailed written statement of such reasons.[202]

(a). Closure Trust Fund

An owner or operator may satisfy the financial assurance for closure requirements by establishing a closure trust fund that conforms to the EPA regulations. The owner or operator must submit an originally signed duplicate of the trust agreement to the regional administrator.[203] The owner or operator of a new facility must submit the originally signed duplicate at least 60 d before the date on which hazardous waste is first received for treatment, storage, or disposal.[204] Owners and operators of existing facilities newly covered by RCRA regulations must submit the originally signed duplicate to the regional administrator by the date that the administrator sets for submitting Part B of the RCRA permit application for that facility.[205]

The trustee of the closure trust fund must be an entity that has the authority to act as a trustee and whose trust operations are regulated and examined by a federal or state agency.[206] The wording of the trust agreement must be *identical* to the wording specified in 40 CFR § 264.151(a)(1). The trust agreement must be accompanied by a formal certification of acknowledgment, an example of which is set forth in 40 CFR § 264.151(a)(2). Schedule A of the trust agreement, which sets forth the facilities and cost estimates to which the trust agreement pertains, must be updated within 60 d after a change in the amount of any current closure cost estimate covered by the trust agreement.[207]

The owner or operator of a permitted facility must make annual payments into the trust fund over the term of the initial RCRA permit for the facility or over the remaining operating life of the facility as estimated in the closure plan, whichever period is shorter.[208] For an interim status facility, such payments must be made annually over either (1) the 20 years beginning with the effective date of the regulations subjecting that facility to RCRA regulation or (2) the remaining operating life of the facility as estimated in closure plan, whichever period is shorter.[209] This period is referred to as the "pay-in period".

For a new facility, the first payment must be made before the initial receipt of hazardous waste for treatment, storage, or disposal. The owner or operator must submit a receipt from the trustee for this payment to the regional administrator before the initial receipt of hazardous waste.[210] For interim status facilities, the first payment must be made by the effective date of the regulations subjecting that facility to RCRA regulation.[211]

The first payment must be at least equal to the current closure cost estimate divided by the number of years in the pay-in period, unless the owner or operator is satisfying its financial assurance requirements through a combination of financial mechanisms.[212] Subsequent payments must be made no later than 30 d after each anniversary date of the first payment. The amount of each subsequent payment must equal the current closure cost estimate minus the current value of the trust fund, divided by the number of years remaining in the pay-in period.[213]

If owner or operator of a interim status facility establishes a closure trust fund, and the value of that trust fund is less than the current closure cost estimate when the facility receives a RCRA permit, the difference must be paid into the trust fund over the term of the RCRA permit or the remaining operating life of the facility, whichever is shorter. The amount of each payment must be determined as set forth above, and payments must continue to be made no later 30 d after each anniversary date of the first payment made when the facility held interim status.[214]

An owner or operator may accelerate payments into the trust fund or deposit the full amount of the current closure cost estimate at the time fund is established. Nonetheless, the owner or operator must maintain the value of the trust fund at no less than the value that the fund would have if annual payments were made as set forth above.[215]

Similarly, if an owner or operator establishes a closure trust fund after having used an alternative financial mechanism, its first payment into the trust fund must be in at least the amount that the fund would contain if the trust fund had been established initially and annual payments made as set forth above.[216]

Whenever the current closure cost estimate changes after the pay-in period is completed, the owner or operator must compare the new estimate with the trustee's most recent annual valuation of the trust fund. If the value of the fund is less than the amount of the new estimate, the owner or operator, within 60 d after the change in the cost estimate, must either deposit an amount at least equal to the deficit into the trust fund or obtain other financial assurance to cover the difference.[217]

If the value of the trust fund is greater than the total amount of the current closure cost estimate, the owner or operator may submit a written request to the regional administrator for release of the excess.[218] Similarly, if an owner or operator substitutes other financial assurance mechanisms for all or part of the trust fund, it may submit a written request to the regional administrator for release of the amount of the trust fund in excess of that portion of the current closure cost estimate covered by the trust fund.[219] Within 60 d after receiving such a request, the regional administrator shall instruct the trustee to release to the owner or operator "such funds as the regional administrator specifies in writing".[220]

Within 60 d after receiving bills for partial or final closure activities from an owner or operator, or other person authorized to conduct partial or final closure activities, the regional administrator will instruct the trustee to make reimbursements in amounts specified by the regional administrator in writing if the regional administrator determines that the performed activities are in accordance with the approved closure plan or otherwise justified. If the regional administrator has reason to believe that the maximum cost of closure over the remaining life of the facility will be significantly greater than the value of the trust fund, he may withhold reimbursements of such amounts as he deems prudent until he determines that the facility has been properly closed. If the regional administrator decides to withhold reimbursements, he will provide the owner or operator with a detailed written statement of reasons.[221]

The Regional Administrator will agree to termination of the trust when the owner or operator substitutes an alternative financial assurance mechanism, or when the Regional Administrator agrees that the facility has been properly closed.[222]

(b). Surety Bond Guaranteeing Payment into a Closure Trust Fund

An owner or operator may satisfy the financial assurance for closure requirements by obtaining a surety bond in an amount that will cover closure costs and establishing a standby closure trust fund into which the bond amount will be paid under certain circumstances. The surety bond, and an originally signed duplicate of the standby closure trust agreement, must be submitted to the regional administrator at least 60 d before the date on which hazardous waste is first received for treatment, storage, or disposal at a new facility.[223] The bond must be effective before the initial receipt of hazardous waste.

The wording of the surety bond must be identical to the wording specified in 40 CFR § 264.151(b).[224] The surety company issuing the bond must, at a minimum, be among those listed as acceptable sureties on federal bonds in Circular 570 of the U.S. Department of Treasury.[225]

Under the terms of the surety bond, all payments made thereunder will be made by the surety directly into the standby trust fund in accordance with instructions from the regional administrator. The standby trust fund must meet the requirements set forth above for trust funds, except that the following are *not required until* the standby trust fund is funded:

1. Payments into the trust fund
2. Updating Schedule A of the trust agreement to show current closure cost estimates
3. Annual valuations as required by the trust agreement
4. Notices of nonpayment as required by the trust agreement[226]

The bond must guarantee that the owner or operator will

1. Fund the standby trust fund in an amount equal to the penal sum of the bond before beginning final closure of the facility; or
2. Fund the standby trust fund in an amount equal to the penal sum of the bond within 15 d after a regional administrator's administrative order to begin final closure becomes final or within 15 d after an issuance of a court order to begin final closure;
3. Provide an alternative financial assurance mechanism, and obtain the regional administrator's written approval of such mechanism, with 90 d after receipt by both the

owner or operator and the regional administrator of a notice of cancellation of the bond from a surety.[227]

Under the terms of the bond, the surety will become liable on the bond obligation if the owner or operator fails to perform as guaranteed by the bond.[228]

The penal sum of the bond must be in an amount at least equal to the current closure cost estimate, unless the owner or operator is utilizing a combination of financial mechanisms to meet the financial assurance requirements.[229] Whenever the current closure cost estimate increases to a sum greater than the penal sum of the bond, the owner or operator, within 60 d after the increase, must either (1) cause the penal sum to be increased to an amount at least equal to the current closure cost estimate and submit evidence of such increase to the regional administrator or (2) obtain other financial assurance to cover the increase. If the current closure cost estimate decreases, the penal sum may be reduced to the amount of the current closure cost estimate following written approval from the regional administrator.[230]

Under the terms of the surety bond, the surety may cancel the bond by sending notice of the cancellation by certified mail to both the owner or operator and the regional administrator. Cancellation may not occur, however, until 120 d after receipt of such notice by both the owner or operator and the regional administrator.[231] The owner or operator may cancel the bond if the regional administrator has given prior written consent based on his receipt of evidence that the owner or operator has obtained an alternative financial mechanism complying with the financial assurance requirements.[232]

(c). Surety Bond Guaranteeing Performance of Closure

The owner or operator of a permitted facility, but not an interim status facility, may satisfy the financial assurance for closure requirements by obtaining a surety bond providing that the surety will either perform final closure at the facility or deposit the sum necessary to fund closure activities in a standby trust fund.

The owner or operator of a new facility must submit the bond to the regional administrator at least 60 d before the date on which hazardous waste is first received at the facility, and the bond must be effective before such initial receipt of hazardous waste. The surety company issuing the bond must, at a minimum, be among those listed as acceptable sureties on federal bonds in Circular 570 of the U.S. Department of the Treasury.[233]

The wording of the surety bond must be identical to the wording specified in 40 CFR § 264.151(c).[234] In addition to the surety bond, the owner or operator must also establish a standby closure trust fund and submit an originally signed duplicate of the trust agreement to the regional administrator with the surety bond. The requirements governing the standby trust fund are identical to those governing the standby trust fund required in combination with a surety bond guaranteeing payment into a closure trust fund.[235]

The bond must guarantee that the owner or operator will either (1) perform final closure in accordance with the facility's closure plan and the facility's permit whenever required to do so, or (2) provide alternate financial assurance and obtain the regional administrator's written approval of such assurance, within 90 d after receipt by the owner or operator and the regional administrator of a notice of cancellation of the bond from the surety.[236]

Under the terms of the bond, the surety will become liable on the bond obligation if the owner or operator fails either to close the facility properly or to provide alternate financial assurance on cancellation of the bond. Following a final administrative determination under RCRA § 3008 that the owner or operator has failed to perform final closure in accordance with the approved closure plan and other permit requirements when required to do so, a surety must either perform final closure or deposit the amount of the penal sum into the standby trust fund.[237]

The penal sum of the bond must be in an amount at least equal to the current closure cost estimate and must be increased or be decreased as required for a bond guaranteeing payment into a closure trust fund.[238] Similarly, a surety may cancel the bond 120 d after the owner or operator and the regional administrator's receipt of a notice of cancellation by certified mail.[239]

The owner or operator may cancel the bond only with the prior written consent of the regional administrator. The regional administrator will provide such written consent when (1) the owner or operator substitutes an alternative financial assurance mechanism or (2) the regional administrator agrees that the facility has been properly closed.[240] Once the regional administrator releases the owner or operator from the financial assurance requirements on the ground that the facility has been properly closed, the surety will not be liable for deficiencies in performance of closure by the owner or operator.[241]

(d). Closure Letter of Credit

An owner or operator may satisfy the financial assurance for closure requirements by obtaining an irrevocable standby letter of credit covering closure costs and establishing a standby closure trust fund. The owner or operator of a new facility must submit the letter of credit, and an originally signed duplicate of the standby trust agreement, to the regional administrator at least 60 d before the date on which hazardous waste is first received at the facility. The letter of credit must be accompanied by a letter from the owner or operator (1) referring to the letter of credit by number, issuing institution, and date; (2) providing the EPA identification number, name, and address of the facility; and (3) stating the amount of funds available for closure of the facility under the letter of credit.[242] The letter of credit must be effective before initial receipt of hazardous waste at a new facility.[243]

The wording of the letter of credit must be identical to the wording specified in 40 CFR § 264.151(d).[244] The issuing institution must be an entity that has the authority to issue letters of credit and whose letter-of-credit operations are regulated and examined by a federal or state agency.[245]

An owner or operator utilizing a standby letter of credit must also establish a standby trust fund. The standby trust fund is subject to the same requirements as a standby trust fund established in conjunction with a surety bond. Under the terms of the letter of credit, all amounts paid pursuant to a draw by the regional administrator will be deposited by the issuing institution directly into the standby trust fund in accordance with instructions from the regional administrator.[246]

The letter of credit must be irrevocable and issued for a period of at least 1 year. The letter of credit must automatically be extended for a period of at least 1 year unless both the owner or operator and the regional administrator receive notice of the decision of the issuing institution not to extend the expiration date by certified mail 120 d before the current expiration date.[247]

The letter of credit must be issued in an amount at least equal to the current closure cost estimate, unless the owner or operator is utilizing a combination of alternative financial mechanisms to cover closure costs.[248] Within 60 d after the increase in the current closure cost estimate, an owner or operator must either (1) increase the amount of the credit to equal the current closure cost estimate (and submit evidence of such increase to the regional administrator) or (2) obtain other financial assurance to cover the increase in the final closure cost estimate. If the current closure cost estimate decreases, the amount of the credit may be reduced to the amount of the current closure cost estimate following written approval from the regional administrator.[249]

The regional administrator may draw on the letter of credit following a final administrative determination pursuant to RCRA § 3008 that the owner or operator has failed to perform final closure in accordance with the closure plan of the facility and other permit requirements when required to do so.[250] Further, the regional administrator will draw on the letter of credit if the owner or operator does not establish (and obtain approval of) an alternative financial assurance mechanism within 90 d after receipt by both the owner or operator and the regional administrator of a cancellation notice from the issuing institution. The regional administrator may delay drawing on the letter of credit if the issuing institution grants an extension of the terms of the credit.[251]

The regional administrator will return a letter of credit to the issuing institution for termination when (a) an owner or operator substitutes alternative financial assurance mechanisms or (b) the regional administrator releases the owner or operator from the financial assurance requirements following final closure of the facility.[252]

(e). Closure Insurance

An owner or operator may satisfy the financial assurance for closure requirements by obtaining closure insurance guaranteeing payment of closure costs. An owner or operator of a new facility submits a certificate of such insurance to the regional administrator at least 60 d before the date on which hazardous waste is first received at the facility, and the insurance must be effective before such initial receipt of hazardous waste.[253]

The wording of the certificate of insurance must be identical to the wording specified in 40 CFR § 264.151(e). The insurer must be licensed to transact the business of insurance, or eligible to provide insurance as an excess or surplus lines insurer, in one or more states.[254]

The closure insurance policy must be issued for a face amount at least equal to the current closure cost estimate, unless the owner or operator is utilizing a combination of financial mechanisms to ensure coverage of closure costs.[255] Whenever the current closure cost estimate increases to an amount greater than the face amount of the policy, the owner or operator, within 60 d after the increase, must either (1) cause the face amount to be increased to cover the deficit (and submit evidence of increase to the regional administrator) or (2) obtain other financial assurance to cover the increase. If the current closure cost estimate decreases, the face amount may be reduced to the amount of the current closure cost estimate upon written approval from the regional administrator.[256]

The policy must guarantee that funds will be available to close the facility whenever final closure occurs. The policy must guarantee that, once final closure begins, the insurer will be responsible for paying out funds, up to an amount equal to the face amount of the policy, on the direction of the regional administrator, to such parties as the regional administrator specifies.[257] The procedure and standards for payment of bills for closure activities is identical to the procedure and standards for such payments from a closure trust fund.[258]

The owner or operator must maintain the policy in full force and effect until the regional administrator consents to the termination of the policy. Failure to pay the premium, without substitution of alternative financial assurance mechanisms, constitutes a significant violation of EPA regulations warranting such remedy as the regional administrator deems necessary.[259]

The policy must provide that the insurer may not cancel, terminate, or fail to renew the policy except for failure to pay the premium.[260] The automatic renewal of the policy must, at a minimum, provide the insured with the option of renewal at the face amount of the expiring policy.[261] If the owner or operator fails to pay the premium, the insured may elect to cancel, terminate, or fail to renew the policy by sending notice by certified mail to the owner or operator and the regional administrator. Cancellation, termination, or failure to renew may not occur until 120 d after receipt of the notice by both the Regional Administrator and the owner or operator.

In any event, cancellation, termination, or failure to renew may not occur, and the policy will remain in full force and effect if, on or before the date of expiration, one of the following occurs:

1. The regional administrator deems the facility abandoned.
2. The permit is terminated or revoked or a new permit is denied.
3. Closure is ordered by the regional administrator or a court.
4. The owner or operator is named as a debtor in a voluntary or involuntary bankruptcy proceeding.
5. The premium due is paid.[262]

The regional administrator will give written consent to the owner or operator to terminate the insurance policy when (1) the owner or operator substitutes alternative financial assurance mechanisms or (2) the regional administrator releases the owner or operator from the financial assurance requirements following final closure of the facility.[263]

(f). Financial Test and Corporate Guarantee for Closure

An owner or operator may satisfy the financial assurance for closure requirements by demonstrating that it, or a parent corporation willing to issue the owner or operator a corporate guarantee, passes a financial test set forth in the RCRA regulations. If the owner or operator, or its guaranteeing parent corporation, meets the financial test, it is permitted to "self-insure".

The owner or operator, or parent corporation, may satisfy either of two prongs of the "financial test". Under the first prong, the owner or operator (or parent corporation) must have all of the following:

1. Two of the following three ratios: a ratio of total liabilities to net worth less than 2.0; a ratio of the sum of net income plus depreciation, depletion, and amortization to total liabilities greater than 0.1; and a ratio of "current assets" to "current liabilities" greater than 1.5[264]
2. "Net working capital" and "tangible net worth", each at least six times the sum of the current closure cost estimates[265]
3. Tangible net worth of at least $10 million
4. Assets located in the U.S. amounting to at least 90% of total assets or at least six times the sum of the current closure cost estimate[266]

In the alternative, an owner or operator may meet the second prong of the financial test by demonstrating all of the following:

1. A current rating for his most recent bond insurance of AAA, AA, A, or BBB as issued by Standard and Poor's or Aaa, Aa, A, or Baa as issued by Moody's
2. Tangible net worth at least six times the sum of the current closure cost estimate
3. Tangible net worth of at least $10 million
4. Assets located in the U.S. amounting to at least 90% of total assets or at least six times the sum of the current closure cost estimate[267]

To demonstrate that he meets either prong of the financial test, the owner or operator must submit the following items to the regional administrator:

1. A letter signed by the owner's or operator's chief financial officer, worded as specified in 40 CFR § 264.151(f)
2. A copy of an independent certified public accountant's report on examination of the owner's or operator's financial statements for the latest completed fiscal year
3. A special report from the owner's or operator's independent certified public accountant to the owner or operator stating that (1) the accountant has compared the data which

the chief financial officer's letter states were derived from the independently conducted year-end financial statements for the latest fiscal year with the amounts in such financial statements and (2) that no matters came to his or her attention that caused the accountant to believe that such data should be adjusted.[268]

The owner or operator of a new facility must submit the items specified above to the regional administrator at least 60 d before the date on which hazardous waste is first received for treatment, storage, or disposal at the facility.[269] In addition, the owner or operator of either a permitted or interim status facility must send updated information meeting the requirements set forth above to the regional administrator within 90 d after the close of each succeeding fiscal year.[270]

If the owner or operator no longer meets the requirements of either prong of the financial test, it must send notice to the regional administrator by certified mail within 90 d after the end of the fiscal year for which the year-end financial data demonstrate the inability to meet such requirements. The notice must inform the regional administrator of the owner's or operator's intent to establish an alternative financial assurance mechanism, and the owner or operator must provide the alternative financial assurance mechanism within 120 d after the end of such fiscal year.[271]

If the regional administrator reasonably believes that the owner or operator cannot meet the requirements of the financial test, the regional administrator may require reports of financial condition from the owner or operator at any time. If the regional administrator finds, on the basis of such reports or other information, that the owner or operator no longer meets the financial test, the owner or operator must provide an alternative financial assurance mechanism within 30 d after notification of such a finding.[272]

The regional administrator may find that the owner or operator fails to meet the financial test based on the qualifications in the opinion expressed by the independent certified public accountant in his report on examination of the owner or operator's financial statements. Use of the financial test will be disallowed based on an adverse opinion or a disclaimer of opinion in such report. The regional administrator will evaluate other qualifications on an individual basis. If use of the financial test is disallowed, the owner or operator must provide an alternative financial assurance mechanism within 30 d after notification of the disallowance.[273]

An owner or operator may meet its financial assurance requirements by obtaining a written guarantee, referred to as a "corporate guarantee", from its parent corporation. The guarantor must meet one or the other prong of the financial test and comply with the requirements for submission of information set forth above. The corporate guarantee must be sent to the regional administrator along with the documents sent to demonstrate compliance with the financial test.[274]

The guarantor must comply with the terms of the corporate guarantee, which must be worded as specified in 40 CFR § 264.151(h). The terms of the corporate guarantee must provide that:

1. If the owner or operator fails to properly perform final closure of the facility covered by the corporate guarantee, the guarantor will either do so or establish a closure trust fund in the name of the owner or operator
2. The corporate guarantee will remain in force unless the guarantor sends notice of cancellation by certified mail to the owner or operator and to the regional administrator. Cancellation will not be effective, however, until 120 days after both the owner or operator and the regional administrator have received the notice of cancellation
3. In any event, if the owner or operator fails to establish (and obtain written approval of) alternative financial assurance mechanisms within 90 days after receipt of the notice

of cancellation, the guarantor is required to provide such alternative financial assurance mechanism in the name of the owner or operator.[275]

iii. Third-Party Liability Coverage Requirements
(a). Liability Insurance

An owner or operator must demonstrate compliance with financial responsibility requirements designed to ensure payments to third parties suffering bodily injury or property damage from operation of a TSD facility. The owner or operator of a TSD facility must demonstrate third-party liability coverage for "sudden accidental occurrences".[276] An owner or operator of a surface impoundment, landfill, land treatment facility, or miscellaneous disposal unit must also procure liability coverage for "nonsudden accidental occurrences".[277]

The owner or operator must have and maintain liability coverage for sudden accidental occurences in the amount of at least $1 million per occurrence with an annual aggregate of at least $2 million exclusive of legal defense costs.[278] An owner or operator may demonstrate liability coverage in any one of six ways.[279]

First, an owner or operator may demonstrate the required liability coverage by having liability insurance complying with the following requirements. The insurance policy must be issued by an insurer that, at a minimum, is licensed to transact the business of insurance, or eligible to provide insurance as an excess or surplus lines insurer, in one or more states.[280] The insurance policy must be evidenced by a "certificate of liability insurance", worded as specified in 40 CFR § 264.141(j), or amended by attachment of the Hazardous Waste Facility Liability Endorsement, worded as specified in 40 CFR 164.151(i).[281]

The owner or operator must submit a signed duplicate original of the endorsement or the certificate of insurance to the regional administrator (or regional administrators if the facilities are located in more than one region). If requested by a regional administrator, the owner or operator must provide a signed duplicate original of the insurance policy.[282]

An owner or operator of a new facility must submit the signed duplicate original of the endorsement or certificate of insurance to the regional administrator at least 60 d before the date on which hazardous waste is first received at the facility, and the insurance must be effective before such initial receipt.[283]

(b). Financial Test or Guarantee for Liability Coverage

Second, an owner or operator may meet the third party liability coverage requirement by passing a financial test *or* obtaining a guarantee for liability coverage.[284] The owner or operator may meet either of two prongs of the financial test. First, the owner or operator may demonstrate that it has (1) net working capital and tangible net worth, each at least six times the amount of liability coverage ($2 million);[285] tangible net worth of at least $10 million; and (3) assets in the U.S. amounting to either 90% of the owner's or operator's total assets or at least six times the amount of the required liability coverage.[286] Second, the owner or operator may demonstrate (1) a current rating on its most recent bond issuance of AAA, AA, A, or BBB as issued by Standard and Poor's, or Aaa, Aa, A, or Baa as issued by Moody's; (2) a tangible net worth of at least $10 million; (3) tangible net worth at least six times the amount of required liability coverage; and (4) assets in the U.S. amounting to 90% of its total assets or at least six times the amount of the required liability coverage.[287]

Further, the owner or operator must submit information to the regional administrator equivalent to (and on the schedule required by) the information required to be submitted to meet the financial test for financial assurance for closure as set forth above.[288] If an owner or operator is using the financial test to demonstrate both financial assurance for closure and liability coverage, he must submit the letter specified in 40 CFR § 264.151(g) to cover both forms of financial responsibility.[289]

Subject to certain conditions,[290] an owner or operator may meet its financial responsibility requirements for liability coverage by obtaining a written guarantee from either (1) its parent corporation, (2) a firm whose parent corporation is also the parent corporation of the owner or operator, or (3) a firm with a "substantial business relationship" with the owner or operator.[291] The guarantor must meet the financial test set forth above and comply with the other requirements imposed on owners and operators meeting their obligations through the financial test.[292]

The wording of the guarantee must be identical to the wording specified in 40 CFR § 264.151(h)(2).

A certified copy of the guarantee must be sent to the regional administrator along with the information required to demonstrate compliance with the financial test. The letter from the guarantor's chief financial officer must describe any value received in consideration of the guarantee where the guarantor is a firm whose parent corporation is also the parent corporation of the owner or operator. If the guarantor is a firm with a "substantial business relationship", the letter must describe the relationship and any value received in consideration of the guarantee.[293]

The terms of the guarantee must provide that, if the owner or operator fails to satisfy a judgment based on a determination of liability for bodily injury or property damage to third parties caused by sudden accidental occurrences arising from the operation of any facility covered by the guarantee, or fails to pay an amount agreed to in settlement of such claims, the guarantor will do so up to the limits of coverage. Further, the terms must provide that the guarantee will remain in force unless the guarantor sends a notice of cancellation by certified mail to the owner or operator and the regional administrator. Nonetheless, the guarantee may not be terminated unless and until the regional administrator approves alternative liability coverage complying with the applicable regulations.[294]

(c). Letter of Credit for Liability Coverage

An owner or operator may satisfy the liability coverage requirements by obtaining an irrevocable standby letter of credit and submitting a copy of such letter to the regional administrator.[295] The wording of the letter of credit must be identical to that specified in 40 CFR § 151(k). The issuer of the letter of credit must have the authority to do so and be regulated and examined by a federal or state agency.[296]

(d). Surety Bond for Liability Coverage

An owner or operator may satisfy the liability coverage requirements by obtaining a surety bond for such coverage and submitting a copy to the regional administrator.[297] The wording on the bond must be identical to the wording specified in 40 CFR § 264.151(1). The issuing surety company must be listed as an acceptable surety on the most recent Circular 570 of the U.S. Department of Treasury.

An owner or operator may use such a surety bond only if the attorneys general or insurance commissioners of both the state in which the surety is incorporated and each state in which a covered facility is located have submitted a written statement to the EPA that such a surety bond is a legally valid and enforceable obligation in that state.[298]

(e). Trust Fund for Liability Coverage

An owner or operator may satisfy the liability coverage requirements by establishing a trust fund for liability coverage.[299] The owner or operator must submit an originally signed duplicate of the trust agreement to the regional administrator. The trustee must be an entity which has authority to act as a trustee and whose trust operations are regulated and examined by a federal or state agency.[300]

The trust fund must be funded for "the full amount of the liability coverage to be provided by the trust fund" to satisfy the liability coverage requirements. If at any time after the trust fund is created the amount of funds in the trust fund is reduced below that amount, the owner or operator, by the anniversary date of the establishment of the fund, must either add sufficient funds to cause the trust funds value to equal that amount or obtain other financial assurance to cover the difference.[301]

The wording of the trust fund must be identical to the wording specified in 40 CFR § 264.151(m).[302]

(f). Combination of Mechanisms

An owner or operator may demonstrate compliance with the liability coverage requirements through a combination of the financial mechanisms noted above.[303] The amount of coverage demonstrated must total at least $1 million per occurrence with an annual aggregate of at least $2 million, exclusive of legal defense costs.

An owner or operator is required to notify the regional administrator in writing within 30 d (1) whenever a claim is made for bodily injury or property damages caused by the operation of the TSD facility and (2) whenever the amount of financial assurance for liability coverage provided by a financial instrument is reduced.[304]

The owner or operator may avoid compliance with the liability coverage requirements if it can demonstrate to the satisfaction of the regional administrator that the level of financial responsibility required is not consistent with the degree and duration of the risk associated with treatment, storage, or disposal of hazardous waste at its facility. The request for such a variance must be submitted to the regional administrator in Part B of the facility's permit application or, if the facility has a permit, pursuant to the procedures for permit modification under 40 CFR § 124.5. If granted, the variance will take the form of an adjusted level of required liability coverage. The level will be based on a regional administrator's assessment of the degree and duration of risk associated with the ownership or operation of the facility.[305]

Similarly, if the regional administrator on his own determines that the levels of liability coverage otherwise required are not consistent with the degree and duration of risk associated with treatment, storage, or disposal of hazardous waste at the facility or group of facilities, the regional administrator may adjust the level of financial responsibility required to an amount as may be necessary to protect health and the environment. Any adjustment of the level or type of coverage for a facility that has a permit will be treated as a permit modification under 40 CFR § 270.41(a)(5) and § 124.5.[306]

Within 60 d after receiving certifications from the owner or operator and an independent registered engineer that the final closure has been completed in accordance with the approved closure plan for the facility, the regional administrator will notify the owner or operator in writing that he is no longer required by this section to maintain liability coverage for that facility unless the regional administrator reasonably believes that the facility has not been properly closed.[307]

iv. Bankruptcy

An owner or operator is required to notify the regional administrator by certified mail of the commencement of a voluntary or involuntary bankruptcy proceeding naming the owner or operator as a debtor within 10 d after commencement of the proceeding. Similarly, a guarantor of a corporate guarantee for closure must also make such notification if it is named as a debtor.[308]

An owner or operator who meets its financial responsibility requirements by obtaining a trust fund, surety bond, letter of credit, or insurance policy will be deemed to be without the required financial assurance or liability coverage in the event of (1) bankruptcy of the trustee or issuing institution or (2) a suspension or revocation of the authority of the trustee institution to act as trustee, or the institution issuing the surety bond, letter of credit, or insurance policy to issue such instruments. In such event, the owner or operator must establish other financial assurance or liability coverage within 60 d.[309]

v. Use of State-Required Mechanisms

If a facility is located in a state where the EPA is administering the RCRA financial responsibility requirements, but the state has hazardous waste regulations including requirements for financial assurance of closure or liability coverage, the owner or operator may use the state-required financial mechanisms to meet the RCRA financial responsibility requirements if the regional administrator determines that the state mechanisms are at least equivalent to the financial mechanisms specified in the RCRA regulations. The regional administrator will evaluate the equivalency of the mechanisms by considering the certainty of the availability of funds under such mechanisms, the amount of funds that will be made available, and other factors that the regional administrator deems appropriate.[310]

To utilize the state-required financial mechanisms, the owner or operator must submit to the regional administrator evidence of the establishment of the state-required mechanism together with a letter requesting that such mechanism be considered acceptable meeting the RCRA financial responsibility requirements. The submission must include the following information: the EPA identification number, name, and address of the facility and the amount of funds for closure or liability coverage assured by the mechanism. The regional administrator will notify the owner or operator of his determination and may require additional information as deemed necessary.[311]

Pending the regional administrator's determination, the owner or operator will be deemed to be in compliance with the RCRA financial responsibility requirements for closure and/ or liability coverage as applicable.[312] If a state-required mechanism is found acceptable except for the amount of funds available, the owner or operator may satisfy the RCRA requirements by increasing the funds available through the state-required mechanism or using additional financial mechanisms as specified in the RCRA regulations. The amount of funds available for the state and federal mechanisms must at least equal the amount required by the RCRA financial responsibility regulations.[313]

vi. State Assumption of Responsibility

If a state either (1) assumes legal responsibility for an owner's or operator's compliance with the closure or liability coverage requirements set forth in the RCRA regulation or (2) assures that funds will be available from state sources to cover those requirements, the owner or operator will be in compliance with the RCRA financial responsibility requirements if the regional administrator determines that the state's assumption of responsibility is at least equivalent to the financial mechanisms specified in the RCRA regulations.[314]

The regional administrator will evaluate the equivalency of state guarantees by considering certainty of the availability of funds for the required closure or liability coverage, the amount of funds that will be made available, and other factors as he deems appropriate.[315]

To utilize a state "guarantee" in lieu of complying with the RCRA regulations, the owner or operator must submit to the regional administrator a letter from the state describing the nature of the state's assumption of responsibility, together with a letter from the owner or operator requesting that the assumption of responsibility by the state be considered acceptable for meeting the RCRA financial responsibility requirements. The letter must include the EPA identification number, name, and address of the facility and the amount of funds for closure and/or liability coverage that are guaranteed by the state.[316]

Pending the regional administrator's determination regarding the acceptability of the state's guarantee in lieu of the financial mechanisms set forth in the RCRA regulations, the owner or operator will be deemed to be in compliance with the RCRA financial responsibility requirements, as applicable.[317] If a state's assumption of responsibility is found acceptable except for the amount of funds available, the owner or operator may satisfy the RCRA financial responsibility requirements by use of both the assurance of the state *and* additional financial mechanisms as specified in the RCRA regulations. The amount of funds available through the state and federal mechanisms must at least equal the amount required by the RCRA financial responsibility requirements.[318]

h. RCRA Standards for Incinerators

The EPA has promulgated regulations specifically applicable to HWIFs at Subpart O of 40 CFR Parts 264 and 265. Distinctions between regulations applicable to a permitted facility and those applicable to an interim status facility will be discussed where applicable.

i. Applicability

Subpart O is applicable to (1) owners or operators of hazardous waste incinerators and (2) owners or operators of boilers or industrial furnaces who burn hazardous waste in boilers or in industrial furnaces (a) to destroy the waste or (b) for any recycling purpose and elect to be regulated under Subpart O.[319] As stated previously, a facility subject to Subpart O will be referred to as a hazardous waste incineration facility (HWIF).

Certain HWIFs may be exempt from all requirements of Subpart O except, in the case of RCRA permitted facilities, those requirements governing waste analysis and closure, and, in the case of an interim status facility, those requirements governing closure.[320] The EPA must exempt a permitted facility if the waste analysis included with Part B of the permitted application demonstrates that the waste contains none of the hazardous constituents listed in 40 CFR Part 261, Appendix VIII, that would reasonably be expected in the waste.[321]

Both RCRA-permitted facilities and interim status facilities may be exempt where the waste to be incinerated meets standards that are specified in the EPA regulations. For RCRA-permitted facilities, the regional administrator must find the waste to be burned meets such standards after consideration of the waste analysis included with Part B of the permit application. In the case of interim status facilities, the owner or operator may simply document, in writing, that the waste would not reasonably be expected to contain the hazardous constituents listed in Part 261, Appendix VIII, and that the waste meets such standards. Under those standards, the wastes must fall into one of the following categories: (1) waste that is listed as hazardous solely because it is ignitable, corrosive, or both; (2) waste designated as hazardous solely because it possesses the characteristic of ignitability, corrrosivity, or both; (3) waste that will not be burned when other hazardous wastes are present in the combustion zone and that is listed as hazardous solely because it is reactive for characteristics other than it (a) generates toxic gases, vapors, or fumes in a quantity sufficient to present a danger to human health or the environment when mixed with water or (b) is a cyanide- or sulfide-bearing waste that generates such toxic gases, vapors, or fumes when exposed to pH conditions between 2 and 12.5; or (4) waste which will not be burned when other hazardous wastes are present in the combustion zone and is classified as hazardous solely because it possesses any of the reactivity characteristics other than those discussed above.[322]

The EPA regional administrator *may*, in establishing permit conditions, exempt an HWIF from all requirements of Subpart O except those requirements governing waste analysis and closure, if, after consideration of the waste analysis included with Part B of the permit application, the regional administrator concludes that the waste to be burned is one of those described above and contains insignificant concentrations of the hazardous constituents listed in Part 261, Appendix VIII. The regional administrator may not so exempt the HWIF if he finds that the waste would pose a threat to human health and the environment when burned in an incinerator.[323]

Finally, the owner/operator of an incinerator may conduct trial burns subject only to the requirements of 40 CFR § 270.62.[324]

ii. Waste Analysis

In either the trial burn plan required by the RCRA permitting regulations or in Part B of the permit application, an HWIF owner/operator must include an analysis of the waste feed sufficient to provide all the information required by the RCRA permitting requirements. Throughout normal operation of the HWIF, the owner/operator must conduct sufficient waste analysis to verify that the waste feed to the incinerator is within the physical and chemical composition limitis specified in his permit.

iii. Performance Standards

An incinerator burning hazardous waste must be designed, constructed, and maintained so that when operated in compliance within the operating requirements of its permit, it will meet certain performance standards. The performance standards are keyed to a destruction and removal efficiency (DRE) for each principle organic hazardous constituent (POHC) designated for each waste feed in the trial burn plan.

One or more POHC will be specified in the facility's permit for each waste feed to be burned during the trial burn. The POHCs specified will be among the hazardous constituents listed in 40 CFR Part 261, Appendix VIII. The specification of POHCs will be based on the degree of difficulty in incinerating the organic constituents in the waste and their concentration or mass in the waste feed, as determined from the results of waste analyses and trial burns or alternative data submitted with Part B of the facility's permit application. Organic constituents most difficult to incinerate will be those most likely to be designated as POHCs. Constituents are more likely to be designated as POHCs if they are present in large quantities or concentration in the waste.[325] For trial burns, POHCs will be designated in accordance with the procedures specified in the EPA permitting regulations.[326]

The EPA regulations specify the formula for calculating the DRE. DRE is determined for each POHC from the following equation:[327]

$$DRE = \frac{(W_{in} - W_{out})}{W_{in}} \times 100\%$$

where W_{in} = mass feed rate of 1 POHC in the waste stream feeding the incinerator and W_{out} = mass emission rate of the same POHC present in exhaust emissions prior to release to the atmosphere.

For RCRA permitted facilities, the performance standards are as follows:

1. Except as provided in paragraph A(2) of this section, an incinerator burning hazardous waste must achieve a DRE of 99.99% for each POHC designated in its permitted waste feed.

2. An incinerator burning hazardous waste designated FO20, FO21, FO22, FO23, FO26, or FO27 in 40 CFR § 261.31 must achieve a DRE of 99.9999% for each POHC designated in its permit. This performance must be demonstrated on POHCs that are more difficult to incinerate than tetra-penta-, and hexachlorodidenzo-cursive *p*-dioxins and dibenzofurans. In addition, the HWIF owner or operator must notify the regional administrator of his intent to incinerate such hazardous wastes.

    a. An incinerator burning hazardous waste and producing stack emissions of more than 1.8 kg/h of hydrogen chloride (HCl) must control HCl emissions such that the rate of the emission is no greater than the larger of either 1.8 kg/h or 1% of the HCl in this gas prior to entering any pollution control equipment.

    b. An incinerator burning hazardous waste must not emit particulate matter in excess of 180 mg per dry standard cubic meter when corrected for the amount of oxygen in the stack gas according to the formula $P^c = P_m \times \left[\dfrac{14}{(21 - Y)}\right]$, where P_c is the corrected concentration of particulate matter, P_m is the measured concentration of particulate matter, and Y is the measured concentration of oxygen in the stack gas, using the orsat method for oxygen analysis of dry flue gas, presented in 40 CFR Part 60, Appendix A (Method 3).[328] The correction procedure indicated above is to be used by all hazardous waste incinerators except those operating under conditions of oxygen enrichment. For those facilities, the EPA regional administrator will select an appropriate correction procedure to be specified in the facility permit.

For purposes of permit enforcement, compliance with the operating requirements specified in the permit will be regarded as compliance with the performance standards described above; nonetheless, evidence that complies with those operating requirements is insufficient to ensure compliance with the performance standards may justify modification, revocation, or reissuance of the facilities permit.[329]

iv. Hazardous Waste Incineration Permits

An HWIF may burn only those wastes specified in its permit and only under the operating conditions specified for those wastes in the permit pursuant to 40 CFR § 264.345, except in approved trial burns under EPA permitting regulations or where otherwise exempt.[330]

An HWIF may burn other hazardous wastes only after operating conditions for such wastes have been specified in a permit or a permit modification. Those operating requirements may be based on either trial burn results or alternative data included with Part B of a permit application.[331] The permit for a new HWIF must establish appropriate conditions for each of the applicable requirements of Subpart O. but not limited to allowable waste feeds and operating conditions that ensure compliance with the performance standards indicated above.[332] The EPA regulations provide for use of operating requirements most likely to ensure compliance with the performance standards during periods immediately before and after a trial burn (*Id.* § 264.344 [c][1][3]). During the trial burn, the operating requirements must be sufficient to demonstrate compliance with the performance standards and must be in accordance with the approved trial burn plan (*Id.* § 264.344[c][2]).

v. Operating Requirements

An incinerator must be operated in accordance with the operating requirements specified in the permit. Such operating requirements will be specified on a case-by-case basis to ensure compliance with the performance standards set forth above. The operating requirements will be based on the results of a trial burn or on alternative data included with Part B of a permit application.[333] Each set of operating requirements will specify the composition of the waste feed, including acceptable variations in the physical or chemical properties of the waste feed that will not affect compliance with the performance standards. For each such waste feed, the permit will specify other operating conditions designed to ensure compliance with the performance standards, including the following conditions: (1) the carbon monoxide (CO) level in the stack exhaust gas; (2) the waste feed rate; (3) the combustion temperature; (4) an appropriate indicator of combustion gas velocity; and (5) allowable variations in incinerator system design or operating procedures.[334]

During startup and shutdown of an incinerator, hazardous waste (except waste exempted as discussed above) must not be fed into the incinerator unless the incinerator is operating within the conditions of operation specified in the permit.[335]

Fugitive emissions from the combustion zone must be controlled by keeping the combustion zone totally sealed against such emissions, maintaining a combustion zone pressure lower than atmosphere, or employing an alternate means of fugitive emissions control equivalent to maintaining the combustion zone pressure lower than atmospheric pressure.[336] The incinerator must be operated with a functioning system to cut off waste feed automatically to the incinerator when operating conditions deviate from limits established as discussed above. An incinerator must cease operation when changes in waste feed, incinerator design, or operating conditions exceed limits designated in its permit.[337]

vi. Monitoring and Inspections

During operation, the owner or operator of a permitted HWIF must continuously monitor the combustion temperature, the waste feed rate, the indicator of combustion gas velocity specified in the permit, and the carbon monoxide concentration at some point downstream of the combustion zone and before release into the atmosphere.[338] On request of the EPA administrator, the HWIF must sample and analyze the waste and exhaust to verify that the operating conditions achieve the performance standards of the operation requirements permit.[339]

The HWIF owner or operator must also inspect, at least daily, the incinerator and associated equipment, including pumps, valves, conveyors, and pipes, for leaks, spills, fugitive emissions, and signs of tampering.[340]

The HWIF owner or operator must also test at least weekly the emergency waste feed cutoff systems and alarms to verify operatability, unless the regional administrator determines that weekly inspections will unduly restrict or upset operations and that less frequent inspection will be adequate. Operational testing must be conducted at least monthly.[341] All monitoring and inspection must be recorded and the records placed in the operating log.[342]

vii. Closure

At closure, the owner or operator must remove all hazardous waste and hazardous waste residues including but not limited to ash, scrubber waters, and scrubber sludges (from the incinerator site).[343] *As throughout the operating period,* unless the owner or operator can demonstrate that the residue removed from the incinerator is not a hazardous waste, the owner or operator becomes a generator of hazardous waste and must manage it in accordance with applicable RCRA requirements.[344]

III. THE PERMIT APPLICATION PROCESS, REQUIREMENTS, AND CONDITIONS

Permitting an incinerator for treatment of hazardous wastes will seldom involve just the incinerator and its air emissions. Section A reviews the permitting process requirements for all environmental disposal permits as an overview for a TSDF facility. Section B then focuses in on the RCRA Part B permit application process and details the requirements for incineration.

A. GENERAL PERMIT APPLICATION PROCESS FOR ALL ENVIRONMENTAL PERMITS

1. Applicable Regulations

General U.S. EPA procedures for issuing, modifying, revoking, and reissuing or terminating the following permits have been issued as 40 CFR Part 124 Subpart A.

- Resource Conservation and Recovery Act (RCRA)
- Safe Drinking Water Act Underground Injection Control (UIC) Program
- Prevention of Significant Deterioration (PSD) under the Clean Air Act (CAA)
- National Pollution Discharge Elimination System (NPDES) under the Clean Water Act (CWA)

2. Overview of the Permitting Process

Subpart 124 allows, after the request of the applicant, the choice of the regional administrator to process the permits above together as consolidated permits (124.1 [d] and

124.4). The advantage to consolidating the permits is to minimize the number of public, evidentiary or judiciary, and nonadversary panel hearings. Any disadvantage is that the permit information should be approximately at equal stages of development throughout the application process for efficient processing and review.

Appendix A of Part 124 was written to assist in reading and understanding the procedural requirements in Part 124, and it contains two flowcharts that are helpful. Figure 2 is based on these charts and includes additional detail from actual permitting activities and a time line that estimates a range of time that can be expected for a permitting effort. This first chart deals with environmental permitting in general. Figure 3 is more specific for the process involved in obtaining a RCRA Part B permit and includes hazardous waste incineration.

The following discussion incorporates Part 124 procedures, Appendix A explanations, and personal experience to the explanation of the permitting process.

3. Submittal of the Application

Any person who requires a permit under the above regulations shall complete, sign, and submit to the regional administrator an application for each permit (see 40 CFR Part 124.3, 270.1 and 270.10, 270.13 and 233.4[404] [RCRA], 144.1 and 144.31 [UIC], 52.21 [PSD], and 122.1 and 122.21 [NPDES]). The regional administrator will give review authority to one of his or her directors of the particular office responsible for the type of facility being permitted.

Permit applications must comply with the signature and certification requirements of the respective regulations (see 40 CFR Part 122.22 [NPDES]; 144.32 [UIC]; 233.6 [404] and 270.11 [RCRA].

4. Agency Review

1. A permit application for a *new* facility will be reviewed by the regional administrator for completeness within 30 d of the first application submittal. A permit application for an *existing* facility will be similarly reviewed within 60 d (124.3[c]). This review will probably be assigned to an engineer or scientist under the director of the applicable EPA regional office, and review may also be subcontracted out to a consultant who reports to the engineer. Completeness is judged in terms of compliance with the application requirements of the applicable regulations, adherence to the various EPA sanctioned guidance manuals, and with any more recently accepted scientific or engineering data or principles as directed by the EPA.

2. After EPA completes the initial review, the applicant will be informed in writing whether or not the application is complete (124.3[c]). A draft permit may be issued at this point for general applications. In applications of moderate or greater complexities such as a RCRA permit application for an HWM facility, it is very rare that the application will be complete after the first or even the second submittal of the application. The EPA will list the information necessary to make the application complete. These lists of additional information required are usually termed notices or lists of deficiencies. In reality, the 30-d review period may not be long enough for the EPA to review a complex application thoroughly. Additional notices of deficiencies may actually occur due to the EPA not "catching" everything in the first review. A date required for supplying the additional information will be supplied. This date will typically be 30 d. Extensions to this time period are occasionally granted for just cause. A company's environmental and technical staff is usually completely involved and extended in supplying the required information within 30 d. Complex applications usually result in up to three notices of deficiencies and required responses before completeness is satisfied. The EPA has been fairly good at providing enough information and guidance so that the application is usually complete after three notices.

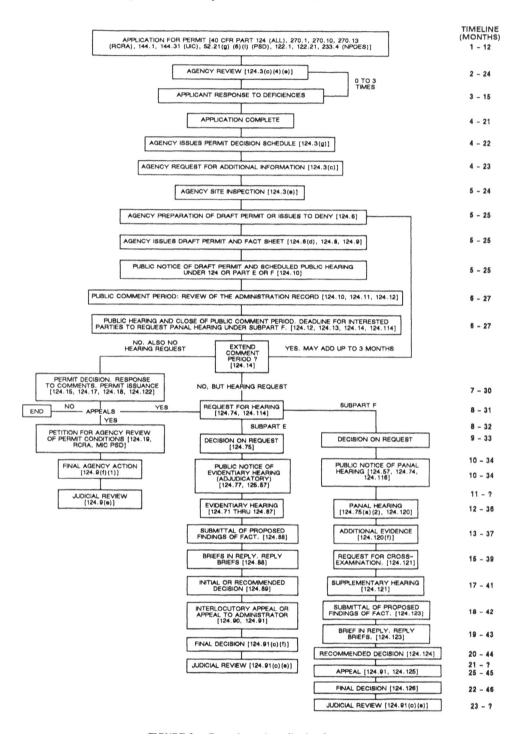

FIGURE 2. General permit application for process.

3. If an applicant fails or refuses to correct application deficiencies, the permit can be denied and enforcement actions taken against existing facilities or facilities under construction that require the permit to operate (124.3[d]) (see #20 Evidentiary Hearings).

4. The EPA usually will conduct at least one site visit in conjunction with the application

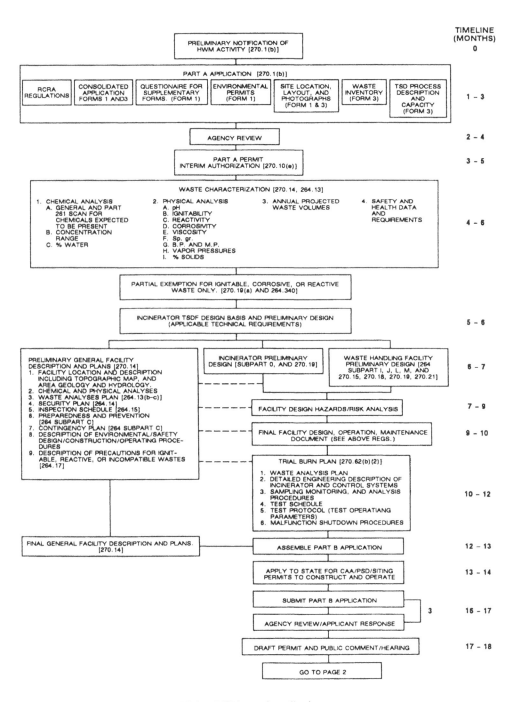

FIGURE 3. RCRA permit application process.

review and/or prior to issuing the permit (124.3[e]). A visit can be counted on before a public hearing occurs.

5. The effective date of the application is the date on which the EPA notifies the applicant that the application is complete (124.3[f]).

6. The EPA may request additional information concerning the application or the facility after the application is considered complete if the information is needed to clarify, modify, or supplement the previously submitted material (124.3[c]). Requests for such

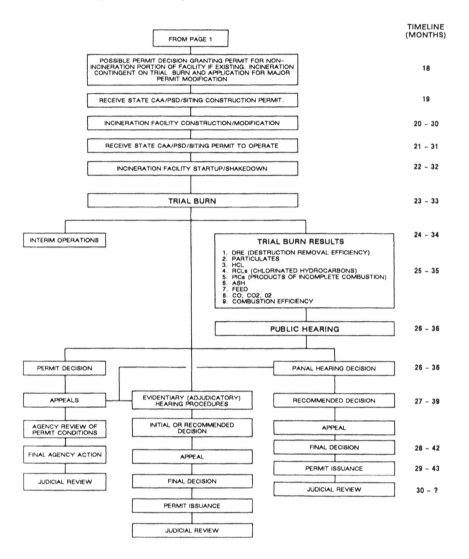

FIGURE   3 (continued).

information do not affect the completeness determination of the application. This information is usually needed to help write the draft permit and prepare the agency for any public hearing. The applicant should expect a request for more information for a major permit application.

7.   On the effective date of the application, the EPA will mail to the applicant a project decision schedule (124.124.3[g]). This decision schedule will specify target dates for the following:

   a. Preparation of a draft permit by the EPA
   b. Giving public notice
   c. Completing the required public comment period, including any public hearing
   d. Issuing a final permit and for NPDES permits, complete any formal proceedings under subparts E or F.

5. Issuance and Review of the Draft Permit
   The next step is the issuance of a draft permit including what is usually termed a fact

sheet (124.6 and 124.8). The draft permit is an extremely important document for the applicant to review carefully. If accepted as is, the applicant must live with all the terms and conditions set forth within. A careful review of the stipulated operating conditions and limitations is necessary. Often, any misunderstandings of either the regulators or the applicant concerning the permit application contents or regulations that did not show up in the permit application or deficiency notices show up in the draft permit. The conditions of the draft permit *can* be changed in the issuance of the final permit by the regulatory agency on showing them evidence of just cause. The applicant must comment on the draft permit by the end of the public comment period for comments to be considered. The agency can also issue at this time a notice of intent to deny the permit. This notice of intent to deny is treated as a draft permit and follows the same subsequent steps as a draft permit (124.6). The contents of the draft permit are based on the administrative record (124.9). This record includes the following:

1. Permit application and any other supporting documentation supplied by the applicant. This would include the notices of deficiencies and the corresponding applicant responses.
2. The draft permit or notice of intent to deny or terminate the permit.
3. The fact sheet.
4. Other documents in the file to support the draft permit.
5. Any environmental assessments and impact statements required of applicants under the National Environmental Policy Act. At present, RCRA, UIC, and PSD permits are not subject to these requirements. It is presently applicable only to NPDES new source draft permits and government projects other than Superfund.

The fact sheet is required by 40 CFR 124.8 and is a short summary of the reason for the permit application, description of the facility or activity being considered for a permit, the type and quantity of wastes to be handled or processed, the basis for the draft permit conditions, discussion on review of alternatives to the facility or activity, and a description of the review process to date and review process subsequent to issuance of the draft permit. The fact sheet will have the beginning and ending dates for the public comment period under 124.10 and the address where the public or interested parties can send their comments. It will detail the procedures for requesting a hearing and the nature of the hearing. Uncontroversial permit applicaitons such as some NPDES permits often do not draw requests for a hearing from the public or interested parties. However, it is rare for a RCRA or other permit dealing with a ''hazardous waste'' not to have a public hearing even if there has been no public comment and request for one. Count on a public hearing if the facility or activity has anything to do with hazardous waste.

6. Public Review

The public must be given notice of the draft permit and given a period to comment (124.10). Notice must be given at least 30 d before closing of public comment and the occurrence of a public hearing (for RCRA the period is 45 d). Notice of a public hearing must also be given and *may* be, and usually is, given at the same time as the notice of the draft permit and public comment period. Determination of the need for a public hearing is sometimes based on public comment during the comment period; therefore, the notice of a public hearing may lag that of the draft permit. Notices of the draft permit and of public hearings must follow the following procedures:

1. Mail
   a. Applicant

b. Other agencies concerned with the permit

c. Users of the facility other than the applicant

d. Persons requesting to be on mailing list

2. A publication notice in a daily or weekly newspaper within the geographic area of facility or activity. For RCRA permits, this notice will also consist of broadcasts over local radio stations.

3. Any other action including press releases that could reasonably be expected to reach potentially affected persons.

The public comment period (40 CFR 124.11) allows any interested person to submit written comments on the draft permit or the permit application and to request a public hearing stating the nature of the issues to be raised in the hearing. Public comments are considered in making the final permit decision. The public hearing, if required, is considered part of the public comment period. An informal public hearing under 124.12 may be held at any time during the public comment period at the discretion of the agency on proper public notice.

7. U.S. EPA Discretionary Review Power

Under 40 CFR 124.14, the EPA at any time before issuing a permit, may do one or more of the following:

1. Reopen or extend the comment period and start the process over either at the draft permit step or at the fact sheet step.

2. Prepare a new draft permit and start the process over from that point.

3. For RCRA, UIC, or initial licensing NPDES facilities, the EPA can elect to begin "panel hearing" proceedings under Subpart F of 40 CFR 124. The most logical time is during or after the public comment period (see Subpart F panel hearing procedures after this discussion on conventional procedures).

4. A public hearing (124.12) is held if the regulator finds it applicable to the decision for granting a permit or there is sufficient interest by parties of concern. Public hearings for applications opposed by environmental groups are interesting. Applicants for RCRA Part B permits (especially if they involve incineration and landfills) should be prepared for organized, and at times emotional, opposition.

8. Permit Issuance

The final permit decision triggers the agency to issue a response to comments (124.17). This response specifies the provisions of the draft permit that have been changed in the final permit decision and the reasons for the change. It contains a description of and response to the significant comments on the draft permit or the permit application.

EPA can decide to take action on their own to change or issue a general permit which would lead directly to issuing a draft permit (124 Appendix A). This occasionally occurs for minor permits, but more frequently, for minor or general type permits, the application, draft permit, fact sheet issuance, and public notice steps can occur within a short period of time if the applicant and regulator are on good terms, the applicant has honestly provided the information requested on any application forms, and the applicant has a good environmental record. It almost never happens for major permits.

9. Administrative Appeals

The appeal of RCRA, UIC, and PSD permit decisions are regulated by 124.19 and evidentiary (adjudicatory) hearings for NPDES and related facilities are regulated by 124.74. Within 30 d after a final permit decision, any person who filed comments on the draft permit or participated in the public hearing on the draft permit may petition for agency review of any permit condition of the decision. During the above review procedures and until final determination by the agency, all contested permit conditions, and those uncontested conditions that are not severable from those contested, are stayed and are identified by the agency (124.16). All other conditions of the permit remain in effect and enforceable. Anyone who did not participate in the public hearing on the draft permit can only petition for review of the changes that occurred in the permit decision since the draft permit. It must be shown that the issues were raised during the public comment period and that, where appropriate, the findings were either clearly erroneous or are indicative of an important policy question or a matter requiring significant discretion. The agency may also decide to act on its own to review a permit but must act within 30 d of the notice of decision. The agency must act within a reasonable time following the petition to grant or deny the petition for review. If review is denied, the final decision becomes the final agency action. The agency may defer consideration of an appeal of a RCRA or UIC permit until completion of formal proceedings under Subpart E or F if an NPDES permit is likely to raise issues relevant to a decision of the RCRA or UIC appeals or if an NPDES permit is likely to be appealed. The agency also has discretion to defer the consideration of an appeal if any adverse effect of the deferment is outweighed by the benefits of a consolidated decision on appeal. Final agency action occurs when a final RCRA, UIC, or PSD permit is issued or denied by the EPA and agency reviews are exhausted. A petition for review to the administrator is a prerequisite to seeking judicial review of the final agency action. Judicial review is not applicable until a final agency determination.

10. Panel F Hearing Procedures

There are advantages and a disadvantage to the hearing procedures. The disadvantage is that the applicant is on the defensive even when the applicant is the party contesting the permit conditions. It is the applicant who always must bear the burden of providing evidence showing why the permit should be granted as requested (124.85 [a][1]). The applicant by this time should have sufficient favorable information in the administrative record that supports the permit issuance. The agency has the burden of presenting a positive case in support of any permit condition (124.85[a][2]). The advantages include: (1) much of the emotionalism is absent and the hearing results should be based on technical and legal arguments, (2) the applicant can further present the case for permit condition modification or permit approval.

a. Subpart F Hearing Procedures

Subpart F Nonadversary Panel Procedures (124.111) incorporate formal hearings for the introduction of evidence for and against the issuance of a permit and are used in the following situations:

1. Initial licensing NPDES permits. These permits include the first decisions on an NPDES permit applied for by any discharger that has not previously held one and the first statutory variance.
2. These procedures will be used for any RCRA, UIC, or PSD permit which has been consolidated with such an NPDES permit.
3. At the discretion of the agency, these procedures may be used for the issuance of individual RCRA or UIC permits.

Steps in these Subpart F procedures are identical to the conventional procedures through the public comment period except that public notice is also regulated under 124.57 and 124.57(c) besides 124.10. Requests for a panel hearing must be received by the end of the public comment period under 124.113 and 124.114.

The public comment period may (and usually does in major permit cases) include an opportunity for a public hearing under 124.12. The agency may also decide before the draft permit is prepared under 124.6 that a panel hearing should be held. In such case, the public notice of a draft permit shall incorporate the notice of the hearing and contain the information in 124.57(c). In practice, for major permits, a public hearing is usually held and followed by the panel hearing the next day or within several days. Persons wishing to participate in the hearing must file a request with the regional hearing clerk before the deadline set forth in the hearing notice as set forth by 124.117. After public notice of the panel hearing the participants must be allowed at least 30 d to submit written comments according to the guidelines of 124.118. These written comments must be filed 30 d prior to the hearing. Parties to any hearing may submit written response material to the comments filed by other parties at the time they appear at the panel hearing.

Either an administering law judge or a lawyer employed by the agency will be named as the presiding officer for the hearing (124.119). According to 124.120, panelists consist of three or more EPA employees having expertise or responsibility in the areas related to the permit but who have not participated in the permit draft. The agency has the discretion to include panelists not employed by the EPA if they are believed to be relevant. The names of the panelists will be included in the hearing notice. Panel members may question anyone participating in the hearing. Cross-examination during the hearing by persons other than panel members is restricted (see 124.120[d] and [e]). Any additional evidence persons desire to submit for consideration must be submitted within 10 d after the conclusion of the hearing. An opportunity for cross-examination is given after the initial hearing by persons submitting a written request following the guidelines of 124.121. The presiding officer will either grant or deny the opportunity for cross-examination. If granted, a supplementary hearing will be held according to the regulations of 124.121, which may include submission of additional written testimony in lieu of cross-examination. Regulations concerning cross-examination procedures include 124.85(d)(2) and 124.84(e). The record for the final permit decision will be based on these proceedings and all previous documentation as outlined in 124.122. Within 20 d after transcripts of all hearings become available, parties may submit proposed findings of fact, and/or proposed modifications to the permit and support brief with references to relevant transcripts and exhibits. Within 10 d parties may submit a brief in reply to the other party's brief containing alternative findings of fact and/or permit modifications (124.123). As soon as practicable after the hearing(s), the agency evaluates the record of the proceedings and files a recommended decision with the regional hearing clerk. A copy of the decision is served on the concerned parties (124.125). The decision may be appealed by either party according to 124.125. As soon as practical after all appeal proceedings are complete, the agency shall issue a final decision (124.126).

b. Subpart E Evidentiary Hearings (Adjudicatory)

Evidentiary hearings are applicable in the following situations:

1. CWA Section 402 proceedings.
2. RCRA permit terminations including interim status termination for failure to submit the required information in the Part B permit application or responses to the notices of deficiencies.
3. In consideration of RCRA and UIC permits which are being issued, modified, or revoked and reissued when the conditions of these permits are closely related with the conditions of an NPDES permit; this situation pertains to many TSDF facility applications that will also have an NPDES permit.
4. Evidentiary hearings may be called by the EPA (as early as notification of the public comment period and public hearing) in cases where the agency believes judicial review

is a possibility (124.74[b][1] Note). The hearing will allow the agency a thorough review of all issues. The hearing may take place immediately after the public hearing or at a later date depending on the date of notification.

5. Evidentiary hearings are not applicable for PSD permits. PSD permits may only be consolidated with other permits under Subpart F hearings (124.74[b][2][iv].

Requests for an evidentiary hearing by any interested person must be made within 30 d of the serving of the final permit decision and must contain the information required by 124.74(b)(1). The agency must decide whether or not to grant the hearing within 30 d after the expiration of the time limit for submitting requests for a hearing (124.75). Unless good cause is shown, evidence or issues presented during the hearing must be evidence or issues that were submitted to the administrative record during the public comment period for the draft permit (124.76). Public notice of the hearing is regulated by 124.77 and 124.57(b). In addition to the party requesting the hearing, any person may be a party to the hearing by requesting admittance within 15 d after notice of the hearing (124.79) with exceptions as noted in 124.79(b). Requests must meet the requirements of 124.74 and 124.76. All submissions for a hearing must be filed with the hearing clerk (124.80).

It is common for the agency to consolidate several of the hearing proceedings to expedite or simplify the review of the issues (124.82).

Prior to the hearing the presiding officer may require participants or their attorneys for one or more conferences to consider the expediting of the hearing procedures, objections to any of the evidence in the record, evidence submission requirements, and the dissemination of expert witnesses identities (124.83).

During the hearing the applicant bears the burden of persuading the agency that the permit should be granted. The agency has the obligation of presenting a positive case supporting any challenged condition of the permit. The applicant needs to have a convincing argument for his permit and/or conditions in the permit he/she disagrees with.

In an evidentiary hearing, testimony of opposing witnesses and cross-examination is allowed based on the discretion of the presiding officer. Direct and rebuttal evidence is given in written form unless the presiding officer determines that oral presentation will facilitate the hearing. Whenever two or more parties have similar interests the presiding officer may limit the number of attorneys who will cross-examine and make and argue motions.

Subsequent to the hearing, and within 45 d of the hearing transcript being filed, any party may file a brief of proposed findings of fact and conclusions of law (124.88). For good cause, the presiding officer may extend the filing time period. Briefs in reply to the above briefs are usually allowed, but are at the discretion of the presiding officer.

The presiding officer issues an initial decision after review of the hearing and administrative record. The regulations do not specify a time period for this review. The applicant should be aware of the political climate. The initial decision of the presiding officer becomes the final decision after 30 d unless a party files a petition of appeals.

Appeals to the initial decision must be made according to 124.91 unless an interlocutory appeal to the agency is made through a certification of such appeal by the presiding officer within 10 d of the decision. The normal process is an appeal to the administrator which must be submitted within 30 d of the initial decision. The appeal is a request for decision review and must include the basis for the request as outlined in 124.91. It should be noted that the agency itself may decide to exercise this review. In such a case, the agency will notify the parties of such a review within 7 d. If the review request is from other parties, the agency must decide on the request within a reasonable time. It must be noted that for any party to proceed to a judicial review, they must first complete the requirement for appeal to the administrator. If the agency denies the review or affirms the initial decision with or without opinion, then the initial decision becomes the final decision. If the review granted is favorable to the requesting party, then the procedure is to rehear the proceeding in question including any appeals.

B. SPECIFIC LAW/REGULATION ADDITIONAL PERMITTING PROCESS REQUIREMENTS

1. The RCRA Application Process

In the following discussion an overview of the general RCRA application process for a treatment, storage, and disposal facility (TSDF) is discussed with emphasis on incineration.

There are three sets of treatment, storage, and disposal (TSD) regulations: (1) interim status standards (Part 265); (2) TSDF management, design, and operation standards (Part 264); and (3) permit program regulations (Part 270).

Both Parts 265 and 264 are divided into two segments categorized as: (1) administrative and nontechnical requirements (management) and (2) technical requirements (design and operation). Part 270 regulates the permitting process and also should be consulted for additional technical requirements. The technical requirements are listed as subparts. Incineration technical requirements are listed as Subpart O in each.

a. Administrative and Nontechnical General Requirements*

i. Preliminary Notification of Hazardous Waste Activity**

Preliminary notification of hazardous waste activity is mandated by Section 3010 of the RCRA Act. This notification form (EPA form 8700-12) is to be filled out by *existing* hazardous waste generators, transporters, and owners/operators of TSDFs. Existing or potential marketers or burners of hazardous waste or used oil burned for energy must also provide notification. The purpose of this notification is to allow the EPA to catalog the facility and keep records. Notification of activity is required no later than 90 d after promulgation or revision of regulations in 40 CFR Part 261. This part lists the applicable wastes to date that trigger the notification requirement.

The primary reason for incorporating the above discussion is that the wastes of Part 261 will increase in number over time, and it will be the waste handlers responsibility to provide notification if the waste handled is added to the regulations. Six months after promulgation or revision of Part 261, the activity must cease unless a RCRA permit is applied for.

A RCRA permit consists of two parts: Part A and Part B. For existing facilities, this requirement is met by submitting Part A of the application until a date is set by the agency for submitting Part B. Submittal of Part A by the applicant and a satisfactory review by the agency qualifies owners/operators for interim status.

New facilities planned do not need to submit a Notification of Activity, and both Part A and Part B are submitted together. Construction cannot begin until a RCRA permit is issued. The consolidated permit application must be submitted at least 180 d prior to construction. In practice, it takes much longer to obtain a permit, and the application should be submitted at least 1 year in advance of construction and up to 2 years in advance if the facility is complex. Since it will take a minimum of 6 months to write the permit application (unless the facility is simple) and the design of the facility is a major portion of the application, the preliminary design should be complete by or should parallel the writing of the permit application.

ii. Contents of Part A***

The RCRA permit effort should be initiated by someone thoroughly familiar with the

---

*   Administrative and Nontechnical General Requirements (270.1, 270.10, 270.13, 270.14, 270.19) (incineration), 264 Subparts A through H, and 265 Subparts A through H, and any other applicable sections of 270.14 through 270.29.

** 270.1(b).

***270.1(b), 270.13, and Consolidated Forms 1 and 3.

regulations and capable of interfacing with agency permit writers. It is critical to establish an effective line of communication with the agency permit writer early in the permitting process. All project personnel involved in the design, construction, and operation of the project should be familiar with the regulations and be capable of interfacing with the application writer.

Part A consists of at least two forms that must be filled out. Form 1 contains questions that delineate the location and nature of the facility and business, ownership/operator relationship the existence of any presently existing environmental permits, and determines if any other forms (supplementary) must be filled out. In all cases involving incineration of hazardous waste and most other hazardous waste activities, Form 3 must also be filled out. If the incineration facility involves wastewater discharges, the applicant will need to request Form 2C or 2D depending if the facility is existing or new (planned). Form 4 needs to be requested if deep well injection is involved. Form 5 must be requested if the facility is a proposed stationary source that is included as one of the 28 industrial categories listed in the Form 1 instructions which would emit 100 ton/year of any pollutant regulated under the CAA and which would affect or be located in a nonattainment area. Form 5 must also be requested and submitted if the facility is one that will emit 250 ton of any air pollutant regulated by the CAA and may affect or be located in an attainment area.

Form 3 consists of questions that further delineates the facility location and description, asks for a description of each hazardous waste and the handling processes involved (TSD), and the annual quantity of each waste processed. A facility layout to scale is required detailing each past, present, and future storage, treatment, and disposal area and operation. Photographs are required of existing facilities and of structures and sites of future facilities.

Completion of Part A may take from 2 weeks to 3 months depending on the number of forms required to submit, the number of different wastes and handling processes involved, current knowledge of the different wastes and processes, and internal review procedures. For existing facilities, the next step is to submit the Part A application to the agency permit writer for review. This review may take 1 to 2 months. Normally, no notices of deficiencies are received by the applicant due to the straightforward nature of the forms, but such an occurrence is possible. Interim status to the facility is granted after a successful review (270.10[e]). Interim status is equivalent to receiving a RCRA permit to operate until the agency makes a final determination of the applicants permit status after submitting the Part B application. Interim status for various hazardous waste units is for a limited time. Interim status termination dates for existing units are as follows:

1. Incineration 270.73(e) — November 8, 1989 unless Part B Application is submitted by November 8, 1986
2. Tank systems 270.73(f) — November 8, 1992 unless Part B is submitted by November 8, 1988
3. Chemical, physical, biological treatment 270.73(f) — November 8, 1992 unless Part B is submitted by November 8, 1988
4. Landfill 270.73(c) — November 8, 1985 or 12 months after the facility becomes subject to the requirement to have a RCRA permit
5. Surface impoundments 270.73(f) — Same as 2 and 3
6. Waste piles 270.73(f) — Same as 2 and 3
7. Containers 270.73(f) — Same as 2 and 3

Facilities with interim status must comply with all the applicable interim status standards in 40 CFR 265 and permitting requirements of 270.70. Incineration is specifically covered by 265 Subpart O. Since all existing hazardous waste incinerators should have submitted Part B by November 8, 1986 (except those that will possibly be included as further hazardous wastes are listed), and the standards are similar to Part 264, interim status is not discussed further.

New facilities do not submit Part A separately, but proceed with Part B and submit the two parts together.

iii. Administrative and Nontechnical Contents of Part B

Administrative and nontechnical contents of Part B include 270.14 and applicable requirements of 270.15 through 270.29, 270.62, and Part 264, with incineration specifically 270.14, 270.19, 270.62, and Part 264, Subpart O.

The information requirements of 270.14 through 270.29 are necessary to show compliance with the general management standards in Part 264. This discussion combines applicable parts of 264 and 270. It is suggested that both parts be referred to while reading this section. General information requirements for HWM facilities (270.14 and Part 264, Subpart B) are as follows.

(a). General Description of the Facility*

Although required by this section, no regulation guidance is given for descriptive items required. The EPA manual "Permit Applicants" Guidance Manual for the General Facility Standards of 40 CFR 264" in Section 5 lists the following minimum required items:

1. Name of company
2. Name of owner
3. Type of facility (on-site, off-site; storage, treatment, disposal)
4. New or existing
5. Location
6. Size (acres, number of units)
7. Activities conducted
8. Waste types and quantities stored, treated, and/or disposed of

This section should remain general in nature; however, each of the items above should be somewhat descriptive and informing. As an example, the location should give street address, longitude and latitude, and a general area map depicting the site in relation to known landmarks. The size of the facility should also contain a general plot of the facility with applicable units (loading and unloading transfer, storage, treatment, disposal) indicated and labeled. The name of the company and the owner should indicate street and mailing addresses and any affiliation with a larger company and their address. Any drawings, however general, should be professional. The description of activities should be complete.

(b). Chemical and Physical Analyses of the Hazardous Waste(s) to Be Handled at the Facility*

This item is where many applicants do not place enough emphasis both for their application and design and operation considerations. It is essential to present the waste analysis data that will allow the applicant in its use to design and operate the facility in a safe and reliable manner. The minimum information needed follows:

1. Name of the waste.
2. Names of waste constituents. (The author recommends reporting identification down to at least 1% by volume and weight of the waste content and identification of detectable amounts of any compounds from a GC/MS scan of compounds that are on the U.S.

---

* 270.14(b)(1).

EPA contract lab program target compound list, which is essentially a pared-down Part 261 Appendix VIII listing [see discussion on waste identification below].) If the wastes and waste constituents to be handled are not known, then the names of the wastes from the Part 261 Appendix VIII list within the design range of the facility should be listed (based on vapor pressure, viscosities, etc.).

3. EPA hazardous waste ID number for the waste(s) and waste constituents.
4. Concentration of the waste constituents if known; if not known, the acceptable design concentration limits.
5. Water content.
6. Ash content.
7. Percent solids.
8. Reactivity.
9. Ignitability (flash point).
10. Vapor pressure(s) (absolutely needed-estimate worst case based on constituents if no data available).
11. Corrosivity.
12. Acidity or alkalinity (pH).
13. Viscosity (absolutely needed-estimate worst case or perform tests if data unavailable. This is mainly needed for pump, valve, agitator, and liquid injection nozzle design and selection).
14. Total organic carbon.
15. Heat of combustion.
16. EP TOX metals or TCLP.
17. Major inorganic constituents.
18. PCB, dioxin, and furans.
19. Heating value in British thermal units per pound (absolutely necessary for incineration).
20. Heavy metals, including concentrations.

If the application is for an off-site facility, the applicant must be able to determine that each receipt of hazardous waste matches the identity of the waste contracted for and that the waste is within key permit parameter limitations. This is called fingerprinting. It is the owner/operator's responsibility to obtain waste analysis data for all wastes received at the facility. The owner/operator should request this data from the generator and verify the information. Recommended information to be obtained from the generator is as follows:

1. Representative waste samples for testing
2. Waste stream data
    a. Volumes (per year, month, week)
    b. Description of process generating the waste
    c. Waste classification
    d. Waste characteristics (see above list)
    e. Known health or environmental effects
    f. Known handling and disposal hazards
    g. Generator site visit for verification
    h. Material safety data sheet

The above two lists should be combined into a standard questionnaire and incorporated into the permit application.

In many instances, the permit applicant will not know the exact composition of wastes to be incinerated (or stored, etc.) prior to the application. What must be decided is what parameters in the above two lists constrain the safe design and operation or economics of the facility. As an example, the applicant may not want to complicate the permit application process or go to the extra expense required for incineration of PCBs (regulated by TSCA, not RCRA). In this case, the permit application chemical and physical analyses section would specify that the incinerator would not process material with a PCB content >50 ppm. Design optimization and economics may restrict the size of the incinerator and, coupled with temperature requirements, will restrict the heat input. Therefore, there will be waste flow rate and waste energy content restrictions. Based on some optimum combination of heat value, incinerator sizing, and flow rate, the chemical analysis section will specify a maximum limit that will be burned. Conversely, there may also be a minimum Btu/lb specified depending on incinerator auxiliary heating capability. Each one of the parameters above should have a maximum and minimum value. The applicant will be constrained by being permitted to accept wastes only within these values. Later in the application, the applicant will have to show that facility is designed, and can be operated, to accept waste within these values.

(c). Waste Analysis Plan*

The applicant must supply a waste analysis plan with the application. This plan gives the written procedures for accomplishing compliance with the waste parameter constraints above. The plan must specify:

1. The parameters for which each waste will be analyzed and the rationale for the selection of these parameters. The trial burn will verify that the incinerator will be able to burn wastes safely within the above parameter boundaries. The applicant must show that wastes received subsequent to the trial burn are within these boundaries. Each waste stream need be characterized only once (with periodic confirmation analysis), however, each time this waste stream is received it must be analyzed to an extent that confirms it is the same waste (fingerprinting).
2. The test methods which will be used to test for these parameters.
3. Sampling methods which will be used to obtain a representative sample. Sampling and testing methodology is expected to conform to Appendix I of Part 261 and the EPA Test Method for Evaluating Solid Waste, SW 846, which includes both sampling and analysis methods. Documentation will be required for equivalency to these methods for proposed methods not included in these documents. Sampling methods may require some ingenuity and cost outlays for safe and representative sampling in many cases. Not all situations are covered by the regulations and guidelines. The number one priority in sampling methodology is worker protection from exposure.
4. Frequency at which the initial sampling and analysis is repeated to verify that the waste handled is consistently the same as in the application or that contracted for within the permit application limits. Minimum frequency that will be accepted is annually. This does not refer to "fingerprinting" which is done every time a waste is received from off site.
5. The sampling, analysis, and handling methods to ensure that ignitable, reactive, or incompatible wastes are handled properly to eliminate the possibility of ignition, reaction, or modification of waste characteristics to that outside the application limits. Prime consideration must be given to the safety of human health and the environment. Samples of all waste material to be handled should be analyzed for ignitability. Laboratory tests should confirm reactivities. All samples should be archived to test the compatibility with other wastes and materials of construction (this is extremely useful

---

* 270.14(b)(3) and 264.13(b), 264.13(c), 264.17, the U.S. EPA "Waste Analysis Plans, A Guidance Manual", and "Sampling and Analysis Methods for Hazardous Waste Combustion", prepared by Arthur D. Little, Inc. for Industrial Research Lab and available through NTIS.

in tank, pump, and valve applications where liners and synthetic parts are sometimes used). *General* guidance on waste intercompatibility is available and detailed in the EPA "Permit Applicants' Guidance Manual for the General Facility Standards of 40 CFR 264" and, specifically, Section 5 of the manual.

6. For facilities receiving wastes from off-site, waste movement must be tracked. Tracking is a methodology similar to the chain of custody concept in sampling and analysis procedures. The following areas are involved in tracking waste through the facility:

   a. Plant entrance — the waste shipment manifest, material safety data sheet (MSDS), and bill of lading is checked against the original waste description from the sender and initial waste analysis prior to entrance to the site. It is also good practice at this point for a safety inspection of entering vehicles and container(s) before entrance is granted. After the proper identification and paperwork inspection, clearance is given to the next checkpoint. Depending on the waste characteristics, specific instructions must be given to the transporter on proceeding to the next applicable checkpoint.

   b. Unloading — unloading may be in a general area for compatible wastes or specific for the waste type to eliminate mixing incompatible or reactive wastes. Some method must be employed to guarantee that the load received here is the same as that expected. This could be in the form of a paper with the waste description, truck identifying number, and the proper unloading area specified. Where it is to be unloaded is important. The number of barrels or truck contents weight is also important for tracking the amount of waste.

   c. Sampling/analysis — a representative sample must be obtained between the unloading area and storage (or treatment/disposal if treatment/disposal is direct) for fingerprinting before any waste is blended, stored, or disposed of. Sample chain of custody procedures should be followed.

   d. Storage/Treatment/Disposal — Following successful fingerprinting, the waste may be transferred to blending, storage, or treatment/disposal if volume or weight amounts are determined and recorded prior to or during transfer. Transfer location and any operations on the material such as blending, storage, or disposal should be documented. Operation logbooks should be kept during the handling and disposal. A record-keeping system for all the above documentation must be in place.

(d). Security Procedures

(264.14): Restricted entry — the owner/operator must prevent the unknowing or unauthorized entry of persons or livestock onto the active portion of the facility. This is to include:

- A 24-h surveillance system consisting of surveillance by guards or facility personnel or television (TV) monitors. In reality, a combination of facility personnel and television monitoring is ideal. TV monitoring has the advantage of constant monitoring of an area (if the monitor is watched) while personnel can also be involved in other duties. However, cameras are unable or unreliable to pick up some safety concerns such as local process instrumentation readouts, minor leaks, and, in some cases, fires in their beginning stages. In the author's opinion an inspection by qualified personnel should be conducted several times in a 24-h period (including off hours unless the facility is not treating, disposing, storing, and handling waste) to verify safe conditions. These inspections should be documented. It is also reasonable to assume that the TV camera views can be circumvented by anyone intent on causing harm to the facility.

- An artificial or natural barrier which completely surrounds the active portion of the facility. This can consist of a properly constructed and maintained fence and can incorporate natural barriers such as a cliff. It is the opinion of the author that dikes should not be included unless they can be shown to be as difficult to scale as a fence.

- Entrance to the active portion of the facility must be controlled through attendants, television monitors, locked entrance, or controlled roadway access. It is the opinion of the author that a hazardous waste facility should have at least one guard attending and patrolling entrance(s) and the facility perimeter. TV monitors can be used to reduce manpower but not used to eliminate it. Controlled access roadways are helpful in reducing chances of an unintentional trespass but useless for prevention of intentional trespass.

- Adequate posting of the site must be erected. As a requirement, the words "Danger — Unauthorized Personnel Keep Out" must be posted at each entrance to the active portion of the facility and at other locations, in sufficient numbers to be seen from any approach to the active portion. The words must be in English and any other language predominant in the area and must be legible from 25 ft. In the author's opinion, signs should be posted at least within every 250 ft and after any perimeter direction change. The signs should also indicate the area contains hazardous waste. Additional signs should be erected warning of any general hazards such as fire or explosion (e.g., "No Smoking" signs).

- Security measures such as perimeter barriers, entrances, surveillance equipment, guard stations, and warning signs should be clearly marked on a facility plot plan in this section of the permit application.

- Security personnel should have communication devices and linkage to outside telephone lines.

(e). General Inspection Requirements

(264.15) — Inspection records must be incorporated into the operating logs and records of the facility according to 264.73. The owner/operator must inspect the facility for malfunctions, deterioration, operator errors, and leaks or discharges from the facility. Inspections must be conducted often enough to identify problems and correct them before they harm human health or the environment. An inspection list(s) of items to be inspected with an inspection schedule(s) for each item must be prepared and be part of the application. The inspection list/schedule must also identify the types of problems which are to be looked for during the inspection. The inspection lists/schedules must include monitoring equipment, safety and emergency equipment, security devices, and operating and structural equipment. Except for areas with great potential for spills, such as loading, unloading, transfer, and storage areas which must be inspected daily, the frequency of inspection of the items can vary depending and based on the rate of estimated or known deterioration of equipment. The probability of an incident involving human health or the environment may also be used in setting the frequency. Minimum required inspection frequencies for the following are

- Container storage areas — weekly. Must inspect for leaking and/or deteriorated containers and containment system integrity/deterioration (264.174).
- Tank systems (264.192, 264.195)
  - New tank systems prior to being put into service. Design and construction must be reviewed/inspected for conformance to code/engineering standards by a qualified registered professional engineer (264.192).
  - Daily inspection of above-ground portions of tank systems to detect corrosion or leakage.
  - Daily inspection of data gathered from monitoring and leak detection equipment to ensure that the tank system is operated according to design.

- Daily inspection of the construction materials and area immediately surrounding the externally accessible portions of the tank system including the secondary containment system to detect erosion or signs of releases of hazardous wastes.
- Annual inspections of cathodic protection systems. New installations must be inspected within 6 months for proper operation. Sources of impressed current must be inspected bimonthly.
- Incinerators (264.347)
  - Continuous monitoring of incinerator combustion temperature, waste feed rate, and combustion gas velocity.
  - Continuous monitoring of incinerator carbon monoxide downstream of the combustion zone and prior to atmospheric release.
  - The incinerator and associated equipment (pumps, valves, conveyors, pipes, etc.) must be thoroughly inspected visually at least daily for leaks, spills, fugitive emissions, and signs of tampering.
  - The incinerator emergency waste feed cutoff system and associated alarms must be tested at a minimum of monthly and must be tested weekly unless shown to be unduly restrictive or likely to upset operations.

As a practical consideration, inspection lists/schedules should be organized by inspection frequency (daily, weekly, monthly, etc.). Each set of records should then be subdivided by system (tank systems, incinerators, etc.). This subsystem then should be subdivided into individual components (Tank 1, Tank 2, Incinerator 1, etc.). Each individual component then is broken down into primary and auxiliary inspection items.

Each inspection sheet must include the date and time of inspection, the name of the inspector, a notation of the observations made, and the date and nature of any repairs or other remedies. The inspection sheets (log) must be kept as records for at least 3 years from the date of inspection (5 years for TSCA). The owner/operator must remedy any deterioration or malfunction of equipment or structures which the inspection reveals according to a time schedule so that the discovered problem does not lead to an environmental or human health hazard. Where a hazard has already occurred or is imminent, the remedial action must be taken immediately. This can be accomplished through a maintenance work order system that is sensitive to the record keeping requirements of the regulations (work order status feedback to the required record keeping) and to the work priority necessary for protecting human health and the environment. An example work order system flow chart is shown in Figure 4.

(f). Personnel Training

(40 CFR 264.16, 40 CFR 265.16, 29 CFR Chap. XVII 1910.120 and the NIOSH/OSHA/ USCG/EPA Occupational Safety and Health Guidance Manual for Hazardous Waste Site Activities) — regulations concerning employee training for working in TSDFs were originally promulgated under RCRA in 1980 (40 CFR 264.16 and 265.16). In 1984, OSHA issued the Hazard Communication Standards (HCS, as 29 CFR 1910.1200) for employees using or manufacturing hazardous materials. SARA in 1986 required OSHA to regulate workers at hazardous waste sites and emergency responders as 29 40 CFR 1910.120.

Subpart Z containing 1910.1200 is the pertinent section of Part 1910 regarding toxic and hazardous substances. Subpart H contains 1910.120 Hazardous waste operations and emergency response.

A pyramid training concept is strongly recommended by the author that incorporates interregulatory required training. This pyramid concept is shown as Figure 5. All corporate officers, facility personnel, including support and administrative personnel such as office workers, should be included in general training concerning the facility. This general training

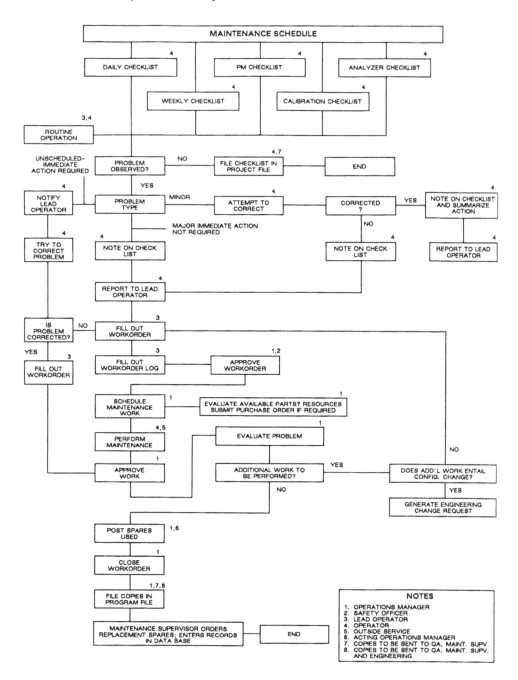

FIGURE 4.   Maintenance schedule.

forms the base of the pyramid and includes the job positions that would be included in the following organizational structure divisions:

- President's office
- Site director or manager's office
- Administrative division
- Operations division
- Maintenance division

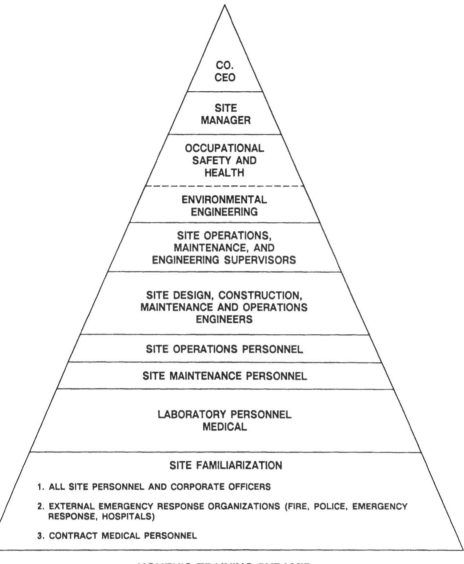

KOKEN'S TRAINING PYRAMID

FIGURE 5. Koken's (training) pyramid.

- Engineering division
- Medical division
- Laboratory division

The base and supporting level of the pyramid, as a minimum, consists of:

- General information and facility familiarization concerning safety, health, and other on-site hazards. Wastes handled at the facility having hazards such as carcinogenicity, toxicity, flammability, and reactivity should be discussed. Personal protective equipment for working within the plant should be discussed. Medical surveillance methods and typical symptoms of over exposure should be given. The emergency response plan should be explained and key features of the plan taught such as the main chain of command for emergency coordination (including the names and phone numbers of personnel and their alternates responsible for site safety and health) and their various responsibilities, evacuation procedures, and routes (why, when, how, where), and emergency communications and alarms.

- The general roles of each main division of labor within the facility with regards to normal operations and emergency response.
- Worker right-to-know laws, regulations, and information should be discussed in enough detail to fulfill regulatory requirements for this group.
- A general test should be given to confirm understanding and document training.
- This facility familiarization should be repeated annually.

Each increase in level of the pyramid above the base becomes increasingly involved in the direct operations and/or maintenance of the active portion of the facility. The decreasing width of the pyramid, in general, signifies the number of people involved. The height of the pyramid signifies the knowledge base required.

The second level of the pyramid consists of medical and laboratory personnel. A medical doctor with industrial medicine credentials with a preference for experience in industrial hygiene and hazardous materials/waste handling is desired to report directly to the plant safety official either on a retainer or as facility staff. Every employee should have a baseline physical at time of hiring. Every employee working with hazardous wastes should have annual comparison medical examinations. Administrative workers should have at least a comparison medical examination at the time of termination and employment. Someone on-site should have intensive first-aid training. This can be a nurse on staff or, in many cases, someone from the safety and health staff such as an industrial hygienist.

Laboratory facilities and personnel are essential. Personnel are required for sampling and analysis of waste samples and are sometimes used in interdisciplinary teams for facility tasks. They must have at least an overview knowledge of the facility and proper waste handling technique for sampling and analysis.

The third level from the bottom is the maintenance personnel directly involved with the active portion of the facility. This training should consist of several modules. The first module consists of normal maintenance skills and training. This module will be different for each maintenance skill (millwright, electrical, instrumentation, pipe fitter, welder, etc.). A training/review program for each module should be in place. The second module is specific to hazardous waste toipics. These include but are not limited to:

- Personal hygiene requirements.
- Personal protective equipment and dress out procedures.
- Hazardous waste areas and contamination zones.
- Decontamination of equipment needing maintenance.
- Personal decontamination and decontamination and storage of tools.
- Required number of workers and supervision for tasks. In general, the guidelines specify a buddy system for safety considerations (most efficient, anyway).
- Emergency rescue procedures.
- Area contamination level monitoring for risk. In potentially contaminated areas, and in general, it is easier to enforce, less confusing, and safer to specify and enforce maximum area personal protection equipment and procedures than to have changing rules depending on differing levels of contamination and risk in an area. Maximum exposure levels for an area should be based on the potential risk if the unexpected happens.

The third module should consist of detailed instruction about the chemicals in which the personnel will be working around. All required worker right-to-know information should be provided. As new chemicals are handled, classes should be scheduled before job assignment.

The fourth module should consist of spill prevention and control procedures and spill cleanup procedures. This training should reinforce the relevant topics above.

Maintenance activities/duties should have job training outlines utilized for classroom and on-the-job training.

The next level of the pyramid is hazardous waste operations personnel. Training should have the same elements as the maintenance level but include necessary process related information such as:

- Process controls and procedures and the reasoning behind such controls and procedures, data accumulation and record keeping, and emergency backup systems.
- Emergency situations and response training pertinent to this position.
- Operations activities/duties should have job training outlines and classroom and on-the-job training.

The next level consists of design and construction engineering and maintenance engineering. Additional training should consist of an overview of all safety and environmental regulations and guidances concerning the facility and in-depth training in the regulations and guidances that directly relate to each engineering or sampling/analysis job. Many engineers still do not realize the critical design, construction, and operation and maintenance constraints, directives, and guidances incorporated into the regulations and policy guidance manuals. It should not be left to an environmental engineering group (or engineer) to take sole responsibility for engineering activities being in compliance. Without a knowledgeable engineering staff concerning regulations, it will be impossible for one responsible individual or group to convince, impose, and implement all the regulatory requirements. A design, construction, maintenance, operations, safety/environmental, and laboratory review team should be implemented to review and accept/reject all major design and construction projects and major activities. This review team must include the environmental/safety engineer or group as a member. The emergency preparedness plan should be taught in depth as it may relate to this level. An in-depth review of the chemicals or type of chemicals used at the facility should be given and would include MSDSs. An in-depth test should be given to confirm understanding and document training.

The next level is the supervisory level. Personnel in this level must have gone through the previous four levels of training. This level must be extremely competent in safe occupational and environmental work practices specific to the wastes. A thorough knowledge of the emergency contingency plan and its implementation is required. Emergency response procedures should be taught in detail. Facility supervisors should be included in design and construction reviews and should be familiar with the design and construction rationale and operating procedures.

The sixth level is occupational safety and health and environmental engineering. Proper emphasis has been placed on environmental health and industrial hygiene in the regulations. Insufficient emphasis has been given to safety engineering and environmental engineering. The proper design, construction, and operation of facilities with regard to engineering controls of hazards (such as conducting hazards analysis) and chemical exposure need to be reevaluated above exposure monitoring, without diminishing the importance of industrial hygiene. Occupational safety and health and environmental engineering departments should be directed by a company engineer with a broad knowledge of safety and environmental engineering and industrial hygiene. This department will be interdisciplinary in nature. The safety and environmental divisions should be staffed with credentialed personnel thoroughly familiar with regulations in their specialty and having a thorough overview of regulations in related areas (safety vs. environmental and vice versa). These divisions should report directly to the plant manager. These divisions should be in charge of training. Any job training only indirectly related to hazardous waste and under other divisions should be monitored by the safety department for adherence to performance and schedule. The safety department, in conjunction with operations and maintenance, should be in charge of emergency response training that consists of employees from the safety and environmental divisions and the operations and maintenance divisions.

The seventh level is the facility manager. The facility manager should be familiar with all the training content in the previous five levels and be extremely knowledgeable concerning the emergency contingency plan and procedures. He/she should have at least an overview of the pertinent environmental/safety regulations so that recommendations from the environmental/safety/engineering/maintenance personnel are not continually in question and criticism. The support for training from this level is imperative. It is also imperative that this level issue an official memorandum expressing the support and policy of obeying all environmental and safety regulations and laws.

The final level of the pyramid is the company president. The company president should have an overview of the pertinent safety and environmental regulations and a concept of associated costs and manpower required to operate one or more facilities according to the regulations and accepted engineering codes and guidelines. It is important for this level to issue a memorandum expecting facility personnel to operate and maintain the facility in accordance with environmental and safety standards set by the regulations, the permit, and other engineering codes and standards.

The regulations require the following as a minimum: 40 C.F.R. 264.16/265.16 — facility personnel must successfully complete a program of classroom and/or on-the-job training that teaches them to perform their duties in a way that ensures the compliance of the facility with the requirements of this part. The applicant must provide a written description of the type and amount of both introductory and continuing training that will be given to each person filling a position having a job title related to hazardous waste management. Each job title must have a job description that must include the required skills for the job, education requirements, other qualifications, and a description of duties assigned to each position. Training records that document the completed training are required.

These records for current employees must be kept on file until facility closure. Former employee records must be kept for at least 3 years from the last date worked. At a minimum, the training program must be designed to ensure that facility personnel are able to respond effectively to emergencies by familiarizing personnel with emergency procedures, equipment, and systems including the following:

- Procedures for using, inspecting, repairing, and replacing facility emergency and monitoring equipment
- Key parameters for automatic waste feed cutoff systems
- Communications and alarm systems
- Response to fire and explosions
- Response to groundwater contamination incidents
- Shutdown of operations

Facility personnel must successfully complete the program required above within 6 months after the date of their employment or their assignment to a facility or to a change of position. Employees must not work in an unsupervised position until they have completed the training. Facility personnel must take part in an annual review of the initial training.

Specific requirements for training personnel handling hazardous wastes are also detailed by 29 CFR Ch. XVII 1910.120. The following information must be incorporated into a training program:

- Establish a safety and health program designed to identify, evaluate, and control safety and health hazards and provide for emergency response for hazardous waste operations.
- Perform a site characterization and analysis to identify site hazards and to determine the appropriate safety and health control procedures needed to protect employees from identified hazards.
- Site control. Procedures must be implemented to control employee exposures to hazardous substances.
- Training. Initial and refresher courses shall be provided to employees before they are permitted to engage in hazardous waste operations. The courses shall include:
  - Names of personnel and alternates responsible for site safety and health
  - Safety, health, and other hazards on site
  - Personal protective equipment use
  - Work practices by the employee to minimize risks from hazards
  - Safe use of engineering controls and equipment on site
  - Medical surveillance requirements and recognition of symptoms
  - Review of the site safety and health plan required by these regulations, and, specifically, review site control measures, decontamination procedures, the site standard operating procedures, the contingency plan, and confined space entry procedures

Hazardous waste workers shall at the time of job assignment receive a minimum of 4 h of *initial* instruction off the site, and a minimum of 3 d of field experience under the direct supervision of a trained, experienced supervisor. If the employer can demonstrate that an employee's previous work experience and/or training is equivalent to the required initial training, then the employee is judged as having satisfied the initial training requirements. However, this only applies to initial training and they are still subject to annual refresher and more specialized training when called for.

On-site management and supervisors directly responsible for or who supervise employees engaged in hazardous waste operations shall receive training of at least 8 additional hours of specialized training on managing such operations at the time of job assignment.

Trainers shall have received a level of training higher than and including the subject matter of the level of instruction that they are providing. Employees responsible for emergency response to hazardous emergency situations shall be trained in how to respond to expected emergencies. Employees and managers shall receive 8 h of refresher training annually.

- Medical surveillance shall be provided for employees exposed or potentially exposed to hazardous substances or health hazards, or who wear respirators.
- Engineering controls, work practices, personnel protective equipment shall be implemented to protect employees from exposure to hazardous substances and health hazards.
- Air monitoring to determine appropriate level of employee protection.
- Informational programs. This program includes the development and use of a site safety and health plan that as a minimum includes the following:
  - Names of key personnel and alternates responsible for site safety and health and appointment of a site safety and health officer
  - A safety and health risk analysis for each site task and operation
  - Employee training assignments

■ Personal protective equipment to be used by employees for each of the site tasks and operations being conducted
■ Medical surveillance requirements
■ Frequency and types of air monitoring, personnel monitoring, and environmental sampling techniques and instrumentation to be used
■ Site control measures
■ Decontamination procedures
■ Standard operating procedures
■ A contingency plan for the safe and effective response to emergencies
■ Confined space entry procedure

(g). Preparedness and Prevention

(264.30) — All TSDF facilities must have the following equipment:

● Internal communications or alarm system for immediate emergency instruction either by voice or signal
● Telephones or hand-held two way radios (pay attention to electrical hazard classification for radios and the hazardous waste being handled) capable of summoning emergency assistance from local police, fire, or emergency response teams
● Portable fire extinguishers, fire control equipment, spill control equipment, and decontamination equipment
● Water at adequate volume and pressure to supply fire control equipment

All facility equipment must be tested and maintained as necessary to ensure its proper operation in time of emergency. As a practical matter the permit application will have to have such a maintenance/testing frequency.

All personnel while working with hazardous waste must have access to emergency communications. The permit application will have to detail h ow this is going to be accomplished.

All hazardous waste areas must be designed, constructed, and operated to have enough access space for emergency response vehicles and activities. This can be shown in the application through a plot plan of the facility.

The applicant must detail the arrangements made with the police, fire department, and emergency response teams to familiarize themselves with the facility layout including entrances and exits, access isles and roads, working areas, and the properties of the hazardous wastes handled. Primary emergency authority agreements between overlapping jurisdictions of police, fire and emergency response teams should be detailed. Local hospitals are required to be informed and familiarized with the hazardous waste properties handled. Affidavits to attest to fulfilling these requirements are required.

(g). Contingency Plan and Emergency Procedures

(264.50) — The owner/operator must furnish a contingency plan as part of the application. The plan must present the measures and design that are or will be implemented to ensure the minimization of hazards (risk) to human health or the environment. Immediate implementation of the plan must be carried out on the occurrences of fire, explosion, or unplanned release of hazardous waste or its constituents. The plan must describe arrangements agreed to by local police and fire departments, hospitals, contractors, and state and local emergency response teams for emergency situation coordination. The plan must list the current names, addresses and phone numbers (office and home) of all persons qualified to act as emergency coordinator at the site. This list and plan must be kept up to date. New facilities applying of a permit submit this information at the time of permit approval rather than with the application. Any list submitted with the application would probably be out of date by the time of permit approval. The application for new facilities should show the anticipated emergency situation chain of command positions. A list of emergency equipment, their physical description, and their actual or proposed location must be included (include a site plot plan). Evacuation procedures and routes must be included in the plan (show primary and secondary evacuation routes on the plot plan). Emergency signal and alarm characteristics and equipment must be detailed and the meaning of each type of signal indicated (if more than one type).

Copies of the plan must be maintained on site and given to the local police, fire departments, hospitals, and state and local emergency response teams that may be called on to assist in an emergency. The plan must be reviewed and immediately amended under the following circumstances:

- The facility permit is revised.
- The plan fails during an emergency.
- The facility design, construction, operation, maintenance, or other circumstance changes in such a way that there is an increased potential for fires, explosions, or hazardous waste or its constituents to be released.
- The list of emergency coordinators changes.
- The list of emergency equipment changes.

Regulatory directives concerning the plan also include the following:

- There must be at lease one employee designated as the emergency coordinator either on the facility premises or on call and available at all times. If the on call option is exercised, this person must be able to respond and reach the facility within a short time. It is recommended that for any relatively large facility or one that handles hazardous waste of a type or quantity that can be extremely detrimental, that an on-site coordinator be present at all times. This coordinatior must be fully familiar with the plan and all operations and activities at the facility. This person must have the authority to commit the resources necessary to carry out the plan.
- Whenever there is an *imminent* or actual emergency the emergency coordinator must immediately do the following (in order of priority):
  - Activate internal facility alarms and communication systems to notify all facility personnel.
  - Identify the character, exact source, amount, and extent of contamination of any released material.
  - Assess possible hazards to human health or the environment of the emergency (e.g., effects of toxic, irritating, or asphyxiating gases or effects of hazardous surface runoff including water or chemical agents used in control of the emergency).
  - The effects of the emergency on human health or the environment outside the facility. If the assessment indicates that evacuation of local areas outside the plant *may* be or is advisable, the emergency coordinator must notify the appropriate local authorities (as listed in the most recent revision of the plan and as part of the application). Immediate notification of the government official designated as the on-scene coordinator for the geographic area (usually a regional EPA or, in some instances with regard to water pollution, a U.S. Coast Guard official, or the National Response Center using their 24-h toll-free number will log and forward the information. It is advised the permittee follow up such a call to the Response Center with one to the on-scene coordinator as the situation allows. Such notification should include all the information from the above assessment and as outlined in 264.56(d)(2)(i).

■ Take all necessary measures to ensure that fires, explosions, and releases do not occur, recur, or spread to other hazardous waste at the facility, including shutting down processes and operations, collecting and containing releases, and removing and isolating containers. Possible leaks, pressure buildup, gas generation, or ruptures in valves, pipes, or other equipment must be monitored for.

■ Provide for the proper (including compatibility) treating, storing, or disposal of waste, contaminated soil, or water arising from or the result of the emergency.

■ All emergency equipment is cleaned and fit for its intended use before processes or operations is resumed.

■ The Regional Administrator (usually a designated officer) and local authorities must be notified prior to the resumption of activities.

■ The details of the incident must be entered into the facility records (operating record), and a written report must be submitted to the Regional Administrator within 15 d. The details entered into the record and the report must include the list of items from 264.56(j)(1-7).

(i). Environmental/Safety Design/Construction/Operating Procedures

(270.14[b][8][i — v]) — The application must detail design, construction, and operating procedures that are or will be used to accomplish the following:

● Prevent hazards in unloading operations
● Prevent runoff and runon from and to the hazardous waste handling and processing areas
● Prevent contamination of water supplies (backflow preventers on all nonelectric utilities and impermeable containment structures correctly sized)
● Prevention of equipment failures or power outages from causing the release of hazardous waste or other unsafe condition
● Prevent exposure of personnel to hazardous waste (proper engineering controls, personal protective equipment, exclusion zones, proper material handling equipment design, and properly designed and supervised operating and emergency procedures are a few examples)

(j). Precautions for Ignitable, Reactive, or Incompatible Waste

(270.14[9]) — The applicant must describe the precautions in the design, construction, and operating procedures to prevent the ignition or reaction of ignitable, reactive, or incompatible waste. As a minimum compliance with 264.17 and 264.17(c) must be shown and documented.

● The waste must be separated and protected from sources of ignition or reaction. These precautions must include as a minimum and not be exclusive of other design, construction, or operating procedures that would be prudent and expected as good engineering practice as warranted by the wastes handled and various engineering guidelines (NFPA, ASME, etc.), separation from open flames, smoking, cutting and welding, hot surfaces, frictional heat, sparks (static, electrical, or mechanical), spontaneous ignition (heat-producing chemical reactions), and radiant heat. Conspicuous "NO SMOKING" signs must be posted wherever there is a hazard from ignitable or reactive waste.
● Whenever the permit application requests permission to treat, store, or dispose of ignitable, reactive, or the mixing of incompatible wastes, the applicant must describe the precautions that are used or will be implemented to prevent reactions that may result in the following:

- Generation of extreme heat or pressure, fire or explosion, or violent reactions
- Production of uncontrolled toxic mists, fumes, dusts, or gases
- Production of uncontrolled flammable fumes or gases
- Damage the structural integrity of the device or facility
- Threaten human health or the environment through the above or other similar means

The owner must document compliance with this section when it is applicable. Documentation must be included in the application and consists of the following:

- Waste analysis (summarize the waste characterization detailing the characteristics of the waste applicable to this section).
- Design, construction, and operating procedures pertinent to the safe handling of these wastes. A suggested such documentation would be a summary of the design rationale incorporated into the design P&IDs, a copy of the pertinent P&IDs, a summary of pertinent physical construction design and associated drawings, and operating procedures that are used or will be implemented specific to these concerns.
- References to published scientific or engineering literature in support of the above rationale.
- Data from trial tests which can include bench-scale and pilot-scale tests.
- The results from the treatment, storage, and/or disposal of the same or similar wastes under the same or similar design, construction methods, and operating procedures.

(k). Vehicular Traffic

The application must have a description of vehicular traffic patterns, estimated volumes, and type of vehicles and their control such as curves, loading/unloading areas, stop signs, speed zones, turning radii, etc. Load-bearing capacities should be indicated. This requirement can be satisfied by incorporating this information on a plot plan and pavement cross-section drawings and a short narrative giving the information. This may be included in the General Facility Description Section or its own short section.

(l). Facility Location Information

(270.14[11][iii]) — This section can be included in the general description of the facility (see item a of this section); however, this application requirement discusses the location with regard to known seismic activity as listed by location in Part 264 Appendix VI. If the requested information of location regarding state, county, township, or election district is not included in Appendix VI, then this information is enough to satisfy this regulation requirement (check state requirements). The entire states of California and Nevada, the Island of Hawaii, a large area of Alaska, Idaho, Montana, New Mexico, Utah, and Washington are included. Smaller areas of Arizona, Colorado, and Wyoming are included. If the area that the facility is or will be located in is included, the design and construction of the facility and its components must meet the applicable seismic standard and the application will have to document conformance with the standard. The documentation information will have to be complete enough to satisfy geologists reviewing the information against the standard. Information needed includes the following:

- No faults within 3000 ft that have had activity (displacement) and/or evidence of such displacement to indicate the possible existence of a fault within Holocene time. Documentation shall consist of:
  - Published geologic studies (to include excerpt and reference)
  - Aerial reconnaissance of the area within a 5-mi radius

■ Analysis of aerial photographs that include the area within 3000 ft of the facility
■ Walking reconnaissance if needed to clarify the above

If faults or possible faults having displacement in Holocene time are present within 3000 ft of the facility, a detailed geologic study must be completed. No faults can pass within 200 ft of a facility designed to seismic standards or not. If such a study is not conclusive as to the absence of faults within 200 ft, subsurface explorations must be made (see 270.14[11][B]).

● The applicant must provide location information concerning the location of the facility in relation to the 100-year floodplain. The source of data must be identified and include a copy of the Federal Insurance Administration (FIA) flood map that includes the site. The 100-year flood level should be indicated in relation to the site location or in relation to other known flood factors (wave action etc.). Design, construction, operating, or maintenance activities affected by such a location must be discussed in relation to the facility withstanding the effects of a 100-year flood. If located within the 100-year floodplain the following information is required:
  ■ Hydrodynamic and hydrostatic force engineering analysis to determine appropriate design and construction rationale and implementation.
  ■ Structural and other engineering studies as needed showing the design of operational units and flood protection devices at the facility and how these will prevent flood damage and protection of human health and the environment.
  ■ If applicable and in lieu of the two points above, the applicant must describe in detail the procedures to remove hazardous waste from the facility before the facility is flooded. This description must include the time needed for waste removal in relation to the warning needed of an impending flood and a description of the location the waste would be relocated to as to its eligibility to receive such waste according to the regulations. Additionally, the description must include the personnel and their availability status to effect such a relocation and a description of the potential for accidental releases or spills of the material.*

(m). Closure/Postclosure Plan
(270.14[13], 264.110 [Subpart G]) — The owner/operator must include as part of the application a written closure plan. The content of the plan must contain the following elements:

● A description of how each hazardous waste management unit at the facility and a description of how the entire facility will be closed, both in accordance with the closure performance standards of 264.111.
● An estimate of the maximum inventory of hazardous wastes ever on site over the active life of the facility and a detailed description of the methods to be used during partial and final facility closure. Identification of methods for removing, transporting, storing, treating, and disposal of the waste is required. Identification of any type(s) of off-site hazardous waste management units to be used is required.
● The methods of partial and/or final closure must be broken down into procedural steps, each of which is fully described. These steps shall include removal or decontamination of all hazardous waste residues and contaminated system components, equipment, structures, and soils. Included will be the procedures for cleaning equipment, sampling

---

*   Author's comment: it is not advisable to locate any hazardous waste TSDF on a fault or in a flood zone unless it is located there for the purpose of site remediation. State laws regulating siting are typically becoming more stringent regarding siting in these areas as well they should. Unfortunately, siting laws are unduly influenced by the not-in-my-backyard syndrome, also making siting extremely difficult in safer areas even if needed.

and testing surrounding soils, and criteria for determining the extent of decontamination required to satisfy the closure performance standards of 264.111.

- Detailed descriptions are required of groundwater monitoring, leachate collection, and runon and runoff control as necessary to meet the closure performance standards.

- The closure plan shall provide a schedule for the above closure activities and steps required for closure of each hazardous waste management unit and for final closure of the facility (each activity and step should have a time to complete associated with it and a total time for the closure that is the sum of the individual steps). An estimate of the expected year of closure should be given (the unit or facility design or economic life projection should be given if actual closure start date is unknown).

- Closure plan modifications, once the application has been reviewed and judged complete, are hard to accomplish. After the permit is issued changes become a major permit modification which requires agency review, public notice, a public comment period, and a public hearing. Although regulations that affect closure are continually changing (especially landfill regulations), it is prudent to plan out closure activities thoroughly using as much foresight as possible, minimizing the severity of future change requests. For a discussion concerning amendment of closure plans, see 264.112(c) and particularly 264.112(C)(4) concerning the procedures of Part 124 and 270 which include permit modification procedures.

- Partial and final closure notification requirements (264.112[D]) — notification of partial or final closure must be given at least 45 d and for some facilities (surface impoundment, waste pile, land treatment, or landfill) 60 d prior to the date the owner/operator expects to begin closure. The expected date to begin closure must be no later than 30 d after the date the unit receives the known final volume of hazardous waste. In some justified cases, an idle unit may be reasonably expected to receive additional hazardous waste. In such cases, closure must begin no later than 1 year after the date of the unit receiving the most recent volume of hazardous waste. In some cases, extensions may be granted based on probability of receiving hazardous waste to the unit in the future. If the permit is terminated or ordered to cease receiving hazardous waste under Section 3008 of RCRA, then closure must proceed as to 264.113 (discussed below).

- Time allowed for closure (264.113)
  - Within 90 d after receiving the final volume of waste, the owner or operator must treat, remove, or dispose of on site all hazardous waste according to the closure plan.
  - A modification to the permit may be requested to extend this time upon showing cause such as treatment, removal, or disposal time necessarily taking longer. The modification request will be reviewed for adherence to the conditions of 264.113(a)(1)(i—ii[C][2]). Such requests are subject to permit modification regulations (Parts 124 and 270). Rather than wait until closure, the closure schedule in the application should consider this eventuality and may be accepted as a permit condition.
  - The owner/operator must complete the partial or final closure within 180 d after receiving the final volume of hazardous waste at the subject unit or facility. Again, a permit modification may be requested extending this time period as in the above point but based on the considerations listed in 264.113(b)(1)(i)—264.113(c).

- Disposal or decontamination of equipment, structures, and soil (264.114) — all contaminated equipment, structures, and soils must be properly disposed of or decontaminated unless directed otherwise by the various Subparts of 264, specific for the type of hazardous waste unit(s) involved. These regulation references are listed in 264.114.

Waste material from or generated by closure is usually considered by the regulatory authorities as a hazardous waste unless shown otherwise through the delisting procedure. At present, it is usually more cost effective to handle and dispose of the waste as hazardous unless it is classified as a land ban waste. If the waste is handled and disposed of as hazardous waste it must be done according to the requirements of Part 262.

• Closure certification (264.115 264.116) — within 60 d of partial and/or final closure, the owner/operator must certify by registered mail that the management unit or facility has been closed according to the approved closure plan. The certification must be signed by the owner/operator and a registered professional engineer with registration documentation. Accompanying the certification should be a survey plat of the location of any present or past hazardous waste units, which has to be certified by a professional land surveyor. A copy of this plat must be sent to the local zoning or land use authority (264.116).

• Postclosure plan and care — in most instances, incinerator facilities will have closed or made plans to close hazardous waste units (impoundments, landfills) that might require postclosure care. Therefore, this chapter merely refers the reader to 264.117 through 264.120 on this topic. The author assumes the clean closure of an incinerator facility which will not require postclosure care.

(n). Financial Requirements and Closure Cost Estimate
(Subpart H, 264.140 and 264.142)

• Cost estimate for closure — the cost estimate must equal the cost of final closure at a time that would be the most expensive (e.g., closure begins with maximum inventory of waste on hand and would include cost of disposal of this waste, based on contracting work to a third party; it cannot incorporate any salvage value or other assets of the facility or include zero costs for wastes that might have an economic value).

The cost estimate must be updated annually to adjust for inflation by multiplying the most recent yearly figure by the inflation factor for the current year. This annual update must be performed within a set period depending on financial assurance options, but in no case is this period greater than 60 d from the anniversary date of the financial assurance instruments.

The closure plan cost estimate must be reviewed and modified as necessary within 30 d of regulatory approval of closure plan modifications and this revision then adjusted for inflation annually as above.

The closure plan must be kept at the facility and incorporate the latest cost estimate.

• Financial assurance for closure — the owner or operator must establish financial assurance that there will be a fund in accordance with the closure cost estimate that will be dedicated to facility closure. There are six financial instrument options.
  ▪ Closure trust fund (264.143[a])
  ▪ Surety bond guaranteeing payment into a closure trust fund
  ▪ Surety bond guaranteeing performance of closure — see 264.143(c)
  ▪ Closure letter of credit — see 264.143(d)
  ▪ Closure insurance — see 264.143(e)
  ▪ Financial test and corporate guarantee for closure
  ▪ Use of multiple financial mechanisms
  ▪ Use of financial mechanism for more than one facility — see 264.143(h)

b. Subpart O Technical Requirements

i. Permit Conditions

Prior to issuing a permit for incineration and usually just prior to the public comment period, the regulating agency will issue a draft permit with permit conditions that are designed to guarantee compliance with the performance standards of 264.343 and operation within design limitations. It is required to supply the operating and design parameters and limits in the Part B application so they can be reviewed and used to establish the permit conditions. Many applicants will tend to skimp on the information required so that competitors do not obtain proprietary information and in an effort to keep the regulators from being too restrictive. What often happens is that the regulators thus put overly restrictive permit conditions in to compensate, and the conditions sometimes are severe enough to make a project economically borderline (such restrictions might be on feed rate, gas flow rates, etc. that severely restrict operation). It is advisable that (1) enough and proper information is given in the Part B application that allows the regulators to make intelligent decisions regarding permit conditions and (2) the permit conditions in the draft permit be reviewed carefully prior to the comment period.

(a). Trial Burn Plan

Applicants must submit with the Part B application a proposed trial burn plan. The trial burn plan must contain the following:

1. An analysis of each waste or mixture of wastes to be burned. The analysis shall include as a minimum:
   a. Heat value of the waste in the form and composition in which it will be burned
   b. Viscosity (absolutely necessary for liquids and semi-liquids)
   c. An identification of any hazardous organic constituents listed in Part 261 Appendix VIII that would be reasonably expected to be present. Reasons for not including specific chemicals from this list must be adequately explained. The waste analysis must be in accordance with SW-846 methods. It is important that communication with the laboratory is sufficient to guarantee that quantification of the compounds is achieved as well as identification even though it is implied by the method. Detection limits should be specified even though the method will have method detection limits. More than one laboratory has given incomplete information on the basis of just specifying the methods of SW-846.
   d. Detailed engineering description of the incinerator, including:

   - Manufacturer's name and model number
   - Type of incinerator
   - Incinerator dimensional data
   - Auxiliary fuel system description
   - Prime mover capacity
   - Automatic waste feed cutoff system description
   - Stack gas monitoring and pollution control equipment
   - Nozzle and burner design
   - Construction materials
   - Location and description of temperature, pressure, and flow indicating and control devices
   - A detailed description of sampling and monitoring procedures, including locations, equipment used, frequency, and the planned analysis methods

- A detailed test schedule for each waste for which the trial burn is planned including dates, duration, and quantity of waste to be burned
- A detailed test protocol including the ranges of temperature, waste feed rate, combustion gas velocity, auxiliary fuel usage, and any other relevant parameters
- A description of, and planned operating conditions for, any emission control equipment that will be used
- Procedures to shut down the incinerator rapidly, stopping the waste feed, and controlling emissions in the case of an equipment malfunction

The regulating agency will review the trial burn plan for sufficiency and will most likely request additional information.

(b). POHCs and PICS

Principal organic hazardous constituents (POHCs) are the hazardous compounds in, or added to characteristic waste, during a trial burn. They are the wastes to be incinerated that are of the most concern from a health and environmental standpoint. They may be easy or difficult to incinerate. Products of incomplete combustion (PICs) are chemical compounds that are formed in the incineration process and detected in the stack gas after incineration. These compounds were formed in the incineration process in either of two ways:

1. Incomplete combustion of the original compound failed to break it completely down into CO_2 and water, leaving parts of the original compound behind.
2. The broken-down compounds reform to make other compounds.

Additionally, PICs arising from sampling contamination, sample recovery contamination, and laboratory contamination may show up in analysis results. Phthalates from rubber gloves are common. It is important that the possible sources are delineated in test reports and these compounds are not counted as PICs. PICs may be less hazardous or more hazardous than the feed POHCs but are usually in smaller concentrations. PICs have been shown to form from merely the combustion of natural gas. In theory, the perfect incinerator would not form PICs due to the 100% breakdown of compounds into CO_2 and water and would prevent reformation into other compounds. In practice, this is not the case and the number and kinds of PICs formed depend on the degree an incinerator approaches the ideal.

The applicant must designate one or more POHCs in the trial burn plan that will be tracked during the trial burn to determine the destruction efficiency of the incinerator. Picking a POHC is not trivial for several reasons.

1. Certain POHCs may generate certain PICs in incomplete combustion.
2. Feed material may contain compounds that are the same as the above PICs. Incomplete combustion, however large or small it might be, will add these compounds to the amount of PICs detected and give a false larger number to the amount of PICs generated. Regulatory agencies and the public are becoming increasingly sensitive to the issue of the possibility of incinerators generating other chemical species that sometimes may be more harmful than the feed POHC(s). It is presently common for the regulatory agencies to request analysis not only for feed and stack sampling for POHC(s) and destruction efficiency, but for a complete SC/MS scan for volatile and semivolatile organics in the feed and stack gases (see volatile and semivolatile organics below).

3. The most important consideration in relation to the success of demonstrating destruction efficiency is that the feed material may contain compounds that as incinerated form PICs that are the same as the chosen POHC(s). This would cause the destruction efficiency for the feed POHC to be artificially low. Unfortunately, at this time, the technical data on what POHCs form what PICs (and vice versa) is incomplete. An incinerator operating near the ideal would minimize these concerns.

4. In addition to these considerations, the regulating agency will want a POHC(s) chosen that is hard to incinerate. The POHC chosen does not have to be present in the feed. The EPA has established an incinerability index based on the heat of combustion. This index is shown with the constituents in alphabetical order as Table 7 and in the order of heat of combustion values in Table 8. Studies have been done since publishing this index that indicate this index may not rank the compounds correctly. Other suggested criteria include heats of formation and chemical kinetics. Table 9 compares incinerability rankings of selected compound based on heats of formation to that based on heats of combustion. Table 10 compares incinerability rankings of selected compounds based on chemical kinetics to that of heat of combustion. To overcome this problem, the EPA has chosen some favorite compounds as a POHC. One of these is carbon tetrachloride. It is one of the harder compounds to incinerate according to several of the incinerability indices and is easy to detect by analytical methods. The disadvantage is that it is highly volatile and toxic. Care must be taken in adding the material to the feed so that it is not lost and in use of the proper personnel protective equipment in handling.

ii. Performance Standards

(a). **Destruction and Removal Efficiency**

An incinerator must demonstrate a destruction and removal efficiency (DRE) of at least 99.99% for each POHC designated by the agency. DRE is designated by the following formula:

$$DRE = [(W_{in} - W_{out})/W_{in}] \times 100\%$$

Where W_{in} = mass feed rate of one POHC in the waste stream feed and W_{out} = mass emission rate of the same POHC present in the stack emissions prior to release to the atmosphere.

For incinerators burning hazardous waste(s) designated as an F020 through F023, F026 and F027, a DRE of 99.9999% must be achieved. This DRE must also be achieved with a POHC(s) more difficult to incinerate than tetra- through hexachlorodibenzo-*p*-dioxins and dibenzofurans. The EPA must be explicitly informed in the application and to the trial burn plan of the owner/operator's intention of burning the above chemicals. Also, for incinerators burning PCBs under TSCA, a 99.9999% DRE is required.

(b). **Volatile and Semivolatile Organics**

There are no current performance standards regarding volatile and semivolatile organics in RCRA. However, due to the public's concern about PICs you will be asked to scan for and quantify these emissions during the trial burn. Proposed regulations may limit these stack emissions on the basis of a risk analysis.

(c). **Hydrochloric Acid**

The current performance criteria for HCl emissions from an incinerator is that either <4 lb/h of HCl is emitted from the incinerator before scrubbing equipment or there is a minimum removal efficiency of HCl entering the scrubber of 99%.

TABLE 7
Heat of Combustion of Organic Hazardous Constituents from Appendix VIII, Part 261

Hazardous constituent	H_c/HW (kcal/g)
Acetonitrile	7.37
Acetophenone	8.26
3-(α-Acetonylbenzyl)-4-hydroxycoumarin and salts (warfarin)	7.00[a]
2-Acetylaminofluorene	7.92[a]
Acetyl chloride	2.77[a]
1-Acetyl-2-thiourea	4.55[a]
Acrolein	6.95
Acrylamide	5.75[a]
Acrylonitrile	7.93
Aflatoxins	5.73[a]
Aldrin	3.75[a]
Allyl alcohol	7.75
4-Aminobiphenyl	9.00
6-Amino-1,1a,2,8,8a,8b-hexahydro-8-(hydroxymethyl)-8a-methoxy-5-methylcarbamate azirino (2′,3′:3,4) pyrrolo(1,2-a)indole-4,7-dione,(ester) (Mitomycin C)	5.41[a]
5-(Aminomethyl)-3-isoxazolol	4.78
4-Aminopyridine	7.37[a]
Amitrole	4.01[a]
Aniline	8.73
Auramine	7.69[a]
Azaserine	3.21[a]
Benz(c)acridine	8.92[a]
Benz(a)anthracene	9.39
Benzene	10.03
Benzenearsonic acid	3.40[a]
Benzenthiol (Thiophenol)	8.43
Benzidine	9.18
Benzo(b)fluoranthene	9.25
Benzo(j)fluoranthene	9.25
Benzo(a)pyrene	9.25
Benzoquinone	6.07
Benzotrichloride	3.90[a]
Benzyl chloride	6.18
Bis(2-chloroethoxy)methane	4.60[a]
Bis(2-chloroethyl)ether	3.38[a]
N,N-Bis(2-chloroethyl)-2-naphthylamine (Chlornaphazin)	6.64[a]
Bis(2-chloroisopropyl)ether	4.93[a]
Bis(chloromethyl)ether	1.97[a]
Bis(2-ethylhexyl)phthalate	8.42[a]
Bromoacetone	2.66[a]
Bromomethane	1.70[a]
4-Bromophenyl phenyl ether	5.84[a]
Brucine	7.42[a]
2-Butanone peroxide	6.96[a]
Butyl benzyl phthalate	8.29[a]
2-sec-Butyl-4.6-dinitrophenol (DNBP)	5.46[a]
Chloral(trichloroacetaldehyde)	0.80[a]
Chlorambucil	5.93[a]
Chlordane	2.71[a]
Chlorinated benzenes, N.O.S.	N/A
Chlorinated ethane, N.O.S.	N/A
Chlorinated fluorocarbons	N/A
Chlorinated naphthalene, N.O.S.	N/A
Chlorinated phenol, N.O.S.	N/A

TABLE 7 (continued)
Heat of Combustion of Organic Hazardous Constituents from Appendix VIII, Part
261

Hazardous constituent	H$_c$/HW (kcal/g)
Chloroacetaldehyde	2.92[a]
Chloroalkyl ethers	N/A
p-Chloroaniline	6.14[a]
Chlorobenzene	6.60
Chlorobenzilate	5.50[a]
p-Chloro-m-cresol	5.08[a]
1-Chloro-2,3-epoxybutane	5.19[a]
2-Chloroethyl vinyl ether	5.19[a]
Chloroform	0.75
Chloromethane	3.25
Chloromethyl methyl ether	3.48[a]
2-Chloronaphthalene	7.37
2-Chlorophenol	6.89
1-(o-Chlorophenyl)thiourea	5.30[a]
3-Chloropropionitrile	4.50[a]
Chrysene	9.37
Citrus red no. 2	—
Coal tars	N/A
Creosote	N/A
Cresol	8.18
Cresylic acid	8.09
Crotonaldehyde	7.73
Cyanogen	6.79
Cyanogen bromide	0.81[a]
Cyanogen chloride	1.29[a]
Cycasin	3.92[a]
2-Cyclohexyl-4,6-dinitrophenol	5.74[a]
Cyclorophosphamide	3.97[a]
Daunomycin	5.70[a]
DDD	5.14[a]
DDE	5.05[a]
DDT	4.51[a]
Diallate	5.62[a]
2,4-D	3.62[a]
Dibenz(a,h)acridine	9.53[a]
Dibenz(a,j)acridine	9.53[a]
Dibenz(a,h)anthracene[dibenzo(a,h)anthracene]	9.40[a]
711-Dibenzo(c,g)carbazaole	8.90[a]
Dibenzo(a,e)pyrene	9.33[a]
Dibenzo(a,h)pyrene	9.33[a]
Dibenzo(a,l)pyrene	9.33[a]
1,2-Dibromo-3-chloropropane	1.48[a]
1,2-Dibromoethane	1.43[a]
Dibromomethane	0.50[a]
Di-n-butyl phthalate	7.34[a]
Dichlorobenzene, N.O.S.	4.57
3,3'-Dichlorobenzidine	5.72[a]
1,4-Dichloro-2-butene	4.27[a]
Dichlorodifluoroethane	0.22[a]
1,1-Dichloroethane	3.00
1,2-Dichloroethane	3.00
trans-1,2-Dichloroethene	3.00
Dichloroethylene, N.O.S.	2.70
1,1-Dichloroethylene	2.70

TABLE 7 (continued)
Heat of Combustion of Organic Hazardous Constituents from Appendix VIII, Part 261

Hazardous constituent	H_c/HW (kcal/g)
Dichloromethane	1.70
Dichloromethylbenzene	5.09[a]
2,4-Dichlorophenol	3.81[a]
2,6-Dichlorophenol	3.81[a]
Dichloropropane	3.99
Dichlorophenylarsine	2.31[a]
1,2-Dichloropropane	3.99
Dichloropropanol, N.O.S.	2.84
Dichloropropene, N.O.S.	3.44[a]
1,3-Dichloropropene	3.44[a]
Dieldrin	5.56[a]
Diepoxybutane	5.74
Diethylarsine	5.25[a]
1,2-Diethylhydrazine	8.68[a]
Diethyl phthalate	6.39
Dihydrosafrole	7.66[a]
3,4-Dihydroxy-α-(methylamino)-methyl benzyl alcohol	6.05[a]
Dimethoate	4.02
3,3'-Dimethoxybenzidine	7.36[a]
p-Dimethylaminoazobenzene	6.97[a]
7,12-Dimethylbenz(a)anthracene	9.61
3,3'-Dimethylbenzidine	8.81[a]
Dimethylcarbamoyl chloride	5.08[a]
1,1-Dimethylhydrazine	7.87
1,2-Dimethylhydrazine	7.87
3,3-Dimethyl-1-(methylthio)-2-butanone-O-(methylamino)carbonyl oxime	5.82[a]
Dimethylnitrosoamine	5.14[a]
α,α-Dimethylphenethylamine	9.54[a]
2,4-Dimethylphenol	8.51
Dimethyl phthalate	5.74
Dimethyl sulfate	2.86
Dinitrobenzene, N.O.S.	4.15
4,6-Dinitro-o-cresol and salts	4.06[a]
2,4-Dinitrophenol	3.52
2,4-Dinitrotoluene	4.68
2,6-Dinitrotoluene di-n-octyl phthalate	6.67[a]
1,4-Dioxane	6.41
Diphenylamine	9.09
1,2-Diphenylhydrazine	8.73[a]
bis-n-Propylnitrosamine	7.83[a]
Disulfoton	5.73[a]
Dithiobiuret	2.12[a]
Endosulfan	2.33[a]
Endrin	3.46[a]
Ethyl carbamate (Urethane)	4.73[a]
Ethylenebisdithiocarbamate	5.70[a]
Ethyl cyanide	4.57
Ethylene imine	7.86[a]
Ethylene oxide	6.86
Ethylenethiourea	5.98[a]
Ethyl methacrylate	7.27[a]
Fluoranthene	9.35
2-Fluoroacetamide	3.24
Formaldehyde	4.47

TABLE 7 (continued)
Heat of Combustion of Organic Hazardous Constituents from Appendix VIII, Part 261

Hazardous constituent	H_c/HW (kcal/g)
Formic acid	1.32
Glycidylaldehyde	5.74
Halomethane, N.O.S.	N/A
Heptachlor	2.96[a]
Heptachlor epoxide	2.71[a]
Hexachlorobenzene	1.79
Hexachlorobutadiene	2.12[a]
Hexachlorocyclohexane (all isomers)	1.12[a]
Hexachlorocyclopentadiene	2.10[a]
Hexachloroethane	0.46
1,2,3,4,10,10-Hexachloro-1,4,4a,5,8,8a-hexahydro-1,4:5,8-endo, endo-dimethanonaphthalene	3.38[a]
Hexachlorophene	3.82[a]
Hexachloropropene	0.70[a]
Hydrazine	4.44[a]
Indeno(1,2,3-c,d)pyrene	8.52[a]
Iodomethane	1.34
Isocyanic acid, methyl ester	4.69[a]
Isobutyl alcohol	8.62
Isosafrola	7.62
Kepone	2.15[a]
Lasiocarpins	—
Maleic anhydride	3.40
Maleic hydrazide	4.10[a]
Malononitrile	5.98
Melphalan	5.21[a]
Methacrylonitrile	8.55[a]
Methanethiol	5.91[a]
Methapyrilene	7.93[a]
Methomyl	5.20[a]
Methoxychlor	5.59[a]
2-Methylaziridine	9.09[a]
3-Methylcholanthrene	9.57[a]
4,4'-Methylene-bis-(2-chloroaniline)	4.84[a]
Methyl ethyl ketone (MEK)	8.07
Methyl hydrazine	6.78[a]
2-Methyl lactonitrile	6.48
Methyl methacrylate	6.52[a]
Methyl methanesulfonate	3.74
2-Methyl-2-(methylthio)propionaldeyde-o-(methylcarbonyl) oxime	5.34[a]
N-Methyl-N'-nitro-N-nitrosoguanidine	4.06[a]
Methylaparathion	4.00[a]
Methylthiouracil	4.79[a]
Mustard gas	4.06[a]
Naphthalene	9.62[a]
1,4-Naphthoquinone	6.97
1-Naphthylamine	8.54
2-Naphthylamine	8.54
1-Naphthyl-2-thiourea	7.50[a]
Nicotine and salts	8.92[a]
p-Nitroaniline	5.50
Nitrobenzene	5.50
Nitrogen mustard and hydrochloride salt	4.28[a]
Nitrogen mustard N-oxide and hydrochloride salt	3.56
Nitroglycerine	3.79

TABLE 7 (continued)
Heat of Combustion of Organic Hazardous Constituents from Appendix VIII, Part 261

Hazardous constituent	H_c/HW (kcal/g)
4-Nitrophenol	4.95
4-Nitroquinoline-1-oxide	5.59
5-Nitro-*o*-toluidine	5.98
Nitrosoamine, N.O.S.	N/A
N-Nitrosodi-*N*-butylamine	8.46[a]
N-Nitrosodiethanolamine	7.02[a]
N-Nitrosodiethylamine	6.86[a]
N-Nitrosodimethylamine	5.14[a]
N-Nitroso-*N*-ethylurea	3.92[a]
N-Nitrosomethylethylamine	6.13[a]
N-Nitroso-*N*-methylurea	2.89[a]
N-Nitroso-*N*-methylurethane	4.18[a]
N-Nitrosomethylvinylamine	7.91[a]
N-Nitrosomorpholine	5.22[a]
N-Nitrosonornicotine	7.07[a]
N-Nitrosopiperidine	7.04[a]
N-Nitrosopyrrolidine	6.43[a]
N-Nitrososarcosine	3.19[a]
7-Oxabicycle(2.2.1)heptane-2,3-dicarboxylic acid	4.70[a]
Paraldehyde	6.30[a]
Parathion	3.61[a]
Pentachlorobenzene	2.05[a]
Pentachloroethane	0.53[a]
Pentachloronitrobenzene (PCNB)	1.62[a]
Pentachlorophenol	2.09
Phenacetin	7.17
Phenol	7.78
Phenylendiamine	7.81
Phenyl dichloroarsine	3.12[a]
Phenylmercury acetate	2.71[a]
N-Phenylthiourea	6.93[a]
Phthalic acid esters, N.O.S.	N/A
Phthalic anhydride	5.29
2-Picoline	8.72
Polychlorinated biphenyl isomers	
Monochloro	7.75[a]
Dichloro	6.36[a]
Trichloro	5.10[a]
Tetrachloro	4.29[a]
Pentachloro	3.66[a]
Hexachloro	3.28[a]
Heptachloro	2.98[a]
Octachloro	2.72[a]
Nonachloro	2.50[a]
Decachloro	2.31[a]
Pronamide	5.72[a]
1,3-Propane sultone	3.67[a]
n-Propylamine	9.58
Propylthiouracil	6.28[a]
2-Propyn-1-ol (Propargyl alcohol)	7.43[a]
Pyridine	7.83
Reserpine	6.70[a]
Resorcinol	6.19
Saccharin	4.49[a]

TABLE 7 (continued)
Heat of Combustion of Organic Hazardous Constituents from Appendix VIII, Part 261

Hazardous constituent	H$_c$/HW (kcal/g)
Safrole	7.68
Strychnine and salts	8.03
2,4,5-TP	5.58[a]
2,4,5-T	2.87[a]
1,2,4,5-Tetrachlorobenzene	2.61[a]
TCDD	3.43[a]
Tetrachloroethane, N.O.S.	1.39
1,1,1,2-Tetrachloroethane	1.39
1,1,2,2,-Tetrachloroethane	1.39
Tetrachloroethene (tetrachloroethylene)	1.19
Tetrachloromethane (carbon tetrachloride)	0.24
2,3,4,6-Tetrachlorophenol	2.23[a]
Tetraethyl lead	4.04[a]
Tetranitromethane	0.41[a]
Thioacetamide	5.95[a]
Thiosemicarbazide	4.55
Thiourea	4.55
Thiram	5.85[a]
Toluene	10.14
Toluene diamine	8.24[a]
o-Toluidine hydrochloride	6.63[a]
Toluene diisocyanate	5.92[a]
Toxaphene	2.50[a]
Tribromomethane	0.13
1,2,4-Trichlorobenzene	3.40[a]
1,1,1-Trichloroethane	1.99
1,1,2-Trichloroethane	1.99
Trichloroethene (trichloroethylene)	1.74
Trichloromethanethiol	0.84[a]
Tricholoromonofluoromethane	0.11[a]
2,4,5-Trichlorophenol	2.88[a]
2,4,6-Trichlorophenol	2.88[a]
Trichloropropane, N.O.S.	2.81
1,2,3-Trichloropropane	2.81
Trypan blue	3.84[a]
Uracil mustard	4.00[a]
Vinyl chloride	4.45[a]

Note: N/A = not applicable; see individual constituents.

[a] Computed by method of Handirck, *Ind. Eng. Chem.*, 48, 1366, 1956.

Sources: Lange's *Handbook of Chemistry*, 11th ed. McGraw-Hill, 1973. Cox and Pilcher, *Thermochemistry of Organic and Organo-metallic Compounds*, Academic Press, London, 1970.

(d). Particulates
At present, particulate emissions are limited to particulate emissions of a maximum of 0.08 gr/dscf when corrected to 7% O$_2$ or as specified in the draft permit. Some state regulations have reduced this allowable amount to 0.05 gr/dscf or less.

iii. Permit and Trial Burn
Only wastes specified in the permit and under the specified operating conditions can be incinerated except in the circumstances of approved trial burns or those listed under the

TABLE 8
Ranking of Incinerability of Organic Hazardous Constituents from Appendix VIII, Part 261 on the Basis of Heat of Combustion

Hazardous constituent	Heat of combustion (kcal/g)
Trichloromonofluoroethane	0.11
Tribromomonoethane	0.13
Dichlorodifluoromethane	0.22
Tetrachloromethane (carbon tetrachloride)	0.24
Tetranitromethane	0.41
Hexachloroethane	0.46
Dibromomonomethane	0.50
Pentachloroethane	0.53
Hexachloropropane	0.70
Chloroform	0.75
Chloral (trichloroacetaldehyde)	0.80
Cyanogen bromide	0.81
Trichloromethenethiol	0.84
Hexachlorocyclohexane	1.12
Tetrachloroethene (tetrachloroethylene)	1.19
Cyanogen chloride	1.29
Formic acid	1.32
Iodomethane	1.34
Tetrachloroethane, N.O.S.	1.39
1,1,1,2-Tetrachloroethane	1.39
1,1,2,2-Tetrachloroethane	1.39
1,2-Dibromomonoethane	1.43
1,2-Dibromo-3-chloropropane	1.48
Pentachloronitrobenzene	1.62
Bromomethane (Methyl bromide)	1.70
Dichloromethane	1.70
Trichloroethene (trichloroethylene)	1.74
Hexachlorobenzene	1.79
Bis(chloromethyl) ether	1.97
1,1,1-Trichloroethane	1.99
1,1,2-Trichloroethane	1.99
Pentachlorobenzene	2.05
Pentachlorophenol	2.09
Hexachlorocyclopentadiene	2.10
Hexachlorobutadiene	2.12
Kepone	2.15
2,3,4,6-Tetrachlorophenol	2.23
Dichlorophenylarsine	2.31
Decachlorobiphenyl	2.31
Endosulfan	2.33
Toxaphene	2.50
1,2,4,5-Tetrachlorobenzene	2.61
Bromoacetone	2.66
Dichloroethylene, N.O.S.	2.70
1,1-Dichloromethylene	2.70
Chlordane	2.71
Heptachlor epoxide	2.71
Phenylmercury acetate	2.71
Octachlorobiphenyl	2.72
Acetyl chloride	2.77
Trichloropropane, N.O.S.	2.81
1,2,3-Trichoropropane	2.81
Dichloropropanol, N.O.S.	2.84
Dimethyl sulfate	2.86

TABLE 8 (continued)
Ranking of Incinerability of Organic Hazardous Constituents from Appendix VIII, Part 261 on the Basis of Heat of Combustion

Hazardous constituent	Heat of combustion (kcal/g)
2,4,5-T	2.87
2,4,5-Trichlorophenol	2.88
2,4,6-Trichlorophenol	2.88
N-Nitroso-N-methylurea	2.89
Heptachlorobiphenyl	2.98
1,1-Dichloroethane	3.00
1,2-Dichloroethane	3.00
trans-1,2-Dichloroethane	3.00
Phenyl dichloroarsine	3.12
N-Nitrososarcosine	3.19
Azaserine	3.21
2-Fluoroacetamide	3.24
Chloromethane	3.25
Hexachlorobiphenyl	3.28
Bis(2-chloroethyl) ether	3.38
1,2,3,4,10,10-Hexachloro-1,4,4,4a,5,8,8a-hexahydro-1,4:5,8-endo, enco-dimethano-naphthalene	3.38
Benzenearsonic acid	3.40
Maleic anhydride	3.40
1,2,4-Trichlorobenzene	3.40
TCDD	3.43
Dichloropropene, N.O.S.	3.44
1,3-Dichloropropene	3.44
Endrin	3.46
Chloromethyl methyl ether	3.48
2,4-Dinitrophenol	3.52
Nitrogen mustard N-oxide and hydrochloride salt	3.56
Parathion	3.61
2,4-D	3.62
Pentachlorobiphenyl	3.66
1,3-Propane sultone	3.67
Methyl methanesulfonate	3.74
Aldrin	3.75
Nitroglycerine	3.79
2,4-Dichlorophenol	3.81
2,6-Dichlorophenol	3.81
Hexachlorophene	3.82
Trypan blue	3.84
Benzotrichloride	3.90
Cycasin	3.92
N-Nitroso-N-ethylurea	3.92
Cyclophosphamide	3.97
Dichloropropane, N.O.S.	3.99
1,2-Dichloropropane	3.99
Methylparathion	4.00
Uracil mustard	4.00
Amitrole	4.01
Dimethoate	4.02
Tetraethyl lead	4.04
4,6-Dinitro-o-cresol and salts	4.06
N-Methyl-N-nitro-N-nitrosoguanidine	4.06
Mustard gas	4.06
Maleic hydrazide	4.10
Dinitrobenzene, N.O.S.	4.15

TABLE 8 (continued)
Ranking of Incinerability of Organic Hazardous Constituents from Appendix VIII,
Part 261 on the Basis of Heat of Combustion

Hazardous constituent	Heat of combustion (kcal/g)
N-Nitroso-*N*-methylurethane	4.18
1,4-Dichloro-2-butene	4.27
Nitrogen mustard and hydrochloride salt	4.28
Tetrachlorobiphenyl	4.29
Hydrazine	4.44
Vinyl chloride	4.45
Formaldehyde	4.47
Saccharin	4.49
3-Chloropropionitrile	4.50
DDT	4.51
Thiourea	4.51
1-Acetyl-2-thiourea	4.55
Thiosemicarbazide	4.55
Dichlorobenzene, N.O.S.	4.57
Ethyl cyanide	4.57
Bis(2-chloroethoxy)methane	4.60
2,4-Dinitrotoluene	4.68
Isocyanic acid, methyl ester	4.69
7-Oxabicyclo(2.2.1)heptane-2,3-dicarboxylic acid (Endothal)	4.70
Ethyl carbamate	4.73
5-(Aminomethyl)-3-isoxazolol	4.78
Methylthiouracil	4.79
4,4′-Methylene-bis-(2-chloroaniline)	4.84
Bis(2-chloroisopropyl) ether	4.93
4-Nitrophenol	4.95
DDE	5.05
Dimethylcarbamoyl chloride	5.08
p-Chloro-*m*-cresol	5.08
Dichloromethylbenzene	5.09
Trichlorobiphenyl	5.10
DDD	5.14
Dimethylnitrosoamine	5.14
N-Nitrosodimethylamine	5.14
Diethylarsine	5.25
Phthalic anhydride	5.29
1-(*o*-Chorophenyl)thiourea	5.30
2-Methyl-2-(methylthio)propionaldehyde-*o*-(methylcarbonyl) oxime	5.34
2-*sec*-Butyl-4,6-dinitrophenol (DNBP)	5.46
p-Nitroaniline	5.50
Chlorobenzylate	5.50
Dieldrin	5.56
2,4,5-TP	5.58
Methoxychlor	5.59
4-Nitroquinoline-1-oxide	5.59
Diallate	5.62
Daunomycin	5.70
Ethylenebisdithiocarbamic acid	5.70
3,3′-Dichlorobenzidine	5.72
Pronamide	5.72
Aflatoxins	5.73
Disulfoton	5.73
4,6-Dinitrophenol	5.74
Diepoxybutane	5.74
Dimethyl phthalate	5.74

TABLE 8 (continued)
Ranking of Incinerability of Organic Hazardous Constituents from Appendix VIII, Part 261 on the Basis of Heat of Combustion

Hazardous constituent	Heat of combustion (kcal/g)
Glycidylaldehyde	5.74
Acrylamide	5.75
3,3-Dimethyl-1-(methylthio)-2-butanone-O(methylamino)carbonyl oxime	5.82
4-Bromophenyl phenyl ether	5.84
Thiram	5.85
Methanethiol	5.91
Toluene diisocyanate	5.92
Chlorambucil	5.93
Thioacetamide	5.95
Ethylenethiourea	5.98
Malononitrile	5.98
5-Nitro-o-toluidine	5.98
Nitrobenzene	6.01
3,4-Dihydroxy-α-(methylamino)methyl benzyl alcohol	6.05
Benzoquinone	6.07
N-Nitrosomethylethylamine	6.13
p-Chloroaniline	6.14
Benzyl chloride	6.18
Resorcinol	6.19
Propylthiouracil	6.28
Paraldehyde	6.30
Dichlorobiphenyl	6.36
Diethyl phthalate	6.39
Dioxane	6.41
2-Methyllactonitrile	6.43
N-Nitrosopyrrolidone	6.43
Methyl methacrylate	6.52
Chlorobenzene	6.60
o-Toluidine hydrochloride	6.63
$N,N,$-Bis (2-chloroethyl)-2-naphthylamine (Chlornaphazine)	6.64
2,6-Dinitrotoluene di-n-octyl phthalate	6.67
Reserpine	6.70
Methyl hydrazine	6.78
Cyanogen	6.79
Ethylene oxide	6.86
N-Nitrosodiethylamine	6.86
2-Chlorophenol	6.89
N-Phenylthiourea	6.93
Acrolein	6.96
2-Butanone peroxide	6.96
p-Dimethylaminoazobenzene	6.97
1,4-Naphthoquinone	6.97
3-(α-Acetonylbenzyl)-4-hydroxycoumarin and salts (warfarin)	7.00
N-Nitrosodiethanolamine	7.02
N-Nitrosopiperidine	7.04
N-Nitrosonornicotine	7.07
Phenacetin	7.17
Ethyl methacrylate	7.27
Di-n-butyl phthalate	7.34
3,3'-Dimethoxybenzidine	7.36
Acetonitrile	7.37
4-Aminopyridine	7.37
2-Chloronaphthalene	7.37
2 Propyn-1-ol	7.43

TABLE 8 (continued)
**Ranking of Incinerability of Organic Hazardous Constituents from Appendix VIII,
Part 261 on the Basis of Heat of Combustion**

Hazardous constituent	Heat of combustion (kcal/g)
1-Naphthyl-2-thiourea	7.50
Isosafrole	7.62
Dihydrosafrole	7.66
Safrole	7.68
Auramine	7.69
Crotonaldehyde	7.73
Allyl alcohol	7.75
Monochlorobiphenyl	7.75
Phenol	7.78
Phenylenediamine	7.81
Di-*n*-propylnitrosoamine	7.83
Pyridine	7.83
Ethyleneimine	7.86
1,1-Dimethylhydrazine	7.87
1,2-Dimethylhydrazine	7.87
N-Nitrosomethylvinylamine	7.91
2-Acetylaminofluorine	7.82
Acrylonitrile	7.93
Methapyrilene	7.93
Strychnine and salts	8.03
Methyl ethyl ketone (MEK)	8.07
Cresylic acid	8.09
Cresol	8.18
Toluene diamine	8.24
Acetophenone	8.26
Butyl benzyl phthalate	8.29
Ethyl cyanide	8.32
Bis (2-ethylhexyl) phthalate	8.42
Benzenethiol	8.43
N-Nitrosodi-*N*-butylamine	8.46
2,4-Dimethylphenol	8.51
Indenol (1,2,3,-*c*,3*d*) pyrene	8.52
Diethylstilbestrol	8.54
1-Naphthylamine	8.54
2-Naphthylamine	8.54
Methacrylonitrile	8.55
Isobutyl alcohol	8.62
1,2-Diethylhydrazine	8.68
2-Picoline	8.72
Aniline	8.73
1,2-Diphenylhydrazine	8.73
3,3′-Dimethoxybenzyldine	8.81
7H-Dibenzo(*c*,*g*)carbazole	8.90
Benz(*c*)acridine	8.92
Nicotine and salts	8.92
4-Amino biphenyl	9.00
Diphenylamine	9.09
2-Methylaziridine	9.09
Benzidine	9.18
Benzo(*b*)fluoranthene	9.25
Benzo(*j*)fluoranthene	9.25
Benzo(*a*)pyrene	9.25
Dibenzo(*a*,*e*)pyrene	9.33
Dibenzo(*a*,*e*)pyrene	9.33

TABLE 8 (continued)
Ranking of Incinerability of Organic Hazardous Constituents from Appendix VIII,
Part 261 on the Basis of Heat of Combustion

Hazardous constituent	Heat of combustion (kcal/g)
Dibenzo(a,i)pyrene	9.33
Fluoranthene	9.35
Benz(a)anthracene	9.39
Dibenz(a,h)anthracene	9.40
Dibenz(a,h)acridine	9.53
Dibenz(a,j)acridine	9.53
αα-Dimethylphenethylamine	9.54
3-Methylcholanthrene	9.57
n-Propylamine	9.58
7,12-Dimethylbenz(a)anthracene	9.61
Naphthalene	9.62
Benzene	10.03
Toluene	10.14

TABLE 9
Incinerability Rankings Based on Heats of Combustion
(H_c) and Heats of Formation (H_f)

Hazardous constituent	H_c (kcal/g)	Relative rank[a]	H_f (kcal/g	Relative rank[a]
Tribromomethane	0.13	1	0.024	5
Tetrachloromethane	0.24	2	−0.22	10
Trichloromethane	0.75	3	−0.26	11
Cyanogen chloride	1.29	4	0.56	1
Iodomethane	1.34	5	−0.14	7
Dibromomethane	1.43	6	−0.005	6
Dibromoethane	1.43	7	−0.10	9
Bromomethane	1.70	8	−0.09	8
Dichloromethane	1.70	9	−0.33	13
Chloromethane	3.25	10	−0.39	14
Formaldehyde	4.47	11	−0.92	18
Thiourea	4.51	12	−0.29	12
Acetaldehyde	6.34	13	−0.90	17
Acrolein	6.95	14	−0.50	16
Acetonitrile	7.37	15	0.31	2
Phenol	7.78	16	−0.40	15
Benzene	10.03	17	0.28	3
Toluene	10.14	18	0.13	4

[a] 1 = most difficult to incinerate.

exemptions of 264.340. In RCRA, a period is stipulated for the owner/operator to incinerate hazardous waste between the date of issuance of a permit and the trial burn. This period is allowed in order for the permit holder to establish operating conditions required to demonstrate compliance with the performance standards above (264.343) and must be in accordance with the trial burn plan. This period is especially applicable to new incinerators with associated shakedown and start-up phases (remember to have your local air pollution district permit for start-up/shakedown and trial burn by this time). This period is limited to a maximum of 720 h and starts as soon as the initial hazardous waste is introduced. It is advisable to do most of the shakedown/start-up on nonhazardous but similar handling char-

TABLE 10
Incinerability Rankings of Some Hazardous Compounds from Appendix VIII, Part 261 on the Basis of Chemical Kinetics (H_k) and Heats of Combustion (H_c)

Rank (H_k)	Compound	Rationale	Rank (H_c)
1	Hexachlorobenzene	All C–H bonds in excess of 110 kcal, C-Cl bonds of 95 kcal.	9
2	Pentachlorobenzene	OH addition is major destruction mode and is dependent on	7
3	Chlorobenzene	substituents on unsaturated structure.	26
4	Benzene		38
5	Napthalene		37
6	Vinyl chloride		20
7	Chloromethane	C-H bonds are in the 90—95 kcal range. All other bonds are	13
8	Ethyldiamine	>80 kcal. Assume that OH abstraction is controlling.	33
9	Dichlorophenol		16
10	Resorcinol		24
11	Chlorotoluene		29
12	Formaldehyde		21
13	Acetaldehyde		25
14	Acrolein		30
15	Dimethylphthalate		23
16	Methylethyl ketone	Bond-breaking processes in the 60—80 kcal range. Unimole-	35
17	Allyl alcohol	cular decomposition is controlling mechanism	34
18	Chloroform		3
19	Bromomethane		8
20	Dinitrobenzene		18
21	Trinitrobenzene		14
22	Tribromomethane		1
23	Hexachloropropene		2
24	Hexachloropentadiene		4
25	Bromoacetone		11
26	Hydrazine		19
27	Methylhydrazine		27
28	1,2-Dichloroethane		12
29	1,2-Dichloropropane	Low-energy pathway through complex fragmentation into sta-	17
30	Hexachlorocyclohexane	ble molecules (A = 10^{12} 10^{14} s E; 45—55 kcal)	10
31	Di-*n*-butylphthalate		32
33	Ethyl carbamate		22
34	Methyliodide		6
35	1,2-Diphenyl hydrazine	Fragmentation into radicals brought about by very weak	36
36	Nitroglycerine	bonds (40—55 kcal range)	15
37	Ni-nitroso-diethyl amine		28
38	2-Butanone peroxide	A-factor 10^{15}/s	31

Note: The compounds are divided into groups purely for present convenience. When one considers all the compounds in the EPA listing there will obviously be many overlaps.

acteristics material. An extension of an additional 720 h is possible on presenting the regulators with evidence of good cause. The trial burn will have to be conducted according to permit conditions and any other requirements necessary to verify compliance with the performance standards of 264.343 and in accordance with the approved trial burn plan. For the period after the trial burn, the owner/operator is again allowed a period up to 720 h that allows incineration of hazardous waste. This period is for the time required for sample analysis, data computation and submission, and review of the trial burn report. Operation must be in accordance with the permit, trial burn plan, and the above performance standards.

After a successful trial burn, the incinerator must be operated according to the requirements of the permit and performance standards of 264.343. The permit will specify the composition and physical/chemical variation range of the waste feed. Operating limits for the following will be specified:

1. Carbon monoxide (CO) level in the stack
2. Waste feed rate
3. Minimum and maximum combustion temperatures
4. Combustion gas velocity
5. Allowable variations in incinerator system design or operating procedures
6. Any other operating procedures necessary to ensure compliance with the performance standards
7. Waste feed cutoff during start-up and shutdown
8. Control of fugitive emissions by maintaining negative pressures in the combustion zone and keeping the combustion zone completely sealed
9. Automatic cutoff system for the waste feed when operating conditions deviate from operating conditions specified in the permit
10. Operation must cease when changes in the waste feed, incinerator design, or operating conditions exceed limits established in the permit

The owner/operator must monitor the following items while incinerating wastes:

1. Continuous monitoring of combustion temperature, waste feed rate, and the indicator of combustion gas velocity. Continuous means continuous, although regulators may vary on this requirement according to their understanding of the intent of the regulation and process control/recording methods. Continuous means a strip chart recording of each of the above parameters plus any other automatic monitor/recording systems employed (such as is common with distributed control capabilities. Most distributed control systems scan values and give time averages and are not truly continuous.).
2. Continuous monitoring of CO downstream of the combustion zone and prior to the release to the atmosphere.
3. The incinerator and auxiliary equipment such as pumps, valves, compressors, fans, piping, regulators, pollution control equipment, heat exchangers, etc., must be inspected at least daily for leaks, spills, fugitive emissions, and tampering. This should be on an inspection form that is filed and linked with a maintenance work order system.
4. The emergency waste feed cutoff system must be tested at least monthly to verify operability. Weekly is required unless it can be shown to be too disruptive to proper operation of the incinerator. All monitoring and inspection data must be entered in the operating log required by 264.73.

Besides the record-keeping requirements of 264.73, Appendix I of Part 264 details additional information requiring incorporation into the operating record.

2. The TSCA Application Process

Disposal of PCBs by either of the two approved methods — incineration or landfill — is regulated by the Toxic Substances Control Act (TSCA) and regulated and approved by procedures contained in 40 CFR 761.7 and 761.75, respectively. The above general application procedures and the above RCRA permit procedures, unless included in these TSCA regulations, and excluding public participation requirements, do not have to be followed by EPA TSCA personnel in their application review. However, experience has shown that a TSCA review will approximate the above general and RCRA procedures. It is believed that the disposal of PCBs will eventually end up being regulated by RCRA. The intent and scope of RCRA would make it a more logical regulatory framework. However, politics being very territorial, PCB regulations may remain where they are. There are many advantages in the permitting process due to the flexibility of TSCA regulators not having to adhere strictly to the RCRA requirements. Flexibility also brings uncertainty, however. At times, TSCA regulators may ask for additional information or impose requirements that RCRA regulators are limited in authority to impose, and many times politics play an even greater role than in RCRA. The peculiarities of TSCA regulations as they apply to PCB incineration (40 CFR Part 761.70) follow.

a. Liquid PCBs

An incinerator for liquid PCBs shall be approved by an EPA Regional Administrator, or if the application is for more than one region, the Assistant Administrator for Pesticides and Toxic Substances in EPA Headquarters, Washington, D.C.

The owner/operator must submit an application and be granted a permit before incineration of PCBs of 50 ppm or greater. The application must have the following information, but in general the same requirements as in RCRA will also be asked for.

1. Incinerator/facility location
2. Detailed description of the incinerator including general site and design drawings
3. Performance specifications
4. Sampling and monitoring equipment
5. Waste volumes to be incinerated
6. Notification of other federal, state, or local approvals needed
7. Schedules and plans for complying with approval requirements

The incinerator design and operation constraints include the following:

1. Combustion efficiency shall be at least 99.9% where combustion efficiency is computed by the following equation:

$$CE = \{[CO_2]/([CO_2] + [CO])\} \times 100\%$$

2. The rate and quantity of PCBs fed to the incinerator shall be measured and recorded at intervals not to exceed 15 min.
3   The temperatures of the incineration process shall be recorded continuously.
4. There must be an automatic cutoff of PCB material feed whenever the combustion temperature drops below permitted levels.
5. Stack emissions will be monitored when the incinerator is first used for the disposal of PCBs. This will be for the trial burn or "demonstration test". TSCA does not have the flexibility to permit operation of the incinerator with $\geq$ 50 ppm PCB material prior to the demonstration test such as RCRA does with hazardous waste. Shakedown and start-up will be with strictly clean material unless state and local authorities grant approval for utilizing < 50 ppm material (remember to have local air polution control authority permits for construction, start-up, and shakedown. Lobby for the local shakedown permit to last through the demonstration test). Parameters that must be monitored for (and quantified) during the demonstration test (trial burn) include O_2, CO, CO_2, NO_x, HCl, total chlorinated organics (RCl), PCBs, and total particulate matter. Additional parameters not in the regulations include the Hazardous Substance List of Volatile and Semivolatile Organics. They will have to be identified and quantified by the approved analysis methods and to the method detection limits. These are being demanded by the regulators based on the public concern about incinerator PIC generation.
6. Whenever the incinerator is combusting PCB material (no distinction of concentrations), monitoring shall be conducted for O_2, CO, and CO_2. Continuous monitoring is required for O_2 and CO. CO_2 might as well be monitored continuously, but not

necessarily recorded continuously. It will be needed to calculate combustion efficiency on a frequent basis.

7. The feed of PCB material must be stopped whenever the following occurs:

    a. Failure of the above monitoring operations.
    b. Failure of the PCB equipment measuring and recording feed rate.
    c. Excess oxygen falls below the percentage specified by the approved demonstration test plan or permit.

8. Water scrubbing or an *approved* alternate method for removing HCl must be used. Any approved method must comply with any other federal and state laws and regulations.

b. Nonliquid PCBs

As for nonliquid PCBs, the following criteria should be noted:

1. Approvals required are identical to liquid PCB incinerators.
2. Mass air emissions shall be no greater than 0.001 g PCB/per kilogram of the PCB introduced into the incinerator. This requirement states that the destruction and removal efficiency shall be demonstrated to meet or exceed 99.9999%.
3. All other performance criteria are identical to the liquid specifications.

c. Maintenance of Data and Records

All data and records shall be maintained in accordance with 40 CFR Part 761.180 (761.80 is a typographical error in the regulations). These records must be kept for a minimum of 5 years.

i. Continuous and Short Interval Data

These data include:

1. Rate and quantity of PCBs fed to the incinerator per 761.70(a)(4).
2. Temperature of the combustion process.
3. Stack emission product including O_2, CO, CO_2, and, very probably, NO_x and SO_2. When a reliable on-line monitor for HCl is established and available, look for this monitoring to be required on a routine basis also. During the demonstration test(s) NO_x, SO_2, HCl, RCl, PCBs, and total particulate matter will also be required with a high probability that volatile and semivolatile organics will be requested.
4. Total weight in kilograms of any solid residues (typically ash) generated by incineration during the calendar year, the total weight in kilograms of any solid residues disposed of by the facility in chemical waste landfills, and the total weight in kilograms of any solid residues remaining at the facility site.
5. Any periodic data specified by the EPA per 40 CFR Part 761.70(d)(4) as being part of any permit condition or approval.

REFERENCES

1. *The American Heritage Dictionary of the English Language,* Houghton Mifflin, Boston, 1970.
2. Publ. L. No. 94-580, 95 Stat. 2795 (1976). *Codified at* 42 U.S.C.A. 6921—6939b (West 1983 7 Suppl. 1989). RCRA replaced the Solid Waste Disposal Act ("SWDA") with nine new subtitles, of which Subtitle C focused on management of hazardous waste. Although Subtitle C of RCRA was codified as Subchapter III of the SWDA, 3001, it continues to be referred to as Subtitle C of RCRA, and that designation will be used herein. The most significant amendments to RCRA were contained in the Hazardous and Solid Waste Amendments of 1984 (HSWA), Publ. L. No. 98-616. 98 State. 3221 (1984)
3. 42 U.S.C.A. § 6902(a) (4).
4. To provide a quick overview of pertinent statutory provisions, the table of contents for RCRA Subtitle C, SWDA Subchapter III, is reprinted in Appendix A to Subpart A.
5. Although the statutory language references Subchapter III of the SWDA, which is RCRA Subtitle C. for all practical purposes that language references RCRA § 3001, 42 U.S.C.A. § 6921.
6. 42 U.S.C.A. § 6924(a).
7. 42 U.S.C.A. § 6924(a).
8. 42 U.S.C.A. § 6924(a).
9. 42 U.S.C.A. § 6925(a).
10. Nonetheless, a RCRA permit shall not be required under 42 U.S.C. § 6925 to *construct* a TSD facility if such a facility is constructed for the incineration of polychlorinated biphenyls pursuant to EPA approval under the Toxic Substances and Control Act, 15 U.S.C.A. § 2605(e). Any person owning or operating such a facility may, at any time after operation or construction of such facility has begun, apply for an RCRA permit authorizing such a facility to incinerate hazardous waste identified or listed under RCRA Subtitle C. 42 U.S.C.A. § 6925(a).
11. 42 U.S.C.A. § 6925(e).
12. 42 U.S.C.A. § 6925(c).
13. 42 U.S.C.A. § 6926(b).
14. Approved state hazardous waste management programs are listed in 40 CFR Part 272. When planning a proposed HWIF, the proponent should examine that part to ascertain whether an authorized state program exists for the state in which the proposed site is located. A proponent may also call the RCRA Hotline, operated by the EPA, at (800)424-9346 (or 382-3000 in the Washington, D.C. metropolitan area) for information.
15. 42 U.S.C.A. § 6926(e).
16. Pursuant to 42 U.S.C.A. § 6930(b). such regulations generally go into effect 6 months after the date such regulations are promulgated (revisions go into effect 6 months after the date of revision). Under certain circumstances, the EPA may provide for such regulations to become effective in less than 6 months.
17. *See* 40 CFR § 264.1 (1988). Such regulations are not generally applicable to TSD facilities holding a "permit by rule"; *see* 40 CFR § 270.60, subject to an authorized state program, or excluded by various specific exclusions contained in 40 CFR § 264.1(g).
18. *See* 40 CFR Parts 264 and 265, Subparts B — H.
19. *See* 40 CFR § 270.32(b) (1). Pursuant to 40 CFR § 270.4, compliance with the terms of a current RCRA permit constitutes compliance for purposes of enforcement with RCRA Subtitle C except for requirements not in the permit that become effective by statute or regulations promulgated under 40 CFR Part 268 regarding placement of hazardous wastes in or on land. Compliance with the permit, however, will not protect a TSD facility owner or operator against suits alleging imminent and substantial endangerment to health or the environment under RCRA 7002, 42 U.S.C.A. § 6972.
20. *See* CFR Parts 264 and 265, Subparts I — O.
21. *See* 40 CFR Part 265, Subpart P.
22. *See* 40 CFR Part 265, Subpart Q.
23. *See* 40 CFR Part 265, Subpart R.
24. *See* 40 CFR Part 264, Subpart X.
25. Because hazardous waste brought to an HWIF for treatment likely will be stored in containers, the owner or operator of an HWIF should review the specific standards for containers set forth in 40 CFR Part 264, Subpart I.
26. A TSD facility for the incineration of polychlorinated biphenyls may be constructed pursuant to EPA approval under the Toxic Substances and Control Act without obtaining a RCRA permit. 42 U.S.C.A. 6925(a); 40 CFR 270.10(f) (3). To incinerate other hazardous wastes identified or listed under RCRA Subtitle C, however, the owner or operator must obtain an RCRA permit.
27. 42 U.S.C.A. § 6925(a).
28. RCRA § 3005(b), 42 U.S.C.A. § 6925(b). Those regulations are set forth at 40 CFR Part 270.
29. 42 U.S.C.A. § 6925(c) (1).

30. 42 U.S.C.A. § 6925(c) (2) (A) (ii).
31. 42 U.S.C.A. § 6925(c) (3).
32. 42 U.S.C.A. § 6925(c) (3).
33. 42 U.S.C.A. § 6925(d).
34. 42 U.S.C.A. § 6925(g) (1).
35. 42 U.S.C.A. § 6925(g) (1) (C) and (2).
36. 42 U.S.C.A. § 6925(g) (1) (A).
37. 42 U.S.C.A. § 6925(g) (4).
38. 42 U.S.C.A. § 6925(g) (1) (B).
39. 42 U.S.C.A. § 6925(e) (1).
40. 42 U.S.C.A. § 6925(c) (2) (A) (ii).
41. 40 CFR 270.60. Such TSD facilities are not generally subject to the requirements of 40 CFR Part 264. *See id.* at § 264.1(c) — (e).
42. The EPA has promulgated regulations governing authorization of state hazardous waste management programs at 40 CFR Part 271.
43. 42 U.S.C.A. § 6924(a); *id.* § 6925(a).
44. 42 U.S.C.A. § 6930(a).
45. For the most part, the proponent of an HWIF will know that he or she expects the facility to handle hazardous waste identified or listed under Subtitle C of RCRA. The definition of hazardous waste in 40 C.F.R Part 261 is relatively self-explanatory. *See also* 40 C.F.R. Part 260, Appendix I. For an exhaustive discussion on the identification of hazardous waste, see Gentry, B. S., Identification of hazardous waste, in *The Law of Hazardous Waste,* Cooke, S. M., Ed., Matthew Bender, 1987, chap. 2.
46. 40 CFR § 260.10.
47. 42 U.S.C.A. § 6903(34); 40 CFR § 260.10.
48. 42 U.S.C.A. § 6903(33); 40 CFR § 260.10.
49. 42 U.S.C.A. § 6903(3); 40 CFR § 260.10.
50. 40 CFR 260.10 (emphasis added).
51. 40 CFR § 264.351 (requiring an owner or operator of an HWIF to remove all hazardous waste and hazardous waste residues from the incinerator site).
52. 40 CFR § 260.10.
53. 40 CFR § 264.340(a) (2).
54. Such regulations are set forth at 40 CFR Part 266, Subpart D.
55. 40 CFR § 260.10.
56. 40 CFR § 260.10.
57. 40 CFR § 260.10.
58. 40 CFR § 262.10.
59. *See* 40 CFR § 264.1.
60. 40 CFR § 264.11; *id.* § 265.11.
61. 40 CFR § 262.12(c).
62. "Regional administrator" means the EPA regional administrator for the EPA region in which the facility is located, or the regional administrator's designee. 40 CFR § 260.10.
63. 40 CFR § 264.12(a);*id.* § 265.12(a).
64. 40 CFR § 264.12(b). *See* 40 CFR § 264.73; *id.* § 265.73.
65. "Operating life" presumably is equivalent to "active life", which means "the period from the initial receipt of hazardous waste of the facility until the regional administrator receives certification of final closure". 40 CFR § 260.10.
66. 40 CFR § 264.12(c); *id.* § 265.12(b).
67. 40 CFR § 264.12(c) at comment; *id.* § 265.12(b) at comment.
68. 40 CFR § 264.13(a) (1); *id.* § 265.13(a) (1). Part 268 addresses land disposal restrictions.
69. 40 CFR § 264.13(a) (2); *id.* § 265.13(a) (2).
70. 40 CFR 264.13(a) (3); *id.* § 265.13(a) (3).
71. 40 CFR § 264.13(a) (2) at comment; *id.* § 265.13(a) (2) at comment.
72. 40 CFR § 264.13(a) (4); *id.* § 265.13(a) (4).
73. 40 CFR 264.13(b); *id.* § 265.13(b).
74. The representative sample may be obtained using either one of the sampling methods described in Appendix I of 40 CFR Part 261 or an equivalent sampling method. 40 CFR § 264.13(b) (3); *id.* § 265.13(b) (3).
75. 40 CFR § 264.13(b) — (c); *id.* § 265.13(b) — (c).
76. 40 CFR § 264.13(b).
77. 40 CFR § 270.14(b) (3).
78. 40 CFR § 264.14(a); *id.* § 265.14(a). "Active portion" means that part of the TSD facility where treatment, storage, or disposal operations are being or have been conducted, and that is not a "closed portion". 40 CFR 260.10.

79. 40 CFR § 264.14(a); *id.* § 265.14(a). An owner or operator that seeks an exemption from the security requirements must demonstrate that it qualifies for the exemption in Part B of its permit application. 40 CFR § 270.14(b) (4).
80. 40 CFR § 264.14(b); *id.* § 265.14(b).
81. 40 CFR § 264.14(b) at comment; *id.* § 265.14(b) at comment.
82. 40 CFR § 264.14(c); *id.* § 265.14(c).
83. 40 CFR § 264.15(a); *id.* § 265.15(a).
84. 40 CFR § 264.15(c); *id.* § 265.15(c).
85. 40 CFR § 270.14(b) (5).
86. 40 CFR § 264.15(b); *id.* § 265.15(b).
87. 40 CFR § 264.15(b) (4); *id.* § 265.15(b) (4).
88. 40 CFR § 264.14(c); *id.* § 265.15(c).
89. 40 CFR § 264.15(d); *id.* § 265.15(d).
90. 40 CFR § 264.16(a) — (c); *id.* § 265.16(a) — (c).
91. 40 CFR § 264.16(a) (2); *id.* § 265.16(a) (2).
92. 40 CFR § 264.52 *id.* § 265.52.
93. 40 CFR § 264.16(a) (3); *id.* § 265.16(a) (3).
94. 40 CFR § 264.16(d); *id.* § 265.16(d).
95. 40 CFR § 264.16(e); *id.* § 2654.16(e).
96. A waste is ignitable if it has any of the characteristics set forth in 40 CFR § 261.21(a). A waste is reactive if it has any of the characteristics set forth in 40 CFR § 261.23(a).
97. 40 CFR § 264.17(a); *id.* § 265.17(a).
98. 40 CFR § 264.17(b); *id.* § 265.17(b).
99. 40 CFR § 264.17(c).
100. 40 CFR § 264.18(a). The EPA defines "fault" to mean "a fracture along which rocks on one side have been displaced with respect to those on the other side". 40 CFR § 264.18(a)(2)(i). "Displacement" is defined as "the relative movement of any two sides of a fault measured in any direction". *Id.* § 264.18(a)(2)(ii). Finally, 'Holocene' means the most recent epic of the Quaternary period, extending from the end of the Pleistocene to the present." *Id.* § 264.18(a)(2)(iii).
101. 40 CFR § 264.18(b)(1). the EPA defines "100-year floodplain" to mean "any land area which is subject to a one percent or greater chance of flooding in any given year from any source". 40 CFR § 264.18(b)(2)(i). "Washout" is defined as "the movement of hazardous wastes from the active portion of the facility as a result of flooding". *Id.* § 264.18(b)(2)(ii). Finally, "100-year flood" is defined to mean "a flood that has a one percent chance of being equalled or exceeded in any given year". *Id.* § 264.18(b)(2)(iii).
102. 40 CFR § 264.18(b)(1) (i).
103 40 CFR § 264.18(b)(1) Comment.
104. 40 CFR § 264.18(b)(1)(ii).
105. 40 CFR § 264.18(c); *id.* § 265.18.
106. 40 CFR § 264.31.
107. 40 CFR § 265.31.
108. 40 CFR § 264.32; *id.* § 265.32.
109. 40 CFR § 264.32; *id.* § 265.32.
110. 40 CFR § 270.14(b)(6).
111. 40 CFR § 264.33; *id.* § 265.33.
112. 40 CFR § 264.34(a); *id.* § 265.34(a).
113. 40 CFR § 264.34(b); *id.* § 265.34(b).
114. 40 CFR § 264.35; *id.* § 265.35.
115. 40 CFR § 270.14(b)(6).
116. 40 CFR § 264.37(a); *id.* § 265.37(a).
117. 40 CFR § 264.51(a); *id.* § 265.51(a).
118. 40 CFR § 264.51(b); *id.* § 265.51(b).
119. 40 CFR § 264.52(a); *id* § 265.52(a). An owner or operator that has already prepared Spill Prevention, Control, and Countermeasures Plan in accordance with 40 CFR Part 112 or Part 1510, or with some other emergency plan, may simply amend that plan to incorporate hazardous waste management provisions complying with the requirements of 40 CFR Part 264. *See* 40 CFR § 264.52(b); *id.* § 265.52(b).
120. 40 CFR § 264.52(c); *id.* § 265.52(c).
121. 40 CFR § 264.52(d); *id.* § 265.52(d). For *new* facilities, this information must be supplied to the EPA Regional Administrator at the time of certification rather than at the time of permit application. *Id.* § 264.52(d).
122. 40 CFR § 264.52(e); *id.* § 265.52(e).
123. 40 CFR § 264.52(f); *id.* § 265.52(f).

124. 40 CFR § 264.53; *id.* § 265.53.

125. 40 CFR § 270.14(b)(7); *id.* § 264.53 at Comment.

126. 40 CFR § 264.54; *id.* § 265.54. Because the contingency plan will be incorporated as a condition of the facility's RCRA permit, amendment of the contingency plan constitutes a modificaiton of the permit. Changes in the list of emergency coordinators or equipment listed in the contingency plan constitutes only a "Class 1 modification" to the facility's permit. 40 CFR § 270.42(a); *see Fed. Regist.*, 53, 37936, 1988.

127. 40 CFR § 264.55; *id.* § 265.55.

128. 40 CFR § 264.56(a); *id.* § 265.56(a).

129. 40 CFR § 264.56(b); *id.* § 265.56(b). The emergency coordinator may do so by observation, by review of facility records, and, if necessary, by chemical analysis. *Id.*

130. 40 CFR § 264.56(c); *id.* § 265.56(c).

131. 40 CFR § 264.56(d)(2); *id.* § 265.56(d)(2).

132. 40 CFR § 264.56(d)(1); *id.* § 265.56(d)(1).

133. 40 CFR § 264.56(e); *id.* § 265.56(e).

134. 40 CFR § 264.56(f); *id.* § 265.56(f).

135. 40 CFR § 264.56(g); *id.* § 265.56(g). In so doing, the owner or operator of the facility becomes a generator of hazardous waste and must follow the applicable requirements of 40 CFR Parts 262, 263, and 264. *Id.* at Comment.

136. 40 CFR § 264.56(h); *id.* § 265.56(g).

137. 40 CFR § 264.56(i); *id.* § 265.56(i).

138. 40 CFR § 264.56(j); *id.* § 265.56(j).

139. 40 CFR § 264.71(a); *id.* § 265.71(a).

140. 40 CFR § 264.71(b); *id.* § 265.71(b).

141. 40 CFR § 264.71(a)(2) at Comment; *id.* § 265.71(a)(2) at Comment.

142. 40 CFR § 264.71(c); *id.* § 265.71(c).

143. 40 CFR § 264.72(a); *id.* § 265.72(a).

144. 40 CFR § 264.72(b); *id.* § 265.72(b).

145. 40 CFR § 264.73(a); *id.* § 265.73(a).

146. 40 CFR § 264.73(b); *id.* § 265.73(b).

147. 40 CFR § 264.74(a); *id.* § 265.74(a).

148. 40 CFR § 264.74(b); *id.* § 265.74(b).

149. 40 CFR § 264.75; *id.* § 265.75.

150. 40 CFR § 264.76; *id.* § 265.76.

151. 40 CFR § 264.76 at Comment; *id.* § 264.76 at Comment.

152. 40 CFR § 264.77(a); *id.* § 265.77(a).

153. 40 CFR § 264.77(b); *id.* § 265.77(c).

154. 40 CFR § 264.77(b) — (c).

155. 40 CFR § 265.77(b).

156. 40 CFR § 264.101(a). The requirements for groundwater monitoring set forth in 40 CFR Part 264, Subpart F do not apply to HWIFs. *See id.* § 264.90(a).

157. 40 CFR § 264.101(b).

158. 40 CFR § 264.101(c).

159. 40 CFR § 260.10.

160. The additional regulations governing postclosure care of disposal facilities are set forth in 40 CFR § 264.116 through 264.120 and *id.* at § 265.116 through 265.120.

161. 40 CFR § 264.111; *id.* § 265.111. As set forth in 40 CFR § 264.351 and § 265.351, at closure of an HWIF, the owner or operator must remove all hazardous waste and hazardous waste residues (including, but not limited to ash, scrubber waters, and scrubber sludges) from the incinerator site.

162. 40 CFR § 264.112(a)(1); *id.* § 265.112(a).

163. 40 CFR § 264.112(a)(2); *id.* § 265.112(a).

164. 40 CFR § 270.14(b)(13).

165. 40 CFR § 270.32; *id.* § 264.112(as)(1).

166. 40 CFR § 264.112(a)(2).

167. 40 CFR § 265.112(a).

168. 40 CFR § 264.112(b); *id.* § 265.112(b).

169. 40 CFR § 264.112(c); *id.* § 265.112(c); as amended by *Fed. Regist.*, 53, 37912/37935, 1988.

170. 40 CFR § 264.112(c)(2); *id.* § 265.112(c)(1).

171. 40 CFR § 264.112(c)(3); *id.* § 265.112(c)(2).

172. The regional administrator may request modifications to the closure plan under such conditions. The owner or operator must submit the modified plan within 60 d of the regional administrator's request or within 30 d if the change in facility conditions occurs during partial or final closure. 40 CFR § 264.112(c)(4); *id.* § 265.112(c)(4).

173. 40 CFR § 264.112(d)(1); *id.* § 265.112(d)(1).
174. 40 CFR § 264.112(d)(2); *id.* § 265.112(d)(2).
175. 40 CFR § 264.112(d)(3). The owner or operator of an interim status facility must submit its closure plan to the regional administrator no later than 15 d after termination of interim status (unless an RCRA permit is issued) or issuance of a judicial decree or final administrative order to cease accepting hazardous wastes. *Id.* § 265.112(d)(3). The regional administrator will review the proposed closure plan in accordance with *id.* § 265.112(d)(4).
176. 40 CFR § 264.112(e); *id.* § 265.112(e).
177. 40 CFR § 264.113(a); *id.* § 265.113(a). An owner or operator of an interim status facility must close the facility within 90 d after approval of its closure plan, if such date is later. *Id.*
178. 40 CFR § 264.113(b); *id.* § 265.113(b). For interim status facilities, the owner or operator must complete partial and final closure 180 d after approval of the closure plan, if such date is later than 180 d after receiving the final volume of hazardous waste. *Id.*
179. 40 CFR § 264.113(a) — (b); *id.* § 265.113(a) — (b).
180. 40 CFR § 264.113(c); *id.* § 265.113(c).
181. 40 CFR § 264.114; *id.* § 265.114.
182. 40 CFR § 264.115; *id.* § 265.115.
183. 42 U.S.C.A. § 6924(t)(1).
184. 42 U.S.C.A. § 6924(t)(1).
185. 40 CFR § 264.140(a),(c); *id.* § 265.140(a),(c).
186. 40 CFR § 264.140(b); *id.* § 265.140(b). Ordinarily, an HWIF will not be a "disposal" facility, which is defined as a facility or part of a facility at which hazardous waste is intentionally placed into or on any land or water, and at which waste will remain after closure. 40 CFR § 260.10.
187. 40 CFR § 264.142(a); *id.* § 265.142(a).
188. 40 CFR § 264.351; *id.* § 265.351. In so doing, the HWIF owner or operator will be a "generator" of hazardous waste subject to the requirements of 40 CFR Part 262.
189. 40 CFR § 264.142(a)(1); *id.* § 265.142(a)(1).
190. 40 CFR § 264.142(a)(2); *id.* § 265.142(a)(2).
191. 40 CFR § 264.142(a)(3); *id.* § 265.142(a)(3).
192. 40 CFR § 264.142(a)(4); *id.* § 265.142(a)(4). In other words, "an owner or operator cannot assume that at closure a third party will take hazardous wastes at no charge." *Fed. Regist.*, 51, 16437, 1986.
193. 40 CFR § 264.142(b); *id.* § 265.142(b).
194. 40 CFR § 264.142(b); *id.* § 265.142(b).
195. 40 CFR § 264.142(c); *id.* § 265.142(c).
196. 40 CFR § 264.142(d); *id.* § 265.142(d).
197. 40 CFR § 264.143(a) — (f).
198. *Fed. Regist.* 50 28702/28734, 1985.
199. 40 CFR § 265.143(a) — (e).
200. 40 CFR § 264.143(g); *id.* § 265.143(f).
201. 40 CFR § 264.143(h); *id.* § 265.143(g).
202. 40 CFR § 264.143(g); *id.* § 265.143(h).
203. 40 CFR § 264.143(a)(1); *id.* § 265.143(a)(1).
204. 40 CFR § 264.143(a)(1).
205. 40 CFR § 270.1(b); *id.* § 270.14(b)(15).
206. 40 CFR § 264.143(a)(1); *id.* § 265.143(a)(1).
207. 40 CFR § 264.143(a)(2); *id.* § 265.143(a)(2).
208. 40 CFR § 264.143(a)(3).
209. 40 CFR § 265.143(a)(3).
210. 40 CFR § 264.143(a)(3)(i).
211. The RCRA regulations state that the first payment must be made by "the effective date of these regulations". 40 CFR § 264.143(a)(3)(i). The regulations do not specifically address the circumstance of a facility burning hazardous waste newly covered by RCRA requirements. The implication of the above provision, confirmed by an EPA representative, is that a financial mechanism to meet the financial requirements should be established and, in the case of a trust fund, the first payment made by the "effective date" of the regulations subjecting the facility to compliance with RCRA regulations applicable to TSD facilities. Regulations generally become effective 6 months after the date of their promulgation. 42 U.S.C.A. § 6930(b).
212. 40 CFR §264.143(a)(3)(i); *id.* § 265.143(a)(3)(i).
213. 40 CFR § 264.143(a)(3)(i); *id.* § 265.143(a)(3)(ii).
214. 40 lCFR § 264.143(a)(3)(ii).
215. 40 CFR § 264.143(a)(4); *id.* § 265.143(a)(4).
216. 40 CFR § 264.143(a)(5); *id.* § 265.143(a)(5).

217. 40 CFR § 264.143(a)(6); *id.* § 265.143(a)(6).
218. 40 CFR § 264.143(a)(7); *id.* § 265.143(a)(7).
219. 40 CFR § 264.143(a)(8); *id.* § 265.143(a)(8).
220. 40 CFR § 264.143(a)(9); *id.* § 265.143(a)(9).
221. 40 CFR § 264.143(a)(10); *id.* § 265.143(a)(10).
222. 40 CFR § 264.143(a)(11); *id.* § 265.143(a)(11).
223. 40 CFR § 264.143(b)(1), (b)(3)(i); *id.* § 265.143(b)(1), (b)(3)(i). Although the regulations do not state when the owner or operator of an interim status facility should submit the surety bond to the Regional Administrator, it probably should be submitted by the date set for submitting Part B of that facility's RCRA permit application.
224. 40 CFR § 264.143(b)(2); *id.* § 265.143(b)(2).
225. 40 CFR § 264.143(b)(1); *id.* § 265.143(b)(1).
226. 40 CFR § 264.143(b)(3);; *id.* § 265.143(b)(3).
227. 40 CFR § 264.143(b)(4); *id.* § 265.143(b)(4).
228. 40 CFR § 264.143(b)(5); *id.* § 265.143(b)(5).
229. 40 CFR § 264.143(b)(6); *id.* § 265.143(b)(6).
230. 40 CFR § 264.143(b)(7); *id.* § 265.143(b)(7).
231. 40 CFR § 264.143(b)(8);; *id.* § 265.143(b)(8).
232. 40 CFR § 264.143(b)(9); *id.* § 265.143(b)(9).
233. 40 CFR § 264.143(c)(1).
234. 40 CFR § 264.143(c)(2).
235. 40 CFR § 264.143(c)(3).
236. 40 CFR § 264.143(c)(4).
237. 40 CFR § 264.143(c)(5).
238. 40 CFR § 264.143(c)(6)—(7).
239. 40 CFR § 264.143(c)(8).
240. 40 CFR § 264.143(c)(9).
241. 40 CFR § 264.143(c)(10).
242. 40 CFR § 264.143(d)(4); *id.* § 265.143(c)(4).
243. 40 CFR § 264.143(d)(1), (3)(i); *id.* § 265.143(c)(1), (3)(i).
244. 40 CFR § 264.143(d)(2); *id.* § 265.143(c)(2).
245. 40 CFR § 264.143(d)(1); *id.* § 265.143(c)(1).
246. 40 CFR § 264.143(d)(3); *id.* § 265.143(c)(3).
247. 40 CFR § 264.143(d)(5); *id.* § 265.143(c)(5).
248. 40 CFR § 264.143(d)(6); *id.* § 265.143(b)(6).
249. 40 CFR § 264.143(d)(7); *id.* § 265.143(c)(7).
250. 40 CFR § 264.143(d)(8); *id.* § 265.143(c)(8).
251. 40 CFR § 264.143(d)(9); *id.* § 265.143(c)(9).
252. 40 CFR § 264.143(d)(10); *id.* § 265.143(c)(10).
253. 40 CFR § 264.143(e)(1); *id.* § 265.143(d)(1). For an interim status facility, EPA regulations provide that the owner or operator must submit to the Regional Administrator a letter from an insurer stating that the insurer is considering issuing closure insurance to the facility "by the effective date of these regulations". *Id.* "Within 90 days after the effective date of these regulations" the owner or operator must submit the certificate of insurance or obtain other financial assurance. *Id.* These requirements should be interpreted to require action by the effective date of the regulations subjecting the facility to RCRA regulation.
254. 40 CFR § 264.143(e)(1); *id.* § 265.143(d)(1).
255. 40 CFR § 264.143(e)(3); *id.* § 265.143(d)(3). The term "face amount means the total amount the insurer is obligated to pay under the policy". *Id.*
256. 40 CFR 264.143(e)(9); *id.* 265.143(d)(9).
257. 40 CFR § 264.143(e)(4); *id.* § 265.143(d)(4).
258. 40 CFR § 264.143(e)(5); *id.* § 265.143(d)(5).
259. 40 CFR § 264.143(e)(6); *id.* § 265.143(d)(6). Such violation will be deemed to begin on receipt by the Regional Administrator of a notice of future cancellation, termination, or failure to renew due to nonpayment of the premium, rather than on the date of expiration. *Id.*
260. 40 CFR § 264.143(e)(8); *id.* § 265.143(d)(8). The policy must permit assignment of the policy to a successor owner or operator. Although such assignment may be conditional on the insurer's consent, such consent may not be unreasonably refused. *Id.* § 264.143(e)(7); *id.* § 265.143(d)(7).
261. 40 CFR § 264.143(e)(8); *id.* § 265.143(d)(8).
262. 40 CFR § 264.143(e)(8); *id.* § 265.143(d)(8).
263. 40 CFR § 264.143(e)(10); *id.* § 265.143(d)(10).

264. For purposes of these regulations, the EPA defines "current assets" to mean cash or other assets or resources commonly identified as those that are reasonably expected to be realized in cash or sold or consumed during the normal operating cycle of the business. The EPA further defines "current liabilities" to mean obligations whose liquidation is reasonably expected to require the use of existing resources properly classifiable as current assets or the creation of other current liabilities. 40 CFR § 264.141(f); *id.* § 265.141(f).

265. For facilities required to prepare postclosure costs estimates as well as plugging and abandonment cost estimates pursuant to 40 CFR § 144.62(a) — (c), net working capital and tangible net worth must each be at least six times the sum of such estimates added to the current closure cost estimate. 40 CFR § 264.143(f)(1)(i)(B); *id.* § 265.143(e)(1)(i)(B). The EPA, for purposes of these regulations, defines "net working capital" to mean current assets minus current liabilities and defines "tangible net worth" to mean tangible assets that remain after deducting liabilities. 40 CFR § 264.141(f); *id.* § 265.141(f).

266. 40 CFR § 264.143(f)(1)(i); *id.* § 265.143(e)(1)(i).

267. 40 CFR § 264.143(f)(1)(ii); *id.* § 265.143(e)(1)(ii).

268. 40 CFR § 264.143(f)(3); *id.* § 265.143(e)(3).

269. 40 CFR § 264.143(f)(4).

270. 40 CFR § 264.143(f)(5); *id.* § 265.143(e)(5).

271. 40 CFR § 264.143(f)(6); *id.* § 265.143(e)(6).

272. 40 CFR § 264.143(f)(7); *id.* § 265.143(e)(7).

273. 40 CFR § 264.143(f)(8); *id.* § 265.143(e)(8).

274. 40 CFR § 264.143(f)(10); *id.* § 265.143(e)(10). A "parent corporation" is defined as a corporation that directly owns at least 50% of the voting stock of the corporation which owns or operates the facility. 40 CFR § 264.141(d); *id.* § 265.141(d).

275. 40 CFR § 264.143(f)(10); *id.* § 265.143(e)(10).

276. 40 CFR § 264.147(a); *id.* § 265.147(a). For purposes of these regulations, the EPA defines "sudden accidental occurrence" to mean an occurrence that is not continous or repeated in nature. *Id.* § 264.141(g); *id.* § 265.141(g). "Accidental occurence" is defined as an accident, including continuous or repeated exposure to conditions, which results in bodily injury or property damage neither expected nor intended from the standpoint of the insured. *Id.* "Bodily injury" and "property damage" are to be given the meanings provided by applicable state law. *Id.*

277. 40 CFR § 264.147(b); *id.* § 265.147(b).

278. 40 CFR § 264.147(a); *id.* § 265.147(a).

279. 40 CFR § 264.147(a); *id.* § 265.147(a); as revised by 53 *Fed. Regist.*, 53, 33938, 1988.

280. 40 CFR § 264.147(a)(1)(ii); *id.* § 265.147(a)(1)(ii).

281. 40 CFR § 264.147(a)(1)(i); *id.* § 265.147(a)(1)(i).

282. 40 CFR § 264.147(a)(1)(i); *id.* § 265.147(a)(1)(i).

283. 40 CFR § 264.147(a)(1)(i).

284. 40 CFR § 264.147(a)(2); *id.* § 265.147(a)(2); as amended by *Fed. Regist.*, 53, 33950, 1988.

285. The regulations define "amount of liability coverage" to mean the annual aggregate amounts for which coverage is required per the financial responsibility requirements governing liability coverage. 40 CFR § 264.147(f)(2); *id.* § 265.147(f)(2). For an HWIF, that amount will be $2 million as set forth in 40 C.F.R. § 264.147(a) or § 265.147(a). An owner or operator of a surface impoundment, landfill, land treatment facility, or miscellaneous disposal unit must obtain additional coverage. *See id.* § 264.147(b); *id.* § 265.147(b).

286. 40 CFR § 264.147(f)(1)(i); *id.* § 265.147(f)(1)(i).

287. 40 CFR § 264.147(f)(1)(ii); *id.* § 265.147(f)(1)(ii).

288. 40 CFR § 264.147(f)(3)—(7); *id.* § 265.147(f)(3)—(7).

289. 40 CFR § 264.147(f)(i); *id.* § 265.147(f)(i).

290. For corporations incorporated in the U.S. a guarantee may be used to satisfy the liability coverage requirements only if the attorneys general or insurance commissioners of the state in which the guarantor is incorporated *and* each state in which a facility covered by the guarantee is located have submitted written statements to the EPA that a corporate guarantee executed as required by the RCRA regulations is a legally valid and enforceable obligation in that state. 40 CFR § 264.147(g)(2)(i); *id.* § 265.148(g)(2)(i). For a corporation incorporated outside the U.S. a guarantee may be used only if (a) the non-U.S. corporation has identified a registered agent for service of process in each state in which a facility covered by the guarantee is located and in the state in which it has its principal place of business and (b) the attorney general or insurance commissioner of each such state has submitted a written statement to the EPA that a guarantee as required by the RCRA regulations is a legally binding and enforceable obligation in that state. 40 CFR § 264.147(g)(2)(ii); *id.* § 265.148(g)(2)(ii).

291. 40 CFR § 264.147(g)(1); *id.* § 265.147(g)(1); as set forth at *Fed. Regist.*, 53, 33938, 1988. The EPA defines a "substantial business relationship" to mean the extent of a business relationship necessary under applicable state law to make a guarantee contract issued incident to that relationship valid and enforceable. The relationship must arise from a pattern of recent or ongoing business transactions. The relationship must be demonstrated to the satisfaction of the regional administrator. 40 CFR § 264.141(h); *id.* § 265.141(h); as set forth at *Fed. Regist.*, 53, 33938, 1988.

292. 40 CFR § 264.147(g)(1); *id.* § 265.147(g)(1).
293. 40 CFR § 264.147(g)(1); *id.* § 265.147(g)(1); as set forth at *Fed. Regist.,* 53, § 3950, 1988.
294. 40 CFR § 264.147(g)(1)(i) — (ii); *id.* § 265.147(g)(1)(i) — (ii).
295. 40 CFR § 264.147(a)(3), (h); *id.* § 264.147(a)(3), (h); as set forth at *Fed. Regist.,* 53, 33938, 1988.
296. 40 CFR § 264.147(h)(2)—(3); *id.* § 264.147(h)(2)—(3); as set forth at *Fed. Regist.,* 53, 33938, 1988.
297. 40 CFR § 264.147(a)(4), (i); *id.* § 265.147(a)(4), (i); as set forth at *Fed. Regist.,* 53, 33938, 1988.
298. 40 CFR § 264.147(i); *id.* § 265.147(i); as set forth at *Fed. Regist.,* 53, 33938, 1988.
299. 40 CFR § 264.147(a)(5); *id.* § 265.147(a)(5); as set forth at *Fed. Regist.,* 53, 33938, 1988.
300. 40 CFR § 264.147(j)(1)—(2); *id.* § 265.147(j)(1)—(2); as set forth at *Fed. Regist.,* 53, 33938, 1988.
301. 40 CFR § 264.147(j)(3); *id.* § 265.147(j)(3); as set forth at *Fed. Regist.,* 53, 33938, 1988. The regulations define "the full amount of the liability coverage to be provided" to mean the amount of liability coverage required to be provided by the owner or operator less the amount of financial assurance for liability coverage provided by other financial assurance mechanisms. *Id.*
302. 40 CFR § 264.147(j)(4); *id.* § 265.147(j)(4); as set forth at *Fed. Regist.,* 33938, 1988.
303. 40 CFR § 264.147(a)(7); *id.* § 265.147(a)(7); as set forth at *Fed. Regist.,* 53, 33938, 1988. An owner or operator, however, may not combine a financial test covering part of the liability coverage requirement with a guarantee unless the financial statement of the owner or operator is not consolidated with the financial statement of the guarantor. *Id.*
304. 40 CFR § 264.147(a)(7); *id.* § 265.147(a)(7); as set forth at *Fed. Regist.,* 53, 33938, 1988.
305. 40 CFR § 264.147(c); *id.* § 265.147(c). The regional administrator may require an owner or operator who requests the variance to provide such technical and engineering information as is deemed necessary by the regional administrator to determine an appropriate level of financial responsibility. *Id.*
306. 40 CFR § 264.147(d); *id.* § 265.147(d).
307. 40 CFR § 264.147(e); *id.* § 265.147(e).
308. 40 CFR § 264.148(a); *id.* § 265.148(a). *See also* 40 CFR § 264.151(h)(terms of corporate guaranty require a notification).
309. 40 CFR § 264.148(b); *id.* § 265.148(b).
310. 40 CFR § 264.149(a); *id.* § 265.149(a).
311. 40 CFR § 264.149(a); *id.* § 265.149(a).
312. 40 CFR § 264.149(a); *id.* § 265.149(a).
313. 40 CFR § 264.149(b); *id.* § 265.149(b).
314. 40 CFR § 264.150(a); *id.* § 265.150(a).
315. 40 CFR § 264.150(a); *id.* § 265.150(a).
316. 40 CFR § 264.150(a); *id.* § 265.150(a).
317. 40 CFR § 264.150(a); *id.* § 265.150(a).
318. 40 CFR § 264.150(b); *id.* § 265.150(b).
319. 40 CFR § 264.340(a); *id.* § 265.340(a).
320. 40 CFR § 264.340(b); *id.* § 265.340(b).
321. 40 CFR § 264.340(b)(2).
322. 40 CFR § 264.340(b)(1)(i) — (iv); *id.* § 264.340(b)(1)—(4).
323. 40 CFR § 264.340(c).
324. 40 CFR § 264.340(d).
325. 40 CFR § 264.342(b)(1).
326. 40 CFR § 264.342(b)(2). Those procedures are specified at 40 CFR § 270.62.
327. 40 CFR § 264.343(a)(1).
328. 40 CFR § 264.343(c).
329. 40 CFR § 264.343(d). Modification or revocation of a permit is governed by *Id.* § 270.41.
330. 40 CFR 264.344(a). Exemptions are set forth at *id.* 264.340.
331. 40 CFR § 264.344(b).
332. 40 CFR § 264.344(c)(4).
333. 40 CFR § 264.345(a).
334. 40 CFR § 264.345(b).
335. 40 CFR § 264.345(c).
336. 40 CFR § 264.345(d).
337. 40 CFR § 264.345(e) — (f).
338. 40 CFR § 264.347(a).
339. 40 CFR § 264. *Id.* § 264.347(a)(3).
340. *Id.* § 264.347(b).
341. *Id.* § 264.347(c).
342. *Id.* § 264.347(d).
343. *Id.* § 264.351.
344. *Id.* § 264.351 at comment.

Chapter 5

SITING ISSUES AND PUBLIC ACCEPTANCE

Martha A. Rozelle

TABLE OF CONTENTS

I.   Introduction 184

II.   Who is the Public? 186

III.   How Is the Public Identified? 187
      A.   Self-Identification 187
      B.   Staff-Identification 188
          1.   Intuitive Information 188
          2.   Lists of Organizations and Individuals 188
          3.   Geographic and Demographic Analysis 188
          4.   Historical Analysis 188
      C.   Third-Party Identification 188

IV.   How Does the Public View Environmental Risk? 189

V.   To What Extent Is the Public Involved? 191

VI.   To What Extent Should the Public Be Involved? 194

VII.   How Can Public Acceptance Be Encouraged? 197
      A.   Understanding the Public's Values 197
      B.   Communicating to the Public About Environmental Risk 198
      C.   Include the Public in the Siting Process 201
          1.   Site Identification 202
          2.   Site Evaluation 202
          3.   Site Selection 203

VIII.   A Step-by-Step Strategy to Involve the Public 204

IX.   Phase I — Candidate Site Identification 205

X.   Phase II — Site Evaluation 205

XI.   Phase III — Site Selection 205
      A.   Issue First News Release 206
      B.   Establish a Point of Public Contact 206
      C.   Begin a Newspaper Clipping Service 206
      D.   Establish a Reading Room 206
      E.   Establish a Mailing List 206
      F.   Prepare Fact Sheet 206
      G.   Conduct Interviews with Opinion Leaders 206
      H.   Establish a Citizen Advisory Committee 208

I.   Develop and Conduct Media Relations Program........................209
J.   Prepare Bulletins 209
K.   Conduct Public Scoping Meeting.......................................209
L.   Conduct Information Fairs/Open Houses 209
M.   Prepare Newspaper Inserts with Response Form.......................209
N.   Distribute Cumulative Brochure or "Fishbowl Planning" 209
O.   Provide Briefings to Community Organizations 209
P.   Conduct Public Workshops...210

XII.   Conclusion...210

References..210

I. INTRODUCTION

With few exceptions, the siting and permitting of new facilities is the most difficult waste management problem facing local governments and industry officials today. In addition to the technical engineering criteria, fiscal constraints, land availability, and environmental regulations compliance, social and political concerns have evolved as equally important considerations in the location of waste disposal facilities. No longer is the "best" site one that is technically adequate. The best site is one that is also acceptable to the public and stands a chance of eventually being built. "Not in my backyard" or "put it in their backyard" have become favorite slogans of the potentially affected public. The "NIMBYs" and the "PITBYs" have become serious stumbling blocks to locating new facilities and to furthering the use of innovative disposal technologies.

In the context of this publication, public involvement is the progressive development of a consensus to take action on a specific issue. When government or industry "assume consent" on the part of the affected public (or impacted party), they often misplace time, money, and priorities. Experience has taught us that public understanding and acceptance (often hard won) are the building blocks which must be in place before technological, financial, and management decisions are implemented.

The modern environmental movement began to evolve with the publication of *Silent Spring* by Rachel Carson in 1962. Carson suggested that mankind was slowly poisoning the environment through the use of pesticides, specifically DDT. The book became a best seller, and by raising public awareness saving the environment became a popular cause. In the past two decades, the public has increasingly demanded more information and accountability from its policy makers and has had more impact on the policy-making process. Most federal and state programs that address cleaning up the environment include citizen participation provisions. In spite of these requirements, both the public and the policy maker are frustrated with the processes of citizen participation. Controversy has erupted over many aspects of science and technology, as decisions once defined as technical have become intensely political. Major projects have been and are being delayed, defeated, or reversed.

The expanded role of citizen participation in decision making that has evolved over the past 20 years is expected to continue and should be recognized as part of a larger effort to reinterpret the meaning and scope of democracy. The demands for participation have forced politicians, seeking to maintain their popular support, to consider technical issues that are beyond their normal scope of activity. A pattern of public, and often negative, reactions has emerged, focused against the system. The system is perceived as unresponsive and unwilling to provide the public with sufficient opportunity to influence decisions. In 1979 Nelkin was one of several researchers who captured this trend when she wrote:

> Thus the past decade has been remarkable for political action directed against science and technology. Issue-oriented organizations have formed to obstruct specific projects; scientists have called public attention to risks; and many groups demand greater accountability and public participation in public decisions.[1]

As we begin the 1990s, this statement is more true than it was 10 years ago. There is a growing trend of dissatisfaction with the role of citizen participation in the policy-making process. Former Energy Secretary James Schlesinger complained, "We have reached the stage of participatory democracy where almost everyone in the society can say 'no' but no one can say 'yes'."[2]

Another result of this trend of dissatisfaction is a growing distrust of the expert opinion. After hearing 120 scientists argue over nuclear safety before the California State Legislature, the hearing officer concluded that the issues were not, in the end, resolvable by expertise. "The questions involved require value judgments and the voter is no less equipped to make such judgments that the most brilliant Nobel Laureate."[3] Resolution of a dispute depends on the nature of underlying perceptions. If the question were merely one of specific interest, direct offsetting compensation measures could reduce conflict. But when more basic ideological principles or attitudes are at stake, no amount of data or dollars will resolve value differences.

Public decisions depend on values, but the values that underlie the decision should reflect some consensus or at least an identifiable aggregation of individual views. In a representative democracy, individuals are entitled to disagree about the importance of their values, and they are entitled to have their values reflected in the final decision. It is not appropriate, however, that they actually make the decision. An accurate portrayal of the public values often permits the decision maker to understand the areas of agreement among various publics. Frequently, the areas of disagreement are loudly voiced and overshadow similarities in values, and the potential for consensus building may not be evident.

This chapter addresses the importance of and effective ways to involve the public in the siting of a hazardous waste incinerator. The following questions are addressed:

1. Who is the public and how does it view the world?
2. What constitutes public acceptance?
3. To what extent should the public be involved in siting decisions? How far does this involvement extend beyond legal requirements?
4. Specifically, how can the public be included in the siting process?
5. Is public involvement in a siting process structured differently for privately and publicly owned facilities?
6. How should we communicate with the public about environmental risk?

Selected events from previous waste disposal facility siting processes are referenced to emphasize the need for public involvement. Not all case studies deal with hazardous waste incinerators, but the issues are similar enough to be illustrative. We can learn from successful (and unsuccessful) siting and permitting experiences of facilities such as municipal incinerators, landfills, radioactive waste repositories, or power plants. Finally, a step-by-step public involvement and information program is presented.

II. WHO IS THE PUBLIC?

The public includes any individual, agency, business, organization, special interest group, or governmental entity that believes it will be affected by the project. They are often referred to in the plural form, publics. The term "stakeholders" suggests those having a stake in the outcome of the project. "Parties at interest," a term borrowed from the legal profession, is frequently used. These terms are used interchangeably in this chapter.

In drawing up a list of publics, one should include not only those who are affected in an obvious and direct way, but also those who are indirectly affected and those who believe they will be affected, even if one is quite sure they will not be. The various publics view the problem from very different contexts, they pursue different purposes or values, and they engage in strikingly different patterns of reasoning. Wedge[4] describes four general groupings of participants in a siting process: the victims and their advocates, the technocrats, the industrialists, and the political leadership.

First, "the victim" is a term disliked by all other stakeholders but is often used by those who are directly affected or threatened. Victims are those who believe their homes and possibly their lives are in danger. Whether the threat is from an existing hazard or a proposed facility, victims may become activists but may have trouble finding a responsive or responsible agency. Victims then seek allies within the community and later in the form of public interest advocates, often turning to the media to make their plight known. Their reasoning tends to be based in deep-seated values of protection of family and property. The perceived threat to their home and children is so self-evident that to them, their demands for immediate response seem justified. The victims believe that they are being asked to take some risk through no fault of their own. When response seems slow or comes in the form of technical studies and formal hearings with long delays, their reaction is one of intense frustration, anger, and increased determination to fight the project.

Advocates who support the victims are drawn from many sources: the general public, public officials, environmental organizations, public health professionals, and the news media. Advocates are typically more detached and less emotional in their reasoning and argument than the victims. Unlike the victim, they are able to communicate with other parties on a more common ground. Often their involvement in a specific issue may further their own organizational agenda.

The technocrats are the scientists and administrators who are authorized to protect the public and support environmental health and welfare. They have an often conflicting responsibility to meet the public need to dispose of potentially dangerous by-products of industrial processes. They have the unenviable task of setting standards of reasonable safety and of monitoring sites for conformity to those standards. But the determination of how safe is safe is not really a scientific question. Scientific thought must rely on probabilities. The scientist can say how unsafe a given agent may be in a given quantity over a given exposure time. But scientific opinion, as distinguished from scientific absolutes, is not uniform, and other scientists or engineers may challenge the findings and often do so. Definitive answers to specific cases are often mistrusted because the facts and circumstances dictate opinions or probabilities and the questioner is seeking simple answers to complex situations.

Wedge's third group of participants, the industrial generators of hazardous waste products, view the issues from a business context.[4] Corporations are responsible to their stockholders for generating profits. Reconciling this with public health concerns that may entail costs or reduce profits can create a strain in their business. While a primary goal is to create products that can be sold at a profit, a secondary goal is to dispose efficiently of the by-products of their industries. The accumulation of stored wastes is a continuing expense and adds to the cost of production. Therefore, virtually all manufacturers who generate wastes, which could possibly be classified as hazardous, eagerly seek solutions. Another option is to contract the problem to waste management firms. These firms largely share the business outlook of the industrial generators but have been forced to become increasingly sensitive to public concerns.

Finally, the political leadership presents still a different context, one of sorting out and prioritizing values and converting those values into workable laws and policies. If the politician is to remain in office, he must remain sensitive to the wishes of his constituency. Once an issue has attracted significant organized force, the politician is forced to move from the position of responsiveness to one of leadership (from collecting input to acting on it). The political reasoning process is largely one of balancing public risks and public good within the conceptual framework of what is considered the public good.

The public can be described in many other ways, but several basic assumptions about the public can be made:

1.   For any proposed decision, the public will consist of people and entities who see themselves as significantly affected by the decision at hand.
2.   The size and composition of the public will be different for each decision.
3.   The size of the public increases with controversy.
4.   The size of the public will increase as the decision becomes closer and the impacts of the decision become clearer. As alternative actions are defined, one group's problems may be solved while another's are just beginning.

Providing each of the stakeholders with enough information to weigh all sides of an issue, evaluate risks, and possibly reach a consensus, is a far more demanding task than simply constructing a mass public information or public relations campaign.

III. HOW IS THE PUBLIC IDENTIFIED?

The importance of compiling a comprehensive list of potentially interested publics cannot be overemphasized. The publics most likely to oppose a project late in its development are those that should have been identified early but were not. It is important to reassess continually who the public is as a project develops.

At the beginning of each public involvement program, a systematic effort should be made to identify those publics that are likely to see themselves as affected by a decision. There are several reasons for this:

1.   To ensure that the active and vocal parties participating in the program represent all potentially affected publics
2.   To establish credibility by including potentially affected publics early, rather than have them discover later that they might be affected
3.   To involve potentially affected publics early in the process when they can have some influence, rather than late in the game when they are forced to become adversaries or reluctant supporters

Three fairly quick ways can be used to identify publics that perceive themselves as potentially affected by a decision: self-identification, staff-identification, and third-party identification.

A. SELF-IDENTIFICATION

Individuals, groups, or governmental entities may show an interest in participating. They may do so in response to news stories, historical development of the project, attendance at a public meeting, or membership in national organizations that follow certain issues.

B. STAFF-IDENTIFICATION

Possible participants may be easily identified by staff in government agencies or by local industry representatives who are active in their community. Some quick methods for staff identification include the following.

1. Intuitive Information

Most agency staff who have worked in an area for a significant period of time can identify individuals and groups that are likely to be involved in a siting study for a hazardous waste incinerator. They can also suggest the concerns that certain publics are likely to have.

2. Lists of Organizations and Individuals

Numerous lists are easily available. Chambers of Commerce keep lists of community clubs and organizations. The League of Women Voters frequently publishes lists of elected officials, and the state Leagues of Cities and Towns frequently print a directory of management and department level employees in every city, town, and county in the state. With a little effort one can often find directories for specific types of organizations such as environmental groups.

3. Geographic and Demographic Analysis

Some people will be interested in a project by virtue of living near potential sites or along transportation routes. They may be particularly concerned if they are elderly, have school age children, or live on fixed incomes.

4. Historical Analysis

Helpful information can be found in past newspaper clippings, files of similar past projects, or agency correspondence files.

C. THIRD-PARTY IDENTIFICATION

In this approach, opinion leaders in the community or people of known interests are interviewed. The purpose of these interviews is to identify public issues and concerns, gauge the public's reaction to the project, and identify ways that people receive information and ways that people want to be involved. Areas of misinformation and the extent of misunderstanding about the project can be identified. Likelihood of support from politicians can be assessed, and supporters as well as opponents can be identified.

Third-party identification begins with developing a list of readily identifiable leaders within the community. Obvious examples include appropriate elected officials, Chambers of Commerce, active environmental group presidents, homeowner associations near potential sites, and waste generators. A trained interviewer then conducts a series of interviews with these individuals, asking questions that will meet the purposes described above. Each individual should be asked to identify other people in the community that they think would be most influential in the decision process. After several interviews have been conducted, the same names will begin to be suggested over and over again. Beyond a certain point this list will not change. If possible, all individuals named during interviews should also be interviewed. Sociologists call this the snowball approach. Depending on the size of the study area, 15 to 20 interviews are usually all that are necessary. These can be done in two or three days.

Experience suggests that the same publics are not necessarily involved in each stage of decision making. Some stages require reveiw by the broadest public possible, while other stages require continuity and an understanding of the technical data base. To decide which publics are most likely to be interested at a specific stage of decision making, consider the following questions:

1. Which publics are capable of providing the decision maker with information needed at this stage? If the information concerns general values and reactions based on those values, aim for the broadest range of publics. If the information is relatively specific, aim for a leadership group.
2. Which publics will be able to understand the information provided at this stage? Do not underestimate the ability of the public to absorb technical information, especially if it is presented clearly and in layman's terms.
3. How much time will be required for participation? Usually, only the leadership publics are able to make an extensive time commitment.
4. How much continuity is required? If participation at this stage requires some form of continuing commitment, such as serving on an advisory committee, participants are usually limited to leadership publics.
5. Whose participation is required for visibility or political acceptability? The public involvement program should have sufficient representation for the effort to be perceived as credible.

IV. HOW DOES THE PUBLIC VIEW ENVIRONMENTAL RISK?

Scientists and policy makers are becoming confused and frustrated by public reactions to environmental risk. Tempers flare at a public meeting concerning a risk that might cause less than one in a million chance of an increase in death by cancer. Yet, people smoke during the meeting break and do not wear seat belts when they drive home. These latter risks are far greater than those discussed at the public meeting. But if scientists point out this contradiction between perception and actual behavior, people become even angrier.

Frequently, experts or agencies dismiss such community reactions as illogical and emotional and conclude that the public is unable to understand the scientific aspects of risk. They may believe that communities are in no position to make decisions about how to deal with the risks that confront them and likewise are in no position to make decisions about how a new incinerator will affect them. But when agencies make decisions that affect communities without involving them first, even angrier responses often result.

Agencies and scientists must begin by recognizing that communities are quite capable of understanding the scientific aspects of risk. Too often, government assumes that because communities do not agree with an agency action, they do not understand it. To reduce the level of hostility between agencies, industry, and the public, those who work within agencies and industry need to understand better how communities perceive risk.

What is the difference between public and expert perceptions of risk? Experts look at risk by measuring how often an event happens and multiplying that number by how serious the consequences of the event are. This measurement is often expressed as "expected annual mortality rates". But this certainly is not how the public looks at risk. Keep in mind that the public can be very smart. People can understand data when they want to. People who play the horses or the tables in Las Vegas understand probability and odds. We are dealing with a motivational problem, not an ignorance problem.

Risk can be defined as hazard plus outrage. The "hazard" will be defined by the experts, while the "outrage" will be defined by the public. What does risk mean to the public? Some of the factors of outrage and risk are described below.[5]

Voluntary vs. involuntary — Research has shown a voluntary risk to be more acceptable to people by three orders of magnitude than a risk to which they are exposed involuntarily. The risk of getting cancer from smoking for 20 years may be statistically greater than the risk from drinking well water contaminated with low levels of trichlorethylene for 20 years. The difference is that a person chooses to smoke in the first instance but may not even know about the contamination in the second.

Natural vs. man-made — Natural risks are less acceptable than voluntary risks but more acceptable than risks created by people. Naturally occurring radon in a person's basement is more acceptable than a radioactive waste disposal facility a mile away from that same person's house.

Not memorable vs. memorable — If people have experience with a specific risk, their reaction will be stronger than if they don't. Flooding is an example. If people suffer severe damage in a flood, they will be less inclined to stay in their houses when the next flood comes along. Symbols can be important in making a risk memorable, too. People may automatically relate 55-gal drums to toxic waste or the skull and crossbones symbol to poison.

Not especially feared vs. very feared — One of the most feared diseases in this country today is cancer. People will be a lot more concerned about exposing their children to the possibility of cancer from hazardous waste incinerator emissions than about exposing them to colds from other children in the classroom.

Diffused in time and space vs. focused in time and space — Many people are more afraid of flying in airplanes than of riding in cars, even though hazard data indicate that cars are more risky. But people in airplanes are clustered in time and space. When an airplane crashes, many people are killed at once. For illustration purposes assume that 5000 people in a given state died from smoking in 1989. This risk is acceptable because those people died one at a time privately. If they all died on the same day and in the same city, smoking would probably be outlawed immediately. For another example, think about which risk is more acceptable: (1) one chance in ten of wiping out a community of 500 people within the next 10 years or (2) having 50 people die each year across the country in household accidents. Most people will accept only the second situation, yet both carry the same risk of killing 50 people per year.

Controlled by the individual vs. controlled by society — The American public consider themselves better than average drivers. They feel in control when they are driving. How many people feel comfortable holding a nail while someone else hammers? The person with the hammer holds the power. A sense of powerlessness can lead to public outrage or political action.

Fair vs. unfair — If the risks and benefits are not distributed evenly, people may perceive the risks as unfair even if the hazard is quite low. For example, when a hazardous waste incinerator is located in a community, the risk is often perceived to be directly proportional to the percentage of waste coming from out of town.

Trusted sources vs. nontrusted sources — Do people trust the entity that is urging them to accept risk? What is the public image of the industry in the community? A project proponent can also suffer from how much or how little trust people have in the industry as a whole. In general, people do not trust government or industry.

Generally, people pay attention when they are told something is a big risk but not when they are told not to worry. People tend to be more trusting when they are told something is dangerous than when they are told something is safe. This is because the public has a greater investment in the dangerous outcome.

In the issue of risk from hazardous waste incinerators, agencies and industry are dealing with perceptions. Perceptions are not necessarily grounded in fact but in people's value systems. What may sound like an irrational view to the scientist is not to the public. It is simply a reflection of those individuals' values and beliefs. These values and beliefs are not easily changed.

The scientist may say, ''If only they would listen to reason; if only they would understand the facts.'' So an expert is brought in to offer an explanation. Does this work? Usually not. Why should the public believe industry's expert any more than they should believe industry or government. Besides, the public will bring in their own expert — someone they trust, perhaps their neighbor who is a doctor, or perhaps someone they have spoken to at the U.S. Environmental Protection Agency. The public will usually get the last word in a battle of experts.

"The greater the number and seriousness of these ["outrage"] factors, the greater the likelihood of public concern about the risk, regardless of the data."[4] The risks that provoke the greatest public concern may not be the same ones that scientists have identified as most significant. When agency officials dismiss the public's concern as misguided or irrational, the result is anger, distrust, and still greater concern. This does not suggest that the public disregards the scientific data and makes decisions solely on Sandman's "outrage" factors. It suggests that outrage also matters and cannot be ignored. If one fails to attend to the outrage factors and people's concerns from the beginning, one will often be forced to attend to them later, after having angered the public — a far more difficult situation.

V. TO WHAT EXTENT IS THE PUBLIC INVOLVED?

In June 1987, the Legislative Commission on Toxic Substances and Hazardous Wastes issued a report on an extensive national survey on hazardous waste facility siting.[6] The survey focused only on commercial hazardous waste treatment and/or disposal facilities and included incineration. It was designed to collect data on siting processes of individual states and their experience with siting applications since 1980 or since the establishment of their siting process.

The survey was divided into groups of questions focusing on five distinct areas of information:

1. Siting statutes and regulations
2. Existing commercial facilities
3. Siting process
4. New siting applications
5. Public participation/education

The fifth section of the survey questioned states on opportunities and funding provided for statewide public participation and education with regard to the siting issue. Unfortunately, lack of response precluded useful interpretation in this area.

The third section, however, examined two fundamental aspects of state siting processes that are frequently at the heart of controversy in the siting issue: the degree of state initiative in siting facilities and the approach of the state to providing for local community input into siting decisions. In addition to being central to the siting issue, these two factors were thought to be easily evaluated and capable of being affected through legislation. The intent of this section was to reveal any correlations between the use and manner of these easily documented factors and success in siting.

To do so, this section of the survey questioned the states' use of various, commonly used siting process elements such as siting boards for reviewing siting applications, negotiation or mediation, economic incentives, and technical assistance for host communities, use of an intrastate regional approach to siting, and state override authority of local zoning laws. Respondents were asked to indicate which parties can initiate siting, what opportunities are provided for public input to the siting process, who makes the final decision on a siting application, and what types of exemptions to the siting regulations are provided for the expansion of existing facilities.

The data summarized from survey responses show that since 1972, 36 states have enacted hazardous waste siting studies. All but five of those states have developed an actual siting process for hazardous waste facilities. In addition to enacting siting statutes, ten states have formulated state hazardous waste management plans.

The authority of a state government to influence siting of hazardous waste facilities within the boundaries of that state proved to be the most important factor in siting approval and issuance of RCRA permits. All seven states judged to be strong in state initiative have received siting applications, and three have issued permits.

Public participation was defined as all non-state-level involvement in the siting process and siting decision. Five elements were identified as indicators of public participation in a siting process:

1. Use of siting board with local representatives for siting application review
2. Technical assistance for host communities
3. Economic incentives for host communities
4. Use of negotiation or mediation with the host community
5. Local decision-making authority

Siting board review — Although survey results indicated that 21 states use siting boards, only 11 states include local host community representatives. Seven of these have received siting applications, and two have successfully sited commercial facilities. Both of these proposals received permits, but to date have not become operational.

Technical assistance to localities — A total of 11 states provide for technical assistance grants to host communities. These grants are generally for a fixed amount of dollars for hiring technical experts to help educate the community. Three of these states have successfully sited and permitted commercial facilities.

Economic incentives — Fourteen states reported having formally instituted some form of economic incentives for encouraging communities to host facilities. These incentives fall primarily into three categories. States such as Kentucky used a direct appropriation to the selected community. Other states, such as Massachusetts, required the developer to provide money to host communities for hiring consultants, expanding emergency response capabilities, or compensation to communities for lowered property values or increased highway maintenance. Several others reported allowing host communities to impose local taxes on the facility's gross receipts. Of the 14 states which provide economic incentives to the host community, 11 have received siting applications and Florida, New York, and North Carolina have successfully permitted facilities.

Negotiation or mediation — Opportunities for negotiation or mediation between facility developers and proposed host communities were reported by 19 states. Exact mechanisms varied. Of the 19 states, 3 have permitted facilities. Only Rhode Island successfully used a negotiation process. Texas and Wisconsin report some success.

Local decision making — In six states the local government hosting the facility may, under certain circumstances, make the final siting decision. Each state (Colorado, Florida, Illinois, Kentucky, Nebraska, and Tennessee) has a unique system for this decision-making process. Five of these six states have received siting applications. Three have successfully permitted operational commercial facilities.

In the analysis of the responses for the five public participation elements, each state was ranked by the number of these elements which occurred in each of their state programs. All elements were weighted equally. Four states — Connecticut, Kentucky, Massachusetts, and Minnesota — possessed four of the five elements. Five more states possessed three out of five, while seventeen states possessed none. A complete listing of all state rankings is contained in Table 1.

The report concludes that extensive public participation programs promote siting activity but, in most cases, have not led to siting approval and issuance of RCRA permits.[7]

TABLE 1[29]
State Ranking in Use of Public Participation in Siting Hazardous Waste Facilities

State	A	B	C	D	E	Score
Alabama						0/5
Alaska				*		1/5
Arizona						0/6
Arkansas						0/6
California						0/5
Colorado					*	1/5
Connecticut	*	*	*	*		4/5
Delaware						N/R
Florida		*			*	2/5
Georgia						0/5
Hawaii						0/5
Idaho	*					1/5
Illinois		*			*	2/5
Indiana	*		*	*		3/5
Iowa	*			*		2/5
Kansas						0/5
Kentucky	*		*	*	*	4/5
Louisiana				*		1/5
Maine	*	*	*			3/5
Maryland				*		1/5
Massachusetts	*	*	*	*		4/5
Michigan	*			*		2/5
Minnesota	*	*	*	*		4/5
Mississippi						0/5
Missouri						0/5
Montana						N/R
Nebraska					*	1/5
Nevada		*				1/5
New Hampshire		*	*	*		3/5
New Jersey		*	*			2/5
New Mexico						0/5
New York	*		*			2/5
North Carolina	*	*	*			3/5
North Dakota						0/5
Ohio	*					1/5
Oklahoma						0/5
Oregon				*		1/5
Pennsylvania				*		1/5
Rhode Island		*		*		2/5
South Carolina						0/5
South Dakota						0/5
Tennessee			*	*	*	3/5
Texas				*		1/5
Utah						0/5
Vermont						0/5
Virginia		*		*		2/5
Washington		*		*		2/5
West Virginia						0/5
Wisconsin			*	*		2/5
Wyoming						N/R
TOTAL	12	11	14	19	6	N/R

Note: Key: A = siting board review; B = technical assistance; C = economic incentives; D = negotiation/mediation; E = local decision making; N/R = no response.

These results indicate that the influence of public participation on siting success may be less dependent on the number of different opportunities offered by a siting process than on the actual manner in which it is incorporated into the process. A single effective element of public participation may be key. For example, a state such as Rhode Island, which was not judged to be particularly strong in public participation (two of five elements) but which did use negotiation and technical assistance, may be more successful in siting than other states whose processes provide for public input in less effective ways.

VI. TO WHAT EXTENT SHOULD THE PUBLIC BE INVOLVED?

The Legislative Commission on Toxic Substances and Hazardous Wastes reported that 81 siting applications for commercial hazardous waste treatment or disposal facilities had been received by 28 states since 1980. Of these, six have resulted in operational facilities (four treatment facilities, one landfill, and one incinerator).[8] The report discusses several reasons for this fact. In this writer's view the public participation elements examined in the survey did not include some of the most important factors for a successful public involvement program. The report did not indicate that primary stakeholders were identified, whether their views were solicited and addressed, or if they were involved in the siting process prior to the identification of final sites.

In this writer's experience, inability to secure a specific site that is acceptable to the public and appropriate from a technical and economic perspective has killed or delayed many projects. Yet, the selection of a site often is not accorded the same level of attention as the selection of a vendor or negotiation of the financial package. In fact, many times, availability of land or a willing seller is the only criterion considered outside of purely technical factors.

To be successful in siting a hazardous waste incinerator, a carefully considered, strategic plan must be developed. *This plan should include a thorough site selection process, a comprehensive public information and public involvement program, and sensitivity to political realities.* A good siting process will use criteria to screen, compare, and evaluate all potential sites. The public should participate in developing the criteria and in designating the relative importance of each. The public needs to feel some ownership in the planning process and the method by which a site is selected. In turn, they will be more likely to accept the final site.

A good example of the importance of developing this ownership can be shown in the following statement made at the final public hearing for an extremely controversial municipal landfill in the northwestern Maricopa County near Phoenix, AZ:

MR. ARTHUR: I think we've been here before. My name is Philip Arthur and my company is in the process of developing 380 mobile home parks within two miles of this landfill. And when this all came up I was very concerned. I went around and worked with the people at Wittmann. We looked at every site. I rode personally on horseback through every site, walked every site, drove every site. There are already people at Wittmann. They [County] were considering some sites within a quarter mile of homes that are actually in, not homes that may be built five or ten years down the road. We determined if this area had to have a landfill, and it definitely needs one, it had to be put in a location that was the least detrimental to the most people who actually have investments and have built their homes. This [site] 106 is the closest of the seven sites to our project, but it still is the least detrimental to the most people. Therefore, we are 100 percent in favor of it . . . So looking from where we stood as far as having to help this [Citizen Advisory] committee, and they have done a magnificent job. It's an awful hard job to pick anything. You just cannot satisfy 100 percent of the people on something like that. But the site that was recommended met every criteria. It went right down the line and turned out to be number one or number two, however you figured your weights. We cannot find any reason [to oppose it] even though it's closest to me. It's practically in my back yard. I cannot see how I can say I don't want it two miles from me, I want it a quarter mile from Mr. Whitaker's home. 106 is definitely the best site under consideration. Thank you.[9]

A hazardous waste incinerator will use somewhat different criteria than a municipal landfill, but many of the public's concerns will be the same: health effects, impact on property values, air quality, noise, visual impacts, heavy truck traffic, and taking someone else's waste. A siting process that includes early and extensive public involvement has a higher probability of adequately addressing public concerns and giving the public a sense of ownership in the final decision. Government and industry frequently lack credibility with the public. More and more people do not trust government at any level to minimize the risks from hazardous waste facilities by enforcing adequate environmental and health standards.[10] Industry's continual assurances that risks will be kept very low are often discounted due to past instances of their carelessness. The public has a long, though selective, memory when it comes to examples of carelessness. Love Canal, Three Mile Island, Times Beach, Mo, to name a few, will often be cited by a distrustful public. Regardless of a company's individual safety record or a state's record on environmental controls, members of the public will often expect them to operate in the same irresponsible manner as some of their worst examples. Many people oppose a new waste incinerator simply because they fail to distinguish these modern facilities from their poorly managed predecessors. Successful development of state and national siting strategies rests on the ability of government and industry to make these distinctions between old and new facilities clear to the public.

Most attempts to site hazardous waste facilities within the past few years have met with vehement opposition in many states. Opposition can be caused or exacerbated by many factors, some of which are illustrated in the following examples provided by Morell and Magorian.[10]

A lack of early and clear communication with the public can cause fear, hostility, and opposition even before a project is actually defined. For example, in July 1980, the Delaware River Basin Commission and the New Jersey Department of Environmental Protection issued draft selection criteria for hazardous waste facility sites in general. This attempt to develop a planning tool prompted 1000 residents of East Windsor Township to meet in the school gymnasium, seeking information from local officials and expressing their opposition to what they assumed was to be a chemical dump.[11]

Communication with the public needs to begin well before siting decisions are made. This is important not only to increase the chance of successful facility siting, but also to avoid wasting industry money and efforts. In New Jersey, for example, residents of Alloway Township discovered that Envirosafe Services, a subsidiary of International Utility Conversion Systems, Inc., planned to build a 400-acre landfill on land known as "Fox Hunt Farms", acquired by Shell Corporation. Apparently, at the time of the land acquisition, many people thought the "developer" planned to build housing in the area. After investing $500,000 in the project, the waste management firm abandoned its plan in the face of rising public opposition.[12]

Elsewhere in New England, residents of Westford, MA successfully resisted an Industrial Tank (IT) Corporation plan for a hazardous waste treatment plant that would have used new and innovative technology. This facility was designed to incinerate, neutralize, and detoxify as much as 500,000 ton of waste per year. Despite new state legislation designed to facilitate successful siting, residents in this community of 14,000, following an emotional town meeting attended by 3000 citizens, eventually were able to persuade the owner of a local granite quarry not to sell his land to the waste disposal company.[13]

The Westford opposition was reportedly based on two objections to the IT proposal: (1) residents felt the siting process had moved so quickly that it was "difficult for amateurs to keep up" with the technical issues involved and (2) they felt it unfair that one small community should have to host a large facility that would process wastes originating from the entire New England region.[14]

Morell and Magorian also discuss two siting cases from New Jersey that illustrate several of the key principles in local opposition to siting new hazardous waste management facilities.[15] While the siting process was essentially the same for both proposals, an on-site industrial landfill was approved, but an off-site regional landfill was rejected due to intense citizen opposition. The level and form of public involvement was a key difference in the two examples.

In 1978 the Monsanto Chemical Company successfully sited a new hazardous waste landfill at its Delaware River Plant near Bridgeport, NJ. The landfill is located on grounds of the plant and is used to dispose of chemical sludges generated at the plant. In contrast, in 1978, the SCA Corporation subsidiary Earthline attempted to establish a commercial hazardous waste landfill in Bordentown, NJ, also on the Delaware River, about 35 mi north of Bridgeport. Citizen opposition rose quickly and led local officials and members of the state legislature to oppose the facility. In 1979, the Earthline application was denied by the New Jersey Solid Waste Administration (SWA), the permit-issuing authority for that state.

Morell and Magorian's comparison of the Bridgeport and Bordentown siting processes makes clear the differences that played a large part in the acceptance of one facility and the defeat of the other. Two primary reasons are responsible for the outcome: *credibility of the parties involved and legitimacy of the siting process.*

In Bridgeport, Monsanto was successful in gaining the trust of the community by providing information and clarification of its application and by being willing to negotiate engineering improvements in its landfill siting proposal. However, in Bordentown, SCA inspired little trust in the community because of its past actions at a site in another state and the actions it took during the actual siting process in Bordentown. SCA adopted a more confrontational approach to the community concern. The credibility of the state decision-making agency was also an issue in Bordentown. Local citizens felt that their fate was being determined by people who were predisposed to approve the landfill request and who, due to a lack of clear criteria and agency resources, did not appear able to provide satisfactory evaluation of the drawbacks of the landfill.

Bridgeport residents did not object to the state-dominated decision-making process, whereas Bordentown critics would have included explicit criteria for local input into judging a landfill application. While the opponents of the SCA proposal complained that their interests would not be satisfactorily taken into account by the state, it is ironic that ultimately their objections were heeded. SWA rejected the application. It is clear from this example and others that, although a siting process may technically establish the power to preempt local opposition to a facility plan, the political impact of this local opposition at the state level should not be underestimated.

Examples of unsuccessful siting of hazardous waste facilities are not confined to the East Coast nor to facilities that have already received appropriate permits. Union Carbide Corporation contracted with the University of North Dakota Energy Research Center at Grand Forks to demonstrate an innovative thermal destruction technology by destroying polychlorinated biphenyls (PCBs) in the Center's existing pilot plant lignite gasifier.[16] After Union Carbide and the Energy Research Center agreed on the scope of the experimental program, the necessary state and federal environmental permits were applied for and received. Shortly after these permits were granted in July 1986, a group of local citizens came out against the PCB program, and the project came under intense public scrutiny.

In July 1986, the Energy Research Center began a series of informational press releases and public meetings to address the concerns expressed by local citizens and the news media. By November 1986, the opposing citizens and their attorney were threatening to file a lawsuit to terminate the project. In the face of this public pressure, the state Department of Health suspended its previously granted research permit, and the President of the University stopped the project pending a public hearing.

The public hearing was held in February 1987. It lasted 3 days and was carried live by the local radio station. During the hearing, the scientific merit of the project was not debated by those speaking in opposition. Most of the testimony opposing the permit appealed to emotional issues, stressing the alleged risk of operating this project at an in-town campus location.[17]

On April 13, 1987, the state Department of Health revoked its permit for the project, stating that the residential location of the Energy Research Center should not be used for the PCB project or for any future research involving hazardous materials.

From this account, it appears that the public had no opportunity to learn about the proposal until the permits had been received and, seemingly, a decision had been made. They were given no alternative but to oppose the proposed facility. It was naive to expect the public to debate the technical benefits of the technology when they were never consulted as to how they might feel about such a facility, the technology, or hazardous waste in general. A good public involvement program deals first with the public's agenda and needs. Until this happens, the public will not deal with the proponent's agenda. Even with a thorough public involvement program and information, there is no guarantee that this facility would have been constructed. However, if opportunities to mitigate some of the public's concerns and negotiate with them had been available, a stronger case could have been made to the university president and the Department of Health that public concerns were being addressed.

VII. HOW CAN PUBLIC ACCEPTANCE BE ENCOURAGED?

Public acceptance of a hazardous waste incinerator can be enhanced if potential project opponents are treated as equals in the siting process. Before developing a public involvement plan it is important to understand the public's values and ways to communicate with them about environmental risk.

A. UNDERSTANDING THE PUBLIC'S VALUES

Understanding the importance of values held by potentially affected publics and the ways in which they perceive environmental risks is a first step toward facilitating public acceptance of a hazardous waste incinerator. Values are the internal standards by which people judge events or behavior to be good or bad, right or wrong, moral or immoral, fair or unfair, just or unjust. They are normative standards by which people judge the way things "ought" to be. People derive their values from upbringing and experiences. Typically, values are implied in people's speech or behavior, rather than explicitly stated. While values play a strong force in shaping people's lives, they are not easily stated. They sound vaguely like "motherhood" and "apple pie" and are difficult to defend. The writers of the Declaration of Independence fell back on the phrase, "we hold these truths to be self-evident" to justify values as fundamental as "life, liberty, and the pursuit of happiness".

The public's opposition or support of a project stems from their basic value and belief system. The more fundamental the value is, the more difficult and resistant it is to change. Examples of fundamental values that are pertinent to any discussion of hazardous waste facilities include individual freedom, consent of the governed, and protection of health, welfare, and property.

It is the nature of values that they are highly personal and, while they have some degree of internal consistency, one person will never have entirely the same perception of reality as another person. A person's values give meaning to events or behavior as good or bad, fair or unfair. Based on this meaning, values are rarely explicitly expressed, and people may find themselves in policy arguments in which the other person appears overly emotional and irrational. As a result, ideas are frequently dismissed. This is a good sign that fundamental value differences exist. This perspective has an important impact on how public comment is received by agencies. The bulk of comment from the general public will usually be general and emotional. It will be an important source of information about values people want applied in a policy decision.

Value judgments are also part of the scientist's or the politician's decision framework. Their value structure cannot help but influence their decisions. Burns agrees when he states:

> Scientists, like all citizens, must play an active role in the discussion of competing values. Their special expertise will inevitably and rightly give them a persuasive voice when issues are discussed in our assemblies and on our streets. But the choice must ultimately be made in a politically responsible fashion.[18]

Once the proponent and the governmental agencies have a good understanding of why the public feels and reacts as they do, a public involvement and public education program can begin.

B. COMMUNICATING TO THE PUBLIC ABOUT ENVIRONMENTAL RISK

In years past when government officials spoke in a public meeting, they were usually believed. People left feeling relieved or alarmed depending on the message, but they believed that the government official was competent and credible. Now the presumption frequently is that government bureaucrats are incompetent, that people have to find the truth themselves, and that the government has a hidden agenda or may even be an obstacle to the truth.

Scientists, by training, tend to focus more on data than on feelings. On the other hand, citizens focus on how they believe a particular situation might affect their lives. Unfortunately, neither side listens very well to the other. Government representatives often feel frustrated with communities that do not seem to listen and that often appear to be frightened of the "wrong" risks. In response, agency decision makers and staff may choose to ignore the public, even though it is likely that they may later face increased hostility. Or they can choose to interact more effectively with the public.

An important challenge for the government or industry representative or the scientist is to place environmental risks in perspective for the public. Effective risk communication can help decision makers to:

1. Understand public perception and more easily predict community response to proposed actions
2. Increase the effectiveness of risk-management decisions by involving concerned publics
3. Improve dialogue and reduce unwarranted tension between communities and agencies
4. Explain risks more effectively

No matter how well it is done, risk communication cannot and should not replace effective risk management or aggressive environmental regulation. The field of risk communication, which explores perception and communication of a variety of risks, is growing rapidly. However, the research literature lags significantly behind the wisdom of the best practitioners, who have been experimenting with a variety of approaches to communicating risk. Most importantly, this literature does not immediately translate to practice. Those who are on the firing line and have to answer the mother's very "simple" question, "Can you guarantee me that my child will not get cancer from this incinerator?" will appreciate risk communication techniques that have a chance at success.

Earlier in this chapter the variables ("outrage" factors) that underlie community perception of risk were described. The greater the number and seriousness of these factors, the greater the likelihood of public concern about risk, regardless of the data. The risks that elicit the greatest public concern may not be the same ones that scientists have identified as most significant. When officials dismiss the public's concern as misguided, the result is controversy, anger, distrust, and greater concern.

It is difficult to explain risks in ways laypeople can understand. But the public's ability to understand the science should not be underestimated. Talking about one in a million increased cancer deaths or comparing 1 ppb to 1 s of time in 32 years or to one sheet in a roll of toilet paper stretching from New York to England leads to two errors: (1) it assumes that low concentration means low risk and (2) it trivializes the risk. Scientists struggle to convey complex technical information, believing that if they could find a way to explain the data more clearly, citizens would understand the risks the way that scientists do and therefore be less concerned.

One must not mistake the need to explain complex data more clearly as a way to placate the public. The scientist must develop sufficient trust and credibility, place a priority on understanding the public's concerns, and involve them in risk decisions that affect their lives. In fact, in this writer's experience, the most successful communications with the public relied more on mutual respect and understanding and improved interaction with the public than on innovative ways to explain the data.

Nonetheless, it is imperative for agency and industry representatives to try to explain risk information as completely and clearly as possible. Some concrete examples may be helpful to the reader. Give the citizens a mental picture of the parts per billion or tons per day. By using a verbal comparison or graphic representation, communicators can help convey a message that is meaningful and memorable. For example, saying that "Pike County produces 125,000 ton of refuse per day, or enough to fill 50 football fields 14 ft deep" will have more effect than simply stating the statistic.

Consider a common question in refuse-to-energy plants, "How much dioxin will come out of the stack?" It may be appropriate to say, "The incinerator will be designed to meet all applicable Resource Conservation and Recovery Act (RCRA) and Toxic Substance Control Act (TSCA) requirements," but that does not explain the possible magnitude of dioxin emissions. Equally unacceptable is the answer, "The design we are currently considering emits 0.11 ng of dioxin or less per cubic meter. We expect to achieve even cleaner emissions with further controls." This statement uses the word "nanogram", which is meaningless to the public, yet it still does not tell them how much dioxin will fall in their backyard. Martha Bean suggests a more satisfactory answer might be:

> We believe this plant will not emit more than 0.11 nanograms of dioxin per cubic meter of exhaust gas. We expect that the incinerator will not emit more than X cubic meters of exhaust gas per day. That works out to about X ounces of dioxin per day falling into the backyards of homes within 500 feet from the stack.[19]

The citizen's question was, "How much?", but an underlying question was, "Will it hurt me?" By answering the question "How much?" with accuracy, the communicator may unwittingly raise the concern of the listener. A possible addendum to the above answer may be:

> You may be wondering what X ounces of dioxin means to your health if you live within 500 ft of the stack. If you breathe the air outside your home 24 h a day for 70 years, you would have slightly less than one chance in a million of contracting cancer from the dioxin that you breathe from the incinerator.

The effective communicator must be an insightful listener to identify the real or underlying questions. Effective communication about the magnitude of risk requires giving the listener information about probability. As an example, the professional is comfortable with the term "excess lifetime cancer risk". But to the layperson, it conveys something unknown and terrible.

Various agencies now require that cancer-causing agents be controlled to the 10^{-6} level at hazardous waste sites. This phrase is unintelligible to the average person even if it is explained by saying, "The excess lifetime risk will be 10^{-6} at this site after cleanup. That means the average incidence of cancer is increased by 1/1 million people exposed over a lifetime."

Martha Bean gives two other examples of how this information may be communicated more clearly:

a. At this site we would expect that among 1 million people drinking 2 liters (about 8 glasses) of water per day for their entire lives, one person may get cancer caused by the contaminant in the water. This person may not die of it.

b. Let me try to explain what we call a 10^{-6} risk of cancer in personal terms. First, let's assume that you know 200 people in this city. Of these 200 people, 50 will probably contract cancer, regardless of where they live or what water they drink. If all 200 people you know drink the water from your well all of their lives, there is a 1 in 5000 chance that one more of the 200 people you know will contract cancer.[19]

Avoid comparisons that seem to minimize or trivialize the risk. If comparisons are used, compare risks that seem similar to the listener. The following guidance from the Chemical Manufacturers Association workshop on risk comparisons may be helpful:

1. Use comparisons of the same risk at two different times — for example, "The risk of X is 40% less than it was before the latest discharge permit" or "We plan to institute new standards next year that will reduce the risk by 10%."

2. Compare with a standard. "The emissions are 10% of what is permitted under the old EPA standard, and slightly under the level established by the new EPA standard."

3. Compare different estimates of the same risk. For example, compare the most cautious "worst case" estimate with the most likely estimate, or compare the agency estimate using one methodology with its estimate using another methodology, or compare the agency estimate with that of industry or an environmental group.

4. Compare with similar data found elsewhere. For example, compare the drinking water data of one community with levels found in other communities in the state.[20]

Thomas Burke of the New Jersey Department of Health says that he tries to put specific risks into context with "the baseline kind of risk . . . the big picture". For example, if you know that the benzene in your air is only slightly higher than the level in the national forests, says Burke, "That's something that's helpful. You're not going to understand whether it's going to cause leukemia in you, but certainly your risks are no higher than anyone else's in the country."[21]

Dealing with the issue of uncertainty is another challenge for the risk communicator. There are at least four types of uncertainty that scientists typically must interact with the public about: (1) the uncertainty of science in general, (2) the inexactness of the risk assessment process, (3) the incompleteness of the information the agency has gathered so far, and (4) differences of opinion as to the implications of the information and the optimum risk management option.

Acknowledge uncertainty by learning to say "I don't know." This may be difficult to do, but it is a better strategy than claiming to know more than is known. It may even enhance the communicator's credibility because people will recognize his or her honesty and forthrightness. However, citizens need to be given enough background so they do not think something is amiss when the agency says it does not know. People need to understand that uncertainty is part of the process. Be specific about what is being done to find the answers and how research is being conducted. It is important to say "I don't know" so that it does not sould like "I don't care."

One of the predictable things that will happen when one explains a risk number is that someone in the audience will point out that it is not zero, following up with the demand that it should be. The agency or industry must explain that zero risk does not exist. In our daily lives people take quite substantial, yet avoidable risks, so it is not possible to demand zero risk for a chemical company. Covello et al. suggest several possible ways to respond to the demand for zero risk:[22]

1. The demand for zero risk may be reasonable. The plant may be able to (in effect) eliminate virtually all the risk through some change in procedure. If so, consider the costs and benefits of doing so.
2. The demand for zero risk may be an exaggerated way of making the point that the risk is too high. The response to this demand may be complicated by the fact that although further risk reduction is entirely feasible, the remaining risk is insignificant and such action makes no economic sense. However, do not point out the economics. The public will generally reject the argument that reduced risk to the community does not justify the added costs to the company. If this happens, the discussion has moved from the inherent correctness of a fact to a discussion of how choices based on that fact compare to other choices. In other words, science is left behind, and the discussion moves into values.
3. The demand for zero risk may be sincere but ill informed. If so, respond gently with some fundamental risk education. Explain that zero risk is nonexistent. Point out that all of life's activities carry some risk, which are usually ignored if the risk is small enough and if the activity is beneficial enough. Be sure to agree that the risk should be made as low as possible, and discuss what is being done to achieve the lowest possible level.
4. The demand for zero risk may be politically motivated and designed to attract the interest of politicians and the media. The agency or industry is unlikely to make much headway in public meetings. Try to arrange a private meeting where frank discussion and negotiation might be possible.
5. Finally, the demand for zero risk may be a reflection of emotional distress stemming from outrage, anger, and distrust. Risk may not be the central issue. Communicators must address themselves to the underlying antagonism and begin the slow, hard work of building bridges.

Much more could be said about risk communication, but, in summary, communicators should empathize with their listeners. When speaking about risk, communicators need to be sensitive to the needs of the public, while, at the same time, know how to present information in a clear and honest way. Risk communicators need to remember that they have a unique perspective on risk because of their occupation and training. By anticipating the community's concerns, the communicator can remain calm and not take attacks personally.

C. INCLUDE THE PUBLIC IN THE SITING PROCESS

Inability to secure a specific site that is acceptable to the public, technically and economically feasible, and able to be permitted is the heart of the issue to which this chapter is dedicated. The bottom-line objective is to find a site that meets the above criteria. Yet many sites are selected with little or no public involvement or consideration of alternatives and without a systematic and defensible site selection process.

Before discussing how to involve the public in a siting process, it is necessary to understand what is meant by a siting process. Usually a site selection program is divided into three major tasks: (1) site identification, (2) site evaluation, and (3) site selection. The basic intent is to conduct a systematic study that is based on accurate data and uses a defensible decision process. A public involvement plan is prepared at the beginning of the study and is implemented throughout as an integral part of the siting process.

1. Site Identification

The goal of the initial site identification stage is to identify a number of possible sites that have a wide range of characteristics in terms of ownership, parcel size, distance from waste generators, proximity to arterials or highways, and so forth. Initial site screening criteria should be developed and could include such topics as avoidance of geologic faults, 100-year floodplains, areas of shallow groundwater, critical habitat for threatened and endangered species, other environmentally sensitive areas, or proximity to existing residences. Sites with similar characteristics may be grouped and screened down to one or two best sites of each group. Variety is important to allow for comparisons among different types of sites and to prevent selection of candidate sites that may all have the same problem. A problem that may not be discovered until later in the study. For example, a local government in the southwestern U.S. had identified five potential sites for a sanitary landfill, and all were on state-owned land. The State Land Department decided that they would not allow their land to be developed for a waste disposal site, so the county had to begin their siting process again.

Next, existing data should be reviewed to determine additional information needed to identify candidate sites. Appropriate information should be collected and mapped by criterion on a uniform base map, so that Mylar^R overlays showing data for each criterion may be used. These data should be interpreted as offering either a siting "opportunity" or "constraint". Site identification criteria will further reduce the areas under consideration by focusing on the more opportune locations, for example, near highways, adjacent to heavy industrial areas, or near populations served. Potential candidate sites of the desired size are delineated, and their advantages and disadvantages are compiled.

2. Site Evaluation

More specific site evaluation criteria should be prepared based on the general characteristics of sites and areas still under consideration, major issues of concern to the public, and availability of site information. These draft criteria should be presented to the public before being finalized. They are likely to include timeliness and cost of land acquisition, potential for groundwater protection, impact to existing land uses and future land uses, visual impacts, noise, air quality concerns, and traffic considerations. Detailed technical studies should be keyed to the criteria identified as appropriate for evaluation of the sites. Concept-level designs for each site should be prepared concurrently with evaluation data studies in sufficient detail to allow cost comparisons. Items such as access road availability or needed construction and availability of utilities should be included. Other studies regarding biological and cultural resource concerns may need to be conducted to address agency concerns.

Once enough information is available to indicate the technical suitability of each site for each criterion, a numerical rating may be assigned by the technical staff. This suitability rating may then be modified by the importance the particular criterion has in the overall evaluation. A citizen advisory committee (if one has been established) or a team of technical people are asked to rate numerically the relative importance of each criterion in relationship to all other criteria. These ratings are combined with the technical suitability ratings to result in overall desirability scores. Interpretation of these scores, in concert with other nonquantifiable considerations and the level of public acceptability for each site, will serve as the basis for site seleciton.

The objective of the criteria weighting process is for an interdisciplinary team of technical people or a citizen advisory committee, as representatives of various constituencies and professional disciplines, to agree to the extent possible on the relative importance of various criteria for the siting of the incinerator. The emphasis should be on evaluating the importance of various siting considerations rather than simply to rank sites according to personal or professional preferences.

This multi-attribute utility technique is based on the concept of utility and has been used successfully by the writer to reach consensus on the best site or alternative in a number of controversial projects. It is designed to focus on criteria rather than alternative sites and is based on the assumption that if stakeholders can agree on the importance of various criteria, they may find it easier to agree on actual sites. Another benefit of this approach is that areas of *agreement* will become evident. It is usually very easy to know which things people disagree about because these factors receive the most attention. In an exercise conducted and reported by the writer, out of 14 criteria weighted by 80 stakeholders, most agreed on the relative importance of nine of them.[23] There was limited disagreement on the importance of three criteria and hearty disagreement on the remaining two. If the study proponents had realized the extent of agreement, they may have concentrated more effort on developing alternatives that addressed the primary areas of disagreement.

By focusing on the importance of the criteria and showing how each site performs on the criteria considered most important by each stakeholder, compromise sites can be identified. In this way the public's values can be incorporated into the decision. People often express surprise that a particular site performs so well because he or she really did not appreciate the technical aspects of the site.

3. Site Selection

The proponents should include their own decision criteria in determining a preferred site. These criteria might include overall cost, ease of acquisition, public sentiment, environmental impacts, and likelihood of receiving permits. Throughout the siting process, the public should be reminded that many factors will be considered in the final decision, only one of which is public acceptability.

A thorough siting process may be viewed by the proponent as too time consuming. In this writer's experience, time spent up front systematically identifying a publicly acceptable site is time better spent than at the other end in litigation. A good siting process should pave the way for a favorable environmental impact study and a smooth permitting process — as well as help win the public's trust and confidence.

Another way to reduce public opposition that has met with some success is the use of compensation and incentives. Site developers provide benefits to the communities or local residents, hoping to gain public acceptance. The benefits may be money, goods, or services. Examples include donating school buses, a community center, or a new road. Likewise, developers can drop those parts of the proposed siting plan that have strong community opposition, such as certain truck routes or acceptance of politically sensitive wastes like PCBs.

In the state of Wisconsin, the law requires anyone who plans to expand or construct a landfill, hazardous waste treatment, storage, or disposal facility to negotiate with the government of the host community. Under the law, anything can be negotiated, but the need for the facility is not a topic for discussion, as this is part of the state licensing review.

More than 90 projects have entered the Wisconsin siting process since 1982. As of mid-1987, all but two of the 21 siting agreements contained compensation provisions. Examples of compensation include reimbursement for the expenses incurred in negotiating the agreement, direct payments to local government, payments for a site impact monitoring committee, road maintenance fees, and reimbursement for lost revenues caused by property tax exemptions. Compensation to private property owners takes three forms in Wisconsin agreements: up-front cash payments, protection against future loss of property value, and protection against future property damages.[24]

If the proposed site would add significantly to the local community in terms of tax base, employment, or facilitating the operation of local industries, these facts should be brought to the public's attention. Every effort should be made to show how the site will benefit the area. Sometimes too much effort is placed on showing how the site will not harm the area, yet the public does not believe the information.

VIII. A STEP-BY-STEP STRATEGY TO INVOLVE THE PUBLIC

Locating "locally unwanted land uses" (LULUs) such as hazardous waste incinerators, landfills, power plants, or freeways, are historically controversial because of potential conflicts with diverse public values. Typically, the greatest concern is voiced by residents, property owners, and developers in the vicinity of potential sites who perceive that such a LULU will have adverse effects on existing land uses, public health, future growth patterns, or their quality of life. The public's confidence cannot be won if they are not informed about the need for the overall management of hazardous waste. However, keeping citizens informed about the process and solutions will not be enough to promote approval. The public must be made an intimate part of the project development.

People may recognize the need for hazardous waste management but are frequently not willing to bear the burden of its disposal, particularly for "industry's garbage". Other public interests may be concerned about increased user costs if remote locations are considered to minimize local impacts, while still others support alternative waste processing technologies. Because of high public sensitivity to hazardous waste and the potential for conflicting public interests, a responsive public information and involvement program is critical to a successful siting process.

Successful siting requires good sound technical studies combined with the ability to obtain enough public support for the process to build consensus around an acceptable site. A public involvement program should be guided by the following objectives:

1.   To establish and maintain the credibility of the overall siting process, including the technical studies
2.   To inform and educate the public as to the need for the incinerator, possible impacts on the environment, and benefits to the public
3.   To identify accurately and consider the values and concerns of the public, agencies, and political entities
4.   To integrate public views and agency policy with technical data into the overall decision-making process

Involving the public includes five general tasks: information dissemination, information gathering, citizen and agency involvement, conflict resolution, and documentation. A comprehensive and well-conceived public involvement program should be integrated into the siting study process. It should be designed to allow for a continual flow of information to and from the public. Early reconnaissance activities should be included to ascertain the public's issues and levels of concern. Refinement of a public involvement plan that is tailored to the project should follow. A thorough public involvement program should provide all interested agencies, organizations, and individuals with opportunities to respond to the project. Staff should remain flexible to adjust the program as necessary to accommodate public concerns. The following steps are suggested as a generic public involvement plan that is integrated with the technical study. More information about specific techniques follows the plan.

IX. PHASE I — CANDIDATE SITE IDENTIFICATION

A. Public involvement tasks

1. Establish point of contact
2. Issue news release
3. Begin newspaper clipping service
4. Establish computerized mailing
5. Prepare fact sheet
6. Identify key constituent groups
7. Conduct opinion leader interviews
8. Analyze issues and prepare public involvement plan
9. Determine willingness of groups to participate
10. Establish citizen advisory committee (CAC)
11. Establish information repository
12. Develop and implement media relations program
13. Conduct CAC meeting No. 1: project orientation, hazardous waste education, discuss and finalize site identification criteria
14. Conduct CAC meeting No. 2: results of site identification studies and candidate sites; develop site evaluation criteria
15. Consider briefing local or regulatory officials
16. Prepare bulletin summarizing Phase I activities
17. Design and conduct public meeting

B. Technical siting tasks

1. Establish study area boundaries
2. Develop initial site identification criteria
3. Inventory and map information on environmental and engineering resources: geology, land use, soils, land ownership, surface hydrology, biology, transportation routes, etc.
4. Identify potential siting areas
5. Apply criteria to identify candidate sites
6. Draft site evaluation criteria
7. Participate in briefings and CAC and public meetings

X. PHASE II — SITE EVALUATION

A. Public involvement tasks

1. Continue established communication mechanisms
2. Conduct CAC meeting No. 3: finalize site evaluation criteria
3. Conduct CAC meeting No. 4: exercise to weigh relative importance of site evaluation criteria
4. Conduct CAC meeting No. 5: discuss evaluation studies, develop mitigation packages to address local concerns, develop consensus on preferred sites
5. Prepare bulletin summarizing Phase II activities focusing on preferred site and announcing public meetings
6. Brief local or regulatory officials
7. Issue news release
8. Design and conduct public meeting

B. Technical siting tasks

1. Prepare generic site designs for use in analysis
2. Prepare cost estimates for facility development, operation, and closure (for purposes of site comparison only)
3. Inventory, map, and analyze candidate site data for environmental and engineering resources
4. Rate sites on technical performance
5. Review and refine mitigation packages
6. Participate in briefings and CAC and public meetings

XI. PHASE III — SITE SELECTION

A. Public involvement tasks

1. Catalog public comments
2. Prepare bulletin announcing final decision
3. Conduct special CAC meeting or luncheon to thank them for participating and tell them of final decision
4. Brief local officials
5. Issue news release

B. Technical siting tasks

1. Review public comment
2. Rate sites on all decision factors
3. Select site
4. Begin permitting process

Throughout the project, responses to public inquiries should be made, individual or small group meetings may be held to address local concerns, and specific techniques might be tried to facilitate consensus or mediate certain issues.

A broad range of techniques can be used to meet specific needs of the public. These techniques, their purpose, and their optimum point of use can be reviewed in Table 2. Techniques that might be particularly appropriate to the siting of an incinerator are described briefly below.

A. ISSUE FIRST NEWS RELEASE

A news release should be prepared and issued by the proponent. It should announce the start of the project, describe the project background, need, scope, schedule, project sponsors, and any consultants who will be performing the work.

B. ESTABLISH A POINT OF PUBLIC CONTACT

It is important to establish a single point of contact for the public. The intent is for the public to place one call with a question and either receive an answer immediately or a study team member will research the information and return the call. If the contact person is not in the same community as the site, a toll-free number could be established.

C. BEGIN A NEWSPAPER CLIPPING SERVICE

Pertinent clippings should be circulated biweekly to the study team. These clippings help sensitize staff to public concerns and issues, as well as document press coverage of public meetings and other project activities.

D. ESTABLISH A READING ROOM

A centrally located information repository may be established so that the public will have access to project information, where they can read project materials and pay for reproduction if desired.

E. ESTABLISH A MAILING LIST

A computerized mailing list should be compiled and include names of individuals and entities identified by the project proponent and agencies. The list must be updated throughout the project.

F. PREPARE FACT SHEET

A short, concise information sheet on the project should be prepared within the first 6 weeks. This fact sheet should include the purpose and need of the project, the participants, a general description of project scope and schedule, a study area map, a listing of the study team, and the address and telephone number of the contact person. It should also include a response form that could be mailed to the project office to record comments and/or additions to the mailing list.

G. CONDUCT INTERVIEWS WITH OPINION LEADERS

An effective, early method of familiarizing key publics with the project and of identifying critical concerns and issues is the opinion leader interview. A series of 10 to 15 interviews with such key individuals as elected officials, special interest group leaders, and community leaders should be conducted. Face-to-face discussions yield important information about issues of concern, historical involvement or positions of various groups, and potential problems, conflicts, opposition, or support. The fact sheet could be mailed to the leaders ahead of the interview.

A summary and analysis of issues should be prepared from these interviews, with specific

TABLE 2
Public Involvement Techniques

Objective	Techniques	Purpose	Optimum point of use
Information dissemination	Media relations program Fact sheets Newsletters Brochures Issues papers Slide show/video Information fairs Open houses Guest speaking Open forums Briefings	To inform public of project goals, objectives, status, and activities	Throughout
	Information contact	To identify a person whom the public can call and either receive an answer or be called back with the information	Throughout
	Mailing list	To inform interested parties of proposed actions and to obtain comments	As needed
Information gathering	Community leader or key informant interviews	To identify reactions to and understanding of project To identify issues of concern To identify other groups or individuals to be contacted or added to the mailing list To assess relationships amoung various interest groups	At beginning of project
	Mailed survey or questionnaires	To assess public awareness of project actions, public issues, and concerns To assess readership of newspapers, radio, etc. for structuring information dissemination	At beginning of project and as needed
Citizen and agency involvement	Advisory groups	To identify issues To review interim results To help find compromises between competing local interests To advise on public involvement approaches To maintain credibility of study process To promote consensus with their constituents	Throughout
	Public workshops	To allow diverse groups to explore solutions to particular problems	During the criteria development and alternative formulation stage of the project After impacts have been assessed

TABLE 2 (continued)
Public Involvement Techniques

Objective	Techniques	Purpose	Optimum point of use
	Public meetings/forums	To obtain local comments regarding a proposed action	At beginning of project At the end of a project, but prior to final decision
Conflict resolution/consensus building	Facilitation	To lead discussions impartially in workshops or other group processes	Any meeting, workshop, or advisory group
	Mediation	To reestablish communication when all positions are polarized and no more communication is occurring	Site-specific conflicts when facilitation attempts have failed
	Nominal group technique	To build consensus on project actions, issues, or mitigation plans	When criteria is being developed When indications of preference or relative importance of issues or mitigation measures, are needed
	Delphi	To build consensus with a group of experts or advisors	When relative importance of criteria is needed to compare alternatives
	Public values assessment	To combine public values with technical facts to result in identification of alternatives that most closely meet what the public has said is important	Toward the end of the project when indication of public acceptability is needed
Analysis and documentation	Computerized comment storage and retrieval system	To summarize objectively and display public comment	Throughout
	Summary and evaluation reports	To provide written documentation of attendees, issues, and comments and to evaluate the public involvement program	At major decision points and at the end of the study

attention to concerns that will help in developing evaluation criteria and the alternative evaluation process. This information should be particularly helpful in determiming how the public perceives hazardous waste and its associated risks.

H. ESTABLISH A CITIZEN ADVISORY COMMITTEE

An advisory committee may be formed to serve as a focal point for public involvement and may be expanded as the study progresses and specific siting areas are identified. Careful selection of the advisors is critical to the success of the siting study because these representatives should serve as liaisons with the general public and provide a vehicle for communicating and incorporating public values into the decision-making process. The committee should remain manageable in size, but be balanced in geographic representation and in interests (residents, property owners, developers, environmental groups, incinerator users, agricultural, commercial, and industrial). In addition, the role and responsibilities of the committee should be clearly defined. They are advisors to the decision makers; they do not make the decision.

Advisory committee members can be appointed by elected officials or by the proponent. The opinion leader interviews can be very helpful in identifying the types of interests that should be represented.

I. DEVELOP AND CONDUCT MEDIA RELATIONS PROGRAM

A list of local newspapers and radio stations may be prepared and contacts made with the editors and news directors, either by telephone or in person. The objective of a media relations program is to establish a positive rapport with appropriate media representatives to ensure accurate coverage of project workshops, meeting announcements, feature stories, and editorial support. A press kit that contains pertinent information and is easily updated should be prepared.

J. PREPARE BULLETINS

Bulletins serve an important function in communicating technical information in an understandable manner, describing project need, announcing upcoming events, describing major issues, and summarizing results of technical studies and the public participation plan.

K. CONDUCT PUBLIC SCOPING MEETING

It may be appropriate to conduct a public scoping meeting early in the project. This will provide the public an opportunity to offer their views on the project, the public participation plan, and the study process, and vent their anger or frustration. Results of these meetings should be documented, analyzed, and incorporated into the study.

L. CONDUCT INFORMATION FAIRS/OPEN HOUSES

Exhibits, displays, or models may be arranged in a school cafeteria or all-purpose community room. Participants may come and go at will. They are encouraged to examine exhibits, discuss them with the study team, and interact informally. The open house may precede a more structured meeting, or it can stand alone.

M. PREPARE NEWSPAPER INSERTS WITH RESPONSE FORM

This technique provides information to the general public and at the same time solicits public comment. Newspaper inserts are costly but reach a much greater percentage of the population than most other public information techniques.

N. DISTRIBUTE CUMULATIVE BROCHURE OR "FISHBOWL PLANNING"

The term "fishbowl" refers to the fact that the technique allows everyone to view all aspects of the planning process and see how the decision is developed. The central element is a cumulative brochure that documents several rounds of public workshops, invites people to submit their own alternatives, and provides space for people to react to various alternatives. As the cumulative brochure develops, it becomes clear what the positions of the different interest groups are, how people feel about the alternatives, and what critical groups have chosen not to participate in the study.

O. PROVIDE BRIEFINGS TO COMMUNITY ORGANIZATIONS

This technique takes advantage of existing community organizations such as neighborhood associations, civic groups, fraternal organizations, parent-teacher associations, or local Chambers of Commerce. It develops a two-way communication channel and working relationship with them. Presentations on the project can be made at their monthly meetings or brief write-ups about the project can be provided to include in their newsletters.

P. CONDUCT PUBLIC WORKSHOPS

Conducting a series of public workshops prior to the selection of the preferred alternative may be appropriate. The intent of these workshops would be to review with the public the siting criteria and the alternatives eliminated from further consideration. They will also highlight those alternatives still under consideration and request additional public comment. Presentation of this information may be followed by a question-and-answer period. In the second part of the meeting, smaller subgroup sessions may be used to facilitate informal, direct discussions between citizens and the study team representatives. Discussions on the types of refinements in locations, design alternatives, and construction techniques that will reduce impacts will help to form the basis for further refinements. Comment sheets should be provided at the meetings to assist in documentation.

XII. CONCLUSION

Well-financed, extensive public involvement programs have failed because no real commitment to reach people or to listen to their concerns was evident. The public's perception frequently is that communication with government or industry is one-sided and that public input is obtained at a point in the process when it is too late to accommodate changes. Based on the writer's experience in controversies over the siting of waste management facilities and other equally unwanted land uses, she has concluded that public involvement in the siting process should occur at the beginning, before decisions have been made and before the public perceives that a project is too far along to be influenced by their input.

Reversing the trend of NIMBY groups to GUMBY (Got To Use Many Backyards)[25] is a difficult challenge. Since the NIMBYs are not going to take the initiative, project proponents must assume a new approach to siting. The siting approach accepts the fact that NIMBYs have a self-interest to protect. Their concerns must be answered appropriately. A well-conceived, thorough public involvement program integrated with a systematic siting process cannot guarantee that a new facility will be sited and permitted. But considering citizen concerns at the start of a siting process can reduce citizen opposition, save facility design costs, and ultimately increase the chances of siting the incinerator (see Table 1).[26]

REFERENCES

1. **Nelkin, D., Ed.,** *Controversy: Politics of Technical Decisions,* Sage Publications, Beverly Hills, CA, 1979, 10.
2. Hostile reception: local opposition halts oil-refinery projects along the east coast, *Wall Street Journal,* (New York), p. 1, June 25, 1979.
3. **Nelkin, D., Ed.,** *Controversy: Politics of Technical Decisions,* Sage Publications, Beverly Hills, CA, 1979, 16.
4. **Wedge, B.,** The NIMBY complex: some psychopolitical considerations, in *Community Reaction to Locally Unwanted Land Use,* Institute for Environmental Negotiation, Charlottesville, VA, 1984, 24.
5. **Sandman, P. M., Caron, C., and Hance, B. J.,** Improving dialogue with communities: a risk communication manual for government, submitted to the New Jersey Department of Environmental Protection, Rutgers University, New Brunswick, NJ, 1987, 6.
6. Legislative Commission on Toxic Substances and Hazardous Wastes, Hazardous Waste Facility Siting, Albany, NY, June 1987, 1.
7. Legislative Commission on Toxic Substances and Hazardous Wastes, Hazardous Waste Facility Siting, Albany, NY, June 1987, 39.
8. Legislative Commission on Toxic Substances and Hazardous Wastes, Hazardous Waste Facility Siting, Albany, NY, June 1987, 23.
9. **Arthur, P.,** Public Hearing Testimony on the Northwest Regional Landfill Siting Study, November 25, 1985.

10. **Morell, D. and Magorian, C.,** *Siting Hazardous Waste Facilities,* Ballinger, Cambridge, 1982, 23.
11. **Carmichael, D.,** Hazardous waste siting in New Jersey, term paper for Engineering 303, Princeton University, 1981, 13; cited by **Morell, D. and Magorian, C.,** in *Siting Hazardous Waste Facilities,* Ballinger, Cambridge, 1982, 24.
12. **Gemperlein, J.,** Firm abandons plans for N.J. chemical dumps, *Philadelphia Inquirer,* p. 7E, December 10, 1980, cited by **Morell, D. and Magorian, C.,** in *Siting Hazardous Waste Facilities,* Ballinger, Cambridge, 1982, 24.
13. **Butterfield, F.,** New England town rises up to block a toxic waste plant, *New York Times,* p. 1 and 7, October 17, 1981; cited by **Morell, D. and Magorian, C.,** in *Siting Hazardous Waste Facilities,* Ballinger, Cambridge, 1982, 24.
14. Local opposition threatens IT expansion into Massachusetts, *Hazardous Materials Intelligence Report,* p. 6, October 23, 1981; quoted by **Morell, D. and Magorian, C.,** in *Siting Hazardous Waste Facilities,* Ballinger, Cambridge, 1982, 24.
15. **Morell, D. and Magorian, C.,** *Siting Hazardous Waste Facilities,* Ballinger, Cambridge, 1982, 27.
16. **Wentz, C. A.,** Risk associated with research of an innovative process for thermal destruction of hazardous substances: a case study, *Hazardous Waste and Hazardous Materials,* 5(2), 155, 1988.
17. Editorial, *Hazardous Waste and Hazardous Materials,* 5(2), v, 1988.
18. **Burns, M. E.,** Striking a reasonable balance, in *Hazardous Waste Management: In Whose Backyard?,* Harthill, M., Ed., Westview Press, Boulder, CO, 1984, 196.
19. Nanograms are not the answer to 'will I be hurt?' questions, *Waste Age,* March, 44, 1988.
20. **Sandman, P. M., Caron, C., and Hance, B. J.,** Improving dialogue with communities: a risk communication manual for government, submitted to the New Jersey Department of Environmental Protection, Rutgers University, New Brunswick, NJ, 1987, 67.
21. **Sandman, P. M., Caron, C., and Hance, B. J.,** Improving dialogue with communities: a risk communication manual for government, submitted to the New Jersey Department of Environmental Protection, Rutgers University, New Brunswick, NJ, 1987, 68.
22. **Covello, V. T., Sandman, P. M., and Slovic, P.,** *Risk Communication, Risk Statistics, and Risk Comparisons: A Manual for Plant Managers,* Chemical Manufacturers Associations, Washington, DC, 1988, 31.
23. **Rozelle, M. A.,** The Incorporation of Public Values into Public Policy, Ph.D. dissertation, Arizona State University, Tempe, 1982, 142.
24. **Shuff, R. G.,** 'Bribes' work in Wisconsin, *Waste Age,* March, 51, 1988.
25. **Goldberg, J. F.,** Gotta use many backyards, *Waste Age,* March, 68, 1988.
26. Legislative Commission on Toxic Substances and Hazardous Wastes, Hazardous Waste Facility Siting, Albany, NY, June 1987, 55.

Chapter 6

ESTABLISHED TECHNOLOGIES

Joseph L. Tessitore and Wendy S. Lessig

TABLE OF CONTENTS

I.   How to Choose the Right Incinerator 214

II.   Incineration Technologies 216

III.   Available Incineration Technologies 218
     A.   Rotary Kiln Incinerators 220
         1.   System Description 220
         2.   System Advantages 223
         3.   System Disadvantages 224
         4.   System Performance 224
         5.   System Operating Experience 224
     B.   Circulating Bed Combustor 226
         1.   System Description 226
         2.   System Advantages 230
         3.   System Disadvantages 231
         4.   System Performance 231
         5.   System Operating Experience 233
     C.   Fluidized Bed Incinerators 235
         1.   System Description 235
         2.   System Advantages 239
         3.   System Disadvantages 239
         4.   System Performance 240
         5.   System Operating Experience 240
     D.   Liquid Injection Incinerators 240
         1.   System Description 240
         2.   System Advantages 243
         3.   System Disadvantages 243
         4.   System Performance 243
         5.   System Operating Experience 243
     E.   Controlled Air Incinerators (i.e., Hearth Type) 243
         1.   System Description 243
         2.   System Advantages 246
         3.   System Disadvantages 246
         4.   System Performance 246
         5.   System Operating Experience 246

IV.   Air Emission Control Systems 246
     A.   Wet Venturi Scrubber 247
         1.   System Description 247
         2.   System Advantages 248
         3.   System Disadvantages 248
         4.   System Performance 248
         5.   System Operating Experience 248

	B.	Dry Chemical Scrubber	248
		1. System Description	248
		2. System Advantages	249
		3. System Disadvantages	249
		4. System Performance	249
		5. System Operating Experience	250
	C.	Hydrosonic Gas Scrubber System	250
		1. System Description	250
		2. System Advantages	251
		3. System Disadvantages	251
		4. System Performance	251
		5. System Operating Experience	251
	D.	Ionizing Wet Scrubbers	251
		1. System Description	251
		2. System Advantages	251
		3. System Disadvantages	252
		4. System Performance	252

V.   Heat Recovery 252

VI. Waste Preparation and Feed Systems 253

I. HOW TO CHOOSE THE RIGHT INCINERATOR

Although there are a wide variety of incinerators, they all rely on thermal oxidation to destroy organic wastes. Figure 1 shows a functional flow diagram of a generic hazardous waste incinerator. It illustrates the fact that the methods of:

1. Feeding air, wastes, coolants, and auxiliary fuel into the incinerator
2. Cooling, filtering, and analyzing the resultant flue gases and ashes

comprise the major process steps and that the incinerator itself is the heart of the system. This is similar to typical chemical process plants and petroleum refineries in which the reactor is part of a much larger facility that includes the auxiliaries that condition the feed and product streams.

It is therefore advantageous to select an incineration system in which the incinerator itself imposes a minimum of constraints on the balance of the system — including both feed and product conditioning. The type of incinerator and the nature of the waste stream have a major impact with these auxiliary requirements.

Several trends are occurring now. One has been toward more versatile incinerators that can handle solids, liquids, or sludges. This allows greater flexibility for future operations.

Yet another trend is toward incinerators that achieve dry scrubbing of acid gases such as HCl and SO_2, thus eliminating the generation of wet scrubber sludges.

After quantifying the composition and mass flow of your waste stream and selecting the types of incinerators that can treat it, the selection of an incinerator system normally

215

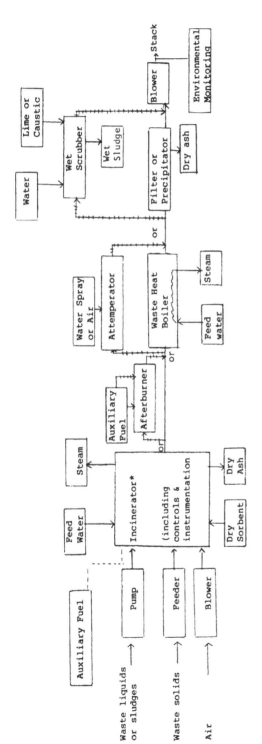

FIGURE 1. Generic functional flow diagram of a hazardous-waste incineration system. *Incinerator: rotary kiln, liquid injection, multiple hearth, fluidized bed, or circulating bed.

proceeds with the issuance of a Request for Proposal (RFP) to those vendors supplying the selected types of equipment.

Figure 2 illustrates the flow of information through this stage and proceeding throughout the entire project. The RFP should be as clear as possible in describing the waste streams and the plant overall operating environment. It is wise to prepare a table of factual information that the equipment supplier must submit as part of a system proposal. This table should include system input and output performance parameters that are warranted by the supplier. Comparisons of bids may then be objectively accomplished. Both minimum and desired criteria for selection should be stated to help obtain the most competitive bids for your incinerator system. Expert consultants are often used to help prepare incinerator RFPs and to evaluate bids.

A typical waste survey form is shown in Figure 3. This form is used by equipment suppliers to gather the information needed to prepare proposals.

II. INCINERATION TECHNOLOGIES

Typical hazardous waste incineration systems are made up of some or all of the following components:

- Waste feed/preparation system
- Primary combustion chamber
- Secondary combustion chamber
- Heat recovery system
- Air pollution control system
- Fan and stack
- Ash handling system

The most significant difference between hazardous and nonhazardous waste incineration systems is the specific technology utilized for the primary combustion chamber. All the incineration technologies discussed herein are actually primary combustion chamber technologies. The specific functions and operating conditions of the other components will differ somewhat depending on the primary combustion chamber technology used. These differences will be identified in the technology discussion.

Incineration is an engineered process using thermal oxidation of a waste material to provide a less bulky, toxic, or noxious material. A waste must be combustible to some extent in order for incineration to be a viable economic disposal option. The three T's of combustion — temperature, residence time, and turbulence — are crucial in controlling operating conditions and incinerator performance. Control of combustion air is also an important operating parameter.

Waste characteristics, including chemical composition and physical form, are important parameters for design of waste feed/preparation equipment. Control of the rate at which hazardous substances are fed into the incinerator, as well as the physical state of the waste or the media containing the waste, are important factors affecting control, operation, and performance of hazardous waste incinerators.

Waste heat recovery systems could be included with virtually any type of incineration system. However, the importance of heat recovery will vary depending on the incineration application, the heat recovery technology utilized, and the need for the energy recovered.

Another important aspect of hazardous waste incineration is emissions control. Hazardous waste incinerators must be equipped with an air pollution control system to ensure that particulates remaining in the flue gas after the incineration process are not allowed to escape to the environment. Control of acid gas emissions caused by the oxidation of chlorinated wastes is also a critical function of an air pollution system.

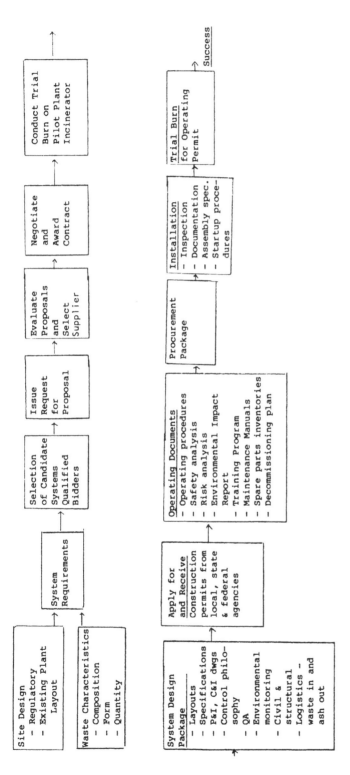

FIGURE 2. Engineering services and software requirements for a typical hazardous-waste incineration project.

WASTE SURVEY FORM

THE INFORMATION PROVIDED IN THIS SURVEY WILL BECOME THE BASIS FOR AN ANALYSIS OF YOUR WASTE, THE VALUE OF ENERGY RECOVERED FROM IT AND THE PAYBACK PERIOD. YOUR CARE IN COMPLETING BOTH SIDES OF THIS FORM WILL ASSURE YOU OF THE MOST COMPREHENSIVE ANALYSIS OF THIS CAPITAL INVESTMENT

NAME OF FIRM _____ DATE _____

ADDRESS _____ CITY _____ STATE _____

YOUR NAME _____ TITLE _____ PHONE _____

1 TYPE OF BUSINESS:

    MANUFACTURING TYPE OF PRODUCT _____

    OTHER DESCRIBE _____

    OPERATING SCHEDULE: HOURS PER DAY _____ ; DAYS PER WEEK _____ ; WEEKS PER YEAR _____ .

    WASTE STORAGE CAPACITY: _____ DAYS

2 TYPE OF WASTES GENERATED:

ID# WASTE DESCRIPTION	HEAT OF COMBUSTION (BTU/LB)	CURRENT/FUTURE AMOUNT (LBS/DAY)	WT% H_2O	WT% SOLIDS	SPECIAL COMMENTS OR HAZARDS
1					
2					
3					
4					

ANALYSIS

WASTE ID#	1	2	3	4	5	6
WT% C						
% H*						
% N						
% O*						
% S						
% Cl						
%						
% INERT						

*EXCLUDE WATER CONTRIBUTION

A

FIGURE 3.   Waste survey form.

This portion of the report outlines the types of incinerator processes and discusses their components. Following the discussions of available and developing technologies, descriptions of applicable air pollution control devices, waste preparation and feed systems, and heat recovery systems are presented.

III. AVAILABLE INCINERATION TECHNOLOGIES

For the purposes of this report, "available" incineration technologies are defined as those technologies which are currently operating under RCRA Part A, Part B, or Research,

METAL ELEMENTS

	1	2	3	4	5	6
% Na						
% Fe						
% Ca						
% P						
% K						
%						
%						
%						

3. DISPOSAL COSTS: CURRENT _____ $/YEAR;   PROJECTED _____ $/YEAR

4. ELECTRICITY COSTS: CURRENT _____ ¢/KWH   PROJECTED _____ ¢/KWH

                         CURRENT _____ $/YEAR;   PROJECTED _____ $/YEAR

5. FUEL COSTS: NATURAL GAS _____ $/HCF/THERM

                                  _____ $/YEAR

                 OIL   _____ $/GALLON

                                    _____ $/YEAR

    OTHER _____   _____

6. ENERGY RECOVERY _____ YES _____ NO

    a. STEAM USE: MAX _____ 1000's LBS/HR AT _____ PSIG _____ °F

                       MIN _____ 1000's LBS/HR AT _____ PSIG _____ °F

        STEAM SOURCE _____ :   COST _____ $/1000's LBS.

    b. HOT WATER USE: _____ GPM; _____ °F FEEDWATER; _____ °F DESIRED.

        HOT WATER SOURCE: _____ ;   COST _____ $/1000 GAL.

    c. HOT AIR USE: _____ CFM AT _____ °F

        HOT AIR SOURCE _____ ;   COST _____ $/1000CFM

7. ELECTRICITY COGENERATED? _____ MW;   PROJECTED _____ MW

8. PROJECT STATUS _____ FORMAL PROPOSAL _____ BUDGET PROPOSAL _____ FEASIBILITY STUDY (PILOT TEST)

    COMMENTS _____

                       _____

                       _____

B

FIGURE 3 (continued).

Development, and Demonstration (RD&D) permits for the treatment of hazardous waste. The available technologies included herein are

1. Rotary kiln incinerators
2. Circulating bed combustors
3. Fluidized bed incinerators
4. Liquid injection incinerators
5. Controlled air incinerators (i.e., hearth type)

The following areas will be discussed for each of the available incineration technologies.

1.   Systems description, consisting of:

    a.   Description of incineration process and incinerator components
    b.   Applicable waste types and throughputs
    c.   Auxiliary burners
    d.   Requirements of waste feed systems
    e.   Ash handling system
    f.   Air pollution control requirements
    g.   Site requirements
    h.   Maintenance history

2.   System advantages
3.   System disadvantages
4.   System performance
5.   System operating experience

The air pollution control, heat recovery, and waste preparation systems available for use with hazardous-waste incinerators are discussed in more detail in Sections IV, V, and VI, respectively, in this chapter.

A. ROTARY KILN INCINERATORS
1. System Description

Rotary kilns have been utilized in both industrial and municipal installations burning solid, liquid, and gaseous wastes. Rotary kilns are currently being used for Comprehensive Environmental Response, Compensation, and Liability Act (CERCLA) site cleanups requiring decontamination of soils and lagoon sediments. In addition, rotary kilns have also been used to destroy obsolete chemical warfare agents and munitions.

Rotary kilns are well suited for waste streams which are highly variable in terms of bulky solids, liquids, and sludges. A typical rotary kiln hazardous waste incineration system is shown schematically in Figure 4.

The rotary kiln design has been in use since the 1920s. At first, it was used primarily in industry as a method for drying solids and continues to be used for these purposes today. Use of rotary kilns for incineration of municipal refuse began in the mid-1950s. Various designs applicable to municipal solid waste (MSW) as well as industrial and hazardous waste have evolved as a result of burning MSW in rotary kilns.

The rotary kiln itself is used as a primary combustion chamber in incineration applications. The kiln is a refractory lined cylindrical metal shell mounted at a slight angle from horizontal, sloping from the solids feed end of the kiln to the discharge end. Rotation of the shell and the slope provides transportation of the solids through the kiln while enhancing mixing of waste with combustion air. Rotational velocity of the kiln is used to control the retention time of solids within the kiln. The combustible portions of the waste are volatilized into gas as the solids travel the length of the kiln. Enhanced mixing and high retention time of solids in the kiln result in a high degree of burnout and increase the chances of producing a nonhazardous ash from the kiln.

Rotary kilns typically have one or two auxiliary burners which can be located at either end of the kiln. The burners are used to preheat the kiln to minimum operating temperatures prior to beginning waste feed and to provide supplemental heat to maintain minimum temperatures during operation. Auxiliary fuel burners can be provided to burn natural gas, propane gas, fuel oil, or high Btu liquid wastes.

Two types of rotary kilns are manufactured in the U.S.: cocurrent and countercurrent. In the cocurrent design, the burner is located at the combustion air inlet end of the kiln.

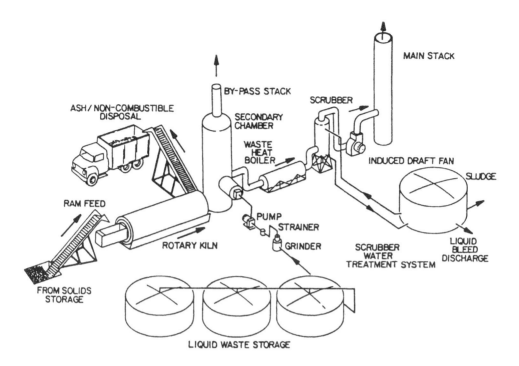

FIGURE 4. Typical rotary kiln hazardous-waste incineration system.

The direction of the fuel/air mixture injection is the same as the gas flow and solids flow through the kiln. A typical cocurrent rotary kiln is shown in Figure 5. In the countercurrent design, the burner is located at the combustion air inlet end of the kiln, but the direction of the fuel/air mixture injection and gas flow through the kiln is opposite to the direction of solids flow. The position of the burner (front or back of the kiln) is arbitrary for waste types which easily sustain combustion. Both configurations will destroy the waste. For wastes of low combustibility (such as low Btu/high moisture sludges: i.e., industrial waste plant sludge), the countercurrent design allows for greater temperature control at both ends of the kiln. The turbulence caused by fuel/air mixture injection at the solids discharge end of the kiln is thought to increase particulate carryover into the secondary combustion chamber, but it also enhances the combustion process by providing thorough mixing of combustion air with volatiles going into the secondary combustion chamber.

Kilns typically have length to diameter (L/D) ratios ranging from 2 to 10 and rotational velocities ranging from 1 to 5 ft/min measured at the kiln periphery. The residence time and temperatures required for proper combustion depend completely on the waste being burned. Combustion temperatures in the kiln will typically range from 1500 to 1900°F (800 to 1600°C). Residence times for solids will range from 30 min to 2 h depending on the kiln dimensions and rotational speeds, while gas residence time will vary from 1 to 5 s depending on the volume of the kiln and the amount of excess air used.

Rotary kilns are typically designed for a volumetric heat release of 20,000 to 25,000 Btu/ft³/h, resulting in typical waste feed capacities of 2000 lb/h to 12,000 lb/h for waste solids, and 50 to 300 gal/h for waste liquids.

Solid wastes are normally batch fed into the primary chamber by a ram feed system but can also be screw fed through a rotating air lock. Liquid wastes are often blended with solids or injected into the primary chamber through atomization with steam or air and/or are pumped through a lance system.

As the combustible portion of the waste is volatilized, the gases formed leave the kiln

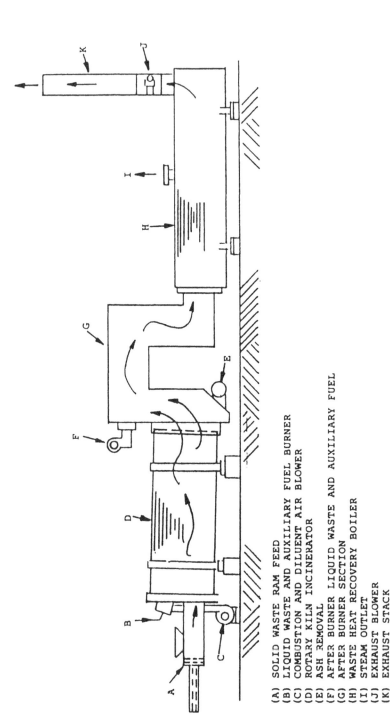

(A) SOLID WASTE RAM FEED
(B) LIQUID WASTE AND AUXILIARY FUEL BURNER
(C) COMBUSTION AND DILUENT AIR BLOWER
(D) ROTARY KILN INCINERATOR
(E) ASH REMOVAL
(F) AFTER BURNER LIQUID WASTE AND AUXILIARY FUEL
(G) AFTER BURNER SECTION
(H) WASTE HEAT RECOVERY BOILER
(I) STEAM OUTLET
(J) EXHAUST BLOWER
(K) EXHAUST STACK

FIGURE 5.   Typical concurrent rotary kiln incinerator.

and enter the secondary combustion chamber. The secondary chamber is equipped with one or more auxiliary burners to maintain temperatures up to 2200°F.

The secondary combustion chamber is a stationary refractory lined metal vessel 30 to 60% of the size of the rotary kiln depending on the air volume and retention times required. Secondary combustion chambers are usually equipped with internal baffles or additional air blowers to maximize turbulence and ensure complete mixing of volatile gases and combustion air. Secondary combustion chambers are typically designed for a minimum of 2-s retention time at 2200°F to satisfy RCRA and TSCA requirements.

The hot gases exiting the secondary combustion chamber may be directed through a heat recovery system or to an air pollution control system. Any of several air pollution control devices can be used with a rotary kiln: Venturi scrubber/packed tower, ionized wet scrubber, or a dry scrubber/baghouse. All of these devices are described in detail in Section IV of this chapter. The most common air pollution control devices used with rotary kilns have been venturi scrubbers for particulate removal followed by a packed tower for acid gas removal.

Ash from the discharge end of the kiln and fly ash from the air pollution control device is collected in an ash collection container and allowed to cool. In some cases, water sprays may be added to help cool the ash.

Analyses for leachable metals and the identified hazardous substances in the ash will be required on a regular basis as set forth in the operating permit.

The actual volume or weight reduction achieved depends entirely on the moisture and volatile content of the waste feed. Maximizing retention time in the primary combustion chamber will increase the total volume reduction. Numerous hazardous wastes, previously disposed of in a potentially harmful manner, are currently being destroyed safely and economically in rotary kiln incinerators combined with the proper air pollution control system. The rotary kiln is generally applicable to the destruction of a wide range of hazardous waste materials.

Unlikely candidates for incineration in a rotary kiln are noncombustibles such as heavy metals, high moisture content sludges, inert material, and inorganic salts, unless there is a need for thorough burnout and detoxification of these materials and unless carefully controlled operating procedures are followed.

2. System Advantages

1.   The system will incinerate a wide variety of liquid and solid hazardous wastes.
2.   It will incinerate some materials passing through a melt phase.
3.   It is capable of receiving liquids and solids independently or in combination.
4.   The feed capability for drums for containerized waste is in very large kilns.
5.   The system is adaptable to a wide variety of waste feed mechanism designs.
6.   It is characterized by high turbulence and air exposure of solid wastes.
7.   Continuous ash removal does not interfere with combustion.
8.   No moving parts are inside the kiln.
9.   The system is adaptable for use with a variety of air pollution control systems.
10. The retention or residence time of the nonvolatile components can be controlled by adjusting the rotational speed of the kiln.
11. The waste can be fed directly into the kiln without preheating or mixing.
12. Rotary kilns can be operated at afterburner temperatures in excess of 2500°F (1400°C), making them well suited for the destruction of toxic compounds that are difficult to thermally degrade.
13. The rotational speed control of the kiln also allows a turndown ratio (maximum to minimum operating range) of about 50%.

TABLE 1
Typical Rotary Kiln Incineration Performance Test Results

Facility type	O$_2$ (ppm)	CO (ppm)	TUHC* (ppm)	DRE (%)	Particulate (mg/m^3)	HCl removal (%)
Commercial rotary kiln	10.5	6.2	1.0	99.999	152	99.4
Commercial rotary kiln	9.4	8.0	0.5	99.999	172	99.9

3. System Disadvantages

1. There is high capital cost for installation.
2. Operating care is necessary to prevent refractory damage.
3. Spherical or cylindrical items may roll through kiln before combustion is completed.
4. The rotary kiln frequently operates with high excess air levels.
5. There is high particulate loading.
6. The system has relatively low thermal efficiency (because of the high excess air).
7. Drying of aqueous sludge wastes or melting of some solid wastes can result in slag formation.

4. System Performance

Rotary kiln incinerators have been tested with TSCA and numerous RCRA wastes. Full-scale, prototype, and pilot units have consistently achieved 99.99 to 99.9999% destruction removal efficiencies with good volume reduction and burnout.

Since 1981, EPA has conducted a substantial program of performance testing at thermal destruction facilities. Testing was conducted to estimate the environmental impact from hazardous waste incinerators and to determine the ability to control particulate emissions. Test facilities, procedures, and results have been published in numerous EPA documents.

Table 1 shows results of rotary kiln incinerator performance tests on organic hazardous wastes. The table shows several parameters which are typical monitoring requirements for hazardous-waste incinerators. The data indicate that the DREs which meet or exceed 99.99% can be consistently achieved in rotary kilns.

Table 2 presents actual performance test data from two fixed facility rotary kiln incinerators. Again, these data show DREs ≥99.99% are achievable with rotary kiln incinerators.

In order to ensure compliance with the EPA permits, measurement of a wide variety of incinerator operating parameters may be necessary to maintain thermal destruction conditions which are equivalent to those obtained during the trial burn. These include combustion temperature, waste feed rate, auxiliary fuel feed rate, oxygen and carbon monoxide concentrations in stack gas, gas flow rates at strategic points and scrubber solution of pH (if a wet scrubber is used). Continuous emission monitors are often used or required in measuring combustion gas components such as carbon monoxide, oxygen, nitrogen oxides, and total unburned hydrocarbons. Requirements for continuous or semicontinuous monitoring of these parameters are not necessarily unique to rotary kiln incinerators and can be expected for all hazardous-waste incineration technologies.

To help minimize conditions which could lead to permit violations, several process parameters are monitored and tied to predetermined set points (derived from trial burn performance), and waste feed is automatically shut down until the condition can be corrected. Table 3 shows typical parameters which may be used to trigger waste feed shutoff.

5. System Operating Experience

Rotary kiln incinerators are in widespread use for hazardous waste disposal. Private industrial plants, TSD facilities, and on-site cleanups using rotary kiln incinerators have accumulated thousands of operational hours of experience.

TABLE 2
Trial Burn Data from Rotary Kiln Incinerators[9]

Compounds[a]	Facility A[a,b]	Facility B[b,c]
Volatile[c]		
Carbon tetrachloride	99.9961[d]	99.9999
Trichloroethylene	99.9976[d]	99.9993
Tetrachloroethylene	99.9985[d]	99.9994
Toluene	99.9985[d]	99.9999
Methyl ethyl ketone	99.9994	
1,1,1-Trichloroethane	99.9992	99.928
1,1,2-Trichloroethane	99.9999[d]	
Semivolatile[e]		
Trans-1, 4-dichloro-2-butene		99.9999[d]
Benzyl chloride		99.9995[d]
Hexachloroethane		
Naphthalene	99.993[d]	
Phenol	99.994[d]	
Cresol(s)	99.9991[d]	
Butyl benzyl phthalate	99.9989[d]	

[a] Ross Incineration Services, Ohio, June 1982, 78 mm-Btu/h rotary kiln.
[b] DuPont Pontchartrain Works, LaPlace, LA, November 1982, 18.5 mm-Btu/h rotary kiln.
[c] By VOST sampling method.
[d] Greater than.
[e] By MM5 sampling method.

TABLE 3
Parameter Typically Employed to Trigger Waste Feed Shutoff for Corrective Action

Parameter	Basis for waste feed shutoff		
	Excess emissions	Work safety	Equipment protection
High CO in stack gas	X		
Low chamber temperature	X		
High combustion gas flow	X		
Low pH of scrubber water			X
Low scrubber P	X		
Low sump levels	X		X
High chamber pressure		X	
High chamber temperature		X	X
Excessive fan vibration			X
Low burner air pressure	X		
Low burner fuel pressure	X		
Burner flame loss	X	X	X

Personnel requirements for a rotary kiln incineration facility are totally dependent on the size of the facility and the number of waste types handled. For most facilities (1 to 3 ton/h), six or seven operations personnel per shift will be required, consisting of one system operator, one operator's assistant, one maintenance foreman, one maintenance worker, one electrician, and one waste feed technician and possibly an assistant waste feed technician to accept, manifest, and transfer wastes to proper storage or blending tank(s). This position could be filled by the equipment mechanic or electrician, depending on the work load.

Operations personnel must, of course, be trained to operate and maintain the equipment, but each staff member must also have hazardous-waste safety training which consists of a 40-h training program with annual refresher courses.

A laboratory staff will also be required in addition to the system operations staff. A separate laboratory and chemist can be provided, or existing base facilities could be utilized. A laboratory will be needed to conduct waste feed analyses, ash analyses, and air pollution control system water analyses.

Operations costs consist of fuel, labor, maintenance, utilities, and other outside costs such as insurance, taxes, and administrative services. Actual operations cost will depend on the size of the unit and the waste types handled.

Operating costs include auxiliary fuel and maintenance. These factors are especially sensitive to the types of wastes handled. Low-Btu, abrasive wastes containing metal drums, large solids, or metal will result in highest operation and maintenance costs due to high auxiliary fuel requirements and excessive wear on the kiln refractory.

An estimate of auxiliary fuel requirements for a rotary kiln incinerator can be made based on individual waste data and quantity projections.

Maintenance costs for rotary kiln facilities are also site specific. The major maintenance concern is the refractory life in the rotary kiln. Life of refractory is dependent on operating conditions, operating schedules, and waste types.

Maintenance costs can be controlled by conducting regularly scheduled inspections and scheduling preventive maintenance for facility equipment according to the manufacturer's recommendations. Outside skin temperatures of the kiln can be monitored to detect any areas of refractory which are experiencing excessive wear. The refractory in a rotary kiln often does not wear uniformly due to the temperature variations and tendency of waste materials to form a slag in the final sections of the kiln. Periodic system downtime should be scheduled to provide physical inspection of kiln refractory and make any necessary repairs.

When properly maintained, rotary kilns will achieve 90% online availability. There have been cases when greater than 90% online time have been achieved for extended periods of time (based on a phone conversation with Mr. Dick Carnes of Rollins Environmental Services of Baton Rouge, LA, who just returned from a tour of a rotary kiln facility in Sweden that has had a 94% online time during the past 6 months while processing about 3000 t per month in a 60×10^6 Btu/h kiln.)

Major repair problems associated with rotary kilns are refractory, kiln end seals, and kiln drive systems. These problems, with the exception of refractory life, can be greatly reduced through proper design and kiln selection. Recent advances in kiln end seal materials have greatly increased life expectancy and seal efficiency. Control of kiln expansion during startup and shutdown will also greatly effect wear of the kiln drive system and supports. Many of these improvements are patented, and selection of the kiln manufacturer will affect maintenance costs.

In general, operation and maintenance costs are difficult to calculate, especially in early stages of project feasibility. The EPA has developed several generic cost curves which can be used to estimate these costs. The curves, based on heat input to the kiln and whether or not the kiln is equipped with a combustion air preheat system and/or waste heat recovery system, are only accurate to within 50% of actual costs. After review of waste data, preliminary mass balance calculations can be made as part of the test burn waste selection and preparation of test burn protocol documents. Preliminary estimates of operating and maintenance costs for a rotary kiln incineration system can be made from the EPA curves and then compared to costs determined by mass and energy balance calculations.

B. CIRCULATING BED COMBUSTOR
1. System Description

The circulating bed combustor (CBC) is an incineration technology that destroys haz-

HOW A CBC WORKS

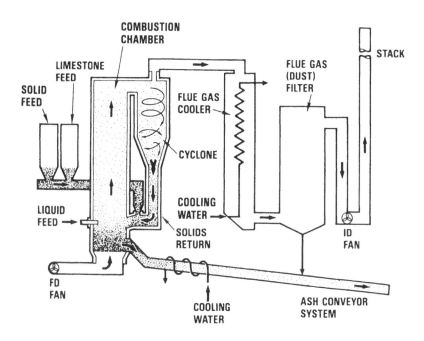

FIGURE 6.   CBC schematics.

ardous waste. It is different from other incinerators because of the turbulence of the combustion bed, the lower operation temperatures, and injection of limestone into the bed along with the waste feed.

High turbulence causes oxygen to circulate among the solids in the combustion bed, so the combustible material burns rapidly at relatively low temperatures (approximately 1600°F). The lower operating temperatures limit the formation of nitrogen oxide acid gases, and the addition of limestone into the bed further neutralizes any halogen- or sulfur-bearing existing acid gases. These properties of the CBC eliminate the need for afterburners and wet scrubbers.

As shown in Figures 6 and 7, CBCs use high-velocity air to entrain circulating solids in a highly turbulent combustion loop. The combustion chamber is typically 30 ft high, diameter, and has a 12-in.-thick ceramic liner.

At the point that solid feed enters the combustor loop, it contacts the hot recirculating solids stream. Liquid and sludge feeds are injected directly into the combustion zone of the CBC. On entering the CBC, hazardous materials are heated and maintained at operating temperatures of 1450 to 1800°F throughout their residence time. Residence times in the combustor range from 2 s for gases to 30 min or longer for larger feed materials.

The feed materials to the CBC system include the contaminated solids, auxiliary fuel, and limestone sorbent (if necessary to control acid gas emissions). The solids feed and limestone are delivered at a controlled rate via the loop seal into the CBC combustion loop. Liquid feeds are introduced directly into the combustion chamber.

The high-combustion air velocity and circulating solids create a uniform temperature around the combustion loop which comprises the combustion chamber, the hot cyclone, and the return leg. During operations, ash is periodically removed from the CBC by means of a water-cooled ash removal system. Hot gas leaving the cyclone at the top of the cyclone is cooled in a flue gas cooler, and particulates that escape the cyclone are collected in a fabric filter baghouse.

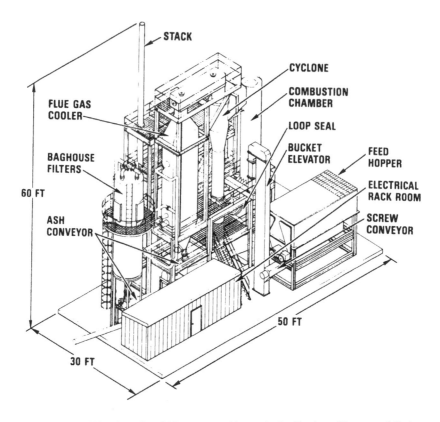

FIGURE 7.   CBC schematics, 36-in. transportable standard soils plant. (Courtesy of Ogden Environmental Services, San Diego, CA.)

The combustion chamber provides a vessel to contain the combustion of the contaminants. The cyclone removes particulates from the combustor flue gas and returns them to the combustion chamber via the loop seal. The loop seal prevents backflow of combustor gases into the bottom of the cyclone. The distributor plate distributes air evenly across the combustor chamber base and prevents ash and solids from entering the windbox below. The refractory lining protects the combustor loop vessel from abrasion and from the high temperatures of the gases and solids in combustor loop.

The purified solids are removed through a valve in the bottom of the combustor and flow into the ash cooler/conveyer system. These bottom solids are cooled from the bed temperature to less than 350°F as they move along a water-jacketed ash cooler provided as part of the conveyer system. Smaller solids flow up the combustor along with the airflow.

The air induction system provides air for fluidization, combustion, combustor loop seal purge, and cooling of the flue gas. In addition, the system provides a means of pressure balance, allowing the system to operate slightly negatively. The flue gas is returned to the atmosphere via the stack.

Combustion air is split into primary air and secondary air. The primary air flows to the distributor through air injection nozzles, and then into the combustion zone. Secondary air can be injected at several different levels above the primary combustion zone to minimize NO_x formation.

From the top of the combustor, the entrained solids, excess air, and flue gas flow horizontally into a hot cyclone, where centrifugal force causes solids larger than about 50 μm to collect against the hot cyclone wall and settle by gravity into the loop seal. These solids are recirculated to the combustor. The very fine solids (fines) pass out of the hot cyclone with the hot flue gas and enter the flue gas cooler (FGC) where they are cooled prior to final filtration.

The FGC lowers the flue gas and fines temperature to about 350 to 400°F. Sonic horns are used intermittently to remove fines which may collect on the tubing.

The FGC consists of three sections. The top section is the flue-gas-to-water heat exchanger. Its function is to remove the heat from the flue gas such that the temperature is reduced sufficiently to eliminate any necessity of using high-temperature metals in the second (air preheat) section. This section operates at nearly constant temperatures and heat load.

The middle section is the combustion air preheat. The waste heat of this section is used to raise the combustion air temperature, thus reducing the fuel requirement and increasing the throughput for wastes having low heats of combustion. The air preheat temperature is typically 645°F.

The third section is the air-to-flue-gas section that controls the flue gas outlet temperature. The flue gas is cooled to about 350°F to prevent damage to the fabric filter located just downstream of the FGC. This section is capable of turndown sufficient to avoid gas temperatures which would approach the dewpoint of the flue gas. The turndown is controlled by regulating the flow rate of the cooling air to the FGC. A dedicated blower services this system.

The flue gas traversing the FGC is at subatmospheric pressure, while the cooling water and airstreams are at positive pressure.

Fines are filtered out by passing the flue gas through a fabric filter (baghouse). Intermittent reverse air pulsing loosens the fine materials collected on the filters. These fines drop through a rotary valve into the ash conveyer system. The pressure drop throughout normal operation is a maximum of 10 in. of water.

The air-to-cloth ratio is based on the type of grain loading and frequency of cleaning operation in order to ensure bag service life of at least 1 year. The air-to-cloth ratio for the CBC ranges up to 3.5 SCFM/ft^2.

The auxiliary fuel system has three functions: (1) to heat the combustion loop from ambient temperatures to operating temperature on a specified temperature ramp, (2) to provide supplemental fuel to maintain combustion loop temperatures during incineration, and (3) to maintain the combustion loop at operating temperatures while idling. When treating contaminated soil, the auxiliary fuel system provides the fuel required to maintain operating conditions.

To accomplish these three functions, the auxiliary fuel system consists of the following subsystems: a startup burner, a set of redundant fuel lances, and independent fuel supply trains supplying each system. Waste feed preparation consists of crushing, shredding, sizing, screening, and drying. The specific procedures involved are dependent on the site and project requirements.

The CBC includes an integrated control and instrumentation system, which provides the four basic functions of automatic or manual process control, alarm, protection, and process monitoring. The instrumentation and control system provides flexibility to control a wide variety of feedstocks. The system monitors and measures pertinent process parameters and uses this information to regulate the performance of the process, initiate required automatic action, and provide displays and records of appropriate information as needed for safe and reliable operation. Direct operator interaction is minimized, consisting primarily of establishment of operating set points, regulation of unit startup, and visual monitoring of performance during operation. The purpose of the control system is to allow safe operation of the unit with a crew of one operator and one assistant.

Control of the system is achieved by adjusting process variables and regulating and/or sequencing the associated equipment in response to operator command and automatic control signals. There are two basic types of control: logic control (interlock) and modulating control.

Logic control initiates and sequences preplanned actions (on/off or open/close). Modulating control regulates the process at the set point of operation for such process parameters as bed pressure differential and combustor temperature.

In the unlikely event of process upset occurring in the CBC system (i.e., fan failure, power outage, etc.) all feeds, air flows, and auxiliary fuel are stopped, and the solids settle together into a slumped packed bed. Feed waste content averages less than 1 wt% of the total bed materials. Hence, the largest part of the packed bed is inert solids and limestone sorbent which surrounds and effectively isolates the waste. Pollutants are thus retained in the stagnant bed of limestone and ash. Experience in full-scale commercial plants shows that the slumped bed is easily and quickly brought back to turbulent circulations, without the release of harmful emissions.

The transportable CBC consists of seven structural steel modules which contain the process equipment and provide the structural framework of the CBC. The module frames are fabricated using wide flanged structural beams. The modules do not exceed measurements of 8 ft 6 in. wide, 10 ft 4 in. high, and 35 ft long; thus, the modules can all be transported on single drop trailers that do not require special transportation permits. The CBC cyclone and combustor are mounted to the top of one of the structural modules.

The transportable 36-in.-I.D. CBC has a nominal throughput of 100 to 150 ton of contaminated soil per day. These numbers can be higher or lower depending on a moisture and energy content of specific soil feed streams. Each CBC is self-contained, including soil feed systems and ash removal systems, and is transportable by truck, rail, or barge, requiring the equivalent of 16 to 18 legal weight tractor-semitrailer shipments. No special permits are required for its transportation.

The CBC operates at ambient temperatures from $-40°F$ to $+110°F$, making it amenable to operation in any environment. The CBC has a footprint of only 30 by 50 ft and a required work area of only 100 by 100 ft, which allows on-site waste destruction in very small areas. The system is inherently safe, operating at subatmospheric pressures that prevent leakage or the escape of hazardous materials in case of an emergency shutdown. Utility requirements average 250 kW electrical power and 9 MBtu natural gas, propane, diesel fuel, or other auxiliary fuels. The cooling loop is closed, so the only water requirements are for personnel needs and decontamination.

2. System Advantages

1. General applicability for the disposal of combustible hazardous solids and liquids
2. Simple design concept requiring no moving parts in the combustion zone
3. Compact design due to high heating rate per unit volume of 100,000 to 200,000 $Btu/h\text{-}ft^3$ which results in smaller units and a lower capital cost
4. Relatively low gas temperatures and excess air requirements which tend to minimize nitrogen oxide formation and contribute to smaller, lower cost emission control systems
5. Capable of receiving liquids and solids independently or in combination
6. Adaptable to a wide variety of waste feed mechanism designs
7. Characterized by high turbulence and air exposure of solid wastes
8. Continuous ash removal which does not interfere with combustion
9. Adaptable for use with a variety of air pollution control systems
10. Large active surface area resulting from fluidizing action enhancing the combustion efficiency
11. Fluctuation in the feed rate and waste composition easily tolerated due to the large quantities of heat stored in the bed
12. Provides for rapid drying of high moisture-content material with combustion taking place in the bed

FIGURE 8. Ogden Environmental Services' circulating bed combustor system burning PCB-contaminated soil in Kenai, AK.

13. Alkaline chemical addition to the bed material suppresses acid gas formation; reduces emission control requirements

3. System Disadvantages

1. Requires bed material preparation and maintenance
2. Feed selection must avoid bed degradation caused by erosion or chemical reactions
3. High particulate loading for fabric filter baghouse with clayey or silty soils
4. High capital cost for installation
5. Temperature excursions can result in agglomeration of the bed.

4. System Performance

OES conducted a polychlorinated biphenyl (PCB)-contaminated soil demonstration test burn at a Swanson River, AK remediation site project in September 1988, in accordance with a test burn plan prepared by OES and approved by the EPA Office of Toxic Substances. The test burn was conducted under witness of the EPA and the Alaska Department of Environmental Conservation (ADEC) at the remote Swanson River Alaska site on the Kenai peninsula. All required performance criteria were met, and in June 1989 OES was granted a nationwide PCB Disposal Operating Permit for its transportable 36-in. CBC unit.

The completion of the Swanson River project is scheduled for the end of 1991. On completion, over 70,000 ton of PCB-contaminated gravel/silt soil will have been treated. Figure 8 is a photograph of the Swanson River site. Soil is fed into the CBC at a rate of 4.5 ton/h. The treated soil is analyzed for PCBs and then released by the on-site laboratory after the PCB level has been verified to be less than 2 ppm. The treated soil is placed into a clean discharge area.

TABLE 4
Swanson River PCB Demonstration Test: Operating
Conditions, Tests 1 through 3

	Test		
Test conditions	1	2	3
Combustor temp (°F)	1620	1606	1620
Residence time (s)	1.68	1.68	1.67
Soil throughput (lb/h)	8217	8602	8603
Soil PCB conc (ppm)	632	615	801
Flue gas oxygen (dry %)	7.1	7.4	6.9
CO emission (ppm)	12	11	17.5
HC emissions (ppm)	2	2	2
SO_2 emissions (ppm)	16	15	13
NO_x emissions (ppm)	89	88	88
Carbon dioxide (%)	8.8	8.7	8.6
HCl emissions (lb/h)	1.49	1.08	1.37
Particulate gr/dscf at 7% O_2	0.0072	0.0065	0.0093
Combustion efficiency (%)	99.980	99.990	99.985
DRE (%)	>99.99993	>99.99992	>99.99997

TABLE 5
Swanson River PCB Demonstration Test: Operating
Conditions, Tests 4 through 6

	Test		
Test conditions	4	5	6
Combustor temp (°F)	1701	1693	1686
Residence time (s)	1.52	1.47	1.53
Soil throughput (lb/h)	8194	9490	9555
Feed PCB conc (ppm)	289	608	625
Flue gas oxygen (dry %)	6.2	6.1	8.1
CO emissions (ppm)	8.7	10	12.5
HC emissions (ppm)	2	2	2
SO_2 emissions (ppm)	27	21	20
NO_x emissions (ppm)	82	90	95
Carbon dioxide (%)	8.8	8.9	8.8
HCl emissions (lb/h)	1.42	1.57	1.21
Particulate gr/dscf at 7% O_2	0.0120	0.0190	0.0182
Combustion efficiency (%)	99.990	99.990	99.990
DRE (%)	>99.99996	>99.99994	>99.99993

The operating conditions during the two demonstration tests are listed in Tables 4 and 5. Two sets of tests were conducted under different process conditions on PCB-contaminated soil in accordance with U.S. Environmental Protection Agency-Toxic Substances Control Act (EPA-TSCA) guidelines and an approved demonstration test plan. Each test consisted of triplicate test burns. A continuous emissions monitoring system measured concentrations of oxygen, carbon monoxide, total hydrocarbons and methane, sulfur oxides, nitrogen oxides, and carbon dioxide in a constant sample of cooled, desiccated flue gas. Monitoring equipment used in OES commercial CBCs is detailed in Table 6. Both tests surpassed all regulatory requirements.

As shown in Tables 4 and 5, the DREs were greater than the required 99.9999% and the purified soil PCB concentrations were significantly less than the regulatory limit of 2 ppm. Combustion efficiency exceeded the regulatory requirement of 99.9%. Dioxin and

TABLE 6
Description of Transportable 36-in. CBC Monitoring Equipment

Name/function	Principal of operation	Range	Accuracy (%) of full scale	Sampling method
Flue gas O_2 probe	Zirconia cell	0—10%	2	*In situ*
Extractive gas analysis				
Oxygen	Paramagnetic	0—25%	1	Extractive
CO	Infrared	0—250 ppm	1	Extractive
CO_2	Infrared	0—25%	1	Extractive
NO/NO_x	Chem lumins	0—500 ppm	1	Extractive
SO_2	Infrared	0—500 ppm	1	Extractive
HC	Flame ionization	0—200 ppm	1	Extractive
Combustor pressure				
Various	Diaphragm	Various	2	*In situ*
Temperatures				
Various	Thermocouple	0—2000°F	0.2	*In situ*
Soil feed rate[a]	Correlation	0—100%	10%	n/a

[a] Soil feed rates are determined from a correlation of motor speed vs. feed rate.

furans were not detected in the purified soil. Stack gas emissions did not contain measurable levels of dioxins and furans above the U.S. EPA Office of Toxic Substances minimum quantitation limit of 10 ng/m^3/cogener. The results demonstrate that the CBC meets or exceeds all EPA-TSCA criteria for incineration of PCB-contaminated soil.

Fuel Oil Site Remediation — For more than 50 years, a leaking underground storage tank at a cannery in Stockton, CA contaminated surrounding clay soil with no. 6 fuel oil. OES was contracted by the site operator to remediate the site using one of its transportable 36-in. CBCs. On completion of the project, over 11,000 ton of contaminated soil will have been treated. Figure 9 is a photograph of the Stockton project site. Table 7 details the February 1989 source test operating conditions performed at Stockton. In July 1989, a demonstration test was performed using naphthalene spiked soil. Table 8 lists the DREs from the three tests, all other results and operating conditions being similar to those recorded during the February source test detailed in Table 7.

5. System Operating Experience

CBC incinerators are in use at several sites for hazardous waste disposal. On-site cleanups using CBC incinerators have accumulated thousands of operational hours of experience.

Personnel requirements for a CBC incineration facility are totally dependent on the size of the facility and the number of waste types handled. For most facilities (3 to 6 ton/h), two or three operations personnel per shift will be required, consisting of one system operator, one operator's assistant, and one waste feed technician to accept, manifest, and transfer wastes to proper storage or blending tank(s).

Operations personnel must, of course, be trained to operate and maintain the equipment, but each staff member must also have hazardous waste safety training which consists of a 40-h training program with annual refresher courses.

A laboratory staff may be required in addition to the system operations staff. A separate laboratory and chemist can be provided or existing local facilities could be utilized. A laboratory will be needed to conduct waste feed analyses, ash analyses, and air pollution control system water analyses.

Operations costs consist of fuel, labor, maintenance, utilities, and other outside costs such as insurance, taxes, and administrative services. Actual operations cost will depend on the size of the unit and the waste types handled.

FIGURE 9.   Ogden Environmental Services' circulating bed combustor system burning hydrocarbon-contaminated soil in Stockton, CA.

TABLE 7
Stockton Source Test: Operating Conditions

Parameter	Test 1	Test 2	Test 3
Combustor temp (°F)	1588	1588	1587
Residence time (s)	1.8	1.8	1.8
Soil throughput (lb/h)	4000	4000	4000
Soil TPH conc (ppm)	2130	1160	3450
Flue gas oxygen (dry %)	13.6	13.6	13.6
CO emissions (ppm at 7% O_2)	28.0	25.4	23.6
HO emissions (ppm at 7% O_2)	<2	<2	<2
SO_2 emissions (lb/d)	16.6	12.0	24.2
SO_2 emissions (ppm at 7% O_2)	84	61	123
NO_x emissions (lb/d)	7.4	7.3	6.7
NO_x emissions (ppm at 7% O_2)	52	52	47
Carbon dioxide (%)	7.0	6.6	6.9
Particulate gr/dscf at 7% O_2	0.045	0.046	0.045
Combustion efficiency (%)	99.989	99.990	99.990

TABLE 8
Stockton Demonstration Preliminary Test Results

Naphthalene conc (ppm)	4314	4703	4106
DRE (%)	>99.99960	>99.99956	>99.99958

These operation and maintenance costs are very site specific, and since regulatory requirements do not require submission of this data, they are not commonly available in current literature.

An estimate of auxiliary fuel requirements for a CBC incinerator can be made based on individual waste data and quantity projections.

Maintenance costs for CBC facilities are also site specific. A maintenance concern is the refractory life in the CBC. Life of refractory is dependent on operating conditions, operating schedules, and waste types.

Maintenance costs can be controlled by conducting regularly scheduled inspections and scheduling preventive maintenance for facility equipment according to manufacturer's recommendations. Outside skin temperatures of the CBC can be monitored to direct any areas of refractory which are experiencing excessive wear. Periodic system downtime should be scheduled to provide physical inspection of refractory and make any necessary repairs.

When properly maintained, CBC incinerators will achieve 80 to 90% on-line availability. There have been cases when greater than 90% on-line time has been achieved for extended periods of time.

In general, operation and maintenance costs are difficult to calculate, especially in early stages of project feasibility. These costs can be determined by CBC incinerator vendors. After review of waste data, preliminary mass balance calculations can be made as part of the test burn waste selection and preparation of test burn protocol documents. Preliminary estimates of operating and maintenance cost for a CBC incineration system can be made from the CBC vendor curves and then compared to costs determined by mass and energy balance calculations.

C. FLUIDIZED BED INCINERATORS
1. System Description
Bubbling fluidized bed incinerators are versatile thermal destruction units which can be used to dispose of a wide variety of solid and liquid wastes. Fluidized beds have been used as industrial process technology for over 60 years. First used in Germany for coal gasification, the technology was subsequently applied to catalytic cracking in the 1930s and in succeeding years for metallurgical roasting, calculating, drying, particle sizing, cooling, and waste combustion applications.

Bubbling fluidized bed incinerators are relatively new to the U.S., with the first commercial applications in the early 1960s, primarily in the petroleum industry, paper industry, wastewater treatment, and for processing nuclear wastes. Currently, bubbling fluidized bed incinerators are in use for combustion of coal, wood chips, refuse derived fuel (RDF), or a combination thereof, in conjunction with power generation and for combustion of dewatered municipal wastewater treatment plant sludge.

A bubbling fluidized bed incinerator has all of the basic components found in a typical incineration system with the exception that the primary and secondary chambers are combined into a single chamber (see Figure 10).

A bubbling fluidized bed incinerator, shown schematically in Figure 11 has the following components:

1. Plenum or windbox
2. Air distribution plate which distributes air from the windbox to the bed
3. The bed media (typically an inert material)
4. A free board area above the fluid bed for further oxidation of volatile gases

The bubbling fluidized bed incinerator consists of a refractory lined vertically oriented cylindrical carbon steel vessel. Typical dimensions of fluidized bed incinerators range from

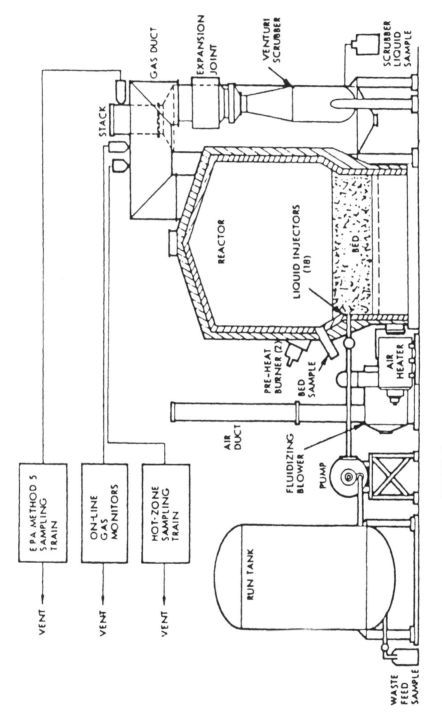

FIGURE 10.   Bubbling fluidized bed incineration system.

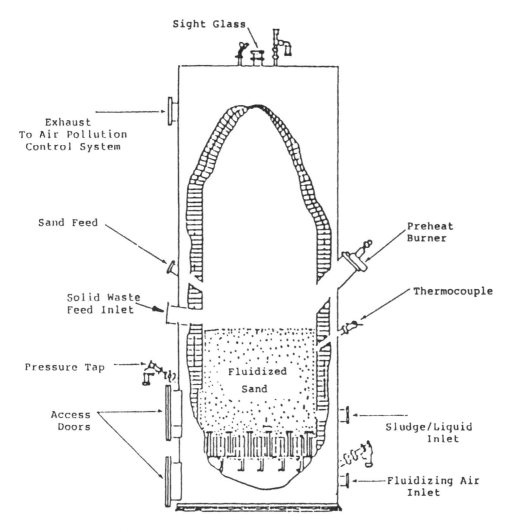

FIGURE 11. Bubbling fluidized bed incinerator (reaction vessel).

10 to 50 ft in diameter and 30 to 50 ft in height. The bottom of the vessel contains an air plenum (or windbox) which receives fluidizing air and combustion air from a blower. Air, under positive pressure, is forced through a metal distribution plate at the top of the plenum. Nozzles or openings in the distribution plate are designed to allow sufficient fluidizing air to pass upward into the bed and at the same time prevent the bed media from falling down into the windbox when the bed is at rest.

Inert material such as silica sand is used as the bed media. Requirements of the bed media are a high melting point and low reactivity with the waste constituents.

The upward flow of air through the bed results in a densely turbulent mass with physical properties similar to a fluid. Waste material is injected near the bottom of the bed at one or more points around the circumference of the vessel. Air passing through the bed produces rapid and turbulent mixing of injected wastes with bed media. Because of the violent mixing achieved in the bed, hazardous waste constituents can be destroyed at lower temperatures than required for other incineration technologies. Lower temperature requirements also result in less excess air than required by rotary kilns.

The mass of the fluidized bed is large in comparison to that of the waste being injected. Bed temperatures are typically maintained at 1400 to 1600°F. Additional auxiliary fuel or water can be injected into the bed to aid in controlling bed temperatures. At these temperatures the heat content of the bed is approximately 25,000 to 35,000 Btu/ft³, thus providing a large heat reservoir. By comparison, the heat capacity of flue gases at similar temperatures is three orders of magnitude smaller than fluidized sand.

Preheating of the bed prior to waste feeding is accomplished through one or more auxiliary burners mounted above the bed. Waste feed is not initiated until the bed has reached minimum operating temperature. Heat is transferred from the bed to the waste material until the waste reaches combustion temperatures. Heat is then transferred back into the bed as the waste combusts. Continuous bed agitation allows large waste particles to remain suspended until completely combusted.

Residual fines (of ash) are carried off the surface of the bed with the flue gases and then controlled in an air pollution control system. In specifying or designing a bubbling fluidized bed incinerator, the primary factors to be considered are gas velocity, bed diameter, bed temperature, and the type and composition of waste incinerated.

Gas velocities range from 5 to 7 ft/s. Higher velocities will lead to attrition of the bed particles which increases the load to the air pollution control system and decreases overall gas residence time in the incinerator.

Fluidized bed depths range from 15 in. to several feet. Beds are expanded up to 6 ft in depth when operational. Variation in bed depth affects both residence time and pressure drop in the system.

After oxidation of the waste in the bed, the combustion gases, ash, and any entrained bed particles rise above the bed into the free board portion of the combustion vessel where further oxidation of volatiles can occur. The free board area of the bed serves as the secondary combustion chamber. Mixing is achieved by the violent boiling action of the bed media as combustion air passes through the bed. Temperatures in the free board can be maintained at 1800 to 2500°F with the use of the auxiliary burners. Total gas residence times are typically 12 to 16 s, and solid residence times will vary depending on the combustion characteristics of the waste.

Waste liquids can be fed at rates of 50 to 300 gal/h depending on the moisture content and Btu value of the wastes. Liquids can be pumped directly from storage containers or trucks after analysis and injected into the bed without atomization.

Incineration of solids in a bubbling fluidized bed reactor is limited to physical waste characteristics. Solids must be injected into the bed mechanically or pneumatically. This requires that waste solids must be homogeneous in nature with a maximum nominal particle size of approximately 2 in. in diameter. This is not a significant problem for waste sludges, but mixed solids containing large items such as drums and pallets must be shredded or pulverized and processed to facilitate injection into the bed. Large differences in specific gravity of the waste materials may cause operating problems.

Sludges with high moisture content must be dewatered prior to injection into the bed. Excess moisture and large amounts of inert material may reduce the bed temperatures, requiring large amounts of auxiliary fuel to maintain minimum operating temperatures. Sludge dewatering can be accomplished by mechanical processes and/or drying with heat recovered from the flue gas in a waste heat boiler.

Removal of the ash from the fluidized bed media is accomplished by continuously removing a portion of the bed media with a bleed auger and disposing of it with the ash from the air pollution control system. Virgin bed media are added to the bed as needed.

Hot gases from the free board portion of the fluidized bed combustion vessel are then routed through an optional heat recovery system or directly to an air pollution control system. The emissions of greatest concern from fluidized bed incinerators are particulates and acid gas (HCl). In some cases, for example, homogeneous waste streams which contain large percentages of chlorinated substances — HCl emissions — can be controlled in the bed by adding lime. The HCl will react with the lime to form calcium salts, effectively removing HCl from the off-gases. The calcium salts are then removed from the bed material along with the ash.

Air pollution control devices for fluidized bed incinerators are typically somewhat smaller than those for rotary kilns because of the relatively low excess air requirements. Common air pollution control devices used with fluidized bed incinerators are venturi scrubbers/packed tower, dry scrubbers, or ionizing wet scrubbers. When required gas velocities through the fluidized bed are such that large amounts of the bed material are carried over to the air pollution control device, a hot cyclone can be used prior to the air pollution control device which will recover large bed particles for return to the incinerator.

Most fluidized bed incinerator applications described in the current literature involve incineration of sludges and slurries. The type and composition of the waste are key design parameters determining the type of feed mechanisms, processing of wastes, and bed specifics. A homogeneous liquid may be immediately injected into the bed, but a nonhomogeneous waste sludge having moderate incineration potential and interspersed with large solid matter will require special feed considerations prior to feeding. Despite the need for pretreatment, the fluidized bed is capable of handling nearly all wastes that the rotary kiln can, depending on the heat limitations of the bed material.

2. System Advantages

1. General applicability for the disposal of combustible hazardous solids, liquids
2. Simple design concept, requiring no moving parts in the combustion zone
3. Compact design due to high heating rate per unit volume 100,000 to 200,000 Btu/h-ft³ (900,000 to 1,800,000 kg-cal/h/m³) which results in smaller units and a lower capital cost
4. Relatively low gas temperatures and excess air requirements which tend to minimize nitrogen oxide formation and contribute to smaller, lower cost emission control systems
5. Capable of receiving liquids and solids independently or in combination
6. Longer incinerator life and lower maintenance costs than other units of the same capacity
7. Large active surface area resulting from fluidizing action, which enhances the combustion efficiency
8. Fluctuation in the feed rate and waste composition easily tolerated due to the large quantities of heat stored in the bed
9. Provides for rapid drying of high-moisture content material with combustion taking place in the bed
10. Alkaline chemical addition to the bed material suppresses acid gas formation; reduces emission control requirements
11. Provides considerable flexibility for shock load of large quantities of highly volatile wastes

3. System Disadvantages

1. Difficult to remove residual materials from the bed
2. Requires bed material preparation and maintenance
3. Feed selection that must avoid bed degradation caused by erosion or chemical reactions
4. May require special operating procedures to avoid bed damage
5. Operating costs possibly high, particularly electric power costs
6. Formation of eutectics a serious problem

4. Systems Performance

Bubbling fluidized bed prototype units have achieved 99.9999% destruction efficiency on TSCA type solid waste, while a full-size private industrial (noncommercial) reactor has achieved 99.99 to 99.9999% destruction efficiency for RCRA type waste (carbon tetra-chloride, chloroform, 1,1,1-trichloroethylene, and perchloroethylene).

5. System Operating Experience

Bubbling fluidized bed incinerators have accumulated significant operating hours in the combustion of municipal and industrial waste, but have accumulated minimal operating hours with RCRA industrial waste. In the incineration of industrial type wastes, bubbling fluidized bed systems have established on-line availability rates as high as 90%.

D. LIQUID INJECTION INCINERATORS

1. System Description

Liquid waste combustors are versatile units which can be used to dispose of almost any combustible waste fluid with a viscosity of less than 10,000 Saybolt seconds units (SSU). Liquid waste incinerators have a wide range of industrial applications. Liquid injection incinerators for hazardous wastes are by far the most common type of hazardous waste incinerator in use today.

High-Btu liquids are easily and inexpensively destroyed by simple liquid injection in-cinerators. Incinerators designed to handle solid hazardous wastes are much more expensive to build and operate than their liquid counterparts. In the past, land disposal facilities for solid waste were readily available and relatively inexpensive; however, realization of the liabilities associated with land disposal of hazardous wastes has resulted in the current efforts to minimize or eliminate these types of facilities and has led to the demand for incinerators capable of destroying both solids and liquids. Therefore, due to waste-type limitations and the need for incinerators to handle a wide range of waste types, liquid injection incinerators are used less frequently for industrial and commercial applications where both solids and liquids must be destroyed.

There are many variations of liquid injection incinerators available today. The two major types are (1) vertically fired units (see Figure 12) which have the waste feed, auxiliary fuel burners, and combustion air blowers at the bottom of the vertically oriented primary com-bustion chamber and (2) horizontally fired units (see Figure 13) which have the waste feed, auxiliary fuel burners, and combustion air blowers located at one end of a horizontally oriented primary combustion chamber.

Liquid injection incinerators have all the components of a typical hazardous waste incineration system. The primary combustion chamber is a refractory lined cylindrical carbon steel vessel. As the name implies, the liquid injection incinerator is limited to liquids, sludges, and slurries, with viscosities less than 10,000 SSU and a low solids content.

The reason for this viscosity limitation is that the waste liquids must be atomized into smaller sized droplets. Liquids with viscosity of greater than 10,000 SSU cannot be atomized to small enough droplet size. As droplet size decreases, the surface area available for heat transfer increases, which minimizes the time needed for the waste to reach its volatization temperature.

There are three stages in liquids combustion: (1) heating to volatilization temperatures, (2) volatilization into gases, and (3) oxidation of the gases in a high temperature atmosphere. The incineration unit must be designed to provide sufficient residence time to accomplish all three stages of combustion prior to the gases reaching the air pollution control system. In order to minimize the combustion chamber volume required, waste preparation and feed (injection) systems are designed to facilitate liquid combustion. Liquid waste injection may be augmented by atomization with steam or compressed gas, which maximizes the surface

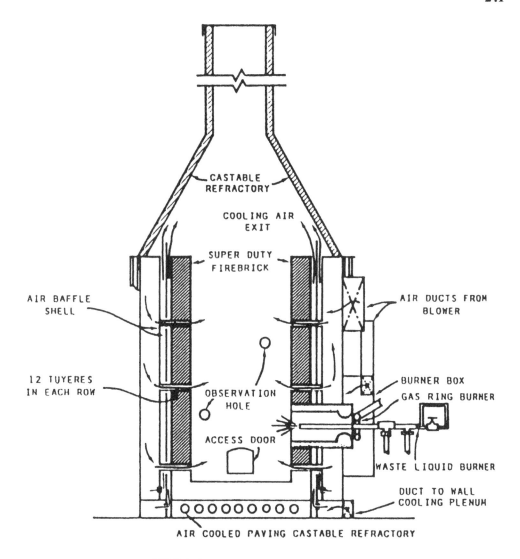

FIGURE 12. Vertical liquid injection incinerator.

area of the waste liquid by minimizing the droplet diameter. The auxiliary burners and combustion air blowers are mounted on the primary chamber to maximize the mixing of waste with combustion air in order to minimize the liquid gas volatilization time. These steps are normally accomplished in the primary combustion chamber.

The time required for volatilization of liquid to gas is a direct function of droplet diameter. Most often, a batch-mixing tank is provided to mix various compatible wastes and supplemental fuel (optional) to optimize Btu value and viscosity of the waste. Proper nozzle selection and maintenance is crucial to incinerator performance. An ideal droplet size is 40 μm in diameter and is attainable mechanically using rotary cups, pressure atomization, or gas-fluid nozzles.

Regardless of the type of liquid injection unit (horizontal or vertical), primary combustion chamber temperatures of 1200 to 1800°F are common depending on the size of the unit, design heat release, and excess air requirements. If good mixing and atomization of wastes are achieved, many organic liquids will combust below 1800°F, typically in the range of 1200 to 1500°F. The autoignition temperature of the waste liquid mix should be determined and the unit maintained several hundred degrees above this point.

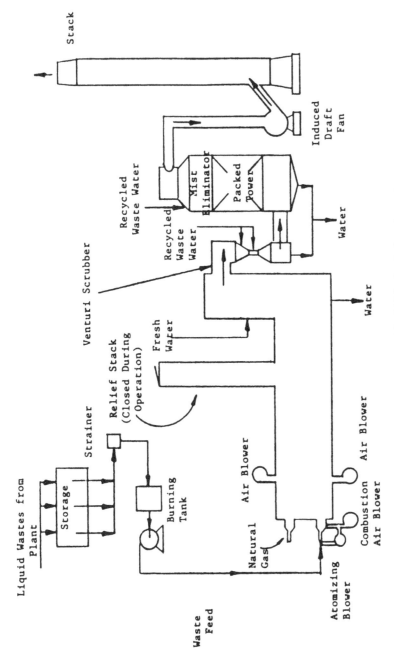

FIGURE 13. Typical horizontal liquid injection incineration system.

Feed rates are dependent on the dimensions of the unit but can be as high as 1500 gal/ h. Hot gases leaving the combustion chamber are routed through an optional heat recovery system or directly to the air pollution control system. Simple gas-to-air or gas-to-organic-fluid heat exchanges are typically used to preheat waste liquids in order to optimize their waste feed characteristics.

Air pollution control devices for liquid injection incinerators typically consist of a quench system (if heat recovery is not included), followed by a venturi scrubber for particulate control and a packed tower for HCl control. In some cases, when waste streams are consistent in Btu value and have a low ash content, particulate control may not be a concern, but for waste streams with high ash content, particulate control will be a requirement.

The only ash produced by a typical liquid injection incinerator is in the form the sludge collected in the scrubber makeup water conditioning tank. These solids are typically removed from the system with a bleed line to a publicly owned treatment works (POTW).

Liquid waste incinerators are capable of handling most combustible liquid hazardous wastes meeting the viscosity and ash content constraints identified above. Unlikely candidates for incineration in liquid injection units are noncombustibles such as heavy metals, high moisture content sludges or liquids, inert materials, inorganic salts, and a general group of liquids having a high inorganic content.

2. System Advantages

1. Capable of incinerating a wide range of liquid hazardous wastes
2. No continuous ash removal system required other than from the air pollution control system
3. Capable of a good high turndown ratio
4. Fast temperature response to changes in the waste fuel feed rate
5. Virtually no moving parts
6. Low maintenance costs

3. System Disadvantages

1. Only wastes which can be atomized through a burner nozzle can be incinerated
2. Heat content of waste burned must maintain adequate ignition and incineration temperatures or a supplemental fuel must be provided
3. Burners are susceptible to plugging by solids

4. System Performance

Liquid waste incinerators have achieved destruction efficiencies of 99.99 to 99.9999% for RCRA- and TSCA-type liquid waste. These efficiencies have been achieved on pilot, prototype, and full-scale operational units.

5. System Operating Experience

Liquid incinerators have been widely used for hazardous waste incineration. It is the most commonly used method for destroying liquid industrial and hazardous wastes. Existing liquid waste incinerators have shown good on-line operational availability (>90%).

E. CONTROLLED AIR INCINERATORS (HEARTH TYPE)
1. System Description

Controlled air incinerators, sometimes referred to as starved air incinerators, first gained popularity in the U.S. in the early 1960s as public attention began to focus on emissions from incineration of solid wastes. Controlled air incinerators are most commonly used for

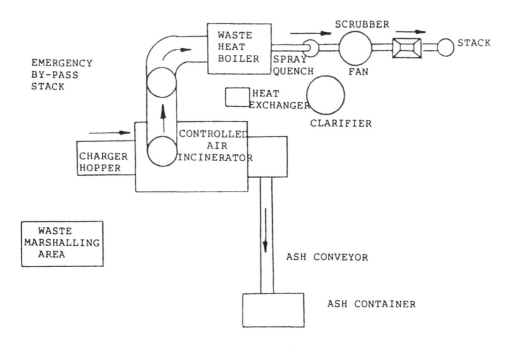

FIGURE 14.   Typical controlled air incineration system. (Not to scale.)

municipal solid waste, infectious wastes, industrial solids wastes, and municipal waste water treatment plant sludge. Controlled air incinerators may also be applicable to gaseous and liquid waste disposal under proper waste feed conditions.

Controlled air incinerators have all of the components of a typical hazardous waste incineration system (see Figure 14). The feature which distinguishes controlled air incinerators from other incinerators is that waste in the primary chamber is volatilized on a stationary hearth in a substoichiometric environment (less than the theoretical amount of oxygen for combustion) rather than in an excess air environment. Controlled air incinerators should not be confused with pyrolysis (thermal decomposition of a compound in the absence of oxygen) which is the basic process utilized in several different emerging or developing technologies.

The primary and secondary combustion chambers consist of refractory lined carbon steel vessels. The shape and dimensions of the chambers are dependent on the orientation of the hearth(s), the secondary retention time required, and the ash collection mechanism. A common type of controlled air incinerator is a stepped hearth incinerator. Figure 15 shows a schematic representation of a typical stepped hearth incinerator.

The stepped hearth incinerator contains two to four stationary hearths in a lineal orientation. Waste is injected onto the first hearth by a ram feeder or an auger screw feeder, depending on the characteristics of the waste. Each successive charge of waste moves previous charges through the unit toward the end of the first hearth. The waste then falls vertically 12 to 24 in. onto the second hearth, which provides for mixing of the waste, exposing new waste surfaces to heat and combustion air. As the waste continues to volatilize, an ash ram is extended from the vertical face of the step between successive hearths. The rams are cycled such that the ram on the lowest hearth is activated prior to receiving waste from the hearth above.

The reduced airflow reduces turbulence in the primary chamber which lessens particulate concentrations in the off-gases. The design heat release for a controlled air incinerator is approximately 25,000 Btu/ft^3/h. Feed capacities range from 100 to 2000 lb/h, depending on Btu value and physical waste characteristics.

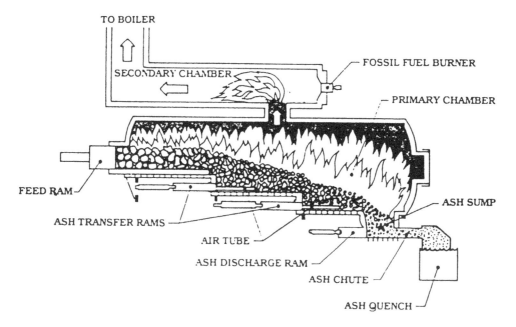

FIGURE 15. Stepped hearth controlled air incinerator.

Volatile gases from the primary combustion chamber enter the secondary combustion chamber. Auxiliary fuel burners and combustion air blowers supply supplemental heat and excess air (typically 150 to 200% of the stoichiometric requirement for the off-gases) to maintain temperatures of 2000 to 2500°F. Combustion air blowers are mounted tangentially and/or baffles are provided in the secondary chamber to provide for turbulent mixing of primary chamber off-gases and combustion air. Residence time in the second combustion chamber is typically 2 s based on 2200°F operating temperature.

The hot gases are then routed to an optional heat recovery system or directly to an air pollution control system. Particulate emissions from controlled air incinerators, when compared with other available technologies (with the possible exception of liquid injection incinerators), are much lower. Low gas velocities in the primary chamber and complete combustion achievable in the secondary combustion chamber result in low particulate carryover to the air pollution control system. Acid gas control may be the primary emissions control concern.

Typically a venturi scrubber/packed tower or dry scrubber/baghouse air pollution control system will provide adequate emissions control. Ash from the primary chamber is collected in ash containers equipped with water spray bars to cool hot ashes. Wet sludge collected from the scrubber water conditioning tank is added to the primary chamber bottom ash. Following analysis of the ash for hazardous constituents, it can be disposed of at an appropriate, secure land disposal facility. Controlled air incinerators are designed for minimal turbulence in the primary chamber, and, therefore, materials requiring turbulence and mixing, such as wet sludges or liquid solvents, are not appropriate candidates for controlled air incineration.

Other unlikely candidates for incineration in a controlled air incinerator include heavy metals, fine organic particles (e.g., power carbon), inorganic salts, inert materials, and a general group of materials having a high inorganic content.

Controlled air incinerators have an on-line time of approximately 80 to 90% depending on the acidity of the off gases. Typically, the majority of downtime for controlled air incineration systems is mostly related to the failure of waste feed/preparation or air pollution control equipment.

2. System Advantages

1. Potential for by-product recovery, (i.e., waste heat recovery)
2. Reduction of waste volume without large quantities of supplemental fuel
3. Thermal efficiency higher than normal incinerators due to the low quantity of air required
4. Reduced uncontrolled air emissions
5. Converts carbonous solids into a gas which is more easily combustible
6. Allows for suppression of particulate emissions
7. Will burn solid wastes with a minimum amount of processing
8. Relatively low capital costs

3. System Disadvantages

1. May have incomplete burnout of carbonous material in the ash due to insufficient retention time in the primary chamber
2. Tends to operate in a batch mode (continuous operation may cause problems with clinkers or scale buildup on their factory surfaces and with burnout)
3. Tendency for push plates for ash removal to buckle over time
4. Not applicable to low Btu and/or contaminated liquid waste
5. Difficult to control operating parameters when the waste stream is highly variable

4. System Performance
   Controlled air incinerators have been extensively used for solid waste destruction. Hazardous waste operational data are limited; therefore, operational data are not readily available.

5. System Operating Experience
   Controlled air incinerators have been extensively used for solid waste destruction. Hazardous waste operational data are limited; therefore, operational data are not readily available.

IV. AIR EMISSION CONTROL SYSTEMS

   Proposed hazardous and toxic waste incineration systems must comply with the following regulations and guidelines:

1. Federal RCRA regulations
2. Federal TSCA regulations
3. Federal prevention of significant deterioration (PSD) criteria
4. National ambient air quality standards

Based on the above regulations and criteria, the control of emissions from incineration systems is especially critical. Considering the above requirements, two basic approaches to the control of air emissions from hazardous and toxic waste incinerators are as follows:

1. A wet venturi scrubber followed by a packed tower scrubber
2. A dry chemical scrubber followed by a fabric filter or an electrostatic precipitator

   Presently, the hazardous waste incinerators at fixed facilities, either industrial or at TSDFs, are equipped with venturi/packed tower systems or wet precipitators. These existing systems are based on wet scrubbers and require the handling and treatment of contaminated wastewater.

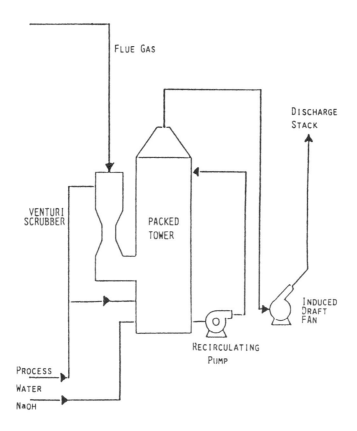

FIGURE 16. Venturi scrubber/packed tower air pollution control system.

Due to the generation of contaminated wastewater and the existence of acidic gases in incinerator combustion gases, the use of a dry scrubbing system is becoming more popular. Recent information derived from testing in municipal solid waste incinerators indicates that the use of a dry scrubber followed by a fabric filter has good particulate and acid gas removal.

It should be noted that any air pollution control systems used with hazardous-waste incinerators must be preceded by a waste heat recovery system or a quench system to reduce the temperature of the combustion exhaust gases. This requirement is due to the high temperatures (>2000°F) typically required in the secondary combustion chamber for the destruction of organics. This temperature is substantially above the design limit for scrubber materials.

In the following sections, the currently available air emission control systems for hazardous waste incinerators are discussed.

A. WET VENTURI SCRUBBER
1. System Description
This control system consists of two separate stages as shown in Figure 16. The first stage, the venturi, removes particulates from the gas stream, while the second stage, the pack tower, removes gaseous contaminants and remaining particulates from the gas stream.

The scrubbing liquid is injected into the high velocity gas stream either at the inlet to the converging section of the scrubber or at the venturi throat. The kinetic energy of the gas stream atomizes the liquid into droplets, maximizing the surface area of the scrubbing liquid. The resulting gas/liquid contact permits removal of particulate contaminants by liquid absorption.

Following scrubbing in the venturi, the solids are separated from the gas stream in a bottom-fed cyclone which is the lower section of the packed tower. Remaining gaseous pollutants are removed in the upper, packed section of the tower. This is generally accomplished through a counterflow packed tower arrangement. The packed tower is a vessel filled with randomly oriented packing material such as saddles and rings. The scrubbing liquid is fed to the top of the vessel and recycled with chemical treatment from the bottom. As the liquid flows down through the tower, it wets the packing material, providing a large surface area for mass transfer with the gas stream. The cleaned gas normally passes through a mist eliminator and is then vented through a stack to the atmosphere. The liquid leaving the absorber is either stripped of the contaminants gas and recycled or passed on for further waste treatment or process use. The packing material (inert chemically) is designed to increase the surface area for gas-liquid contact.

2. System Advantages

1. Established technology
2. Provides high removal efficiencies for particulate and gaseous pollutants

3. System Disadvantages

1. High capital cost
2. High operating cost due to substantial pressure drop across venturi
3. Results in a liquid waste and sludge by-product

4. System Performance
The venturi scrubber/packed tower system has provided removal efficiencies of particulate and gaseous pollutants which are consistent with RCRA and TSCA requirements.

5. System Operating Experience
The venturi scrubber/packed tower system has extensive operational experience in industrial waste incineration and RCRA/TSCA incineration facilities. SCA in Chicago, IL, Eastman Kodak in Rochester, NY, Upjohn in Kalamazoo, MI, and EPA/Combustion Research Facility (CRF) in Pine Bluff, AR are some of the major existing facilities utilizing this concept.

B. DRY CHEMICAL SCRUBBER
1. System Description
A dry chemical scrubber system consists of two units: a spray dryer followed by a baghouse or an electrostatic precipitator (see Figure 17). The spray dryer consists, in turn, of a reactor vessel and a slurry atomizing system.

The incineration flue gas is first passed through a heat exchanger system to reduce the gas temperature to approximately 400°F, and then introduced into the spray dryer. It flows through the dryer in either an upward (upflow reactor) or downward (downflow reactor) direction depending on the equipment design. Upflow reactors are often preceded by cyclones which remove a portion of the fly ash. Downflow reactors incorporate a bottom hopper where large solids (fly ash, reaction products, and unreacted reagent) settle out of the gas stream.

A calcium or sodium slurry is injected into the flue gas in the reaction vessel. The slurry is atomized into fine droplets either by compressed air in spray nozzles or centrifugally in rotary atomizers. The choice of reactor type and atomizer depends on the selection of the air pollution control equipment vendor.

In the reactor, each droplet is comprised of water with suspended lime particles and

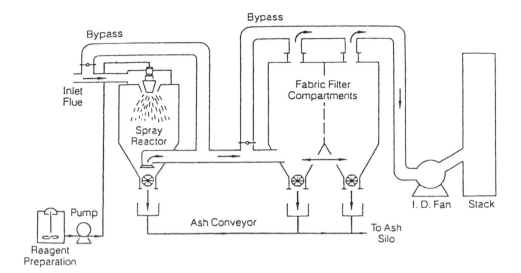

FIGURE 17. Dry chemical scrubber/baghouse air pollution control system.

dissolved lime. The SO_2 and acid gases (HCl, HF, SO_3/H_2SO_4) contained in the flue gas contact the atomized droplets and are absorbed.

Neutralization of SO_2 and the acid gases occurs as the droplets evaporate prior to exiting the spray dryer. Evaporating water cools the flue gas stream. Maintaining a low flue gas temperature out of the spray dryer (approximately 250°F) results in a gas with a high relative humidity. This keeps the droplets wet and moist as long as possible, thereby increasing the rate of reaction. The droplets exiting the spray dryer are dry particles comprised of open-pored reaction salts and unreacted reagent.

The gas is then passed to a baghouse or electrostatic precipitator. Baghouses have generally shown higher removal efficiencies than precipitators. In the baghouse, fabric filters consisting of cylindrical filter bags are vertically suspended in a baghouse compartment. The filter bags are porous which allows the gas to pass through them while capturing a high percentage of the particles suspended in the gas stream. The fly ash, reaction salts, and unreacted reagent are captured in the baghouse as the solids collect on the filter bags. Further SO_2 and acid gas removal takes place on the filter bags as the flue gas passes through the filter cake containing the unreacted reagent. As the particulate cake builds on the bags, the pressure drop across the bags also increases and bags must be cleaned. Cleaning the bags dislodges the particulate cake which falls into a hopper for removal disposal.

2. System Advantages

1. High removal efficiency of particulates and acid gases
2. Good removal of trace organics and heavy metals
3. No liquid wastewater produced

3. System Disadvantages

1. High capital cost
2. High operating cost due to high cost of reagent
3. Extensive additional material handling and mixing requirements for scrubbing reagent

4. System Performance

Extensive testing of full-scale operating systems shows particulate and acid gas removal

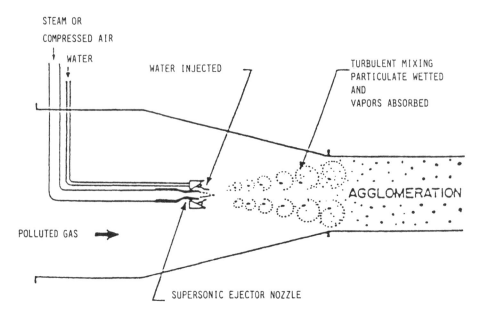

FIGURE 18.   Hydrosonic gas scrubber air pollution control system.

efficiencies which exceed RCRA and TSCA requirements. In addition, the system has shown high removal efficiencies for trace organics and heavy metals.

5. System Operating Experience

The dry scrubber system has been used in municipal solid waste incineration facilities, and is currently planned for use at several proposed facilities.

C. HYDROSONIC GAS SCRUBBER SYSTEM
1. System Description

The hydrosonic gas scrubber is a wet scrubbing system in which the energy for cleaning and pumping of polluted gases is provided by the flow of a compressible fluid from a supersonic ejector nozzle. Steam or compressed air at 30 psig (minimum) is commonly employed as the driving fluid. Absence of fans or other moving parts makes this system attractive when abrasion, corrosion, or vibration preclude use of in stream mechanical equipment.

There are five basic processes involved in the operation of this system and are defined as follows and shown in Figure 18.

Droplet formation — Polluted gas is shown into the system by ejector nozzle, which is fitted with a water injector ring. A water spray is injected onto the exhaust of the expanding jet of the supersonic nozzle. When the water spray strikes the high-speed gas stream, it is shattered into many small droplets. It is desirable to create water droplets 10 to 30 times the size of the particulate to be captured, since the finer dust particles tend to be deflected around a droplet that is too large.

Wetting process — The wetting process consists of the initial collision of the water droplets with the dust particles and acid gas molecules in the exhaust gas stream (see Figure 18).

Mixing process — The mixing process immediately follows and overlaps the wetting process. It insures complete wetting of any particles not already wetted by collision. Mixing is accelerated and enhanced by the recirculation and subsequent turbulence created by the low pressure areas in the mixing tube caused by the ejector. The optimum mixing of the gas stream and water occurs by the combination of these effects.

Growth process — The growth process follows and partly overlaps the mixing process so that a single water drop may contain hundreds of micron and submicron dust particles. As a result of the growth of droplets containing particulates into increasingly larger size, the initial size and shape of the particulate in the gas stream has only a small effect on its removal.

Separation — Separation of cleaned gas from the water-encapsulated particulates is accomplished in a specially designed low-pressure-drop cyclone. Water and particulate are gravity drained from the cyclone bottom while the cleaned gases exit through the top. No demisters are required for this process.

2. System Advantages

1. Low energy consumption
2. Low capital cost
3. Good particulate removal efficiency

3. System Disadvantages

1. Generates wastewater by-products
2. Acid gas removal not demonstrated

4. System Performance
Performance data for RCRA- and TSCA-type waste is not readily available.

5. System Operating Experience
The hydrosonic gas scrubber has had considerable operational experience with the EN-SCO mobile incinerator and the Stablex fixed incinerator. However, the experience has been mainly with soil-type waste and not with the variety of waste available on this project.

D. IONIZING WET SCRUBBERS
1. System Description
The ionizing wet scrubber (IWS) combines the principles of electrostatic particle charging, image force attraction, inertial impaction, and absorption to collect submicron solid particles, liquid particles, and noxious and odoriferous gases simultaneously. The IWS utilizes Tellerette packing as its collection surface to achieve particulate removal. Scrubbing liquid droplets that are formed also act as collection surfaces. Particles of any size or composition are collected by the IWS. The noxious gases are removed concurrently with particles through physical absorption and/or absorption that is accompanied by chemical reaction.

The IWS as shown in Figure 19 utilizes high voltage ionization to charge particulates in the gas stream before the particles enter a Tellerette packed scrubber section where they are removed either by inertial impaction or by attraction of the charged particles to a neutral surface. Particles of 3 to 5 μm and larger are collected through inertial impaction within the packing bed. As the smaller charged particles flow through the scrubber, they pass very close to the surface of either a Tellerette or a scrubbing liquid droplet. The particles become attracted to, and attached to, one of these surfaces by image force attraction. They are eventually washed out of the scrubber in the exit liquor. Noxious and odoriferous gases are absorbed and reacted in the same scrubbing liquor.

2. System Advantages

1. Low energy consumption

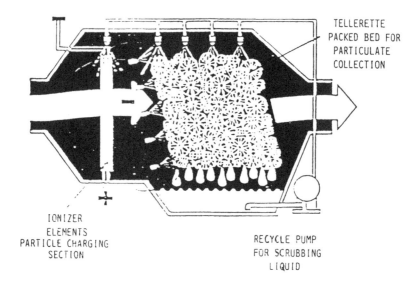

FIGURE 19.   Ionizing wet scrubber air pollution control system.

2.   Collection efficiency independent of system load
3.   Impervious to acid gases because of inert material construction

3. System Disadvantages

1.   Produces wastewater by-products

4. System Performance
Currently, there are several ionizing wet scrubbers installed on existing hazardous waste incinerators. The Chemical Waste Management incinerator in Chicago has a rotary kiln incinerator followed by an ionizing wet scrubber control system.

V. HEAT RECOVERY

Recovery of waste heat is possible with virtually any incineration system. There are three basic types of waste heat recovery systems: (1) gas to water, (2) gas to air, and (3) gas to organic fluid.

Gas-to-water systems produce steam and are the most common heat recovery system used with incineration system. Gas-to-air systems generally use recuperators or air heaters to preheat incoming combustion air. Gas-to-organic-fluid systems use heat exchangers (similar to those used to produce steam) to preheat waste liquids to improve feed characteristics and combustion properties.

Waste heat recovery systems on industrial waste incinerators are typically used to produce steam for other on-site processes. Often, for economic reasons, these incineration systems must be able to replace existing fossil fuel boilers and, therefore, must be capable of producing a constant Btu output. To reduce quantities of auxiliary fuel needed, waste streams must be consistent in terms of Btu values and qualities for the heat recovery system to be economically feasible.

Gas to air and gas to organic fluid waste heat recovery systems are most often used to recover heat of use in the incineration process. Preheating combustion air and/or waste liquids reduces the time needed for the waste to reach its oxidation temperature, thus increasing the effective oxidation time in the primary combustion chamber. Preheating combustion air and waste liquids also greatly reduces the amount of auxiliary fuel needed to maintain minimum temperatures which is the largest portion of operating costs for industrial solid/hazardous waste incinerators. These types of waste heat recovery systems have the advantage of being low pressure systems and, therefore, are much less expensive and less complex than gas to water systems generating steam.

The primary goal of hazardous waste incineration systems is to destroy hazardous wastes. Due to the complexity of hazardous waste incinerators, heat recovery systems which do not directly benefit the combustion process should be a secondary consideration and should not be included unless a significant economic benefit can be identified. The addition of the heat recovery system to recover steam for use in other on-site processes will increase the time and expense for permitting and testing, as well as creating an additional set of operation and maintenance considerations.

Simple gas to air or gas to organic fluid heat recovery systems are the most common types utilized with hazardous waste incinerators. The heat recovery system to recover heat for use in the incineration process itself can benefit combustion efficiency and, therefore, aid in the destruction of hazardous waste as well as reduce auxiliary fuel costs. These types of systems are relatively inexpensive and easy to operate because they are low pressure systems and have little effect on the performance of the incinerator. To design any heat recovery system properly, consideration of the following parameters is required:

1. Chemical nature, temperature, and corrosiveness of the exhaust gases
2. Quantity, specific gravity, size, and nature of the fly ash
3. Available draft
4. Type of exhaust gas system (pressure or vacuum)
5. Space available
6. Requirements for supplemental firing for: startup, preheating, emergency use, stabilizing furnace conditions, or other uses
7. Equipment redundancy requirements (if producing steam)
8. Other special requirements of the individual processes for which the energy is being produced

VI. WASTE PREPARATION AND FEED SYSTEMS

Current waste feed techniques available for hazardous waste incinerators include: (1) screw feeders, (2) ram feeders, (3) lances, and (4) atomizing nozzles. The use of a particular feed mechanism is highly dependent on the type of waste, its physical state, and the specific requirements of the incineration system. Table 9 summarizes the applicability of various waste feed techniques.

In general, most solids are fed with ram or screw feeders, while liquids are usually lanced or injected by atomization with steam or air. Materials that have physical properties between solids and liquids may require pretreatment to facilitate their use with a specific waste feed system. For example, some sludges which are high in solids content may be screw fed, while sludges with low solids content would be applicable to lance feeding. Table 10 identifies possible pretreatment alternatives for wastes.

TABLE 9
Applicability of Various Waste Feed Techniques to Wastes

Feeding technique	Applicable Waste Type				
	Liquid	Solid	Sludge	Foam	Emulsion
Screw feed	No	Yes[a]	No	No	No
Ram feed	No	Yes[b]	No	Yes[b]	No
Lance	Yes	No	Yes[c]	No	Yes[c]
Atomizing nozzle	Yes[c]	No	No	No	No

[a]   Solids must be processed to meet maximum size requirements of screw auger.
[b]   Ram feeders typically require solids to be prepackaged in bags, boxes, or other containers.
[c]   Sludges can be pumped through a lance providing the viscosity of the waste is suitable for size of the lance, desired capacity, and the type of nozzle chosen.

TABLE 10
Pretreatment Options for Wastes

Pretreatment options	Type of Material			
	Liquid	Solid	Sludge	Foam
Dewatering	X	X	X	
Filtration	X			
Homogenization	X			
Containerization		X	X	X
Mixing/blending	X	X	X	

Chapter 7

INNOVATIVE THERMAL DESTRUCTION TECHNOLOGIES*

C. C. Lee, George L. Huffman, Gregory Ondich, and Stephen C. James

TABLE OF CONTENTS

I. Introduction 256

II. Technology Descriptions 257
   A. Oxygen-Enriched Incineration 257
      1. Process Description 257
      2. Case Study Results 258
         a. Throughput Increase 258
         b. Fuel Savings 259
         c. Kiln Puff Reduction 259
      3. Current Technology Status 261
      4. Technology Advantages 261
      5. Technology Disadvantages 261
   B. Westinghouse/O'Connor Combustor 261
      1. Process Description 261
      2. Case Study Results 266
      3. Current Technology Status 267
      4. Technology Advantages 267
      5. Technology Disadvantages 267
   C. Fluidized Bed Combustion 267
      1. Process Description 267
      2. Case Study Results 267
      3. Current Technology Status 268
      4. Technology Advantages 268
      5. Technology Disadvantages 269
   D. Circulating Bed Combustion 270
   E. Molten Salt Combustion 270
      1. Process Description 270
      2. Case Study Results 270
      3. Current Technology Status 270
      4. Technology Advantages 270
      5. Technology Disadvantages 272
   F. Infrared Systems 272
      1. Process Description 272
      2. Case Study Results 272
         a. Destruction and Removal Efficiency 273
         b. Particulate Emissions 273
         c. Acid Gas Emissions 274
         d. Metals Fixation/Ash Leaching 275
         e. Organics in Ash or Scrubber Effluent Solids 275
         f. Metals Emissions from the Stack 276
         g. Dioxins and Furans 276
      3. Current Technology Status 277
      4. Technology Advantages and Disadvantages 277

---

* This chapter was taken from a report originally prepared for the EPA in 1988.

G.   Advanced Electrical Reactor...277
     1.   Process Description 277
     2.   Case Study Results 277
     3.   Current Technology Status.....................................280
     4.   Technology Advantages..280
     5.   Technology Disadvantages....................................280
H.   Plasma Arc...280
     1.   Process Description 280
     2.   Case Study 1: Mobile Plasma Arc Unit.........................281
     3.   Case Study 2: Plasma Arc for Unopened Capacitor
          Destruction..283
     4.   Current Technology Status.....................................283
     5.   Technology Advantages..284
     6.   Technology Disadvantages....................................284
I.   Wet Air Oxidation...285
     1.   Process Description 285
     2.   Case Study Results 286
     3.   Current Technology Status.....................................287
     4.   Technology Advantages..288
     5.   Technology Disadvantages....................................288
J.   Supercritical Fluid..288
     1.   Process Description 288
     2.   Case Study Results 289
     3.   Current Technology Status.....................................289
     4.   Technology Advantages and Disadvantages....................289

III.   Conclusion...289

References...291

I. INTRODUCTION

One of the functions of the U.S. Environmental Protection Agency (EPA) in their Resource Conservation and Recovery Act (RCRA) and Superfund programs is to assist industry in developing innovative technologies for waste destruction.[1] Their main activities in this area are described below.

During 1980 to 1984, EPA's Hazardous Waste Engineering Research Laboratory in Cincinnati, OH sponsored a research program under RCRA to evaluate the performance of innovative thermal technologies for hazardous waste destruction. The program selected and tested the following six technologies:[2]

1.   Fluidized bed
2.   Molten salt
3.   Advanced electrical reactor
4.   Plasma arc
5.   Wet air oxidation
6.   Supercritical water

TABLE 1
Innovative Technologies

Oxygen-enriched incineration	High-temperature oxidation (1500—2000°F)	High-purity O_2 is used as a replacement for combustion air (so gas flows and N_2 content are low).
Westinghouse/O'Connor combustor	Same	Reportedly can be used for both hazardous waste and municipal waste incineration
Fluidized bed combustion (FBC)	Low-temperature (1000—1500°F) incineration coupled with simultaneous scrubbing of products by bed materials	Bed materials are generally lime or limestone powders.
Circulating-bed combustor (CBC)	Now fully commercialized	Now fully commercialized
Molten salt combustion (MSC)	Same	Bed materials are generally molten carbonate salts.
Infrared system	High-temperature pyrolysis or oxidation (1500—2000°F)	Can be operated under starved-air or excess air conditions
Advanced electrical reactor (AER)	High-temperature pyrolysis (~3000°F)	Radiative heat transfer
Plasma arc	High-temperature pyrolysis (18,000°F)	Ionized gas flow
Wet air oxidation (WAO)	Multiphase oxidation	Oxidation occurs in a saturated zone.
Supercritical fluid (SCF)	Same	Oxidation occurs in a supercritical zone.

As a result of the 1980 Superfund Act, industry began to develop transportable incineration systems aimed at destroying both industrial and dump-site remedial waste. In 1986, the Superfund Amendments and Reauthorization Act (SARA) required EPA to establish a Superfund Innovative Technology Evaluation (SITE) program to support specifically innovative technology demonstrations each year. So far, EPA has selected 21 technologies for funding support under the SITE program. Of the 21 technologies, five technologies are thermal processes. They are

Technology	Testing status
Infrared system	Testing results are included in this paper.
Circulating bed combustor	This technology has been fully commercialized and is described more fully in Chapter 6.
Westinghouse plasma arc	No testing data are available yet.
Westinghouse electric pyrolyzer	No testing data are available yet.
Centrifugal reactor	No testing data are available yet.

This paper describes selected innovative technologies either supported by EPA RCRA and SARA programs or developed by industry since 1980. Two of the important criteria used in selecting these technologies are that they are (or had been) at least at the stage of pilot-scale demonstration and appear to be promising, in terms of destruction effectiveness. The technologies selected for this paper are listed in Table 1.

II. TECHNOLOGY DESCRIPTIONS

A. OXYGEN-ENRICHED INCINERATION
1. Process Description

Successful hazardous waste incineration requires an intensive, complete destruction/oxidation of waste molecules with oxygen. Current incinerators require about 150% excess air to provide enough oxygen for oxidation and require that the so-called 3-T factors (temperature, turbulence, and residence time) be adhered to to insure efficient destruction.

Because 79% of air is nitrogen, the majority of any excess air used will not contribute to the effectiveness of incineration and will only result in extra energy required to raise the nitrogen to combustion temperature and in additional product-gas handling and cleaning requirements. As a matter of fact, two of the 3-Ts (turbulence and residence time) are essentially the physical parameters used to promote the contact of hazardous waste particles with oxygen. Therefore, it is quite intuitive that increased oxygen concentration should improve incineration or destruction efficiency.

In late 1985, under the EPA SITE Program, the Union Carbide Corporation (UCC) tested its oxygen-enrichment burner, the Linde Oxygen Combustion System (OCS), on EPA's Mobile Incineration System (MIS). The burn took place using the dioxin-contaminated solids and liquid waste located at the Denney Farm in McDowell, MO. The MIS was built in 1981.[3] It originally consisted of four heavy-duty, over-the-road semitrailers. They were

1. Trailer 1: rotary kiln (4.9 MMBtu/h, i.e., million Btu/h)
2. Trailer 2: secondary combustion chamber (3.9 MMBtu/h)
3. Trailer 3: scrubber and air pollution controls (APC)
4. Trailer 4: combustion and stack gas monitoring equipment

In order to test the oxygen-enrichment concept, the MIS unit was modified. The modified system consisted principally of (1) a rotary kiln, (2) an added refractory-lined cyclone, (3) a secondary combustion chamber (SCC), (4) a wetted throat quench elbow with sump, (5) a wet electrostatic precipitator (WEP) to replace the particulate filter, (6) a mass transfer scrubber, and (7) an induced draft (ID) fan. Ancillary support equipment consisted of bulk fuel storage, waste blending, feed equipment for both liquids and solids, scrubber solution feed equipment, ash receiving drums, and an auxiliary diesel power generator.[4]

As part of the system modifications, the conventional air burner system in the kiln was replaced by the Linde OCS. The Linde OCS consists of the patented "A" Burner,[5,6] an oxygen flow control piping skid, and a control console. The control console utilizes a programmable controller to integrate optimally all system components into a flexible combustion package while providing for easy operator interfacing and safety interlocking. The unique design of the "A" Burner allows the use of up to 100% oxygen in place of air for incineration without creating high flame temperatures, high NO_x, poor mixing, and nonuniform heat distribution. However, for incinerators under even a slight vacuum, close to 100% oxygen enrichment would be very difficult to achieve due to the inevitable air infiltration.

2. Case Study Results

Operation of the modified MIS from early June to mid-September 1987 confirmed that the system had achieved the following major research goals:[4]

1. Throughput increase
2. Specific fuel savings over 60%
3. Kiln puff reduction

Each of these achievements is described below.

a. Throughput Increase

The maximum contaminated soil throughput of the MIS during its operation with air burners had been 2000 lb/h. However, this maximum rate was not sustainable. For example, the average throughput rate of four test runs in the spring of 1985 had been only 1478 lb/h. With the Linde OCS, the MIS achieved a sustainable soil throughput rate of 4000 lb/h,

TABLE 2
EPA Mobile Incineration System: Oxygen Demonstration
Result Summary[4]

	Air case[a]	Oxygen case	Percent change (%)
Contaminated soil (lb/h)	1478	4000	+171%
Firing rate (MMBtu/h)			
Kiln	4.9	3.9	
SCC (secondary combustion chamber)	3.9	5.4	
Total	8.8	9.3	+6%
Specific Fuel Use (MMBtu/ton soil)	11.9	4.7	−61%
Pure oxygen input (lb/h)			
Kiln[b]		574	
SCC		0	
Total		574	
Specific oxygen consumption (ton O_2/ton soil)		0.14	
Fuel savings (MMBtu/ton O_2)		50	
Kiln superficial velocity (ft/s)[c]	8.1	3.3	−59%
SCC residence time (s)[c]	2.6	3.2	+21%
Quenched gas volume (DSCFM)	3250	2250	−31%

[a] Average of four runs. Maximum throughput was 2000 lb/h.
[b] 58% of total oxygen entering the kiln (equivalent to 40% O_2 enrichment).
[c] Calculated data.

as confirmed by a certified verification test. The comparison of the two sustainable conditions is shown in Table 2.

b. Fuel Savings

Without oxygen enrichment, supplemental fuel was added to provide the heat required to operate the rotary kiln at 1500 to 1600°F and the secondary combustion chamber at about 2100°F, because the waste materials did not have a sufficient heating value to sustain self-combustion.

Specific fuel savings of over 60% was achieved during operation of the EPA/MIS with the Linde system. This result can also be expressed as 50 MMBtu saved per ton of oxygen used, as shown in Table 2.

The economics of using oxygen to save fuel, of course, depend on the relative cost of fuel and oxygen. With No. 2 fuel oil costing $0.70/gal (or $5.50/MMBtu) and a fuel savings of about 50 MMBtu for every ton of oxygen used, the break-even oxygen cost is $275/ton of oxygen. The cost of oxygen depends on the method of oxygen generation, the size of the plant, and the location. For example, it ranges from about $50/ton of oxygen produced by a large on-site facility to about $120/ton for delivered liquid.

c. Kiln Puff Reduction[4]

When high-Btu wastes are fed into rotary kiln incinerators in an intermittent mode, the transient combustion behavior of these materials creates unsteady releases of combustible gases which may momentarily deplete the oxygen supply to the incinerator. These temporary oxygen-deficient conditions could cause the release of products of incomplete combustion (PICs) and often are called kiln "puffs".[7,8]

In the field operation of the EPA/MIS, large quantities of plastic materials were burned periodically. These materials were ram fed in the rotary kiln every 1 to 2 min. To respond to the transient oxygen demand as a result of burning these materials, a unique oxygen feed-

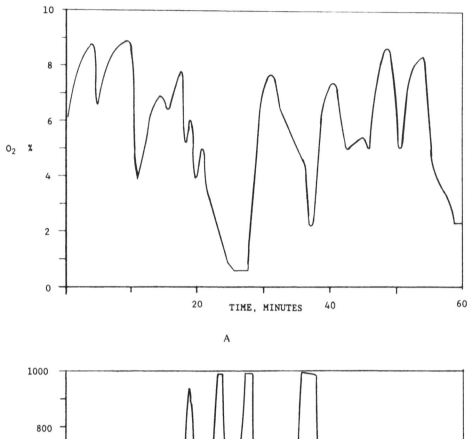

A

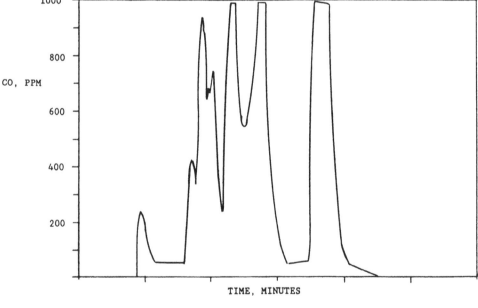

B

FIGURE 1.   Kiln puffs.

forward/feedback control logic was designed into the Linde system. Automatic water spray was used to modulate kiln temperature when required.

Before the implementation of this O_2 control feature, the MIS had difficulty in burning these plastic materials smoothly, partly due to its relatively small capacity. Shown in Figure 1 are examples obtained during the early shakedown period. Even though the normal excess oxygen level was high, occasional feeding-practice upsets caused puffs to occur as evidenced by the drop in the O_2 concentration (to close to 0%) and the CO spikes. Although the MIS is designed so that the waste feed is automatically cut off whenever the O_2 level is below 4% and/or the CO level is above 100 ppm, the waste materials already in the kiln can continue to release combustible gases for a few minutes, during which time the complete destruction of hazardous materials may not be assured. Extreme caution by operators to limit the waste feed rate and to adjust the airflow rate was used to avoid these upset conditions. This was a significant operational constraint.

Immediately after the implementation of the oxygen control feature, the transient upset conditions associated with the release of the combustible gases were virtually eliminated in the operation of the MIS. As shown in Figure 2, the oxygen level of the gas entering the SCC was controlled to be within $\pm 1\%$ from the setpoint of 9% O_2, while the O_2 level at the SCC is maintained at about 6% (dry). Carbon monoxide spikes were not detected. Note also in Figure 2 that the oxygen flow rate responded promptly to the transient oxygen demand. This can be attributed to the fast response (short lag time) of the pure oxygen system and the Thermox® WDG-III *in situ* O_2 analyzer. In addition, the high-momentum oxygen jets in the Linde burner also enhanced mixing in the kiln to eliminate any pockets of unburned combustibles.

3. Current Technology Status

Although Union Carbide Corporation (UCC) completed its burn evaluation with the MIS, EPA is continuing to use the UCC burner to test-burn other wastes. Test results relative to destruction and removal efficiency (DRE) will be available later.

In addition to UCC's oxygen-enrichment tests, the American Combustion Company (ACC) also tested its Pyretron oxygen-enrichment burner at the EPA Incineration Research Facility (IRF) in Jefferson, AAR. Testing results are being evaluated at this writing.

4. Technology Advantages[4]

1. When using oxygen enrichment, the waste throughput of the incinerator, which is normally limited by the air blower capacity, and the gas residence time can be significantly increased.
2. The fuel consumption, if supplemental fuel is required, is lowered primarily due to the reduced sensible heat loss to the flue.
3. The DRE should be improved due to the higher oxygen concentration in the fuel-oxidant mixture and the longer residence time.
4. Pollution control of the reduced flue gas is less costly and more effective.

5. Technology Disadvantages

1. Because of less diluents such as N_2 in the combustion off-gas system, oxygen-enriched incineration, in general, has higher flame temperatures compared to the conventional air-only incineration. The higher flame temperatures often result in higher NO_x formation and local overheating.
2. Oxygen supply (and cost) is the major problem associated with oxygen-enriched incineration.

B. WESTINGHOUSE/O'CONNOR COMBUSTOR

1. Process Description

The heart of the system is the Westinghouse/O'Connor combustor, a water-cooled rotary barrel constructed of alternating longitudinal water tubes and flat perforated steel plates welded together to form the perimeter. The combustor is installed on a slight incline and is

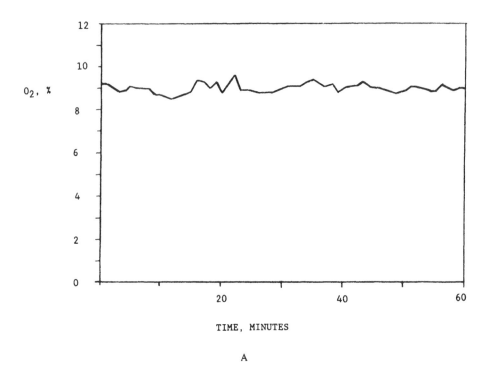

A

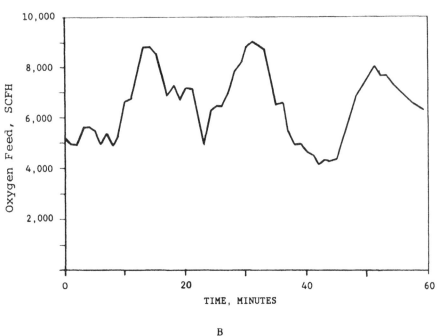

B

FIGURE 2.   Effects of O_2 control.

slowly rotated by a chain and roller drive. The perforations between the water tubes provide controlled distribution of combustion air, while the water-cooled walls remove heat and protect the barrel from overheating.

Waste is fed directly from the receiving area into the upper end of the tilted combustor. As the waste tumbles down the length of the rotating barrel, it dries and then progressively burns. Ash dropping out of the lower end is essentially totally incinerated; remaining unburned material is more completely combusted in an afterburner grate. The remaining residue ash represents only one tenth of the original waste volume.

Combustion gases transmit heat to the water tubes and the water-walled boiler to produce steam. The steam, in turn, is used to drive a turbine generator to produce electricity, or can be piped directly to neighboring industries for direct use.

Exhaust gases pass through air pollution control equipment which, depending on local requirements, may include a dry scrubber to neutralize acid gases released during combustion and bag filters or electrostatic precipitators to remove particulate matter before the gases reach the stack.

The city of Panama City, FL constructed two Westinghouse/O'Connor units at its Bay County Resource Management Facility to incinerate 510 ton/d of municipal solid waste (MSW) in 1987.[9] Heat generated by the combustion of waste produces steam to drive a turbine generator. A process flow diagram of the Bay County facility is shown in Figure 3. The plant consists of two combustor/boiler units, a turbine-generator truck scale, tipping floor, front-end loaders, conveyers, air pollution control equipment, a stack, ash handling equipment, a central control room, and all required ancillary equipment. The facility also includes administration offices, change rooms, parking areas, roadways, and security fencing.

All MSW received at the plant enters through an automatic gate system and is unloaded on the tipping floor. Large items are separated from the MSW; the large combustible items are processed through a shear shredder; the large noncombustible items are removed and stored temporarily for landfill disposal. After sorting, the MSW is thoroughly mixed and then pushed onto the horizontal apron conveyer by the front-end loader. The horizontal apron conveyer transfers the MSW to the inclined apron conveyer and then into the combustor charging chute. From the combustor charging chute, the MSW is pushed into the combustor by the hydraulic ram feeders. The speed of the ram feeders and, consequently, the amount of solid waste fed to the combustor are controlled by the automatic combustion control system. The combustion process begins when the MSW is pushed into the combustor. The slightly inclined combustor barrel rotates slowly, causing the waste to tumble and advance as combustion proceeds. A forced-draft fan draws combustion make-up air from the tipping area to reduce odor and dust levels in the tipping hall and to prevent odor and dust from escaping the building. The air is preheated before entering the multiple zone windbox located beneath each combustor barrel. Figure 4 shows the cross-section of the rotary combustor and the flow of underfire and overfire air into the combustor. The combustor barrel has a diameter of 10 ft and is constructed by alternating steel tubes with carbon steel perforated webs and welding them together. The steel webs have a width of $1\frac{1}{2}$ in. with $\frac{3}{4}$-in.-diameter holes used to bring in combustion air. The tubes direct cooling water through the outside wall of the combustor barrel which on heating is delivered to the boiler through the rotary joint. Hot gases, produced during the combustion process, flow from the combustor barrel through the radiant, superheater, and convection sections of the boiler. The combustion gases exiting the convection section pass through a heat exchanger that preheats the incoming combustion air.

The flue gases from the air heater enter the electrostatic precipitator (ESP) to remove particulate matter before exiting the stack. The flue gas is drawn from the ESP by an induced-draft fan before being discharged to the atmosphere through a separate flue in the common stack. The stack is made of precast concrete with two 4-ft, 6-in. diameter flues that are constructed of 4-in.-thick, acid-resistant bricks. The stack is 125 ft tall and has air emissions monitoring ports located 60 ft from the stack base.

Three types of ash by-products are produced by the process: fly ash, siftings, and bottom ash. Fly ash is collected in hoppers under the convection, superheater, air heater, and ESP

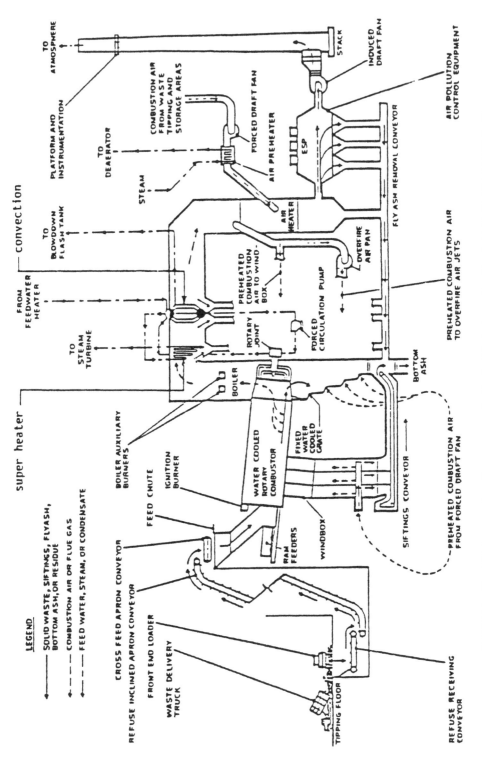

FIGURE 3.   Simplified process flow diagram, gas cycle, for the Westinghouse-Bay County Resource Management Center.

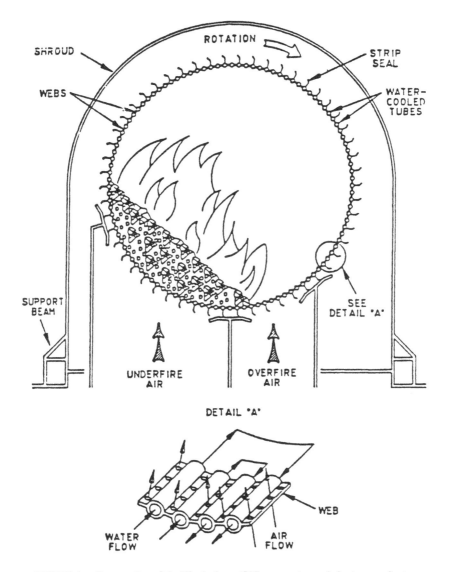

FIGURE 4.   Cross section of the Westinghouse/O'Connor water-cooled rotary combustor.

sections of each combustor/boiler train and is conveyed pneumatically to the bottom ash conveyer. Siftings are collected underneath the combustor by the siftings conveyer and are transferred by an ash drag system to the bottom ash conveyer. Bottom ash falls from the rotary combustor onto a fixed afterburner grate located beneath the combustor outlet. The afterburning grate provides additional time for the remaining combustibles to be consumed prior to their discharge through a bifurcated chute into one of two submerged wet drag conveyers. The fly ash, siftings, and bottom ash mixture are water-quenched, dewatered, and removed by one of two redundant bottom ash drag conveyers into trucks, and then disposed of in a landfill.

Heat from the combustion of MSW produces steam to drive the turbine generator. Boiler feedwater moves through the boiler tubes by natural circulation as it is transformed into a mixture of saturated steam and water. Pumps circulate water through the rotary combustor by drawing water from the lower drum of the boiler through the rotary joint and into one of the combustor barrel's ring headers. The water passes through the combustor tubes and returns to the boiler steam drum as a mixture of saturated water and steam. Steam leaves

TABLE 3

Emission Compliance Test Results from the Bay County Resource Management Center

Date	Time	Flue gas flow (kdscfm)	Flue gas flow (kacfm)	Stack temp (°F)	Steam flow (klb/h)	Percent of rated capacity	Particulate matter or/dscf @ 12% CO_2
Bay County Compliance Test Results — Unit 1							
6/5	959	25.8	52.4	425.0	71.1	104.5	0.0140
6/5	1140	27.9	55.1	429.0	66.5	97.8	0.0240
6/5	1307	25.8	52.8	427.0	65.0	95.6	0.0200
Average					67.5	993	0.0193
Bay County Compliance Test Results — Unit 2							
6/4	945	27.7	52.6	429.0	69.7	102.5	0.0250
6/4	1310	28.4	58.1	449.0	62.7	92.2	0.0190
6/4	1525	29.2	59.0	451.0	62.3	91.6	0.0290
Average					64.9	95.4	0.0243

the drum and passes through the primary and secondary tubes of the superheater section where the steam is heated to the design steam condition for the turbine (750°F). The steam flows from the superheater to the multiple-extraction condensing turbine generator where a portion of its energy is converted to electricity. The generator produces three-phase, 60-Hz electrical power. Transformers provide power at reduced voltage for inplant use, and at increased voltage for distribution to the utility grid. Turbine exhaust steam is condensed in a shell-and-tube condenser that is cooled by an external cooling tower. Steam condensate is pumped back to the boiler through feedwater heaters and a deaerator.[9]

2. Case Study Results

Particulate emission was tested on the Panama City units from April 22 through June 5, 1987.[9] The results indicate that both Units 1 and 2 are in compliance with the particulate and visual emission levels required by the State of Florida Department of Environment's Prevention of Significant Deterioration (PSD) regulations. The PSD regulations require particulate matter emissions to be less than 0.03 gr/dscf corrected to 12% CO_2 and limits plume opacity to no more than 10%. The particulate matter concentration levels at the design capacity of 255 ton of MSW per day per unit averaged 0.0193 gr/dscf at 12% CO_2 for Unit 1 and 0.0243 gr/dscf at 12% CO_2 for Unit 2. The EPA Method 5 particulate measurements conducted for determining compliance on June 4 and 5 are given in Table 3. The EPA Method 9 opacity measurements were consistently at or less than 10% for both units during the test runs.

Gaseous emissions tests were also conducted during April through June 1987 to determine the stack gas concentration of SO_2, NO_x, and HCl. The testing was conducted to verify the emission factors used to project the annual emission rates in the PSD permit application. EPA Method 8 was used to determine SO_2 emissions. The average SO_2 concentration from nine tests performed on 5 d was 111 ppm_{dv} corrected to 12% CO_2 where dv = dry volume. A continuous emissions monitor (CEM) was used to measure NO_x emissions over a 9-d test program. EPA Method 7 was also used to verify the NO_x levels measured by the CEM during an 8-h period by simultaneous sampling. The average of the NO_x emissions data from the CEM and EPA Method 7 were 180 and 157 ppm_{dv} corrected to 12% CO_2, respectively, for the 8-h test. NO_x emissions measured by the CEM during the 9-d test period were in the range of 150 to 200 ppm_{dv} with a maximum of 300 ppm_{dv} during boiler excursions.

NIOSH Method 112B was used to determine HCl concentrations. Twenty samples were taken on seven different days with an average HCl concentration of 467 ppm_{dv} corrected to 12% CO_2.

3. Current Technology Status

Currently, Westinghouse has constructed 29 modular units at 12 different facilities. It appears that the technology has good potential for the future.[9]

4. Technology Advantages

1. The technology can reportedly handle a variety of wastes including municipal wastes, hazardous wastes, and, possibly, even hospital wastes. It can burn liquids, semisolids, solids, sewage sludges, and residual oils or refinery bottoms.
2. The technology can be quite flexible in that the modular design concept can be applied to it. Individual units can be as large as 500 ton/d of MSW, and, therefore, four units would allow for over 2000 ton/d capacity. Units can be added as required.
3. The technology has a high energy recovery efficiency, because the water tubes are used as an integral part of the reactor wall.

5. Technology Disadvantages

1. Siftings from the combustion chambers may contain residuals which are not completely incinerated.
2. The technology has not been fully demonstrated yet on hazardous wastes.

C. FLUIDIZED BED COMBUSTION

1. Process Description[2]

Fluidized bed combustion (FBC) has been used for coal combustion and municipal sludge incineration for the last 10 years. Because of its special character, simultaneous combustion, and scrubbing in the bed, FBC has recently been investigated for possible hazardous waste disposal. In general, an FBC hazardous waste incinerator is not significantly different from one that burns coal or municipal sludge. Its major components include a plenum chamber (or windbox), an orifice place (or air distributor), and a bed (or combustion zone). The bed is usually composed of either inert materials (such as sand) or limestone, depending on the particular application. A cross-section of an FBC is shown in Figure 5.

During operation, a chemical waste, along with fuel oil if necessary, is injected into a fluidized bed combustor operating at approximately 750°C. Air is forced up through the bed material at such a velocity that the mixture behaves like a boiling liquid, and limestone is continuously added to the bed to replenish that expended by reaction with combustion products. Also during operation, bed material is periodically drained from the vessel to maintain the appropriate bed depth and fluidity.

2. Case Study Results[10]

In the 1970s, incineration tests of selected chemical wastes were conducted at the fluidized bed incinerator facility operated by Black Clawson Fibreclaim, Inc., in Franklin, OH. These tests were performed under contract with the Systems Technology Corporation to determine the effectiveness of thermal destruction of two different industrial liquid wastes: (1) an aqueous phenolic sludge and (2) an aqueous solution of methyl methacrylate monomer. Each waste was burned at two different conditions to determine the effects of normal operating and equipment variables. The testing results are shown in Table 4.

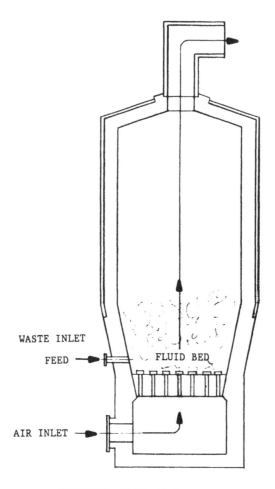

WASTE INLET

FEED →

FLUID BED

AIR INLET →

FIGURE 5.   Fluidized bed reactor.

3. Current Technology Status

Black Clawson discontinued using its FBC for hazardous waste destruction after these early tests. In 1986, the EPA SITE Program selected an FBC technology being developed by Waste-Tech Services, Inc. for Superfund waste destruction testing. According to Waste-Tech, its FBC has a unique feature in that the bed material can be recycled without shutting down the combustor. Material that is dense enough falls out the bottom of the vessel, and bed material can be reinserted. An internal secondary combustion chamber, operating at the same temperature as the bed, provides additional residence time for the reportedly complete destruction of the waste material.

The Waste-Tech combustor can reportedly handle all types of waste, including solids up to 3 in. in size. Residuals from the process include liquid, wastewater, air emissions, ash, and used bed material. However, as of this writing, Waste-Tech has not tested its FBC at any Superfund site.

4. Technology Advantages[2]

1.   Because of its intimate contact with heated bed particles, wastes can be combusted at lower temperatures than that of conventional incinerators.
2.   Temperatures in the vessel are reportedly high enough to destroy wastes, but low enough to prevent formation of significant amounts of NO_x.

TABLE 4
Testing Results of FBC

	Phenol waste	Methyl methacrylate
Bed temperature, average (°C)	740—757	774—788
Freeboard temperature (°C)	813—899	824—843
Residence time (s)	12—14	12
Waste feed rate (l/min)	34—50	30—36
Waste-to-auxiliary fuel[a] (l/l)	2.3—3.0	2.0—2.6
Quality of stack emissions		
Particulate (mg/m³)	1280—1430	560—630
Trace metals (mg/m³)	0.44—0.87 Pb	0.55—2.2 Pb
Quality of combustion gas (before scrubbing)		
Total organics (mg/m³)	7.0—7.6	7.5
Waste content (mg/m³)	Not detected (<0.03)	Not detected (<0.16)
Trace metals (mg/m³)	1.0—1.2 Pb	0.85—4.7 Pb
Quality of scrubber water		
Total organics (mg/l)	Not detected (<0.03)	Not detected (<0.2)
Trace metals (mg/l)	0.50—2.7 Pb	0.45—1.8 Pb
Quality of ash (fluidizing sand)		
Waste content (mg/kg ash)	Not detected (<0.2)	Not detected (<1.0)
Destruction efficiency		
Total organics (%)	99.7—99.98	99.99
Waste constituents (%)	>99.999	>99.999

[a]   No. 2 oil utilized as auxiliary fuel for all tests.

3.   The bed material acts as a scrubber to capture acid gas from the process, reportedly creating a nontoxic solid residue.

5. Technology Disadvantages[2]

1.   Disposal of the supposedly inert residual bed material may represent a problem (if the removed bed materials are considered to be hazardous, they would have to be disposed of at a secure landfill site).
2.   Removal of the residual materials from the bed is a potential problem area. As waste combustion proceeds, noncombustible ash accumulates in the bed. Beds with higher than 20 wt% ash are undesirable, since at ash levels of this magnitude, defluidization can occur.[10]
3.   Relatively large amounts of fine particulate matter entrained in the exhaust gases may require elaborate pollution control devices.
4.   Waste feed particle size must be controlled to maintain a uniform feed rate.
5.   Accurate control is needed to ensure that retention time in the bed is sufficient for complete combustion and that radical increases in the heat value of the waste will not drastically boost bed temperatures and adversely affect bed operation.
6.   Sophisticated analytical methodologies for the analysis of hazardous residues in the spent bed materials may be required. The analysis is normally required to determine how much unburned waste remains in the residues and to examine how much waste residue is leachable to the environment. It is quite possible that milligram amounts of hazardous materials may remain in the spent bed materials undetected. Disposal of

the spent materials could then recycle the hazardous material back into the environment. This situation would be especially dangerous in the case of chemical warfare agents.[10]

7.   Other problems include corrosion of the combustor and mechanical devices for feeding wastes to the combustor.[10]

D. CIRCULATING BED COMBUSTION

Because this technology has been fully commercialized, it is described in Chapter 6.

E. MOLTEN SALT COMBUSTION

1. Process Description[2]

Molten salt combustion (MSC) has long been used in the metallurgical industry to recover metals, especially aluminum. More recent research has been conducted on possible applications of the technology to organic waste destruction, coal gasification, and flue gas desulfurization.

Greenberg[11,12] developed much of the theoretical basis for MSC of hazardous waste. He also made an extensive survey of salt combinations which can be used in the process. The salts used in the process should be stable at temperature required for combustion of specific hazardous substances.

Since beginning the development of MSC technology in 1969, Rockwell International Corporation has constructed four units of varying sizes. Two of these units were bench-scale combustors with feed rates of 0.25 to 1.0 kg/h. They were used for feasibility and optimization studies. A pilot-plant unit had a feed rate of 15 to 100 kg/h. The fourth combustor was portable and was used for disposal of empty pesticide containers. All tests were successful in that the objectives of achieving the required extent of destruction were met.

Molten salt destruction is a method of combusting organic material while, at the same time, scrubbing *in situ* objectionable by-products of the combustion and, thus, preventing their emission in the effluent gas stream. This process of simultaneous combustion and scrubbing is accomplished by injecting the material to be burned with air, or oxygen-enriched air, under the surface of a pool of (generally) molten carbonate salts. The melt is maintained at temperatures on the order of 1650°F (900°C) to promote rapid oxidation of organic waste. Halogen species in the waste form halide salts, and phosphorus, sulfur, arsenic, or silicon (from glass or ash in the waste) form the oxygenated salts such as sodium phosphate, sulfate, arsenate, or silicate, respectively. These products are retained in the melt as inorganic salts rather than released to the atmosphere as volatile gases.

2. Case Study Results[13]

Rockwell conducted many tests in Canoga Park, CA during the development of their molten salt reactor. Table 5 shows their typical bench-scale data on the destruction of hazardous waste.

In 1981, under their contract with EPA, Rockwell tested hexachlorobenzene and chlordane in their pilot-scale molten salt reactor. The results showed that the destruction and removal efficiency of the reactor exceeded 99.99% for these chemicals.

3. Current Technology Status

Since those early tests, Rockwell has decided not to pursue the hazardous waste disposal business. Instead, Rockwell is now trying to use MSC to convert black liquor from the pulp and paper industry to a combustible gas to power a gas turbine.

4. Technology Advantages

The primary advantage of MSC is that the molten material acts as a scrubber to capture the halogens liberated during the destruction of the waste material.

TABLE 5
Rockwell's Testing Data

Type of waste	Waste	Temperature (°C)	Concentration of unreacted waste in off gas (mg/m³)	Concentration of unreacted waste in melt (wppm)	Destruction efficiency (%)
Chemical	PCB	750—900	0.5	0.8	99.9994
	Perchloroethylene bottoms	850—950	<0.5		
	Chloroform	818	<0.5	<0.1	>99.999
	Trichloroethane	840	<1.7		>99.999
	Diphenylamine HCl	922	<0.4	<0.1	>99.999
	Nitroethane	892	<4.4	<1	>99.993
	Aqueous slurry of TBP	939	0.1		>99.99
	para-Arsanilic acid	924	<0.8	<0.1	>99.999
Pesticides	DDT Powder	900	0.3	<0.05	99.998
	Malathion	900	0.06	<0.01	99.9998
Chemical warfare agents	VX	930	<0.00003	<3	>99.99999
	GB	925	<0.0003	<0.1	>99.999999
	Mustard	925	<0.03	<0.03	>99.99998

5. Technology Disadvantages

In general, MSC has the same disadvantages as stated earlier for FBC. However, MSC may not be applicable to dilute acidic-type wastes because rapid contact between water and the salt melt may result in an explosion.[2]

F. INFRARED SYSTEMS

1. Process Description

In general, an infrared system has the following major components:

Primary combustion chamber (PCC) — The electric-powered PCC utilizes a high-temperature alloy, wire-mesh belt for waste feed conveyance and is capable of achieving temperatures up to 1850°F by exposure to infrared radiant heat provided by horizontal rows of electric-powered silicon carbide rods located above the conveyer belt.

Secondary combustion chamber (SCC) — The gas-fired SCC is capable of reaching temperatures of 2300°F; it provides residence time, turbulence, and supplemental energy, if required, to destroy gaseous volatiles emanating from the waste.

Air pollution control equipment (APCE) — The equipment includes an emissions control system where particulates are removed generally in a venturi section, acid vapor is neutralized in a packed tower scrubber, and an induced-draft blower draws the cleaned gases from the scrubber into a free-standing exhaust stack.

Process management and monitoring control center — The center contains all process and mechanical controls and monitors, including control panels, emissions monitors, motor control units, and office space.

2. Case Study Results[14]

Beginning in the 1950s, Peak Oil Company operated a used-oil processing facility at a site on Reeves Road in Brandon, FL. Various liquid and sludge waste streams from the rerefining operation which included toxic organics and heavy metals were dumped into an existing natural lagoon located on the property. The lagoon has some 7000 ton (and approximately the same number of cubic yards) of material in it and is about one quarter to one half of an acre in size.

The EPA Region IV office in Atlanta initiated actions for a cleanup program. The regional office contracted with the firm of Haztech, Incorporated, an emergency response contractor, to proceed with the cleanup program. The EPA Hazardous Waste Engineering Research Laboratory (HWERL) in Cincinnati, through its Superfund Innovative Technology Evaluation (SITE) program, also participated in the cleanup for this particular lagoon. The HWERL effort primarily involved on-site monitoring and the performance evaluation of the infrared system.

Initial efforts required that the lagoon be drained of water; the remaining sludge was mixed with sand, soil, and lime to form a waste/soil solid matrix that could be handled by earth-moving equipment. (The lime, in addition to providing binder to the moisture-laden matrix, also counteracted and neutralized the acidic properties of the waste.) This approach greatly facilitated the use of the infrared system (with a high-temperature afterburner). The SITE demonstration of this innovative incineration system was carried out between July 31, 1987 and August 5, 1987.

The waste/soil matrix was excavated from the lagoon area and screened with a power screen. This device effectively broke up the lumpy waste, blended and aerated the feed, and prepared it for further processing. The screening system allowed the 1-in. and smaller clumps to be separated from the rock, roots, and other extraneous material found in the lagoon. From the power screen, the waste feed soil was weighed and conveyed to the infrared system.

The infrared system was brought to the site in four individual units consisting of (1) a

67-ft-long electrically heated infrared primary chamber capable of gas temperatures up to 1850°F, (2) a 72-ft-long gas-fired secondary combustion chamber (afterburner) capable of reaching temperatures of 2300°F, (3) an emissions control system with a 60-ft-tall stack, and (4) a process management and monitoring control center.

Wastes were metered, spread to the proper width, leveled to optimum process thickness and were fed into the PCC at a rate of 3.6 to 4.0 ton/h in this nominal 100-ton/d unit. The waste was exposed to infrared radiant heat provided by horizontal rows of electric-powered silicon carbide rods located above the conveyer belt. Typically, the waste was retained on the primary chamber conveyer belt for 18 to 19 min. The processed feed material dropped off the end of the belt into a discharge module and was quenched with water sprays prior to being discharged by the screw conveyer system to an outside receiving container.

Exhaust gases from the primary combustion chamber were introduced into the secondary combustion chamber which provided over 3.0 s of residence time, turbulence, and supplemental energy to destroy any gaseous organics from the primary furnace. The gases leaving the secondary combustion chamber were quenched, particulates were removed in a venturi section, and acid vapor was neutralized in a packed tower scrubber using a sodium hydroxide solution. An induced-draft fan drew the cleaned gases from the scrubber into the exhaust stack which was equipped with sample ports.

Radian Corporation was employed to conduct the sampling and analytical phases of the Peak Oil SITE test program. During a series of three replicate test runs, samples of the solid waste feed, liquid streams, furnace ash, and stack gas emissions were obtained under rigorous, EPA-approved sampling protocols and quality assurance criteria. The most intricate sampling procedures involved the stack gas emissions collection. Sampling methods included a Method 5 train for HCl and particulates, and a source assessment sampling system (SASS) for polychlorinated biphenyls (PCBs), dioxins, furans and semivolatile priority pollutants. A modified Method 5 train for soluble chromium, a volatile organic sampling train (VOST) for volatile priority pollutants, and various continuous emission monitors (for O_2, CO_2, CO, THC, and NO_x) were employed. Ambient air monitoring stations were also deployed both upwind and downwind of the unit to monitor any air contamination from the operation of the unit beyond the defined site perimeter.

Table 6 shows the typical range of organic and metallic contaminants in the waste feed excavated from the lagoon after the material was blended into a matrix with sand, soil, and lime in order to facilitate its handling. The primary organic of concern was the PCB, while the primary metal of concern was lead. Other characteristics such as heating value and moisture are also shown.

Presented below is a summary of the results related to each of the objectives for the test program:[14]

a. Destruction and Removal Efficiency

As shown in Table 7, the unit successfully achieved a destruction and removal efficiency (DRE) in excess of 99.99% for the PCB contaminants, thus satisfying the required RCRA standard. TSCA regulations, which require 99.9999% DRE, were not applicable since the contaminated material at the site contained less than 50 ppm of PCBs. Operational temperatures of approximately 1900°F in the secondary combustion chamber and more than 3 s of residence time, instead of the TSCA-mandated 2200°F and 2-s level, accounted for the unit's satisfactory operation in this situation. Residual PCBs in the ash were below the 1-ppm target level mandated by Region IV, thereby achieving one of the most important test objectives.

b. Particulate Emissions

As shown in Table 7, the infrared system at the Peak Oil site demonstrated problems with particulate emissions and failed to meet the regulatory limit of 0.08 gr/dscf at 7% O_2

TABLE 6
Waste Feed Solid Matrix Properties

PCB (total)	3480 to 5850 ng/g
Heptachlorobiphenyl	940 to 2200 ng/g
Hexachlorobiphenyl	1100 to 1700 ng/g
Pentachlorobiphenyl	200 to 490 ng/g
Tetrachlorobiphenyl	400 to 830 ng/g
Trichlorobiphenyl	570 to 820 ng/g
Dichlorobiphenyl	120 to 190 ng/g
Ethyl benzene	80 to 140 ng/g
Methylene chloride	80 to 120 ng/g
Toluene	130 to 440 ng/g
Xylenes	260 to 780 ng/g
Lead	0.44 to 0.59%
Antimony	2.1 to 3.6 µg/g
Arsenic	2.0 to 2.9 µg/g
Cadmium	3.9 to 4.6 µg/g
Chromium	20 to 24 µg/g
Copper	44 to 55 µg/g
Strontium	50 to 62 µg/g
Vanadium	7 to 11 µg/g
Zinc	950 to 1100 µg/g
Moisture	14.2 to 16.6%
Carbon	7.0 to 7.8%
Sulfur	1.8 to 2.5%
Chlorine	<0.1%
Ash	70—75%
Btu value (HHV)	1640—2065 Btu/lb

TABLE 7
Stack Emissions Data

Date of run	DRE for PCB (%)	Particulates gr/DSCF @ 7% O_2	HCl (g/h)	SO_2 (g/h)
8/1/87	99.99967	0.1590	<0.8	27.24
8/2/87	99.9988	0.0939	8.6	1070.0
8/3/87	99.99972			
8/4/87	99.99905	0.0768	2.9	22.0
8/4/87 (duplicate)		0.0761	2.7	20.6

(by volume) on two of the three days they were measured. Emission control system modifications and maintenance before the second test run appear to have been responsible for lowering particulate emissions and eventually meeting RCRA standards in the last two runs.

Interpretation of the data indicates that the inability of the unit's emissions control system to meet particulate emissions requirements was probably due to the lead content in the waste feed and the subsequently high lead oxide loading in the feed to the emissions control equipment.

c. Acid Gas Emissions

The emissions of HCl and SO_2 are also shown in Table 7. The HCl emissions were relatively low since the chlorine concentration in the waste feed was below the 0.1 wt% detection limit. Therefore, an actual HCl removal efficiency could not be determined. The more difficult to remove SO_2 constituent was reduced by an average of 99.1 wt%. With

TABLE 8
Leaching Test Results

Metal	Reg level (mg/l)	Ash EP toxicity (avg mg/l)	Ash TCLP test (avg mg/l)	Effluent solids EP test (avg mg/l)	Effluent solids TCLP test (avg mg/l)
Arsenic	5.0	0.020	0.007	ND	<0.22
Barium	100	1.35	0.25	<0.56	<0.80
Cadmium	1.0	0.98	0.008	0.5—1.9	<0.22
Chromium	5.0	0.57	0.037	<0.65	<0.08
Lead	5.0	31	0.011	3.1—40.0	0.12—0.38
Mercury	0.2	0.0015	ND	ND	ND
Selenium	1.0	ND	0.031	ND	ND
Silver	5.0	0.031	0.059	<0.04	<0.06

Note: ND = not detectable.

sulfur at approximately 2% in the waste feed or a sulfur input rate of about 150 lb/h SO_2 emissions were under 2.5 lb/h at a maximum measured rate of 1070 g/h.

d. Metals Fixation/Ash Leaching

One of the objectives of this test program was to determine the fate of heavy metal contaminants in the waste feed and whether any metals remaining in the ash were significantly reduced in concentration or would be stabilized in the ash residue and other solid effluents, rendering the ash potentially suitable for delisting as a hazardous waste. The solid waste feed, furnace ash, and scrubber effluent solids were subjected to both the current toxicity characteristic leaching procedure (TCLP) and the EP toxicity leaching tests.

The results on the waste feed for both TCLP and EP testing showed that the waste feed exceeded regulatory limits only for lead. The TCLP regulatory limit for lead of 5.0 mg/l was exceeded in one sample of waste feed at 8.8 mg/l, while other samples showed values of 2.5 to 3.5 mg/l. However, the EP regulatory limit (also 5.0 mg/l) was exceeded for lead in the waste feed by larger amounts, namely, values of 24.0 to 29.0 mg/l and one sample investigation (though not within acceptable spike recovery limits) showed 57.0 mg/l. All other metals tested for their TCLP and EP values in the waste feed (eight metals total) were found to be below regulatory levels.

For the ash residue, the results of TCLP and EP testing are summarized in Table 8. Here, TCLP results were lower than regulatory limits in every case. However, the EP results again showed a regulatory limit exceedence for lead which would probably prevent delisting of the ash residue even though the other metals appeared well below their regulatory limits.

Similarly, for the scrubber water effluent solids, Table 8 shows that, again, the EP testing values for lead would prevent the material's delisting, while other metals were found to pass the regulatory levels. Thus, the hoped-for result that lead and other heavy metals would somehow be inactivated by some process of conversion into nonleachable forms was not met (if EP test results are used and not the TCLP ones).

e. Organics in Ash or Scrubber Effluent Solids

TCLP testing for organic materials other than PCB was done on the waste feed, the ash residue, and the scrubber solid samples. The results, which involved testing for some 36 organic hazardous compounds, will be reported in the upcoming final EPA report. However, other than those compounds reported in Table 6 for the waste feed material, all 36 compounds were generally found to be either nondetectable or far below any TCLP regulatory levels.

TABLE 9
Metals Analysis

Parameter	Solid waste feed (μg/g)	Ash (μg/g)	Average stack emission rate (g/h)
Antimony	2.15	3.3	0.13
Arsenic	2.55	2.6	0.065
Cadmium	4.15	4.1	3.4
Chromium	22	27	0.54
Copper	49	64	0.37
Lead	4800	6400	1150
Mercury	ND	ND	ND
Sodium	5550	5600	33
Strontium	57	76	0.017
Vanadium	9	13	<0.036
Zinc	1030	1060	16

Note: ND = not determined.

f. Metals Emissions from the Stack

The largest mass release of metals measured in the stack emissions was that of lead. The specific amounts varied from a release of 2000 g/h of lead for the first test run (on 8/1/87) down to a range of 780 to 1000 g/h in subsequent runs where changes were made to the operation of the pollution control system. It is interesting to note that the calculated net input feed rate of lead from the waste ranged between 17,000 to 19,000 g/h for the test runs made. Thus, the amount of lead exiting the stack represented only about 11.2% of the feed rate of lead for the first run and between 4.2 to 5.5% for subsequent runs. Also, it should be noted that lead (or lead compounds) represented 54 to 60% of the mass of particles from the stack.

In total, 30 different metals were sampled for and analyzed in the stack particulate material. Other than lead, the next highest amounts of metals emitted as particulate were measured to be zinc (16 to 17 g/h), sodium (31 to 34 g/h), and chromium (0.51 to 0.53 g/h). Stack sampling and analysis for chromium VI concentrations yielded results of less than 160 μg/m^3.

A partial listing of the data for metals is presented in Table 9. Presented there are 11 metals in terms of their levels in the waste feed, in the ash residue, and their levels emitted via the stack gas particulate material.

g. Dioxins and Furans

Since PCB was in the waste, it was desirable to investigate whether there may have been any significant levels of dioxins and/or furans in the lagoon and, thus, in the waste matrix processed by the infrared system. Sampling and analysis for these compounds was conducted for the waste matrix, the ash, the scrubber waters, scrubber solids, and the stack gas. All of the above analyses resulted in less-than-detectable levels of these compounds as summarized below for TCDD, TCDF, PCDD, and PCDF (tetra- and polychlorinated dibenzodioxins and -furans):

Sample	TCDD	PCDD	TCDF	PCDF	Detection limit
Waste feed matrix	ND[a]	ND	ND	ND	1.1 ppb
Ash after treatment	ND	ND	ND	ND	1.4 ppb
Scrubber water	ND	ND	ND	ND	0.022 ng/l
Stack gas	ND	ND	0.47[b]	ND	0.34 $\mu g/m^3$

[a] ND = Nondetectable.
[b] Reported for one out of four tests.

3. Current Technology Status

The infrared system is, so far, probably the most used technology of all the innovative thermal technologies for both RCRA waste destruction and Superfund waste remediation.

4. Technology Advantages and Disadvantages

The technology has the ability to control residence time and temperature more accurately in the PCC compared to other technologies. However, the waste feed should be pretreated so as not to exceed recommended size limitations. If the waste is a liquid, it should be mixed with sand, or other solid material, in order to destroy it effectively in the PCC.

G. ADVANCED ELECTRICAL REACTOR
1. Process Description[2]

The high-temperature fluid-wall (HTFW) reactor was originally designed for the continuous dissociation of methane into carbon black and hydrogen. The process requires the generation of stable temperatures above 1600°C and the prevention of precipitate formation on the reactor wall.

The key within the reactor is to keep the reacting stream out of physical contact with the reactor wall by using a gaseous blanket to keep them separated. The gaseous blanket is formed by flowing an inert gas radially inward through a porous reactor tube (a core). The inert gas is also used as heat transfer media to carry high rates of heat to accomplish the heating of the porous carbon core to incandescence so that the predominant mode of heat transfer is by radiation from the core to the stream. By avoiding physical contact of the stream with the core, the problem of chemical compatibility of the two (stream and core) becomes a nonproblem. Additionally, the heating of particles is accomplished without the necessity of bringing the entire system to the same temperature.

While successfully pyrolyzing methane to carbon and hydrogen at rates of commercial interest, the reactor also demonstrates its usefulness in other applications including the pyrolysis of organic material to carbon and hydrogen or to carbon, carbon monoxide, and hydrogen (depending on the composition of the starting material), and the production of high-melting-point refractories and Portland cement.

A partial horizontal cross-section of a typical fluid-wall reactor is shown in Figure 6 and a vertical cross-section is shown in Figure 7. The reactor is heated electrically with the six carbon resistance heaters. Because of the extreme temperatures encountered in operation of the device, the insulation package consists not of refractory brick but of a radiation shield made of multiple layers of graphite paper backed up with carbon felt.

During operation, the waste material to be pyrolyzed is finely ground to a 20 mesh and introduced into the top of the reactor. As the material falls through the tubular space it is exposed through radiative coupling to power densities of over 1200 W/in^2. The finely divided reactants are heated through the direct impingement of electromagnetic radiation. Examples of the reactions that take place are shown in Table 10.

2. Case Study Results[3]

J. M. Huber Corporation owns the advanced electrical reactor (AER) patent and per-

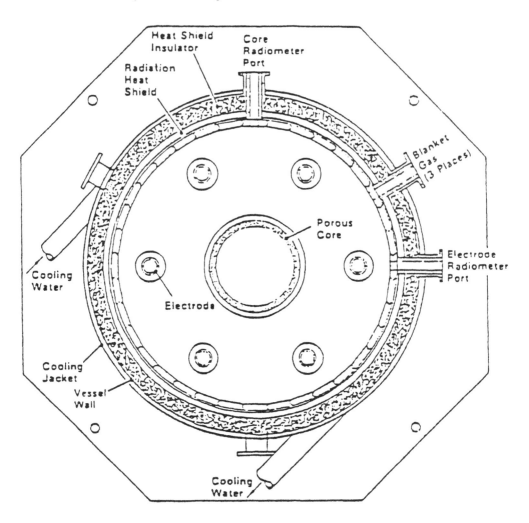

FIGURE 6.   Horizontal cross section of a typical fluid-wall reactor.

formed tests on both a 3-in. (36 lb/h capacity) and a 12-in.-diameter (2500 lb/h capacity, 3.4 million Btu/h) reactor. The purpose of conducting these tests was to determine (1) the performance of the AER to destroy PCBs, CCl_4, and dioxin mixed with soils and (2) the suitability of the tested solids for landfill under the RCRA guidelines. The dioxins tested were octachlorodibenzo-*p*-dioxin (OCCD) and 2,3,7,8 tetra-chlorodibenzo-*p*-dioxin (TCDD).

PCB (Aroclor 1260) was mixed with sand to form a solid feed containing approximately 3000 μg/g PCB. Carbon black was added to the feedstock at approximately a 6.25:1 ratio to the PCB oil to simulate the organic carbon content of soil. Carbon tetrachloride was mixed with screened, dried soil (less than 35 mesh) with CCl_4 concentration approximately 0.37 to 13.76%. Activated carbon was added to the feedstock at 94:1 ratio (w/w) to CCl_4 to reduce the volatility of CCl_4 and to prevent its loss before reaching the reactor.

PCB test results reportedly show that in all test cases, DREs exceeded 99.99999%, at least an order of magnitude greater than the requirement of the Toxic Substances Control Act (TSCA) regulations. Maximum PCB concentrations in the treated feed and baghouse filter catch were 0.001 and 0.53 μg/g, respectively. These values are well below the TSCA limit of 50 μg/g set for solids to be treated as hazardous wastes. Although results for the scrubber liquid were variable, ranging from 0.29 to 2.7 μg/l, all were well below the TSCA limit of 50 mg/l set for liquids to be treated as hazardous wastes. The results also reportedly

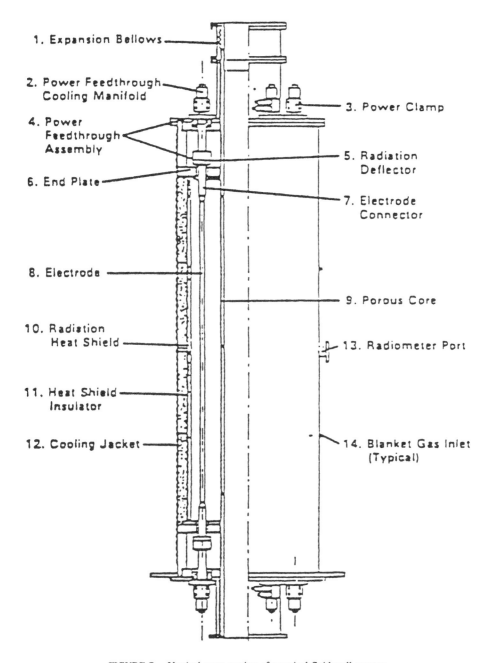

FIGURE 7. Vertical cross section of a typical fluid-wall reactor.

show that PCDDs (dioxins) and PCDFs (furans) at the cyclone outlet were below analytical detection limits.

For the CCl_4 testing, its DRE results reportedly show values greater than 99.9999%. These results are at least two orders of magnitude better than RCRA requirements for hazardous waste incinerators.

Huber conducted triplicate tests in Borger, TX on their 12-in.-diameter reactor with OCDD (OCDD mixed with clean soils) in late October 1984. Feed concentrations up to 18,000 ppb, by weight, and feed rates up to 15 lb/min were used. In all cases, no OCDD or products of incomplete pyrolysis (PIPs) were reportedly detected.

TABLE 10
HTFW Reaction Pattern

Feedstock	Example reaction	Product
Hydrocarbon: oil or gas	$CH_4 \rightarrow C + 2H_2$	Carbon black
Chlorinated hydrocarbons	$C_6Cl_6 + 3Ca(OH)_2 \rightarrow 3CaCl_2 + 6CO + 3H_2$	Inert material
Hydrocarbon oil emulsions or coal/water slurries	$Hydrocarbon + H_2O \rightarrow CO + H_2$	Synthesis gas
High-sulfur oils or synthesis gases	$Oil + Ca(OH)_2 + Air + H_2O \rightarrow CO + H_2 + CaS$	Sulfur-free Syngas
Calcium carbonate and silica	$CaCO_3 + SiO_2 \rightarrow Cement$	Cement
Coal and calcium hydroxide	$Coal + Ca(OH)_2 \rightarrow CaC_2 + CO + H_2$	Calcium carbide
Nuclear waste	$Waste + Ca(OH)_2 + clay \rightarrow glass$	Inert glass beads

Triplicate tests of TCDD-contaminated soils were conducted with the 3-in.-diameter reactor at Times Beach, MO on November 13, 1984. Again, no dioxin or PIPs were reportedly detected.

3. Current Technology Status

After these early tests, the J. M. Huber Corporation decided not to pursue the use of their process for hazardous waste destruction. Consequently, development work on the AER technology for this application is currently not active.

4. Technology Advantages[2]

1. The high operating temperature provides high radiant energy densities that should assure rupture of the chemical bonds, thus breaking the wastes into their simplest constituents, and eliminating the formation of intermediate products.
2. The high temperature appears sufficient to destroy virtually any chemical bond and produce clean off-gases, thus minimizing the needs of off-gas cleanup systems.
3. The solid residue is a glasslike particle that appears to be nonleachable. If this is indeed the case, the residue would not have to be disposed of in a secure chemical landfill.
4. The configuration of the reactor and the nature of the process make scaling from pilot units a relatively simple procedure.
5. The simplicity of the design should reduce capital outlay and maintenance costs.

5. Technology Disadvantages[2]

1. Solid wastes need to be ground to about 20-mesh size before processing.
2. The reactor core can only accept solid materials. Therefore, if a waste is gaseous or liquid, it must be blended with nonhazardous solid materials to form a solid powder suitable for feeding.

H. PLASMA ARC
1. Process Description[2]

The plasma arc technology has been used in the U.S. space program. The evaluation of heat shields that protect space vehicles on reentry required an intense heat source with plasma characteristics. Plasmas have been referred to as the fourth state of matter since they do not always behave as a solid, liquid, or gas. A plasma may be defined as a conductive gas flow consisting of charged and neutral particles, having an overall charge of approximately zero and all exhibiting collective behavior.

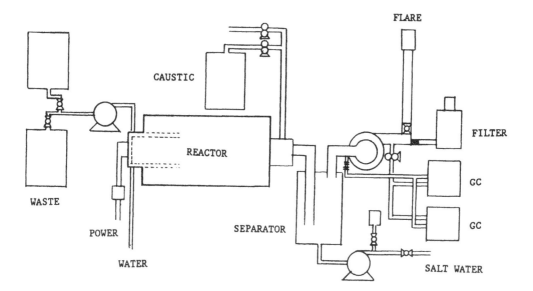

FIGURE 8.   Process schematic of plasma pyrolysis system.

The most common method of plasma generation is electrical discharge through a gas. A low pressure gas is used as a medium through which an electrical current is passed. The type of gas used is relatively unimportant in creating the discharge, but will ultimately affect the products formed. In passing through the gas, electrical energy is converted to thermal energy by the absorption of gas molecules, which are activated into ionized atomic states, losing electrons in the process. Arc temperatures up to 10,000°C may be achieved along the centerline recirculation vortex. Ultraviolet radiation is emitted when molecules or atoms relax from the highly activated states to lower energy levels.

The plasma, when applied to waste disposal, can best be understood by thinking of it as an energy conversion and energy transfer device. The electrical energy input is transformed into a plasma with a temperature equivalent of up to 10,000°C at the centerline of the reactor. As the activated components of the plasma decay, their energy is transferred to waste materials exposed to the plasma. The wastes are then broken into atoms, ionized, pyrolyzed, and, finally, destroyed as they interact with the decaying plasma species. The heart of this technology is that the breakdown of the wastes into atoms occurs virtually instantaneously, and no large molecular intermediary compounds are produced during the kinetic recombination.

2. Case Study 1: Mobile Plasma Arc Unit[15]

The New York State Department of Environmental Conservation (NYSDEC) Division of Solid and Hazardous Waste and the U.S. Environmental Protection Agency (USEPA) Hazardous Waste Engineering Research Laboratory established a cooperative agreement in 1982 for the construction and testing of a mobile plasma arc system for the high-efficiency destruction of hazardous wastes.

Figure 8 shows a process diagram of the plasma pyrolysis system which consists of a liquid waste feed system, plasma torch, reactor, caustic scrubber, on-line analytical equipment, and flare. The system is rated at 4 kg/min or approximately 55 gal/h of waste feed. Product gas production rates are about 5 to 6 m³/min prior to flaring. For the purposes of this program, a flare containment chamber and 10-m stack were constructed to facilitate testing. After flaring, the stack gas flow rate is approximately 30 to 40 standard cubic meters per minute.

TABLE 11
Carbon Tetrachloride Test Results

	Run		
	1	2	3
Date	2/18/85	2/26/85	2/26/85
Sample time (min)	60	60	60
Feed rate (l/min)			
CCl_4	0.63	0.63	0.63
MEK mixture	2	1.6	2
Total mass fed, CCl_4 (kg)	60.0	60.6	60.6
Chlorine loading, mass (%)	35	40	35
Reactor operating temp. (°C)	974	1008	1025
Plasma torch power (kW)	280	298	300
Avg stack gas flow rate (dscfm)	1346.3	1048.2	1052.7
Avg stack gas flow rate (dscmm)	38.13	29.69	29.81
Average temp (°C)	893.3	807.1	667.3
NO_x conc. (ppm [v/v])	106	92	81
Emission rate (kg/h)	0.46	0.31	0.28
CO conc. (ppm [v/v])	48	57	81
Emission rate (kg/h)	0.13	0.12	0.17
O_2 (%)	12.7	14.4	15.1
CO_2 (%)	6.0	5.7	4.9
HCl (mg/dscm)	[a]	137.7	247.2
Emission rate (kg/h)	[a]	0.25	0.44
CCl_4 conc. (ppb)	<2[b]	<2[b]	<2[b]
Emission rate (mg/h)	<29.3	<22.8	<22.9
Scrubber effluent flow rate (l/min)	30.0	30.0	30.0
CCl_4 conc. (ppb)	1.3	5.5	3.3
Discharge rate (mg/h)	2.3	9.9	5.9
Destruction removal efficiency[c]	99.99995	99.99996	99.99996
CCl_4 (% DRE)			

[a] Invalid data.
[b] The detection limit of 2 ppb CCl_4 in the stack gas was used to calculate the CCl_4 mass emission rate for each run.
[c] The DRE is based on stack emissions and excludes scrubber effluent.

A gas chromatograph is installed in the system's mobile trailer to provide information on bulk gas composition. Preflare gas samples are analyzed for hydrogen, carbon monoxide, carbon dioxide, water, nitrogen, methane, ethylene, ethane, acetylene, propane, propylene, and 1-butene. A Hewlett-Packard Model 5792A gas chromatograph is also coupled to a Hewlett-Packard Model 5970A mass selective detector and used to analyze preflare gas samples for waste feed residuals.

One-hour carbon tetrachloride (CCl_4) tests were conducted in Canada to demonstrate the destruction of a simple chlorinated compound and to demonstrate effective HCl removal by the scrubber. The CCl_4 was introduced into the system in a blend of methyl ethyl ketone (MEK), ethanol, and water at a rate of 1 kg CCl_4 per minute. The stack testing was conducted by GCA Corporation. Results of the three tests are presented in Table 11. The results indicate that the system is capable of achieving destruction and removal efficiencies much greater than the 99.99% required under RCRA for this listed waste.

Three 1-h tests using PCBs were also conducted to determine the DRE for PCBs and to acquire operating data for this environmental contaminant. On reaching a product gas temperature of 1100°C at the reactor exit while using MEK/MeOH (methanol) as the feed material, the feed was then switched to a blend of PCBs, MEK, and MeOH. The scrubber water was retained before discharge to sewer to ensure that the concentration of BaP, PCBs,

TABLE 12
Operating Data for PCB Run No. 1

Elapsed Operating Time: 70 min at Operating
Temperature

Feed rate	
Total feed	3.09 l/min
	2.83 kg/min
PCB feed	0.40 kg/min
Feed composition (mass)	14.1% PCB
	11.0% TCB[a]
	74.9% MEK/MeOH
Reactor operating temperature	1136°C
Plasma torch power	327 kW

[a]   Trichlorobenzene.

dioxins, and furans did not exceed criteria established by Canadian regulatory authorities. Operating data are presented in Table 12.

The stack gases were monitored for particulates, PCBs, dioxins, furans, NO_x, HCl, flow rate, and temperature. For these runs the stack monitoring was conducted by IMET, Inc. (Markham, Ontario, Canada). The sampling and analysis program included the necessary quality assurance/quality control protocols. The procedures and equipment were standard. A summary of the results is presented in Table 13.

3. Case Study 2: Plasma Arc for Unopened Capacitor Destruction[16]

A schematic of this system is shown in Figure 9.[16] The research on this system is sponsored by the Electric Power Research Institute (EPRI) and Arc Technologies Company (ATC) which is a joint venture of Chemical Waste Management Inc. and Electro-Pyrolysis, Inc. The purpose is to destroy whole, unopened PCB capacitors at the rate of 3000 to 5000 lb/h. The sponsors have obtained RCRA permits for the construction of the system in Model City, NY and are in the process of constructing the unit.

Basically, the process involves four steps as follows (as coded on Figure 9):

1. A whole, unopened PCB capacitor is fed into a molten metal (iron) bath.
2. The capacitor shells melt and the internal organic components (PCBs) are subjected to intense radiation from the plasma arc and heat from the molten metal. As a result, the PCB fluids evaporate and decompose.
3. The PCB and/or decomposition products are further directed through the plasma arc in the vicinity of the high current (DC) arc for complete destruction.
4. The products are channeled to a scrubber system where any inorganic materials in capacitors are converted to inert solid residues. Metals may be recovered if desired from both the metal bath and scrubber ash.

4. Current Technology Status

Since the Canadian tests, Westinghouse has built a mobile plasma unit (Pyroplasma) of 3-gal/min (1-ton/h) capacity. Also, for the same high-temperature application, Westinghouse constructed a "Westinghouse Electric Pyrolyzer" for contaminated soil detoxification.

The Westinghouse Electric Pyrolyzer uses electric energy to heat the waste feed to about 3000°F. At this temperature, the organic contaminants are destroyed by decomposition to carbon monoxide and water, and the inorganics melt to form a glass-like fluid which is removed. Both the Pyroplasma and the Pyrolyzer units have been selected for EPA's SITE program tests. However, no Superfund site has yet been identified for testing.

TABLE 13
PCB 1-h Test Results

	Run		
	1	2	3
Date	12/5/85	12/17/85	1/16/86
Sample time (min)	50	60	60
Stack gas parameters			
Flow rate (dscmm)	37.9	45.0	38.1
Temperature (°C)	836	678	962
NO_x (ppm)	117	N/A	139
HCl (mg/dscm)	N/A	43	68
O_2 (%)	14	14.5	16.5
CO_2 (%)	5.5	5.0	3.0
CO (%)	0.01	0.01	0.01
Total PCB[a]	<0.013	0.46	3.0
μg/dscm[b]	<0.013	0.32	<0.011
Total dioxins (μg/dscm)	0.076[c]	<0.43	<0.13
Total furans (μg/dscm)	0.26	1.66	<0.30
Total BaP (μg/dscm)	0.18	0.45	2.8
Scrubber effluent parameters			
Effluent flow rate (l/min)	41	36	33
Total PCB (ppb)[a]	1.56	2.15	9.4
Total PCB (ppb)[b]	0.06	4.7	<0.01
Total dioxins, (ppt)	5.8	<259	<1.05
Total furans (ppt)	1.5	399	<1.05
Total BaP (mg/l)	0.04	0.92	2.0
Destruction removal efficiency			
PCB (% DRE)			
[a]	99.999999	99.99999	99.999999
[b]	99.999999	99.99997	99.999999

[a] These values are based on monodecachlorobiphenyl.
[b] These values are based on tridecachlorobiphenyl.
[c] No tetra- or pentadioxins were detected at 0.05 ng on a GC column, except for Run No. 1, where 0.06 ng tetradioxin was reported.

5. Technology Advantages[16]

1. Since radiative heat transfer proceeds as a function of the fourth power of temperature, a plasma system has very intense radiative power and therefore is capable of transferring its heat much faster than a conventional flame.
2. Organic chlorides are known to dehydrogenate when excited by the ultraviolet radiation which is abundant from thermal plasmas.
3. Because the plasma arc for waste destruction is a pyrolytic process, it virtually does not need oxygen at all. Compared to conventional incinerators which normally require about 150% excess air to ensure proper combustion, the plasma arc will save the energy required to heat the excess air to the combustion temperatures and will thereby produce significantly fewer oxygenated by-products that would otherwise need to be treated in downstream equipment.
4. The process has a very short on/off cycle.
5. Because of its compactness, a plasma arc system has potential for use in a mobile trailer for movement of the system from site to site.

6. Technology Disadvantages[16]

1. Because the temperatures are so high (about 10,000°C at the arc's centerline), the

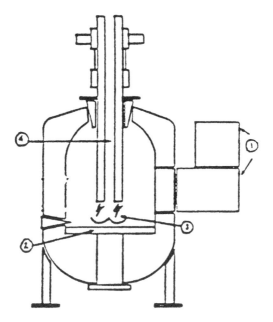

FIGURE 9. Process concept of plasma arc for unopened capacitor destruction.

    durability of the arc and the refractory materials could be a potential problem.
2.   Because the arc is very sensitive to many factors such as sudden drops in voltage, the operation of the system requires highly trained professionals.

I. WET AIR OXIDATION
1. Process Description[2]

    Wet air oxidation (WAO) has been used worldwide to treat both industrial and municipal waste streams for the last 30 to 40 years. Operating designs vary from "low oxidations", for sludge conditioning, to "high oxidations", for chemical and power recovery and for carbon regeneration. During recent years, bench-scale and pilot testing has been performed by various companies to determine the applicability of wet oxidation for the destruction of various hazardous organic wastes. Wet oxidation units currently detoxify specific waste streams at several generators.

    WAO refers to the aqueous phase oxidation of dissolved or suspended organic substances at elevated temperatures and pressures. Water, which represents the aqueous phase, serves to catalyze the oxidation reactions so that they proceed at relatively low temperatures (175 to 345°C), and at the same time it serves to moderate the oxidation rates removing excess heat by evaporation. Water also provides an excellent heat transfer medium which enables the WAO process to be thermally self-sustaining with relatively low organic feed concentration.

    The oxygen required by the WAO reactions is provided by an oxygen-containing gas, usually air, bubbled through the liquid phase in a reactor used to contain the process: thus, the commonly used term "wet air oxidation". The process pressure is maintained at a level high enough to prevent excessive evaporation of the liquid phase, generally between 200 and 3000 psi.

    A temperature-volume diagram, Figure 10, shows the operating zone of WAO. The water-waste mixture is maintained at the saturated zone within EBCF (Figure 10). For the particular case shown, total pressure is maintained at 2000 psi, and temperature varies from 175 to 320°C.

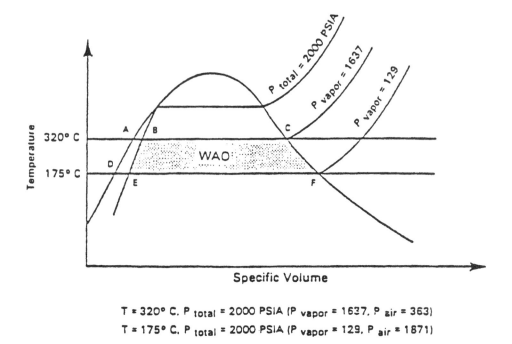

$T = 320° C. P_{total} = 2000 PSIA (P_{vapor} = 1637, P_{air} = 363)$

$T = 175° C. P_{total} = 2000 PSIA (P_{vapor} = 129, P_{air} = 1871)$

FIGURE 10.   Typical wet air oxidation zone.

2. Case Study Results[2]

In 1976, the EPA published a list of 65 compounds or classes of compounds having top priority for elimination from effluents. Based on chemical structure, Zimpro Inc. chose a variety of these compounds for study. Their results are shown on Tables 14 and 15.

The solutions resulting from the WAO of these compounds at 320 and 275°C and 1 h were tested for toxicity using *Daphnia magna*. These results were compared with the toxicity of the starting compound. Toxicity reduction was quite good for all compounds studied. The data also indicated that compound destruction or toxicity reduction is nearly as good at 275°C as that obtained at 320°C.

These studies provided some empirical observations concerning the susceptibility of a particular compound to conventional (with or without a copper catalyst) WAO based on its molecular structure. These observations were

1. Aliphatic compounds, even with multiple halogen atoms, can be destroyed within conventional WAO conditions. Formation of oxygenated compounds, such as low molecular weight alcohols, aldehydes, ketones, and carboxylic acids result, but these are readily biotreatable.
2. Aromatic hydrocarbons such as toluene, acenaphthene, or pyrene are easily oxidized.
3. Halogenated aromatic compounds can be oxidized provided there is at least one non-halogen functional group present on the ring, i.e., pentachlorophenol (–OH) or 2,4,6-trichloroaniline (–NH₂).
4. Halogenated aromatic compounds such as 1,2-DCB or Aroclor 1254, a PCB, are resistant to conventional conditions, but can be oxidized in the presence of a copper catalyst.
5. Halogenated condensed-ring compounds, such as the pesticides aldrin, dieldrin, or endrin, are expected to be resistant to conventional WAO, but should be oxidizable in the presence of a copper catalyst.

TABLE 14
Wet Air Oxidation

Compound	Starting conc. (g/l)	Percent starting material destroyed after 1-h oxidation		
		320°C	275°C	275°C[a]
Acenaphthene	7.0	99.96	99.99	
Acrolein	8.41	>99.96[b]	>99.05[b]	
Acrylonitrile	8.06	99.91	99.00	99.50
2-Chlorophenol	12.41	99.86	94.96	99.88
2,4-Dimethylphenol	8.22	99.99	99.99	
2,4-Dinitrotoluene	10.0	99.88	99.74	
1,2-Diphenylhydrazine	5.0	99.98	99.88	
4-Nitrophenol	10.0	99.96	99.60	99.97
Pentachlorophenol	5.0	99.98	81.96	97.30
Phenol	10.0	99.97	99.77	99.93

[a] Cu used as the catalyst.
[b] The concentration remaining was less than the detection limit of 3 mg/l.

TABLE 15
Wet Air Oxidation

1 h With or Without Catalyst

Compound	Temp (°C)	Catalyst	Percent Destroyed
Sodium cyanide (NaCN)	275	0	99.96
Potassium thiocyanate (KSCN)	275	0	99.98
Formic acid (CO_2H_2)	275	0	99.3
N-Nitrosodimethylamine	275	0	99.56
($C_2H_6N_2O$)	275	+	99.38
Toluene	275	0	99.73
(C_7H_6)	275	+	99.96
Pyrene	275	0	99.995
($C_{16}H_{10}$)	275	+	99.997
Malathion	200	+	99.87
($C_{10}H_{19}O_6PS_2$)	250	0	99.85
	300	0	99.97
1,2-Dichloroethane	275	0	99.8
($C_2H_4Cl_2$)	275	+	99.9
Chloroform	275	0	99.92
($CHCl_3$)	275	+	99.94
Carbon tetrachloride	275	0	99.99
(CCl_4)	275	+	99.99

6.   DDT, although destroyed, results in intractable oil formation using conventional WAO.
7.   Heterocyclic compounds containing O, N, or S atoms provide a point of attack for oxidation and are expected to be destroyed by WAO.

3. Current Technology Status

Zimpro, Inc. installed its first WAO unit in Casmallia, CA for commercially treating hazardous wastes in 1982. Currently, Zimpro has about 20 units around the world for using WAO for both hazardous waste and chemical process applications.

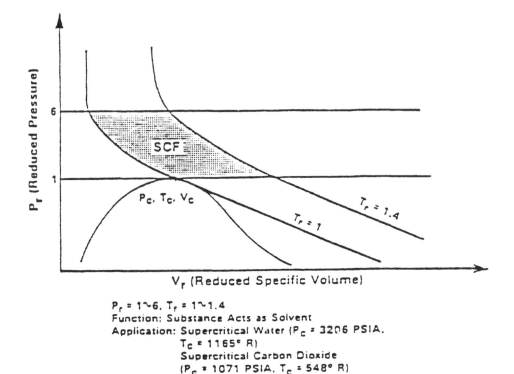

P$_r$ = 1∼6, T$_f$ = 1∼1.4
Function: Substance Acts as Solvent
Application: Supercritical Water (P$_c$ = 3206 PSIA.
          T$_c$ = 1165° R)
          Supercritical Carbon Dioxide
          (P$_c$ = 1071 PSIA, T$_c$ = 548° R)

FIGURE 11.   Typical supercritical fluid (SCF) operation zone.

4. Technology Advantages

1. Since oxidation takes place in the liquid state, it is not necessary to evaporate the water content of the waste. The process, therefore, is most useful for wastes which are too dilute to incinerate economically, yet too toxic to treat biologically.
2. Because wet oxidation reactions take place in the liquid phase, the process can operate without the consumption of heat for vaporizing water and organics. The sensible heat required to heat the waste to reaction temperature is recovered from the effluent by heat exchangers.

5. Technology Disadvantages

1. At elevated temperatures and pressures, WAO may require the use of high-alloy materials such as titanium to prevent corrosion. This use of high alloys increases capital costs.
2. Its application is generally limited to aqueous wastes.

J. SUPERCRITICAL FLUID
1. Process Description[2]

Supercritical fluid (SCF) is characterized as a form of matter in which the liquid and gaseous states are indistinguishable from one another. It is formed when both temperatures and pressures to which the fluid is subjected exceed the critical point (T$_c$ and P$_c$) as shown in Figure 11. Under these supercritical states, the character of the fluid becomes very unusual compared to that under ambient conditions. For example, if water is under the supercritical conditions, the density, dielectic constant, hydrogen bonding, and certain other physical properties are so altered that water behaves much as a moderately polar organic liquid. Thus, *n*-heptane or benzene could become miscible in all proportions with supercritical water (SCW) which cannot happen with water under subcritical conditions. Even some types of wood fully dissolve in SCW. On the other hand, the solubility of sodium chloride (NaCl) could be as low as 100 ppm and that of calcium chloride can be less than 10 ppm. This is the reverse of the solubilities in water that are encountered under subcritical conditions — under which the two salts solubilities are about 37 wt% and up to 70 wt%, respectively.

The SCF region can best be visualized with the aid of a reduced pressure and volume diagram, as in Figure 11. The reduced pressure, temperature and volume are defined as:

$$P_r = P/P_c, \ T_r = T/T_c, \ V_r = V/V_c$$

where the subscripts r and c denote a reduced property and the property at the critical point, respectively.

In a study of the use of supercritical CO_2 for the regeneration of activated carbon, Arthur D. Little, Inc. suggested that the SCF region be in the range of 1 to 1.4 of reduced temperatures and 1 to 6 of reduced pressures as shown in Figures 7 to 11.[17]

2. Case Study Results[2]

Modar Inc. has conducted an extensive series of bench-scale tests to establish the technical feasibility of SCW oxidation. A summary of the test results is given in Tables 16 and 17.

3. Current Technology Status

In 1986, Modar tested its 50-gal/d unit at the CECOS site in Niagara Falls, NY. Two waste streams — isopropyl alcohol and transformer fluids — were tested. Test results indicated that both wastes were destroyed to levels below detectable limits. Currently, Modar is designing a commercial unit with 10,000- to 20,000-gal/d capacity. It is expected that the unit will be operational in Texas.

4. Technology Advantages and Disadvantages

In general, SCF has the same technology advantages and disadvantages as stated for WAO. However, there are two basic differences:

1. WAO requires longer residence times of 20 min to 1 h, compared to SCF which requires residence times of generally less than 1 min, thus reducing the volume of the oxidizer.
2. The SCF vessel generally requires higher temperatures and pressures to maintain the mixture at the supercritical conditions.

III. CONCLUSION

The authors believe that the above ten technologies are the most worthy of compilation for illustrating the development of innovative thermal destruction/incineration technologies. The authors have summarized the ten technologies in accordance with their testing sites and wastes tested (Table 18). In most cases, these technologies can meet either RCRA or TSCA destruction requirements, depending on which compound or chemical (waste) is to be destroyed.

TABLE 16
Composition of Feed Mixtures for Runs 11 to 15

		Wt%	Wt% Cl
Run 11			
DDT	$C_{14}H_9Cl_5$	4.32	2.133
MEK	C_4H_8O	95.68	—
Total		100.0	2.133
Run 12			
1,1,1-Trichloroethane	$C_2H_3Cl_3$	1.01	0.806
1,2-Ethylene dichloride	$C_2H_2Cl_2$	1.01	0.739
1,1,2,2-Tetrachloroethylene	C_2Cl_4	1.01	0.866
o-Chlorotoluene	C_7H_7Cl	1.01	0.282
1,2,4-Trichlorobenzene	$C_6H_3Cl_3$	1.01	0.591
Biphenyl	$C_{12}H_{10}$	1.01	—
o-Xylene	C_8H_{10}	5.44	—
MEK	C_4H_8O	88.48	—
Total		100.0	3.284
Run 13			
Hexachlorocyclohexene	$C_6H_6Cl_6$	0.69	0.497
DDT	$C_{14}H_9Cl_5$	1.00	0.493
4,4'-Dichlorobiphenyl	$C_{14}H_8Cl_2$	1.57	0.495
Hexachlorocyclopentadiene	C_5Cl_6	0.65	0.505
MEK	C_4H_8O	96.09	—
Total		100.0	1.99
Run 14			
PCB1242	$C_{12}H_xCl_{4-6}$	0.34	0.14
PCB 1254	$C_{12}H_xCl_{5-8}$	2.41	1.30
Transformer oil	C_{10}–C_{14}	29.26	—
MEK	C_4H_8O	67.99	—
Total		100.0	1.44
Run 15			
4,4'-Dichlorobiphenyl	$C_{12}H_8Cl$	3.02	0.96
MEK	C_4H_8O	96.98	—
Total		100.0	0.96

TABLE 17
Summary of Results: Oxidation of Organic Chlorides

Run No.	11	12	13	14	15
Residence time (min)	1.1	1.1	1.1	1.1	1.3
Carbon analysis					
Organic carbon in (ppm)	26,700	25,700	24,500	38,500	33,400
Organic carbon out (ppm)	2.0	1.0	6.4	3.5	9.4
Destruction efficiency (%)	99.993	99.996	99.975	99.991	99.97
Combustion efficiency (%)	100	100	100	100	100
Gas composition					
O_2	25.58	32.84	37.10	10.55	19.00
CO_2	59.02	51.03	46.86	70.89	70.20
CH_4	—	—	—	—	—
H_2	—	—	—	—	—
CO	—	—	—	—	—
Chloride analysis					
Organic chloride in (ppm)	876	1266	748	775	481
Organic chloride out (ppm)	0.023	0.037	<0.028	0.032	0.036
Organic chloride conversion (%)	99.997	99.997	99.996	99.996	99.993

TABLE 18
Summary of Testing Results

	Process	Testing sites	Major compounds (wastes) tested
1.	Oxygen-enriched incineration (EPA's mobile incineration system)	Denney Farm, McDowell, MO	Dioxin-contaminated soils
2.	Westinghouse/O'Connor combustor	Panama City, FL	Municipal waste
3.	Fluidized bed combustion (FBC)	Franklin, OH	Phenol waste
4.	Circulating bed combustor (CBC)	San Diego, CA	Oily water sludges, chemical plant wastes, chlorinated organic sludges, aluminum pot linings, and PCB-contaminated soil
5.	Molten salt combustion (MSC)	Canoga Park, CA	Hazardous chemicals, pesticides, warfare agents
6.	Infrared system	Brandon, FL	Superfund waste
7.	Advanced electric reactor (AER)	Times Beach, MO	Superfund waste
8.	Plasma Arc	Kingston, Canada	CCl_4a and PCB
9.	Wet air oxidation (WAO)	Rothschild, WI	RCRA wastes
10.	Supercritical fluid (SCF)	Niagara Falls, NY	Isopropyl alcohol, transformer fluids (PCB)

REFERENCES

1. **Lee, C. C., Huffman, G. L., and Oberacker, D. A.,** An overview of hazardous/toxic waste incineration, *J. Air Pollut. Control Assoc.,* 36, 8, 1986.
2. **Lee, C. C.,** A comparison of innovative technology for thermal destruction of hazardous waste, in Proc. 1st Annu. Hazardous Materials Management Conf., Philadelphia, July 12 to 14, 1983.
3. **Lee, C. C., and Huffman, G. L.,** An overview of pilot-scale research in hazardous waste thermal destruction, in Proc. 1st Int. Conf. New Frontiers for Hazardous Waste Management, Pittsburgh, PA, September 15—18, 1985.
4. **Ho, M. D. and Ding, M. G.,** Field testing and computer modelling of an oxygen combustion system at the EPA mobile incinerator, presented at the American Flame Research Committee Int. Symp. Incineration of Hazardous Waste, Palm Springs, CA, November 2 to 4, 1987 (not included in the Conference Proceedings).
5. **Anderson, J. E.,** Oxygen Aspirator Burner and Process for Firing a Furnace, U.S. Patents 4,378,205 and 4,541,796, March 29, 1983 and September 17, 1985.
6. **Anderson, J. E.,** A low NO_x, low temperature oxygen-fuel burner, in Proc. Symp. on Industrial Combustion Technologies, Chicago, IL, American Society of Matals, April 29, 1986.
7. **Linak, W. P., Kilgroe, J. D., McSorley, J. A., Wendt, J. O. L., and Durn, J. E.,** On the occurrence of transient puffs in a rotary kiln incinerator simulator. I, *J. Air Pollut. Control Assoc.,* 37, 8, 1987.
8. **Linak, W. P., Kilgroe, J. D., McSorley, J. A., Wendt, J. O. L., and Durn, J. E.,** On the occurrence of transient puffs in a rotary kiln incinerator simulator. II, *J. Air Pollut. Control Assoc.,* 36, 8, 1987.
9. **Beachler, D. S., Weldon, J., and Pompelia, D. M.,** Bay County, Florida waste-to-energy facility air emission test results, presented at the Air Pollution Control Association Specialty Conf. on Thermal Treatment of Municipal, Industrial, and Hospital Wastes, Pittsburgh, PA, November 3 to 6, 1987.
10. **Ackerman, D. G.,** Destroying Chemical Wastes in Commercial Scale Incinerators, Facility Rep. No. 3, Systems Technology, PB-265-540, National Technical Information Service, Springfield, VA, 1977.
11. **Greenberg, J.,** Method of Catalytically Inducing Oxidation of Carbonaceous Materials by the Use of Molten Salts, U.S. Patent 3,647,358, assigned to Anti-Pollution Systems, Inc., Pleasantville, NJ, 1972.
12. **Greenberg, J. and Whitaker, D. C.,** Treatment of Sewage and Other Contaminated Liquids With Recovery of Water by Distillation and Oxidation, U.S. Patent 3,642,583, assigned to Anti-Pollution Systems, Inc., Pleasantville, NJ, 1972.

13. **Yosium, S. J. et al.,** Disposal of hazardous wastes by molten salt combustion, paper presented at the American Society of Mechanical Engineers Meeting, Hawaii, August 1979.
14. **Oberacker, D. A., Wall, H. O., and Rosenthal, S.,** EPA's incineration of an industrial refinery lagoon sludge using infrared furnace technology, presented at the 2nd Annu. Natl. Symp. on Incineration of Industrial Wastes, San Diego, CA, May 9 to 11, 1988.
15. **Kolak, N. P., Barton, T. G., Lee, C. C., and Peto, E. F.,** Trial burns — plasma arc technology, in Proc. 12th Annu. Research Symp., Publ. No. 600-9-86/022, U.S. Environmental Protection Agency, Washington, D.C., August 1986.
16. Arc Pyrolysis Project, SCA Chemical Services, Inc., Model City, NY, 1986.
17. **De Filippi, R. P.,** Supercritical Fluid Regeneration of Activated Carbon for Adsorption of Pesticides, Publ. No. 600/2-80-054, U.S. Environmental Protection Agency, Washington, D.C., March 1980.

Chapter 8

COMBUSTION CALCULATIONS

Joseph L. Tessitore and Dennis Engler

TABLE OF CONTENTS

I.   Introduction 294

II.   Waste Quantities and Characteristics 294
        A.   Material Properties Calculations 294

III.   Combustor Feed Characteristics 295

IV.   Combustor Mass Balance for Rotary Kiln and Circulating Bed
        Combustor 296
        A.   Definitions for Combustor Mass Balance 296
        B.   Combustor Waste Mass Flow Calculations for Rotary Kiln 298

V.   Heat Balance for Rotary Kiln and Circulating Bed Combustor 299
        A.   Definitions for Combustor Heat Balance 300
        B.   Combustor Heat Balance Calculations for Rotary Kiln 301

VI.   Design of Additional Equipment Only Required for Rotary Kiln 302
        A.   Afterburner Heat and Mass Balance 302
                1.   Definition for Afterburner Heat and Mass Balance 302
                2.   Afterburner Calculations 303
        B.   Kiln Design 305
        C.   Afterburner Design 305
        D.   Heat Available for Recovery 305
        E.   Precooler Design 306
        F.   Venturi Scrubber Design 306
        G.   Packed Tower Scrubber Design 306

VII.   Ash Quantity Produced for Rotary Kiln and Circulating Bed
        Combustion 307

References 307

I. INTRODUCTION

Equipment sizing is determined by conducting mass and energy balances around each unit operation, using actual waste quantities and characteristics. Examples of this sizing process are presented for two markedly different technologies: (1) rotary kiln with afterburner and wet scrubber and (2) circulating bed with dry filters.

Each technology is applied to the same large remediation site.

II. WASTE QUANTITIES AND CHARACTERISTICS

Based on the waste quantities and characteristics in Table 1, additional material properties (i.e., density, specific gravity) were calculated to establish the incinerator feed rate (Section III). These calculations are based on textbook densities and specific gravities for each waste type's composition and are presented in the following sections. The results are summarized in Table 2.

A. MATERIAL PROPERTIES CALCULATIONS

Lagoon surface oil

Specific gravity	$= 0.91 \text{ lb/yd}^3 = (0.91)(1685 \text{ lb/yd}^3)$
	$= 1533 \text{ lb/yd}^3$
Density	$= (1685 \text{ lb/yd}^3)(0.36) + (1533 \text{ lb/yd}^3)(0.64)$
	$= 1588 \text{ lb/yd}^3$
Weight	$= (4200 \text{ yd}^3)(1588 \text{ lb/yd}^3) = 6.67 \times 10^6 \text{ lb}$

Nonaqueous-phase liquids (NAPL)

Specific gravity	$= 1.5 \text{ lb/yd}^3$
	$= (1.5)(1685 \text{ lb/yd}^3)$
	$= 2528 \text{ lb/yd}^3$
Density	$= (1685 \text{ lb/yd}^3)(0.15) + (2528 \text{ lb/yd}^3)(0.85)$
	$= 2402 \text{ lb/yd}^3$
Weight	$= 90,000 \text{ yd}^3 \times 2402 \text{ lb/yd}^3$
	$= 216.18 \times 10^6 \text{ lb}$

Contaminated lagoon sediment

Specific gravity	$= 1.69 \text{ lb/yd}^3$
	$= (1.69)(1685 \text{ lb/yd}^3)$
	$= 2848 \text{ lb/yd}^3$
Density	$= (1685 \text{ lb/yd}^3)(0.44) + (2848 \text{ lb/yd}^3)(0.56)$
	$= 2336 \text{ lb/yd}^3$
Weight	$= 18,000 \text{ yd}^3 \times 2336.1 \text{ lb/yd}^3$
	$= 42.05 \times 10^6 \text{ lb}$

Surface soil

Specific gravity	$= 2.65 \text{ lb/yd}^3$
	$= (2.65)(1685 \text{ lb/yd}^3)$
	$= 4465 \text{ lb/yd}^3$
Density	$= (1685 \text{ lb/yd}^3)(0.15) + (4465 \text{ lb/yd}^3)(0.85)$
	$= 4048 \text{ lb/yd}^3$
Weight	$= 8,000 \text{ yd}^3 \times 4048 \text{ lb/yd}^3$
	$= 32.38 \times 10^6 \text{ lb}$

TABLE 1
Waste Quantities and Characteristics

Materials	Quantity[a]	Heat value (Btu/lb)	H₂O content (% wt)	Ash content (% wt)
Lagoon surface oil	4,200 cy	10,000	36	1
NAPL	90,000 cy	5,000	15	20
Contaminated lagoon sediment	18,000 cy	2,700	44	39
Tank oil	500,000 gal	18,000	—	1
Spent solvent	150,000 gal	18,000	—	1
Surface soil	8,000 cy	500	15	80

[a] cy = cubic yards.

TABLE 2
Waste Quantity Summary

Material	Quantity (yd³)	Density (lb/yd³)	Specific gravity	Weight (× 10⁶ lb)
Lagoon surface oil	4,200	1588	0.94	6.67
NAPL	90,000	2402	1.43	216.18
Contaminated lagoon sediment	18,000	2336	1.39	42.05
Surface soil	8,000	4048	2.40	32.38
Total				297.28

Total waste quantity $= (6.67 + 216.18 + 42.05 + 32.38) \times 10^6$
$= 297.28 \times 10^6$ lb

Thus, the following waste summary along with Table 1 provide sufficient information about the waste stream to determine incinerator feed rate and design.

III. COMBUSTOR FEED CHARACTERISTICS

From the above waste quantities, weight fractions of each type were calculated in order to determine overall waste properties as follows:

Waste input (weight fractions)

$$\text{Lagoon surface oil} = \frac{6.67 \times 10^6 \text{ lbs}}{297.28 \times 10^6 \text{ lbs}} = 0.0224$$

$$\text{NAPL} = \frac{216.18 \times 10^6 \text{ lbs}}{297.28 \times 10^6 \text{ lbs}} = 0.7272$$

$$\text{Surface soil} = \frac{32.38 \times 10^6 \text{ lbs}}{297.28 \times 10^6 \text{ lbs}} = 0.1086$$

$$\text{Contaminated lagoon sediment} = \frac{42.05 \times 10^6 \text{ lbs}}{297.28 \times 10^6 \text{ lbs}} = 0.1414$$

From these data and Tables 1 and 2 a composite value for waste BTU content, moisture content, and ash content have been determined as follows:

Waste Btu content	= (10,000 Btu/lb)(0.0224) + (5000 Btu/lb)(0.7272) + (2700 Btu/lb)(0.1414) + (500 Btu/lb)(0.1086)
	= 4296.1 Btu/lb

Waste moisture content	= % H_2O
	= [(0.0224)(0.36) + (0.7272)(0.15) + (0.1414)(0.44) + (0.1086)(0.15)] × 100
	= 19.56%
Ash content	= % Ash
	= [(0.0224)(0.01) + (0.7272)(0.20) + (0.1414)(0.39) + (0.1086)(0.80)] × 100
	= 28.75%

Having established the incineration properties of the waste, the design parameters of the kiln system can be evaluated. The following assumptions are necessary to establish the kiln feed rate and capacity: incineration cleanup period of 2 years and incinerator availability = 75%.

Waste input rate

$$= \frac{297.28 \times 10^6 \text{ lbs}}{(2000 \text{ lb/ton}) (24 \text{ h/d}) (7 \text{ day/wk}) (52 \text{ wk/yr}) (2 \text{ yr}) (0.75)}$$

$$= 11.34 \text{ ton/h}$$

(Assume 2 units @ 6-ton/h capacity each.)

The following mass balance, heat balance, and design calculations are based on one 6-ton/h unit.

IV. COMBUSTOR MASS BALANCE FOR ROTARY KILN AND CIRCULATING BED COMBUSTOR

To determine the required incinerator size, air flow, and fuel requirements, a mass balance was performed for the proposed incinerator systems. The inputs are waste quantities, airflow (including humidity entrained within the air), and any supplemental fuel flow. The effluent gas from the incinerator includes moisture, dry exhaust gas, and ash. The parameters involved in the mass balance are presented in Table 3.

A. DEFINITIONS FOR COMBUSTOR MASS BALANCE
Wet feed — As received, charging rate is 12,000 lb/h.

Moisture — Percent moisture, by weight, of the wet feed is 19.56%. The moisture rate is 19.56% of the wet feed rate; 0.1956 × 12,000 lb/h = 2347 lb/h.

Dry feed — Total wet feed less moisture; 12,000 lb/h − 2347 lb/h = 9653 lb/h.

Ash — Percent of feed that remains after combustion. Ash is 28.75% of total wet feed; 0.2875 × 12,000 lb/h = 3450 lb/h.

Volatile — Portion of dry feed that is combusted. It is found by subtracting the ash from the dry feed: 9653 lb/h − 3450 lb/h = 6203 lb/h.

Heating value, wet feed — Btu value of the waste per pound of wet feed = 4296.1 Btu/lb.

Heating value, volatile — Btu value of the waste per pound of volatile matter. It is calculated as wet feed × heating value: 6203 lb/h = 8311 Btu/lb.

TABLE 3
Combustor Mass Balance Table

	Rotary kiln	CBC
Wet feed rate	12,000 lb/h	,,
% Moisture	19.56%	,,
Moisture content flow	2,347 lb/h	,,
Dry feed rate	9,653 lb/h	,,
% Ash	28.75%	,,
Ash flow rate	3,450 lb/h	,,
Volatile flow rate	6.203 lb/h	,,
Heating value, wet feed	4,296.1 Btu/lb	,,
Heating value, volatile	8,311 Btu/lb	,,
Input heating value	51.55 MBtu/h	,,
Dry gas production rate	8.55 lb/10,000 Btu	,,
Dry gas flow	44,075 lb/h	,,
Combustion H_2O produced	0.65 lb/10,000 Btu	,,
Comb. H_2O Flow	3,351 lb/h	,,
Dry gas + comb. H_2O	47,426 lb/h	,,
100% Air required	41,223 lb/h	,,
Total air fraction	2.5	1.25
Total air intake	103,058 lb/h	51,529
Excess air intake	61,835 lb/h	10,306
Humidity/dry gas (air)	0.01	,,
Humidity mass flow rate	1,031 lb/h	515
Total H_2O mass flow (exit)	6,729 lb/h	6,213
Total dry gas flow (exit)	105,910 lb/h	54,381
Total gas flow (exit)	112,639 lb/h	60,594

Input heating value — Total heating value as charged; 12,000 lb/h × 4296.1 Btu/lb = 51.55 MBtu/h where M = 10^6.

Dry gas — Produced from combustion of the waste is assumed to be 8.55 lb/10,000 Btu. This figure is multiplied by the Btu released; 8.55 lb/10,000 Btu × 51.55 MBtu/h = 44,075 lb/h.

Combustion H_2O — Moisture generated from burning the waste is assumed to be 0.65 lb/10,000 Btu. Therefore, the combustion moisture flow rate is 0.65 lb/10,000 Btu × 51.55 MBtu/h = 3351 lb/h.

Dry gas + comb. H_2O — Sum of the dry gas and the moisture products of combustion; 44,075 lb/h + 3351 lb/h = 47,426 lb/h. This figure is used in obtaining the air required for combustion.

100% Air — Dry gas and moisture weights that are produced by combustion of the volatile component, which equals the weight of the volatile component plus the weight of the air provided. Likewise, the air requirement is equal to the sum of the dry gas and moisture of combustion less the volatile component. This air requirement is the stoichiometric air requirement, that amount of air necessary for complete combustion of the volatile component of the waste. The value for the rotary kiln is 100% air, which is calculated as follows: 47,426 lb/h − 6,203 lb/h = 41,223 lb/h.

Total air fraction — The air required for effective combustion. It is assumed that 150% excess air is required in order to achieve complete burning in the rotary kiln, providing 1.50 + 1.00 = 2.5 total air fraction. By contrast, only 25% excess air is required for the CBC system.

Total air — The stoichiometric requirement multiplied by the total air fraction; 41,223 lb/h × 2.5 = 103,058 for the rotary kiln. CBC calculations indicate a total of 51,529 lb/h.

Excess air — Total air less the stoichiometric requirement; 103,058 lb/h − 41,223 lb/h = 61,835 for the rotary kiln or 10,306 for the CBC.

Humidity/dry gas (air) — Humidity of the air entering the system is assumed to be 0.01 lb H_2O/lb of dry air.

Humidity — Flow of moisture into the system with the air supply. When the airflow and the fractional humidity are known, the humidity can be calculated as 103,058 lb/h air × 0.01 lb H_2O/lb of air = 1031 lb/h moisture for the rotary kiln or 515 lb/h for the CBC.

Total H_2O — Moisture exiting the system. Equal to the sum of the three moisture components, moisture in the feed plus moisture of combustion plus humidity; 2347 lb/h + 3351 lb/h + 1031 lb/h = 6729 lb/h total H_2O for the rotary kiln or 6213 lb/h for the CBC.

Total dry gas — Dry gas exiting the system is equal to the sum of the dry gas generated by the combustion of the volatiles, with stoichiometric air, plus the flow of excess air into the system. Thus, 44,075 lb/h dry gas + 61,835 lb/h excess air = 105,910 lb/h total dry gas for the rotary kiln or 54,381 lb/h for the CBC.

Total gas — Exiting the system is the sum of the total H_2O and the total dry gas; 6729 lb/h total H_2O + 105,910 lb/h total dry gas = 112,639 lb/h total gas for the rotary kiln or 60,594 lb/h for the CBC.

B. COMBUSTOR WASTE MASS FLOW CALCULATIONS FOR THE ROTARY KILN

Wet feed
6 ton/h × 2000 lb/ton = 12,000 lb/h

Moisture
12,000 lb/h × 19.56% H_2O = 2347 lb/h

Dry feed
12,000 lb/h − 2347 lb of H_2O/h = 9653 lb/h

Ash
12,000 lb/h × 28.75% ash = 3450 lb/h

Volatile
Dry feed − ash = 9653 lb/h − 3450 lb/h
            = 6203 lb/h

Heating value, volatile

$$\frac{\text{Wet feed} \times \text{wet feed Btu Value}}{\text{Volatile}} = \frac{12,000 \text{ lb/h} \times 4296.1 \text{ Btu/lb}}{6203 \text{ lb/h}}$$
$$= 8311 \text{ Btu/lb}$$

Waste Btu loading
4296.1 Btu/lb × 12,000 lb/h = 51.55 MBtu/h

Dry gas flow
(8.55 lb/10,000 Btu) × (51.55 MBtu/h) = 44,075 lb/h

Combustion H_2O Flow
(0.65 lb/10,000 Btu) × (51.55 MBtu/h) = 3351 lb/h

Dry gas + comb. H₂O

44,075 lb/h + 3351 lb/h = 47,426 lb/h

100% Air (stoichiometric)

(Dry gas + comb. H_2O) − volatile
= 47,426 lb/h − 6203 lb/h
= 41,223 lb/h

Total air

Total air fraction = 100% + 150% excess air = 250% = 2.5
100% air × total air fraction = 41,223 lb/h × 2.5
= 103,058 lb/h

Excess air

Total air − 100% air = 103,058 lb/h − 41,223 lb/h
= 61,835 lb/h

Humidity

Total air × 0.01 lb H_2O/lb air
= 103,058 × 0.01
= 1031 lb/h

Total H₂O

Moisture + comb. H_2O + humidity
= 2347 lb/h + 3351 lb/h + 1031 lb/h
= 6729 lb/h

Total dry gas

Dry gas flow + excess air
= 44,075 lb/h + 61,835 lb/h
= 105,910 lb/h

Total gas flow

Total H_2O + total dry gas = 6729 lb/h + 105,910 lb/h
= 112,639 lb/h

V. HEAT BALANCE FOR ROTARY KILN AND CIRCULATING BED COMBUSTOR

A heat balance was also performed for the proposed incineration system. This balance assumes that the heat entering a system minus losses (due to radiation, heat lost to ash, etc.) equals the heat exiting the system. The result of a heat balance is a determination of the incinerator outlet temperature, outlet gas flow, supplemental fuel requirement, and total air requirement. By determining how much of the total heat in the incinerator is present in the exhaust gas, the exhaust gas temperature can be calculated. If the calculated exhaust gas temperature is greater than or equal to the desired exhaust gas outlet temperature, supplemental fuel is not required. If the exhaust gas temperature is less than the desired exhaust gas temperature, supplemental fuel must be added. The products of combustion for the system must include the products of combustion of the supplemental fuel. A heat balance analysis for the rotary kiln is presented in Table 4.

TABLE 4
Combustor Heat Balance

	Rotary kiln	CBC
Cooling air loss	0.85 MBtu/h	"
Ash loss	0.449 MBtu/h	"
Radiation loss	0.773 MBtu/h	"
Humidity correction	−1.0 MBtu/h	"
Total heat loss	1.072 MBtu/h	"
Input heat value	51.55 MBtu/h	"
Outlet heat value	50.48 MBtu/h	"
Dry gas flow (exit)	105,910 lb/h	54,381
H_2O flow (exit)	6,729 lb/h	6,213
Total gas flow (exit)	112,639 lb/h	60,594
Actual temperature	1,483°F	1,600
Desired temperature	1,200	1,600
Volumetric gas flow (exit)	94,767 ACFM	50,980
Heat recovery from combustion chamber	0	32.6 MBtu/h

A. DEFINITIONS FOR COMBUSTOR HEAT BALANCE

Cooling air wasted — Assume the incinerator is cooled by a flow of 2000 SCFM of air. A standard cubic foot of air weighs 0.075 lb/ft^3 = 9000 lb/h air. It is also assumed that this flow is discharged to the atmosphere. The temperature of the air at the discharge point is assumed to be 450°F is 94 Btu/lb. The total heat lost due to the wasted cooling air is 9000 lb/h × 94 Btu/lb = 0.85 MBtu/h.

Ash heat loss — The heat content of the ash is assumed to be 130 Btu/lb. The heat loss due to the ash is 3450 lb/h × 130 Btu/lb = 0.449 MBtu/h.

Radiation heat loss — The heat lost by radiation from the incinerator can be approximated as a percentage of total heat of combustion. For a heat release of 51.55 MBtu/h a radiation loss of 1.5% is used; 51.55 MBtu/h × 0.015 = 0.773 MBtu/h.

Humidity correction — The heat of vaporization of humidity at 60°F, 960 Btu/lb, is added to the total heat capacity of the flue gas; 960 Btu/lb × 1031 lb/h = 1.0 MBtu/h.

Total losses — Sum of the heat discharged as cooling air plus the heat lost in the ash discharge plus the radiation loss minus the humidity correction; 0.85 + 0.449 + 0.773 − 1.0 = 21.072 MBtu/h.

Input heat value — Heat generated from the combustible portion of the feed is 51.55 MBtu/h.

Outlet heat value — Heat content of the flue gas exiting the system is equal to the input heat value less the total heat loss; 51.55 MBtu/h − 1.072 MBtu/h = 50.48 MBtu/h.

Dry gas — Dry gas content in the flue gas is 105,910 lb/h for the rotary kiln and 54,381 lb/h for the CBC.

H_2O — Water content in the flue gas is 6729 lb/h for the rotary kiln and 6213 lb/h for the CBC

Temperature (exhaust gas) — The heat content of the dry gas flow plus the heat content of the moisture flow exiting the system equals the outlet heat content of 50.48 MBtu/h. At 1450°F the heat content is 51.0 MBtu/h. By interpolation the outlet temperature of the exhaust gas is equal to 1483°F for the rotary kiln. The CBC has sufficient heat exchange in the combustion chamber and flue gas cooler to allow the recovery of 32,625,000 Btu/h as steam or hot water, while maintaining the combustion chamber at a constant temperature of 1600°F. The much lower excess air requirement of the CBC, which is due to the higher degree of turbulence and mixing, results in this large amount of recoverable energy, with no auxiliary fuel.

Desired temperature — The favored temperature of the rotary kiln exhaust gas is 1200°F. Therefore, no supplemental fuel is needed in the rotary kiln itself; however, substantial supplemental fuel is required for the rotary kiln's afterburner (24×10^6 Btu/h to be purchased). The CBC does not require an afterburner, having completed all combustion in the main combustion chamber.

Gas flow — Gas flow of the exhaust gas was calculated to be 94,767 ACFM at 1483°F for the rotary kiln, or 50,980 ACFM for the CBC at the same temperature.

B. COMBUSTOR HEAT BALANCE CALCULATIONS FOR ROTARY KILN
Cooling air wasted
Assume incinerator shell is cooled by a flow of 2000 SCFM of air and that it is discharged into atmosphere:

$$2000 \text{ ft}^3/\text{min} \times 60 \text{ min/h} \times 0.075 \text{ lb/ft}^3 = 9000 \text{ lb/h}$$

Assume temperature of air discharge is 450°F.
Enthalpy of air at 450°F is 94 Btu/lb.
Total heat lost = 9000 lb/h × 94 Btu/lb = 0.85 MBtu/h.

Ash
Assume heating value of ash is 130 Btu/lb:

Ash heat loss = 3450 lb/h × 130 Btu/lb = 0.449 MBtu/h

Radiation
Radiation loss % is 1.5 (from Table 8-3 Brunner[1]):

Radiation loss = 0.015 × 51.55 MBtu/h
              = 0.773 MBtu/h

Humidity
Correction − heat of vaporization at 60°F = 970 Btu/lb:

$$1301 \text{ lb/h} \times 970 \text{ Btu/lb} = 1.0 \text{ MBtu/h}$$

This is added to total heat capacity of flue gas.

Total losses
Cooling air loss + ash loss + radiation loss − humidity correction
= (0.85 + 0.449 + 0.773 − 1.0) MBtu/h

Outlet heat value
Input heat value − total losses = (51.55 − 1.072) MBtu/h = 50.48 MBtu/h

Outlet and desired temperatures
Dry gas = 105,910 lb/h
H_2O = 6729 lb/h

Temp (°F)	Dry gas (Btu/lb)	H_2O (Btu/lb)	Dry gas (MBtu/h)	H_2O (MBtu/h)	Total (MBtu/h)
1450	355.58	1747.70	37.7	11.8	49.5
1500	369.37	1775.52	39.1	11.9	51.0

1450°F	49.5 MBtu/h
X	50.48 MBtu/h (outlet)
1500°F	51.0 MBtu/h

TABLE 5
**Afterburner Mass/Heat Balance (Required Only for Rotary
Kiln)**

Dry gas flow (inlet)	105,910 lb/h
H_2O flow (inlet)	6,729 lb/h
Temperature (inlet)	1,483°F
Btu Value available	50.48 MBtu/h
Desired temperature	2200°F
Desired Btu value	74.90 MBtu/h
Net Btu value	24.42 MBtu/h
Available heat @ 2200°F from no. 2 fuel oil	49.294 Btu/gal
Fuel required	495.4 gal/h (3,765 lb/h)
Air required	61,956 lb/h
Dry gas produced	62,191 lb/h
H_2O produced	4,335 lb/h
Total dry gas flow (exit)	168,101 lb/h
Total H_2O flow (exit)	11,064 lb/h
Total gas flow (exit)	179,165 lb/h
Fuel Btu value	69.2 MBtu/h
Total heat value	119.63 MBtu/h
Heat loss (10%)	11.96 MBtu/h
Outlet heat value	107.67 MBtu/h
Volumetric gas flow (exit)	207,610 ACFM

$$X = 1400 + (1500 - 1450)\ \frac{50.48 - 49.5}{51.0 - 49.5}$$

$$= 1483°F = \text{outlet temperature}$$

Desired temperature = 1200°F
Therefore, no auxiliary fuel needed in the kiln

Total heat outlet
Outlet heat value = 50.48 MBtu/h.

$$\frac{(105,910\ \text{lb/h})\ (48.7\ \text{ft}^3/\text{lb})\ +\ (6729\ \text{lb/h})\ 78.5\ \text{ft}^3/\text{lb})}{60\ \text{min/h}}$$

$$= 94,767\ \text{ACFM}$$

VI. DESIGN OF ADDITIONAL EQUIPMENT ONLY REQUIRED FOR ROTARY KILN

A. AFTERBURNER HEAT AND MASS BALANCE

The information obtained from the mass and heat balance of the rotary kiln was used to conduct a heat and mass balance for the afterburner. This was done to determine how much supplemental fuel will be required in the afterburner to raise the kiln exhaust gases to 2200°F and the resulting gas flows. The results of the heat and mass balance are presented in Table 5.

The CBC does not require an afterburner, since all combustion is efficiently completed in the combustion chamber.

1. Definition for Afterburner Heat and Mass Balance

Desired Btu value — Given the dry gas and moisture flows from the kiln, the desired

Btu value at 2200°F can be calculated as (105,910 lb/h × 567.52 Btu/lb) + (6729 lb/h × 2189.92 Btu/lb) = 74.90 MBtu/h.

Net Btu value — This is the amount of heat that must be added to the flue gas to raise its heat content to the desired level. It can be calculated as 74.90 MBtu/h − 50.48 MBtu/h = 24.42.

Fuel required — Using the available heat from no. 2 fuel oil at 2200°F with 20% excess air, the fuel required to raise the temperature of the gas in the afterburner to 2200°F can be calculated as 24.42 MBtu/gal − 49,294 Btu/gal = 495.4 gal/h of no. 2 fuel oil.

Air required — For combustion of no. 2 fuel oil, air required is 125.062 lb/gal of fuel oil; 125.062 lb/gal × 495.4 gal/h = 61,956 lb/h of air required.

Dry gas produced — From the combustion of no. 2 fuel oil, this is 125.537 lb/gal of fuel oil; 125.537 × 495.4 gal/h 62,191 lb/h of dry gas produced.

H_2O produced — From the combustion of the no. 2 fuel oil, this is 8.751 lb/gal of fuel oil; 8.751 lb/gal × 495.4 gal/h = 4,335 lb/h of H_2O produced.

Total dry gas — Total dry gas exiting the system is the sum of the dry gas produced from the combustion of the fuel oil plus the dry gas from the kiln; 62,191 lb/h + 105,910 lb/h = 168,101 lb/h.

Total H_2O — Total water exiting the system is the sum of the H_2O produced from the combustion of the fuel oil plus the H_2O from the kiln; 4335 lb/h + 6729 lb/h = 11,064 lb/h.

Total heat value — Total heat value the exhaust gas exiting the afterburner is equal to the sum of the heat content of the gas prior to the added fuel oil plus the heat content of the fuel oil. The heat content of the fuel oil is 139,703 Btu/gal × 495.4 gal/h = 69.2 MBtu/h. The total heat value is 69.2 MBtu + 50.48 MBtu/h = 119.6 MBtu/h.

Heat loss — In the afterburner, this is assumed to be 10%; 119.6 MBtu/h × 0.10 = 11.96 MBtu/h.

Outlet heat value — Total heat value less the heat loss; 119.6 − 11.96 = 107.6 MBtu/h.

2. Afterburner Calculations
Required Btu Value
@ 2200°F, H_{air} = 567.52 Btu/lb

H_{water} = 2189.92 Btu/lb

Dry gas = 105,910 lb/h

H_2O = 6729 lb/h

(105,910 lb/h) (567.52 Btu/lb) + (6729 lb/h) (2198.92 Btu/lb)
= 74.90 MBtu/h

Additional heat required
@ 1475°F − 50.48 MBtu/h

74.90 MBtu/h − 50.48 MBtu/h
= 24.42 MBtu/h

Auxiliary fuel required
49,294 Btu/gal available from no. 2 fuel oil @ 2200°F with 20% excess air

$$\frac{24.42 \text{ MBtu/h}}{49.294 \text{ Btu/gal}} = 495.4 \text{ gal/h}$$

Air required
125.062 lb/gal × 495.4 gal/h
= 61,956 lb/h

Dry gas produced
125.537 lb/gal × 495.4 gal/h
= 62,191 lb/h

H₂O produced
8.751 lb/gal × 495.4 gal/h
= 4335 lb/h

Total dry gas
62,191 lb/h + 105.910 lb/h
= 168,101 lb/h

Total H₂O
6729 lb/h + 4335 lb/h
= 11,064 lb/h

Total air required
61,959 lb/h + 103.058 lb/h
= 165,017 lb/h

Heat value of fuel
495.4 gal/h × 139.703 Btu/gal
= 69.2 MBtu/h

Outlet heat value
24.42 MBtu/h + 50.48 MBtu/h
= 74.90 MBtu/h

Total gas flow

$$\frac{(168,101 \text{ lb/h}) (67 \text{ ft}^3/\text{lb}) + (11,064 \text{ lb/h}) (107.9 \text{ ft}^3/\text{lb}}{(60 \text{ min/h})}$$

= 207,610 ACFM

Total gas
Total dry gas + total H₂O = 168.101 lb/h + 11,064 lb/h
                           = 179,165 lb/h

Fuel Btu input
139,703 Btu/gal × 495.4 gal/h
= 69.2 MBtu/h

Total heat value
(168,101 lb/h)(567.52 Btu/lb) + (11,064 lb/h)(2189.92 Btu/lb)
= 119.63 MBtu/h

Heat loss from radiation and conduction
(119.63 MBtu/h) × (10% heat loss)
= 11.96 MBtu/h

Outlet heat value
119.63 MBtu/h − 11.96 MBtu/h
= 107.67 MBtu/h

B. KILN DESIGN

From the heat release rate calculated in Section V. (50.48 MBtu/h), the required kiln volume and dimensions can be calculated. Assuming a kiln heat release of 25,000 Btu/h-ft^3,[2] the following kiln size is needed:

$$\frac{50.48 \times 106 \text{ Btu/h}}{25,000 \text{ Btu/h-ft}^3} = 2019.2 \text{ ft}^3$$

Assuming length (L) = 4 × diameter (D),[3]

L = 4D
V = D^3 = 2019.2 ft^3
D = (V/π)$^{1/}_3$ = (2019.2 ft^3/π)$^{1/}_3$
D = 8.63 ft ≈ 9 ft

Hence, L = 4D = 4 ft × 9 ft = 36 ft.

After adding 3 ft of refractory to the diameter and 1 ft of refractory to the length, O.D. = 12 ft and L = 37 ft.

C. AFTERBURNER DESIGN

Knowing the gas flow in the afterburner from Table 5 and the required 2-s retention time for PCBs (under TSCA), the afterburner volume can be calculated:

Gas flow = 207,610 ft^3/min
Retention time = 2 s
207,610 ft^3/min × (1 min/60 s)(2 s) = 6920 ft^3

To provide adequate mixing, a velocity of 20 ft/s is needed, so (20 ft/s) × (2 s) = 40 ft = afterburner height.

Area = volume/height = 6920/40 = 173 ft^2
Diameter = (4 × area/π)$^{1/}_2$ = (4 × 173/π)$^{1/}_2$ = 14.8 ft

Final dimensions with refractory are 16-ft O.D. and 40-ft height.

D. HEAT AVAILABLE FOR RECOVERY

From the outlet heat value obtained in Section VI.A (107.67 MBtu/h) and the heat loss due to temperature reduction, the heat available for recovery in the rotary kiln is calculated:

Outlet heat value from afterburner = 107.67 MBtu/h @ 2200°F
At 600°F, H$_{air}$ = 131.69 Btu/lb
H$_{water}$ = 1307.12 Btu/lb
Dry gas flow from afterburner = 168,101 lb/h
H$_2$O flow from afterburner = 11,064 lb/h

Heat lost to temperature reduction = (168,101 lb/h) (131.69 Btu/lb) + (11,064 lb/h) (1307.12 Btu/lb)
= 36.60 MBtu/h

So, the net heat available for recovery = 107.67 MBtu/h − 36.60 MBtu/h
= 71.07 MBtu/h

E. PRECOOLER DESIGN

The exhaust gases from the rotary kiln's afterburner must be cooled to 600°F prior to entering the precooler equipment. The heat content of the exhaust gases at 600°F is 36.6 MBtu/h (see Section VI.C). Assuming a temperature reduction in the exhaust gases from 600 to 150°F in the precooler system, the following heat loss would occur:

Heat content of exhaust gases @ 600°F = 36.6 MBtu/h @ 150°F

H_{air} = 21.61 Btu/lb

H_{water} = 1091.92 Btu/lb

Heat loss due to temperature reduction = (168,101 lb/h) × (21/61 Btu/lb) + (11,064 lb/h) (1091.92 Btu/lb)

= 15.71 MBtu/h

So, heat content remaining in exhaust gas is 36.6 MBtu/h − 15.71 MBtu/h.

The amount of water required to achieve this temperature reduction (600 to 150°F) is as follows:

$$\text{Water consumption} = \frac{(20.89 \text{ MBtu/h})}{970 \text{ Btu/lb}}$$

$$= 2.15 \times 10^4 \text{ lb/h}$$

$$= 2578 \text{ gal/h}$$

$$= 42.97 \text{ gal/min/unit}$$

F. VENTURI SCRUBBER DESIGN

While the CBC uses a dry filtration system, rotary kilns typically use a wet scrubber.

Table 5 gives the volumetric gas flow exiting the afterburner of the rotary kiln as 207,610 ACFM. At the exit temperature of the precooler (105°F), the gas flow becomes:

$$207,610 \text{ acfm} \times \frac{150 + 460}{2200 + 460} = 47,609.8 \text{ acfm}$$

Assuming a throat velocity of 400 ft/s,[4] the throat area is

47,609 ft³/min × (1/400 ft/s) (1/60 s/min) = 1.98 ft² ≈ 2.0 ft²

Also, assuming the scrubber water required is 6 gal/1000 ft³,[5] then 6 gal/1000 ft³ × (47,609.8 ft³/min) = 285.65 gal/min of water required. (Use 350 gal/min/unit for design purposes.)

B. PACKED TOWER SCRUBBER DESIGN

From Section VI.E, at 150°F the rotary kiln volumetric gas flow rate is 47,609 ACFM. Assuming the required scrubber water is 2.5 gal/1000 ft³ of gas, then

$$\text{H}_2\text{O required} = \frac{2.5 \text{ gal}}{1000 \text{ ft}^3} (47,609 \text{ ft}^3/\text{min}) = 119 \text{ gal/min}$$

(Use 150 gal/min/unit for design purposes.)

The chemical reaction which takes place to neutralize the acid gas is HCl + NaOH → NaCl + H₂O. The HCl produced can be estimated as follows: H + Cl → HCl. NAPL is 0.13% Cl, so 216.2 × 106 lb NAPL × (0.0013) − 2.81 × 105 lb Cl = 8.0286 × 103 lb-mol Cl. Thus,

HCl produced   = (8.0286 × 10³ lb-mol) (36 lb/lb-mol)

= 2.89 × 10⁵ lb or 21.99 lb/h

$$\text{NaOH required} = 21.99 \text{ lbs/h} \times \frac{40 \text{ lb/lb-mol}}{36 \text{ lb/lb-mol}}$$

$$= 24.43 \text{ lb/h/unit}$$

$$\text{NaCl produced} = 21.99 \text{ lb/h} \times \frac{23 + 35}{36}$$

$$= 35 \text{ lb/h} \times \frac{23 + 35}{36} = 35 \text{ lb/h/unit}$$

So, scrubber sludge produced (NaCl) is 35 lb/h/unit × (24 h/d) (365 d/yr)(2 yr) × 0.75 = 459,900 lb/unit. Total sludge = 919,800 lb = 459.9 or 460 ton for the rotary kiln and none for the CBC.

VII. ASH QUANTITY PRODUCED FOR ROTARY KILN AND CIRCULATING BED COMBUSTION

From Table 8-3, the ash quantity produced per unit is 3450 lb/h. So, 3450 lb/h-unit × 2 unit = 6900 lb/h. 6900 lb/h (24 h/d) (7 d/wk) (52 wk/yr) (2 yr) (0.75)

$$= 90.42 \times 10^6 \text{ lb of ash}$$

$$= 45,209 \text{ ton of ash for either the CBC or rotary kiln}$$

REFERENCES

1. **Brunner, C. R.,** *Incinerator Systems: Selection and Design,* Van Nostrand Reinhold, New York, 1984.
2. **Brunner, C. R.,** *Incinerator Systems: Selection and Design,* Van Nostrand Reinhold, New York, 1984, 243.
3. **Brunner, C. R.,** *Incinerator Systems: Selection and Design,* Van Nostrand Reinhold, New York, 1984, 350.
4. **Brunner, C. R.,** *Incinerator Systems: Selection and Design,* Van Nostrand Reinhold, New York, 1984, 331.
5. **Brunner, C. R.,** *Incinerator Systems: Selection and Design,* Van Nostrand Reinhold, New York, 1984, 332.

Chapter 9

TRIAL BURN*

P. Gorman, R. Hathaway, D. Wallace, and A. Trenholm

TABLE OF CONTENTS

I.   Introduction 311

II. Overview of a Trial Burn 311
     A.   What Does a Trial Burn Involve? 311
          1.   Regulatory Limits 312
          2.   Permit Conditions 312
          3.   Sampling and Analysis Activities 312
          4.   Trial Burn Time Requirements 313
          5.   Assessing Potential Performance Problems 313
     B.   What Types of Sampling and Analysis Are Typically Required? 314
          1.   Selecting the S&A Matrix 314
          2.   Identifying S&A Methods 314
          3.   Adverse Stack Sampling Conditions 315
          4.   Sample Train Sealing Problems 316
          5.   Need for Specialized Methods 316
     C.   What Skills, Equipment, and Facilities Are Needed to Conduct
          a Trial Burn? 316
          1.   Facilities and Equipment 319
          2.   Staffing Needs 319
          3.   Selecting a Contractor 319
     D.   What Are the Major Cost Factors Associated with a Trial
          Burn? 320
          1.   Planning and Preparation 320
          2.   Sampling and Analysis 320
          3.   Quality Assurance 320
          4.   Estimating the Costs 320

III. Planning for a Trial Burn 321
     A.   What Equipment/Instrumentation Is the Incinerator Required to
          Have? 321
     B.   How Should the Operating Conditions Be Selected? 321
          1.   Operating Parameters that Affect Permit Conditions 322
          2.   Use of Pretest or Miniburns 322
     C.   How Should Trial Burn POHCs Be Selected? 323
     D.   What Types and Quantities of Waste Are Needed, and How
          Can They Be Prepared? 324
          1.   Quantities of Waste 324
          2.   Types of Waste 324
          3.   Waste Preparation 324
          4.   Mixing 325

* Contents of this chapter are from a report entitled Practical Guide — Trial Burns for Hazardous Waste Incinerators, prepared under EPA Contract No. 68-03,3149 in 1985. The EPA Project Officer was Donald Oberacker.

		5.	Time Requirements	325
	E.		How Many Runs Are Necessary, and How Long Is Each Run?	325
	F.		How Many People Will Be Needed, with What Experience?	325
	G.		How Are POHC Stack Sampling Methods Selected?	326
	H.		What Detection Limits Are Required for the Sampling and Analysis Methods?	326
		1.	Waste Feed Detection Limits	326
		2.	Stack Gas Detection Limits	326
		3.	High Concentration of Volatile POHCs	328
	I.		What QA/QC Needs to Be Done?	328
	J.		How Is It Best to Plan for the Possibility that Trial Burn Results Are Outside RCRA Requirements?	328
IV.			Conducting the Trial Burn	329
	A.		What Is Involved in Preparing for the Tests?	329
		1.	Schedule	329
		2.	Sampling Crew	330
		3.	Equipment	330
		4.	Facility Readiness	330
		5.	Process Data	333
		6.	Data Sheets and Labels	333
		7.	Safety Precautions	336
		8.	Observers	336
	B.		What Is Involved in Conducting the Actual Sampling, and What Are the Problems that May Occur?	336
		1.	Equipment Setup	337
			a. Sampling Train Setup	337
			b. Setup for Drummed Waste Sampling	337
		2.	Preliminary Testing	338
			a. Velocity Traverse	338
			b. Cyclonic Flow	338
			c. Moisture Measurement	338
		3.	Actual Testing	338
	C.		What Is Involved in Analysis of Samples, and What Are the Problems that May Occur?	339
		1.	Sample Check-in	340
		2.	Analysis Directive	340
		3.	Sample Inhomogeneity	340
		4.	Analytical Interferences	340
		5.	Saturation of GC/MS Data System	340
		6.	High Blanks	341
		7.	Poor Precision	341
		8.	Recovery Efficiency	341
		9.	Actual vs. Expected Results	341
	D.		How Are the Sampling Data and Analysis Data Converted to Final Results?	341
		1.	Blank Correction	341
		2.	Significant Figures and DRE	343
		3.	Rounding Off DRE Results	343
		4.	Reporting DRE with a "<" or ">" Sign	343
	E.		How Are the Data and Results Usually Reported?	344

References | 348 |

I. INTRODUCTION

On May 19, 1980, the U.S. Environmental Protection Agency (EPA) published regulations under the authority of the Resource Conservation and Recovery Act (RCRA) for hazardous waste incinerators. These regulations require that new and existing incinerators adequately destroy hazardous organic compounds and maintain acceptable levels of particulate and chloride emissions. Owners and operators of incinerators are required to demonstrate the performance of the facility by means of a trial burn. As a consequence, industry and control agency personnel have become involved in planning for, conducting, and interpreting the results from trial burns. This manual is written to assist those individuals in their efforts.

The manual concentrates on those aspects of a trial burn that are the most important and those that are potentially troublesome. The manual contains practical explanations based on experience of Midwest Research Institute (MRI) and others in conducting trial burns and related tests for EPA. It includes the comments of several industrial plant owners and operators. It is directed mainly to incinerator operators, those who may conduct the actual sampling and analysis, and those who must interpret trial burn results. It will also be useful for regulatory personnel and others that need to understand trial burns.

One of the major objectives was to make this guide readily usable. For that reason, the discussion is brief and avoids dwelling on detail. A question and answer format is used to relate the material to operator concerns. Each subsection begins as a question that could well be posed by an incinerator operator who needs to conduct a trial burn. The narrative following each question provides answers to the question or provides information pertinent to the question. For each question, the most important considerations are discussed, and potential trouble spots are identified.

This guide addresses multiple components of the trial burn process including planning and preparation, sampling and analysis for the trial burn, process monitoring during the trial burn, and data reduction reporting. The guide does not directly address the preparation of the trial burn plan, but it does address some planning aspects that affect trial burn plan preparation and subsequent interpretation of the trial burn results.

The remainder of the guide is divided into three sections. Section II presents an overview of the trial burn process and requirements. Section III discusses planning for the trial burn. Section IV discusses conducting the trial burn and reducing and reporting data from the trial burn.

II. OVERVIEW OF A TRIAL BURN

This section summarizes different aspects of the trial burn. It describes the trial burn process and requirements for the trial burn. Basic information is provided to help answer four questions:

1. What does a trial burn involve?
2. What types of sampling and analyses are typically involved?
3. What skills, equipment, and facilities are needed to conduct a trial burn?
4. What are the major cost factors associated with a trial burn?

A. WHAT DOES A TRIAL BURN INVOLVE?

When an incinerator operator is faced with the need to perform a trial burn, the first questions that come to mind are "what do I do for a trial burn?" and "what does the trial burn do to me?" From the operator's perspective, the key trial burn considerations are the regulatory limits that must be achieved, the permit conditions that result from the burn, and the extent of sampling and analysis activities required. Potential trouble spots that have been

encountered are (1) trial burns frequently take more time and effort than an operator anticipates and (2) failure to meet the trial burn requirements. Each of these considerations is
discussed below.

1. Regulatory Limits

The trial burn provides regulatory agencies with data that will allow them to issue an
operating permit. Consequently, the trial burn is directed to testing the plant to show that
it achieves the RCRA limits, under the desired plant operating conditions. Those RCRA
limits are

1. Destruction and removal efficiency (DRE) >99.99% for all subject principal organic
    hazardous constituents (POHCs)
2. Particulate emission <180 mg/dscm (corrected to 7% O_2)
3. Hydrogen chloride (HCl) emissions <4 lb/h, or >99% removal efficiency

In addition to the above standards, state permit officials may add their own individual trial
burn and permit conditions to the federal standards.

2. Permit Conditions

From the operator's standpoint, operating conditions imposed by a permit need to allow
the plant to incinerate the types and quantities of waste they expect to handle, at the necessary
feed rates and within an acceptable range of operating conditions. That is, the permit
conditions need to provide the plant with the desired flexibility, within limits that are
reasonably achievable. Based on the trial burn results, the operating permit may specify
certain criteria such as:

1. No wastes may be incinerated which contain any Appendix VIII* compound having
    a higher heating value (HHV) below that of the most difficult to incinerate POHC
    used in the trial burn.
2. Maximum concentration of certain principal organic hazardous constituents (POHCs)
    in waste feed.
3. Maximum waste feed rate and/or maximum total heat input rate.
4. Maximum air feed rate, and/or maximum total heat input rate.
5. Minimum combustion temperature.
6. Maximum carbon monoxide (CO) content of stack gas.
7. Maximum chloride (Cl) and ash content of waste feed.

Additional criteria are discussed in Reference 1.

The trial burn involves testing at conditions that meet the operating needs of the plant
while meeting the three RCRA limits. It may be necessary to test at more than one operating
condition in order to satisfy all those needs. For example, it might be difficult to achieve a
high heat input rate (i.e., design heat input rate) with a waste feed that contains desired
high levels of Cl and ash. These factors are discussed more fully as a part of planning
activities in Sections III.B and III.C.

3. Sampling and Analysis Activities

Each test run in the trial burn includes sampling of the waste feeds and the stack effluent.
These samples are then split into a series of subsamples to be analyzed for POHCs, Cl,
HHV, ash, etc. The subsamples are then analyzed for the subject POHCs by rather complex
methods that include analyses by gas chromatography/mass spectrometry (GC/MS). Analysis

---

* *Fed. Regist.*, 46(97), 27477, 1981; 40 CFR 261, Appendix VIII.

TABLE 1
Time Factors Involved in a Trial Burn

Receive notification to submit Part B application.
Evaluate all conditions at which plant desires to be permitted (1 month).
Prepare trial burn plan and submit to EPA (required 6 months after notification).
Prepare responses to EPA on any questions or deficiencies in the trial burn plan (1 month).
Make any additons or modifications to plant that may be necessary (1 to 3 months).
Prepare for trial burn.
   Prepare for all S&A, or select S&A contractor (2 to 3 months).
   Select date for trial burn, in concert with S&A staff or contractor (completed 1 month prior to test).
   Notify all appropriate regulatory agencies (1 month).
   Obtain required quantities of waste having specified characteristics.
   Calibrate all critical incinerator instrumentation (2 weeks).
Conduct trial burn sampling (1 week).
Sample analysis (1—$1^{1}/_{2}$ months).
Calculate trial burn results ($^{1}/_{2}$ month).
Prepare results and requested permit operating conditions for submittal to EPA ($^{1}/_{2}$—1 month).

results, along with waste feedrates and stack gas flow rates measured during each run, are used to calculate the destruction and removal efficiencies (DREs). Usually, samples of ash and scrubber waters are also taken and analyzed for the subject POHCs. Although not required by RCRA, regulatory agencies may impose other additional sampling and/or analysis requirements. More detail on sampling and analysis procedures is included in Section II.B.

For any trial burn, at any one set of operating conditions (and waste feed characteristics), EPA documents recommend three replicate runs. However, it may be acceptable to make three or more runs with each run done at different conditions or with different waste feed characteristics. In this regard, there appear to be differences in what is acceptable from case to case, so plans must be approved by the responsible regulatory agencies before the trial burn. The trial burn plan should specify the number of runs and the test conditions for each run.

An important thing to remember in planning for three or more runs is that the quantities of waste required are substantial. Each run may require 4 to 8 h of plant operating time. It is probably best also to burn the same, or very similar, wastes during nontest periods (i.e., at night) in order to maintain reasonably steady conditions over the test period. The total trial burn period can require a rather large quantity of the specified waste(s). Those quantities are also specified as part of the trial burn plan.

4. Trial Burn Time Requirements

A major factor in performing a trial burn is time. Many steps are involved in the trial burn sequence of events listed in Table 1. Some of the steps have time limits dictated under RCRA. For others, adequate time must be allowed. For example, the many samples obtained in a trial burn and the complexity of POHC analysis, make it desirable to allow $1^{1}/_{2}$ months to complete the analyses and another half month to prepare a detailed report of all results. General guidelines for time requirements are included in Table 1. In specific instances, greater amounts of time may be required. If an operator is unfamiliar with trial burns, consultation with other operators, consultants, or agency personnel early in the process can provide more exact estimates of time requirements for specific situations.

In addition to the time required to prepare and conduct the trial burn adequately, time is also required for preparation of the Part B Permit Application. Frequently, the applicant will be working on trial burn preparations and responding to letters and comments on the RCRA permit simultaneously.

5. Assessing Potential Performance Problems

Probably the most important question faced by the operator is "will I pass?" (i.e., meet the RCRA requirements). The trial burn can be designed to include several different operating conditions including some where potential incinerator performance problems are minimized. Another alternative selected by some plants is to conduct an unofficial preliminary "miniburn" (i.e., one run) prior to the actual trial burn. This miniburn provides some indication of the results that can be expected, but it must be done at least 2 months before the scheduled trial burn in order to complete all analyses, evaluate the results, and make whatever changes are required.

B. WHAT TYPES OF SAMPLING AND ANALYSIS ARE TYPICALLY REQUIRED?

The primary objectives of the sampling and analysis (S&A) program are (1) to quantify POHC input and output rates to determine whether DRE requirements are met, (2) to measure input and output rates of chloride, and (3) to determine stack effluent particulate concentrations. The two most important considerations are selecting the S&A "matrix" (i.e., selecting the streams to be sampled and analytes to be measured) and identifying appropriate S&A methods. Specific problems which can be encountered are adverse stack conditions, sample train sealing problems, and the need for specialized S&A methods. Each of these factors is discussed briefly in this section and in more detail in Sections III.F and III.G and IV.A and IV.B.

1. Selecting the S&A Matrix

The main focus of the sampling activities is collection of the waste feed and the stack effluent samples, the latter being the most complex. Usually, the ash and scrubber waters are also sampled and analyzed. The main focus of the analysis activities is on the POHCs. The stack S&A also includes determination of HCl and particulate emissions, but these methods are relatively simple compared to those for POHCs. A discussion of sampling and analysis needs can also be found in References 1, 2, and 3.

Overall, the S&A typically required consists of the following, as a minimum.

1. *Obtain representative samples of each waste feed stream to the incinerator. Analyze those samples for the selected POHCs and for HHV, Cl, and ash.* (Remember that the input rate of each waste feed must also be determined in order to compute the POHC input rate which is used in the calculation of DRE). To achieve a "representative" waste feed sample, liquid waste feeds are often sampled once every 15 min and composited in each run. Solid waste feeds must also be sampled using the best practical method of obtaining representative samples of each type of solid waste used in the trial burn.

2. *Sample stack emissions to determine stack gas flow rate, HCl, particulate concentration, and to determine concentration of POHCs.*

2. Identifying S&A Methods

An example of S&A methods that could be specified for a trial burn is shown in Tables 2 and 3. These tables identify the main references that are available on recommended S&A methods, particularly References 2 and 3. These documents contain valuable information but do take considerable time to understand. They are best utilized by personnel experienced in S&A methods. These references are the best sources to identify the methods that can be used in a trial burn plan. However, experience helps a great deal in selecting the most appropriate of the available recommended methods.

Determination of stack gas flow rate and particulate emissions is done according to the conventional stack sampling method commonly referred to as Method 5 (M5). This method encompasses EPA Methods 1 to 5 and is defined in detail in 40 CFR Part 60, Appendix A. HCl emissions are sampled by modifying the M5 train to include caustic impingers.

TABLE 2
Sampling Methods and Analysis Parameters

Sample	Sampling frequency for each run	Sampling method	Analysis parameter
Liquid waste feed	Grab sample every 15 min	S004	V&SV-POHCs, Cl⁻, ash, ult. anal. viscosity, HHV
Solid waste feed	Grab sample of each drum	S006, S007	V&SV-POHCs, Cl⁻, ash, HHV
Chamber ash	Grab 1 sample after all 3 runs are completed	S006	V&SV-POHCs, EP toxicity
Stack gas	Composite	MM5 (3 h)	SV-POHCs,
	Composite	M5 (3 h)	particulate, H₂O, HCl
	Three pair of traps, 40 min each pair	VOST (2 h)	V-POHCs
	Composite in Tedlar® gas bag	S011	V-POHCs[a]
	Composite in Mylar® gas bag	M3 (1—2 h)	CO₂ and O₂ by Orsat
	Continuous (3 h)	Continuous monitor	CO (by plant monitor)

Note: VOST denotes volatile organic sampling train; MM5 denotes EPA Modified Method 5; M3 denotes EPA Method 3; S0XX denotes sampling methods found in "Sampling and Analysis Methods for Hazardous Waste Combustion[3]," V-POHCs denotes volatile principal organic hazardous constituents (POHCs); SV-POHCs denotes semivolatile POHCs; HHV denotes higher heating value.

[a] Gas bag samples may be analyzed for V-POHCs only if VOST samples are saturated and not quantifiable.

Sampling of stack effluent for POHCs, in order to determine DRE, may require from one to three separate methods (or more), depending on the number of POHCs to be quantified and their characteristics (e.g., volatile or semivolatile) and on the detection limits that are required to prove a DRE of 99.99%. These methods are

1. Modified Method 5 (MM5) for semivolatile (SV) POHCs
2. Volatile organic sampling train (VOST) for volatile POHCs
3. Gas bags for volatile POHCs
4. Special methods for certain POHCs which cannot be sampled with any of the above methods

Semivolatile POHCs commonly require use of MM5. This one sampling train, shown in Figure 1, provides for determination of the SV-POHCs. Components can be extracted, and combined with extracts of the XAD® resin and the condensate, for analysis by GC/MS to determine SV-POHCs. A diagram of the VOST train commonly used for sampling volatile POHCs is shown in Figure 2. This train, unlike M5 or MM5, does not involve traversing the stack. However, the VOST preparation and analysis procedures are quite complex. Those interested in the detailed procedures should refer to Reference 6.

3. Adverse Stack Sampling Conditions

Adverse stack sampling conditions are frequently encountered at hazardous waste incinerators. Problems that have been encountered include cyclonic flow, very high temperature stack (1600 to 1800°F), and high moisture content (saturated with H₂O at 150°F with droplet carryover). These potential problems should be considered during planning, and appropriate actions should be taken. More complete discussions of cyclonic flow and moisture are included in Section IV.B.2.

TABLE 3
Example Analytical Procedures

Sample	Analysis parameter	Sample preparation method	Sample analysis method
Liquid waste feed	V-POHCs	8240	8240
	SV-POHCs	8270	8270
	Cl⁻	—	E442-74
	Ash	—	D482
	HHV	—	D240
	Viscosity	—	A005
Solid waste feed	V-POHCs	8240	8240
	SV-POHCs	8270	8270
	Cl⁻	—	D-2361-66 (1978)
	Ash	—	D-3174-73 (1979)
	HHV	—	D-2015-77 (1978)
Ash	V-POHCs	8240	A101
	SV-POHCs	P024b, P031	A121
	Toxicity	—	C004
Stack gas			
M5 train filter and probe rinse	Particulate	M5	M5
MM5 train components	SV-POHCs	P024b, P031	A121
Condensate	Cl⁻	—	325.2
	SV-POHCs	P021a	A121
XAD® resin	SV-POHCs	P021a	A121
Caustic impinger	Cl⁻	—	325.2
VOST	V-POHCs	—	A101
Tedlar® gas bag	V-POHCs	—	A101ᵃ
Gas bag	CO₂, O₂	—	M3 (Orsat)
Continuous monitor	CO	—	Continuous monitor

Note: Four-digit numbers denote methods found in "Test Methods for Evaluating Solid Waste", SW-846.[2] Numbers with prefixes of A, C, and P denote methods found in "Sampling and Analysis Methods for Hazardous Waste Combustion."[3] Method No. 325.2 (for Cl⁻) is from "Methods for Chemical Analysis of Water and Wastes". EPA-600/4-79-020, March 1979.[4] Numbers with prefixes D and E denote methods established by the American Society for Testing and Materials Standards (ASTM). M3, M5 refer to EPA testing methods found in Fed. Regist. 42(160), Thursday, August 18, 1977.[5]

ᵃ   Tedlar® gas bag samples will be analyzed for V-POHCs only if VOST samples are saturated and not quantifiable.

4. Sample Train Sealing Problems

Both the VOST method and available guidance of the MM5 method state that no grease be used on any of the connections in the train (i.e., ball joints). Teflon® or Viton O-rings have been used in VOST, and in MM5, to provide adequate seals without use of grease. Added care must be taken to ensure leak-check integrity of the sampling trains, with some added risk that a test may have to be repeated if any sampling train fails the post-test leak check.

5. Need for Specialized Methods

Although the majority of POHCs are sampled with either VOST or MM5, specialized sampling methods must be used for some POHCs. Those POHCs which require specialized methods are identified in Appendix B of Reference 3. This reference should be consulted during the planning stage to assure that proper methods are used.

C. WHAT SKILLS, EQUIPMENT, AND FACILITIES ARE NEEDED TO CONDUCT A TRIAL BURN?

The incinerator facility operator is responsible for conducting the trial burn. The facility must provide the types and quantities of waste needed and operate the plant during the trial

317

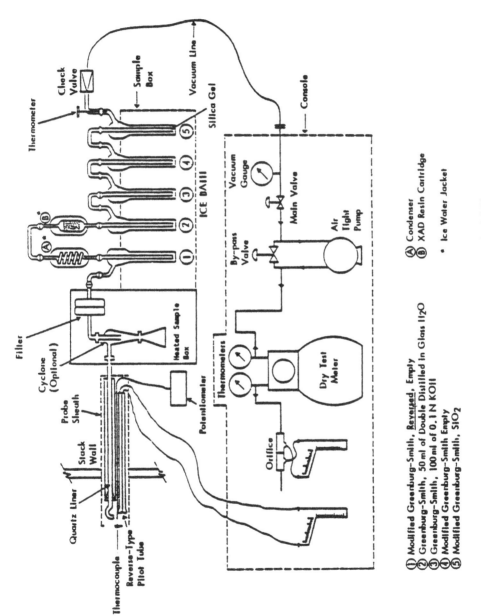

FIGURE 1. Modified Method 5 sampling train (MM5).

① Modified Greenburg-Smith, Reversed, Empty
② Greenburg-Smith, 50 ml of Double Distilled in Glass H₂O
③ Greenburg-Smith, 100 ml of 0.1 N KOH
④ Modified Greenburg-Smith Empty
⑤ Modified Greenburg-Smith, SiO₂

Ⓐ Condenser
Ⓑ XAD Resin Cartridge

* Ice Water Jacket

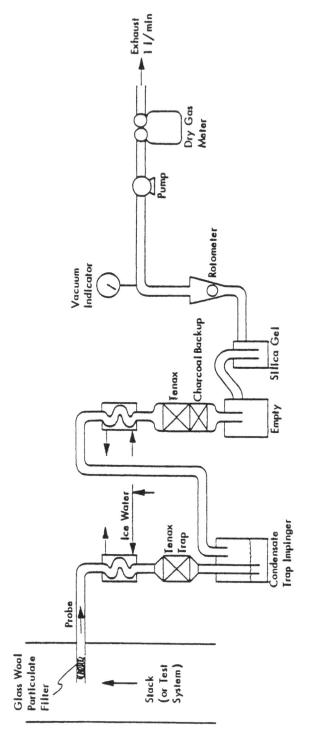

FIGURE 2.   Volatile organic sampling train (VOST).

TABLE 4
Capabilities Necessary for Trial Burn Sampling and Analysis

Sampling equipment for solid waste feeds (especially drummed wastes)
Stack sampling equipment, usually including the following:
  EPA Method 5 equipment and all associated test equipment (e.g., EPA Methods 1, 2, and 3)
  Method 5 equipment adaptable to Modified Method 5 (greaseless) and associated XAD resin preparation,
  extraction, and analysis facilities
  Volatile organic sampling train (VOST) equipment with at least 18 pairs of VOST traps; also, all facilities
  needed for preparing, checking, and analyzing the traps
  Field laboratory equipment for sample recovery
Facilities for analyzing all samples, including:
  Laboratories containing relevant safety equipment such as hoods and equipped with sample preparation equip-
  ment including Soxhlet extractors, separatory funnels, continuous extractors, blenders, Sanifiers or other
  equipment to homogenize waste feed samples, sodium-sulfate drying tubes, Kuderna-Danish glassware, etc.
  Equipment for preparing VOST traps to allow simultaneous heating and purging of the traps; ideally the traps
  should be prepared and stored in an organic-free laboratory
  All required compounds to prepare calibration standards and surrogate recovery spiking solutions
  Computerized GC/MS instrumentation
  Established QA procedures for assessing precision and accuracy of analytical methods
Knowledge and preferably experience in all of the sampling and analysis methods and calculation/reporting of
results
Process monitoring experience, especially quantification of waste feedrates and documentation of plant operating
conditions

burn at the conditions under which they desire to be permitted. However, the specialized sampling and analyses required in a trial burn are beyond the capability of most facilities. A facility that has most of the necessary capabilities still may decide to use an experienced contractor because of the specialized nature of the methods and the fact that the trial burn may be only a one time need. The important considerations in the decision are facilities and equipment requirements and staffing needs. Each of these factors, plus a brief consideration of contractor selection, are addressed below.

1. Facilities and Equipment
Whether the operator uses a contractor or his own staff for the trial burn S&A, certain capabilities are required. The facilities and equipment that are usually necessary are shown in Table 4.

2. Staffing Needs
Decisions regarding use of in-house or contractor expertise to conduct a trial burn also depend on staffing needs. At a minimum the trial burn staff should be knowledgeable in stack sampling methods, have experience in analysis of low concentrations of organics in complex matrices, and be familiar with calculating and reporting trial burn results. Knowledge of process monitoring is also helpful. A detailed discussion of the number and capability of S&A personnel required is included in Section III.E.

3. Selecting a Contractor
Trial burn procedures are relatively new, and are much more sophisticated than a normal EPA M5 test for particulate emissions. There are about 10 to 20 organizations in the U.S. that have trial burn S&A experience, and probably several more are capable of doing so. If a facility would like to know who to contact for S&A service, inquiries should be made at state and federal regulatory agencies, or other incinerator facilities should be contacted that may have already conducted trial burns.

D. WHAT ARE THE MAJOR COST FACTORS ASSOCIATED WITH A TRIAL BURN?

The three major cost components of the trial burn are planning and preparation, sampling and analysis, and quality assurance (QA). Each of the components must be included in trial burn budgets.

1. Planning and Preparation

One of the first cost factors for a trial burn is preparation of the trial burn plan, including preparing responses to questions and additional information requested by the regulatory agencies.

The second cost factor is plant additions and modifications needed to comply with RCRA regulations (see Section III.A). These may include CO monitors, waste feed flow monitors, etc., and stack sampling ports and scaffolding needed for the trial burn.

A third cost factor is associated with acquiring/storing the types and quantities of wastes necessary for the trial burn (see Section III.C).

2. Sampling and Analysis

The major cost factor is the sampling and analysis required by the trial burn. In general, this cost usually ranges from $30,000 to $150,000, depending on the number of runs, the number of samples to be taken in each run, and the analysis required on each sample.

The number of runs to be conducted in the trial burn is one of the major cost factors. EPA recommends three replicate runs at each operating condition to be tested. Thus, a minimum of three runs are usually done. If two operating conditions are involved, then six runs may be necessary. In some cases, an array of operating conditions are tested, with only one run at each condition.

In each run, all influent and effluent streams are usually sampled and analyzed as described earlier (see Table 2). The number of these samples and the complexity of their analysis obviously affect the cost. Ordinarily, the field sampling activity, including all preparation for sampling, accounts for one fourth to one third of the S&A cost. Analysis of samples usually accounts for one third to one half the cost, with the remaining costs for data reduction, calculations, and reporting of results.

3. Quality Assurance

Analysis costs, especially for the POHCs, are a rather large cost factor, partly because analytical QA activities typically include:

1.   Replicate analysis of some samples
2.   Analysis of samples spiked with POHCs or surrogates
3.   Analysis of blanks
4.   Analysis of calibration standards
5.   Analysis of blind audit samples

All of the above can easily involve analysis of twice the number of samples actually taken. All these samples then must be multiplied by the number of analyses to be performed on each sample. Some QA is an essential part of a trial burn, but excessive QA can rapidly increase the cost.

4. Estimating the Costs

Determining the cost for a trial burn is highly site dependent. In general, the trial burn costs will depend on the following factors, as discussed in the preceding sections:

1.   Number of runs

2.   Number and type of waste feed samples
3.   Number of effluent samples
4.   Number of different analyses performed on each sample
5.   Complexity of the QA/QC plan
6.   Modifications required to prepare facility for trial burn

The normal range for a trial burn sampling and analysis program conducted by an outside contractor is $30,000 to $150,000+. This range does *not* include plant modifications, preparations on-site for the test, or preparation of the permit application and trial burn plan.

The breakdown of costs for a trial burn is roughly: one third for the field sampling program, one third for sample analysis and project QA/QC, and one third preparation, engineering calculations, and reporting. These are rough estimates, and frequently the analysis portion of the program can involve as much as half of the total cost.

In summary, the S&A part of a trial burn is costly, and each time another run, another sample, or another analysis is added to the test plan the cost will rise. Each such addition needs to be carefully considered in order to hold cost at the lower end of the range.

III. PLANNING FOR A TRIAL BURN

The probability for success of a trial burn is enhanced by good planning. The major objectives of the planning process are (1) to select trial burn conditions that provide the plant adequate operating flexibility, (2) to assure that the trial burn will be conducted in a manner acceptable to regulatory agencies, and (3) to make the trial burn cost-effective. Key questions addressed during planning are

1.   What equipment/instrumentation is the incinerator required to have?
2.   How should operating conditions be selected?
3.   How should trial burn POHCs be selected?
4.   What types and quantities of waste are needed, and how can they be prepared?
5.   How many runs are necessary, and how long is each run?
6.   How many people are needed and with what experience?
7.   How are POHC sampling methods selected?
8.   What detection limits are required for the sampling and analysis method?
9.   What QA/QC needs to be done?
10.   How is it best to plan for the possibility that trial burn results may be outside RCRA requirements?

A. WHAT EQUIPMENT/INSTRUMENTATION IS THE INCINERATOR REQUIRED TO HAVE?

The incinerator is required by RCRA to have the equipment/instrumentation shown in Table 5. The regulatory agencies also may require monitoring of other important operating parameters (e.g., scrubber water flow rates, venturi scrubber ΔP, etc.). Minimum or maximum levels for each parameter may be specified in the operating permit. Analysis of waste feeds may also be required if the operating permit stipulates limitations on HHV, Cl, or ash content of waste feed.

B. HOW SHOULD THE OPERATING CONDITIONS BE SELECTED?

Operating conditions for the trial burn are selected to provide the plant with operating flexibility. Important considerations are the key operating parameters that affect permit conditions and the use of pretests or miniburns to help establish those conditions while meeting RCRA requirements (e.g., DRE).

TABLE 5
Incinerator Equipment/Instrument Requirements for Trial Burn

Equipment to maintain particulate emissions below 0.08 gr/dscf
Equipment to maintain 99% HCl removal or HCl emissions below 4 lb/h
Equipment that provides 99.99% DRE on POHCs
Stack test ports and scaffolding
Valves, taps, etc., for sampling all waste feeds, liquid effluents, ash, etc.
Equipment to maintain noncyclonic flow in stack when testing
Continuous CO monitor
Continuous waste feed flow monitor
Continuous monitor for combustion gas velocity or air input rate
Continuous combustion temperature monitor
Automatic interlock system to shut off waste feed under the following situations:
  Low combustion temperature
  High CO concentration[a]
  High combustion airflow to incinerator or high combustion gas velocity

[a] Established based on trial burn results or state statutory limitations.

1. Operating Parameters that Affect Permit Conditions

The operating conditions selected for the trial burn must represent the worst case conditions under which the incinerator may expect to operate and, therefore, needs to be permitted to operate. The conditions selected may include any or all of the following:

1. Waste containing hardest-to-burn POHC (lowest HHV)
2. Highest concentrations of all POHCs selected
3. Maximum waste feed rates
4. Maximum combustion airflow rate (minimum residence time)
5. Maximum CO level in stack gas
6. Minimum combustion temperature
7. Minimum HHV of waste
8. Maximum thermal input (Btu/h)
9. Minimum O_2 level in stack gas
10. Maximum Cl content of waste feed
11. Maximum ash content of waste feed
12. Minimums of maximums on other operating conditions (e.g., venturi scrubber ΔP, scrubber water flow rate, and pH).

Obviously, it is very difficult to achieve all of the above at any one set of operating conditions. In fact, some of the conditions are almost direct opposites (e.g., maximum airflow rate but minimum O_2 in stack gas).

The first six items in the above list are probably the most important and may be achievable in one set of operating conditions that also include some of the other conditions. If so, one trial burn (three runs) at those conditions may suffice. If not, additional runs that include the other conditions may be necessary. Of course, operating conditions which result in permit conditions most favorable to each individual facility will have to be determined on a case-by-case basis.

The major problem with the worst-case conditions is that they maximize the chance of failure (not meeting RCRA requirements). Since the plant wants to pass, the exact conditions must be carefully selected, balancing operating needs against increasing chance of failure. Plant operating experience is very important in these decisions.

2. Use of Pretest or Miniburns

Preliminary testing and miniburns can be extremely valuable in helping to select operating conditions for the actual trial burn. The following types of miniburns may be useful.

The hardest to burn POHC, at high concentration, can be used in a miniburn that is conducted at the lowest temperature and the highest CO level. If the results show a DRE exceeding 99.99%, then it is likely that 99.99% DRE will be achieved regardless of any other operating conditions.

At high Cl input rates, a well-designed scrubber will not usually fail 99% removal even at minimum conditions. A pretest could verify that presumption.

Achieving the particulate limit causes problems more frequently than does achieving DRE. A pretest with EPA M5 will help identify any problems and help in selecting conditions for the trial burn. The pretest can also uncover specific sampling and analysis problems that may not be readily apparent.

Mist carried over from a recirculating scrubber solution or alkaline scrubbers can have a drastic impact on particulate emission measurements, especially if the scrubbers are not equipped with efficient mist eliminators. It may be advisable to conduct a preliminary particulate test, well in advance of the actual trial burn, to identify possible problems.

For existing plants, any of the above pretesting could be done prior to submitting the trial burn plan for approval. For new plants, pretesting will have to be part of the approved trial burn plan.

C. HOW SHOULD TRIAL BURN POHCs BE SELECTED?

POHCs for the trial burn should be selected during development of the trial burn plan. The selection conforms with the regulatory approach laid out in the *Guidance Manual for Hazardous Waste Incinerator Permits*.[1] In addition to the regulatory criteria, the following two considerations should be taken into account: (1) maximum flexibility of operating conditions under the permit and (2) ease of sampling and analysis during the trial burn. Currently, the regulation requires that a DRE of 99.99% be demonstrated for the selected POHCs. In addition, the following limits will result from the selection of POHCs for the trial burn:

1. Appendix VIII compounds in any subsequently burned waste feed must be present in concentrations lower than the POHC in highest concentration during the trial burn.
2. Appendix VIII compounds in any subsequently burned waste must have a heat of combustion ranking higher than the POHC with the lowest heat of combustion during the trial burn. (Heats of combustion for all Appendix VIII compounds have been determined and can be found in Reference 1).

Because of these limits, the POHCs chosen for trial burn testing must include the Appendix VIII compounds in the waste feed, usually the compounds in the highest concentration and with the lowest heat of combustion. "Appendix VIII" refers to the appendix of the hazardous waste regulations which lists compounds that are considered hazardous (see 40 CFR 261 Appendix VIII).

It is important that the Appendix VIII compound present in highest concentration in any proposed waste feed be present in the feed during the trial burn at the maximum concentration expected, in order to obtain the necessary permit conditions. Likewise, it is important that the compounds with the lowest heat of combustion be present in the waste feed used during the trial burn at sufficient levels to determine 99.99% DRE (see Section II.H).

In selecting POHCs for a trial burn, sampling and analysis implications also must be considered. From this point of view, Appendix VIII compounds fall into three categories:

1. Volatiles: compounds which can be sampled using the VOST (in certain cases other methods may be more appropriate, as discussed in Section III.H)
2. Semivolatiles: compounds which can be sampled using the MM5 train
3. Other compounds which must be sampled using different techniques: special trains,

calorimetric methods, etc., including compounds which degrade easily in water or which have special interferences or are otherwise difficult to quantify using GC/MS analysis

Ideally, all trial burn POHCs could be selected from either the volatile or semivolatile group. This minimizes the number of sampling trains used in the field and simplifies the analysis. If possible the "other" category should be avoided, because more specialized equipment may be needed, which will have to be cleared by permit reviewers in advance of the test and may result in higher sampling and analysis costs. An additional consideration is to avoid POHCs which also might show up as products of incomplete combustion from the burning of the waste (e.g., chlorinated benzenes, ethanes, and methanes).

All of these considerations must be taken into account when selecting POHCs for a trial burn. One solution, which has been used at incinerators in the hope of burning a wide variety of wastes, is to spike a low heat of combustion compound (e.g., carbon tetrachloride or perchloroethylene) in significant concentrations (5—10%) into the waste feed.

D. WHAT TYPES AND QUANTITIES OF WASTE ARE NEEDED, AND HOW CAN THEY BE PREPARED?

The response addresses calculation of waste quantities, assuring adequate supplies of waste by type, and preparation of wastes. Specific problems addressed include mixing of synthetic or spiked wastes and time requirements.

1. Quantities of Waste

The quantity of waste required is dependent on the waste feed rate to be used during each run, the number of runs, and the duration of each run. Waste feed rate and the number of runs are selected by the incinerator operator and are specified in the trial burn plan. The sampling time required in each run is usually 3 to 4 h plus 1 h to line out the unit before the start of testing and 1 to 2 h for contingencies (plant operating problems or sampling problems). Considering these, a quantity of waste sufficient for 8 h of operation should be available for each run. If the trial burn involved only three runs at one set of operating conditions, then waste sufficient for 24 h of operation should be available.

2. Types of Waste

Sufficient quantities of waste must be available for each type of waste feed that is used. Each type of waste must have all the specific characteristics that are required to meet selected operating conditions. For example, the waste to be burned during the trial burn might include both continuous feeding of an organic liquid and intermittent feeding of drummed solids. Each of these wastes must meet certain specifications selected for the trial burn, including POHC concentrations, heating value, Cl and ash content, etc.

3. Waste Preparation

Three methods can be used to prepare the required quantities of wastes possessing the correct characteristics. These three methods pertain mainly to POHC characteristics but may be used to achieve any of the necessary charactcristics. The three methods are

1. Use actual wastes
2. Use synthetically prepared wastes
3. Continuously spike POHCs into the waste during the trial burn

Method 1 usually is desirable, if it is possible to acquire actual wastes that have the necessary characteristics or to achieve those characteristics by blending of actual wastes.

Method 2 usually involves using actual wastes mixed with purchased chemical compounds (i.e., POHCs). Method 3 is similar to method 2 except that it applies mainly to continuous liquid feeds, with the purchased chemical(s) continuously pumped into this feedline.

4. Mixing

All of the above three methods require that the waste feeds be well mixed, but mixing is especially important for methods 2 and 3. Lack of good mixing for any waste feed can, and has, caused problems in trial burns. For method 3 the trial burn may involve continuous spiking of POHCs (pure components or mixtures thereof) into a liquid waste feed line. When this is to be done, a connection ($^1/_4$ or $^1/_2$-in. valve) must be provided, with another sample tap located further downstream. It is also highly advisable for the plant to install an in-line mixer between these two connections to help ensure that the "spiked" components are well mixed with the waste and that the samples collected are representative of this mixture.

5. Time Requirements

Another important factor in waste preparation is time. The quantities of waste involved can be rather large, and it may take several weeks to acquire sufficient quantities of wastes to prepare a homogeneous batch with the proper characteristics. Storage space for these "special" wastes, over some time period, can impact normal plant operations. Finally, some additional time may be needed to sample and analyze the wastes to be sure they have the necessary characteristics.

Adequate time also must be allowed for numbering, weighing, and sampling of drummed solids before the trial burn. Since the number of drums may exceed 300, the problem of weighing and sampling initially may not be realized. Also, samples of drummed solids must be representative of those drums used in each run. Representative samples may be obtained by sampling each drum during each run or by "staging" the drums to be used in each run and sampling them prior to the trial burn.

E. HOW MANY RUNS ARE NECESSARY, AND HOW LONG IS EACH RUN?

This question was discussed in Section II.A.3 and II.D.2. Additional points offered as guidance are

1. Each run will require at least 2 to 4 h. It is best to plan only one run per day, except in special cases when sampling is less complex than usual. Quite often, when an incinerator operator hears that the sampling time required for each run is 3 to 4 h, it is assumed that the sampling crew can do two runs each day. However, the sampling crew has about 2 h of work in preparing for each run and at least 2 h of work after each run is completed to recover, label, and package each sample. In many instances, a variety of problems do occur, both in plant operations and in sampling, so that one 3- to 4-h run may involve a 12- to 16-h day for the sampling crew. The most reasonable assumption is that one run can be completed each day.
2. EPA *recommends* three runs at each set of operating conditions to be tested.
3. If several sets of operating conditions are to be tested, regulatory agencies may allow fewer than three runs at each condition.
4. Conducting more than six runs may not be cost effective.

F. HOW MANY PEOPLE WILL BE NEEDED, WITH WHAT EXPERIENCE?

Personnel required for sampling during the trial burn usually number between 5 and 10, depending on the complexity of the sampling. A typical example, reflecting the sampling plan shown earlier in Table 2, is presented in Table 6.

The personnel list in Table 6 is only an example. In some cases, one person can do multiple jobs depending on sampling frequency and complexity, and physical layout of

TABLE 6
A Typical Example of Sampling Personnel Required

Job	Number of personnel	Experience required
Sample liquid feed (once every 15 min)	1	Technician with sampling experience and safety training
Drum solid sampling and recording (once every 5—10 min)	1	Technician with sampling experience and safety training
Sampling ash and scrubber waters every $^1/_2$—h	1	Technician with safety training
Stack sampling MM5	2	Experienced console operator and technician for probe pushing
VOST	1	Experience with VOST operation
Process monitor to record operating data every $^1/_4$—$^1/_2$ h and determine waste feed rates	1	Engineer or other person experienced in plant oeprations and trial burn requirements
Field laboratory	1	Experienced chemist for check-in and recovery of all samples and preparation of sampling equipment for each run
Crew chief	1	Person experienced in all aspects of trial burn sampling to direct all activity and solve problems that may occur
Total	9	

sampling locations. Also, quite often the crew chief performs one of the sampling activities, again depending on complexity of the sampling activity. Plant personnel may perform the process monitoring. However, the date should be separate from any normal plant operating log and usable in the trial burn test report.

G. HOW ARE POHC STACK SAMPLING METHODS SELECTED?

A general procedure to identify the appropriate POHC stack sampling methods is outlined in Table 7. When both volatile and semivolatile POHCs are present, both MM5 and VOST are needed. Analyses performed on these samples must provide the necessary detection limits for the POHCs, as mentioned in Tables 7 and 8.

H. WHAT DETECTION LIMITS ARE REQUIRED FOR THE SAMPLING AND ANALYSIS METHODS?

1. Waste Feed Detection Limits

Analyses of POHCs in waste feeds must be capable of detecting the expected concentrations, which usually are above 10,000 ppm (1%). But it is desirable that the detection limit be 100 ppm (commonly achieved by recommended analytical techniques). A POHC at this concentration or above may be considered (under RCRA) to be "significant".

2. Stack Gas Detection Limits

Detection limits required for POHCs in stack gases are discussed in Table 8. The rule-of-thumb that can be used in most cases is

100 ppm in waste feed $= 1$ $\mu g/m^3$ in stack gas at 99.99% DRE

This equation can be used to estimate stack gas concentrations for any waste feed concentration (i.e., 2000 ppm in waste $= 20$ $\mu g/m^3$ in stack gas, at 99.99% DRE).

Some POHCs may require special sampling/analysis methods or may show low recovery efficiencies for MM5 samples. Therefore, each case must be considered separately to ensure

TABLE 7
Procedure for Identifying Necessary Stack Sampling Methods

Step 1. *Determine whether each POHC is a volatile or semivolatile compound*
Volatile compounds are generally those that have boiling points below 130°C. Most can be sampled with VOST or gas bags. The best way of determining the sampling method needed is to refer to Appendix B of Reference 2. If the POHC shows sampling by "particulate/sorbent", then MM5 is required. If it shows "sorbent" or "gas bulb", then VOST or gas bags will be the sampling method. Regardless of whether a POHC is a volatile or semivolatile, some POHCs require special sampling methods as indicated in Appendix B of Reference 2 (e.g., formaldehyde).

Step 2. *Estimate concentration of each POHC in the stack gas, assuming a DRE of 99.99%*
Estimation of the concentration of each POHC requires some knowledge of approximation of POHC concentration in waste feeds, waste feed rates, and stack gas flow rates. Using that information, concentrations of each POHC in the stack gas can be estimated, for an assumed DRE of 99.99% (see sample calculation in Table 8 of Chapter 4).
For semivolatile POHCs, MM5 is suitable for any stack gas concentration above 1 μg/m³.
For volatile compounds, VOST should be used when stack gas concentrations fall with the range of 1—100 ng/l. However, if the estimated stack gas concentration exceeds 100 ng/l, the gas bags should also be used and analyzed in the event that the VOST sample concentrations saturate the GC/MS data system.

TABLE 8
Sample Calculation for Estimating POHC Concentration in Stack Gas at DRE of 99.99%

Basis
  Waste feed flow rate approximately 4 gal/min
  Density of waste approximately 9 lb/gal
  POHC concentration in waste feed is near minimum significant level of 200 ppm (200 ppm = 0.0002000 g POHC/g feed)
  Stack gas flow rate unknown, but total heat input to incinerator is approximately 30×10^6 Btu/h with 100% excess air
Calculation
  Rule of thumb (applies in most, but not all cases):
  Each 100 Btu of heat input produces about 1 dscf of flue gas at 0% excess air or 2 dscf of dry flue gas at 100% excess air

1. Flue gas flow rate:

   $(30 \times 10^6$ Btu/h) (1h/60 min) (2 dscf/100 Btu) = 5000 dscf/min

   (5000 dcsf/min) (1 m³/35.3 ft³) = 142 m³/min

2. Waste feed rate:

   (4 gal/min) (9 lb/gal) (454 g/lb) = 16,300 g/min

3. POHC input rate:

   (16,300 g feed/min) (0.000200 g POHC/g feed) = 3.26 g POHC/min

4. POHC stack output rate, at 99.99% DRE:

   (3.26 g/min) (1.0 − 0.9999) = 0.000326 g/min

5. POHC concentration in stack gas (at 99.99% DRE):

   $\dfrac{0.000326 \text{ g/min}}{142 \text{ m}^3/\text{min}}$ = 0.0000023 g/m³ = 2.3 g/m³ = 2.3 ng/l

TABLE 9
Sample QA for a Trial Burn

All equipment used in S&A activities should have written calibration procedures. Procedures and documentation of the most recent calibration should be available.

Traceability procedures (not necessarily chain of custody) should be established to ensure sample integrity.

A GC/MS performance check sample should be analyzed each day prior to sample analysis. If results are outside acceptable limits, samples should not be run.

All samples from a least one run should be analyzed in triplicate to assess precision.

Blank samples should be analyzed to assess possible contamination and corrective measures should be taken as necessary. Blank samples include:

  Field blanks — These blank samples are exposed to field sampling conditions and are analyzed to assess possible contamination from the field (a minimum of one for each type of sample preparation or the number specified by the appropriate method).

  Method blanks — These blank samples are prepared in the laboratory and are analyzed to assess possible laboratory contamination (one for each lot of samples analyzed).

  Reagent and solvent blanks — These blanks are prepared in the laboratory and are analyzed to determine the background of each of the reagents or solvents used in an analysis (one for each new lot number of solvent or reagent used).

Field audits and laboratory performance and systems audits may be included in some cases. Cylinders of audit gases for volatile POHCs are available from EPA.

A minimal level of calculatiom checks (e.g., 10%) should be established.

that the detection limit for the methods used are low enough to quantify those specific POHCs at the concentrations expected in the stack of 99.99% DRE. Consultation with analytical chemists or with the authors and EPA project officer given in Reference 3 can be most valuable in this regard.

3. High Concentration of Volatile POHCs

Since the GC/MS analytical techniques for MM5 and VOST samples can easily achieve a detection limit of 1 μg/m^3, there is usually no problem. However, high stack concentrations of some volatile POHCs may exceed the range for VOST samples (i.e., that saturate the GC/MS). Those samples require use of gas bags in order to determine if 99.99% DRE was or was not achieved. The gas bags are analyzed by transferring a small volume of sample onto a VOST trap prior to GC/MS analysis. For example, 5 l may be taken for analysis. This quantity is four times less than the quantity sampled by VOST under normal sampling conditions.

I. WHAT QA/QC NEEDS TO BE DONE?

It is important in planning the trial burn to stipulate exactly what QA will be done and to know why it is needed. Some QA activities may be desirable but are not essential in specific cases. Blanket statements that "full QA" will be employed in the trial burn are not definitive, and excessive QA can drastically increase costs. An example list of basic QA for a trial burn is given in Table 9. Preliminary discussion of QA procedures with the responsible regulatory agency is recommended prior to submittal of the trial burn plan.

One example of the QA activity that may be specified without adequate thought is "chain of custody". The number of samples collected in a trial burn normally numbers 100 to 300. Adherence to chain-of-custody procedures for all of these samples requires considerable time and effort, with its associated cost impacts. Unless there is reason to believe that sample results will be a part of some judicial proceedings, chain-of-custody procedures on the samples may be an unnecessary added cost when traceability procedures would suffice.

J. HOW IS IT BEST TO PLAN FOR THE POSSIBILITY THAT TRIAL BURN RESULTS ARE OUTSIDE RCRA REQUIREMENTS?

There is always the haunting possibility that the trial burn results may show failure to meet one or more of the RCRA requirements. This result is more likely when the trial burn is conducted under "worst case" conditions at which the plant wants to be permitted to operate.

A miniburn and other preliminary testing (e.g., M5) can help identify problems before the trial burn and, after modifications, avoid failure during the trial burn. Another alternative is to conduct runs at two sets of operating conditions. One set would be worst case, while the other set could be conditions that increase the chance of passing and that the plant could tolerate for continued operation. The latter would be less desirable and would not be cost efficient if the plant passed under worst case conditions. Therefore, it may be possible to test only under worst case conditions. If the plant fails the RCRA requirements, then perhaps the contingency plan could be to request a variance and retest under the other set of operating conditions, as soon as possible after the results from the first test are available.

IV. CONDUCTING THE TRIAL BURN

Many questions arise when preparing for and conducting a trial burn. Some questions which may be asked are

1. What is involved in preparing for the tests?
2. What is involved in conducting the actual sampling?
3. What is involved in analysis of samples?
4. How are sampling and analysis data converted to final results?
5. How are the data and results usually reported?

Each of the above questions are broad and cover many specific items. Subsequent sections of this manual will attempt to address these questions by discussing specific areas that are most important and areas that may cause problems. In formulating answers, it has been assumed that the incinerator operator has obtained all necessary approvals of the trial burn plan and is preparing to implement that plan. At that point, the realities of the test come to the forefront and answers are needed to many questions like those discussed below.

A. WHAT IS INVOLVED IN PREPARING FOR THE TESTS?

Preparations for the test are numerous; several of the most important items are scheduling, sampling crew activities, equipment preparation and calibration, facility readiness, process data, data sheets and labels, and safety precautions. One potential problem that should be addressed during preparation is how to coordinate with observers during the trial burn.

1. Schedule

Many scheduling problems can occur if the lead time required is not anticipated. These include:

Item	Time often required
Make additions or revisions to trial burn plan	Varies
Acquire all wastes needed	Varies
Select test contractor	3 months before test
Pretest site visit by contractor	1 month before test
Notify all regulatory agencies	1 month before test
Install necessary sampling access	Complete 1 week before test
Test contractor begins preparation	2 to 3 weeks before test
Numbering, weighing all drum solid feed	2 to 3 d before test
Conduct test	3+ days
Sampling equipment tear down	1 d after last test
Analysis of samples	1 to $1^1/_2$ months after test
Report of results from contractor	2 months after test
Submit results to EPA	3 months after test

If a miniburn or other preliminary testing is involved, it should be done at least 2 months before the trial burn, which would increase some of the required time intervals shown above.

2. Sampling Crew

If a sampling/analysis contractor is used, that contractor will have to make crew selections and assignments at least 2 to 3 weeks prior to the test. Many of those crew members and the analytical personnel will perform the complex activities for equipment preparation and calibration, preparation of all absorbent traps (MM5 and VOST), and special cleaning of all sampling containers. Other logistical arrangements for transporting equipment and personnel to the site must also be made. For these reasons a firm date for the test should be established, in concert with the S&A contractor, at least 1 month prior to the test.

3. Equipment

The large amount of sampling equipment needed for a trial burn is usually surprising to the operator. A list of some of this equipment is shown in Table 10. What that list does not show is the detailed preparation procedure for much of the equipment. For example:

1. Calibrate all MM5 consoles.
2. Condition and check all VOST traps.
3. Clean and pack all MM5 resin traps (XAD).
4. Preweigh all M5 filters.
5. Clean all glassware and sample bottles.
6. Purchase special reagents and solvents.
7. Modify equipment for special sampling situations.
8. Reserve time for use of analytical instrumentation.
9. Collect and pack all necessary field laboratory equipment.

In many cases, the complexity of the sampling plan or lack of available plant facilities requires provision of a large field trailer for the samples and sampling equipment. This trailer also serves as the field laboratory.

4. Facility Readiness

Facility readiness is critical to conducting the test as scheduled. Checking operational readiness of the incinerator and its components, including critical instrumentation (especially flow meters), is vital and should be done early enough to correct any problems identified. When the plant is operated under worst case conditions, unanticipated problems often occur. The plant should be operated under test planned conditions prior to the tests to minimize costly delays during the test.

Other facility readiness needs are identified during the pretest site survey. This survey will identify most of the sampling needs, especially those related to the stack sampling ports and sampling platform, which can require some installation work by the plant. Frequently, the contractor will need to rent a trailer to be used on site for sample workup and storage. A suitable location for the trailer should be identified during the survey. The survey also will identify other needs such as electrical supply requirements. (These requirements are usually much larger than the plant expects.) The survey should be conducted at least 1 month prior to the test to allow time for modification of the facility.

Facility readiness also includes preparation of all the wastes to be used in the tests. Waste preparation is especially important for drummed solid waste. It is highly desirable to have all drummed waste on site at least a week prior to the test and have the drums arranged in a staging area in the order that they will be used. These preparations will facilitate numbering, weighing, and sampling of each drum. Drum preparation can require considerable time and effort on the part of the plant operating staff and the sampling crew.

TABLE 10
List of Sampling Equipment and Supplies Typically Used

Method 5 — particulate train
  Console with pump heaters
  Sample box with pump heaters
  Umbilical-sampler hookup (gooseneck)
  Umbilical cords and adapter (elec.)
  Probe tips (4/set)
  Probes (type SS, glass)
  Extra quartz inserts
  Extra glass inserts
  "S"-type pitot (ft-5)
  Rails — dexangle
  Port-rail clamps (collar size 4 in.)
  Probe-support
  Sample box guide attachment
  Potentiometer
  Spare thermocouples
  Intercom (with cable)
  Manometer, inclined
  Manometer, "U"
  5-impinger foam inserts (for sample box)
  Digital pyrometer
  Umbilical thermocouple adapter
  Submersible pumps
  Latex tubing for condensers
  Glass tape (high temp)
  Console supply briefcase
Method 5 — glassware
  Aluminum cases
  Impingers
  "U" connector
  90° connector
  Cyclone
  Filter holder with frit
  Cyclone bypass, 90°
  Teflon® sleeves (45/50)
  2-l impinger bottle
  Condenser
  XADs
  U-tubes
  2-l bottle foam inserts
  Socket flask, 500 ml
  Plastic caps for probe ends
  Miscellaneous clamps, gaskets, stoppers
Method 4 — moisture train
  Probe
  Midget impinger
  Midget connectors
  Glass wool
  Silicone grease
  Micrometer valve
  Vacuum pump
  Vacuum gauge
  Impinger box/ice bath
  Dry gas meter with thermometer
  Spring clips
  Vacuum tubing
  Stopwatch
Integrated gas train

Grab sample
  Probe
  Squeeze blub
  Gas bags
Integrated gas train
  Probe
  Midget impingers
  Micrometer valve
  Pump
  Rotameter
  Box with bag insert
  Bag
  Pitot tube (S-type)
  Inclined manometer
  "U" manometer (H_2O)
  Vacuum gauge
  Purge fitting
  Miscellaneous tubing
  Glass wool
  Sealant
  Analysis
    Orsat analyzer
    Spare Orsat parts
VOST equipment
  Teflon® line
  Rotameter
  Spare 602 polymer
  Probes
  Dry gas meter
  Manifold
VOST — glassware and fittings
  Condenser
  Fittings
  Impinger
  Teflon® tubing
  Glass stopcock
  Glass tapered joint
  Spare glassware
  Teflon® sleeves for glassware
  Tenax® traps
  Clean coolers
Fyrite O_2 and CO_2
  O_2 Fyrite (21% scale)
  CO_2 Fyrite (20% scale)
  Sampling pumping line consisting of
    Hose from probe to filter
    Filter
    Aspirator bulb with check valve
    Rubber connector tip
  Spare parts
    Filtering yarn
    Diaphragm
    Gaskets
  Spare chemicals
  Orsat, stand
  O_2 buret
  CO_2 buret
  Manifold

TABLE 10 (continued)
List of Sampling Equipment and Supplies Typically Used

Graduated buret
Leveling bottle with tubing
Orsat sampler
Mylar® bags
Conditioning manifold
THC analyzer
CO analyzer
O_2 analyzer
Recorder
Nitrogen cylinder
Air cylinder (THC)
H_2/N_2 cyclinder (THC)
Calibration gases
Gas regulators
Teflon® tubing
Ascarite trap
Silica gel trap
Silica gel
Ascarite
Heated lines
Controller
Variacs
Probe filter
Parts box
Safety equipment item
Hard hats
Hard-hat liners
Safety shoes
Safety glasses, goggles
Slip on shields for glasses
Ear plugs
Neutralizer
Ear protectors
Face masks (full face respirators)
Dust respirators
Climbing belt and lanyard
Gloves (high temperature)
Rain gear
First-aid kit
Water jug
Saf-T-Lok
Fire extinguisher
Face shield
Viton gloves
Jumpsuits
Restricted area sing
Black and yellow ribbon
Tarps
"No smoking" sign
Respirators
Cartridges — organic vapors and acids
Eyewash bottle
Fire blanket
Computer and associated items
  Hardware
  Pocket computer
  Printer cassette interface
  Cassette recorder

Cassette to interface connecting cable
Ribbon cartridge
Printer paper
Level II basic reference manual
Extra blank cassette
Software
Program printout copies
Laboratory — general equipment
  Laboratory toolbox
  Oven
  Kimwipes®
  Chix wipes
  Wash bottles
  Glass wool
  Barometer
  Plastic for lab floor
  Graduated cylinders
  Pipettes
  Funnel glass
  Beakers
  Thermometer (0—125°C)
  Bulbs for disposable pipettes
  Brushes and soap
  pH paper
  Ultrasonic cleaner
  Triple beam balance with weights
  Ramrods with brushes
  Washtub
Laboratory chemicals
  General
    Orsat chemicals:
    3 oz oxorbent (O_2)
    3 oz cosorbent (CO)
    3 oz disorbent (CO_2)
    3 oz burette solution
    Distilled-deionized H_2O double-distilled
    B and J acetone (1 part/part. test gal)
    Silica gel (2 run/lb) (large can)
    Spare Fyrite chemicals
  Particulate (Method 5)
    Filter paper (glass fiber)
    Sample bottles (glass)
    B and J methanol
    60 ml poly bottles with caps
    0.1 N KOH
General items
  Spare hardware equipment
  Electrical toolbox
  FM 2-way radio
  Heat gun
  Sample labels
  Icebags
  Laboratory notebooks
  Air tank for blow-back
  Portable welding unit
  Label tape and dispenser
  Data sheets
  SOPs

TABLE 10 (continued)
List of Sampling Equipment and Supplies Typically Used

Traceability sheets
Timers for consoles
Rubber bands
Clipboards
Paper clips
Scotch tape
Masking tape

5. Process Data

Process data recorded during a trial burn is of equal importance with the sampling activity, for two reasons:

1. Process data are necessary for computation of DRE.
2. Some process readings recorded during the trial burn may, and probably will, become the limits specified in the operating permit (e.g., minimum combustion chamber temperature).

The average feedrate of each waste and fuel input stream must be determinable for each run by flow meter, tank level change, drum weights, etc. Those feed rates, and analysis results for waste feed samples are used to compute thermal input rate (Btu/h) and POHC input rate. These parameters are used to calculate DRE.

Three steps for process monitoring are recommended:

1. Before the test, determine what process data must be taken and reported to EPA.
2. Record all possible data, to help identify any problems during the test, or in the test results.
3. Before the test, establish the acceptable range for each critical operating parameter.

Item 3 above is not as simple as it may appear. For example, combustion chamber temperatures for the trial burn might be 2000 ± 50°F. Questions then arise as to whether the temperature range is maintained at all times during the test:

1. Is sampling to be interrupted if the temperature goes outside the established range?
2. How long can the temperature be outside the range, or how far outside the range, before ordering an interruption in sampling?
3. If sampling is interrupted, how long must the temperature be back within range before sampling can be restarted?

This one example demonstrates the complexity of questions that frequently arise. These questions should be anticipated and guidance developed before the trial burn to assure trial burn operating conditions that meet plant needs. There is often precious little time to make those decisions when the questions are faced during a test.

6. Data Sheets and Labels

Preparation of all data sheets and sample labels that will be needed for the test is important.

Many different data sheets are needed for a trial burn as shown in Table 11. The units of measure must be shown for every item on every data sheet. Too often, data are taken (i.e., number recorded) without showing the units of measure (e.g., °F, gal/min, etc.).

TABLE 11
Sample List of Data Forms

Traverse point locations
Preliminary velocity traverse
Method 5 data sheets
Isokinetic performance work sheet
M5 sample recovery data
Integrated gas sampling data (bag)
Orsat data sheet
VOST sampling data

Drum weighing record
Drum sampling record
Liquid waste feed sampling record
Fuel oil sampling record
Drum feed record

Process data (control room)
Miscellaneous process data (in-plant)
Tank level readings
Log of activities

Ash sampling record
Scrubber waters sampling record
Sample traceability sheets
GC/MS data calculation sheets

Note: Units of measure must be shown for each item
on each data sheet.

Instrument factors (e.g., Rdg $\times$ 100 = °F) should be noted during trial burn preparation to assure that data are accurately compiled. Data sheets may be a better record than copies of strip charts, since the latter do not show units of measure or multiplication factors and are often difficult to interpret for other reasons.

All data sheets should be prepared before the test to ensure that all necessary data are recorded. Specific assignment should be made as to who is responsible for each sheet. The data to be recorded can be identified during the pretest survey and the data forms prepared thereafter.

Sample labels should also be prepared prior to the test. Labels can be prepared most efficiently using computer printed labels like the sample shown in Figure 3. Replicate labels which each contain all essential information including a unique sample number for each sample are printed. Replicate labels are needed in order to place one on the sample container, one on the outside of the final packaged sample, one in the field laboratory log book, and the fourth left on the sheet of labels. The last label provides a quick way of checking which samples were taken, since some changes may be made in the field. Blank labels are also always provided for such changes or additional samples that may be taken.

Preparing the computerized labels requires careful thought to identify each and every specific sample that will be taken during each run, including all replicates and blanks. Label preparation also helps identify all the sizes and types of sample containers that will be needed, how they must be prepared, and the number of each that is required (including spares). This activity often shows that 50 to 100 individual samples (and labels) will be involved in each run. Given this magnitude of samples, preprinted labels with specific sample names and a consistent numbering system should be prepared before the trial burn to help avoid confusion and errors that can occur if labels are prepared later in the field.

RUN # 1　101 Liq. Organic Waste Feed Proj. #　DATE: Plant Name	RUN # 1　101 Liq. Organic Waste Feed Proj. #　DATE: Plant Name	RUN # 1　101 Liq. Organic Waste Feed Proj. #　DATE: Plant Name
RUN # 1　102 Aqueous Waste Feed Proj. #　DATE: Plant Name	RUN # 1　102 Aqueous Waste Feed Proj. #　DATE: Plant Name	RUN # 1　102 Aqueous Waste Feed Proj. #　DATE: Plant Name
RUN # 1　103 Kiln Ash Sluice Water Proj. #　DATE: Plant Name	RUN # 1　103 Kiln Ash Sluice Water Proj. #　DATE: Plant Name	RUN # 1　103 Kiln Ash Sluice Water Proj. #　DATE: Plant Name
RUN # 1　104 Liquid Scrubber Effluent Proj. #　DATE: Plant Name	RUN # 1　104 Liquid Scrubber Effluent Proj. #　DATE: Plant Name	RUN # 1　104 Liquid Scrubber Effluent Proj. #　DATE: Plant Name
RUN # 1　105 VOST Trap Pair #1, Tenax Proj. #　DATE: Plant Name	RUN # 1　105 VOST Trap Pair #1, Tenax Proj. #　DATE: Plant Name	RUN # 1　105 VOST Trap Pair #1, Tenax Proj. #　DATE: Plant Name
RUN # 1　106 MM5 Caustic Solution Proj. #　DATE: Plant Name	RUN # 1　106 MM5 Caustic Solution Proj. #　DATE: Plant Name	RUN # 1　106 MM5 Caustic Solution Proj. #　DATE: Plant Name

FIGURE 3.　Example of computer labels.

7. Safety Precautions

Most plants and sampling crews utilize common safety equipment such as safety glasses, steel-toed shoes and hard hats. However, an outside contractor's sampling crew needs to be made aware of all plant safety requirements and any special hazards that may exist, especially with regard to particularly toxic components in the feed streams or the stack effluent (e.g., high CO levels).

Sampling personnel need to be instructed on any special safety equipment and procedures for liquid waste sampling or sampling of any other hazardous waste feeds. The need for protective equipment such as specific types of gloves, goggles, respirators, chemical resistant suits, etc., should be established early enough in the planning stage so that the sampling crew can prepare for their use. Once at the test site, plant personnel must ensure that the sampling personnel are informed of the plant's safety procedures, especially if they could impact on the test program (e.g., evacuation of the sampling area caused by a process upset in an adjoining portion of the plant).

One special note about safety for sampling of drummed wastes is encountering "bulging" drums. A bulging drum can, of course, indicate pressure buildup in the drum. If a bulging drum is encountered in the course of drum sampling, only experienced plant personnel should attempt to open it. Furthermore, a drum can be under pressure, even if it is not bulging. It is recommended that plant personnel be assigned to assist the sampling crew and be responsible for opening all drums to be sampled.

8. Observers

The operator frequently does not realize until a test starts that trial burns bring out everyone with a vested interest and even those who are just interested. Observers may include regulatory authorities, extra operating and maintenance personnel, responsible plant management, and many others who otherwise are seldom present. They all usually congregate in the control room. These "observers" usually have reason to be present, but their numbers can create problems.

Observers often want to ask questions and to have discussions with the plant operators. Some of this may be necessary, but it can divert the operators' attention from their primary function and responsibility. Similarly, the observer may want to ask questions of the sampling crew at times when they must give their full attention to their sampling responsibilities. Also, suggestions made by "observers" to operators or samplers are sometimes interpreted as a directive to change how they are doing something.

To help avoid the above problems, the following should be done prior to the trial burn:

1. Instruct all operating personnel and samplers not to make any changes unless directed to do so by their supervisor or other designated individuals.
2. Require that each observer minimize discussion or interference with operators or samplers during busy periods, especially during test periods.
3. Assign one plant person as the primary contact for all observers, and request that the observers direct questions and comments to that person first.
4. Since most observers are interested in what is being done and how it is being done, have descriptive material available, and, if needed, make arrangements for them to discuss the test plan and sampling/analysis methods with appropriate personnel at appropriate times before or after the actual test periods.

B. WHAT IS INVOLVED IN CONDUCTING THE ACTUAL SAMPLING, AND WHAT ARE THE PROBLEMS THAT MAY OCCUR?

The main factors involved in the actual sampling for a trial burn are

1. Equipment setup
   Sampling train setup
   Set up waste feed sampling
2. Preliminary testing
   Velocity traverse
   Cyclonic flow check
   Moisture measurements
3. Actual testing
   Waste feed sampling
   Process monitoring and determination of waste feed rates
   Sampling of ash and scrubber waters, etc.
   Stack sampling
   Sample recovery
   Labeling and sample packaging/storage
   Preparation of equipment for next run
4. Equipment dismantling and packing

Brief discussions of the above items and procedures to avoid problems that may occur are presented below.

1. Equipment Setup

a. *Sampling Train Setup*

The first job of the sampling crew after arriving at the facility is unloading and setting up of equipment (usually in a field laboratory trailer) including setup for the stack sampling. Setup on the stack for MM5 is usually the most difficult step. First, relatively heavy equipment (40 to 80 lb) must be moved up to the sampling platform. Second, support rails or a monorail must be installed to allow the MM5 train to traverse the stack. These rails often must extend outward from the port, a distance equal to the stack diameter plus about 2 ft. The rails must be rigidly secured to support the sampling train over its entire length (8 ft) and allow free movement with no interfering objects (e.g., guardrails). The problem usually encountered is the lack of means to support the rail, especially at the outer end which may extend 4 to 8 ft further out than the platform. (Some platform designers assume that a 6-ft-diameter stack can be sampled from a 2-ft-wide platform). Also, the sampling ports are almost always at about the same level as the platform guardrail, so part of the guardrail must be removed to provide clearance for the sample box, if not done earlier.

Stack samplers have necessarily developed various means of supporting the rails, but each test site always requires something slightly different. A pretest survey may identify some modifications the plant can make to facilitate the stack setup, consistent with the design of the rail system to be used. But ideally, the width of the platform would be at least equal to the diameter of the stack plus 2 ft.

Another common problem encountered during stack setup is inadequate electrical outlets. As a minimum, at least four 110-V, 20-A electrical outlets should be available on the stack sampling platform. If at all possible, these circuits shoudl be dedicated to the test, without interferences from other plant equipment.

b. *Setup for Drummed Waste Sampling*

The equipment setup period may include numbering, weighing, and sampling of drummed wastes. Plant personnel will be needed to assist in this activity. Adequate time must be set aside for this work, and the plant needs to preplan the work so that the wastes are properly staged and equipment provided (scales, forklift, etc.). Preparation of drummed waste is done most efficiently with two- or three-person teams. Each team can process a drum in 2 to 5 min.

2. Preliminary Testing

a. Velocity Traverse

One of the first preliminary tests is a preliminary velocity traverse of the stack and determination of stack moisture content. Values obtained are used to determine sampling train conditions, so the preliminary measurements need to be made at test conditions. This preliminary test will require additional time on the first test day.

b. Cyclonic Flow

Another preliminary test is a check for cyclonic flow in the stack, as required by EPA Method 1. Sampling under cyclonic flow conditions requires special equipment and methods. Alternatively, a flow straightener can be installed in the stack, as far upstream of the test ports as possible. This installation could involve considerable delay to design, fabricate, and install the flow straightener. Installation may require a plant shutdown. If there is any reason to suspect that the stack flow may be cyclonic, then the plant should either have the flow checked or install a flow straightener well in advance of the tests to avoid the possibility of having to delay testing at the last minute.

c. Moisture Measurement

Another possible problem is high moisture content of the stack gases from wet scrubbers. For scrubber stack test, crews should determine moisture content at saturated conditions prior to the particulate run to ensure that the runs are conducted under isokinetic conditions.

3. Actual Testing

Sampling usually consists of taking representative samples of all influent and effluent streams, especially waste feeds and stack effluent, during each of the three runs. Process monitoring, including collection of data needed to determine all waste feed rates, is also done during each run. When recording process data, common practice is to read and record the instrument readings once every 15 min, even if they are also continuously recorded. These manual readings provide a good written record for inclusion in test reports. The results reported should include notation of momentary excursions. Otherwise, the operating permit might not contain any allowance for those types of occurrences that are a part of normal plant operations.

During the test one person who knows the conditions under which sampling should be interrupted, and who is in radio contact with the stack test crew at all times should be responsible for process monitoring. That person must notify the crew to interrupt sampling whenever deemed necessary, but especially when a serious process upset or shutdown occurs during a test. Such transient conditons could have a drastic impact on the samples, so all sampling must be stopped immediately. In that event, sampling can be resumed after the desired plant operating conditions have been reestablished.

Immediately after each run, the sampling crew must recover all samples, properly label each, and package them for storage and shipment. Samples are usually double-bagged with protective wrapping and stored in coolers, many of which must be iced each day. After that work has been completed, all sampling equipment must be prepared for the next run. All this may sound relatively simple, but there are numerous problems that can occur. Some of these problems are listed in Table 12 and are briefly explained below.

Plant operating problems during a run are not uncommon and are due in part to the fact that the plant is being operated under "worst case" conditions. This scenario may cause operating conditions to go outside the specified range for the test, or upsets may occur to cause a plant shutdown. Such situations require interruption of all sampling until desired conditons are reestablished.

Any interruptions in sampling, for whatever reason, can have an impact on proper

TABLE 12
Potential Problems that May Occur
during Tests

Plant operating problems
Determination of waste feed rates
Weather
Sampling equipment problems
High filter vacuum due to filter loading

deterination of waste feed rates, especially liquid wastes. This may not be much of a problem if the plant is equipped with reliable waste flow meters. However, if tank level changes are the basis for determining feed rate, then levels have to be read whenever the sampling is interrupted and again when it is restarted.

Another problem that can often occur during testing is bad weather. This can alter plant operating conditions, but usually it impacts the sampling activity. Heavy rain, lightning, or high winds are common conditions that will require interruption in sampling, since it is unsafe for the sampling crew to remain up on the stack unless it is enclosed.

Most of the other types of problems that occur during a test relate to the stack sampling equipment. Equipment used to conduct stack sampling per M5 include pumps, heaters, thermocouples, etc., all of which are subject to mechanical failure. Another type of problem is high pressure drop across the sampling train, usually caused by material on the particulate filter used in the train. When this occurs, sampling must be interrupted for 30 to 60 min to change filters.

The most serious problems related to the stack sampling are failure to achieve prescribed sampling rate (i.e., isokinetic sampling) or failure to pass final leak checks. Either situation can invalidate a run, probably requiring that it be repeated. The plant must be aware that this can happen, requiring additional quantities of waste and additional time.

C. WHAT IS INVOLVED IN ANALYSIS OF SAMPLES, AND WHAT ARE THE PROBLEMS THAT MAY OCCUR?

Even in a relatively simple trial burn, 100 to 200 samples may be acquired for analysis. Each sample will need to be analyzed for several different parameters (HHV, Cl, ash) and several analytes (POHCs). Numerous problems can occur in the analysis phase of a trial burn due to the complexity of sample analysis, the variety of sample types, and the detection limits required for some samples. Most frequently, these problems occur in the analysis of samples for POHCs. Some of the most common problem areas are

1. Sample check-in
2. Analysis directive
3. Sample inhomogeneity
4. Analytical interferences
5. Saturation of GC/MS detection instrumentation

Usually, the analytical results are reported to the project leader who is responsible for using them to calculate final results (DRE) and to prepare a test report. The project leader first examines the analytical results. Potential problem areas which may appear in the data are

1. High blanks
2. Poor precision for analysis of replicate samples
3. Poor accuracy for recovery of surrogate recovery components spiked into the sample prior to analysis
4. Nonconformance with "expected" results

Each potential problem area is discussed below.

1. Sample Check-in

Samples taken during a trial burn are usually brought to the analytical laboratory for transfer to an analytical task leader. At that point, when the samples are checked in and tranferred, the project leader needs to cross-check that each sample taken in the field has arrived and is intact. Any missing samples can in this way be immediately identified and hopefully located. Also, any extra samples that have been taken can be identified. The sample information should be recorded in a bound laboratory record book (LRB) and sample traceability sheets should be used, even if strict chain of custody sheets are not used.

2. Analysis Directive

After completing sample check-in, the project leader should prepare a written directive for sample analysis specifying the following for each and every sample:

1. Sample number and type of sample (e.g., 141 — waste feed)
2. Notation of any safety or hazard considerations in handling/analyzing samples
3. Analysis parameters and analytes
4. Analysis methods for each parameter or analyte
5. Analysis sequence (where necessary)
6. Indication of whether sample is to be analyzed in duplicate (or triplicate)
7. Samples to be spiked with analytes and surrogate recovery compounds for determination of percent recovery efficiency

The directive should also indicate expected concentrations, analysis priorities, and the schedule for reporting of analysis results.

Preparation of the above directive is complex but very important. It documents the analysis needs and will be very helpful to those who must perform the analyses. Most importantly, it helps assure that all the necessary analyses will be done as needed to satisfy the trial burn requirements.

3. Sample Inhomogeneity

When analyses of samples are initiated, a problem often encountered is nonhomogeneity of samples, especially waste feed samples. Liquid waste feeds often separate during shipment/storage. Steps must be taken to homogenize samples before portions or aliquots are removed for analysis. Solid waste samples often present even more difficult problems that should be discussed with the analysis task leader.

4. Analytical Interferences

Waste feed samples are organically complex and usually contain many consitutents other than POHCs. These other consitutents may interfere with the POHC analysis, especially if the POHC concentrations are low in comparison to the other constituents. Some type of sample cleanup may be required.

5. Saturation of GC/MS Data System

Saturation of the GC/MS data system may occur for any type of sample, but especially VOST traps. VOST samples are analyzed by thermal desorption and cannot be prescreened or diluted. If saturation occurs, gas bag samples must be analyzed. Saturation can be a complex problem if there are several volatile POHCs with a wide range in their anticipated concentrations. VOST results may be essential for determining DRE of some POHCs present at low concentrations in waste feeds, but those POHCs may be masked by high concentration of other POHCs in the VOST samples. Such problems are best avoided by careful preplanning, including selection of POHCs and/or their concentrations in the waste feeds.

6. High Blanks

High blanks may occur for any parameter or analyte, but probably more frequently for POHCs. Blank levels can be used to "correct" sample levels, but this is not possible if blank levels are near to or exceed sample levels. If both the blanks and sample levels are high, causing the uncorrected sample value to yield a DRE below 99.99%, no useful information is obtained for that sample. Every precaution must be taken in the laboratory and in the field to prevent contamination of samples.

7. Poor Precision

The QA protocol usually requires triplicate analysis of critical samples from at least one run. Wide variability in these triplicate analysis results may occur if samples were not homogeneous, if the samples contain some interfering component, or if other problems occur with the analytical technique. In any case, the precision obtained provides an indication of the possible variability in results reported for each sample and analyte. For POHCs, knowledge of this variability may be quite important if the calculated DREs are close to 99.99% (e.g., 99.989%).

8. Recovery Efficiency

Another normal part of the QA/QC is spiking of samples with known amounts of POHCs and/or surrogates to determine recovery efficiency for the analysis method. The desired recovery range is normally 70 to 130%, but actual results may in some cases be much lower or much higher for any of several reasons. Sample results usually are not corrected based on recovery efficiency results, but knowledge of the recovery efficiency is important for the same reason as knowledge of the precision. Also, poor precision or poor recovery efficiency may indicate a need to reanalyze the samples.

9. Actual vs. Expected Results

In certain cases the project leader has some idea of what the analytical results should be. For example, the amount of POHC added into a waste feed tank might be known, so there is some expected concentration of the POHC in those samples. Usually the analytical results are in good agreement with expected results, but in some instances the analytical results may disagree with the expected value. The analyses and calculations then must be rechecked, and the QA/QC data scrutinized more closely for a clue. Mathematical errors are the most common cause of such problems, but other possible causes, such as a poorly mixed feed tank, cannot be overlooked.

D. HOW ARE THE SAMPLING DATA AND ANALYSIS DATA CONVERTED TO FINAL RESULTS?

This section addresses the calculation methods used to convert laboratory data on organics, and field data on flow rates, into DRE numbers. The values necessary to calculate DREs, and how they are obtained, are listed in Table 13. A brief review of the method used to calculate DRE is presented at the end of this section. First, however, it is necessary to give attention to the areas of blank correction, significant figures, rounding of DREs, and the need to use "<" and ">" signs in reporting DRE data.

1. Blank Correction

Because achievement of 99.99% DRE often results in stack concentrations that are at or below ambient or laboratory levels for POHCs, contamination of samples can be a significant problem. The purpose of blank correction procedures is to account for any portion of the sample results that represent contamination, or something other than the value intended to be measured (e.g., stack emissions).

TABLE 13
Data Necessary for Calculating DRE

Measured value	Example units	How value is obtained
Mass flow rate of feed	g/min	Measured during test or calculated from flow and density
Volumetric flow of feed	l/min	Measured during test
Density of feed	g/ml	Density analysis from lab
Concentration of POHC in feed	µg/g	Analysis of waste feed samples
Total quantity of POHC in stack sample	µg	Reported by lab for each sample taken during test
Volume sampled of stack gas	Nm³	For VOST this is found in the sample train data
		For M5 this is found with the M5 train data
		For gas bags this is reported as the volume analyzed by the lab
Total quantity of POHC in blank samples	µg	Analysis of "blanks"
Volumetric stack flow rate	Nm³/min	Reported as result of pitot traverse with M5 train

Note: $1 \text{ µg} = 10^{-6} \text{ g}$; $1 \text{ ng} = 10^{-9} \text{ g}$; $1 \text{ µg/m}^3 = 1 \text{ ng/l}$; $\text{Nm}^3 = $ normal cubic meters $=$ dry standard cubic meters.

The underlying philosophy of the procedure is based on a paper prepared by the American Chemical Society (ACS) Committee on Environmental Improvement[7] and on experience in conducting and interpreting trial burn data. The ACS paper assumes that blank values are random samples that vary because of preparation, handling, and analysis activities. Under this assumption, blank values can be treated statistically. The "best estimate" for the blank for any particular sample is the mean of the available blanks. The ACS procedure also enables determination of whether a sample is "different from" the blank. If the sample value is not significantly different from the blank value, a sample cannot be blank corrected. Even so, the measured sample value does provide an upper bound for the emission value and may still provide sufficient information for determining if the required DRE of 99.99% was met.

The blank correction procedure applies mainly to stack emission samples and consists of the following:

1. Assemble data for each POHC from all of the field and trip blanks. An example of such data for VOST might be

	Run 1	Run 2	Run 3
POHC A			
Field blank	0.008 µg	<0.002 µg	0.004 µg
Trip blank	0.005 µg	0.004 µg	0.003 µg

2. Determine whether or not the field blanks are statistically different from the trip blanks by using the paired t-test (consult a statistics text).
    If the field blanks are significantly different than the trip blanks, use the field blank data only. If the blanks are not significantly different, use all of the blank values.
3. Calculate the average and standard deviation of the blanks. (Many calculators have statistics functions which allow you to do this easily.)
4. Determine whether or not each measured sample value is "different from" the blank value by using the following test for each sample:

s = sample value (μg)

b = (blank average) + 3 (standard deviation of blanks)

If s is greater than b, then the sample is "different" from the blanks.

5.   If the measured sample value is different from the blank value, then the blank correction procedure is applied:

Blank corrected emission value (μg) = measured sample value (μg) − average blank value (μg).

6.   If measured sample value is not different from blank value, then the measured sample value is used as an upper bound emission value and the emission rate is considered less than or equal to the measured value. This results in the reporting of emission concentration, and mass emission rate with a "<" sign. As a consequence, DRE would be reported with a ">" sign.

2. Significant Figures and DRE

DRE is usually reported with one or two significant figures, depending on the accuracy of the measured values which go into the calculation of DRE. It is improtant to note that a reported DRE of 99.99 or 99.999% has only one significant figure. The reason for this is that what is actually being measured is the penetration, which is the amount of a compound which is *not* destroyed. That is, DRE = 100% − penetration. For a DRE of 99.99%, the penetration is 0.01% (one significant figure). For a DRE of 99.9916% the penetration is 0.0084% (two significant figures).

The DRE is reported with the same number of significant figures as the least accurately measured value used in the calculations. The controlling measurement that determines the number of significant figures is usually the stack concentration. GC/MS methods can normally only report concentrations with one or two significant figures unless another measured value (waste feed concentration, waste feed flow rate, or stack gas flow rate) has fewer significant figures.

3. Rounding Off DRE Results

The rules on this are stated in the Guidance Manual for Hazardous Waste Incineration Permits: "if the DRE was 99.9880 percent, it could not be rounded off to 99.99 percent." In other words, your calculated value, after rounding to the proper number of significant figures, must equal or exceed 99.99% to be acceptable. (Note: This same rule applies to rounding HCl results to 99%.)

4. Reporting DRE with a "<" or ">" Sign

As mentioned in the section on blank corrections, if the sample is not "different" from the blank (greater than the average blank plus three standard deviations), then it cannot be blank corrected. As a consequence, the DRE will be reported with a ">" sign. This reported ">" value will also occur when the POHC in the sample is undetected (below detection limit of the analysis method). But, as long as the DRE is >99.99%, this is not a problem.

In cases where both the blanks and samples have high values, a DRE below 99.99% may be preceded by a ">" sign (i.e., >99.96%). Such a number is useless in evaluating achievement of 99.99%. Experience in using the recommended sampling methods and avoiding contamination is the only way to minimize this possibility.

Occassionally, a sample may saturate the GC/MS with the POHC in question. This will result in an emission rate with a ">" sign and a DRE with a "<" sign. If such a DRE is below 99.99% the incinerator clearly fails. If it is above 99.99% (i.e., <99.9964%), the number is useless. To avoid such problems, alternate sampling methods should be used, based on preliminary estimates of the stack concentrations that may exist.

TABLE 14
Incinerator Operating Conditions[a]

Parameter	Run 1 (11/3/82)	Run 2 (11/4/82)	Run 3 (11/4/82)
Organic waste flow rate,[b] kg/min (lb/h)	3.76 (497)	4.01 (542)	4.50 (595)
Aqueous waste flow rate,[b] kg/min (lb/h)	6.13 (811)	5.38 (712)	4.90 (648)
Heat input rate, GJ/h (10^6 Btu/h)	8.37 (7.93)	9.01 (8.54)	10.50 (9.96)
Combustion chamber temp. °C (°F)	1053 (1925)	1066 (1950)	1094 (2000)
Calculated residence time,[c] sec	2.5	2.4	2.2
Stack height, m (ft)	11.6 (38)	11.6 (38)	11.6 (38)
Stack exit velocity, m/s (fpm)	10.7 (2110)	10.3 (2030)	11.3 (2230)
Stack temperature, °C (°F)	810 (1490)	749 (1380)	766 (1410)

[a]   Data collected by reading plant monitoring instruments at regular intervals. Values shown are averages from each run.
[b]   Determined by measuring storage tank liquid levels at start and finish of each run.
[c]   Calculated from chamber volumes and stack flow rates.

TABLE 15
Concentrations of POHCs in Waste Feeds (µg/g)

	Aqueous waste			Organic waste		
	Run 1	Run 2	Run 3	Run 1	Run 2	Run 3
Volatile POHCs						
Carbon tetrachloride	<2[a]	<2	<2	6,400	6,000	4,700
Trichloroethylene	<1	<1	<1	5,700	5,500	4,300
Benzene	<3	<3	<3	2.7	260	140
Toluene	94	110	100	1,800	2,400	1,900
Semivolatile POHCs						
Phenol	42,000	34,000	—[b]	4,200	1,000	—[b]
Naphthalene	<100	<100	—[b]	510	350	—[b]

[a]   Results reported as less-than values represent limits of detection.
[b]   MM5 sample voided for this run due to equipment problems. Therefore, the waste feed samples were not analyzed for semivolatile POHCs.

The conclusion of this section is to always design the sampling and analysis so that passage/failure of the 99.99% criterion is determinable. This can best be done by preliminary estimates of POHC concentrations in the stack (assuming 99.99% DRE) and with selection of sampling and analysis methods having appropriate upper and lower limits of detection. Experience in use of these methods to avoid contamination is also a key factor.

E. HOW ARE THE DATA AND RESULTS USUALLY REPORTED?

The results should be reported in a format which includes all information and data necessary to calculate final results, presented in as clear and succinct a format as possible. This will include: a description of the operating system; the operating conditions during the test; the measured quantities of POHCs, HCl, and particulate in all samples; and the calculated results. Example formats for presentation of these data are presented in Tables 14 through 25. Using part of the data in these tables, a sample calculation of DRE is shown in Table 26.

TABLE 16
Calculated Input Rates for POHCs in Waste Feeds

	Input rates (g/min)[a]		
	Run 1	Run 2	Run 3
Volatile POHCs			
Carbon tetrachloride	24	25	21
Trichloroethylene	22	22	19
Benzene	0.010	1.1	0.63
Toluene	7.3	10	9.0
Semivolatile POHCS			
Phenol	270	180	—[b]
Naphthalene	1.9	1.4	—[b]

[a] Combined input rates for both waste feeds.
[b] Samples not analyzed for semivolatiles since MM5 sample was voided for this run.

TABLE 17
Concentrations of Volatile POHCs by VOST in Stack Effluent (Not Blank Corrected) (ng/l)

POHC	1st pair	2nd pair	3rd pair	Average
Run 1				
Carbon tetrachloride	2.3	0.47	0.57	1.1
Trichloroethylene	20	1.8	1.6	7.8
Benzene	2.2	2.3	2.2	2.2
Toluene	6.2	0.99	2.1	3.1
Run 2				
Carbon tetrachloride	2.3	1.7	1.7	1.9
Trichoroethylene	17	1.8	1.0	6.6
Benzene	2.0	7.4	2.6	4.0
Toluene	21	7.5	4.3	11
Run 3				
Carbon tetrachloride	3.1	0.58	0.45	1.4
Trichloroethylene	4.8	0.95	0.66	2.1
Benzene	6.0	7.1	6.2	6.4
Toluene	15	9.7	5.7	10

TABLE 18
VOST Blank Correction Values

	Average blank value (ng)	Standard deviation (ng)
POHCs		
Carbon tetrachloride	<2	0
Trichloroethylene	<1	0
Benzene	<3	0
Toluene	3.7	1.8

TABLE 19
VOST Sample Volumes (Dry Standard Liters)

Run no.	Pair no.	Volume (l)
1	1	18.4
1	3	18.1
1	6	17.5
2	1	18.4
2	3	18.5
2	6	18.5
3	1	18.9
3	3	18.9
3	6	19.0

TABLE 20
Blank Correction Values for Semivolatile POHCs

Compound	Blank correction value (μg)
POHCs	
Phenol	3.4
Naphthalene	6.0

TABLE 21
Destruction and Removal Efficiencies (DREs)

POHCs	Run 1	Run 2	Run 3
VOST			
Volatile			
Carbon tetrachloride	99.99966	99.99942	99.99946
Trichloroethylene	99.9975	99.9977	99.99906
Benzene	a	99.972	99.914
Toluene	>99.9973	99.9926	99.9916
Semivolatile compounds			
MM5			
Semivolatile			
Phenol	99.9985	>99.99996	>99.9996
Napthalene	99.96	99.98	99.986

ᵃ Waste feed concentration <100 μg/g in this run.

TABLE 22
Modified Method 5 Test Data

Volume of gas sampled (Nm³, dry)	2.277	2.101
Sampling time (min)	140	140
Percent isokinetic	95.0	96.8
Moisture content (%)	15.8	13.0
Percent O_2 (dry)	10.5	10.8
Percent CO_2 (dry)	7.8	7.7
Stack flow rate (actual m³/min)	355	341
Stack temperature (°C)	809	749
Stack flow rate (Nm³/min, dry)	73	76
Particulate concentration		
(mg/dscm)	842	523
(gr/dscf)	0.367	0.228
(mg/dscm corrected to 7% O_2)	1125	719
(gr/dscf corrected to 7% O_2)	0.491	0.313
(lb/h)	8.1	5.2
Chloride emissions		
(g/min)	31.1	37.1
(lb/h)	4.1	4.9

Note: Run 3 was voided by a broken probe liner. Nm³ = normal cubic meters.

TABLE 23
Continuous Monitoring Data[a]

	Run 1 (11/3/82)	Run 2 (11/4/82)
Oxygen (%)		
Range	7.1—11.0	8.3—11.0
Average	9.4	10.5
Carbon dioxide (%)		
Range	7.2—10.6	7.2—8.1
Average	8.5	7.6
Carbon monoxide (ppm_v)		
Range	<1—5.8	<1—5.3
Average	1.4	1.8
Total hydrocarbons (ppm_v)[b]		
Range	<1	<1
Average	<1	<1

[a] Concentration on dry gas basis.
[b] Total hydrocarbons reported as propane.

TABLE 24
General Analysis of Aqueous Waste

Parameter	Run 1	Run 2	Run 3
Heating value, kJ/kg	1800	1720	1550
(Btu/lb)	780	730	660
% Chlorides	0.39	0.36	0.26
% Water	88.54	93.89	89.67
% Ash	0.70	0.78	0.74
Saybolt viscosity	28.9	28.2	29.5

TABLE 25
General Analysis of Organic Waste

Parameter	Run 1	Run 2	Run 3
Heating value, kJ/kg	34,140	34,390	37,110
(Btu/lb)	14,680	14,800	15,960
% Chlorides	1.03	1.26	0.72
% Water	2.15	3.18	5.73
% Ash	1.53	2.13	2.36
Saybolt viscosity (s)	32.1	30.1	30.3

TABLE 26
Example DRE Calculation

The following is a sample calculation showing the method used to convert the analytical results to DREs for trichloroethylene in Run 2 using the VOST sample.

$DRE = [(W_{in} - W_{out})/W_{in}] \times 100$

Determine input rate (W_{in})

W_{in} = (organic waste flow rate × TCE concentration) + (aqueous waste flow rate × TCE concentration)
        (Table 14)   (Table 15)   (Table 14)   (Table 15)

W_{in} = (4010 g/min) (5500 μg/g) + (5380 g/min) (<1 μg/g) = 22 × 10^6 μg/min = 22 g/min (Table 16.)

Calculate output rate (W_{out})

Stack flow rate = 76 Nm³/min (Table 22.)
VOST avg concent = (20 + 1.8 + 1.6)/3 = 7.8 ng/l (not blank corrected)
(concentration values taken from Table 17.)

Blank correction

VOST = (<1 ng/sample)/(18.5 l/sample) = <0.05 ng/l (Table 18 and Table 19.)

Blank corrected value = 7.8 ng/l − <0.05 ng/l = <7.8 ng/l = <7.8 μg/m³

VOST output rate

Mass flow = (<7.8 μg/m³) (76 Nm³/min) (1 × 10^{-6} g/μg)
        = <0.00059 g/min (corrected)
Calculated DRE

DRE = {[(22 g/min) − (<0.00059 g/min)]/(22 g/min)} × 100
= >99.9973% (Table 21.)

REFERENCES

1. Guidance Manual for Hazardous Waste Incinerator Permits, Office of Solid Waste, U.S. Environmental Protection Agency, Washington, D.C., March 1983.
2. Test Methods for Evaluating Solid Waste — Physical/Chemical Methods, Office of Solid Waste, U.S. Environmental Protection Agency, SW-846 (1980), SW-846 Revision A (August 1980), and SW-846 Revision B, July 1981.

3. **Harris, J., Larsen, D., Rechsteiner, C., and Thurn, K.,** Sampling and Analysis Methods for Hazardous Waste Combustion, 1st Ed., prepared for U.S. Environmental Protection Agency, Contract No. 68-02-3211 (124) by Arthur D. Little, Inc., December 1983.
4. Methods for Chemical Analysis of Water and Wastes, EPA-600/4-79-020, U.S. Environmental Protection Agency, Washington, D.C., March 1979.
5. *Fed. Regist.,* 42(160), August 18, 1977.
6. Protocol for the Collection and Analysis of Volatile POHCs Using VOST, EPA-600/8-84-007, Industrial Environmental Research Laboratory, U.S. Environmental Protection Agency, Washington, D.C., March 1984.
7. Guidelines for data acquisition and data quality evaluation in environmental chemistry, *Anal. Chem.,* 52(14), 2242, 1980.

CASE HISTORIES

A.1 Site Remediation

A.1.a PYRETRON BURNER SYSTEM

United States Environmental Protection Agency	EPA/540/S5-89/008 May 1989

SUPERFUND INNOVATIVE
TECHNOLOGY EVALUATION

Technology Demonstration Summary

The American Combustion Pyretron Thermal Destruction System at the U.S. EPA's Combustion Research Facility

Under the auspices of the Superfund Innovative Technology Evaluation, or SITE, program, a critical assessment was made of the American Combustion Pyretron™ oxygen enhanced burner system during eight separate tests at the United States Environmental Protection Agency's Combustion Research Facility (CRF) in Jefferson, Arkansas. The report includes a description of the Pyretron and of the facilities used at the CRF, the tests conducted as part of this demonstration, the data obtained, and an overall performance and cost evaluation of the system.

Results show that Destruction and Removal Efficiencies (DREs) of 99.99 percent were achieved for a series of polycyclic aromatic hydrocarbons found in decanter tank tar sludge, RCRA listed waste K087, the organic waste tested during this demonstration. Particulate emissions of less than 180 mg/dscm at 7 percent O_2 were measured for all tests. The use of oxygen enhancement with the Pyretron enabled the feed rate of the waste to be doubled. All solid and liquid residues generated during these tests were contaminant free.

The costs associated with using the Pyretron in place of an air-only burner depend upon the relative costs of oxygen and fuel and to some extent the capital costs of the burners themselves. For this demonstration, operating the Pyretron with oxygen enhancement used oxygen worth between $3250 and $3870 (it was provided free of charge) and roughly $2672 worth of propane. Operation without oxygen enhancement consumed $4000 worth of propane. During this period 1820 kg of waste were treated using oxygen and 1180 kg were treated without oxygen. The Pyretron burners used in this demonstration had an estimated cost of $150,000 and involved $50,000 of design and development effort.

This Summary was developed by EPA's Risk Reduction Engineering Laboratory, Cincinnati, OH, to announce key findings of the SITE program demonstration that is fully documented in two separate reports (see ordering information at back).

Introduction

The SITE demonstration of the American Combustion, Inc. Pyretron

PYRETRON BURNER SYSTEM (continued)

oxygen-enhanced burner system was conducted from November 16, 1987 to January 29, 1988 at the U.S. Environmental Protection Agency's Combustion Research Facility (CRF) in Jefferson, Arkansas. The Pyretron was installed on the CRF's Rotary Kiln Incinerator System (RKIS). This demonstration was conducted using a mixture of decanter tank tar sludge from coking operations (RCRA listed waste KO87) and waste soil excavated from the Stringfellow Superfund site near Riverside, California. These two wastes were mixed together to provide a feed stream that had high levels of organic contamination and was in a soil matrix. This was determined to be the best feed material to use to evaluate the performance of the Pyretron. The purpose of the demonstration tests was to provide the data to evaluate three ACI claims regarding the Pyretron system. These claims are as follows:

- The Pyretron system with oxygen enhancement reduces the magnitude of the transient high levels of organic emissions, CO, and soot ("puffs") that occur with repeated batch charging of waste to a rotary kiln.

- The Pyretron system with oxygen enhancement is capable of achieving the RCRA mandated 99.99 percent destruction and removal efficiency (DRE) of principal organic hazardous constituents (POHCs) in wastes incinerated at a higher waste feedrate than conventional, air-only, incineration.

- The Pyretron system is more economical than conventional incineration.

Process and Facility Description

Two Pyretron burners were installed on the RKS. One was installed on the kiln and one on the afterburner. Valve trains for supplying these burners with controllable flows of auxiliary fuel, oxygen, and air; and a computerized process control system were also provided. A schematic of the system as it was installed at the CRF is shown in Figure 1. The Pyretron burners use the staged introduction of oxygen to produce a hot luminous flame which efficiently transfers heat to the solid waste fed separately to the kiln. Oxygen, propane and oxygen-enriched air enter the burner in three separate streams each concentric to one another. A stream of pure oxygen is fed through the center of

the burner and is used to burn propane in a substoichiometric manner. This produces a hot and luminous flame. Combustion is completed by mixing these hot combustion products with the stream of oxygen-enriched air.

All tests were performed in the RKS at the CRF. A simplified schematic of this system is given in Figure 2. The system consists of an 880 KW (3MM BTU/hr) rotary kiln incinerator, a transition section, a fired afterburner chamber, a venturi-scrubber, and a packed-column scrubber. In addition, a backup air pollution control system consisting of a carbon-bed adsorber and a HEPA filter is installed downstream of the previously mentioned air pollution control devices. With the exception of the carbon bed and HEPA filter, the system is typical of what might exist on an actual commercial or industrial incinerator. The carbon bed and HEPA filter are installed to ensure that organic compound and particulate emissions to the atmosphere are negligible.

The waste incinerated during the SITE demonstration was a mixture of 60% decanter tank tar sludge from coking operations, RCRA listed waste KO87, and 40% contaminated soil from the Stringfellow Superfund site. The KO87 waste was included in the test mixture to provide high levels of several polynuclear automatic hydrocarbon compounds. Six of these, naphthalene, acenaphthylene, fluorene, phenanthrene, anthracene, and fluoranthene were selected as the Principal Organic Hazardous Constituents (POHCs) for the test program. The Stringfellow soil was included in order to make the resulting feedstream more closely resemble the type of waste that would be incinerated using this technology. For all tests the waste was packed into 5.7 L (1.5 gal) fiber pack drums. Each drum contained between 4.1 and 7.9 kg (9 and 17 lb) of waste.

Eight tests were performed. These tests were designed to compare oxygen enhanced incineration to air-only incineration using the Pyretron. Table 1 summarizes the test conditions for the eight tests conducted.

During each test the feed and all effluent streams were sampled and analyzed to determine levels of contamination. In addition levels of carbon monoxide, carbon dioxide, oxygen, total unburned hydrocarbons, and nitrogen oxides in the exhaust gas were continuously measured and recorded. Comparison of the stripchart recordings of the oxygen-enhanced and air-only operation made it possible to

determine whether or not the controlled introduction of oxygen reduced transient emissions.

Results and Discussion

A detailed summary of the SITE demonstration test results is presented in Figures 3 and 4 that follow and in Tables 2 and 3. Based on the test objectives outlined in the Introduction, the following results were obtained and conclusions drawn.

Transient Emissions

American Combustion claimed that the Pyretron with oxygen enhancement could reduce the levels of transient emissions that occur when solid waste is batch charged to a rotary kiln. Transient emissions occur when organic contaminants originally present in the solid waste are volatilized in the hot kiln environment. The rapid volatilization and reaction of these organic contaminants depletes the kiln environment of oxygen. Pyrolysis occurs in the oxygen-depleted kiln environment. This results in the production and emission of soot and other pyrolysis products.

The basis for American Combustion's claim of reduced transient emissions is based upon the belief that the timed addition of oxygen would provide sufficient oxygen to the kiln atmosphere to oxidize the volatilized organic matter, thus reducing pyrolysis and the resulting emissions of pyrolysis products. The demonstration tests were planned to evaluate this claim by deliberately feeding the kiln in a way that would produce transient emissions and then by measuring and recording those emissions as carbon monoxide "spikes" on continuous emission monitor stripcharts. The stripcharts produced under air-only and oxygen-enhanced operation would be compared in order to determine any statistically significant differences in transient emissions between air-only and oxygen-enhanced operation.

Figures 3 and 4 show the stripchart data for tests 2 (air-only operation at 105 lb/hr feedrate) and 5 (oxygen enhancement at 210 lb/hr feedrate). These two figures are presented to indicate the frequency and level of transient emissions during the demonstration. All of the continuous emission monitor stripcharts obtained during the demonstration are presented in the Technology Evaluation Report. Comparison of the carbon monoxide emissions indicates that no significant differences in transient emissions could

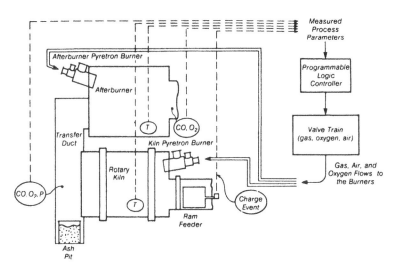

Figure 1. Pyretron thermal destruction system process diagram.

be readily observed. Statistical analysis of carbon monoxide peak height indicated that test-to-test variation was greater than the variation observed between air-only and oxygen-enhanced operation. Thus, it was not possible to state as a result of these tests whether the timed addition of oxygen reduced transient emissions.

There are two possible explanations for the inconclusive results of these tests. First, in order to achieve throughput increases with the Pyretron, water was injected into the kiln. This was not part of the original test plan, but was later deemed necessary in order to demonstrate throughput increases with high heating value waste. Even though water injection is a reasonable way to achieve throughput increases with high heating value wastes, the injection of water made it impossible to assess the ability of the Pyretron to reduce transient emissions through the timed injection of oxygen. An explanation of this is as follows.

Without the water, temperature excursions and other operational problems would have resulted from feeding high heating value waste (24.61

MJ/kg) to the kiln at elevated feed rates. Higher kiln temperatures are believed to increase transient emissions by driving more organic material off of the solid waste fed to the kiln. The use of oxygen often results in higher kiln temperatures. The added water reduced kiln temperatures and may have reduced them enough to reduce transient emissions over what they would have been had the Pyretron been used without water. Thus, had there been any statistically significant reduction in transient emissions, it would have been the result of the addition of water and not the timed injection of oxygen into the kiln atmosphere.

Secondly, Superfund wastes are very heterogeneous in nature. Even though these wastes were mixed prior to the start of testing, it is likely that significant variation in organic content existed from batch to batch during feeding. Variation in waste feed organic content may have also affected the variations observed in transient emissions. It was difficult to separate variations of this nature from variations resulting from the performance of the Pyretron. Studies of transient emissions are best carried out in a

laboratory setting using specially prepared, and therefore uniform, surrogate wastes. Such a study was conducted on a smaller version of the Pyretron at the U.S.EPA's Air and Energy Engineering Research Laboratory. Results indicate that the elevated temperatures that result from the use of oxygen enhancement result in an increase in transient emissions despite the additional oxygen present in the kiln atmosphere. This study is described in more detail in the Applications Analysis Report.

Destruction and Removal Efficiencies (DREs) and Particulate Emissions at Elevated Feed Rates

Even at double the feedrate, no organic contamination was detected in all but one of the twelve stack samples taken. Table 2 summarizes the DREs achieved. It should be noted that for the waste treated in this demonstration, 136 L/hr of water had to be added to the system at the elevated feed rates in order to achieve an increase in throughput. This is because the waste feed had a heating value of

PYRETRON BURNER SYSTEM (continued)

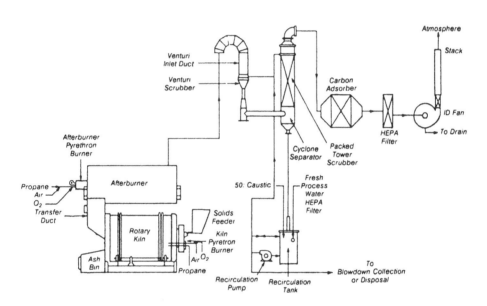

Figure 2. *CRF rotary kiln system.*

Table 1. *Summary of Demonstration Test Conditions*

Test		Feed Rate		Kiln			Afterburner		
No.	Mode*	kg/hr	(lb/hr)	Temp. °F	(C)	02. %	Temp. °F	(C)	02. %
1	A	47.7	(105)	954	(1750)	13.3	1121	(2050)	7.7
2	A	47.7	(105)	921	(1690)	12.8	1121	(2050)	7.4
3	02	47.7	(105)	1035	(1895)	17.6	1121	(2050)	15.2
4	02	47.7	(105)	963	(1765)	14.5	1121	(2050)	15.0
5	02	95.5	(210)	979	(1795)	13.9	1121	(2050	14.0
6	02	95.5	(210)	979	(1795)	14.6	1121	(2050)	15.3
7	02	55	(120)	1010	(1850)	13.5	1121	(2050)	13.5
8	A	55	(120)	1010	(1850)	8.8	1121	(2050)	11.4

*A = Air only
02 = Oxygen enhanced

24.61 MJ/kg (10,400 Btu/lb). At the feed rates obtained with oxygen enhancement, this resulted in a total heat input of 640 KW (2.2MMBtu/hr). Without all of the nitrogen provided by air-only operation, an additional heat sink had to be provided. This problem would be alleviated when treating a lower heating value waste.

As indicated by Table 2, the stack gases were virtually free of organic contamination. The solid and liquid residues produced from these tests were also free of contamination. The composite scrubber blowdown liquor and kiln ash samples from each test were analyzed for the POHCs and other Method 8270 semivolatile organic hazardous constituents. No POHC was detected in any blowdown sample at a detection limit of 20 µg/L; no other

PYRETRON BURNER SYSTEM (continued)

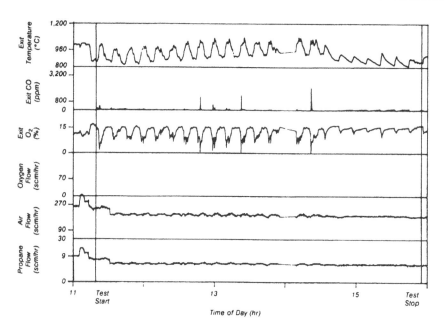

Figure 3. *Kiln data, Test 2.*

semivolatile organic hazardous constituent was detected at detection limits ranging from 100 µg/L (nitrophenols and pentachlorophenol) to 20 µg/L (all other Method 8270 constituents). No POHC analyte was detected in any kiln ash sample at a detection limit of 0.4 mg/kg ash. No other semivolatile organic hazardous constituent was detected at detection limits ranging from 2.0 mg/kg (nitrophenols and pentachlorophenol) to 0.4 mg/kg (all other Method 8270 constituents). This high level of decontamination is understandable given that all tests were performed at relatively high kiln and afterburner temperatures.

Particulate concentrations in the flue gas at the two locations sampled are summarized in Table 3. The two locations sampled were the scrubber discharge, and the CRF stack. Between the scrubber discharge and the stack are the carbon adsorber and the HEPA filter. Particulate levels were measured in the stack for all

tests. Sampling port availability limitations precluded measuring scrubber discharge flue gas particulate emissions for the tests during which simultaneous MM5 sampling was performed (Tests 1, 2, 5, and 6).

The data in Table 3 show that particulate levels in the scrubber discharge flue gas for three Pyretron tests and one conventional incineration test were in the 20 to 40 mg/dscm at 7 percent O_2 range. All levels measured were below the RCRA incinerator performance standard of 180 mg/dscm at 7 percent O_2.

Costs

Since the Pyretron is a burner and, therefore, only one of many components of an incineration system, the use of the Pyretron can be expected to affect cost only incrementally. Since the capital cost for any burner is only a fraction of the capital cost for the entire incinerator, the

majority of the costs associated with the use of the Pyretron will be associated with the costs of fuel and oxygen. Table 4 summarizes the costs for fuel and oxygen during the SITE demonstration. A range of costs is presented to give the reader an estimate of the variability in costs associated with using this technology. More information on costs is provided in the Applications Analysis Report on this demonstration.

The capital costs for the Pyretron system used in the SITE demonstration was $150,000. In addition $50,000 was spent in design and development work on the system.

Since this demonstration was done at a research facility and not under actual field conditions, the incremental effect that using the Pyretron has on the cost of incinerating a ton of hazardous waste cannot be directly determined. It is likely that the major factor in determining the cost effectiveness of the Pyretron will remain the oxygen and fuel. These costs

PYRETRON BURNER SYSTEM (continued)

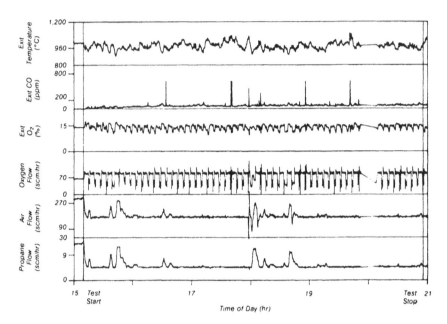

Figure 4. Kiln data. Test 5

vary widely depending upon location and scale of operation. More discussion of costs is provided in the Applications Analysis Report.

Unit Problems

Three problems were identified with the Pyretron during the course of the SITE demonstration. First, EPA is not certain to what extent the Pyretron's process controller reacts to conditions within the incinerator. EPA was not provided with documentation on the control system. EPA's knowledge of how the control system operated during the demonstration is based on conversations with American Combustion personnel during the course of the demonstration. EPA's understanding of how the control system worked during the demonstration is as follows.

While the Pyretron allows for variation in the amount of oxygen fed into the incinerator during the course of a test run, the process controller requires that

adjustments in the flowrate of oxygen be preset prior to the initiation of feed. During the SITE demonstration, the Pyretron's control system increased the oxygen level in a stepwise fashion to a series of preset levels if any one of the following three things happened:

1. Thirty seconds elapsed since the initiation of a batch feed cycle (which was indicated by activation of the ram feeder)

2. Carbon monoxide levels in the kiln exhaust reached a preset level. (undisclosed to the EPA)

3. Oxygen levels in the kiln exhaust reached a preset level (undisclosed to the EPA)

In the event that kiln pressure suddenly increased, the combustion air flowrate was reduced in a stepwise manner and the flow of oxygen was increased in order to keep the overall level of oxygen in the

kiln constant. The initial and final levels of oxygen fed to the system were the same regardless of whether the stimulus was an elapsed time of 30 seconds or a carbon monoxide spike. Further, these levels were preset by the operator prior to the initiation of incineration and were based on the operators judgment as to the likely combustion behavior of the waste. This requires some prior knowledge about the way in which a given waste stream is likely to ignite and burn in the incinerator. This is difficult to ascertain unless that particular waste has been incinerated before.

Second, high heating value wastes are difficult to incinerate at elevated feed rates with oxygen enhancement. This is because when oxygen is added to the combustion air stream it displaces nitrogen. Without that nitrogen to act as a heat sink, the practical heat release limitations of the incinerator are soon reached when high heating value waste is treated. Additional heat absorption

PYRETRON BURNER SYSTEM (continued)

Table 2. Destruction and Removal Efficiencies (DREs)

Test No.	Naphthalene	Acenaphthylene	Fluorene	Phenanthrene	Anthracene	Fluoranthene
1a	> 99.9988	> 99.9955	> 99.9900	> 99.9961	> 99.9868	> 99.9944
*	> 99.9989	> 99.9962	> 99.9915	> 99.9970	> 99.9898	> 99.9957
2a	> 99.9989	> 99.9954	> 99.9905	> 99.9971	> 99.9904	> 99.9955
*	> 99.9940	> 99.9739	> 99.947	> 99.956	> 99.985	> 99.931
3o	> 99.9986	> 99.9941	> 99.9918	> 99.9961	> 99.987	> 99.9926
4o	> 99.99970	> 99.9987	> 99.9974	> 99.99922	> 99.9974	> 99.9983
5o	> 99.99985	> 99.99942	> 99.9988	> 99.99968	> 99.99896	> 99.99932
*	> 99.99987	> 99.99952	> 99.9990	> 99.99972	> 99.99909	> 99.99941
6o	> 99.99989	> 99.99956	> 99.9991	> 99.99976	> 99.99922	> 99.99944
*	> 99.99987	> 99.99946	> 99.9989	> 99.99970	> 99.99901	> 99.99929

	Hexachloroethane	1,3,5 Trichlorobenzene
7o@	99.9951	> 99.9922
8a@	> 99.9926	> 99.9865

> indicates that the DREs are based on the analytical method detection limit
a = air, o = oxygen
* the second set of DREs are from duplicate samples
@ tests 7 and 8 were done at the request of Region 9 and involve spiking Stringfellow soil with the two POHCs listed.

Table 3. Particulate Emission Summary

	Particulate Concentration (mg/dscm at 7 percent O_2)[a]	
	Scrubber Discharge	
Test No.	Flue Gas	Stack Gas
1 (12-9-87)	b	8
2 (12-11-87)	b	9
3 (12-17-87)	21	99
4 (1-14-88)	26	59
5 (1-20-88)	b	63
6 (1-21-88)	b	21
7 (1-27-88)	27	37
8 (1-29-88)	38	38

[a] Measured particulate concentration directly corrected to 7 percent O_2 using flue gas O_2 level. RCRA standard is 180 mg/dscm corrected to 7 percent O_2. This does not provide a direct comparison for tests with O_2 enhancement (Tests 3, 4, 5, 6, and 7).
[b] Denotes measurements not performed. Particulate levels increased at the stack partly because of particulate entrainment downstream of the scrubber discharge.

capacity must be provided if throughput is to be increased with this kind of waste. During this demonstration water was used. This was sufficient for the 24 16MJ/kg (10,400 Btu/lb) waste treated during the demonstration. In some situations, however, water injection may not provide sufficient heat absorbing capacity. In these cases throughput increases may be difficult to achieve.

Third, levels of NO_x produced by the Pyretron were elevated over those that occurred without oxygen enhancement. The high flame temperatures that result when the Pyretron is used with oxygen enhancement are responsible for this. Air-only operation resulted in average NO_x levels of 92 ppm. Use of the Pyretron with oxygen enhancement resulted in average NO_x levels of 1073 ppm. Appendix C of the Technology Evaluation Report contains all of the NO_x data obtained during the demonstration. The Applications Analysis Report discusses the implications of the Pyretron's high NO_x levels.

Conclusions and Recommendations

Based on the results and experience obtained from this SITE demonstration, the following conclusions and recommendations can be made concerning the operation and performance of the American Combustion Pyretron oxygen-enhanced burner.

1. Overall, the Pyretron may be useful in increasing the efficiency of incinerators that are treating many of the wastes that are found at Superfund sites.

2. With respect to the first claim made about the Pyretron, we were unable to conclusively determine whether the Pyretron system with oxygen enhancement was able to reduce the magnitude of transient emissions produced when waste is batch charged to a rotary kiln. Part of the reason for this is that the waste feed was not uniformly contaminated with high levels of organic waste. Because of this, variations in the levels of transient emissions observed could not be solely attributed to the action of the Pyretron. Further, there was not a clear difference in the frequency or level of transient emissions produced by the Pyretron with oxygen enhancement over conventional incineration.

3. As for the second claim made about the Pyretron, the demonstration clearly showed that thorough waste decontamination can be obtained at throughput rates double those achievable without oxygen enhancement provided that sufficient heat absorption capacity is provided when high heating value wastes are treated.

4. As for the third claim made for the Pyretron, the results of the demonstration indicate that the incremental cost of operating an incinerator equipped with the Pyretron will vary depending on the size and

PYRETRON BURNER SYSTEM (continued)

Table 4. *Utility Costs Incurred During the Pyretron Site Demonstration*

Mode	Total Feed kg (lb)	Total Cost Oxygen $ HIGH LOW ACTUAL	Total Cost Propane $ HIGH LOW ACTUAL	Water Injection Cost $* HIGH LOW ACTUAL	Total Utility Cost $ HIGH LOW ACTUAL	Unit Cost $ kg ($ lb) HIGH LOW ACTUAL
Air	1180 (2596)	--	6000	--	6000	5.08 (2.31)
		--	3000	--	3000	2.54 (1.15)
		--	4008	--	4008	3.39 (1.54)
O_2	1820 (4004)	3870	4000	6.12	7876	4.32 (1.97)
		3250	2000	6.12	5256	2.89 (1.31)
		3560 ⁺	2672	6.12	6238	3.43 (1.56)

* only needed for high heating value wastes
⁺ average value

location of the application as well as on the magnitude of the throughput increases achievable and is predominantly influenced by the costs of oxygen and fuel.

5. The NO_x levels observed during the demonstration may limit the applicability of the Pyretron in situations requiring stringent control of these emissions. Further development of the Pyretron may alleviate this problem.

The EPA Project Manager, **Laurel J. Staley**, is with the Risk Reduction Engineering Laboratory, Cincinnati, OH 45268 (see below).
The complete report, entitled Technology Evaluation Report, Site Program Demonstration Test: The American Combustion Pyretron Thermal Destruction System at the U.S. EPA's Combustion Research Facility," (Order No. PB89-167894/AS; Cost $28.95. subject to change) will be available only from:
National Technical Information Service
5285 Port Royal Road
Springfield, VA 22161
Telephone: 703-487-4650
A related report, which discusses application and costs, is under development.
The EPA Project Manager can be contacted at:
Risk Reduction Engineering Laboratory
U.S. Environmental Protection Agency
Cincinnati, OH 45268

United States
Environmental Protection
Agency

Center for Environmental Research
Information
Cincinnati OH 45268

BULK RATE
POSTAGE & FEES PAID
EPA
PERMIT No. G-35

EPA/540/S5-89/008

Appendix A.1.b

SHIRCO INCINERATION SYSTEM

United States
Environmental Protection
Agency

EPA/540/S5-89/007
April 1989

SUPERFUND INNOVATIVE
TECHNOLOGY EVALUATION

Technology Demonstration Summary

Shirco Pilot-Scale Infrared Incineration System at the Rose Township Demode Road Superfund Site

Under the Superfund Innovative Technology Evaluation or SITE Program, an evaluation was made of the Shirco Pilot-Scale Infrared Incineration System during 17 separate test runs under varying operating conditions. The tests were conducted at the Demode Road Superfund site in Rose Township, Michigan using 1,799 kg (3,967 lb) of soils contaminated with poly-chlorinated biphenyls (PCBs) and other organics, and lead and other heavy metals. The report includes a process description of the unit, unit operations data, sampling and analytical procedures and data, and an overall evaluation of performance and energy consumption.

The Shirco Infrared Incineration System uses electricity for infrared heating rods which heat the soil and desorb or incinerate the organic contaminants, followed by a conventional, propane-fired combustion chamber to complete the destruction of gaseous organic compounds. The system was evaluated for effectiveness in removing and destroying organic contaminants and reducing the mobility of metal contaminants under both standard and varied operating parameters. The achievement of applicable regulatory standards and the effect of operating conditions on energy consumption were also assessed.

The results show that the unit achieved destruction and removal efficiencies (DREs) for PCBs exceeding 99.99%, based on detection limits. Several semivolatile and volatile organic compounds were measured in the stack gas at very low levels (parts per billion) and may be products of incomplete combustion (PICs). The unit achieved regulatory standards for acid gas removal and particulate emissions. Levels of residual PCBs in the furnace ash were less than 0.2 ppm under most unit operation conditions. The majority of the heavy metals remained in the furnace ash, but there was no evidence of a decrease in the mobility of lead as a result of treatment. Also, residual heavy metals were measured in the scrubber water effluent. The optimization of the heat content of the waste, retention time, and primary combustion chamber temperature can significantly reduce energy consumption and cost.

SHIRCO INCINERATION SYSTEM (continued)

This Summary was developed by EPA's Risk Reduction Engineering Laboratory, Cincinnati, OH, to announce key findings of the SITE Program demonstration that is fully documented in two separate reports (see ordering information at back).

Introduction

In response to the Superfund Amendments and Reauthorization Act of 1986 (SARA), the Environmental Protection Agency's Office of Research and Development (ORD) and Office of Solid Waste and Emergency Response (OSWER) have established a formal program to accelerate the development, demonstration, and use of new or innovative technologies as alternatives to current containment systems for hazardous wastes. This new program is called Superfund Innovative Technology Evaluation or SITE. The principal goal of the SITE program is to demonstrate new technologies in the field and develop reliable economics and performance information.

The SITE program demonstration of the Shirco Pilot-Scale Infrared Incineration System for thermal treatment developed by Shirco Infrared Systems, Inc. of Dallas, Texas, was conducted at the Demode Road Superfund Site in Rose Township, Michigan. The Demode Road site is a 12-acre waste site previously used to bury, dump, and store industrial wastes such as paint sludges, solvents, and other wastes containing PCBs, oils and greases, phenols, and heavy metals . PCBs and lead are the principal contaminants in the soil used for the test of the Infrared System.

The test was conducted from November 2-13, 1987 and treated 1,799 kg (3,967 lb) of contaminated soil under various test conditions. The major objectives of this demonstration were to determine the following:

- DRE levels for PCBs and the presence of PICs in the stack gas. The regulatory standards are 99.99% DRE under the Resource Conservation and Recovery Act (RCRA) and 99.9999% DRE under the Toxic Substances and Control Act (TSCA).

- Level of hydrogen chloride (HCl) and particulates in the stack gas. The RCRA standard for HCl in the stack gas is 1.8 kg/hr (4 lb/hr) or 99 wt% HCl removal efficiency. The RCRA standard for particulate emissions in the stack gas is 180 mg/dscm (0.08 gr/dscf).

- Level of residual PCBs in the furnace ash at normal and varied operating conditions.

- Mobility of heavy metals, particularly lead, in the furnace ash as compared to the feed.

- Mobility of heavy metals in the furnace ash as compared to the RCRA Extraction Procedure Toxicity (EP Tox) Characteristic (as measured by the EP Tox test) and the proposed Toxicity Characteristic (as measured by the Toxicity Characteristic Leaching Procedure (TCLP)).

- Level of residual heavy metals and organic compounds, and other physical and chemical characteristics in the scrubber water discharged from the unit .

- The operating conditions that reduce energy consumption without decreasing soil decontamination effectiveness.

- Effect of varying operating conditions on residual levels of heavy metals and organics in the furnace ash versus the levels in the feed.

- Adherence of the quality assurance (QA) procedures to the requirements of the RREL approved QA Project Plan (Category II), as defined by the Document No. PA QAPP-0007-GFS, "Preparation Aid for HWERL's Category II Quality Assurance Project Plans", June, 1987.

Feed Preparation

The demonstration test used soil from an area of the site that was highly contaminated with PCBs and lead, as determined in the original remedial investigations performed at the site. Pretest sampling and analysis further identified those sectors within the area most highly contaminated with PCBs and lead for excavation. Other organics and heavy metals were also present in these sectors. Soil from these sectors to be used as feed for the test was excavated and mixed into a pile using a front-end loader, and then screened to remove aggregate and debris greater than one inch in diameter. The screened soil was drummed and transferred to a designated zone adjacent to the test unit. Two drums of soil were blended with 3 wt% fuel oil to be used for several of the test runs to investigate the effect of increased feed heating value on overall unit performance and energy consumption at varying operating conditions.

Process Description

The Shirco Pilot-Scale Infrared Incineration System consists of a waste feed system, an (electric) infrared primary combustion chamber, a supplemental propane-fired secondary combustion chamber, a venturi scrubber emissions control system, an exhaust system, and a data collection and control system, all enclosed in a 45-ft trailer. The system process flow and the overall 250 ft x 100 ft test site layout are presented schematically in Figures 1 and 2, respectively.

During the test, the feed material was transferred from the drums to pails, weighed, and then fed manually to a hopper mounted over a metering conveyor belt. The waste was fed at a controlled rate through a sealed feed chute onto the incinerator conveyor (a tightly woven wire belt which moved waste material through the primary combustion chamber). The conveyor belt speed can be adjusted to achieve feed residence times in the PCC from 6 to 60 min. Typically residence times range from 10 to 25 min. The depth of the waste on the conveyor belt ranged from one to one and a half inches.

The primary combustion chamber (PCC) is a rectangular box insulated by layers of ceramic fiber. Combustion air is supplied to the primary combustion chamber through a series of air ports at points along the length of the chamber . The gas flow in the incinerator is countercurrent to the conveyed feed material. Electric infrared heating elements installed above the conveyor belt heat the waste to the designated temperature (nominally 1600°F), which results in desorption or incineration of organic contaminants from the feed. Rotary rakes gently turn the material to ensure adequate mixing and complete desorption. When the thermally treated soil, now referred to as furnace ash, reaches the discharge end of the PCC, it drops off the belt through a chute and into an enclosed hopper and discharge storage drum. The drums of furnace ash are then stored for final disposal.

Exhaust gas containing the desorbed contaminants exits the primary combustion chamber into a secondary combustion chamber (SCC) or afterburner, where a propane-fired burner combusts residual organic compounds into CO_2, CO, HCl, and H_2O. The SCC is typically operated at 2200°F and a gas residence time exceeding 2 sec. Secondary air is supplied to ensure adequate excess oxygen levels for

SHIRCO INCINERATION SYSTEM (continued)

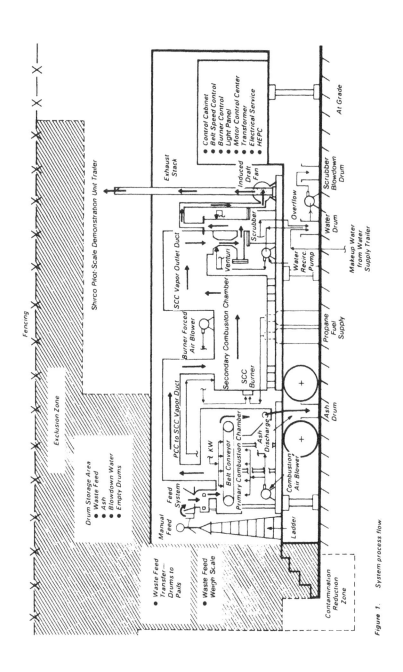

Figure 1. System process flow

SHIRCO INCINERATION SYSTEM (continued)

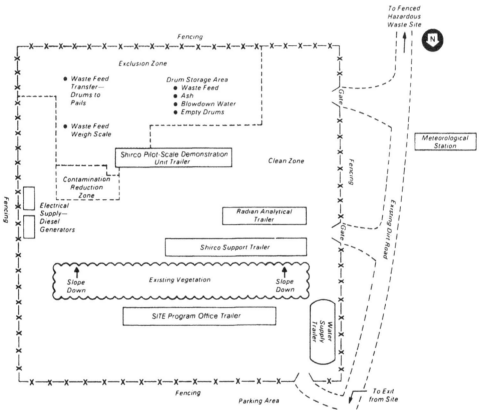

Figure 2. *Overall test site.*

complete combustion. Exhaust gas from the secondary combustion chamber then is quenched by a water-fed venturi scrubber emissions control system to remove particulate matter and acid gases. An induced draft fan transfers the gas to the exhaust stack for discharge to the atmosphere.

The same trailer housing the thermal system also contains the control panel for the main unit, and data collection indicators and recorders. Safety interlocks also are integrated into the trailer-mounted unit to automatically correct abnormal operating conditions, maintain system performance, and if necessary, shut down feed and heat input to the unit.

Test Procedure

In order to meet the objectives of the demonstration test (see Introduction), a total of 17 test runs were conducted. Three runs were performed under design operating conditions to assess overall unit operation and system performance (Phase I), and 14 runs were conducted under varying operational parameters to evaluate their effect on system performance and energy consumption (Phase II).

The Phase I runs were conducted at 1600°F PCC temperature, a 2200°F SCC temperature, and a PCC residence time of 20 min. Each of the three runs was sufficiently long (six to ten hours) to

SHIRCO INCINERATION SYSTEM (continued)

gather a large enough sample of stack gas to analyze it for PCBs. An additional run was conducted at the same operating conditions to obtain specific stack samples that had not been successfully collected during two of the previous runs.

The Phase II runs were conducted for approximately one hour under varied operating conditions that included the PCC temperature (900, 1200, 1400, and 1600°F), SCC temperature (1800 and 2200°F), PCC feed residence time (10, 15, 20, and 25 minutes), and PCC combustion air flow (on, off to simulate oxidizing or non-oxidizing (pyrolytic) PCC atmosphere).

For the Phase I runs, samples were taken of the feed, scrubber makeup water, furnace ash, scrubber water, and scrubber solids. These streams were analyzed for PCBs, dioxins and furans, metals, organics, and other physical and chemical properties and components specific to the characterization of each sampled matrix. In addition, the EP Tox and TCLP leaching tests were performed on these sampled streams (exclusive of scrubber water makeup) and the extracts were evaluated for metals. Samples were also taken of the upwind and downwind ambient air, PCC offgas, and stack gas. Ambient air upwind and downwind of the unit was monitored for PCBs and heavy metals by high volume samplers. For the stack gas and PCC offgas, several sampling methods were employed, including an EPA Method 5 for particulate matter (and subsequent metals analysis of particulates) and hydrochloric acid; a Source Assessment Sampling System (SASS) for semivolatile organic pollutants, PCBs, dioxins and furans; a Volatile Organic Sampling Train (VOST) for volatile organic pollutants; a Modified Method 5 for soluble chromium; continuous emission monitoring for oxygen, carbon dioxide, carbon monoxide, oxides of nitrogen, and total hydrocarbons; and an experimental method for vaporous lead emissions in the PCC offgas.

For the Phase II runs, samples were taken of the feed, furnace ash, scrubber water, scrubber solids, PCC offgas, stack gas, and upwind and downwind ambient air. EPA Method 5 was used again to sample stack gas and PCC offgas but was analyzed only for particulate matter. In addition, throughout the Phase II tests, the stack gas and PCC offgas were continuously monitored for oxygen, carbon dioxide, carbon monoxide, oxides of nitrogen, and total hydrocarbons. All of the remaining sampled streams were analyzed for PCBs, dioxins and furans,

metals, and other physical and chemical properties and components specific to the characterization of each sampled matrix. In addition, the EP Tox and TCLP leaching tests were performed on these samples and the extracts were evaluated for metals.

All of the sampling and analytical work was conducted in accordance with QA/QC Category II and include data quality credibility statements for the precision and accuracy of the data reported.

Results and Discussion

A detailed summary of the SITE demonstration test results is presented in Table 1. Based on the test objectives outlined in the Introduction, the following results were obtained.

• Characteristics of the Feed

Based on data from the previous remedial investigation of the site, a specific area within the site was identified with the highest concentrations of both PCBs and lead, the major soil contaminants of concern. The remedial investigation also described the soil as a dry, brown, sandy, and silty clay topsoil which upon excavation proved to be an accurate observation. Subsequent pretest sampling and analysis of the specific area of the site identified particular sectors with the highest contamination of PCBs and lead. A composite sample of all the sectors within the area indicated a 7.8 pH, 9.0 wt.% moisture, 81 wt.% ash, less than 1000 Btu/lb high heating value, and a 0.95 g/cc density. The composite sample contained 570 ppm of total PCBs and 580 ppm lead (elemental lead after digestion and conversion to inorganic form). A composite sample of the 10 sectors chosen for excavation contained 626 ppm PCBs, 560 ppm of lead, 55 ppb of tetrachlorodibenzo-p-dioxin (TCDD), and 4.2 ppb of tetrachlorodibenzofuran (TCDF). Once the feed excavation was begun, it became evident that the front-end loader could not confine its large-scale activities to the 10 specific sectors and an area comprising 14 specific sectors was excavated for the unit's feed source

Table 1 summarizes the PCB and lead contaminant concentrations measured in the soil from the composite of the grab samples of feed taken during each of the test runs. In

addition to lead, where concentrations ranged from 290 to 3000 ppm and averaged 778 ppm, several other metals were present at average concentrations exceeding 50 ppm including barium (591 ppm), zinc (301 ppm), and chromium (85 ppm). Total PCBs concentration ranged from 10.2 to 669 ppm and averaged 272 ppm.

Several samples of the feed contained small quantities of TCDFs ranging from 0.04 to 0.1 ppb. Volatile and semivolatile organic compounds including methyl ethyl ketone, trichloroethene, and bis(2-ethyl-hexyl)phthalate were measured in feed samples at concentrations less than 50 ppm. Methyl ethyl ketone and trichloroethene were also detected in solvent blanks and are attributed to analytical laboratory contamination.

• Characteristics of the Furnace Ash

Table 1 summarizes the PCB and lead contaminant concentrations measured in the furnace ash from the composited grab samples taken at the conclusion of each test run. In addition to lead, where concentrations ranged from 420 to 2000 ppm and averaged 1173 ppm, several other metals were present at average concentrations exceeding 50 ppm including barium (1061 ppm), zinc (410 ppm), and chromium (81 ppm). Total PCBs concentration ranged from 0.004 to 3.396 ppm. Two samples of furnace ash contained 0.07 and 0.3 ppb of TCDF during two runs conducted at a 900°F PCC operating temperature; the normal PCC operating temperature is 1600°F. These runs were also conducted without the input of PCC combustion air to simulate non-oxidizing or pyrolytic combustion conditions. The low PCC temperature and pyrolytic environment could have led to the incomplete desorption or incineration of TCDF present in the feed or to the production of TCDF from the incomplete combustion of PCBs in the feed. Volatile compounds including methylene chloride, methyl ethyl ketone, tetrachloroethene, and trichloroethene were also measured in the furnace ash samples in concentrations ranging from 3.9 to 64 ppm with one sample containing 980 ppm of methylene chloride. Methyl ethyl ketone and trichloroethene were also detected in solvent blanks and methylene chloride is commonly employed in laboratory procedures; therefore these compounds may be

SHIRCO INCINERATION SYSTEM (continued)

Table 1. Site Demonstration Test Results Summary

Operating Conditions PCC		Waste Feed Characteristics				Furnace Ash Characteristics			
Temp. °F	Residence Time min.	PCB ppm	Pb ppm	EP Tox (Pb) ppm	TCLP (Pb) ppm	PCB ppm	Pb ppm	EP Tox (Pb) ppm	TCLP (Pb) ppm
900[a,b]	20	327	590	0.29	0.81	2.079	1,000	0.38	2.90
900[b]	20	20.2	660	0.67	0.88	3.396	1,400	0.89	6.20
900[b]	25	367	290	0.32	7.00	0.168	0.860	0.88	3.80
1200	20	297	640	0.05	0.56	0.115[d]	1,100	4.10	1.60
1200	15	27.6	870	0.20	0.44	0.077	1,000	0.38	3.60
1200[b]	25	456	590	0.12	0.53	0.108[d]	1,200	0.14	0.05
1200[b]	20	669	610	0.20	0.71	0.066[d]	1,200	0.06	4.10
								4.90[g]	2.80[g]
1200[b]	15	602	470	0.18	0.53	0.025[d]	2,000	h	h
1200[a,b]	15	309	370	0.21	0.96	0.066[d]	1,000	0.46	0.82
1400	20	56.0	740	0.07	0.89	0.087[d]	1,600	ND	0.15
1600	20	10.2	3000	0.15	0.67	0.037	1,100	0.05	ND
1600	20	35.2	1400	0.20	0.35	0.112	1,300	ND	ND
1600	20	20.4	550	0.23	1.30	0.003	1,100	0.13	0.05
1600	20	f	1100	0.14	0.49	f	0.420	0.28	1.80
1600	10	391	620	0.25	0.73	0.045[d]	1,700	ND	1.00
1600[a]	15	451	620	ND	0.66	0.117[d]	0.840	0.43	0.17
1600[a]	15[c]	271	390	0.53	1.80	0.004	1,500	0.27	0.23
1600[a,b]	15	311	500	0.07	0.55	0.061[d]	0.800	1.10	2.40
				3.00[g]	1.40[g]				

[a] Waste feed blended with 3 wt.% fuel oil.
[b] Non-oxidizing atmosphere.
[c] PCC bed depth at 1 inch. All other tests at 1-1/2 inches.
[d] PCB levels below analytical detection limits. Total shown is sum of detectable limits indicated in analyses.
[e] ND - nondetectable value.
[f] Run was conducted to make up for incomplete semivolatile organics, PCDD-PCDF, soluble chromium and stack gas particulate samplings on other runs.
[g] Data from additional EP Tox and TCLP tests.
[h] ND due to broken sample container.

products of incomplete combustion and/or the result of laboratory contamination.

• Residual PCBs in Furnace Ash

During the demonstration test, a total of 17 runs were conducted at varying operating conditions. In addition to the DRE levels, which are an indication of the performance of the Shirco Pilot-Scale Infrared Incineration System and its ability to meet RCRA and/or TSCA regulatory standards, the reduction of PCB concentration from the feed to the furnace ash is also a measure of the unit's ability to effectively destroy PCBs and produce a furnace ash with a PCB concentration below the TSCA guidance level of 2 ppm.

Based on the data presented in Table 1, two samples of furnace ash

exceeded the TSCA guidance levels and contained 3.396 and 2.079 ppm of total residual PCBs. The samples were produced during two runs conducted at a 900°F PCC operating temperature (20 minutes residence time), which is significantly lower than the normal PCC operating temperature of 1600°F. These runs were also conducted without the input of PCC combustion air to simulate non-oxidizing or pyrolytic combustion conditions. At this low PCC temperature and pyrolytic condition, these higher total residual PCB levels in the furnace ash may be the result of the incomplete combustion of PCBs in the feed. This is further substantiated by the residual TCDF present in the furnace ash samples from these same two runs, as discussed previously. The remaining

runs conducted at 1200, 1400, and 1600°F resulted in total residual PCB concentrations in the furnace ash ranging from 0.003 to 0.117 ppm. A third run, which was conducted at a 900°F PCC operating temperature but with an increased PCC residence time of 25 minutes resulted in a total furnace ash PCB concentration of 0.168 ppm with no detectable TCDF. It is possible that the increased residence time in the PCC may have offset the low 900°F PCC operating temperature and provided the additional processing time for the satisfactory destruction of the PCBs in the feed.

• Mobility of Heavy Metals - Feed and Furnace Ash

In order to determine whether heavy metals, particularly lead, would leach

SHIRCO INCINERATION SYSTEM (continued)

from the furnace ash produced in the Shirco Pilot-Scale Infrared Incineration System, EP Tox and TCLP tests were conducted to determine the mobility of heavy metals from the furnace ash as compared to the feed.

The initial EP Tox analyses for lead in the leachate ranged from 0.05 to 0.67 ppm for the feed and 0.05 to 4.1 ppm for the furnace ash. The initial TCLP analyses ranged from 0.35 to 1.80 ppm (with one sample at 7.0 ppm) for the feed and 0.05 to 4.1 ppm (with one sample at 6.2 ppm) for the furnace ash.

A comparison of the EP Tox and TCLP analyses conducted on the furnace ash and the feed do not show any trend or evidence that indicate reduced mobility of lead from the furnace ash versus the feed as a result of the thermal treatment. The comparison did reveal that the concentrations of lead in the TCLP leachates from both the feed and the furnace ash were consistently higher than the corresponding EP Tox tests on the same samples.

When several samples were retested to verify the results, the concentrations of lead in the EP Tox leachates (4.9 ppm feed, 3.0 ppm furnace ash) were higher than during the initial tests and, in direct reversal to the original data, exceeding corresponding TCLP leachate concentrations (2.8 ppm feed, 1.4 ppm furnace ash). The results of the retest again did not indicate reduced mobility of lead from the furnace ash versus the feed as a result of the thermal treatment.

- Mobility of Heavy Metals - EP Tox and Proposed TCLP Toxicity Characteristic Standards

EP Tox and TCLP tests were conducted on the feed, furnace ash, scrubber water, and scrubber solids. All of the results were below the EP Tox and proposed TCLP toxicity characteristic standards of 5 ppm arsenic, 100 ppm barium, 1 ppm cadmium, 5 ppm chromium, 5 ppm lead, 0.2 ppm mercury, 1 ppm selenium, and 5 ppm silver except for one feed sample at 7.0 ppm lead (TCLP) and one furnace ash sample at 6.2 ppm lead (TCLP). A comparison of the EP Tox and TCLP analyses on all the sampled streams to the above mentioned standards do not show any trend or evidence that indicate reduced mobility of heavy metals as a result of the thermal treatment.

- Destruction and Removal Efficiency (DRE) of PCBs

The DRE of PCBs for the first three runs (Phase I) is greater than 99.99%. In contrast, the regulatory standard for incineration under the RCRA is 99.99% DRE and under TSCA is 99.9999% DRE. The low PCB concentrations in the feed resulted in PCB levels in the stack gas that were less than the analytical detection limits for two of the runs. Therefore for these runs, DRE is calculated based on the sum of the detection limits of the PCB congenors in order to compare the DRE for the runs on the same basis. Stack gas measurements conducted during the third run did detect trichlorobiphenyl and tetrachlorobiphenyl congenors and a DRE is shown based on this measurement. The less rigorous sampling in Phase II of the test was not designed to allow calculation of DRE.

- Other Organic Stack Gas and PCC Offgas Emissions

Several volatile and semivolatile organic compounds were detected in the stack gas at concentrations less than 100 ppb and established standards for direct inhalation. Low levels of several phthalate compounds were also detected in blank samples and may be traced to contamination from plastic components in the process, sampling equipment, or laboratory apparatus. Several volatile organic compounds including benzene and toluene were detected in the stack gas and the scrubber makeup water and may be attributable to contamination from the makeup water although PIC formation is a possibility. Other volatile and semivolatile organic compounds, which probably represent PICs, were detected. They include halomethanes, chlorinated species including chlorobenzene and methylene chloride, other volatile organics including xylenes, styrene and ethylbenzene, oxygenated hydrocarbons including acetone and acrolein, carbon disulfide, and p-chlor-m-cresol. Dioxins and furans were not detected in the stack gas samples.

The majority of the organic compounds present in the PCC off gas samples at levels less than 500 ppb were also present in the stack gas. The additional destruction of organics that take place in the SCC and emissions scrubbing system reduced the concentration of these organic compounds in the corresponding stack gas samples.

- Acid Gas Removal

During the Phase I Runs 1-3, HCl emissions ranged from 0.181 to 0.998 g/hr, which were significantly below the RCRA performance standard of 1800 g/hr that would require a 99 wt.% HCl removal efficiency. HCl removal efficiencies ranged from 97.23 to 99.35 wt.%. Acid gas removal was not measured in Phase II.

- Particulate Emissions

Particulate emissions were measured throughout the test and ranged from 7 to 68 mg/dscm, well below the RCRA standard of 180 mg/dscm.

- Analysis of Scrubber Makeup Water, Scrubber Water, and Scrubber Solids

Scrubber makeup water was transported to the site in a tank truck that may have contained some residual contamination prior to fill up. Samples of scrubber makeup water were taken at the end of each run. No PCBs, dioxins, furans, or semivolatile organic compounds were detected. Several volatile organics including benzene, toluene, and trichloroethene were measured at concentrations less than 15 ppm. The concentrations of heavy metals were all less than 0.2 ppm.

Samples of the water recirculation through the venturi scrubber system, referred to as scrubber water, were also taken at the end of each run. PCB concentrations were less than 200 ppt and no dioxins, furans, or semivolatile organic compounds were detected. Small quantities of benzene (2 ppm) and toluene (5.7 to 11 ppm) were measured in several of the samples and are attributable to the similar contaminants in the scrubber makeup water. The concentrations of heavy metals in the scrubber water were all less than 1 ppm except for barium, which ranged from 0.2 to 2.2 ppm, and lead, which ranged from 0.12 to 1.8 ppm.

Insufficient quantities of scrubber solids in the scrubber water were available for analysis.

- Overall Disposition of Metals

Total metals analyses of the feed, furnace ash, PCC offgas and stack gas particulates, scrubber makeup water, scrubber water, and scrubber solids showed that the majority of the detectable metals, including lead, that

SHIRCO INCINERATION SYSTEM (continued)

entered the unit with the feed remained in the furnace ash. An overall mass balance of lead through the unit was calculated based on the analysis of lead in the samples, the measured feed rate as weighed during the runs' operating periods, the calculated furnace ash flow rate based on the ultimate analysis of ash in the feed sample, and the measured particle mass and gas volume obtained from the gas' EPA Method 5 sampling trains. Phase I results indicate an average lead mass flow rate of 28.3 g/hr in the feed, 37.0 g/hr in the furnace ash, 0.206 g/hr in the PCC offgas particulates, and 0.109 g/hr in the stack gas particulates. The quantity of lead leaving the unit with scrubber water effluent is approximately 0.204 g/hr based on the maximum measured concentration of 1.8 ppm lead in the scrubber water and an overall approximate water flow rate of 30 gph. The PCC offgas particulates sampled during the Phase I runs contained an average of 5364 ppm of lead as compared to stack gas particulates, which contained an average of 15,830 ppm of lead. By contrast, the average concentration of lead in the feed was 1550 ppm. Although the concentration of lead in the particulate matter increases as the process flow progresses through the unit, the actual mass flow of lead decreases as the gas stream is cooled and treated through the emissions control system.

For the Phase I runs sampling and analysis procedures were conducted to evaluate vaporous lead concentrations in the PCC offgas and soluble chromium concentrations in the PCC offgas and stack gas particulates. The special sampling for vapor phase lead and soluble chromium were unable to detect any of either metal at levels less than 2.7 ppb and 264 ppb, respectively; therefore the evaluations were inconclusive.

Other heavy metals, particularly barium and zinc, with average concentrations exceeding 100 ppm in the feed (barium 591 ppm, zinc 301 ppm) were also present in high concentrations, relative to other heavy metals, in the furnace ash (barium 1061 ppm, zinc 410 ppm) and scrubber water (barium 0.8 ppm, zinc 0.3 ppm).

• Optimum Operating Conditions

Phase II was designed to examine the effect on energy consumption and changes in the residual levels of heavy

metals and organics in the furnace ash versus the levels in the feed by varying operating conditions.

Based on the data obtained an analysis was conducted to compare energy consumption in the unit at operating conditions that did not affect the performance of the unit. A reduction in the PCC operating temperature from 1600°F to 1200°F reduced the average PCC power usage 48% from 0.2294 to 0.1200 kwhr.lb feed. A reduction in the SCC operating temperature from 2200°F to 1800°F reduced the average propane fuel consumption by 51% from 3997 to 1952 Btu/lb feed. The use of 3 wt.% fuel oil to supplement the fuel value of the feed further decreased PCC power usage by 26 to 67% at PCC operating temperatures of 1600°F and 1200°F, respectively, with accompanying increases in overall feed rate of 32% and 26%. The costs for fuel oil and its attendant facilities still must be examined for specific applications to determine the cost effectiveness of a fuel oil additive to the waste feed.

As discussed in previous sections the results did not provide any trend or change in the residual levels of the heavy metals and organics in the furnace ash versus the levels in the feed as the operating conditions were varied and PCC operating temperatures were maintained at 1200°F to 1600°F. At an abnormally low PCC operating temperature of 900°F, without the input of combustion air to simulate non-oxidizing or pyrolytic combustion conditions, total PCB and TCDF concentrations in the furnace ash increased. The increases may indicate that these PCC conditions led to incomplete desorption or incineration of PCB and TCDF and to the production of TCDF from the incomplete combustion of PCBs in the feed.

• QA Summary

The Phase I and II runs had a well-defined quality assurance/quality control program to ensure the collection of accurate data. This program was developed as part of the test program preparation activities and was formalized in the RREL approved QA Project Plan (Category II). All of the sampling and analytical work was conducted in accordance with this QA Project Plan and the results include data quality credibility statements and information that confirm the satisfactory

precision and accuracy of the data reported.

Conclusions

Based on the above data and discussion, the following conclusions can be made concerning the operation and performance of the Shirco Pilot-Scale Infrared Incineration System.

1. The PCC equipped with infrared heating rods reduced PCBs from an average of 272 ppm and a maximum of 669 ppm in the feed to less than 0.2 ppm PCBs in the furnace ash when PCC temperature was 1200°F or higher. PCB levels in the ash were well below the TSCA guidance level of 2 ppm of PCBs in treatment residuals.

2. The majority of the lead and other heavy metals present in the feed remained in the furnace ash, regardless of operating conditions. However, the scrubber water contained levels of lead and barium (up to 1.8 to 2.2 ppm, respectively), and metals also concentrated to some extent in the furnace ash. Both residual streams may require further treatment when metals are present in the feed.

3. In most cases concentrations of metals in the extract of the furnace ash did not exceed their respective EP Tox and TCLP toxicity characteristic standards. The need for further treatment of the furnace ash to reduce or immobilize the metals is site specific, and will depend on the cleanup standards for the site.

4. Based on two leaching tests, the EP Tox and TCLP, the mobility of lead and other heavy metals was similar in the feed and the furnace ash, and there was no evidence that treatment affected metals leaching.

5. The unit achieved DREs of PCBs greater than 99.99%, based on one actual calculation and in two cases on detection limits. PCB concentrations in the feed and analytical detection limits did not allow the demonstration of 99.9999% DRE required under TSCA. However, this unit achieved greater than 99.9999% DRE in other tests, and at this time at least one full-scale infrared system has demonstrated greater than 99.9999% DRE for PCBs and is permitted under TSCA to process PCB waste. The upcoming Appli-

SHIRCO INCINERATION SYSTEM (continued)

cations Analysis Report will incorporate this additional data.

6. The unit achieved regulatory standards for acid gas removal and particulate emissions. These data apply to the operation and performance of the air pollution control system installed on this unit. Additional data on the performance of air pollution control systems on full-scale Shirco Infrared units will be discussed in the Applications Analysis Report.

7. Several semivolatile and volatile organic compounds measured in the stack gas in the parts per billion may be PICs. These levels are much lower than established standards for direct inhalation of these compounds.

8. The unit was able to reduce the PCBs in the feed using less power when fuel oil was added to the waste and when PCC temperature was reduced. The addition of fuel oil also increased the feed rate. Cost savings in specific applications will depend on local fuel and electrical costs, and a minimum PCC temperature must be maintained to avoid inadequate desorption of the organics in the feed and the production of PICs.

United States
Environmental Protection
Agency

EPA 540 S5-88 002
January 1989

SUPERFUND INNOVATIVE
TECHNOLOGY EVALUATION

Technology Demonstration Summary

Shirco Electric Infrared Incineration System at the Peak Oil Superfund Site

Under the auspices of the Superfund Innovative Technology Evaluation or SITE Program, a critical assessment is made of the performance of the transportable Shirco Infrared Thermal Destruction System* during three separate test runs at an operating feedrate of 100 tons per day. The unit was operated as part of an emergency cleanup action at the Peak Oil Superfund site in Brandon, Florida. The report includes a process description of the unit, unit operations data and a discussion of unit operations problems, sampling and analytical procedures and data, and an overall performance and cost evaluation of the system.

The results show that the unit achieved destruction and removal efficiencies (DREs) of polychlorinated biphenyls (PCBs) exceeding 99.99% and destruction efficiencies (DEs) of PCBs ranging from 83.15% to 99.88%. Acid gas removal efficiencies were consistently greater than 99%. Particulate emissions ranged from 171 to 358 mg/dscm, exceeding 180 mg/dscm during two of the four tests. The

Extraction Procedure (EP) Toxicity Test on the furnace ash exceeded the RCRA EP Toxicity Characteristic standard for lead. Small quantities of tetrachlorodibenzofuran (TCDF) were detected in one of the four stack gas samples. Also detected were low levels of some semivolatile organics and a broader range of volatile organics, which can be considered products of incomplete combustion (PICs). Ambient air monitoring stations detected quantities of PCBs, which appear to be caused by the transport of ash from the ash pad to the ash storage area. Waste feed and ash samples were not mutagenic according to the standard Ames Salmonella mutagenicity assay. Unit costs are estimated to range from $196 to $795 per ton with a normalized cost per ton of $425 for the Peak Oil cleanup.

This Summary was developed by EPA's Risk Reduction Engineering Laboratory, Cincinnati, OH, to announce key findings of the SITE Program demonstration that is fully documented in three separate reports (see ordering information at back).

SHIRCO INCINERATION SYSTEM (continued)

Introduction

The SITE Program demonstration test of the Shirco infrared incineration system was conducted from July 1, 1987 to August 4, 1987 at the Peak Oil Superfund site in Brandon, Florida during a removal action by EPA Region IV. The Region had contracted with Haztech, Inc., an emergency removal cleanup contractor, to incinerate approximately 7,000 tons of waste oil sludge contaminated with PCBs and lead after determining that high temperature thermal destruction of the nonrecyclable sludge was capable of destroying the PCBs in a cost-effective and environmentally sound manner. Metals that concentrated in the ash residue would be dealt with after the thermal destruction of the sludge. The removal action offered an ideal opportunity for the SITE program to obtain specific operating, design, analytical, and cost information to evaluate the performance of the unit under actual operating conditions. Also, the SITE program studied the feasibility of utilizing the Shirco transportable infrared incinerator as a viable hazardous waste treatment system at other sites throughout the country. To this end, specific test objectives of the Shirco system were:

- To determine the system's destruction and removal efficiency (DRE) for PCBs.

- To report the unit's ability to decontaminate the solid material being processed and to determine the destruction efficiency (DE) for PCBs based on the PCB content of the furnace ash.

- To evaluate the ability of the unit and its associated air pollution control scrubber system to limit hydrochloric acid and particulate emissions.

- To determine whether heavy metals contaminants in the waste feed are chemically bonded or fixated to the ash residue by the process.

- To determine the effect of the thermal destruction process in producing combustion byproducts or products of incomplete combustion (PICs).

- To determine the impact of the unit operation on ambient air quality and potential mutagenic exposure.

- To provide unit cost data for effective development of a cost-economic analysis for the unit.

- To document the mechanical operations history of the unit and analyze and provide potential solutions to chronic mechanical problems.

Facility and Process Description

Solid waste processed at the Peak Oil site was incinerated in a transportable infrared incinerator, designed and manufactured by Shirco Infrared Systems, Inc. of Dallas, Texas and operated by Haztech, Inc. of Decatur, Georgia. The overall incineration unit consists of a waste preparation system and weigh hopper, infrared primary combustion chamber, supplemental propane-fired secondary combustion chamber (afterburner), emergency bypass stack, venturi scrubber system, exhaust system, and data collection and control systems, all mounted on transportable trailers. The system process flow and the overall test site layout are presented schematically in Figure 1.

Solid waste feed material is processed by waste preparation equipment designed to reduce the waste to the consistency and particle sizes suitable for processing by the incinerator. After transfer from the waste preparation equipment, the solid waste feed is weighed and conveyed to a hopper mounted over the furnace conveyor belt. A feed chute on the hopper distributes the material across the width of the conveyor belt. The feed hopper screw rate and the conveyor belt speed rate are used to control the feedrate and bed depth.

The incinerator conveyor, a tightly woven wire belt, moves the solid waste feed material through the primary combustion chamber where it is brought to combustion temperatures by infrared heating elements. Rotary rakes or cakebreakers gently stir the material to ensure adequate mixing, exposure to the chamber environment, and complete combustion. When the combusted feed or ash reaches the discharge end of the incinerator, it is cooled with a water spray and then is discharged by a screw auger/conveyor to an ash hopper.

The combustion air to the incinerator is supplied through a series of overfire air ports located at various locations along the incinerator chamber; combustion air flows countercurrent to the conveyed waste feed material.

Exhaust gas exits the primary combustion chamber and flows into the secondary combustion chamber where

propane-fired burners combust any residual organics present in the exhaust gas. The secondary combustion chamber burners are set to burn at a predetermined temperature. Secondary air is supplied to ensure adequate excess oxygen levels for complete combustion. Exhaust gas from the secondary combustion chamber is quenched by a water-fed venturi scrubber to remove particulate matter and acid gases, the exhaust gas is then transferred to the exhaust stack by an induced draft fan, and finally discharged to the atmosphere.

The main unit controls and data collection indicators comprising the data collection and control system are housed in a specially designed van.

An emergency bypass stack is mounted in the system directly upstream of the venturi scrubber for the diversion of hot process gases under emergency shutdown conditions.

Results and Discussion

A detailed summary of the SITE demonstration test results is presented in Table 1. Based on the test objectives outlined in the Introduction, the following results and conclusions were obtained.

PCB Destruction and Removal Efficiency

PCBs were analyzed in the solid waste feed, furnace ash, scrubber effluent solids, stack gas, scrubber liquid effluent, and scrubber water inlet. The DRE calculation for PCBs is based on the following:

$$DRE = \frac{W_{in} - W_{out}}{W_{in}} \times 100$$

where: W_{in} = mass rate of PCBs fed to incinerator

W_{out} = mass emission rate of PCBs in stack gas

The unit achieved a DRE for PCBs of 99.99%.

It should be noted that the unit was operated to produce an ash that contained 1 ppm or less of PCB. The PCB concentration in the waste feed to the unit varied from 5.85 to 3.48 ppm during the tests. These low PCB concentrations in the waste feed were the result of mixing the original oily waste having up to 100 ppm of PCBs with the PCB-free surrounding soil, lime, and sand so that the resulting material could

SHIRCO INCINERATION SYSTEM (continued)

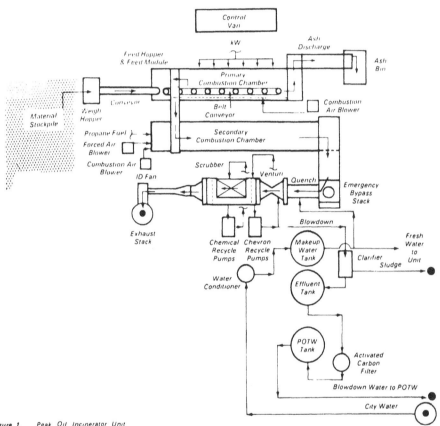

Figure 1   Peak Oil Incinerator Unit

be handled and processed as a solid waste It was not possible to calculate the DRE beyond two decimal places because of the detection limits associated with the analytical procedures employed

Decontamination of Solid Waste and Destruction Efficiency

Residual PCBs in the furnace ash were below the 1 ppm operating standard. ranging from 0 007 ppm on August 1 to 0 900 ppm on August 3. DE was determined by the formula

$$DE = \frac{W_{in} - W_{out}}{W_{in}} \times 100$$

where W_{in} = mass rate of PCBs fed to incinerator

W_{out} = mass rate of PCBs in stack gas, furnace ash. and scrubber effluent

A basis for calculating DE was based on the PCB concentrations in the waste feed and the furnace ash. The DE or removal of the PCBs from the waste feed ranged from 99 88 wt% (August 1) to 83.15 wt% (August 3).

Acid Gas Removal

Measured HCl emission rates ranged from less than 0.8 to 8.6 g/hr. Since the chlorine concentration in the solid waste feed was below the 0.1% detection limit, it was impossible to determine actual HCl removal efficiency. However, SO_2

SHIRCO INCINERATION SYSTEM (continued)

Table 1. Site Demonstration Test Results Summary

	8/1/87	8/2/87	8/3/87	8/4/87
Waste Feed Characteristics				
Moisture, wt. %	16.63	16.06	14.24	14.37
Ash, wt. %	69.77	69.80	72.40	75.21
HHV, Btu/lb	2064	1639	1728	2018
PCB, ppm	5.850	3.850	5.340	3.480
Pb, ppm	5900	4900	5000	4400
Chlorine, ppm	< 1000	< 1000	< 1000	< 1000
Sulfur, ppm	25300	17800	18900	16700
Chlorine (as HCl), kg/hr	< 5	< 5	< 5	< 5
Sulfur (as SO_2), kg/hr	200	132	138	125
EP Tox (Pb), mg/L, ppm	27.00	29.00	- -	24.00
TCLP (Pb), mg/L, ppm	8.60	2.50	3.00	3.50
Stack Gas				
HCl, ppmv	< 0.051	0.60	0.22	0.20
SO_2, ppmv	0.99	41.80	0.96	0.91
HCl, g/hr	< 0.8	8.60	2.90	2.70
SO_2, g/hr	27.40	1070.0	22.0	20.6
Particulates (@ 7% O_2), mg/dscm	358	211	173	171
PCB, µg/hr	57.70	174.50	58.10	126.20
Ash				
PCB, ppm	0.01	0.240	0.900	0.540
Pb, ppm	7100	6000	6400	6200
EP Tox (Pb), mg/L, ppm	25.0	28.0	36.0	36.0
TCLP (Pb), mg/L, ppm	0.01	0.01	0.02	0.01
Operating Conditions				
Waste Feedrate (avg. daily), kg/hr	3328	3287	3626	3600
DRE (PCB), wt. %	99.99	99.99	99.99	99.99
DE (PCB), wt. %	99.88	93.77	83.15	84.48
Primary Combustion Chamber				
Exhaust Temperature (avg.), F	1797	1836	1922	1885
Residence Time, min.	19	19	18	19
Secondary Combustion Chamber				
Chamber Temperature (avg.), F	1886	1887	1889	1907
Residence Time, sec.	> 3	> 3	> 3	> 3
Acid Gas Removal Efficiency, wt. % SO_2	> 99.9	> 99.1	> 99.9	> 99.9

emissions were less than 1100 g/hr, with an average 149 kg/hr SO_2 feedrate giving an average removal of SO_2 in excess of 99%. SO_2 is more difficult to remove than HCl in a caustic scrubber, and the tests show that HCl removal should be in excess of the 99% determined for SO_2 removal.

Particulate Emissions

The particulate emissions during the first day were 358 mg/dscm. The unit was cleaned and mechanical adjustments were made resulting in an emission rate of 211 mg/dscm during the second day. The emissions during the third day were 172 mg/dscm (average of duplicate measurements). These values exceeded the RCRA standards during two of the four sampling periods. Particulate emissions were about 60% lead, when analyses of all samples were averaged.

Leaching Characteristics

The solid waste feed, furnace ash, and scrubber effluent solids were subjected to the EP Toxicity and proposed TCLP tests to evaluate the toxicity characteristics of these materials.

The EP Toxicity and the TCLP data present a contradictory picture regarding leaching of metals. The EP Toxicity data did not indicate that the process "encapsulates" or ties up heavy metals (lead) in the ash to prevent leaching. The EP Toxicity data show that lead content in the ash was 30 ppm and exceeded the 5 ppm toxicity characteristic standard. The measured lead content of leachates for feed material and ash are almost equal, indicating that the process appears not to affect leaching characteristics for lead.

In contrast to the EP Toxicity data, the TCLP data show that the lead content for both the feed and ash were less than the proposed toxicity characteristic standard

of 5 ppm. Measured lead concentrations were an order of magnitude lower in the TCLP leachate (about 2 ppm compared to about 30 ppm for EP Toxicity).

The significant differences in results from these two analytical techniques have been documented in a recent Oak Ridge National Laboratory report (ORNL, "Leaching of Metals from Alkaline Wastes by Municipal Waste Leachate," ORNL/TM-11050, March, 1987). It appears that the differences in the test procedures and alkalinity of the matrix provide a difference in the pH environment that is sufficient to affect the solubility and leachability of heavy metals, particularly lead.

Products of Incomplete Combustion

Small quantities of products of incomplete combustion (PICs) were identified in the sampled streams from

SHIRCO INCINERATION SYSTEM (continued)

the unit. No polychlorinated dibenzo-dioxins (PCDDs) or polychlorinated dibenzofurans (PCDFs) were identified in any of the sampled streams above detection limits with the exception of trace quantities (2.1 ng) of tetrachlorodibenzofuran (TCDF) found in the stack gas sampled on August 2.

Low levels of some semivolatile organic compounds were identified in all streams. These compounds were primarily phthalates, which may be the result of contamination from plastic components in the process, sampling equipment, or laboratory apparatus. Other semivolatile compounds included aromatic, polyaromatic, and chlorinated aromatic hydrocarbons. Low levels of pyrene, chrysene, anthenes, naphthalenes, and chlorinated benzene were identified in the waste feed stream; although possible PICs, their presence must be discounted to some extent, because they were originally introduced into the unit with the waste feed.

Low concentrations of volatile organics were measured in the stack gas and included halogenated methanes, chlorinated organics, and aromatic hydrocarbons including BTX compounds. No volatile organics were identified in the water streams. Low levels (ppb) of chlorinated hydrocarbons and BTX compounds were measured in all solid streams. Low levels of BTX compounds, carbon disulfide, chloroform, ditri-chlorofluoromethane, and trichloro-fluoromethane, dichloroethane, and trichloroethane, and methylene chloride were identified in the waste feed. Methylene chloride, a solvent used during testing, was also detected in laboratory and field blanks. These compounds, although possible PICs, must also be discounted to some extent based on their introduction to the unit from an external source and because of possible contamination.

Ambient Air Sampling and Mutagenic Testing

Ambient air monitoring stations placed upwind and downwind of the Shirco unit were designed to collect airborne PCB contaminants. Based on the downwind sampler data, it appears that the Peak Oil site boundaries limited the location of the downwind sampler to an area that was significantly exposed to fugitive emissions during the transport of ash from the ash pad to the ash storage area.

Samples of the waste feed and ash were collected on August 2 and forwarded to the EPA Health Effects Laboratory, Research Triangle Park, North Carolina for mutagenic testing. The results of these tests indicate that although the samples contain hazardous contaminants, they are not mutagenic based on the standard Ames Salmonella mutagenicity assay.

Cost/Economic Analysis

Several cost scenarios examined were based on a model for a Shirco unit operation equivalent in processing capacity to the unit that operated at Peak Oil, and on cost data available from Shirco and other sources. The economic analysis concludes that in using currently available Shirco transportable infrared incineration systems, commercial incineration costs will range from an estimated $196 per ton for a Shirco unit operation at an 80% on-stream capacity factor to an estimated $795 per ton for the operation at the Peak Oil site at a 19% on-stream capacity factor. A normalized total cost per ton of $425 represents a more realistic interpretation of the costs accrued to the Peak Oil cleanup action based on a 37% on-stream capacity factor.

Unit Problems

A review of the Haztech, EPA Technical Assistance Team (TAT), and EPA logbooks and progress reports, plus discussions with unit and project personnel, provided a summary of mechanical and operating problems encountered in this first application of a full-scale commercial Shirco incineration system at a Superfund site. These problems were categorized by unit operating sections, and a profile of the major problem areas within the unit were defined and analyzed to ascertain the reasons for and possible solutions to these specific operational difficulties. The review revealed that materials handling and emissions control were the most significant problem areas affecting operation of the unit. Prior to the operation of such a unit, extensive pretest analysis should be conducted on the waste feed matrix. The characteristics of the feed, including the nature of contaminants plus the feed's effect on incineration system chemistry, must be defined to allow appropriate assembly of the unit. The unit must be equipped with the proper feed preparation system and materials handling capabilities and adequate emissions control capacity and effectiveness. At the Peak Oil site, the solidified sludge feed continually

agglomerated, clogged, bridged, and jammed feed preparation and handling equipment. The high levels of lead contaminant and the excessive carryover of calcium and magnesium salts were a continuous source of problems for the emissions control system, which had difficulty in meeting stack emissions criteria.

Conclusions and Recommendations

Based on the above data and discussions, the following conclusions and recommendations can be made concerning the operation and performance of the transportable Shirco infrared thermal destruction system.

1. The unit achieved DREs of PCBs greater than 99.99%. Detection limits were used for this calculation so actual DREs were greater.

2. The unit achieved DEs of PCBs ranging from 83.15 to 99.88%. The unit was operated to produce an ash that contained 1 ppm or less of PCB.

3. Acid gas removal efficiencies were consistently greater than 99%. Particulate emissions during two days of testing were 358 mg/dscm and 211 mg/dscm, which contained 60% lead. The unit's emissions control system experienced particulate removal problems due to a combination of excessive fines carryover from the waste feed matrix and scrubber-washer and an overall emissions control system design that was not able to operate efficiently at abnormally high particulate loadings. As a result, two of the four samples taken exceeded the 180 mg/dscm RCRA standard.

   Pretest analysis of the waste feed and its combustion and emissions control chemistry and mechanisms must be performed to identify potential emissions control problems. A more flexible and adaptable emissions control system should be developed that can respond to and control a wider range of particulate and stack gas flows.

4. The furnace ash failed to meet the toxicity characteristic standard for lead for the EP Toxicity Test Procedure. Although the ash passed the similar standard for the proposed TCLP, its failure under EP Tox indicates that the unit did not immobilize lead in the ash product.

SHIRCO INCINERATION SYSTEM (continued)

5. Small quantities of PICs were identified in the sampled streams from the unit. In addition to trace quantities of TCDF on one sample, low levels of semivolatile compounds, including aromatic, polyaromatic, and chlorinated aromatic hydrocarbons were identified. Low concentrations of a broader range of volatiles including halogenated methane, chlorinated organics, and BTX compounds were also identified.

6. Ambient air monitoring stations detected quantities of PCBs, which appear to be caused by the wind transport of ash resulting from the nearby roadway. Waste feed and ash samples were not mutagenic based on the standard Ames Salmonella mutagenicity assay.

7. Overall costs ranged from $196 per ton with the unit operating at an 80% on-stream capacity (292 days per year) to $795 per ton with the unit operating at a 19% on-stream capacity (70 days per year). A normalized cost per ton for the Peak Oil cleanup was estimated at $425.

8. In addition to the particulate emissions control system problems, waste feed handling and materials handling problems consistently affected the unit's ability to treat the waste feed at design capacity. Pretest analysis of the waste feed and its handling characteristics must be performed to identify and design for any potential materials handling or feeding problems that the waste matrix may present at a specific site.

Appendix A.1.c

SUPERFUND SITE APPLICATIONS

Experience in Incineration Applicable to Superfund Site Remediation

Notice

The information in this document has been funded wholly or in part, by the U.S. Environmental Protection Agency (EPA) under Contract No. 68-03-3312, Work Assignment No. 1-03, to PEER Consultants, P.C. This document has been subject to the Agency's peer and administrative review and has been approved for publication as an EPA document. Mention of trade names or commercial products does not constitute endorsement or recommendation for use.

Acknowledgments

This document was prepared under Contract No. 68-03-3312 by PEER Consultants, P.C., under the sponsorship of the U.S. Environmental Protection Agency. Clarence A. Clemons of the U.S. EPA, Office of Research and Development, Center for Environmental Research Information was the project officer responsible for the preparation of this document. Special acknowledgement is given to H. Douglas Williams, Donald A. Oberacker, Frank Freestone, Robert Mournighan, Ivars Licis, Gregory Carroll and Robin Anderson of the U.S. EPA for their assistance and comments and to Fred Hall of PEI Associates, Inc., Joseph Santoleri of Four Nines, Inc., Wayne Westbrook of Research Triangle Institute and Lawrence Doucet of Doucet and Mainka, P.C. for the technical review of early drafts of this report. Participating in the development of this document for PEER Consultants, P.C. were Richard E. Frounfelker, Edward J. Martin and Joseph T. Swartzbaugh.

CHAPTER 1

Introduction

The Superfund Amendment and Reauthorization Act (SARA) of 1986 reauthorized the Comprehensive Environmental Response, Compensation and Liability Act (CERCLA), and together they are popularly known as the Superfund program. When SARA was enacted, it specifically emphasized the use of *permanent* remedies in effecting site cleanups. As a result, technologies which achieve detoxification of waste through destruction are being chosen for Superfund actions at an increasing frequency. Incineration, which is technology utilizing an integrated system of components for waste preparation, feeding, combustion and emissions control, has been proven to achieve acceptable levels of destruction of the organic portion of hazardous wastes. Therefore, incineration is being proposed and adopted for the remediation of several Superfund sites. A review of the Superfund Records of Decision (ROD) indicated that as of February 1988, 34 National Priority List (NPL) sites had selected thermal destruction as a possible remediation alternative.

Clearly, the application of incineration in the Superfund program is in the planning stage. Full scale field applications of this technology on-site at Superfund sites is now in the planning and initial implementation stages. However, off-site incineration of waste from Superfund actions and from analogous RCRA Corrective Actions are occurring. In addition, the EPA's Superfund Innovative Technology Evaluation (SITE) program has demonstrated the application of infrared technology. The EPA's mobile test incinerator has been used for large scale pilot evaluations of on-site incineration, and the EPA Combustion Research Facility has demonstrated the application of rotary kiln technology to the remediation of specific types of wastes found at Superfund sites. From these early applications and from the broad experience gained in the destruction of hazardous wastes at commercial incinerators for purposes of RCRA compliance, some useful information and guidance can be obtained which is directly applicable to the use of incineration for the remediation of Superfund sites.

It should always be remembered that the application of new technology, as well as the utilization of established technology for new purposes always exposes new and unexpected problems. It is common for engineers and technologists to speak of

this period of learning about, and addressing new problems, as "being on the learning curve." However, appropriate application of as much information as possible about analogous problems will serve to minimize the extent and duration of this learning curve.

Various types of incineration systems are commercially available and are well demonstrated technologies for the disposal of hazardous wastes in the RCRA context. Considerable design and operating experience exists, and both design and operating guidelines are available for the proper implementation and use of such systems. The most common incinerator designs incorporate one of four major combustion chamber design concepts: liquid injection, rotary kiln, fixed hearth, or fluidized bed.

If off-site incineration is chosen as a treatment option for the remediation of a Superfund site, then the commercially accessible, RCRA permitted facilities are candidates which are presently available for accepting wastes removed from Superfund sites. The operators of these facilities know their facilities well and are competent in their operation. It can be expected that the operators will carefully analyze the characteristics of any candidate wastes to be received prior to their agreeing to accept such wastes for destruction in their units. Note that the acceptance of a new waste by a commercial facility may require modification of their permit or other "nontechnical" concerns that the proposed facility is unwilling to deal with. Since there are, at present, a limited number of permitted incinerator facilities, a "sellers market" condition prevails and operators may pick and choose which waste streams they will accept. Thus, for sites proposing off-site incineration, the most important point of guidance that can be offered is that a clear-cut understanding of the nature of the waste and of the commercial facility's ability, as well as their willingness, to accept that waste must be agreed upon before final remedy selection.

On-site incineration will most often utilize a mobile or transportable technology rather than a unit to be permanently constructed at a site. The distinction between mobile and transportable units is that mobile units are truck mounted (on one or more semi-trailers) and can be rapidly connected and assembled in the field, often without removal from the truck beds. Thus, the design of mobile units is limited by the size of road-usable trailers, and this limiting restriction may not always be consistent with optimum incinerator design practice. In contrast, transportable units consist of a set of compatible modules which can be readily disassembled at one site, transported to a new site and reassembled on a concrete pad or other stable base and utilized for several weeks or months as necessary for site remediation activities. Design limits on a transportable unit are more forgiving and, in general, the transportable units can be constructed of stronger materials and can be designed to more closely meet optimum incinerator design requirements. Capacities of mobile and transportable technologies presently available are in the 2-10 tph range, although some of the operating experience reported herein was gained in first generation units having capacities of approximately 500 pounds per hour. Such units can, and have demonstrated, adequate destruction of hazardous wastes in tests and trials performed to date.

Conditions for effective disposal of hazardous wastes in incinerators have been well established and are presented in Table 1-1. Note that tests on many different types of combustors being fed different types of wastes have all shown consistent results (see Table 1-2). For incineration systems, the destruction and removal efficiencies observed have been between 99.99% and 99.9999%, as required for destruction of specific toxic materials under RCRA and TOSCA. In addition, control efficiency requirements for HCl emissions also have been met. In fact, one of the most difficult requirements to meet in any test burns of Superfund wastes was the particulate emission standard of 0.08 gr/dscf. However, this difficulty was probably related more to the fact that tests reported herein were performed in

an attempt to gain a range of design information rather than to actually utilize specific incinerators for long term remediation of Superfund sites. Properly designed air pollution control devices can meet current state-of-the-art emission levels of 0.03 to 0.04 gr/dscf. It should also be noted that it is not always possible to demonstrate that solid residues from hazardous waste incineration would pass tests for delisting purposes.

Although only limited information exists about the specific application of incineration technology to Superfund wastes, guidance does exist about the selection, operation and implementation of incineration technology, and this guidance is directly applicable to utilization of such technology in Superfund site remediation. In general, this guidance exists in the areas of identifying the type of data and information necessary to properly select and evaluate incinerator technologies; probable best operating practice for incinerators receiving wastes (such as contaminated soils) which could be expected from Superfund site remediation; and specific regulatory and safety concerns which will effect the timing and cost of the implementation of incineration technology to Superfund site cleanups. These topic areas are the foci of the following chapters of this document.

Table 1-1. General Combustion Chamber Conditions Favorable to the Destruction of Hazardous Waste

Combustion chamber temperature level:	820°C - 1500°C (1500°F − 2700°F)
Residence time:	0.2 - 6.5 sec for gases and liquid/vapors several minutes to 1/2 hour for solids/sludges
Excess combustion air:	60 - 130%

(Oppelt, 1986)

Table 1-2. Stack Emission Characteristics Observed During Successful Combustion of Hazardous Waste

Stack gas characteristics:	
Oxygen:	8 - 15%
Carbon dioxide:	6 - 10%
Carbon monoxide:	0 - 50 ppm (rare exceptions up to 500 ppm or more)
Total hydrocarbons:	0 - 5 ppm (rare exceptions up to 75 ppm or more)
Stack particulate emissions:	
Units without control devices:	60 - 900 mg/dscm (0.03 - 0.39 gr/dscf)
Units with control devices:	20 - 400 mg/dscm (0.01 - 0.17 gr/dscf)

(Oppelt, 1986)

CHAPTER 2
Background Information on Incineration Technology

This chapter presents brief descriptions of the types of incinerator technology presently in use (or planned to be used in the near futu:a) for CERCLA site remediation. It is not the intent of this document to serve as a design guide or a textbook on incineration. There are several excellent texts and reference documents presently available which supply a wealth of information about different incineration technologies. The technologies described herein include the most widely used technology for hazardous waste incineration, namely the rotary kiln incinerator. In addition, brief descriptions of liquid injection incinerators, of fluidized bed and circulating bed combustors and of infrared incinerators are presented.

An incinerator is a complex system of interacting pieces of equipment and is not just a simple furnace. A generic concept diagram indicating the type of operations which make up an incineration system is presented in Figure 2-1. In considering whether an incinerator can combust a specific hazardous waste stream, one must take into account the waste feed mechanism of the incinerator, the size and configuration of the furnace itself, the nature of the furnace's refractory material and the design of its ash handling mechanism. Many operators and hazardous waste incineration experts consider that the feed mechanism is the most critical aspect of an incinerator. This is because experience has shown that the feed mechanism is one of the major sources of problems in actual operation of incinerators. It is obvious that, if a waste is to be burned, there must be a viable mechanism for introducing the waste into a combustion system. The limitations of the feed mechanism as well as limitations of the ash removal mechanism set the requirements for any preprocessing of the waste. In fact, because Superfund sites often contain such a varied mix of waste types and matrices, preprocessing requirements may be so extreme that the preprocessing could turn out to be the most extensive (and possibly the most expensive) operation in the entire remediation process. Feed mechanisms for the different types of incinerator systems are presented in the discussion of each specific incinerator type.

Any discussion of incineration as an option for treatment of wastes, usually includes mention of the three T's of incineration: time, temperature and

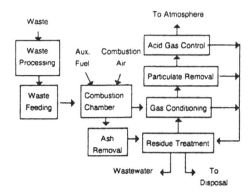

Figure 2-1.   Incineration system concept flow diagram.

turbulence. Specifically, these are the temperature at which the furnace is operated, the time during which the combustible material is subject to that temperature and the turbulence required to ensure that all the combustible material is exposed to oxygen to ensure complete combustion. Different combustion systems utilize different mechanisms for addressing the three T's.

Rotary Kiln Incineration

A rotary kiln incinerator is essentially a long inclined tube that is mounted in such a way that it can be slowly rotated. Rotary kilns are intended primarily for combustion of solids but liquids, sludges and gases may be co-incinerated with solids. Wastes and auxiliary fuels are introduced into the high end of the kiln and the kiln's rotation constantly agitates any solid materials being burned to expose the solids to oxygen and to improve heat transfer. However, because the solids are being agitated, there is a high likelihood that solids will be entrained in the gas stream and this will require post-combustion control. Any ash residues from the combustion process are discharged and collected at the low end of the kiln. Exhaust gases from a kiln typically pass to a secondary combustion chamber or afterburner for further oxidation. Exhaust gases typically require acid gas removal and particulate removal. Note that the ashes and air pollution control process residues from an incinerator may require further treatment before being deposited in the land. Most types of solid, sludge and liquid organic wastes can be combusted with this technology. Solids are usually introduced into the rotary combustor by a ram feeder or an auger conveyer feeder. Sludges are introduced into the combustion chamber by a cooled tube injected into the flame zone and liquids are injected either into the primary chamber or the secondary chamber by means of liquid injection nozzles.

Fluidized Bed Incineration

Fluidized bed incinerators utilize a very turbulent bed of inert granular material to improve the transfer of heat to the waste streams to be incinerated. Air is blown through the granular bed material until the granules are suspended and able to move and mix in a manner similar to a fluid. The bed is preheated using an auxiliary burner and wastes and auxiliary fuel are introduced in a manner that will facilitate their intimate contact with the heated bed particles. The technology is particularly appropriate for liquids and sludges and can be used for solids having small particle sizes. However, fluidized bed combustion is not appropriate for bulky or for viscous wastes. Wastes having high sodium levels or wastes which are not uniformly mixed can cause bed agglomeration problems (i.e., particles in the bed will stick together and become too large to be able to be fluidized). Fluidized beds require frequent attention for maintenance and for cleaning of the bed. Typically these units operate at low combustion temperatures (e.g., 750-1000°C) and therefore, organic wastes that are especially refractory may not be fully destroyed.

The circulating bed combustor (CBC) is an off-shoot of the fluidized bed combustor. Circulating bed combustor systems operate at higher velocities and with finer particles in the bed than fluidized bed systems. Bed materials are carried up through the first combustion chamber and typically are passed through to a second, more quiescent chamber, where they are collected for reinjection into the first chamber. The use of higher velocities and finer bed materials permits the construction of a unit that is more compact while still achieving great turbulence. Manufacturers claim that the high turbulence achieved allows especially efficient destruction of all types of halogenated hydrocarbons including PCB's and other aromatic materials at temperatures lower than 850°C. The required 99.9999% DRE has been demonstrated in a trial burn of a commercial size CBC unit (Vrable, 1985). Wastes must be fairly homogeneous in composition when fed to a CBC since most designs utilize only a single feed point.

Liquids and sludges can be introduced into fluidized bed combustors by injection nozzles or cooled wands into locations within the bed. Solid

materials can be introduced pneumatically or through an air lock into the freeboard above the bed in the fluidized bed combustor or at a point approximately 1/3-1/2 of the height of the first chamber in the circulating bed combustor.

Liquid Injection Incineration

Liquid materials with solids contents below 10-12 percent can be combusted in a relatively simple furnace which usually has no provisions for ash capture other than fly ash removal. Typically, combustible liquid waste material is introduced to the combustion chamber by means of specially designed nozzles. Different nozzle designs result in various droplet sizes which then mix with air and auxiliary fuel as needed. Pre-treatment such as blending and filtration, may be required for feeding some wastes to specific nozzles in order to achieve efficient mixing with the oxygen source and to maintain continuous homogeneous flow of waste into the combustion chamber. Liquid injection incineration can be applied to all pumpable organic wastes, including wastes with high moisture content. However, it is absolutely necessary that care be taken in matching the waste characteristics to specific nozzle design. Solids in the waste stream can easily clog nozzles and it should be noted that abrasive solids (even at low concentrations) can cause pump wear, even in progressing cavity pumps which are specifically designed to handle high solids content liquids.

Infrared Incineration

Infrared radiators can be used as a heat source to destroy hazardous wastes. The infrared incineration system recently tested within the Superfund Innovative Technology Evaluation (SITE) Program was made up of a primary chamber consisting of a rectangular carbon steel box lined with layers of a lightweight ceramic fiber blanket. Energy is provided to the system as infrared radiation from silicon carbide resistance heating elements. The wastes to be destroyed are conveyed through the furnace on a open wire belt. This technology is intended to be used to treat solids, sludges and contaminated soils. Any solid materials and soils must be crushed or shredded and then screened to a size less than approximately 1 inch. Liquid and gaseous injection feed systems may be added to this technology. Solid waste is carried through the furnace on the belt until reaching the discharge end of the furnace where it drops off into a hopper. These incinerators typically require further treatment of the gaseous emissions, such as secondary combustion, acid gas control, and particulate removal.

CHAPTER 3
The Effect of Waste Characteristics on Technology Selection

The evaluation and selection of a technology to be used in a CERCLA remedial action must be based on complete and accurate analyses of the contaminants that exist at the site. The effectiveness of any technology, including incineration, is dependent upon the waste's chemical and physical characteristics and often upon the uniformity of those characteristics. Therefore in order to assess whether incineration is a valid treatment technology to be utilized in the remediation of a Superfund site, it is necessary to have a thorough chemical analysis of the characteristics of the wastes to be disposed in the incinerator. It is also necessary to have a complete characterization of the physical form of the waste and to have a sampling plan adequate to determine the uniformity of the waste material to be fed to the incinerator.

Incineration is applicable to the full range of organic waste constituents. Information published in the RODs for the 34 sites at which incineration is an alternative indicated that VOCs and PCBs were contaminants in 50% of the sites. Other major contaminants were reported as metals (41%), organics (35%), PAH compounds (27%) and inorganics (18%). No single contaminant or contaminated media was common to all sites. Note that wastes containing organics but which also contain metals may be burned, but in such cases, care must be taken to properly control stack emissions as well as to properly dispose of all residues.

Characterizing Waste Constituents

There are several purposes for which the analytical data will be utilized in selecting and evaluating incineration. An understanding of all of these purposes will better enable the site engineer to plan the sampling and analytical program necessary to allow an appropriate evaluation of the incineration option. Obviously the first and foremost purpose of any analytical plan is to assess whether or not the waste present is amenable to destruction or detoxification by thermal processes. Further, information is necessary to determine the incinerator design to be utilized, as well as to determine the appropriate feed mechanism, stack gas cleaning

equipment, and solid residue removal equipment that will be necessary. Thus, it is necessary to know:

- The organic hazardous constituents of the waste, especially whether the waste contains highly toxic materials such as PCBs or chlorinated dioxins and furans.

- The more refractory of the organic hazardous constituents, in order to properly design a test burn protocol if and when the assessment of incineration as an option proceeds to that point.

- The toxic heavy metal constituents of the waste stream, especially the more volatile of these metals such as lead, in order to assess the need for appropriate air pollution control devices should the incineration option be chosen. Certain other inorganic constituents must be identified. For example, high levels of sodium can cause serious problems of bed agglomeration in a fluidized bed combustor.

- The presence and concentration of any acid gas precursors, such as sulfur or halogens in the waste material, so that appropriate air pollution control measures can be incorporated into the incineration option.

- The moisture content of the material to be incinerated. High moisture content wastes can be incinerated, but the amount of moisture present has a significant impact on furnace volume requirements and especially on auxiliary fuel requirements. Furthermore, the impact of any high moisture levels in the exhaust gas stream must be taken into consideration in determining the proper downstream air pollution control equipment.

It is always important that any remediation option be evaluated on the basis of *actual* analytical data for the waste as it exists at the site. There have been situations where inappropriate or less than optimum treatment strategies have been utilized because improper or incomplete analytical programs were established and the true nature of the waste to be remediated was not known. One example of such a situation is the Peak Oil Site in Florida at which the decision was made to perform field-scale evaluation of an infrared incineration system based on knowledge of the site's history. Since some PCB-laden oils had been disposed at the site, it was believed that this was the most significant constituent in the waste stream. Therefore, the system vendor proposed the use of the infrared destruction technology as part of the Superfund Innovative Technology Evaluation (SITE) program and proceeded to install and implement equipment prior to a full evaluation of the waste material. Analyses of the waste feed performed at the time of incinerator process evaluation showed that the total PCB content in the waste material averaged about 5 mg per gram and other toxic organic materials were also present. Incineration would usually be considered appropriate for such wastes. However, the lead content in the waste was approximately 0.5 percent. The system was operated without proper planning for, or control of, volatile metal emissions. Measurement of the lead in the stack emissions from the process showed releases of 0.78 to 2 kilograms per hour of lead from the system during operation. Zinc and mercury emissions were also detected, although at much lower levels than the lead emissions. (Oberacker, 1988)

At the Laskan Poplar Site in Ashtabula, Ohio (Engineering Science, 1987), PCB-contaminated oils and oily sludges were evaluated as being appropriate candidates for incineration. However, the lead content seen in the oil ranged from 30-543 mg per kilogram and the lead content in the oily sludges ranged from 69-12,400 mg per kilogram. Although it was originally believed by the site engineer that the high lead content would not require special incineration considerations, it was found that off-site commercial incineration facilities would not accept the waste sludges because of limiting condition in their existing RCRA permits or because of concerns that lead concentration in their scrubber water effluent would exceed the requirements of the incinerator facility's NPDES permit. Therefore, the decision was made to incinerate the waste sludges on site using a commercially available, transportable incinerator. Note that this decision was made for regulatory concerns and not on a strictly technical basis.

Determining the Physical Characteristics of the Waste

Determination of the form of the waste to be treated is much more involved than a more visual assessment that the waste is a solid, liquid or a sludge. The physical characteristics of a waste determine whether and how a waste can be moved and fed into any treatment system, whether the treatment be chemical, biological or thermal treatment. The most common types of Superfund wastes which will be candidates for treatment by incineration are contaminated soils and liquid wastes. Different problems exist when characterizing such different wastes.

Characterization of the contaminants in the soil across a Superfund site is complex because the waste and waste matrix are probably not homogeneous (either horizontally or vertically) in the soil at the site. Conversely, liquid organics in tanks, drums, and lagoons are easier to identify and are more homogeneous, at least within each container. Liquids can be pumped and mixed to produce a nearly homogeneous feedstock prior to injection into an incinerator. Liquid feedstocks are therefore easier to control than soil feedstocks and result in more uniform operation.

Both liquids and soils can be expected to have some "trash" intermingled. In this context, "trash" is any major extraneous material (even if inert) which can complicate the handling and feeding of the waste. Examples include rocks or small pieces of wood in liquids, or wire, scrap metal, pieces of rubber or large rocks and boulders in contaminated soils.

As shown in Figure 3-1, 80 percent of the Superfund sites recommending incineration have contaminated soils to treat. The material handling of soils can cause significant problems to the incinerator (although many of these problems are common to other remedial actions). Of the sites reporting solid wastes, forty percent have less than 10,000 cy of solid waste, 32% have between 10,000 and 50,000 cy and only 12% have more than 100,000 cy to incinerate.

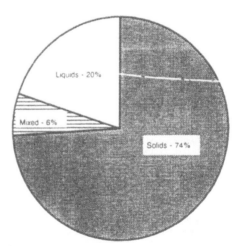

Figure 3-1. Waste matrices as reported in Records of Decision for Superfund sites recommending incineration.

Soil characteristics affect the method of excavation, the possible requirement for and selection of pretreatment technologies, and the selection of the incineration technology itself.

1. Soils with a high moisture and clay content can agglomerate into balls or clumps. Reduction of the size of this agglomerated material is very difficult. A large percentage of this type soil must be screened out of the feed stream for many types of incinerator technologies.

2. The contaminants are usually not uniformly spread throughout the soil. Therefore, the concentration of contaminant in the feed to the incinerator will vary as the excavation moves across the site. Overloading of the incinerator feed with organic contaminant can reduce the DRE of the contaminant, while feeding very low concentrations of contaminant would waste time and money.

3. Moist or "sticky" soils affect the waste handling system. Wet soil can stick and jam numerous types of conveying, screening and feeding equipment. Highly variable moisture content can affect the attainable DRE in the combustor.

4. Gravel and rocks in the soil can cause problems with material handling equipment. Screw and belt conveyors are affected by the abrasive, sharp edges as well as by the size of different types of gravel. Screens can be plugged or damaged by a combination of gravel, rocks, and other debris.

5. High loam content soils can produce large amounts of contaminated fugitive dust and VOC emissions during handling and conveyance to the incinerator.

Sampling to Characterize the Variability of the Wastes

It is probable that the characteristics of any waste at a Superfund site will be highly non-uniform. Liquids and sludges in surface impoundments will have stratified, solids will have settled and some liquids will have seeped into soils. Wastes disposed in the ground were probably not applied in a uniform manner, and the liquid and soluble portions of such wastes will migrate from the original point of deposit with the passage of time.

It is a critical need in determining waste feed rate to an incinerator to attempt to maintain relatively constant heat input. Non-uniform feeding can result in carbon monoxide spikes and insufficient oxygen to attain proper destruction of the organic constituents. Thus, it is necessary for proper incinerator operation that some measure of uniformity of the waste in the waste matrix be known. If the wastes to be fed to an incinerator are fairly uniform, then uniform feed and heat input rates can be attained through operator training. However, if the waste is highly variable, then there is probably some need for mixing, blending,

shredding, or even more extensive pretreatment of the waste material to achieve higher uniformity prior to feeding it into the incineration system.

Non-uniformity of wastes to be fed to an incinerator is an especially important concern for fluidized bed and circulating bed combustors. Unexpected changes in the heat input rate, as well as unexpected changes in the inorganic constituents in a waste stream, can result in bed agglomeration or in bed material degradation. Bed agglomeration problems were observed on a fairly regular basis during extended test burnings of hazardous waste at the fluidized bed combustor system at the Franklin, Ohio, Solid Waste Resource Recovery Demonstration Project facility. Although the waste being fed was a blended liquid waste (more typical of a RCRA waste), slight changes in waste composition and especially in moisture content of the input waste stream resulted in several bed agglomeration episodes. Finally, the plant operator decided to accept no more of the liquid industrial waste for incineration in the system (Ackerman, 1978).

CHAPTER 4

Operating Experience

Operating Concerns Related to Waste Feed

The incinerator system itself must be selected to "fit" the waste type and as-found matrix of the waste in the field. For example, a rotary kiln incinerator can be charged with hazardous wastes which are widely variable in nature. The size of kiln-fed wastes is only limited by the size of the opening in the unit. However, other types of incinerators may be very limited; for example, the waste fed to a fluidized bed type must be very homogeneous. In such a system, some form of a pretreatment is necessary to assure that the feedstock is homogeneous and compatible with the incinerator technology.

Pretreatment, including screening, shredding, metal separation and size separation, dewatering, or other processing, is often costly. Pretreatment costs can even equal or exceed the cost of the primary treatment itself. Therefore, pretreatment costs must be considered when making the project cost estimates. Failures in pretreatment equipment have caused significant system downtime. Almost continuous attention to these devices is required, especially when the original feed material is nonhomogeneous.

Numerous problems have been identified with waste feed systems:

- Soil has been fed into rotary kilns by means of screw conveyors. Screw conveyors work well for uniform sandy soils but they have experienced problems when conveying gravel-bearing soil and soil mixed with scrap. Excessive wear of the screws can require weekly shutdown for replacement. Jams are often caused by pieces of metal wire or other scrap.

- Soils have sometimes been fed into an incinerator by means of a ram feeder. The fines in some soils (e.g., sandy river loam) can spill around the ram. The volume of spilled material can become so great that it can prevent the ram from fully retracting past the feed hopper. To correct this, a small tubular conveyor (screw) can be installed beneath the loading hopper to remove the spillage

buildup from behind the ram and carry the fines into the incinerator.

- Dump trucks used to haul excavated soil to an incinerator feeding system are frequently lined with large sheets of plastic film. The plastic is then dumped with the soil. During pretreatment processing, the plastic film has been found to cause jamming of the feed system. The plastic does not shred into small pieces but rather shreds into strips and then rolls up into balls. These become entangled in screens and around belt pulleys. Introduction of large pieces of plastic into the incinerator causes a rapid increase in temperature and decrease in excess oxygen. This could trigger a system shutdown.

- The use of shearing-type shredders can result in long strings of plastic, wire or cloth as well as large wood splinters. These stringers jam the system conveyor, screens and feeders.

- Sludges, oils, and tars typically are fed continuously to combustors by means of progressing cavity pumps. Some debris (sticks, rocks, and waste) can be handled by such pumps. However, abrasive debris can cause rapid wear in progressing cavity pumps. Soils and nonpumpable gels have been packed into fiber packs or plastic drums and then successfully fed one at a time through a ram into the combustor. Steel drums of waste have also been successfully fed one at a time into rotary kilns.

Sometimes, unique wastes present unique problems in feeding the waste to an incinerator. For example, the U.S. Army Toxic and Hazardous Materials Agency (USATHAMA) used a transportable rotary kiln to successfully burn explosive-contaminated soil dredged from red and pink water lagoons (Noland, 1984). The technology (developed by Therm-All, Inc.) had been used in industry for destruction of solid wastes. The normal screw feed system was not used due to fear of an explosion during the extruded plug feed process. Therefore, the soil was placed in combustible buckets and individually fed by a ram into the incinerator. The feed rate was 300 to 400 lb/hr and the operational temperature was 1200 to 1600°F in the kiln and 1600 to 2000°F in the secondary chamber. A DRE of 99.99

percent was reported. Analysis of the soil contaminants is presented in Table 4-1.

Table 4-1.   Analysis of Soil Matrices at a Site Contaminated with Explosive Waste

Descriptor		
Soil type	Sand	Clay
Moisture content	12-26%	25-30%
Ash content (as received)	44-83%	54-66%
Explosives content (dry basis)		
− TNT	9-41%	5-14%
− RDX	0.02%	3-10%
− HMX	Not detected	0.6-1.4%
− Other	0.03%	0.06%
− Total explosives	9-41%	10-22%
Heating value (as received)	50-2400 Btu/lb	600-1200 Btu/lb

(Noland, 1984)

Experience Gained in the Operation of Incinerators

General process control systems and strategies exist to control incinerator performance. However, none of the real time monitoring performance indicators appear to correlate with actual organic compound DRE. No correlation between indicator emissions of CO or unburned hydrocarbons and DRE has been demonstrated for field-scale incinerator operations, although CO may be useful as an estimator of a lower boundary of acceptable DRE performance. It may be that combinations of several potential real-time indicators (CO, THC, surrogate compound destruction) may be needed to more accurately predict and assure DRE performance on a continuous basis. In any case, even the approach of using real-time indicators to control DRE suffers a response time lag due to the time required to sample, analyze and transmit the sample analysis to the controller.

Typically, a high volume of excess air is passed through the system to control the chamber temperature and to assure sufficient oxygen for the combustion of the waste. However, large volumes of excess air reduces the system waste processing capacity and increases the need for auxiliary fuel to heat the air. Better design to control air leakage and better operator control of air levels will yield improved system effectiveness. The use of a burner that uses oxygen, rather than air, can reduce the effective volume of gases to be heated. This volume reduction effectively increases the throughput capacity of the system and lowers the auxiliary fuel cost. However, oxygen production systems can be very costly, and care must be exercised in properly performing cost analyses prior to their selection.

In the use of an incinerator for soil treatment, much of the energy produced is used for heating and/or drying the soil, since the amount of con-

tamination is usually much smaller than the mass of soil. Once the soil is dried and its contaminants volatilized, soil particles can readily be lofted to become particulate emissions which must be controlled. High solids processing in rotary kilns is accomplished using a downstream cyclone separator, or by using indirect calcining type kilns to reduce the solids carryover in the flue gases. The approach of using a cyclone separator has been used for wastes which are predominantly soil.

Sites with contaminated soil of 5000 tons or less can be considered small sites. At such sites, units with throughput capacities of less than 5 tons per hour (tph) can be used, even though the cost per unit processed is high: $1000 to $1500/ton. A site with soil volume of 5000 to 30,000 tons to be treated is a medium-sized site. System with capacities between 5 and 10 tph, can keep the time for remedial action short. The cost per ton for this size unit can range from $300 to $800 per ton. Sites with more than 30,000 tons of soil to be processed are considered large sites, and high-capacity technologies (greater than 10 tph) may be considered for application under these conditions. Costs of cleanups for such units could range from $100 to $400 per ton (Cudahey, 1987). The above cost information compares closely with actual bid prices for incineration projects as published in various news reports appearing in Hazardous Waste News, Waste Age and the Journal of the Air Pollution Control Association.

Compounds that are difficult to burn, such as PCBs and dioxin-contaminated soils, have been successfully incinerated to a DRE of 99.9999 percent. However, some tests have been conducted which have failed various regulatory requirements. Early tests of incinerators indicated less than 99.99 percent DRE, and in addition, some particulate emissions exceeded permit limits. Some recent testing of contaminated (PCB) soil showed 99.999 percent rather than 99.9999 percent DRE. Possible reasons for these apparent failures include faulty or contaminated equipment (Weston, 1988). In other tests, insufficient concentrations of input POHCs have resulted in failure to demonstrate the required DREs because of high detection limits in stack emissions analyses. Dioxin wastes, including dioxin-containing soils, often contain only low levels of dioxins and furans. Nonetheless, it is necessary to treat the entire waste matrix in order to manage these toxic compounds. Technology which is applicable to dioxin decontamination is expected to be applicable to Superfund site wastes, which may be mixed solids, sludges and liquids containing a range of organics. The summary of treatment processes in Table 4-2 also shows that potentially contaminated air stream cleaning residues are produced. These thermal treatment by-product streams require further treatment and thus additional costs are incurred.

Table 4-2. Technologies Applicable to Dioxin Decontamination

Process Name	Applicable Waste Streams	Stage of Development	Performance/Destruction Achieved	Cost	Residuals Generated
Stationary Rotary Kiln Incineration	Solids, liquids, sludges	Several approved and commercially available units for PCBs; not yet used for dioxins	Greater than six nines DRE for PCBs; greater than five nines DRE demonstrated on dioxin at combustion research facility	$0.25-$0.70/lb for PCB solids	Treated waste material (ash), scrubber wastewater, particulate from air filters, gaseous products of combustion
Mobile Rotary Kiln Incineration	Solids, liquids, sludges	EPA mobile unit is permitted to treat dioxin wastes; ENSCO unit has been demonstrated on PCB waste	Greater than six nines DRE for dioxin by EPA unit; process residuals delisted	NA*	Treated waste material (ash), scrubber wastewater, particulate from air filters, gaseous products of combustion
Liquid Injection Incineration	Liquids or sludges with viscosity < ssu (i.e., pumpable)	Full scale land-based units permitted for PCBs; only ocean incinerators have handled dioxin wastes	Greater than six nines DRE on PCB wastes; ocean incinerators only demonstrated three nines on dioxin containing herbicide orange	$200-$500/ton	Same as above, but ash is usually minor because solid feeds are not treated
Fluidized-bed Incineration	Solids, sludges	GA Technologies mobile circulating bed combustor has a TSCA permit to burn PCBs anywhere in the nation; not tested yet on dioxin	Greater than six nines DRE demonstrated by GA unit on PCBs	$60-$320/ton for GA unit	Treated waste (ash), particulates from air filters
High Temperature Fluid Wall (Huber AER)	Primarily for granular contaminated soils, but may also handle liquids	Huber stationary unit is permitted to do research on dioxin wastes; pilot scale mobile reactor has been tested at several locations on dioxin-contaminated soils	Pilot scale mobile unit demonstrated greater than five nines DRE on TCDD-contaminated soil at Times Beach (79 ppb reduced to below detection)	$300-$600/ton	Treated waste solids (converted to glass beads), particulates from baghouse gaseous effluent (primarily nitrogen
Infrared Incinerator (Shirco)	Contaminated soils/sludges	Pilot scale portable unit tested on waste containing dioxin; full scale units have been used in other applications; not yet permitted for TCDD	Greater than six nines DRE on TCDD-contaminated soil	Treatment costs are $200-$1,200/ton	Treated material (ash), particulates captured by scrubber (separated from scrubber water)
Molten Salt (Rockwell Unit)	Solids, liquids, sludges; high ash content wastes may be troublesome	Pilot scale unit was tested on various wastes – further development is not known	Up to eleven nines DRE on hexachlorobenzene; greater than six nines DRE on PCB using bench scale reactor	NA	Spent molten salt containing ash, particulates from baghouse
Supercritical Water Oxidation	Aqueous solutions or slurries with less than 20 percent organics can be handled	Pilot scale unit tested on dioxin-containing wastes – results not yet published	Six nines DRE on dioxin-containing waste reported by developer, but not presented in literature; lab testing showed greater than 99.99% conversion of organic chloride for wastes containing PCB	$0.32-$2.00/gal $77-480/ton	High purity water, inorganic salts, carbon dioxide, nitrogen
Plasma Arc Pyrolysis	Liquid waste streams (possibly low viscosity sludges)	Prototype unit (same as full scale) currently being field tested	Greater than six nines destruction of PCBs and CCl_4	$300-$1,400/ton	Exhaust gases (H_2 and CO) which are flared and scrubber water containing particulates
In Situ Vitrification	Contaminated soil – soil type is not expected to affect the process	Full scale on radioactive waste; pilot scale on organic-contaminated wastes	Greater than 99.9% destruction efficiency (DE) (not offgas treatment system) on PCB-contaminated soil	$120-$250/m³	Stable/immobile molten glass; volatile organic combustion products (collected and treated)

(Breton, 1987)

There are significant quantities of brominated wastes requiring disposal. Testing has been done using the EPA mobile incinerator, but DREs were not determined. Particulate results were consistently below RCRA standards of 0.08 gr/dscf. Rotary kiln temperatures were maintained within the operating range of the testing protocol. The waste burned was a brominated sludge, with bromine content between 1 and 12 percent. Moisture varied between 35 and 42 percent and the heating value between 3200 and 4600 Btu/lb. By-product streams were tested for delisting purposes, and the purge water, cyclone ash, and kiln ash failed the initial testing. Subsequent process modifications and operation changes resulted in successful delisting of these by-product streams.

Test burning of ethylene dibromide wastes was performed at a commercial hazardous waste incinerator for the purpose of assessing an innovative approach to bromine emissions control. Sulfur was added to the combustion chamber to force the formation of hydrogen bromide which is readily scrubbed from stack gases. Successful destruction of all POHCs were observed with no significant bromine stack emission (Oberacker, 1988).

Experience Related to Incinerator Emissions and Residues

Testing of incinerators has shown them capable of performing at or above RCRA and TSCA requirements for organic destruction or removal. Although hazardous wastes may be burned in incinerators, boilers and industrial furnaces, other thermal processes have also been shown to be highly effective. Limited data on incinerator ash and air pollution control residues suggest that organic compound levels are low and that destruction is the primary reason for high DREs, not removal. Metal concentrations in ash and residues vary widely, depending upon metal input rate and process operation (e.g., scrubber water recycle and make-up rates). Data in Tables 4-3 (organics) and 4-4 (metals) illustrate that the residual concentrations of toxic metals after leach testing are very low. These concentrations are acceptable values according to the EP Toxicity test requirements.

Incinerators equipped with appropriate air pollution control devices (APCD) can meet HCl and particulate emission requirements. Typical test results are given in Table 4-5. Air cleaning systems significantly affect the air flow balance and vacuum balance of the incineration system as a whole. These systems are important to meet air pollution permit requirements and must be carefully designed to operate within system constraints and design factors.

High solids content in wastes is of concern because of the potential for solids entrainment and carryover to downstream devices. As much as 10 to 25 percent of the soil solids fed can be carried over into the secondary combustion chamber and then into the APCD. The effect of such carryover can significantly reduce on-stream availability of the entire system.

In systems utilizing a horizontal, secondary combustion chamber with tangential firing burners, solids accumulating in the chamber can effectively "reshape" the gas flow path. This can result in poor gas mixing in the chamber, as well as increased load on the induced draft fan. Such systems need frequent cleanout to ensure operation as designed.

Lead concentrations in the kiln ash and blowdown solids for the tests on combustion of soil and soil plus sludge performed at the EPA's Combustion Research Facility (CRF) clearly reveal partitioning of this metal to the fly ash rather than the kiln ash. The partitioning is not evident for other metals. Kiln leachate analyses and blowdown liquids indicate concentrations typically less than 1 mg/L, or well below the EP toxicity test requirement. Antimony was evenly distributed between the particulate and vapor phases in flue gas samples taken at the afterburner exit. The same was generally true for arsenic, although a somewhat greater fraction was found in the particulate phase for a greater number of tests. Most, if not all, of the antimony and arsenic was found in the particulate catch of the sampling trains run at the scrubber exit. Comparable fractions of the input antimony and arsenic were accounted for in the scrubber exit flue gas and the scrubber blowdown. No clear dependence of the distribution of arsenic and antimony discharges with the primary test variables was apparent. When chlorine was present in the waste, the distribution of metals increased in the flue gases. The metals are reported to partition more into the vapor phase than into particulate in these situations. (Actually, the test protocol called for analysis of the impinger catch and of filterable particulates from EPA Method 5 sampling trains.) (Waterland, 1987b)

Often, air cleaning systems are not "matched" to system removal requirements. Contaminated soil or soil-laden wastes are likely to result in high solids in the off gases and may require high efficiency flue gas cleaning systems to remove particulate emissions. Air cleaning systems, especially scrubbers, generate large quantities of wastewater. Wastewater treatment facilities must be designed to "match" other system capacities.

Table 4-3. Concentration of Volatile and Semivolatile Organics in Incinerator Ash Residuals and their EP III Leachates

Site Number Stream Description	1 Kiln ash Concentrations[a,b]		2 Kiln ash Concentrations[a,b]		3 Kiln ash Concentrations[a,b]		3 Boiler ash Concentrations[a,b]		4 Cyclone ash Concentrations[a,b]	
	Ash (mg/kg)	EP III Leachate (µg/L)	Ash (mg/kg)	EP III Leachate (µg/L)	Ash (mg/kg)	EP III Leachate (µg/L)	Ash (mg/kg)	EP III Leachate (µg/L)	Ash (mg/kg)	EP III Leachate (µg/L)
Volatile Organics[c]										
Nominal Detection Limit	1	–	100	–	0.5	–	0.1	–	0.5	–
Chloromethane	–	–	–	–	–	–	–	–	–	–
trans-1,2-Dichloroethane	–	–	–	–	–	–	–	–	1.1	–
1,1,1-Trichloroethane	–	–	–	–	–	–	–	–	3.7	–
Trichloroethane	–	–	–	–	2.5	–	–	–	5.4	–
Tetrachloroethene	–	–	–	–	–	–	–	–	16	–
Toluene	–	–	–	–	–	–	–	–	6.4	–
Chlorobenzene	–	–	–	–	–	–	–	–	–	–
Ethylbenzene[d]	–	–	–	–	0.5	–	–	–	–	–
Carbon Disulfide[d]	–	–	–	–	2.8	–	–	–	–	–
2-Butanone[d]	34	–	–	–	–	–	–	–	–	–
4-Methyl-2-Pentanone[d]	–	–	–	–	–	–	–	–	–	–
Styrene[d]	–	–	–	–	4.3	–	–	–	–	–
Total Zylene[d]	–	–	–	–	1.5	–	–	–	–	–
Semivolatile Organics[c]										
Nominal Detection Limit	0.1	2	0.1	2	0.1	2	0.1	2	0.01	2
Acenaphthene	–	–	–	–	–	–	–	–	–	–
1,2,4-Trichlorobenzene	–	–	–	–	–	–	–	–	–	–
Fluoranthene	–	–	0.61	–	–	–	–	–	–	–
Isophorone	–	–	–	–	–	–	–	–	–	–
Naphthalene	–	–	–	–	0.17	–	–	–	–	–
2-Nitrophenol	–	–	–	–	–	–	–	–	–	–
4-Nitrophenol	–	–	–	–	–	–	–	–	–	–
N-Nitrosodiphenylamine	–	–	–	–	–	–	–	–	–	–
Phenol	–	–	0.35	–	3	116	–	–	–	–
Bis(2-ethylhexyl) phthalate	–	56	0.31	24	4.4	–	0.4	30	–	–
Benzyl butyl phthalate	–	–	–	–	0.28	–	–	–	–	–
Di-n-butyl phthalate	–	–	–	–	0.41	–	–	–	–	–
Di-n-octyl phthalate	–	–	–	–	0.76	–	–	–	–	–
Diethyl phthalate	–	–	–	–	–	6	–	–	–	–
Dimethyl phthalate	–	–	–	–	–	–	–	–	–	–
Benzo(b) fluoranthene	–	–	0.2	–	–	–	–	–	–	–
Chrysene	–	–	0.2	–	–	–	–	–	–	–
Anthracene[d]	–	–	–	–	–	–	–	–	–	–
Phenanthrene[d]	–	–	0.48	–	–	–	–	–	–	–
Pyrene[d]	–	–	0.66	–	–	–	–	–	–	–
Benzoic Acid[d]	–	–	–	–	–	–	–	–	–	–
2-Methylphenol[d]	–	–	–	–	–	–	–	–	–	–
2-Methylnaphthalene[d]	–	–	–	–	.12	–	–	–	–	–
Aniline[d]	–	–	–	–	–	–	–	–	–	–
Comments: Wet or Dry Ash	Wet		Wet		Wet		Wet		Dry	

(Van Buren, 1988)

[a] means not detected, hence less than nominal detection limit.
[b] Volatile organics data for EP III leachate not available.
[c] RCRA Appendix VIII unless otherwise noted.
[d] Not RCRA Appendix VIII compound.

Table 4-3.   (Continued)

Site Number	5		6		7		8		8	
Stream Description	Small Incinerator Bottom Ash Concentrations[a,b]		Incinerator Bottom Ash Concentrations[a,b]		Incinerator Bottom Ash Concentrations[a,b]		Kiln Ash Concentrations[a,b]		Incinerator Bottom Ash Concentrations[a,b]	
	Ash (mg/kg)	EP III Leachate (µg/L)	Ash (mg/kg)	EP III Leachate (µg/L)	Ash (mg/kg)	EP III Leachate (µg/L)	Ash (mg/kg)	EP III Leachate (µg/L)	Ash (mg/kg)	EP III Leachate (µg/L)
Volatile Organics[c]										
Nominal Detection Limit	100	–	100	–	100	–	0.5	–	0.5	–
Chloromethane	–	–	–	–	–	–	1.7	–	–	–
trans-1,2-Dichloroethane	–	–	–	–	–	–	–	–	–	–
1,1,1-Trichloroethane	–	–	–	–	–	–	6.2	–	–	–
Trichloroethane	–	–	–	–	–	–	5.3	–	–	–
Tetrachloroethene	–	–	–	–	–	–	3.6	–	–	–
Toluene	–	–	–	–	–	–	120	–	2.1	–
Chlorobenzene	–	–	–	–	–	–	2.5	–	–	–
Ethylbenzene	–	–	–	–	–	–	7.6	–	–	–
Carbon Disulfide[d]	–	–	–	–	–	–	–	–	–	–
2-Butanone[d]	–	–	–	–	–	–	–	–	–	–
4-Methyl-2-Pentanone[d]	–	–	–	–	–	–	29	–	–	–
Styrene[d]	–	–	–	–	–	–	–	–	–	–
Total Zylene[d]	–	–	–	–	–	–	15	–	–	–
Semivolatile Organics[c]										
Nominal Detection Limit	0.1	2	0.1	2	0.1	2	0.5	2	0.1	2
Acenaphthene	–	–	0.26	–	–	–	–	–	–	–
1,2,4-Trichlorobenzene	–	–	10	10	–	–	–	–	–	–
Fluoranthene	–	–	0.23	–	–	–	–	–	–	–
Isophorone	–	–	1.1	20	11	62	2.5	–	–	–
Naphthalene	–	–	6.8	8	0.75	–	2.3	–	–	–
2-Nitrophenol	–	–	–	60	–	–	–	–	–	–
4-Nitrophenol	–	–	–	90	–	–	–	–	–	–
N-Nitrosodiphenylamine	–	–	–	–	1.5	–	–	–	–	–
Phenol	–	–	1.7	30	–	6	400	1800	–	120
Bis(2-ethylhexyl) phthalate	–	–	500	–	150	–	–	–	–	–
Benzyl butyl phthalate	–	–	5	–	7	–	–	–	–	–
Di-n-butyl phthalate	–	–	39	14	7	–	–	–	–	–
Di-n-octyl phthalate	–	–	2.5	–	–	–	–	–	–	–
Diethyl phthalate	–	–	–	–	–	–	–	–	–	–
Dimethyl phthalate	–	–	31	580	–	–	–	–	–	–
Benzo(b) fluoranthene	–	–	–	–	–	–	–	–	–	–
Chrysene	–	–	–	–	–	–	–	–	–	–
Anthracene[d]	–	–	0.15	–	–	–	–	–	–	–
Phenanthrene[d]	–	–	1.3	–	–	–	0.9	–	–	–
Pyrene[d]	–	–	0.34	–	–	–	1.3	–	–	–
Benzoic Acid[d]	–	46	2.4	–	–	–	–	–	–	–
2-Methylphenol[d]	–	–	–	–	–	–	1	–	–	–
2-Methylnaphthalene[d]	–	–	6.2	4	0.3	–	15	–	–	–
Aniline[d]	–	–	–	–	–	20	–	–	–	–
Comments: Wet or Dry Ash	Dry		Dry		Wet		Wet		Dry	

(Van Buren, 1988)

[a] means not detected, hence less than nominal detection limit.
[b] Volatile organics data for EP III leachate not available.
[c] RCRA Appendix VIII unless otherwise noted.
[d] Not RCRA Appendix VIII compound.

Table 4-4. Concentration of Priority Metals in Incinerator Residuals

Site Number Stream Description	1 Kiln Ash Concentrations			2 Kiln Ash Concentrations			3 Kiln Ash Concentrations		
	Sample (mg/kg)	EP Leachate (mg/L)	EP III Leachate (mg/L)	Sample (mg /kg)	EP Leachate (mg/L)	EP III Leachate (mg/L)	Sample (mg/kg)	EP Leachate (mg/L)	EP III Leachate (mg/L)
Antimony	2	<0.05	0.04	6	<0.01	<0.01	18	0.06	<0.01
Arsenic	4	0.23	<0.01	2	<0.01	<0.01	3	<0.01	<0.01
Beryllium	<1	<0.01	<0.01	<2	<0.01	<0.01	<7	<0.01	<0.01
Cadmium	<2	<0.01	<0.01	<1	<0.01	<0.01	<1	<0.01	<0.01
Chromium	120	0.10	0.22	110	0.09	0.10	660	0.03	0.06
Copper	6900	8.6	16	840	3.7	7.9	400	0.02	0.09
Lead	220	2.3	3.5	100	<0.01	<0.01	610	0.04	<0.01
Mercury	<0.05	<0.001	<0.001	1.5	<0.001	<0.001	<0.1	<0.001	<0.001
Nickel	190	0.49	0.45	7300	6.9	6	240	0.79	13
Selenium	<1	<0.05	0.02	6	0.2	0.05	13	0.17	1.4
Silver	11	<0.01	<0.01	8	0.05	<0.01	4	0.02	0.05
Thallium	<1	<0.01	0.02	<1	0.01	<0.02	7	<0.01	<0.02
Zinc	160	0.14	0.42	640	1.8	2	21000	27	300
Comments Wet or Dry Ash	Wet			Wet			Wet		

Site Number Stream Description	3 Boiler Ash Concentrations			4 Cyclone Ash Concentrations			5 Large Incinerator Bottom Ash Concentrations		
	Sample (mg/kg)	EP Leachate (mg/L)	EP III Leachate (mg/L)	Sample (mg/kg)	EP Leachate (mg/L)	EP III Leachate (mg/L)	Sample (mg/kg)	EP Leachate (mg/L)	EP III Leachate (mg/L)
Antimony	190	<0.01	<0.01	<1	<0.01	0.01	3	<0.01	<0.01
Arsenic	14	<0.01	<0.01	<1	0.01	<0.01	9	<0.12	<0.10
Beryllium	6	<0.01	<0.08	<2	<0.01	<0.01	<2	<0.01	<0.01
Cadmium	61	8.6	6.7	<1	<0.01	<0.01	2	<0.01	<0.01
Chromium	1800	0.03	0.36	7	0.03	0.03	520	0.98	0.20
Copper	780	31	21	<4	<0.01	0.02	500	<0.01	0.11
Lead	5000	4.4	4.5	<1	<0.01	<0.01	1800	<0.01	<0.01
Mercury	0.2	<0.001	<0.001	<0.1	<0.001	<0.001	<0.1	<0.001	<0.001
Nickel	4700	20	13	25	0.18	0.22	34	0.03	0.02
Selenium	13	<0.1	1.4	<1	<0.01	<0.01	8	<0.01	0.03
Silver	190	0.09	0.05	120	<0.01	<0.01	16	<0.01	<0.01
Thallium	9	0.7[a]	<0.02	<1	<0.01	0.02	<1	<0.01	<0.02
Zinc	32000	1400	1200	200	2.2	2.6	1300	0.14	0.17
Comments Wet or Dry Ash	Wet			Dry			Dry		

[a]Thallium EP leachate concentration for Site 3 boiler ash measured as 0.7 but probably less due to interference

Table 4-4.   (Continued)

Site Number Stream Description	5 Small Incinerator Bottom Ash Concentrations			6 Incinerator Bottom Ash Concentrations			7 Incinerator Bottom Ash Concentrations		
	Sample (mg/kg)	EP Leachate (mg/L)	EP III Leachate (mg/L)	Sample (mg/kg)	EP Leachate (mg/L)	EP III Leachate (mg/L)	Sample (mg/kg)	EP Leachate (mg/L)	EP III Leachate (mg/L)
Antimony	<1	<0.01	0.10	<1	0.07	0.06	49	<0.01	0.02
Arsenic	<1	0.12	0.54	8	<0.01	<0.01	12	<0.06	<0.01
Beryllium	<2	<0.01	<0.01	<2	<0.01	<0.01	<1	<0.01	<0.01
Cadmium	<1	<0.01	<0.01	<1	0.04	<0.01	<1	<0.01	<0.01
Chromium	100	0.03	2.7	110	0.03	<0.02	120	<0.03	<0.02
Copper	40	0.02	0.07	120	1.9	0.64	2000	13	11
Lead	<1	<0.01	<0.01	1300	3.3	12	160	0.11	0.50
Mercury	<0.1	<0.001	<0.001	0.1	<0.001	<0.001	0.25	<0.001	<0.001
Nickel	3	0.04	0.27	22	0.33	0.49	650	13	4.0
Selenium	<1	<0.1	0.12	12	0.03	0.02	19	<0.05	0.02
Silver	54	<0.01	<0.01	21	<0.01	<0.01	9	<0.01	<0.01
Thallium	6	<0.01	<0.02	<1	0.05	<0.02	4	<0.01	0.02
Zinc	200	0.31	0.17	810	16	9.5	850	65	98
Comments Wet or Dry Ash	Dry			Dry			Wet		

Site Number Stream Description	8 Kiln Ash Concentrations			8 Incinerator Bottom Ash Concentrations			9 Kiln Ash Concentrations		
	Sample (mg/kg)	EP Leachate (mg/L)	EP III Leachate (mg/L)	Sample (mg/kg)	EP Leachate (mg/L)	EP III Leachate (mg/L)	Sample (mg/kg)	EP Leachate (mg/L)	EP III Leachate (mg/L)
Antimony	240	0.49	0.36	32	<0.05	<0.01	<0.8	<0.01	<0.02
Arsenic	11	0.06	0.02	27	0.22	<0.01	2	<0.06	<0.01
Beryllium	<1	<0.01	<0.01	<1	<0.01	<0.01	<1	<0.01	<0.01
Cadmium	36	0.12	0.19	3	0.03	<0.01	<1	<0.01	<0.01
Chromium	250	<0.03	<0.02	110	0.63	0.28	29	0.08	<0.02
Copper	2900	0.33	1.8	14	0.09	0.05	120	<0.02	0.67
Lead	1600	0.11	<0.01	280	<0.07	<0.01	490	<0.07	<0.50
Mercury	0.1	<0.001	<0.001	<0.05	<0.001	<0.001	<0.05	<0.001	<0.001
Nickel	100	0.42	0.71	15	<0.03	<0.01	21	2	0.49
Selenium	<40	<0.05	0.04	8	<0.05	<0.01	<4	<0.05	<0.01
Silver	3	<0.01	<0.01	<1	<0.01	<0.01	9	0.09	<0.01
Thallium	3	<0.01	0.18	4	<0.01	<0.02	6	<0.01	<0.02
Zinc	2500	12	35	2200	8.5	20	44	0.67	1.9
Comments Wet or Dry Ash	Wet			Dry			Wet		

(Van Buren, 1988)

Table 4-5.   Summary of Data on Particulate Concentration and HCl Removal Efficiency for Incinerators Burning Wastes Containing Chlorinated Compounds

Plant	Scrubbing System	Alkaline Scrubbing Media	Mist Eliminator	Avg HCl Removal Efficiency and pH of Scrubber Effluent	Avg Particulate Emissions mg/Nm³ (gr/dscf) at 7% O_2
A	75-cm water column (w.c.) ΔP venturi – sieve tray tower	Caustic	See last column	>99% at pH 4; <99% at pH 1	1520 (0.666) (no mist eliminator) 220 (0.095) (with mist eliminator)
B	Packed tower – 1-stage electrified scrubber (ES)	Lime	Yes	>99% at pH 2	78 (0.034)
C	Packed tower – packed tower – 2-stage (ES)	Caustic	Yes	>99% at pH 5	150 (0.066)
D	300 cm w.c. ΔP venturi – 3 packed towers in series	H_2O only	Yes	>99%*	23 (0.010)
E	1 packed tower	Once-through H_2O	Yes	>99% at pH 1	200 (0.088)
F	1 packed tower	Once-through H_2O	Yes	>99% at pH 1.5	4 (0.002)

(PEER, 1984; Gorman, 1988)

* pH of final effluent was 2 but is not comparable to other systems since this plant used three scrubbers in series with feed-forward water flow.

CHAPTER 5

Issues Affecting the Implementation of Incineration

Permitting Requirements

The steps required to conform to environmental protection regulations are different for on-site and off-site incinerators. Off-site incineration is typically performed in a commercial incinerator which should already be permitted under a RCRA Part B to accept a wide variety of hazardous wastes. In order to accept a waste stream different from those for which it is permitted, any such incinerator will probably be required to perform a trial burn (i.e. a test burn utilizing a (spiked) sample of the candidate waste with sampling and analyses performed to demonstrate a DRE of 99.99% for the identified POHCs).

Preparation of the trial burn plan, including the sampling and analysis plan, and the quality assurance plan, can take the majority of predesign time. Time required to plan and execute trial burns varies from 4 to 9 months. Steps in the trial burn procedures are presented in Table 5-1. It may be possible to meet TSCA and RCRA requirements simultaneously in the case of burning PCBs. Applicants should be aware of this and should consider business ramifications. Technical requirements for successful trial burns at off-site incinerators are shown in Table 5-2.

Table 5-1.   Steps in Trial Burn Procedures

- Prepare trial burn plan and submit to appropriate state or federal agency (required 6 months after notification).
- Prepare responses to EPA on any questions or deficiencies in the trial burn plan (1 month).
- Make any additions or modifications to the incinerator that may be necessary (1 to 3 months).
- Prepare for trial burn.
  - ▸ Prepare for all sampling and analysis (S&A) (2 to 3 months).
  - ▸ Select date for trial burn, in concert with S&A staff or contractor (completed 1 month prior to test).
  - ▸ Notify all appropriate regulatory agencies (1 month).
  - ▸ Obtain required quantities of waste having specified characteristics.
  - ▸ Calibrate all critical incinerator instrumentation (2 weeks).
- Conduct trial burn sampling (1 week).
- Conduct sample analysis (1 to 1-1/2 months).
- Calculate trial burn results (1/2 month).
- Prepare results for submittal to EPA (1/2 to 1 month). Include requested permit operating conditions.
- Obtain permit to accept candidate waste.

(U.S. EPA, 1987)

During the planning process it is necessary to select trial burn conditions that (1) provide the plant adequate operating flexibility, (2) assure the trial burn will be conducted in a manner acceptable to regulatory agencies, (3) make the trial burn cost-effective, and (4) represent worst case conditions under which the incinerator may be expected to operate. Failure to meet the trial burn requirements on the first series of tests is common and this is just one of many reasons why trial burns frequently take more time and effort than an operator anticipates. Note that it is often necessary to "spike" trial burn wastes with significantly higher concentrations of some POHCs in order to ensure adequate measurement of DREs. If a POHC is not detected, then the DRE is calculated using the analytical detection limit and this results in misleadingly poor, apparent DREs.

On-site incineration has somewhat different requirements. Superfund site remediation is exempt from specific compliance with other regulations. However, the requirement that actions at a CERCLA site comply with all applicable or relevant and appropriate requirements (ARARs) means that any technology (including incineration) must meet the technical requirements (but not the administrative requirements) of federal and state laws. Thus, a trial burn will be necessary even at a Superfund site. The technical requirements are the same as described above and so the time requirements, including planning, are approximately the same.

The Use of Risk Assessment as a Tool in Evaluating Waste Management Alternatives

If hazardous organic waste is fed to an incinerator, the RCRA regulations require the achievement of "four-nines" (99.99%) minimum destruction and removal efficiency (DRE). This technical requirement constitutes an ARAR for incineration at a CERCLA site. This means that for every 10,000 pounds of a POHC fed, no more than one (1) pound may be released from the stack. Using the amount of waste fed, the DRE for the incinerator, and the available toxicity data for compounds in the waste, estimates can be made of risk of exposure to organic emissions. Similarly, estimates of risk to human health can be made for products of

Table 5-2.   Regulatory Requirements for Off-Site Incineration Permits

	RCRA	TSCA
POHC Destruction and Removal Efficiency (DRE)	99.99%	99.9999%
Temperature	per trial burn or per the permit	1200°C min (PCB liquids) per trial burn (non-liquids)
Residence Time	per trial burn or per the permit	2 sec min (1200° C) 1.5 sec min (1600° C) non-liquid per trial burn
Combustion Efficiency	none	99.9%
Stack O$_2$	per trial burn or per the permit	3% min (PCB Liquids) 2% min (1600° C) non-liquid per trial burn
HCl Control	4 lb/hr max or 99% control*	per regional administrator
Particulates	180 mg/dscm*	per regional administrator

(Oberacker, 1988)

* May be subject to more stringent state or local standards

incomplete combustion (PICs) and for inorganic emissions, including heavy metals.

There are risks to human health and the environment associated with any approach for remediating a contaminated site (even the "no-action" alternative). One way of comparing remediation alternatives is to perform a risk assessment of each alternative prior to remedy selection. Risk assessment can also be a useful tool for risk management when used to identify the source of greatest risk in any action.

There are uncertainties in the risk assessment process. These arise from a number of sources including toxicity data, source emissions estimates or measurements, and dispersion modeling. To deal with such uncertainties, conservative assumptions can be made during the risk analysis process. Such assumptions provide large factors of safety in the process. Being conservative in the risk estimating process means that overestimating risk is encouraged. Risk estimates are usually made for a 70 year average total lifetime exposure, even though persons living near disposal facilities are exposed for much less time. The values used for compound toxicity are derived from conservative assumptions about chemical dose resulting in specific effects. Atmospheric dispersion models have been developed so as to overestimate ambient air concentrations to which possible receptors are exposed. For example, continuous exposure to ambient outdoor air concentrations is assumed, while most people spend a large portion of time indoors.

Risks may occur for a remediation activity from any one of three different mechanisms: 1) normal operations, 2) abnormal operations (upsets), and 3) accidents. Normal operations refer to the day-to-day operational procedures when systems are working properly. Upsets and accidents are unexpected conditions leading to higher concentrations of toxics, but only for relatively short periods of time.

Partitioning in this way helps to characterize scenarios in the risk estimating process.

The results of application of accepted risk analysis includes three steps, i.e., estimating the amount of material escaping from the source, analysis of the movement (direction and amount) of the toxics from the source to a potential point of exposure (say a human "receptor"), and estimating the toxicity of the compound(s) once a receptor is reached.

The Regulatory Impact Analysis (RIA) developed to support RCRA hazardous waste incineration regulations, contained a "worst case" risk assessment for hazardous waste incineration performed for a wide range of possible scenarios (PEER, 1984). The range included normal and upset conditions. Calculations were performed assuming DREs for organics of 99 percent, 99.9 percent and 99.99 percent (the required DRE), and also assuming zero removal of metals by air pollution control devices (APCDs). These "worst case" assumptions simulate incinerator upsets. The results of these analyses are presented in Table 5-3. As can be seen, the risks of metals are likely to be higher than those from POHCs and PICs. When the risks from metals are added to the risk from POHCs and PICs, the total risk did not exceed one in 100,000 (1 X 10^{-5}).

Table 5-3.   Total Excess Lifetime Cancer Risk to the Maximum Exposed Individual

Emission Item	Risk Range	Probability Statement
POHCs	10^{-7} to 10^{-10}	1 in 10,000,000 to 1 in 10 billion
PICs	10^{-7} to 10^{-11}	1 in 10,000,000 to 1 in 100 billion
Metals	10^{-5} to 10^{-8}	1 in 100,000 to 1 in 100,000,000
Total	10^{-5} to 10^{-8}	1 in 100,000 to 1 in 100,000,000

(PEER, 1984)

Health and Safety Considerations

On-site incineration actions often involve excavation of contaminated soils, special materials handling of wastes, crushing or other size reduction operations, other pretreatments, the combustion process itself and finally, disposal (and possibly treatment) of residues. All of these actions require special consideration of the volatility and flammability (or even explosiveness) of the wastes which are candidates for incineration. As with any remedial activity, a zone of decontamination is needed to store soils or wastes prior to processing. Furthermore, the residues also require zone of decontamination storage prior to final treatment and/or disposal.

There is a need for contractual commitments between contractors and subcontractors regarding health and safety requirements. The site health and safety plan which is developed early in the program does not always have input from the site subcontractors. Making the plan available during contract negotiations would make the subcontractors aware of the requirements earlier. Personnel turnover also makes enforcement of the plan difficult. There is a continuing need for health and safety training throughout remediation activities.

The health and safety plan should take into account the wide variations in temperature and other weather conditions which could be encountered at a site. As an example, respirators have been found to fog during operations. They can require modifications, and this in turn can cause significant project delays. Similarly, heat stress during the summer months, caused by required heavy protective clothing can be a major concern and can cause significant delays. This problem can be exacerbated when work near a hot combustion chamber is required.

The general health and safety plan developed during the early phases of site remediation needs to be upgraded as the work proceeds. It is necessary to do this to meet changing conditions and changing project demands. A flexible plan which takes into account the need for possible changes is very useful and eliminates, or at least mitigates, significant project delays.

References

Ackerman, D., J. Clausen, A. Grant, R. Johnson, C. Shih, R. Tobias, C. Zee, J. Adams, N. Cunningham, E. Dohnert, J. Harris, P. Levins, J. Stauffer, K. Thrun and L. Woodland. *Destroying Chemical Wastes in Commercial Scale Incinerators*, NTIS No. DB 278816. TRW Defense and Space Systems Group, Redondo Beach, CA, 1978.

Breton, M., M. Arienti, P. Frillici, M. Kravett, S. Palmer, A. Shayer, and N. Suprenant. *Technical Resource Document: Treatment Technologies for Dioxin-Containing Wastes*, EPA/600/S2-86-096, February 1987.

Carpenter, B. and D. Wilson. "Technical/Economic Assessment of Selected PCB Decontamination Processes," *J. of Hazardous Materials*, 17 (1988), Elsevier Science Publishers, B.V., Amsterdam, p. 125-148.

Carpenter, B. "PCB Sediment Decontamination Processes - Selection for Test and Evaluation," Research Triangle Institute for the U.S. EPA Contract No. 68-02-3992, HWERL, Cincinnati, Ohio, September 1987.

Code of Federal Regulations, "RCRA Regulations on Permitting," Office of Federal Register National Archives and Records Administration, Vol 40, Part 260, July 1986.

Cudahey, J., S. DeCicero, and W. Tofler. "Thermal Treatment Technologies for Site Remediation," In: *Proceedings of the International Congress on Hazardous Materials Management*, Chattanooga, TN, June 1987.

Code of Federal Regulations, "RCRA Regulations on Permitting," Office of Federal Register National Archives and Records Administration, Vol 40, Part 261, July 1986.

Daily, P. "Performance Assessment of Portable Infrared Incinerator," In: *Management of Uncontrolled Hazardous Waste Sites Proceedings*, Washington, D.C., 1985, p. 383.

Engineering Science, Inc., *Remedial Action Work Plan for the Laskin/Poplar Oil Company Site in Jefferson, Ohio*, Vol. 1, Atlanta, Georgia, in fulfillment of Administrative Order V-W-86-C-015, March 1987.

Engineering Science, Inc., *Results of Soil Sampling, for the Laskin/Poplar Oil Company Site in Jefferson, Ohio*, Vol. 2, Atlanta, Georgia, in fulfillment of Administrative Order V-W-86-C-015, March, 1987.

Federal Register, "Burning of Hazardous Waste in Boilers and Industrial Furnaces," Office of Federal Register, National Archives and Records Administration, Vol 52, May 6, 1987, p. 16982.

Frank, J., M. Dinkel, and D. Chari. "Use of Mobile Incineration to Remediate the Lenz Oil Site", In: *Superfund 1987: Proceedings of the 8th National Conference*, The Hazardous Materials Control Research Institute, p. 459, 1987.

Freeman, H. "Innovative Thermal Processes for the Destruction of Hazardous Wastes," In: *Incinerating Hazardous Wastes*, ed. Freeman, H.M., Technomic Publications, Lancaster, Pennsylvania, 1988.

Freeman, H. and R. Olexsey. "A Review of Treatment Alternatives for Dioxin Wastes," In: *Land Disposal, Remedial Action, Incineration and Treatment of Hazardous Waste: Proceedings of the Thirteenth Annual Research Symposium*, EPA/600/9-87/015, July 1987, p. 285.

Fuhr, H. "Hazardous Waste Incineration at Bayer, AG," *Hazardous Waste & Hazardous Materials*, Vol 2, No. 1, 1985, p. 1-5.

Gorman, P. and D. Oberacker. "Practical Guide to Trial Burns at Hazardous Waste Incinerators," In: *Incinerating Hazardous Wastes*, ed. Freeman, H.M., Technomic Publications, Lancaster, Pennsylvania, 1988.

Gupta, G., et al., MIS (Mobile Incineration System) Modifications, Trial Burn Operations, February, 1986, to September, 1987, Enviresponse and Foster Wheeler for the U.S. EPA, HWERL, Edison, NJ, EPA Contract No. 68-03-3255, April, 1988.

Hatch, J. and E. Hayes. "State-of-the-Art Remedial Action Technologies Used for the Sydney Mine Waste Disposal Site Cleanups," In: *Management of Uncontrolled Hazardous Waste Sites Proceedings*, Washington, D.C., 1985, p. 285.

Hazaza, D., S. Fields and G. Clemons. "Thermal Treatment of Solvent Contaminated Soils," In: *Management of Uncontrolled Hazardous Waste Sites Proceedings*, Washington, D.C., 1984, p. 404.

Janssen, J., R. Munger, J. Noland, N. McDevitt, and L. Velazquez. "Utilization of Mobile Incineration at the Beardstorm Lauder Salvage Yard Site," In: *Superfund 1987: Proceeding of the 8th National Conference*, The Hazardous Materials Control Research Institute, p. 453.

Kristensen, A. "Operating the Rotary Kiln Incinerators at Komsmunekeni," *Hazardous Waste & Hazardous Materials*, Vol 2, No. 1, 1985, p. 7-21.

Lee, J., T. Backhouse, R. Ross, and L. Waterland. *PCB Trial Burn Report for the U.S. EPA Combustion Research Facility Liquid Injection Incinerator System*, EPA/600-S2-87/051, U.S. EPA, HWERL, Cincinnati, Ohio, September 1987.

Martin, E. "Data and Information on Incineration Permits and Status," Personal Communication, E.J. Martin with Sonya Stelmack, Office of Solid Wastes, U.S. EPA, Washington, D.C., data current December 10, 1987, from Hazardous Waste Data Management System (HWDMS), February, 1988.

Martin, E., L. Weinberger, J. Swartzbaugh, A. Mathews, and C. Lee. "Practical Limitations of Waste Characteristics for Effective Incineration," In: *Incinerating Hazardous Wastes*, ed. Freeman, H.M., Technomic Publications, Lancaster, Pennsylvania, 1988.

Noland, J. and W. Sisk. "Incineration of Explosives Contaminated Soils," In: *Management of Uncontrolled Hazardous Waste Sites Proceedings*, Washington, D.C., 1984, p. 203.

Oberacker, D. "Hazardous Waste Incinerator Performance Evaluations by the U.S. EPA," In: *Incinerating Hazardous Wastes*, ed. Freeman, H.M., Technomic Publications, Lancaster, Pennsylvania, 1988.

Office of Technology Assessment, "Are We Cleaning Up?" A Special Report of OTAs Assessment on Superfund Implementation, Congress of the United States, OTA-ITE-362, June, 1988.

Olexsey, R., G. Huffman, and G. Evans. "Emission and Control of Byproducts from Hazardous Waste Combustion Processes," In: *Incinerating Hazardous Wastes*, ed. Freeman, H.M., Technomic Publications, Lancaster, Pennsylvania, 1988.

Oppelt, E. "Incineration of Hazardous Wastes; A Critical Review," *J. Air Pollution Control Assoc.*, Vol 37, No. 5, May 1987.

Oppelt, E. "A Profile of Hazardous Waste Thermal Destruction Facilities: Performance and Prospects," Vol 20, No. 4, *Environmental Science & Technology*, April 1986.

PEER Consultants, Inc., Supporting Documentation for the RCRA Incinerator Regulations, 40 CFR 264, Subpart O, Incinerators, Rockville, Maryland, for the U.S. EPA Office of Solid Waste, Washington, D.C., October 1984.

Rowe, W. and E. Martin. *Risk Analysis of the Proposed Erieway, Inc., Waste Processing Plant*, Rowe Research and Engineering Associates for Erieway, Inc., Bedford, Ohio, April, 1988.

Toxler, W., R. Miller, and C. Pfrommer. *Destruction of Dioxin-Contaminated Solids and Liquids by Mobile Incineration*, IT Corp., for the U.S. EPA, Edison, New Jersey, April, 1987.

Trenholm, A., R. Hathaway, and D. Oberacker. "Products of Incomplete Combustion for Hazardous Waste Incinerators," In: *Incinerating Hazardous Wastes*, ed. Freeman, H.M., Technomic Publications, Lancaster, Pennsylvania, 1988.

U.S. EPA. *Total Mass Emission from a Hazardous Waste Incinerator*, EPA/600/S2-87/064, HWERL, Cincinnati, Ohio, Nov 1987.

U.S. EPA. *PCB Trial Burn Report for the US EPA Combustion Research Facility Liquid Injection Incinerator System*, EPA/600/S2-87/051, HWERL, Cincinnati, Ohio, Sept 1987.

U.S. EPA. *Practical Guide - Trial Burns for Hazardous Waste Incinerators*, EPA/600/S2-86/50, HWERL, Cincinnati, Ohio, July 1986.

Unterberg, W., R. Melvold, S. Davis, F. Stephens, and F. Bush. *Reference Manual of Countermeasures for Hazardous Substances Releases*, EPA/600/S2-87/069, U.S. EPA, HWERL, Cincinnati, Ohio, November 1987.

Van Buren, D., G. Poe, and C. Castaldini. "Characterization of Hazardous Waste Incineration Residuals," In: *Incinerating Hazardous Wastes*, ed. Freeman, H.M., Technomic Publishers, Lancaster, Pennsylvania, 1988.

Vogel, G., A. Goldfarb, R. Zier, A. Jewell, and I. Licis. Incinerator and Cement Kiln Capacity for Hazardous Waste Treatment, In: *Incinerating Hazardous Wastes*, ed. Freeman, H.M., Technomic Publishers, Lancaster, Pennsylvania, 1988.

Vrable, D. and D. Engler. "Transportable Circulating Bed Combustor for the Incineration of Hazardous Waste," In: *Management of Uncontrolled Hazardous Waste Sites Proceedings*, Washington, D.C., 1985, p. 378.

Waterland, L., J. Lee, R. Ross II, J. Lewis, and C. Castaldini. "Incineration of Cleanup Residues from the Bridgeport Rental and Oil Services Superfund Site," In: *Land Disposal, Remedial Action, Incineration and Treatment of Hazardous Waste Proceedings of the Thirteenth Annual Research Symposium*, EPA/600/9-87/015, July 1987, p. 318.

Waterland, L., Operations and Research at the U.S. EPA Combustion Research Facility: Annual Report for FY '87, Acurex Corp., for the U.S. EPA, HWERL, Cincinnati, OH, EPA Contract No. 68-03-3267, December, 1986.

Waterland, L., Operations and Research at the U.S. EPA Combustion Research Facility: Annual Report for FY '87, Acurex Corp., for the U.S. EPA, HWERL, Cincinnati, OH, EPA Contract No. 68-03-3267, December, 1987.

Westbrook, C., C. Tatsch, and L. Cottone. "Control of Air Emission from Hazardous Waste Combustion Sources," In: *Field Evaluation of Pilot-Scale Air Pollution Control Devices*, EPA/600/S2-86/011, Hazardous Waste Engineering Laboratory, Cincinnati, Ohio, July 1986.

Weston, Roy F., Inc., *Demonstration Test Report: "PCB Destruction Unit Executive Review and Presentation of Results,"* Vol. 1, West Chester, PA, for Office of Toxic Substances, U.S. EPA, Washington, D.C., February 1988.

A.2 In-House Wastes

A.2.a CHEMICAL INCINERATION AT THE KODAK PARK SITE*

Providing Environmentally Sound Treatment for Combustible Chemical Wastes

New York State

WHY INCINERATE

Incineration is a key technology used by Eastman Kodak Company to effectively treat combustible chemical waste which cannot be reused, recycled, or recovered. It is a readily available method of treatment which is environmentally sound for the destruction of combustible chemical wastes generated at Kodak Park.

Most chemical wastes are made up principally of carbon, hydrogen, and oxygen in various combinations. Properly controlled high-temperature incineration converts these materials into carbon dioxide and water, the same products as human respiration. Kodak Park's chemical incinerator utilizes modern technology and well-trained operators, meeting stringent regulations and testing requirements to ensure that chemical wastes are treated safely and properly.

ALTERNATIVES TO INCINERATION

Elimination of waste is a key element in Kodak's dedication to the production of quality products for its customers. Therefore, the company's waste management plan begins with minimization of the production of waste.

Recycling, reuse, and recovery of materials are the next best alternatives to nonproduction of waste. Most of the potential waste chemicals at Kodak are recycled in some manner. Certain waste products which cannot be reused by Kodak are sold for use by other companies. Only a small fraction of chemicals used by Kodak end up as waste.

The only widely used disposal techniques for chemical wastes are incineration and land disposal. Government regulators are placing severe restrictions on the future use of land disposal due to long-term risks associated with it.

Modern chemical incineration at Kodak Park effectively destroys chemical waste without affecting the environment.

THE INCINERATOR — HISTORY

Kodak Park's chemical incinerator was built in 1976 to replace an older incinerator. In 1983 a new high-intensity vortex secondary burner was added to make use of the most advanced technology. This incinerator is recognized as one of the most technologically advanced in the country. Representatives from other companies and government agencies frequently visit the facility to study its design and operation.

INCINERATION PROCESS

COMBUSTION PROCESS

The two main parts of the incinerator are the combustion processes and the gas cleaning and handling processes.

---

* Courtesy of Ron Bastian and Roy Wood, Eastman Kodak Company, Rochester, New York.

Rotary Kiln

Solid waste is charged into a rotary kiln for combustion. Liquid waste or fuel is fired to control the combustion temperature assuring destruction of the solid waste prior to ash discharge.

Ash Quench

Noncombustible materials, such as glass, salts, and steel drums, drop out of the rotary kiln into an ash quench. This material is shipped to a reclaiming operation for silver recovery.

Secondary High-Intensity Vortex Burner

Other liquids are burned in a high-intensity burner at an average temperature of 2100°F.

Secondary Combustion Chamber

The combustion gases from the vortex burner mix with the combustion gases from the rotary kiln and enter the secondary combustion chamber. Combustion is completed in this chamber forming carbon dioxide and water at an average temperature of 1750°F. A minimum temperature of 1550°F is required to ensure complete combustion of the waste feeds.

GAS CLEANING AND HANDLING

Gas Cleaning

Gases are cooled and cleaned with high-pressure, high-velocity water sprays in the quench, venturi scrubber, and demister section. Hydrochloric acid, a by-product from the combustion of chlorinated solvents, is removed from the gas stream by the water sprays and converted to sodium chloride by the addition of sodium hydroxide to the liquid.

Solid ash particles are removed from the gas stream by the high-energy venturi scrubber to meet stringent emission limits. The waste water from this system is treated at Kodak Park's industrial waste water treatment plant.

Fans and Silencer

The combustion air is drawn through the combustion and gas cleaning system by two induced-draft fans. The noise generated by the fans is minimized by a silencer.

Stack

The gases are discharged through a 200-ft stack. A white plume exiting the stack is caused by water vapor from the gas cleaning system condensing when it hits cooler outside air.

COMPOSITION OF WASTE

Kodak treats common liquid, solid, and semisolid chemical wastes at its Kodak Park facility. Nonrecoverable simple organic ketones, acids, alcohols, and hydrocarbon solvents used in chemical manufacture make up the bulk of the liquid wastes. Filter cake sludge, chemically contaminated rags and glass, distillation still bottoms, paint cans, storage tank cleanings, and nonrecoverable plastic are typical solid wastes. Kodak's incinerator burns only wastes which have received prior approval by the New York State Department of Environmental Conservation and U.S. Environmental Protection Agency as defined by the permit to operate issued by these agencies. Compounds such as PCBs, PBBs, dioxins, and mercury are not fed to the incinerator.

All materials are identified and labeled according to composition and hazard type before being shipped to the incinerator. Material which does not meet specifications is sent outside of Kodak for disposal.

THE INCINERATION CONTROL SYSTEM

A sophisticated automatic control system ensures that the incinerator operates properly and meets permit conditions in accordance with state and federal regulations. Waste feed rate is modulated automatically to control temperature. Carbon monoxide, a measure of incomplete combustion, is monitored continuously in the stack gas and limited to less than 90 ppm. Water flow rates to the gas cleaning system and venturi pressure drop are monitored continuously to insure proper gas cleaning. Many other parameters are monitored, and an audible alarm sounds if they approach the specification limits. If the incinerator fails to meet any permit requirements such as temperature, carbon monoxide, or gas cleaning water flow rate limits, waste feeds will be automatically stopped and replaced with fuel oil. A trained operator monitors the process in the control room at all times.

PERSONNEL

There are five trained operators at the incinerator at all times. A solids handler and a liquids handler insure all wastes are properly inspected, handled, and prepared for the incinerator. An incinerator operator is in the control room at all times to ensure the incineration process operates properly. A senior operator inspects equipment, reviews records, and provides assistance to other operators. A shift supervisor manages the operation, provides guidance, and ensures all operators are properly trained. A general supervisor, a laboratory technician, a records clerk, a maintenance supervisor, electricians, pipe fitters, instrumentation technicians, and mechanics work the day shift and are on call to handle problem situations. An engineering assistant, Environmental Services, Chemical Quality Laboratory Services, and specialized engineering services support the facility. All personnel receive specialized hazardous waste handling training exceeding specifications in the Federal Resource Conservation and Recovery Act. All personnel must have specific knowledge and meet minimum qualifications before receiving a job at the incinerator. An "Analytical Methods Training System" insures their knowledge is current and complete. Personnel also participate in quality teams and receive specialized safety training.

TESTING THE INCINERATOR STACK GASES

Kodak has spent over a million dollars testing the Kodak Park incinerator to ensure that it operates in an environmentally safe manner. Kodak determined relative destructibility of the compounds fed into the incinerator through laboratory testing. To receive the U.S. Environmental Protection Agency permit to operate the incinerator, Kodak had to show that greater than 99.99% of the most difficult-to-incinerate combustible compound fed to the unit was destroyed. To receive the New York State Department of Environmental Conservation operating permit, Kodak also had to demonstrate that no significant level of dioxins was emitted from the facility. Continuous monitoring of carbon monoxide insures the incinerator operates in the same efficient manner as it did during the testing programs.

REGULATORY COMPLIANCE

Kodak submitted extensive test results, equipment descriptions, waste descriptions, operating procedures, instrument calibration procedures, and training procedures to EPA before the EPA operating permit was issued. NYS Department of Environmental Conservation asked for additional data before issuing their permit. Both agencies were convinced that the incinerator and its stack gas will pose no health or environmental threat to the community. Public notification and comments were part of the permitting process.

The incinerator operating permit includes limitations of waste feed rates, gas flow rates, combustion chamber temperatures, gas cleaning water flow rates, carbon monoxide levels, and other parameters.

Hazardous waste feeds are automatically stopped before any permit requirements can be exceeded. Records of waste received and burned, instrument calibrations, equipment inspection results, operating parameters, and operator training are kept at the incinerator. The NYS Department of Environmental Conservation can inspect the incinerator equipment, its operating parameters, and records at any time.

EASTMAN KODAK COMPANY CHEMICAL WASTE INCINERATION ORGANIC COMPOUND DESTRUCTION

Ronald E. Bastian and Roy W. Wood

Eastman Kodak Company
Rochester, New York 14650

INTRODUCTION

Incineration technology is a key element of Eastman Kodak Company's waste management strategy for the treatment of chemical residuals which cannot be reused, recovered, or recycled. A chemical waste incineration unit consisting of a rotary kiln followed by secondary combustion chamber and a venturi scrubber has been in successful operation since 1976 in Rochester, N.Y. This unit was designed primarily to treat a variety of combustible liquid and solid wastes generated at the Kodak Park manufacturing site in Rochester. Upgrades to the original design have enabled demonstration of compliance with applicable federal and New York State regulations.

RCRA Trial Burn Testing was completed in November 1984. EPA issued the RCRA Part B operating permit[1] in April 1986. Further testing for the potential emission of products of incomplete combustion was conducted for New York State in March 1985 and June 1986 as a requirement of the Certificate to Operate issued by New York State Department of Environmental Conservation (NYSDEC).

All process emissions have been shown to be within the limits required by New York State and EPA standards of performance. Operation of the process within the limits established by the RCRA Part B permit has been effective in minimizing emissions of organic compounds, particulate and hydrochloric acid to the atmosphere.

DESTRUCTION AND REMOVAL EFFICIENCY (DRE) TESTING

The RCRA Part B permitting process officially started for Kodak Park with the call for and submission of the complete permit application during 1983.[2] As a result of investigation and research done by Kodak and discussions with the EPA, tetrachloromethane and acetonitrile were selected as principle organic hazardous constituents (POHCs) for DRE testing. The Trial Burn Plan[3] established the concepts and methods by which each permit condition would be established. Trial Burn Plan approval was received and testing was conducted in October and November 1984. The Trial Burn Report[4] submitted in February 1985 was an 1800-page, eight-volume document. Expected permit conditions, based on the results of the Trial Burn Testing, were proposed in the report. Following discussion, these recommendations were substantially accepted by the EPA permit writer. The final RCRA permit was received in April 1986.[1]

MODES OF OPERATION

Permit limits were established for five modes of operation. Three of the five modes were tested. The other two were agreed by EPA to be submodes of those tested. Table 1 describes operations in each of these modes. Mode 1 represents full-system operation with both solid and liquid-wastes feed. Liquid waste only is fed in Mode 2, operating one of the two induced draft fans resulting in a reduced venturi pressure drop. Mode 3 is a kiln-only operating mode, feeding both liquid and solid waste.

* Courtesy of Ron Bastian and Roy Wood, Eastman Kodak Company, Rochester, New York.

TABLE 1
Incinerator Operating Modes

Mode	Kiln feeds	Secondary feeds	Kiln minimum temp (°F)	Secondary minimum temp (°F)	Venturi ΔP rolling hour av (H_2O)	Combustion gas flow maximum (K-ACFM)	Heat input MBtu/h	Carbon monoxide rolling hour (Av ppm)	Carbon monoxide instantaneous (max ppm)
1	Solids & liquids	Liquids	≥1400	≥1600	≥78	≤105	≤120	≤90	1000
2	Liquids	Liquids	≥1400	≥1600	≥42	≤75	≤80	≤90	1000
3	Solids & liquids	None	≥1600	—	≥78	≤90	≤100	≤90	1000
4	None	Liquids	—	≥1600	≥42	≤75	≤80	≤90	1000
5	Liquids	None	≥1600	—	≥42	≤75	≤80	≤90	1000

TABLE 2
Composition of Waste Streams Mode 1

	Heating value (Btu/lb)	Ash (wt%)	Chloride (wt%)	Tetrachloromethane (wt%)	Acetonitrile (wt%)
Fuel-grade liquid	7,600	1.9	4.7	4.3	9.5
	(7,200—7,900)	(0.9—2.9)	(4.2—5.0)	(2.8—5.7)	(7.8—10.7)
Non-fuel-grade liquid	2,600	2.3	1.7	ND	10.0
	(2,500—2,700)	(1.2—2.9)	(1.4—1.8)		(8.6—12.6)
Synthetic solid waste packs	3,000	42	30	26	20
		(41—44)	(28—31)	(24—27)	(19—21)
High-heat-content solid waste packs	14,000	—	—	—	—
Low-heat-content solid waste packs	4,000	—	—	—	—

WASTE FEED

Two grades of liquid feed and three categories of solid-waste feed were used during the testing. POHCs (tetrachloromethane and acetonitrile), sodium chloride (ash), water, and organic chlorine were added to these wastes to establish agreed-upon "worst case" waste feed characteristic (see Table 2). Fuel-grade waste liquid was normally fed to two burners in the kiln and two burners in the secondary combustion chamber. Non-fuel-grade liquid was normally fed to one burner in the kiln and four peripheral burner nozzles in the secondary combustion chamber. Containerized solids were fed continuously, alternating high heat value, low heat value, and "synthetic", which contained ash, organic chlorine, and POHC.

Mixtures of liquid waste were prepared to specification prior to each test. Fuel-grade liquid used a methanol waste as a base. Acetonitrile and tetrachloromethane were added to the waste in a 10,000-gal agitated tank. In a separate tank the desired amount of water to lower the heat value to approximately 7500 Btu/lb was mixed with sodium chloride salt to adjust the ash to approximately 3% in Modes 1 and 3 and 1% in Mode 2. The water and salt mixture was transferred to the 10,000-gallon waste feed tank containing the methanol waste with POHC.

POHC SELECTION

Ranking parameters considered when selecting the Trial Burn POHCs were

1. Heat of combustion
2. Temperature of 99.99% DRE at 2 s of residence time by flameless oxidation (T99.99/2)
3. Degree of water solubility
4. DREs in full-scale preliminary tests of the incinerator
5. Concentration in the waste

Heat of combustion was an important ranking parameter because EPA had proposed its use in their Guidance Manual for Hazardous Waste Incinerator Permits (1982). Full-scale testing conducted in 1982 at the Kodak incinerator showed little correlation between the heat of combustion of individual POHCs and their DREs; however, there was a strong correlation between the heat of combustion of the aggregate waste mixture and the constituent DREs.[5]

Flameless oxidation destruction efficiency was the most important ranking parameter, because Kodak felt it most nearly indicated the actual difficulty of destruction of a POHC. The University of Dayton conducted a series of flameless vapor phase thermal oxidation tests using waste mixtures containing candidate POHCs.[6] They calculated the temperature required to give 99.99% destruction efficiency at 2 s of residence time (T99.99/2) using pseudo first-order reaction kinetics and computing the rate constant using Arrhenius equation.[6] Earlier testing indicated this model was quite accurate over the (1200—2000°F) temperature range.[7]

The degree of water solubility was included in the criteria because earlier full-scale testing of this unit showed that waste mixtures containing water or water-soluble compounds were much more difficult to destroy than those containing water-insoluble compounds.[5] This testing was conducted before the installation of a new vortex secondary burner in 1983. It appeared that poor atomization, resulting in delayed vaporization in the original secondary burners, was the mode of failure in the earlier tests. Water-soluble compounds in aqueous mixtures have a lower vapor pressure and vaporize more slowly than water-insoluble compounds, so they have less vapor-phase residence time. The vortex secondary burner eliminated the poor atomization and added additional turbulence and residence time and a higher temperature zone. Therefore, it was not clear whether water solubility should have been an important ranking parameter in the existing incinerator.

Full-scale testing results on the incineration unit were also considered when selecting POHCs. Most of the full-scale data were obtained before the installation of the new vortex secondary burner, so its applicability was unclear.

Another ranking factor listed in the EPA guidance document was concentration in the waste. POHCs which might be present at greater than 1000 ppm in our waste were ranked on each of the five parameters previously discussed. These rankings are listed in Table 3. In addition, two compounds which had the highest T99.99/2 rankings and were present at 100—1000 ppm in our waste are included in Table 3.

Acetonitrile ranks as most difficult to destroy using the T99.99/2 criteria which was the most important criteria for selecting POHCs, and it was also one of the most difficult-to-destroy compounds in full-scale tests. It is water-soluble and one of Kodak's highest-volume POHCs. Tetrachloromethane ranks most difficult to destroy based on heat of combustion and ranks third using T99.99/2. Since these two compounds ranked high on each of the ranking criteria, they were selected for the trial burn plan and accepted for the trial burn by the EPA.

PROCEDURE

Pretrial burn testing had shown that the failure mode for DRE at >1600°F was related to the introduction of high-heat-value containerized solids, the combustion air supplied to the kiln, and the kiln rotation speed. Each of these affected the average carbon monoxide (CO) level. High-heat-value containerized solids can cause spikes of CO in the event the heat release rate of the container is too great for the oxygen available. Duration of these CO spikes is not normally over 60 s.

The amount of combustion air available in the kiln directly affects the duration and magnitude of CO spikes from containerized solids. In addition, normal average CO concentration can be affected by the amount of oxygen available. Kiln rotation speed affects the rate of combustion of solids burning in the kiln. An increase in rotation speed can increase CO where a decrease could reduce CO.

System capacity and combustion-chamber temperature(s) above 1500°F were not significant parameters influencing CO. Thus, the trial burn was designed to determine the maximum CO level which correlated to an acceptable DRE. The CO parameter used was a rolling 1-h average. Three tests of three 1-h runs each were conducted in Mode 1 (full system

TABLE 3
Incinerability Difficulty Rankings

Compound	Present[a]	Heat of combustion (cal/g)	Heat of combustion rank	Flameless oxidation T99.99/2 (°C)	Flameless oxidation rank	Autoignition temp (°C)	Autoignition temperature rank	Percent solubility in water	Water solubility rank	Qualitative assessment of destruction from full-scale tests[b]
Acetonitrile	1	7.73	10	951	1	524	7		1	1
Chlorobenzene	1	6.60	9	726	6	638	5	0	11	2
Chloroform	1	0.75	2	629	10	—	2	0.8	8	3
1,2-Dichloroethane	1	3.00	7	634	9	413	13	0.9	7	—
Dichloromethane	1	1.70	4	821	5	662	4	2	6	2
1,4-Diethylene dioxide	1	6.41	8	—	—	180	14		1	1
Isobutyl alcohol	1	8.62	13	—	—	416	12		1	2
Methyl ethyl ketone	1	8.07	12	674	8	515	8	35	5	1
Pyridine	1	7.83	11	840	4	482	10		1	1
Tetrachloromethane	2	0.24	1	894	3	—	1	0.1	10	2
1,1,2,2-Tetrachloro-ethane	2	1.19	3	928	2	—	3	0.3	9	—
Toluene	1	10.14	14	700	7	482	11	0	11	3
1,1,1-Trichloroethane	1	1.99	5	600	11	560	6	0	11	—
1,1,2-Trichloroethane	1	1.99	5	600	11	513	9	0	11	—

a Compound present at >1000 ppm = (1) or 100—1000 PPM = (2).
b (1)-Difficult (3)-Easy.

operation) and one test in Mode 3 (kiln only operation). The carbon monoxide aim was increased from 50 ppm rolling-hour average in the first test in Mode 1 to 100 ppm in subsequent tests by manipulating containerized solid feed rates, damper settings for oxygen control, and kiln rotation rate. Operating conditions in each mode are shown in Table 1.

The stack emissions were drawn into an evacuated bag and analyzed for acetonitrile and tetrachloromethane using a gas chromatograph/flame ionization detector.

RESULTS

DRE results from the trial burn are presented in Table 4. Since Mode 1 is full-system operation at maximum capacity, three separate tests were made, taking the system to "failure". Failure to achieve 99.99% DRE for acetonitrile occurred in Run 4 on the test conducted 11/5/84. All other runs in both modes tested demonstrated 99.99% DRE. Slightly better DRE results were achieved with full system (Mode 1) vs. kiln-only operation (Mode 3). Similar DREs were demonstrated for both acetonitrile and tetrachloromethane; however, acetonitrile was more sensitive to failure, based on the 11/5/84 results.

Log (100% DRE) of acetonitrile is plotted against the maximum rolling-hour average CO in each run (Figure 1). The plot shows a correlation between DRE and CO for the 11 test runs, with expected failure at about 180—200 ppm. The rolling-hour average CO accounts for any combination of spikes of CO which occurred during the test run. As shown by the data, significant spiking associated with the feeding of high-heat-value containerized waste occurred. Rolling-hour average CO is thus a measure which can be used to limit the frequency and duration of CO spikes.

The reported CO concentration is an actual concentration in the stack gas measured using an on-line *in situ* CO monitor during each test. The stack gas is typically about 35% moisture and 14% oxygen. No correction for oxygen and moisture has been made to the data presented. When corrected to dry gas conditions at 7% oxygen, the 90 ppm rolling-hour average permit limit would be equivalent to about 270 ppm. The estimated corrected CO concentration at 99.99% DRE would be about 500 ppm.

AMBIENT AIR IMPACT

The mass of acetonitrile emissions ranged from a low of 0.000642 lb/h to a high of 0.264 lb/h. The corresponding range for tetrachloromethane was 0.00179 to 0.0723 lb/h. Data for runs which represent the "worst case" in each mode of operation are presented in Table 5.

The estimated maximum, 1-h, and annual average concentrations of tetrachloromethane and acetonitrile associated with this data are:[8]

Predicted Maximum 1-h Average Concentrations (μg/m³)

	11/05/84	10/17/84
Tetrachloromethane	0.092	0.027
Acetonitrile	0.122	0.336

Predicted Maximum Annual Average Concentrations (μg/m³)

	11/05/84	10/17/84
Tetrachloromethane	9.92×10^{-4}	2.83×10^{-4}
Acetonitrile	3.63×10^{-3}	1.16×10^{-3}

The recommended interim ambient air level (AAL) for tetrachloromethane is 100 μg/m³.[9] The predicted maximum *1-h* concentrations of tetrachloromethane are 1000 to 3700

TABLE 4
DRE Results USEPA Trial Burn Test

					Peak CO # >		POHC balance					
							Acetonitrile			Tetrachloromethane		
Test date	Run	Comb. temp (°F)	Heat input (MBtu/h)	CO max rolling h av	300	1000	In (lb/h)	Out (Stack) (lb/h)	DRE %	In (lb/h)	Out (lb/h)	DRE %
Mode 1												
10/10/84	1	1625	122	52	2	0	2233	0.00146	99.999935	1357	0.00422	99.99969
	2	1625	123	59	5	0	1911	0.00156	99.999918	1149	0.0103	99.99940
	3	1625	117	NA	2	1	1991	0.00141	99.999929	1061	0.00199	99.99981
	4	1600	116	45	2	0	1912	0.00238	99.99988	1163	0.00179	99.99985
	Avg	1620	120	52	3	0	2012	0.00168	99.999916	1182	0.00458	99.99961
10/14/84	1	1630	119	87	10	0	2152	0.0355	99.9984	1034	0.0146	99.9986
	2	1700	144	107	13	1	2340	0.000642	99.999973	976	0.00330	99.99966
	3	1700	122	84	9	2	2026	0.002442	99.99988	817	0.00349	99.99957
	Avg	1675	128	93	11	1	2173	0.0129	99.99942	942	0.00713	99.99928
11/05/84	2	1650	109	147	18	3	1991	0.0813	99.9959	1018	0.00361	99.99965
	4	1680	98	198	23	8	1613	0.264	99.984	910	0.0723	99.9921
	Avg	1675	105	173	20	5	1802	0.1725	99.98995	964	0.03796	99.9959
Mode 3												
10/17/84	1	1650	103	59	2	0	1733	0.00784	99.99955	1142	0.00588	99.99948
	2	1650	95	76	1	1	1466	0.0918	99.9937	998	0.0203	99.9980
	3	1650	98	85	4	0	1492	0.00461	99.99969	1103	0.00373	99.99966
	Avg	1650	99	73			1564	0.0348	99.9976	1081	0.00997	99.99905

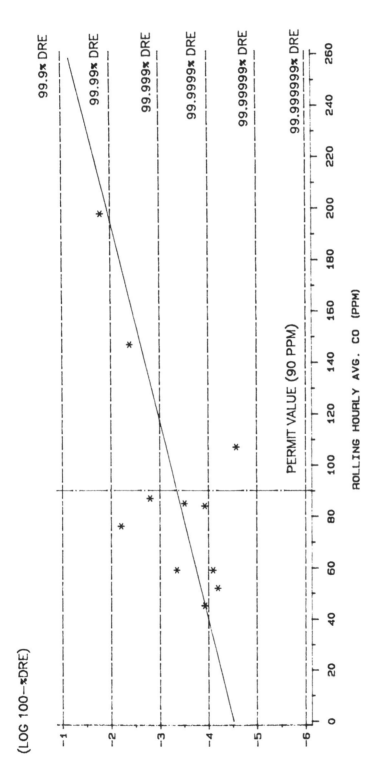

FIGURE 1.   USEPA trial burn using acetonitrile — log (100 - % DRE) vs. maximum rolling hour average CO (ppm).

TABLE 5
Emission Data for Ambient Air Impact

Date	Mode	Stack temp (°F)	Stack moisture (%)	Stack gas flow (ACFM)	Tetrachloromethane (lb/h)	Acetonitrile (lb/h)
10/17/84	3	150	25	69,400	0.0203	0.0918
11/05/84	1	158	35	80,100	0.0723	0.264

times less than DEC's *annual* AAL. The predicted maximum annual concentrations of tetrachloromethane are 100,800 to 353,300 times less than DEC's *annual* AAL.

The recommended interim AAL for acetonitrile is 1400 $\mu g/m^3$.[8] The predicted maximum *1-h* concentrations of acetonitrile are 4100 to 11,400 times less than DEC's *annual* AAL. The predicted maximum annual concentrations of acetonitrile are 385,600 to 1,111,100 times less than DEC's *annual* AAL.

PERMIT CONDITIONS

Permit conditions related to DRE which were established by the trial burn include combustion temperature, carbon monoxide (CO) concentration in the stack gas, and waste feed rate. The conditions established are shown in Table 6. The rationale for establishing these conditions was presented and agreed to in the trial burn plan.[3] For example, an aim point for combustion zone temperature of 1600°F, with automatic shutdown if the temperature dropped 100°F below aim for 1 min, was suggested in the trial burn plan. The actual permit condition established was "the combustion chamber temperature shall be maintained above 1600°F. All hazardous waste feed shall shutdown automatically if this temperature is less than 1550°F for 1 minute." This was established by the data presented in Table 4 plus continuous strip-chart recording, which showed variations establishing the minimum temperature shutdown condition.

Two permit limits for CO were suggested based on the trial burn data: one based on the rolling 1-h average CO concentration affecting all hazardous-waste feeds and a second based on instantaneous CO concentration, which only affects containerized solid-waste feed. The proposed limit in the Trial Burn Report for Mode 1 operation was 120 ppm and for Mode 3 operation 85 ppm. The permit writer's specification was 90 ppm for both modes, primarily based on the 11/14/84 test for Mode 1, where the DREs were greater than 99.999% for both POHCs at an average of 93 ppm maximum rolling-hour average for the three runs. The proposed and acceptable limit for instantaneous CO value was 1000 ppm.

Two stages of hazardous-waste feed cutoff were proposed and accepted by the USEPA. Containerized waste fed is automatically shut down if the rolling-hour average CO exceeds 90 ppm at any time. In addition, containerized waste cannot be fed in the event the instantaneous CO is above 1000 ppm. All hazardous waste shall automatically shut down if the rolling-hour average CO exceeds 90 ppm for 15 min. This time delay is to avoid unnecessary system upsets. Based on trial burn testing (Mode 2) and supplementary data submitted, the CO concentration without containerized solid-waste feed would be expected to average about 20 ppm, with no peaks exceeding 300 ppm. Therefore, the waste-feed cutoff strategy purposely prevents the feeding of solid waste as the primary measure to control CO, and hence assures compliance with DRE demonstrated during the trial burns.

The waste-feed rate permit conditions included total waste feed in MBtu/h measured by stack gas-flow rate (I.D. fan amps) and combustion zone temperature (maximum 1-h average). In addition, a pack feed rate of 40/h and a non-fuel-grade liquid feed of 10 gpm to the kiln and 5 gpm to the secondary were established.

TABLE 6
Permit Limit Summary

Operating parameter	Permit limit	Permit condition
Mode 1		
Secondary combustion chamber (SCC) temperature	1600°F	The secondary combustion chamber temperature shall be maintained above 1600°F. All hazardous waste feed shall shut down automatically if this temperature is less than 1550°F for 1 min.
Kiln temperature	1400°F (minimum)	The kiln temperature shall be maintained above 1400°F. No containerized solid waste may be fed any time the kiln temperature drops below 1400°F for 1 min. Containerized waste feed may not resume until the kiln temperature is above 1400°F. All hazardous waste feed to the kiln shall automatically shut down if the kiln temperature drops below 1400°F for 5 min. Hazardous waste feed may not resume until the kiln temperature is above 1400°F.
Combustion gas flow	220 A (aim point) 232 A (maximum)	Combustion gas flow shall be maintained less 105,000 ACFM as measured by induced draft fan amperage at 232 A. An alarm shall be set at 227 A and all hazardous waste feed shall be shut down automatically if the permit limit of 232 A is exceeded of 5 min.
Carbon monoxide (CO)	90 ppm rolling 1-h average	Two stages of hazardous waste feed shut down shall occur based on the rolling 1-h average limit of 90 ppm. 1. The containerized solid waste system shall shut down automatically if the rolling hour average CO exceeds 90 ppm at any time and remain shut down until the rolling hour average CO has dropped below 90 ppm for 15 min. 2. All hazardous waste shall automatically shut down if the rolling hour average CO exceeds 90 ppm for 15 min. Liquid hazardous waste feed may be resumed once the rolling hour CO has dropped below 90 ppm for at least 15 min. Solid waste may be resumed 15 min after the liquid waste feed is resumed.
	1000 ppm instantaneous	No containerized solid waste may be fed at any time the instantaneous CO concentration is above 1000 ppm.
Waste feed rate	133 million Btu/h (MBtu/h)	Total waste feed rate shall be below 133 MBtu/h by maintaining the combustion gas flow (measured by induced draft fan amperage) below 232 A and the secondary combustion chamber temperature below 2000°F. All hazardous waste feed shall automatically shut down if the induced draft fan amperage exceeds 232 A for 5 min OR the secondary combustion chamber temperature exceeds 2000°G for a 1-h average.
Containerized solid waste	40 packs/h feed rate	The rate of containerized solid waste feed shall not exceed 40 packs per hour.
Nonfuel grade liquid feed rate	10 gpm (kiln) 5 gpm (SCC)	The rate of nonfuel grade liquid waste shall not exceed 85 lb/min to the kiln or 45 lb/min to the secondary combustion chamber.

CONCLUSION

System performance was demonstrated while operating the incinerator under "worst case" conditions. Normal operating feed and operating conditions seldom approach the conditions tested. As a result of the test, permit conditions could be set which allow a flexibility in operation under five operating modes while always complying with the performance standard of 99.99% DRE required by 40 CFR 264.343.

PRODUCTS OF INCOMPLETE COMBUSTION (PICs) TESTING

Both Eastman Kodak Company and the New York State Department of Environmental Conservation (NYSDEC) were interested in the possible emission of products of incomplete combustion (PICs) from Kodak's Building 218 Chemical Waste Incinerator. During the planning of the USEPA Trial Burn, it was agreed between Kodak and NYSDEC that separate PIC testing, if required, would be done as a condition of the NYSDEC Certificate to Operate. Of primary concern were the emission of polychlorinated dibenzodioxins and dibenzofurans (PCDD/PCDF). As a "worst case" test, a potential precursor to the formation of these PICs not normally present in Kodak waste, 1,2,4-trichlorobenzene, was chosen as the POHC for these tests. Tests measured the DRE of 1,2,4-trichlorobenzene and the emission of selected PICs — including benzo(a)pyrene, chrysene, PCDD and PCDF. The PCDD and PCDF results were used to determine the "2,3,7,8-TCDD equivalent" concentration using both the New York State and USEPA methods of calculation. The impact of these emissions on ground-level concentrations was assessed using mathematical dispersion modeling procedures.

PIC FORMATION THEORETICAL DISCUSSION

Dr. Barry Dellinger provided Kodak with a report entitled "Theoretical Discussion of Potential Emissions of Products of Incomplete Combustion from the Kodak Park Building 218 Chemical Waste Incinerator"[10] in April 1985. The purpose of that report was to review hazardous organic emissions data available in the literature at that time, assess the various mechanisms of PIC formation, and utilize this information along with results of laboratory thermal decomposition testing on a simulated waste stream to predict the type and level of organic emissions from the Kodak Park Chemical Waste Incinerator. Nonflame laboratory thermodegradation testing of two waste mixtures typical of Kodak's waste-liquid profile was conducted. One waste was a mixture of 10 listed compounds and two additional solvents; the second a process till-bottom tar sample. Acetonitrile and chlorobenzene were shown to be the most stable POHCs of the mixture. PICs were identified; all were eliminated at 775°C and 2 s of residence time. Similar results were achieved on the tar residue with elimination of PICs at 927°C. No 2,3,7,8-TCDD or 2,3,7,8-TCDF were observed in this laboratory study.

Based on the laboratory data and the Kodak incinerator flame (1100°C) and postflame temperatures of 800—900°C for 2—3 s, it would be expected that liquid wastes would be destroyed very efficiently in the flame zone (i.e., process tar) and only that waste escaping the flame zone would potentially cause POHC or PIC emissions.

Kodak's EPA Trial Burn[4] verified that DRE of acentonitrile approached 99.999% for the permit condition established (<90 ppm rolling-hour average CO). Thus, the actual incinerator POHC emissions were low and PICs should be corresponding low.

Dellinger's discussion[10] concluded that laboratory thermal decomposition testing suggested emission levels of any single PIC resulting from the partial destruction of the waste feed (including 2,3,7,8-TCDD and 2,3,7,8-TCDF) may be as much as 100,000 times less than the acetonitrile or other refractory POHC emission rate. He also concluded, based on the laboratory testing, identified PICs should be of lesser thermal stability than acetonitrile and comparable to the laboratory determined stability of tetrachloromethane.

PERFORMANCE AND EMISSION TESTS FOR NYSDEC PRE-USEPA TRIAL BURN TESTING

Although not indicative of current performance, since modifications to the process made in 1983 improved organic destruction of liquid waste significantly, results of testing accom-

TABLE 7
B-218 Organic Emissions and Destruction Removal Efficiencies 1980
Testing

| Test no. | Organic[a] feed, lb/h | Location | Organic emissions (lb/h) | | | DRE[e] |
			VOC[b]	Semivolatile[c]	Nonvolatile[d]	
1	9,640	Stack	1.3	0.023	1.0	99.976
		Scrubber	0.53	0.0063	ND[f]	
2	6,926	Stack	0.29	0.019	0.79	99.983
		Scrubber	0.25	0.019	0.25	
3	10,793	Stack	1.0	0.18	3.2	99.959
		Scrubber	0.50	0.0075	ND[f]	

[a]   Organic feed = total mass − (water + ash).
[b]   Boiling points <100°C (total of 19 specific compounds).
[c]   100°C < boiling points < 300°C.
[d]   Boiling points >300°C.
[e]   DRE = destruction removal efficiency.
[f]   ND = none detected.

plished in the Fall of 1980[11] can be used to substantiate the theory that PIC emissions from efficient incinerator processes, as measured by DRE, do not significantly increase the total emission from the process over that which would be predictable by 99.99% DRE of all organic matter fed.

In the testing, the main organic emissions from the incinerator were totally characterized, in that the volatile, semivolatile, and nonvolatile fractions were quantified. Volatile organics were collected in evacuated Tedlar bags and analyzed by GC/FID and GC/MS. Nonvolatiles and semivolatiles were collected on XAD-2 resin in a modified EPA Method 5 Sampling Train, and analyzed using the EPA Level 1 environmental 5 Sampling Train, and analyzed using the EPA Level 1 environmental procedure.[12] As shown in Table 7, total organic mass feed was approximately 7000 to 11,000 lb/h while total emissions ranged from 1.1 to 4.4 lb/h. The DREs of 19 volatile organic compounds were determined. Most individual DREs were 99.99% or greater. The overall volatile organic compound (VOC) DRE was 99.99% for two of the three test runs and 99.97% for the third. When considering the total DRE of all organics in and out of the process, two of the three runs were approximately 99.98% while the third and lowest DRE was 99.96%.

If one assumed all the semivolatile and nonvolatile emissions were PICs, which they probably were not, they would represent approximately 0.01% of the feed, or approximately equivalent to the mass of total POHC emission when POHCs DREs are 99.99%. With the process improvements made since 1980 and operating under conditions of the USEPA permit, predicted total emissions today would be an order of magnitude less (99.999% DRE and 0.001% total PICs). With an organic feed rate of 10,000 lb/h, expected total PIC emissions based on this extension of test data would be approximately 0.1 lb/h, and expected total organic emissions would be approximately 0.2 lb/h.

POST-USEPA TRIAL BURN TESTING

Testing in June 1986[12] was designed to assess organic process emissions to include full evaluation of polychlorinated dibenzodioxins and dibenzofurans (PCDD/PCDF) and to substantiate the theory supported by Dellinger's discussion.[10] Waste feed prepared to specification introduced a dioxin precursor, 1,2,4-trichlorobenzene, not normally present in Kodak's waste profile but mentioned in discussion with the NYSDEC and chosen as the POHC for demonstration of DRE for this trial burn. The DRE of 1,2,4-trichlorobenzene and the emission

413

TABLE 8
Emissions Summary 2,3,7,8-TCDD Equivalents (ng/m³)[a]

Run	PCDD	NYS method PCDF	Total	EPA method PCDD	PCDF	Total
1	0.038	0.75	0.79	0.019	0.23	0.25
2	0.02	0.70	0.72	0.011	0.22	0.23
3	0.02	0.60	0.62	0.011	0.18	0.19
Average	0.026	0.68	0.71	0.014	0.21	0.22

[a] Actual cubic meter based on stack exit conditions of 166°F and 35% moisture.

of selected PICs — including benzo(a)pyrene, chrysene, PCDD and PCDF — were measured. The PCDD and PCDF results were used to determine the "2,3,7,8-TCDD equivalent" concentration using both the New York State and USEPA methods of calculation. The impact of these emissions on groundlevel concentrations was addressed using mathematical dispersion modeling procedures. A modified Method 5 sampling train containing XAD-2 resin was used to collect the organic. High-resolution gas chromatography/low-resolution mass spectroscopy and USEPA method 8270 of SW-846 were used for analysis. The American Society of Mechanical Engineers protocol "Analytical Procedures to Assay Stack Effluent Samples and Residual Combustion Products of Polychlorinated Dibenzo-p-Dioxins (PCDD) and Polychlorinated Dibenzofurans (PCDF)" — as issued September 18, 1984 and revised December 31, 1984 — was used for isolation and purification of the PCDD and PCDF.

The system was operated near capacity, with all POHC in the specially prepared, containerized solid-waste ("synthetic packs"). These packs also contained diatomaceous earth as an ash source and tetrachloromethane as a source of organic chlorine. Other waste feeds were representative of those normally treated. They included the complex tarry waste which had been subjected to laboratory testing by Dellinger and introduced chlorinated organics and other constituents that could contribute to PIC formation. This combination of waste feed simulated a "worst case" combination for the potential of PIC formation, especially TCDD and TCDF. The carbon monoxide concentration was purposely elevated, as it had been for the EPA Trial Burn, by the controlled feeding of high-heat-value containerized solid waste.

The average DRE for trichlorobenzene was 99.9953%, with associated average emission rate of 0.018 lb/h. Table 8 presents PCDD and PCDF emissions measured in ng/m³, including the average isomer emissions and "TCDD equivalents". PIC/POHC ratios for individual PICs again were shown to be low, ranging from 1/600 to 1/250,000 (Table 9).

Applying the PIC/POHC ratio range from this test to typical POHC emissions from the USEPA Trial Burn ranging from 0.001 to 0.05 lb/h would result in theoretical emission of 0.000000004 to 0.00008 lb/h for individual PICs. As discussed earlier, based on the total assessment of all process emissions made during 1980 testing and on improvements to the process made subsequently, the current estimate of total PIC emission during routine operation would be approximately 0.1 lb/h. For the aggregate of individual PICs to reach this estimated total normal mass emission, 1000 to 10,000,000 individual compounds would need to be identified.

Applying this same logic to actual PCDD and PCDF emission in the order of magnitude measured, i.e., 0.000001 to 0.0000001 lb/h, 100,000 to 1,000,000 individual PICs of this magnitude would need to be identified. It is very unlikely this many individual compounds would be identified if a full PIC assessment were made. In addition, it is unlikely that the impact of any other compound on human health or the environment would be greater than that of any compound measured in these PIC tests.

TABLE 9
PIC/POHC Ratio

PIC	Range (ng/m³)[a]	Average (ng/m³)[a]	PIC/POHC[b]
PCDD	<0.059—0.59	0.23	1/220,000
PCDF	<1.11—3.02	2.2	1/250,000
TCDD equivalent			
EPA	0.19—0.25	0.22	1/250,000
NYS	0.62—0.79	0.71	1/70,000
Benzo(*a*)pyrene	14—15	15	1/3,500
Chrysene	60—110	85	1/600

[a] Actual cubic meter.
[b] Based on average trichlorobenzene emission of 51,000 ng/m³.

TABLE 10
Air Quality Impact Worst Case Annual
Impact Test Average (μg/m³)

Contaminant	Modeled ground-level concentration
Total TCDD	$<3.4 \times 10^{-10}$
2,3,7,9-TCDD Equivalent	
NYS method	3.4×10^{-9}
EPA method	1.1×10^{-9}
Benzo(*a*)pyrene	$<7.1 \times 10^{-8}$
Chrysene	$<4.3 \times 10^{-7}$
1,2,4-Trichlorobenzene	2.5×10^{-4}

AMBIENT AIR IMPACT

Worst-case annual air quality impact data using GALSON dispersion model GRSGTS[13] are presented in Table 10. Ground level concentrations of total TCDD and "2,3,7,8-TCDD equivalents" were predicted to be significantly less than the previous NYSDEC "Air Guides" acceptable ambient level (AAL) for total TCDD of 9.2×10^{-8} μg/m³. Ambient levels for total TCDD were less than 3.4×10^{-10} μg/m³ with NYSDEC and USEPA "equivalents" of 3.5×10^{-9} μg/m³ and 1.1×10^{-9} μg/m³, respectively. The ambient impact of 1,2,4-tri-chlorobenzene was determined to be 2.5×10^{-4} μg/m³. Benzo(*a*)pyrene and chrysene were modeled to have ambient concentrations of 7.1×10^{-8} and 4.3×10^{-7} μg/m³, respectively.

CONCLUSIONS

1. A well-designed and well-operated incinerator can perform with insignificant quantities of stack emissions.
2. Existing performance standards adopted by EPA in 40 CFR 264.343 adequately control all emissions, including PICs, from hazardous waste incinerators.
3. Laboratory thermal stability data of POHCs are a good measure of difficulty of incineration and of potential for PIC formation.

REFERENCES

1. U.S. Environmental Protection Agency, Part B Permit for Kodak Park Division EPA ID. No. NYD980592497, April 5, 1986.
2. Eastman Kodak Company, Kodak Park Division Hazardous Waste Management Facility EPA ID. No. NYD980592497 RCRA Part B Application, October, 1983.
3. Eastman Kodak Company, Kodak Park Division Hazardous Waste Management Facility EPA ID. No. NYD980592497 RCRA Part B Application Trial Burn Plan Revision TB-1, July 16, 1984.
4. Eastman Kodak Company, Kodak Park Division Trial Burn Report for Chemical Waste Incinerator EPA ID. No. NYD980592497, February 13, 1985.
5. **Austin, D. S., Bastian, R. E., and Wood, R. W.,** Factors Affecting Performance in a 90 Million Btu/HR Chemical Waste Incinerator: Preliminary Findings, May 1982.
6. **Dellinger, B., Hall, D. L., Graham, J. L., and Rubey, W. A.,** Destruction Efficiency Testing of Selected Compounds and Wastes, University of Dayton, Dayton, OH, September 20, 1984.
7. **Dellinger, B., Torres, J. L., Rubey, D. L., Graham, J. L., and Carnes, R. A.,** Determination of the thermal stability of selected hazardous organic compounds, *Hazardous Waste,* July 1984.
8. **Moses, S.,** Modeling of B-218 Tetrachloromethane and Acetonitrile Emissions, June 22, 1987.
9. New York State Department of Environmental Conservation, Air Guide — 1.
10. **Dellinger, B.,** Theoretical Discussion of Potential Emissions of Products of Incomplete Combustion from the Kodak Park Building 218 Chemical Waste Incinerator, April 29, 1985.
11. Eastman Kodak Company, Emission Test Report Building 218 Chemical Waste Incinerator, submitted to NYSDEC, November 1980.
12. **Lentzen, et al.,** IERL-RTP Procedures Manual: Level 1 Environmental Assessment, National Technical Information Service, Research Triangle Park, NC, 1978.
13. Eastman Kodak Company, Building 218 Chemical Waste Incinerator Products of Incomplete Combustion Test Burn, June 1986.

Appendix B

DATA BASE

Mark White

B.1   Physical Properties of Materials..418

B.2   Combustion Parameters 505

B.3   Equipment Performance...547

Appendix B.1

PHYSICAL PROPERTIES OF MATERIALS

TABLE 1
Density of Various Solids

The approximate density of various solids at ordinary atmospheric temperature.
In the case of substances with voids such as paper or leather the bulk density is indicated rather than the density of the solid portion.

(Selected principally from the Smithsonian Tables.)

Substance	Grams per cu cm	Pounds per cu ft	Substance	Grams per cu cm	Pound per cu ft	Substance	Grams per cu cm	Pounds per cu ft
Agate	2.5–2.7	156–168	Glass, common	2.4–2.8	150–175	Tallow, beef	0.94	59
Alabaster, carbon-			flint	2.9–5.9	180–370	mutton	0.94	59
ate	2.69–2.78	168–173	Glue	1.27	79	Tar	1.02	66
sulfate	2.26–2.32	141–145	Granite	2.64–2.76	165–172	Topaz	3.5–3.6	219–223
Albite	2.62–2.65	163–165	Graphite*	2.30–2.72	144–170	Tourmaline	3.0–3.2	190–200
Amber	1.06–1.11	66–69	Gum arabic	1.3–1.4	81–87	Wax, sealing	1.8	112
Amphiboles	2.9–3.2	180–200	Gypsum	2.31–2.33	144–145	Wood (seasoned)		
Anorthite	2.74–2.76	171–172	Hematite	4.9–5.3	306–330	alder	0.42–0.68	26–42
Asbestos	2.0–2.8	125–175	Hornblende	3.0	187	apple	0.66–0.84	41–52
Asbestos slate	1.8	112	Ice	0.917	57.2	ash	0.65–0.85	40–53
Asphalt	1.1–1.5	69–94	Ivory	1.83–1.92	114–120	balsa	0.11–0.14	7–9
Basalt	2.4–3.1	150–190	Leather, dry	0.86	54	bamboo	0.31–0.40	19–25
Beeswax	0.96–0.97	60–61	Lime, slaked	1.3–1.4	81–87	basswood	0.32–0.59	20–37
Beryl	2.69–2.7	168–169	Limestone	2.68–2.76	167–171	beech	0.70–0.90	43–56
Biotite	2.7–3.1	170–190	Linoleum	1.18	74	birch	0.51–0.77	32–48
Bone	1.7–2.0	106–125	Magnetite	4.9–5.2	306–324	blue gum	1.00	62
Brick	1.4–2.2	87–137	Malachite	3.7–4.1	231–256	box	0.95–1.16	59–72
Butter	0.86–0.87	53–54	Marble	2.6–2.84	160–177	butternut	0.38	24
Calamine	4.1–4.5	255–280	Meerschaum	0.99–1.28	62–80	cedar	0.49–0.57	30–35
Calcspar	2.6–2.8	162–175	Mica	2.6–3.2	165–200	cherry	0.70–0.90	43–56
Camphor	0.99	62	Muscovite	2.76–3.00	172–187	dogwood	0.76	47
Caoutchouc	0.92–0.99	57–62	Ochre	3.5	218	ebony	1.11–1.33	69–83
Cardboard	0.69	43	Opal	2.2	137	elm	0.54–0.60	34–37
Celluloid	1.4	87	Paper	0.7–1.15	44–72	hickory	0.60–0.93	37–58
Cement, set	2.7–3.0	170.190	Paraffin	0.87–0.91	54–57	holly	0.76	47
Chalk	1.9–2.8	118–175	Peat blocks	0.84	52	juniper	0.56	35
Charcoal, oak	0.57	35	Pitch	1.07	67	larch	0.50–0.56	31–35
pine	0.28–0.44	18–28	Porcelain	2.3–2.5	143–156	lignum vitae	1.17–1.33	73–83
Cinnabar	8.12	507	Porphyry	2.6–2.9	162–181	locust	0.67–0.71	42–44
Clay	1.8–2.6	112–162	Pressed wood			logwood	0.91	57
Coal, anthracite	1.4–1.8	87–112	pulp board	0.19	12	mahogany		
bituminous	1.2–1.5	75–94	Pyrite	4.95–5.1	309–318	Honduras	0.66	41
Cocoa butter	0.89–0.91	56–57	Quartz	2.65	165	Spanish	0.85	53
Coke	1.0–1.7	62–105	Resin	1.07	67	maple	0.62–0.75	39–47
Copal	1.04–1.14	65–71	Rock salt	2.18	136	oak	0.60–0.90	37–56
Cork	0.22–0.26	14–16	Rubber, hard	1.19	74	pear	0.61–0.73	38–45
Cork linoleum	0.54	34	Rubber, soft			pine, pitch	0.83–0.85	52–53
Corundum	3.9–4.0	245–250	commercial	1.1	69	white	0.35–0.50	22–31
Diamond	3.01–3.52	188–220	pure gum	0.91–0.93	57–58	yellow	0.37–0.60	23–37
Dolomite	2.84	177	Sandstone	2.14–2.36	134–147	plum	0.66–0.78	41–49
Ebonite	1.15	72	Serpentine	2.50–2.65	156–165	poplar	0.35–0.5	22–31
Emery	4.0	250	Silica, fused trans-			satinwood	0.95	59
Epidote	3.25–3.50	203–218	parent	2.21	138	spruce	0.48–0.70	30–44
Feldspar	2.55–2.75	159–172	translucent	2.07	129	sycamore	0.40–0.60	24–37
Flint	2.63	164	Slag	2.0–3.9	125–240	teak, Indian	0.66–0.88	41–55
Fluorite	3.18	198	Slate	2.6–3.3	162–205	African	0.98	61
Galena	7.3–7.6	460–470	Soapstone	2.6–2.8	162–175	walnut	0.64–0.70	40–43
Gamboge	1.2	75	Spermaceti	0.95	59	water gum	1.00	62
Garnet	3.15–4.3	197–268	Starch	1.53	95	willow	0.40–0.60	24–37
Gas carbon	1.88	117	Sugar	1.59	99			
Gelatin	1.27	79	Talc	2.7–2.8	168–174			

* Some values reported as low as 1.6.

TABLE 1 (continued)
Density of Various Solids

The approximate density of various solids at ordinary atmospheric temperature.
In the case of substances with voids such as paper or leather the bulk density is indicated rather than the density of the solid portion.

(Selected principally from the Smithsonian Tables.)

Substance	Grams per cu cm	Pounds per cu ft	Substance	Grams per cu cm	Pound per cu ft	Substance	Grams per cu cm	Pounds per cu ft
Agate	2.5–2.7	156–168	Glass, common	2.4–2.8	150–175	Tallow, beef	0.94	59
Alabaster, carbon-			flint	2.9–5.9	180–370	mutton	0.94	59
ate	2.69–2.78	168–173	Glue	1.27	79	Tar	1.02	66
sulfate	2.26–2.32	141–145	Granite	2.64–2.76	165–172	Topaz	3.5–3.6	219–223
Albite	2.62–2.65	163–165	Graphite*	2.30–2.72	144–170	Tourmaline	3.0–3.2	190–200
Amber	1.06–1.11	66–69	Gum arabic	1.3–1.4	81–87	Wax, sealing	1.8	112
Amphiboles	2.9–3.2	180–200	Gypsum	2.31–2.33	144–145	Wood (seasoned)		
Anorthite	2.74–2.76	171–172	Hematite	4.9–5.3	306–330	alder	0.42–0.68	26–42
Asbestos	2.0–2.8	125–175	Hornblende	3.0	187	apple	0.66–0.84	41–52
Asbestos slate	1.8	112	Ice	0.917	57.2	ash	0.65–0.85	40–53
Asphalt	1.1–1.5	69–94	Ivory	1.83–1.92	114–120	balsa	0.11–0.14	7–9
Basalt	2.4–3.1	150–190	Leather, dry	0.86	54	bamboo	0.31–0.40	19–25
Beeswax	0.96–0.97	60–61	Lime, slaked	1.3–1.4	81–87	basswood	0.32–0.59	20–37
Beryl	2.69–2.7	168–169	Limestone	2.68–2.76	167–171	beech	0.70–0.90	43–56
Biotite	2.7–3.1	170–190	Linoleum	1.18	74	birch	0.51–0.77	32–48
Bone	1.7–2.0	106–125	Magnetite	4.9–5.2	306–324	blue gum	1.00	62
Brick	1.4–2.2	87–137	Malachite	3.7–4.1	231–256	box	0.95–1.16	59–72
Butter	0.86–0.87	53–54	Marble	2.6–2.84	160–177	butternut	0.38	24
Calamine	4.1–4.5	255–280	Meerschaum	0.99–1.28	62–80	cedar	0.49–0.57	30–35
Calcspar	2.6–2.8	162–175	Mica	2.6–3.2	165–200	cherry	0.70–0.90	43–56
Camphor	0.99	62	Muscovite	2.76–3.00	172–187	dogwood	0.76	47
Caoutchouc	0.92–0.99	57–62	Ochre	3.5	218	ebony	1.11–1.33	69–83
Cardboard	0.69	43	Opal	2.2	137	elm	0.54–0.60	34–37
Celluloid	1.4	87	Paper	0.7–1.15	44–72	hickory	0.60–0.93	37–58
Cement, set	2.7–3.0	170.190	Paraffin	0.87–0.91	54–57	holly	0.76	47
Chalk	1.9–2.8	118–175	Peat blocks	0.84	52	juniper	0.56	35
Charcoal, oak	0.57	35	Pitch	1.07	67	larch	0.50–0.56	31–35
pine	0.28–0.44	18–28	Porcelain	2.3–2.5	143–156	lignum vitae	1.17–1.33	73–83
Cinnabar	8.12	507	Porphyry	2.6–2.9	162–181	locust	0.67–0.71	42–44
Clay	1.8–2.6	112–162	Pressed wood			logwood	0.91	57
Coal, anthracite	1.4–1.8	87–112	pulp board	0.19	12	mahogany		
bituminous	1.2–1.5	75–94	Pyrite	4.95–5.1	309–318	Honduras	0.66	41
Cocoa butter	0.89–0.91	56–57	Quartz	2.65	165	Spanish	0.85	53
Coke	1.0–1.7	62–105	Resin	1.07	67	maple	0.62–0.75	39–47
Copal	1.04–1.14	65–71	Rock salt	2.18	136	oak	0.60–0.90	37–56
Cork	0.22–0.26	14–16	Rubber, hard	1.19	74	pear	0.61–0.73	38–45
Cork linoleum	0.54	34	Rubber, soft			pine, pitch	0.83–0.85	52–53
Corundum	3.9–4.0	245–250	commercial	1.1	69	white	0.35–0.50	22–31
Diamond	3.01–3.52	188–220	pure gum	0.91–0.93	57–58	yellow	0.37–0.60	23–37
Dolomite	2.84	177	Sandstone	2.14–2.36	134–147	plum	0.66–0.78	41–49
Ebonite	1.15	72	Serpentine	2.50–2.65	156–165	poplar	0.35–0.5	22–31
Emery	4.0	250	Silica, fused trans-			satinwood	0.95	59
Epidote	3.25–3.50	203–218	parent	2.21	138	spruce	0.48–0.70	30–44
Feldspar	2.55–2.75	159–172	translucent	2.07	129	sycamore	0.40–0.60	24–37
Flint	2.63	164	Slag	2.0–3.9	125–240	teak, Indian	0.66–0.88	41–55
Fluorite	3.18	198	Slate	2.6–3.3	162–205	African	0.98	61
Galena	7.3–7.6	460–470	Soapstone	2.6–2.8	162–175	walnut	0.64–0.70	40–43
Gamboge	1.2	75	Spermacéti	0.95	59	water gum	1.00	62
Garnet	3.15–4.3	197–268	Starch	1.53	95	willow	0.40–0.60	24–37
Gas carbon	1.88	117	Sugar	1.59	99			
Gelatin	1.27	79	Talc	2.7–2.8	168–174			

* Some values reported as low as 1.6.

From Weast, R. C., Ed., *Handbook of Chemistry and Physics*, 69th ed., CRC Press, Boca Raton, FL, 1988.

TABLE 2
Viscosity of Liquids

Viscosity of liquids in centipoises (cp) including elements, inorganic and organic compounds and mixtures.

Liquid	Temp. °C	Viscosity cp	Liquid	Temp. °C	Viscosity cp
Acetaldehyde	0	.2797		700	1.26
	10	.2557		800	1.08
	20	.22		850	1.05
Acetanilide	120	2.22	Benzaldehyde	25	1.39
	130	1.90	Benzene	0	.912
Acetic acid	15	1.31		10	.758
	18	1.30		20	.652
	25.2	1.155		30	.564
	30	1.04		40	.503
	41	1.00		50	.442
	59	.70		60	.392
	70	.60		70	.358
	100	.43		80	.329
anhydride	0	1.24	Benzonitrile	25	1.24
	15	.971	Benzophenone	55	4.79
	18	.90		120	1.38
	30	.783	Benzyl alcohol	20	5.8
	100	.49	Benzylamine	25	1.59
Acetone	−92.5	2.148	Benzylaniline	33	2.18
	−80.0	1.487		130	1.20
	−59.6	.932	Benzyl ether	0	10.5
	−42.5	.695		20	5.33
	−30.0	.575		40	3.21
	−20.9	.510	Bismuth	285	1.61
	−13.0	.470		304	1.662
	−10.0	.450		365	1.46
	0	.399		451	1.280
	15	.337		600	.998
	25	.316	Bromine, liq	−4.3	1.31
	30	.295		0	1.241
	41	.280		12.6	1.07
Acetonitrile	0	.442		16	1.0
	15	.375		19.5	.995
	25	.345		28.9	.911
Acetophenone	11.9	2.28	o-Bromoaniline	40	3.19
	23.5	1.59	m-Bromoaniline	20	6.81
	25.0	1.617		40	3.70
	50.0	1.246		80	1.70
	80.0	.734	p-Bromoaniline	80	1.81
Air, liq	−192.3	.172	Bromobenzene	15	1.196
Alcohol. See *Ethyl, Methyl,*				30	.985
etc.			Bromoform	15	2.152
Allyl alcohol	0	2.145		25	1.89
	15	1.49		30	1.741
	20	1.363	Butyl acetate	0	1.004
	30	1.07		20	.732
	40	.914		40	.563
	70	.553	n-Butyl alcohol	−50.9	36.1
Allylamine	130	.506		−30.1	14.7
Allyl chloride	15	.347		−22.4	11.1
	30	.300		−14.1	8.38
Ammonia	−69	.475		0	5.186
	−50	.317		15	3.379
	−40	.276		20	2.948
	−33.5	.255		30	2.30
n-Amyl acetate	11	1.58		40	1.782
	45	.805		50	1.411
alcohol	15	4.65		70	.930
	30	2.99		100	.540
ether	15	1.188	sec-Butyl alcohol	15	4.21
Aniline	−6	13.8	n-Butyl bromide	15	.626
	0	10.2	n-Butyl chloride	15	.469
	5	8.06	Butyl chloride, tertiary	15	.543
	10	6.50	n-Butyl formate	0	.940
	15	5.31		20	.689
	20	4.40	Butyric acid	0	2.286
	25	3.71		15	1.81
	30	3.16		20	1.540
	35	2.71		40	1.120
	40	2.37		50	.975
	50	1.85		70	.760
	60	1.51		100	.551
	70	1.27	Cadmium, liq	349	1.44
	80	1.09		506	1.18
	90	.935		603	1.10
	100	.825	Carbolic acid. See *Phenol.*		
Anisol	0	1.78	Carbon dioxide, liq., pres-	0	.099
	20	1.32	sure that of saturated	10	.085
	40	1.12	vapor	20	.071
Antimony, liq	645	1.55		30	.053

TABLE 2 (continued)
Viscosity of Liquids

Liquid	Temp. °C	Viscosity cp	Liquid	Temp. °C	Viscosity cp
disulfide	−13	.514		−40	.461
	−10	.495		−20	.362
	0	.436		0	.2842
	5	.380		17	.240
	20	.363		20	.2332
	40	.330		25	.222
Carbon tetrachloride	0	1.329		40	.197
	15	1.038		60	.166
	20	.969		80	.140
	30	.843		100	.118
	40	.739	Ethyl acetate	0	.582
	50	.651		8.96	.516
	60	.585		10	.512
	70	.524		15	.473
	80	.468		20	.455
	90	.426		25	.441
	100	.384		30	.400
Cetyl alcohol	50	13.4		50	.345
Chlorine, liq	−76.5	.729		75	.283
	−70.5	.680	Ethyl alcohol	−98.11	44.0
	−60.2	.616		−89.8	28.4
	−52.4	.566		−71.5	13.2
	−35.4	.494		−59.42	8.41
	0	.385		−52.58	6.87
Chlorobenzene	15	.900		−32.01	3.84
	20	.799		−17.59	2.68
	40	.631		−30	1.80
	80	.431		0	1.773
	100	.367		10	1.466
Chloroform	−13	.855		20	1.200
	0	.700		30	1.003
	8.1	.643		40	.834
	15	.596		50	.702
	20	.58		60	.592
	25	.542		70	.504
	30	.514	Ethyl alcohol, anh.	−148	8,470
	39	.500		−146	5,990
o-Chlorophenol	25	4.11		−130	467
	50	2.015	Ethyl aniline	25	2.04
m-Chlorophenol	25	11.55	Ethylbenzene	17	.891
p-Chlorophenol	50	4.99	Ethyl benzoate	20	2.24
Copper, liq	1,085	3.36	Ethyl bromide	−120	5.6
	1,100	3.33		−100	2.89
	1,150	3.22		−80	1.81
	1,200	3.12		0	.487
o-Cresol	40	4.49		10	.441
m-Cresol	10	43.9		15	.418
	20	20.8		20	.402
	40	6.18		30	.348
p-Cresol	40	7.00	n-Ethyl butyrate	15	.711
Creosote	20	12.0	Ethyl carbonate	15	.868
Cycloheptane	13.5	1.64	Ethylene bromide	0	2.438
Cyclohexane	17	1.02		17	1.95
Cyclohexanol	20	68		20	1.721
Cyclohexene	13.5	.696		40	1.286
	20	.66		67.3	.922
Cyclooctane	13.5	2.35		70	.903
Cyclopentane	13.5	.493		82.2	.750
n Decane	20	.92		99.0	.648
Diethylamine	25	.346	chloride	0	1.077
	25	.367		15	.887
Diethylaniline	5	3.84		19.4	.800
	20.0	2.18		40	.652
	25.0	1.95		50	.565
Diethylcarbinol	15.0	7.34		70	.479
Diethylketone	15	.493	glycol	20	19.9
Dimethylaniline	10	1.69		40	9.13
	20	1.41		60	4.95
	25	1.285		80	3.02
	30	1.17		100	1.99
	40	1.04	oxide	−49.8	.577
	50	.91		−38.2	.488
Dimethyl-α-naphthylamine	130	.868		−21.0	.394
Dimethyl-β-naphthylamine	130	.952		0	.320
Diphenyl	70	1.49	Ethyl formate	20	.402
	100	.97	iodide	0	.727
Diphenylamine	130	1.04		15	.617
Dodecane	25	1.35		20	.592
Ether (diethyl-)	−100	1.69		40	.495
	−80	.958		70	.391
	−60	.637	malate	24.7	3.016
			oxalate	15	2.31

TABLE 2 (continued)
Viscosity of Liquids

Liquid	Temp. °C	Viscosity cp	Liquid	Temp. °C	Viscosity cp
propionate	15	.564	Isopropyl alcohol	15	2.86
Eugenol	0	29.9		30	1.77
	20	9.22	Isoquinoline	25	3.57
	40	4.22	Isosafrol	25	3.981
Fluorobenzene	20	.598	Lead, liq	350	2.58
	40	.478		400	2.33
	60	.389		441	2.116
	80	.329		500	1.84
	100	.275		551	1.70
Formamide	0	7.55		600	1.38
	25	3.30		703	1.349
Formic acid	7.59	2.3868		844	1.185
	10	2.262	Menthol, liq	55.6	6.29
	20	1.804		74.6	2.47
	30	1.465		99.0	1.04
	40	1.219	Mercury	-20	1.855
	70	.780		-10	1.764
	100	.549		0	1.685
Furfural	0	2.48		10	1.615
	25	1.49		19.02	1.56
Glucose	22	9.1×10^{11}		20	1.554
	30	6.6×10^{11}		20.2	1.55
	40	2.8×10^{11}		30	1.499
	60	9.3×10^{7}		40	1.450
	80	6.6×10^{5}		40.8	1.45
	100	2.5×10^{4}		41.86	1.44
Glycerin	-42	6.71×10^{8}		50	1.407
	-36	2.05×10^{8}		60	1.367
	-25	2.62×10^{8}		70	1.331
	-20	1.34×10^{8}		80	1.298
	-15.4	6.65×10^{4}		90	1.268
	-10.8	3.55×10^{4}		100	1.240
	-4.2	1.49×10^{4}		150	1.130
	0	12,110		200	1.052
	6	6,260		250	.995
	15	2,330		300	.950
	20	1,490		340	.921
	25	954	Methyl acetate	0	.484
	30	629		20	.381
Glycerin trinitrate	10	69.2		40	.320
	20	36.0	Methyl alcohol	-98.30	13.9
	30	21.0	(Methanol)	-84.23	6.8
	40	13.6		-72.55	4.36
	60	6.8		-44.53	1.98
Heptane	0	.524		-22.29	1.22
	17	.461		0	.82
	20	.409		15	.623
	25	.386		20	.597
	40	.341		25	.547
	70	.262		30	.510
n-Heptyl alcohol	15	8.53		40	.456
Hexadecane	20	3.34		50	.403
Hexane	0	.401	Methyl amine	0	.236
	17	.374	aniline	25	2.02
	20	.326		30	1.55
	25	.294	chloride	20	.1834
	40	.271	Methylene bromide	15	1.09
	50	.248		30	0.92
Hydrazine	1	1.29	chloride	15	.449
	10	1.12		30	.393
	20	.97	Methyl iodide	0	.606
Hydrogen, liq		.011		15	.518
Iodine, liq	116	2.27		20	.500
Iodobenzene	15	1.74		30	.460
Iron, 2.5% carbon, liq	1,400	2.25		40	.424
Isoamyl acetate	8.97	1.030	Naphthalene	80	.967
	19.91	.872		100	.776
alcohol	10	6.20	Nitric acid	0	2.275
amine	25	.724		10	1.770
Isobutyl alcohol	15	4.703	Nitrobenzene	2.95	2.91
amine	25	.553		5.69	2.71
Isobutyric acid	15	1.44		5.94	2.71
	30	1.13		9.92	2.48
Isoeugenol	25	26.72		14.94	2.24
Isoheptane	0	.481		20.00	2.03
	20	.384	Nitromethane	0	.853
	40	.315		25	.620
Isohexane	0	.376	o-Nitrotoluene	0	3.83
	20	.306		20	2.37
	40	.254		40	1.63
Isopentane	0	.273		60	1.21
	20	.223	m-Nitrotoluene	20	2.33

TABLE 2 (continued)
Viscosity of Liquids

Liquid	Temp. °C	Viscosity cp	Liquid	Temp. °C	Viscosity cp
	40	1.60	n-Propyl alcohol	0	3.883
	60	1.18		15	2.52
p-Nitrotoluene	60	1.20		20	2.256
n-Nonane	20	.711		30	1.72
n-Octane	0	.706		40	1.405
	16	.574	n-Propyl alcohol	50	1.130
	20	.542		70	760
	40	.433	Propyl aldehyde	10	.47
Octoderane	40	2.86		20	.41
n-Octylalcohol	15	10.6		40	.33
Oil, castor	10	2,420	bromide	0	.651
	20	986		20	.524
	30	451		40	.433
	40	231	chloride	0	.436
	100	16.9		20	.352
cottonseed	20	70.4		40	.291
cylinder, filtered	37.8	240.6	n-Propyl ether	15	.448
	100	18.7	Pyridine	20	.974
cylinder, dark	37.8	422.4	Salicylic acid	10	3.20
	100	24.0		20	2.71
linseed	30	33.1		40	1.81
	50	17.6	Salol	45	.746
	90	7.1	Sodium bromide	762	1.42
machine, light	15.6	113.8		780	1.28
	37.8	34.2	chloride, liq	841	1.30
	100	4.9		896	1.01
machine, heavy	15.6	660.6		924	.97
	37.8	127.4	nitrate, liq	308	2.919
Oil, olive	10	138.0		348	2.439
	20	84.0		398	1.977
	40	36.3		418	1.828
	70	12.4	Stearic acid	70	11.6
rape	0	2,530	Sucrose (cane sugar)	109	2.8×10^8
	10	385		124.6	1.9×10^5
	20	163	Sulfur (gas free)	123.0	10.94
	30	96		135.5	8.66
soya bean	20	69.3		149.5	7.09
	30	40.6		156.3	7.19
	50	20.6		158.2	7.59
	90	7.8		159.2	9.48
sperm	15.6	42.0		159.5	14.45
	37.8	18.5		160.0	22.83
	100.0	4.6		160.3	77.32
Oleic acid	30	25.6		165.0	500.0
Pentadecane	22	2.81		171.0	4,500.0
Pentane	0	289		184.0	16,000.00
	20	240		190.5	19,700.0
o-Phenetidine	0	16.5		197.5	21,300.0
	20	6.08		200.0	21,500.0
	30	4.22		210.0	20,500.0
m-Phenetidine	30	12.9		217.0	19,100.0
p-Phenetidine	20	12.9		220.0	18,600.0
	30	8.3	Sulfur dioxide, liq	-33.5	5508
Phenol	18.3	12.7		-10.5	4285
	50	3.49		0.1	3936
	60	2.61	Sulfuric acid	0	48.4
	70	2.03		15	32.8
	90	1.26		20	25.4
Phenylcyanide	28	1.96		30	15.7
	20.0	1.33		40	11.5
Phosphorus, liq	21.5	2.34	Sulfuric acid	50	8.82
	31.2	2.01		60	7.22
	43.2	1.73		70	6.09
	50.5	1.60		80	5.19
	60.2	1.45	Tetrachloroethane	15	1.844
	69.7	1.32	Tetradecane	20	2.18
	79.9	1.21	Tin, liq	240	2.12
Potassium bromide, liq	745	1.48		280	1.678
	775	1.34		300	1.73
	805	1.19		301	1.680
nitrate, liq	334	2.1		400	1.43
	358	1.7		450	1.270
	333	2.97		500	1.20
	418	2.00		600	1.08
Propionic acid	10	1.289		604	1.045
	15	1.18		750	.905
	20	1.102	Toluene	0	772
	40	.845		17	.61
Propyl acetate	10	66		20	.590
	20	.59		30	.526
	40	.44		40	.471

TABLE 2 (continued)
Viscosity of Liquids

Liquid	Temp. °C	Viscosity cp	Liquid	Temp. °C	Viscosity cp
	70	.354		70	.728
o-Toluidine	20	4.39	Turpentine, Venice	17.3	1.3×10^5
m-Toluidine	20	3.81	n-Undecane	20	1.17
p-Toluidine	50	1.80	o-Xylene (xylol)	0	1.105
Triacetin	17	28.0		16	.876
Tributyrin	20	11.6		20	.810
Trichlorethane	20	1.2		40	.627
Tridecane	23.3	1.55	m-Xylene (xylol)	0	.806
Triethylcarbinol	20	6.75		15	.650
Tripalmitin	70	16.8		20	.620
Tristearin	75	18.5		40	.497
Turpentine	0	2.248	p-Xylene (xylol)	16	.696
	10	1.783		20	.648
	20	1.487		40	.513
	30	1.272	Zinc, liq	280	1.68
	40	1.071		357	1.42
				389	1.31

From Weast, R. C., Ed., *Handbook of Chemistry and Physics*, 69th ed., CRC Press, Boca Raton, FL, 1988.

TABLE 3
Viscosity of Gases

Gas or vapor	Temp. °C	Viscosity micro- poises	Gas or vapor	Temp. °C	Viscosity micro- poises
Acetic acid, vap	119.1	107.0	Benzene, vap.	14.2	73.8
Acetone, vap	100	93.1		131.2	103.1
	119.0	99.1		194.6	119.8
	190.4	118.6		252.5	134.3
	247.7	133.4		312.8	148.4
	306.4	148.1	Bromine, vap.	12.8	151
Acetylene	0	93.5		65.7	170
Air	−194.2	55.1		99.7	188
	−183.1	62.7		139.7	208
	−104.0	113.0		179.7	227
	−69.4	133.3		220.3	248
	−31.6	153.9	Bromoform, vap	151.2	253.0
	0	170.8	Butyl alcohol, n, vap	116.9	143
	18	182.7	tert, vap	82.9	160
	40	190.4	chloride, n, vap	78	149.5
	54	195.8	iodide, vap	130	202
	74	210.2	β-Butylene	18.8	74.4
	229	263.8		100.4	94.5
	334	312.3		200	119.2
	357	317.5	Butyric acid, vap	161.7	130.0
	409	341.3	Carbon dioxide	−97.8	89.6
	466	350.1		−78.2	97.2
	481	358.3		−60.0	106.1
	537	368.6		−40.2	115.5
	565	375.0		−21	129.4
	620	391.6		−19.4	126.0
	638	401.4		0	139.0
	750	426.3		15	145.7
	810	441.9		19	149.9
	923	464.3		20	148.0
	1034	490.6		30	153
	1134	520.6		32	155
				35	156
Alcohol. See *Ethyl*, *Methyl*, etc.				40	157
Ammonia	−78.5	67.2		99.1	186.1
	0	91.8		104	188.9
	20	98.2		182.4	222.1
	50	109.2		235	241.5
	100	127.9		302.0	268.2
	132.9	139.9		490	330.0
	150	146.3		685	380.0
	200	164.6		850	435.8
	250	181.4		1052	478.6
	300	198.7	disulfide, vap	0	91.1
Argon	0	209.6		14.2	96.4
	20	221.7		114.3	130.3
	100	269.5		190.2	156.1
	200	322.3		309.8	196.6
	302	368.5	monoxide	−191.5	56.1
	401	411.5		−78.5	127
	493	448.4		0	166
	584	481.5		15	172
	714	525.7		21.7	175.3
	827	563.2		126.7	218.3
Arsenic hydride (Arsine)	0	145.8		227.0	254.8
	15	114.0		276.9	271.4
	100	198.1	tetrachloride, vap	76.7	195.0

TABLE 3 (continued)
Viscosity of Gases

Gas or vapor	Temp. °C	Viscosity micropoises	Gas or vapor	Temp. °C	Viscosity micropoises
Carbon tetrachloride,				−252.5	8.5
vap	127.9	133.4		−198.4	33.6
	200.2	156.2		−183.4	38.8
	314.9	190.2		−113.5	57.2
Chlorine	12.7	129.7		−97.5	61.5
	20	132.7		−31.6	76.7
	50	146.9		0	83.5
	100	167.9		20.7	87.6
	150	187.5		28.1	89.2
	200	208.5		129.4	108.6
	250	227.6		229.1	126.0
Chloroform, vap	0	93.6		299	138.1
	14.2	98.9		412	155.4
	100	129		490	167.2
	121.3	135.7		601	182.9
	189.1	157.9		713	198.2
	250.0	177.6		825	213.7
	307.5	194.7	bromide	18.7	181.9
Cyanogen	0	92.8		100.2	234.4
	17	98.7	chloride	12.5	138.5
	100	127.1		16.5	140.7
Ethane	−78.5	63.4		18	142.6
	0	84.8		100.3	182.2
	17.2	90.1	iodide	20	165.5
	50.8	100.1		50	201.8
	100.4	114.3		100	231.6
	200.3	140.9		150	262.7
Ether (diethyl), vap	0	67.8		200	292.4
	14.2	71.6		250	318.9
	100	95.5	phosphide	0	106.1
	121.8	98.3		15	112.0
	159.4	107.9		100	143.8
	189.9	115.2	sulfide	0	116.6
	251.0	130.0	Hydrogen sulfide	17	124.1
	277.8	135.8		100	158.7
Ethyl acetate, vap	0	68.4	Iodine, vap	124.0	184
	100	94.3		170.0	204
	128.1	101.8		205.4	220
	158.6	109.8		247.1	240
	192.9	119.5	Isobutyl acetate, vap	16.1	76.4
	212.5	126		116.4	155.0
alcohol, vap	100	108	alcohol, vap	108.4	144.5
	130.2	117.3	bromide, vap	92.3	179.5
	170.7	129.3	butyrate, vap	156.9	167.0
	191.8	135.5	chloride, vap	68.5	150.0
	212.5	140	iodide, vap	120	204.7
	251.7	151.9	Isopentane, vap	25	69.5
	308.7	167.0		100	86.0
bromide, vap	38.4	186.5	Isopropyl alcohol, vap	99.8	109
butyrate, vap	119.8	160.0		120.3	103.1
chloride, vap	0	93.7		198.4	124.8
Ethylene	−75.7	69.9		293.1	148.8
	−44.1	76.9	bromide, vap	60	176.0
	−38.6	78.5	chloride, vap	37.0	148.5
	0	90.7	iodide, vap	89.3	201.5
	13.8	95.4	Krypton	0	232.7
	20	100.8		15	246
Ethylene	50	110.3	Mercury, vap	273	494
	100	125.7		313	551
	150	140.3		369	641
	200	154.1		380	654
	250	166.6	Methane	−181.6	34.8
bromide, vap	131.6	221.0		−78.5	76.0
chloride, vap	83.5	168.0		0	102.6
Ethyl formate, vap	99.8	92		20	108.7
iodide, vap	72.3	216.0		100.0	133.1
Helium	−257.4	27.0		200.5	160.5
	−252.6	35.0		284	181.3
	−191.6	87.1		380	202.6
	0	186.0		499	226.4
	20	194.1	Methyl acetate, vap	99.8	98
	100	228.1		100	100
	200	267.2		143.3	113.9
	250	285.3		218.5	134.8
	282	299.2	alcohol, vap	66.8	135.0
	407	343.6		111.3	125.9
	486	370.6		217.5	162.0
	606	408.7		311.5	192.1
	676	430.3	chloride	−15.3	92
	817	471.3		0	96.9
Hydrogen	−257.7	5.7		15.0	104

TABLE 3 (continued)
Viscosity of Gases

Gas or vapor	Temp. °C	Viscosity micropoises	Gas or vapor	Temp. °C	Viscosity micropoises
	99.1	137		829	501.2
	182.4	168	n-Pentane, vap	25	67.6
	302.0	211		100	84.1
iodide, vap	44	232	Propane	17.9	79.5
Neon	0	297.3		100.4	100.9
	20	311.1		199.3	125.1
	100	364.6	n-Propyl alcohol, vap	99.9	93
	200	424.8		121.7	102.5
	250	453.2		209.7	126.7
Neon	285	470.8		273.0	143.4
	429	545.4	bromide, vap	99.8	119
	502	580.2	Propylene	16.7	83.4
	594	623.0		49.9	93.5
	686	662.6		100.1	107.6
	827	721.0		199.4	133.8
Nitric oxide (NO)	0	178	Propyl iodide, vap	102	210.0
	20	187.6	Sulfur dioxide	-75.0	85.8
	100	227.2		-20.0	107.8
	200	268.2		0	115.8
Nitrogen	-21.5	156.3		0	117
	10.9	170.7		18	124.2
	27.4	178.1		20.5	125.4
	127.2	219.1		100.4	161.2
	226.7	255.9		199.4	203.8
	299	279.7		293	244.7
	490	337.4		490	311.5
	825	419.2	Trimethylbutane. (2,2,3-), vap	70.3	73.4
Nitrosyl chloride	15	113.9		132.2	82.7
	100	150.4		262.1	104.8
	200	192.0	Trimethylethylene, vap	25	70.1
Nitrous oxide (N$_2$O)	0	135		100	86.9
	26.9	148.8	Water, vap	100	125.5
	126.9	194.3		150	144.5
n-Nonane, vap	100.3	63.3		200	163.5
	202.1	78.1		250	182.7
n-Octane, vap	100.4	67.5		300	202.4
	202.2	84.8		350	221.8
Oxygen	0	189		400	241.2
	19.1	201.8	Xenon	0	210.1
	127.7	256.8		16.5	223.5
	227.0	301.7		20	226.0
	283	323.3		127	300.9
	402	369.3		177	335.1
	496	401.3		227	365.2
	608	437.0		277	395.4
	690	461.2			

From Weast, R. C., Ed., *Handbook of Chemistry and Physics*, 69th ed., CRC Press, Boca Raton, FL, 1988.

TABLE 4
Heat Capacity of Organic Liquids and Vapors at 25°C

William E. Acree, Jr.

The values in this table are expressed as J/mol·K at 1 atmosphere pressure. To convert to calories/mol·K multiply at 0.2390057. Compounds in this table are arranged in order of increasing number of carbon atoms in the molecules.

Formula	Name	C$_p$(l)	C$_p$(g)	Ref.
H$_4$N$_2$	Hydrazine	92.9	51.0	1
CCl$_4$	Carbon tetrachloride	132.6	83.3	1
CS$_2$	Carbon disulfide	76.1	45.6	1
CHCl$_3$	Chloroform	115.5	65.7	1
CHN	Hydrogen cyanide	70.7	36.0	1
CH$_2$O$_2$	Formic acid	98.7	45.2	1
CH$_3$NO$_2$	Nitromethane	106.3	57.7	1
CH$_3$NO$_3$	Methyl nitrate	156.9	—	1
CH$_4$O	Methanol	81.6	43.9	1
C$_2$Cl$_4$	Tetrachloroethene	146.5	—	2
C$_2$H$_2$Cl$_2$	1,1-Dichloroethylene	111.3	67.4	1
C$_2$H$_3$Cl$_3$	1,1,1-Trichloroethane	143.9	93.7	1
C$_2$H$_4$Br$_2$	1,2-Dibromoethane	134.7	76.1	1
C$_2$H$_4$O$_2$	Methylformate	119.7	—	3
C$_2$H$_4$O$_2$	Acetic acid	123.0	66.9	1
C$_2$H$_6$O	Ethanol	113.0	65.7	1
C$_2$H$_6$S	Ethanethiol	118.0	72.8	1
C$_2$H$_6$S	Dimethyl sulfide	118.4	74.5	1
C$_2$H$_6$S$_2$	Dimethyl disulfide	146.0	94.6	1
C$_3$H$_3$N	Acrylonitrile	110.9	62.3	1
C$_3$H$_6$O	Acetone	126.4	75.3	1
C$_3$H$_6$O	Propionaldehyde	137.2	—	4
C$_3$H$_6$O$_2$	Ethyl formate	144.3	—	3
C$_3$H$_6$O$_2$	Methyl acetate	143.9	—	3
C$_3$H$_6$S	(CH$_2$)$_3$S (Trimethylene sulfide)	113.4	69.9	1
C$_3$H$_8$O	Propanol-1	144.4	87.4	5,1
C$_3$H$_8$O	Propanol-2	155.2	89.1	1
C$_3$H$_8$S	Methyl ethyl sulfide	144.8	95.4	1
C$_3$H$_8$S	Propanethiol-2	145.2	96.6	1
C$_3$H$_{10}$N$_2$	Tetramethylhydrazine	186.2	—	1
C$_4$H$_4$O	Furan	114.6	—	1
C$_4$H$_5$N	Pyrrole	127.7	—	6
C$_4$H$_6$	Butadiene-1,3	123.4	79.5	1
C$_4$H$_6$	Butyne-2	123.0	77.8	1
C$_4$H$_6$O$_2$	γ-Butyrolactone	141.4	—	3
C$_4$H$_6$O$_2$	Methyl acrylate	161.5	—	3
C$_4$H$_8$O	Butanone-2	159.2	—	7
C$_4$H$_8$O$_2$	p-Dioxane	150.7	—	8
C$_4$H$_8$O$_2$	Ethyl acetate	167.4	—	3
C$_4$H$_8$S	(CH$_2$)$_4$S (Tetrahydrothiophene)	140.2	91.6	1
C$_4$H$_9$N	Pyrrolidine	156.5	—	1
C$_4$H$_{10}$O	Butanol-1	177.1	—	5
C$_4$H$_{10}$O	Diethyl ether	172.0	107.9	1
C$_4$H$_{10}$S	Butanethiol-1	172.8	118.8	1
C$_4$H$_{10}$S	Methyl n-propyl sulfide	171.5	118.0	1
C$_4$H$_{10}$S	Diethyl sulfide	171.1	117.6	1
C$_4$H$_{10}$S	Methyl isopropyl sulfide	172.4	120.5	1
C$_4$H$_{10}$S	2-Methylpropanethiol-2	174.9	121.3	1
C$_4$H$_{10}$S	2-Methylpropanethiol-1	172.0	118.8	1
C$_4$H$_{10}$S	1-Methylpropanethiol-1	171.1	119.7	1
C$_4$H$_{10}$S$_2$	Diethyl disulfide	203.8	141.8	1
C$_4$H$_{12}$Si	Tetramethylsilane	208.4	140.2	1
C$_5$H$_8$	Pentadiene-1,4	145.6	105.0	1
C$_5$H$_8$	2-Methyl butadiene-1,3	153.6	104.6	1
C$_5$H$_{10}$	Pentene-1	155.2	109.6	1
C$_5$H$_{10}$	cis-Pentene-2	151.9	101.7	1
C$_5$H$_{10}$	trans-Pentene-2	157.3	108.4	1
C$_5$H$_{10}$	2-Methyl butene-1	156.9	111.7	1
C$_5$H$_{10}$	3-Methyl butene-1	156.1	118.8	1
C$_5$H$_{10}$	2-Methyl butene-2	152.3	105.0	1
C$_5$H$_{10}$	Cyclopentane	127.2	82.8	1
C$_5$H$_{10}$O	Pentanone-2	185.5	—	4
C$_5$H$_{10}$O	Pentanone-3	190.1	—	4
C$_5$H$_{10}$O	3-Methyl butanone-2	180.6	—	4
C$_5$H$_{10}$O	Valeraldehyde	189.2	—	4
C$_5$H$_{10}$O	Trimethylacetaldehyde	186.2	—	4
C$_5$H$_{10}$S	(CH$_2$)$_5$S (Thiane)	163.2	108.8	1
C$_5$H$_{12}$	2-Methylbutane	164.8	119.2	1
C$_5$H$_{12}$O	Pentanol-1	209.0	—	5
C$_5$H$_{12}$S	Methyl n-butyl sulfide	200.8	141.4	1
C$_5$H$_{12}$S	Ethyl n-propyl sulfide	198.3	139.7	1
C$_5$H$_{12}$S	Pentanethiol-1	201.7	141.8	1
C$_6$H$_3$Cl$_3$	1,2,4-Trichlorobenzene	194.6	—	9
C$_6$H$_5$Br	Bromobenzene	155.2	101.3	1
C$_6$H$_5$Cl	Chlorobenzene	146.0	98.7	1
C$_6$H$_5$F	Fluorobenzene	146.9	95.0	1
C$_6$H$_6$	Benzene	135.6	81.6	1
C$_6$H$_6$S	Benzenethiol	163.6	105.4	1
C$_6$H$_7$N	Aniline	193.3	109.2	1
C$_6$H$_{10}$O	Cyclohexanone	179.5	—	4
C$_6$H$_{12}$	Hexene-1	183.3	132.2	1
C$_6$H$_{12}$	2,3-Dimethyl butene-2	174.5	127.6	1
C$_6$H$_{12}$	Cyclohexane	152.3	106.3	1
C$_6$H$_{12}$	Methylcyclopentane	159.0	109.6	1
C$_6$H$_{12}$O	3,3-Dimethyl butanone-2	210.0	—	4
C$_6$H$_{12}$O	4-Methyl pentanone-2	215.8	—	4
C$_6$H$_{12}$O	Hexanone-2	214.8	—	4
C$_6$H$_{12}$O	Hexanone-3	216.9	—	4
C$_6$H$_{12}$O$_2$	Methyl pentanoate	229.3	—	3
C$_6$H$_{12}$O$_2$	n-Butyl acetate	228.4	—	3
C$_6$H$_{12}$O$_2$	t-Butyl acetate	231.0	—	3
C$_6$H$_{14}$	n-Hexane	195.0	143.1	1
C$_6$H$_{14}$	2-Methyl pentane	193.3	144.3	1
C$_6$H$_{14}$	3-Methyl pentane	190.8	143.1	1
C$_6$H$_{14}$	2,3-Dimethyl butane	188.3	140.6	1
C$_6$H$_{14}$	2,2-Dimethyl butane	188.3	141.8	1
C$_6$H$_{14}$O	Hexanol-1	241.3	—	5
C$_6$H$_{14}$O	2-Methyl pentanol-1	247.6	—	10
C$_6$H$_{14}$O	3-Methyl pentanol-2	275.9	—	10
C$_6$H$_{14}$O	4-Methyl pentanol-2	273.0	—	10
C$_6$H$_{14}$O	3-Methyl pentanol-3	293.4	—	10
C$_6$H$_{14}$S	Di-n-propyl sulfide	225.9	161.9	1
C$_6$H$_{18}$OSi$_2$	Hexamethyl disiloxane	311.3	—	1
C$_7$H$_6$O	Benzaldehyde	177.9	—	4
C$_7$H$_8$	Toluene	156.1	103.8	1
C$_7$H$_8$O	Benzyl alcohol	218.0	—	1
C$_7$H$_9$N	Benzyl amine	207.2	—	11
C$_7$H$_{12}$O	Cycloheptanone	211.8	—	4
C$_7$H$_{14}$	Heptene-1	211.3	155.2	1
C$_7$H$_{14}$	Methyl cyclohexane	184.9	135.1	1
C$_7$H$_{14}$	Ethyl cyclopentane	186.2	131.8	1
C$_7$H$_{14}$	1,1-Dimethyl cyclopentane	186.6	133.5	1
C$_7$H$_{14}$	cis-1,2-Dimethyl cyclopentane	188.7	134.3	1
C$_7$H$_{14}$	trans-1,2-Dimethyl cyclopentane	188.7	134.3	1
C$_7$H$_{14}$	trans-1,3-Dimethyl cyclopentane	188.7	134.3	1
C$_7$H$_{14}$O	Heptanone-2	243.6	—	4
C$_7$H$_{14}$O	Heptanone-4	246.3	—	4
C$_7$H$_{14}$O	2,4-Dimethyl pentanone-3	233.7	—	4
C$_7$H$_{14}$O	2-Methyl hexanone-3	241.5	—	4
C$_7$H$_{16}$	n-Heptane	224.7	166.1	1
C$_7$H$_{16}$	2-Methyl hexane	222.2	—	1
C$_7$H$_{16}$	3-Ethyl pentane	219.7	—	1
C$_7$H$_{16}$	2,2-Dimethyl pentane	220.9	—	1
C$_7$H$_{16}$	2,3-Dimethyl pentane	216.7	—	1
C$_7$H$_{16}$	2,4-Dimethyl pentane	223.8	—	1
C$_7$H$_{16}$	3,3-Dimethyl pentane	215.1	—	1
C$_7$H$_{16}$	2,2,3-Trimethyl butane	213.4	164.4	1
C$_7$H$_{16}$O	Heptanol-1	273.7	—	5
C$_8$H$_8$	Cyclooctatetraene	184	—	1
C$_8$H$_8$O	Acetophenone	204.6	—	4
C$_8$H$_8$O$_2$	Methyl benzoate	221.3	—	3
C$_8$H$_{10}$	Ethylbenzene	186.2	128.4	1
C$_8$H$_{10}$	p-Xylene	182.4	126.8	1
C$_8$H$_{10}$	o-Xylene	187.6	—	12
C$_8$H$_{10}$	m-Xylene	181.8	—	12
C$_8$H$_{10}$O	2-Phenyl ethanol	252.6	—	11
C$_8$H$_{11}$N	2-Phenyl ethylamine	239.2	—	11
C$_8$H$_{14}$O$_4$	Diethyl succinate	330.5	—	3
C$_8$H$_{16}$	Ethyl cyclohexane	209.2	159.0	1
C$_8$H$_{16}$	Octene-1	241.4	178.2	1
C$_8$H$_{16}$	1,2-Dimethyl cyclohexane	210.9	156.5	1
C$_8$H$_{16}$	cis-1,3-Dimethyl cyclohexane	209.2	157.3	1
C$_8$H$_{16}$	trans-1,3-Dimethyl cyclohexane	212.5	157.3	1
C$_8$H$_{16}$	cis-1,4-Dimethyl cyclohexane	211.3	157.3	1
C$_8$H$_{16}$	trans-1,4-Dimethyl cyclohexane	210.5	157.7	1
C$_8$H$_{16}$O	Octanone-2	276.9	—	4
C$_8$H$_{18}$	n-Octane	254.0	188.7	1
C$_8$H$_{18}$	2,2,4-Trimethyl pentane	236.4	—	1
C$_8$H$_{18}$	2,2,3,3-Tetramethyl butane	233.9	192.5	1
C$_8$H$_{18}$O	Octanol-1	305.6	—	5
C$_8$H$_{18}$S	Di-n-butyl sulfide	284.5	207.9	1
C$_9$H$_7$N	Quinoline	199.2	—	1
C$_9$H$_{10}$O	Propiophenone	243.8	—	4
C$_9$H$_{10}$O$_2$	Ethyl benzoate	246.0	—	3
C$_9$H$_{12}$	1,2,3-Trimethyl benzene	216.7	154.4	1
C$_9$H$_{12}$	1,2,4-Trimethyl benzene	215.9	155.2	1
C$_9$H$_{12}$	1,3,5-Trimethyl benzene	209.6	150.2	1
C$_9$H$_{12}$O	3-Phenyl propanol-1	280.7	—	11
C$_9$H$_{13}$N	3-Phenyl propylamine-1	265.6	—	11
C$_9$H$_{18}$O	Nonanone-5	305.9	—	4
C$_9$H$_{20}$	n-Nonane	284.5	211.7	1

TABLE 4 (continued)
Heat Capacity of Organic Liquids and Vapors at 25°C

Formula	Name	$C_p(l)$	$C_p(g)$	Ref.	Formula	Name	$C_p(l)$	$C_p(g)$	Ref.
$C_{10}H_{12}$	Tetralin	218	—	1	$C_{12}H_{12}O$	4-Methyl benzophenone	325.1	—	4
$C_{10}H_{12}O$	4-Ethyl acetophenone	260.7	—	4	$C_{12}H_{14}O_4$	Diethyl phthalate	357.7	—	3
$C_{10}H_{18}$	cis-Decalin	231.8	—	1	$C_{12}H_{24}$	Dodecene-1	360.7	269.4	1
$C_{10}H_{18}$	trans-Decalin	228.0	—	1	$C_{12}H_{26}$	n-Dodecane	375.3	280.3	1
$C_{10}H_{20}$	Decene-1	300.4	223.8	1	$C_{13}H_{28}$	n-Tridecane	406.7	303.3	1
$C_{10}H_{22}$	n-Decane	314.6	234.7	1	$C_{14}H_{30}$	n-Tetradecane	438.5	325.9	1
$C_{10}H_{22}$	2-Methyl nonane	316.3	—	1	$C_{15}H_{32}$	n-Pentadecane	469.4	348.9	1
$C_{11}H_{22}$	Undecene-1	330.1	246.9	1	$C_{16}H_{32}$	Hexadecene-1	484.9	361.1	1
$C_{11}H_{24}$	n-Undecane	345.2	257.3	1	$C_{16}H_{34}$	n-Hexadecane	501.7	372.0	1
$C_{12}H_{10}O$	Diphenyl ether	268.6	—	1					

REFERENCES

1. Shaw, R., J. Chem. Eng. Data, 14, 461, 1969.
2. Grolier, J.-P. E., Inglese, A., and Wilhelm, E., J. Chem. Thermodyn., 14, 523, 1982.
3. Fuchs, R., J. Chem. Thermodyn., 11, 959, 1979.
4. Fuchs, R., Can. J. Chem., 58, 2305, 1980.
5. Zegers, H. C. and Somsen, G., J. Chem. Thermodyn., 16, 225, 1984.
6. Scott, D. W., Berg, W. T., Hossenlopp, I. A., Hubbard, W. N., Messerly, J. F., Todd, S. S., Douslin, D. R., McCullough, J. P., and Waddington, G., J. Phys. Chem., 71, 2263, 1967.
7. Grolier, J.-P. E., Benson, G. C., and Picker, P., J. Chem. Eng. Data, 20, 243, 1975.
8. Grolier, J.-P. E., Inglese, A., and Wilhelm, E., J. Chem. Thermodyn., 16, 67, 1984.
9. Wilhelm, E., Inglese, A., Quint, J. R., and Grolier, J.-P. E., J. Chem. Thermodyn., 14, 303, 1982.
10. Bravo, R., Pintos, M., Baluja, M. C., Paz Andrade, M. I., Roux-Desgranges, G., and Grolier, J.-P. E., J. Chem. Thermodyn., 16, 73, 1984.
11. Nichols, N. and Wadso, I., J. Chem. Thermodyn., 7, 329, 1975.
12. Fortier, J.-L. and Benson, G. C., J. Chem. Eng. Data, 25, 47, 1980.

From Weast, R. C., Ed., *Handbook of Chemistry and Physics*, 69th ed., CRC Press, Boca Raton, FL, 1988.

TABLE 5
Emissivity of Total Radiation, ϵ_{tot}, for Various Materials

Material	Temperature (°C)	ϵ_{tot}
Alloys		
Nickel-Chromium		
20 Ni—25 Cr—55 Fe, oxidized	200	0.90
	500	0.97
60 Ni—12 Cr—28 Fe, oxidized	270	0.89
	560	0.82
80 Ni—20 Cr	100	0.87
	600	0.87
	1300	0.89
Aluminum		
Polished	50—500	0.04—0.06
Rough surface	20—50	0.06—0.07
Strongly oxidized	55—500	0.2—0.3
	25	0.022
	100	0.028
	500	0.060
Oxidized	200	0.11
	600	0.19
Asbestos board	20	0.96
Bismuth		
Unoxidized	25	0.048
Brass		
Dull tarnished	200	0.61
Oxidized at 600°C	200	0.61
	600	0.59
Unoxidized	25	0.035
	100	0.035
Polished	200	0.03
Rolled sheet	20	0.06
Bronze		
Polished	50	0.1
Carbon		
Filament	1000—1400	0.53
Graphite	0—3600	0.7—0.8
Lamp black	20—400	0.96
Soot applied to solid	50—1000	0.96
Soot with water glass	20—200	0.96
Unoxidized	100	0.81
Chromium		
Polished	50	0.1
	500—1000	0.28—0.38
Cobalt		
Unoxidized	500	0.13
	1000	0.23
Columbium		
Unoxidized	1500	0.19
Copper		
Calorized	100	0.26
Calorized, oxidized	200	0.18
	600	0.19
Commercial, scoured to a shine	20	0.07
Oxidized	50	0.6—0.7
	500	0.88
Polished	50—100	0.02
Unoxidized	100	0.02
Unoxidized, liquid	—	0.15
Fire brick	1000	0.75
Glass	20—100	0.94—0.91
	250—1000	0.87—0.72
	1100—1500	0.7—0.67
Gold		
Carefully polished	200—600	0.02—0.03
Unoxidized	100	0.02
Enamel	100	0.37
Graphite	0—3600	0.7—0.8
Gypsum	20	0.93
Iron		
Cast		
Oxidized	200	0.64
	600	0.78
Strongly oxidized	40	0.95
	250	0.95
Unoxidized	100	0.21
Unoxidized liquid	—	0.29
Oxidized	100	0.74
	500	0.84
	1200	0.89
Rusted	25	0.65
Wrought, dull	100	0.05

Material	Temperature (°C)	ϵ_{tot}
	25	0.94
Lamp black	20—400	0.96
Lead		
Oxidized	200	0.05
Unoxidized	200	0.63
Mercury		
Unoxidized	25	0.10
	100	0.12
Molybdenum	600—1000	0.08—0.13
	1500—2200	0.19—0.26
Monel metal		
Oxidized	200	0.43
	600	0.43
Nichrome		
Wire		
Clean	50	0.65
	500—1000	0.71—0.79
Oxidized	50—500	0.95—0.98
Nickel		
Industrial, polished	200—400	0.07—0.09
Oxidized	200	0.37
Oxidized at 600°C	200—600	0.37—0.48
Unoxidized	25	0.045
	100	0.06
	500	0.12
	1000	0.19
Platinum		
Clean, polished	200—600	0.05—0.1
Unoxidized	25	0.037
	100	0.047
	500	0.096
	1000	0.152
	1500	0.191
Wire	50—200	0.06—0.07
	500—1000	0.1—0.16
	1400	0.18
Porcelain		
Glazed	20	0.92
Rubber		
Hard	20	0.95
Soft, gray, rough	20	0.86
Silica brick	1000	0.80
	1100	0.85
Silver		
Clean, polished	200—600	0.02—0.03
Unoxidized	100	0.02
	500	0.035
Soot applied to a solid surface	50—1000	0.94—0.91
Soot with water glass	20—200	0.96
Steel		
Alloyed (8% Ni, 18% Cr)	500	0.35
Aluminized	50—500	0.79
Dull nickel plated	20	0.11
Flat, rough surface	50	0.95—0.98
Cast, polished	750—1050	0.52—0.56
Sheet, ground	50	0.56
	950—1100	0.55—0.61
Oxidized	200—600	0.8
Calorized, oxidized	200	0.52
	600	0.57
Sheet with shiny layer of oxide	20	0.82
Strongly oxidized	50	0.88
	500	0.98
Unoxidized	100	0.08
Unoxidized, liquid	—	0.28
Tantalum		
Unoxidized	1500	0.21
	2000	0.26
Tungsten		
Unoxidized	25	0.024
	100	0.032
	500	0.071
	1000	0.15
	1500	0.23
	2000	0.28
Varnish	40—100	0.8—0.95
Dull black	40—100	0.96—0.98
Glossy black sprayed on iron	20	0.87
Zinc		
Polished	200—300	0.04—0.05
Unoxidized	300	0.05

From Weast, R. C., Ed., *Handbook of Chemistry and Physics, 69th ed.*, CRC Press, Boca Raton, FL, 1988.

TABLE 6
Selected Values of Chemical Thermodynamic Properties

(From National Bureau of Standards Technical Notes 270-3, 270-4, 270-5, 270-6, 270-7 and 270-8

D. D. Wagman, W. H. Evans, V. B. Parker, R. H. Schumm, S. M. Bailey, I. Halow, K. L. Churney, and R. L. Nuttall

The compounds listed in the table represent only a small fraction of those for which data are given in the six Technical Notes referenced above. Copies of these Technical Notes may be purchased from the Superintendent of Documents, U.S. Government Printing Office, Washington, D.C. 20402. Conversion of units used in the table to SI units or other units may be accomplished by employing the following factors.

CONVERSION FACTORS FOR UNITS OF MOLECULAR ENERGY

	J mol^{-1}	cal mol^{-1}	cm^3 atm mol^{-1}	kWh mol^{-1}	Btu lb^{-1} mol^{-1}	cm^{-1} molecule^{-1}	eV molecule^{-1}
J mol^{-1} =	1	2.390057×10^{-1}	9.86923	2.77778×10^{-7}	0.429923	8.35940×10^{-2}	1.036409×10^{-5}
cal mol^{-1} =	4.18400*	1	41.2929	1.162222×10^{-6}	1.798796	3.49757×10^{-1}	4.33634×10^{-5}
cm^3 atm mol^{-1} =	0.1013250*	2.42173×10^{-2}	1	2.81458×10^{-8}	4.35619×10^{-2}	8.47016×10^{-3}	1.050141×10^{-6}
kWh mol^{-1} =	3,600,000	860,421	3.55292×10^7	1	1,547,721	300,938	37.3107
Btu lb^{-1} mol^{-1} =	2.32600*	5.55927×10^{-1}	22.9558	6.46111×10^{-7}	1	1.944396×10^{-1}	2.41069×10^{-5}
cm^{-1} molecule^{-1} =	11.96258	2.85912	118.0614	3.32294×10^{-6}	5.14299	1	1.239812×10^{-4}
eV molecule^{-1} =	96487.0*	23060.9	952,252	2.68019×10^{-2}	41482.0	8065.73	1

* These numbers represent the fundamental values used in deriving the data in the table. The remaining factors were obtained by applying the relationships:

$$n_{ij} = n_{ik} \cdot n_{kj} \qquad n_{ii} = n_{ik} \cdot n_{ki} = 1$$

INTRODUCTION

Substances and Properties Included in the Tables

The tables contain values of the enthalpy and Gibbs energy of formation, enthalpy, entropy and heat capacity at 298.15 K (25°C), and the enthalpy of formation at 0 K, for inorganic substances and organic molecules containing not more than two carbon atoms.

No values are given in these tables for metal alloys or other solid solutions, fused salts, or for substances of undefined chemical composition.

Physical States

The physical state of each substance is indicated in the column headed "State" as crystalline solid (c), liquid (l), glassy or amorphous (amorp), or gaseous (g). For solutions, the physical state is that normal for the indicated solvent at 298.15 K. Isomeric substances or various crystalline modifications of a given substance are designated by a number following the letter designation, as c2, g2, etc.

Definition of Symbols

The symbols used in these tables are defined as follows: P = pressure; V = volume; T = absolute temperature; S = entropy; H = enthalpy (heat content); G = H − TS = Gibbs energy (formerly the free energy); $C_p = (dH/dT)_p$ = heat capacity at constant pressure.

Conventions Regarding Pure Substances

The values of the thermodynamic properties of the pure substances given in these tables are for the substances in their standard states (indicated by the superscript * on the thermodynamic symbol). These standard states are defined as follows:

1. For a pure solid or liquid, the standard state is the substance in the condensed phase under a pressure of one atmosphere.*
2. For a gas the standard state is the hypothetical ideal gas at unit fugacity, in which state the enthalpy is that of the real gas at the same temperature and zero pressure.

The values of ΔHf° and ΔGf° given in the tables represent the change in the appropriate thermodynamic quantity when one gram-formula weight of the substance in its standard state is formed isothermally at the indicated temperature from the elements, each in its appropriate standard reference state. The standard reference state at 298.15 K for each element except phosphorus has been chosen to be the standard state that is thermodynamically stable at that temperature and one atmosphere pressure. For phosphorus the standard reference state is the crystalline white form; the more stable forms have not been well characterized thermochemically. The same reference states have been maintained for the elements at 0 K except for the liquid elements bromine and mercury, for which the reference states have been chosen as the stable crystalline forms.

The value of H°$_{298}$ − H°$_0$ represents the enthalpy difference for the given substance between 298.15 K and 0 K. If the indicated standard state at 298.15 K is the gas, the corresponding state at 0 K is the hypothetical ideal gas; if the state at 298.15 K is solid or liquid, the corresponding state at 0 K is the thermodynamically stable crystalline solid, unless otherwise specifically indicated.

The values of S° represent the virtual or "thermal" entropy of the substance in the standard state at 298.15 K, omitting contributions from nuclear spins. Isotope mixing effects, etc., are also excluded except in the case of the hydrogen-deuterium system. Where data have been available only for a particular isotope, they have been corrected when possible to the normal isotopic composition.

The values of the enthalpies of formation of gaseous ionic species are computed on the convention that the value of ΔHf° for the electron is zero. Conversions between 0 and 298.15 K are claculated using the value of H°$_{298}$ − H°$_0$ = 1.481 kcal per mole of electrons, and assuming that the values of H°$_{298}$ − H°$_0$ for the ionized and nonionized molecules are the same.

Conventions Regarding Solutions

For all dissolved substances the composition of the solvent is indicated following the chemical formula of the solute. In most instances the number of moles of solvent associated with 1 mole of solute is stated explicitly. In some cases the concentration of the solute cannot be specified. For aqueous solutions this is indicated in the State column by "au" (aqueous, unspecified). Such solutions may be assumed to be "dilute".

* One standard atmosphere equals 101325 pascal.

TABLE 6 (continued)
Selected Values of Chemical Thermodynamic Properties

The standard state for a strong electrolyte in aqueous solution is the ideal solution at unit mean molality (unit activity). For a non-dissociating solute in aqueous solution the standard state is the ideal solution at unit molality.

The value of $\Delta Hf°$ for a solute in its standard state is equal to the apparent molal enthalpy of formation of the substance in the infinitely dilute solution, since the enthalpy of dilution of an ideal solution is zero. At this dilution the partial molal enthalpy is equal to the apparent molal quantity. At concentrations other than the standard state, the value of $\Delta Hf°$ represents the apparent enthalpy of the reaction of formation of the solution from the elements comprising the solute, each in its standard reference state, and the appropriate total number of moles of solvent. In this representation the value of $\Delta Hf°$ for the solvent is not required. The experimental value for an enthalpy of dilution is obtained directly as the difference between the two values of $\Delta Hf°$ at the corresponding concentrations. At finite concentrations the partial molal enthalpy of formation differs from the apparent enthalpy. In some instances the partial molal enthalpy of formation is given in the Tables. In this case the concentration designation is preceded by "D:".

The values for the thermodynamic properties for an individual ion in aqueous solution are for that undissociated ion in the standard state and are based on the convention that $\Delta Hf°$, $\Delta Gf°$, $S°$, and $Cp°$ for $H^+(a)$ are zero. The properties of the neutral strong electrolyte in aqueous solution in the standard state are equal to the sum of these values for the appropriate number of ions assumed to constitute the molecule of the given electrolyte. By adopting the above convention with respect to $H^+(a)$, it follows that for an individual ionic species the G–H–S relation becomes

$$\Delta Gf° = \Delta Hf° - T\{\Delta Sf° + (n/2)S°[H_2(gs)]\}$$

with n = the algebraic value of the ionic charge. For neutral electrolytes and gaseous ions the normal consistency relation holds (see below).

Unit of Energy and Fundamental Constants

All of the energy values given in these tables are expressed in terms of the thermochemical calorie. This unit, defined as equal to 4.1840 joules, has been generally accepted for presentation of chemical thermodyanmic data. Values reported in other units have been converted to calories by means of the conversion factors for molecular energy given in the table which precedes this discussion.

The following values of the fundamental physical constants have been used in these calculations:

- R = gas constant = 8.3143 ± 0.0012 J/deg mol = 1.98717 ± 0.00029 cal/deg mol
- F = Faraday constant = 96487.0 ± 1.6 coulombs/mol = 23060.9 ± 0.4 cal/V equivalent
- Z = Nhc = 11.96258 ± 0.00107 J/cm^{-1} mol = 2.85912 ± 0.00026 cal/cm^{-1} mol
- c_2 = second radiation constant = hc/k = 1.43879 ± 0.00015 cm deg 0°C = 273.15 K

These constants are consistent with those given in the Table of Physical Constants, recommended by the National Academy of Sciences — National Research Council.* The formula weights listed in the tables have been calculated for the empirical molecular formula given in the Formula and Description column.

Internal Consistency of the Tables

The various aspects of internal consistency are specified below:

1. Subsidiary and auxiliary quantities used. All of the values given in these tables have been calculated from the original articles, using consistent values for all subsidiary and auxiliary quantities. The original data were corrected where possible for differences in energy units, molecular weights, temperature scales, etc. Thus we have sought to maintain a uniform scale of energies for all substances in the tables.

2. Physical and thermodynamic relationships for the tabulated properties of a substance. The tabulated values of the properties of a substance satisfy all the known physical and thermodynamic relationships among these properties. The quantities $\Delta Hf°$, $\Delta Gf°$, and $S°$ at 298.15 K satisfy the relation (within the assumed uncertainty)

$$\Delta Gf° = \Delta Hf° - T \Delta Sf°$$

Substance		0 K		298.15 K (25°C)				
Formula and Description	State	Formula weight	ΔHf_0 kcal/mol	$\Delta Hf°$	$\Delta Gf°$ kcal/mol	$Hf_{98} - Hf_0$	$S°$	$Cp°$ cal/deg mol
ACTINIUM								
Ac	c	227.0280	0	0	0	—	13.5	6.5
	g	227.0280	—	97.0	87.6	1.481	44.92	4.98
ALUMINUM								
Al	c	26.9815	0	0	0	1.094	6.77	5.82
	g	—	77.44	78.0	68.3	1.654	39.30	5.11
Al⁺	g	—	215.476	217.517	—	—	—	—
Al²⁺	g	—	649.663	653.185	—	—	—	—
Al³⁺	g	—	1305.70	1310.70	—	—	—	—
std. state, m = 1	aq	—	—	−127.	−116.	—	−76.9	—
Al₂	g	53.9630	116.	116.14	103.57	2.33	55.7	8.7
AlO₂⁻								
std. state, m = 1	aq	58.9803	—	−219.6	−196.7	—	−5.	—
Al₂O₃								
α, corundum	c	101.9612	−397.59	−400.5	−378.2	2.394	12.17	18.89
δ	c	—	—	−398.	—	—	—	—
ϱ	c	—	—	−391.	—	—	—	—
κ	c	—	—	−397.	—	—	—	—
γ	c	—	—	−395.	—	—	—	—
	amorp	—	—	−390.	—	—	—	—
Al₂O₃·H₂O								
boehmite	c	119.9765	—	−472.0	−436.3	—	23.15	31.37
diaspore	c	—	—	−478.	−440.	—	16.86	25.22
Al₂O₃·3H₂O								
gibbsite	c	156.0072	—	−612.5	−546.7	—	33.51	44.49
bayerite	c	—	—	−610.1	—	—	—	—
AlH₃	c	30.0054	—	−11.	—	—	—	—
AlOH²⁺								
std. state, m = 1	aq	43.9889	—	—	−165.9	—	—	—
Al(OH)₃	amorp	78.0036	—	−305.	—	—	—	—
AlF₃	c	83.9767	−358.02	−359.5	−340.6	2.778	15.88	17.95
AlF₃	g	—	−287.01	−287.9	−284.0	3.37	66.2	14.97
un-ionized; std. state, m = 1	aq	—	—	−363.	−338.	—	−6.	—
ionized; std. state, m = 1	aq	—	—	−366.	−316.	—	−86.8	—

* NBS Technical News Bulletin, October 1963. See also Report of the CODATA Task Group on Fundamental Constants, CODATA Bulletin 11, December 1973.

SELECTED VALUES OF CHEMICAL THERMODYNAMIC PROPERTIES (Continued)

Formula and Description	State	Formula weight	0 K ΔHf$_0$ kcal/mol	298.15 K (25°C) ΔHf° kcal/mol	ΔGf° kcal/mol	H$_{298}$ − H$_0$	S° cal/deg mol	C$_p$
AlCl$_3$	c	133.3405	−168.02	−168.3	−150.3	4.104	26.45	21.95
	g	—	—	−139.4	—	—	—	—
std. state, m = 1	aq	—	—	−247.	−210.	—	−36.4	—
AlCl$_3$ 6H$_2$O	c	241.4325	—	−643.3	—	—	—	—
Al$_2$Cl$_6$	g	266.6810	—	−308.5	−291.7	—	117.	—
AlBr$_3$	c	266.7085	—	−126.0	—	—	—	24.3
	g	—	—	−101.6	—	—	—	—
std. state, m = 1	aq	—	—	−214.	−191.	—	−17.8	—
Al$_2$Br$_6$	g	533.4170	—	−232.0	—	—	—	—
AlI$_3$	c	407.6947	—	−75.0	−71.9	—	38.	23.6
	g	—	—	−49.6	—	—	—	—
	aq	—	—	−165.8	—	—	—	—
Al$_2$I$_6$	g	815.3894	—	−123.5	—	—	—	—
Al$_2$S$_3$	c	150.1550	—	−173.	—	—	—	—
Al$_2$(SO$_4$)$_3$	c	342.1478	—	−822.38	−740.95	—	57.2	62.00
std. state, m = 1	aq	—	—	−906.	−766.	—	−139.4	—
Al$_2$(SO$_4$)$_3$ 6H$_2$O	c	450.2398	—	−1269.53	−1104.82	—	112.1	117.8
Al$_2$Se$_3$	c	290.843	—	−135.	—	—	—	—
Al$_2$Te$_3$	c	436.763	—	−78.	—	—	—	—
AlN	c	40.9882	−74.80	−76.0	−68.6	0.925	4.82	7.20
Al(NO$_3$)$_3$								
std. state, m = 1	aq	212.9962	—	−276.	−196.	—	28.1	—
Al(NO$_3$)$_3$ 6H$_2$O	c	321.0882	—	−681.28	−526.74	—	111.8	103.5
AlCl$_3$ NH$_3$	c	150.3711	—	−212.6	—	—	—	—
AlCl$_3$ 3NH$_3$	c	184.4323	—	−283.0	—	—	—	—
AlCl$_3$ 6NH$_3$	c	235.5242	—	−363.4	—	—	—	94.
AlBr$_3$ NH$_3$	c	283.7391	—	−177.6	—	—	—	—
AlBr$_3$ 3NH$_3$	c	317.8003	—	−252.7	—	—	—	—
AlBr$_3$ 6NH$_3$	c	368.8922	—	−343.0	—	—	—	—
AlI$_3$ NH$_3$	c	424.7253	—	−119.0	—	—	—	—
AlI$_3$ 3NH$_3$	c	458.7865	—	−205.9	—	—	—	—
AlI$_3$ 6NH$_3$	c	509.8784	—	−312.9	—	—	—	—
NH$_4$Al(SO$_4$)$_2$	c	237.1433	—	−562.2	−487.2	—	51.7	54.12
std. state, m = 1	aq	—	—	−593.	−491.	—	−40.2	—
NH$_4$Al(SO$_4$)$_2$ 12H$_2$O	c	453.3274	—	−1420.26	−1180.21	—	166.6	163.3
AlP	c	57.9553	—	−39.8	—	—	—	—
AlPO$_4$								
berlinite	c	121.9529	−411.40	−414.4	−386.7	3.528	21.70	22.27
AlAs	c	101.9031	—	−27.8	—	—	—	—
Al$_4$C$_3$	c	143.9594	−48.71	−49.9	−46.9	3.936	21.26	27.91
Al(CH$_3$)$_3$	c	72.0867	−29.76	—	—	—	—	—
	liq	—	—	−32.6	−2.4	8.114	50.05	37.19
	g	—	—	−17.7	—	—	—	—
Al$_2$(CH$_3$)$_6$	g	144.1734	—	−55.19	−2.34	—	125.4	—
Al$_2$Si$_2$O$_7$ 2H$_2$O								
kaolinite	c	258.1615	—	−979.6	−903.0	—	48.5	58.62
halloysite	c	—	—	−975.0	−898.5	—	48.6	58.86
Al$_6$Si$_2$O$_{13}$								
mullite	c	426.0532	—	−1632.8	−1541.2	—	61.	77.94
ANTIMONY								
Sb								
III	c	121.75	0	0	0	1.410	10.92	6.03
IV, explosive	amorp	—	—	2.54	—	—	—	—
	g	—	62.63	62.7	53.1	1.481	43.06	4.97
Sb⁺	g	—	261.91	263.46	—	—	—	—
Sb^{2+}	g	—	643.1	646.1	—	—	—	—
Sb^{3+}	g	—	1227.1	1231.6	—	—	—	—
Sb^{4+}	g	—	2245.4	2251.4	—	—	—	—
SbO	g	137.749	48.	47.67	—	2.122	—	—
SbO⁺								
std. state, m = 1	aq	—	—	—	−42.33	—	—	—
SbO$_2$⁻								
std. state, m = 1	aq	153.749	—	—	−81.32	—	—	—
Sb$_2$O$_3$	aq	291.498	—	−164.9	—	—	—	—
Sb$_2$O$_4$	c	307.498	—	−216.9	−190.2	—	30.4	27.39
Sb$_2$O$_5$	c	323.497	—	−232.3	−198.2	—	29.9	—
SbH$_3$	g	124.774	36.625	34.681	35.31	2.502	55.61	9.81
HSbO$_2$								
undissoc.; std. state, m = 1	aq	154.757	—	−116.6	−97.4	—	11.1	—
Sb(OH)$_3$	c	172.772	—	—	−163.8	—	—	—
undissoc.; std. state, m = 1	aq	—	—	−184.9	−154.1	—	27.8	—
H$_3$SbO$_4$	aq	188.772	—	−216.8	—	—	—	—
SbF$_3$	c	178.745	—	−218.8	—	—	—	—
SbCl$_3$	c	228.109	—	−91.34	−77.37	—	44.0	25.8
SbCl$_3$	g	—	−74.57	−75.0	−72.0	4.269	80.71	18.33
SbCl$_5$	liq	299.015	—	−105.2	−83.7	—	72.	—
SbOCl	c	173.202	—	−89.4	—	—	—	—
	g	—	−25.	−25.5	—	—	—	—
SbBr$_3$	c	361.477	—	−62.0	−57.2	—	49.5	—
	g	—	−41.03	−46.5	−53.5	4.727	89.09	19.17
in CS$_2$	—	—	—	−58.4	—	—	—	—
SbI$_3$	c	502.463	—	−24.0	—	—	—	—
	aq	—	—	−23.6	—	—	—	—
Sb$_2$S$_3$								
black	c	339.692	—	−41.8	−41.5	—	43.5	28.65
orange	amorp	—	—	−35.2	—	—	—	—
Sb$_2$(SO$_4$)$_3$	c	531.685	—	−574.2	—	—	—	—
ARGON								
Ar	g	39.948	0	0	0	1.481	36.9822	4.9679
std. state, m = 1	aq	—	—	−2.9	3.9	—	14.2	—
Ar⁺	g	—	363.42	364.90	—	—	—	—
Ar^{2+}	g	—	1000.5	1003.5	—	—	—	—
Ar^{3+}	g	—	1943.9	1948.3	—	—	—	—
ARSENIC								
As								
α, gray, metallic	c	74.9216	0	0	0	1.226	8.4	5.89
γ, yellow, cubic	c	—	—	3.5	—	—	—	—
β	amorp	—	—	1.0	—	—	—	—
	g	—	72.04	72.3	62.4	1.481	41.61	4.968
As⁺	g	—	298.38	300.12	—	—	—	—
As^{2+}	g	—	764.42	767.64	—	—	—	—

SELECTED VALUES OF CHEMICAL THERMODYNAMIC PROPERTIES (Continued)

Formula and Description	State	Formula weight	ΔHf₀ kcal/mol (0 K)	ΔHf° kcal/mol	ΔGf° kcal/mol	H°₂₉₈ − H°₀	S° cal/deg mol	C°p cal/deg mol
As²⁺	g	—	1417.44	1422.14	—	—	—	—
As⁴⁺	g	—	2573.58	2579.76	—	—	—	—
As₂	g	149.8432	53.30	53.1	41.1	2.251	57.2	8.366
AsO	g	90.9210	16.88	16.72	—	2.101	—	—
AsO⁺								
undissoc.; std. state, m = 1	aq	—	—	—	−39.15	—	—	—
AsO₂⁻;								
std. state, m = 1	aq	106.9204	—	−102.54	−83.66	—	9.7	—
AsO₄								
std. state, m = 1	aq	138.9192	—	−212.27	−155.00	—	−38.9	—
As₂O₃	c	229.8402	—	−221.05	−187.0	—	25.2	27.85
As₂O₃·4H₂O	c	301.9016	—	−503.0	—	—	—	—
AsH₃	g	77.9455	17.70	15.88	16.47	2.438	53.22	9.10
HAsO₂								
undissoc.; std. state, m = 1	aq	107.9284	—	−109.1	−96.25	—	30.1	—
HAsO₄²⁻;								
undissoc.; std. state, m = 1	aq	139.9272	—	−216.62	−170.82	—	−0.4	−−
H₂AsO₃⁻								
undissoc.; std. state, m = 1	aq	124.9357	—	−170.84	−140.35	—	26.4	—
H₂AsO₄⁻								
undissoc.; std. state, m = 1	aq	140.9351	—	−217.39	−180.04	—	28.	—
H₃AsO₄	c	141.9431	—	−216.6	—	—	—	—
	aq	—	—	−216.2	—	—	—	—
AsF₃	liq	131.9168	—	−196.3	−185.04	—	43.31	30.25
	g	—	−186.82	−187.80	−184.22	3.413	69.07	15.68
AsCl₃	liq	181.2806	—	−72.9	−62.0	—	51.7	—
	g	—	−62.12	−62.5	−59.5	4.137	78.17	18.10
AsBr₃	c	314.6486	—	−47.2	—	—	—	—
	g	—	−25.55	−31.	−38.	4.569	86.94	18.92
AsI₃	c	455.6348	−13.91	−13.9	−14.2	5.964	50.92	25.28
	g	—	—	—	—	4.834	92.79	19.27
As₂S₂	c	213.9712	—	−34.1	—	—	—	—
As₂S₃	c	246.0352	—	−40.4	−40.3	—	39.1	27.8
NH₄H₂AsO₃								
std. state, m = 1, (NH₄⁺ + H₂AsO₃⁻)	aq	143.9743	—	−202.51	−159.32	—	53.5	—
(NH₄)₂HAsO₄	c	176.0043	—	−282.4	—	—	—	—
std. state, m = 1, (2NH₄⁺ + HAsO₄²⁻)	aq	—	—	−279.96	−208.76	—	53.8	—
BARIUM								
Ba	c	137.34	0	0	0	1.65	15.0	6.71
	g	—	43.2	43.	35.	1.481	40.663	4.968
Ba⁺	g	—	163.39	164.67	—	—	—	—
Ba²⁺	g	—	394.09	396.86	—	—	—	—
std. state, m = 1	aq	—	—	−128.50	−134.02	—	+2.3	—
BaO₂	c	169.339	—	−151.6	—	—	—	16.0
BaO₂·H₂O	c	187.354	—	−222.3	—	—	—	—
BaO₂·8H₂O	c	313.462	—	−718.6	—	—	—	—
BaH	g	138.348	53.6	53.	47.	2.082	52.29	7.19
BaH₂	c	139.356	—	−42.7	—	—	—	—
Ba(OH)₂	c	171.355	—	−225.8	—	—	—	—
Ba(OH)₂·H₂O	c	189.370	—	−298.4	—	—	—	—
Ba(OH)₂·3H₂O	c	225.401	—	−442.0	—	—	—	—
Ba(OH)₂·8H₂O	c	315.477	—	−798.8	−667.6	—	102.	—
BaF₂	c	175.337	−288.19	−288.5	−276.5	3.452	23.03	17.02
	g	—	−195.0	−195.5	−198.0	3.262	71.91	12.85
std. state, m = 1	aq	—	—	−287.50	−267.30	—	−4.3	—
BaCl₂	c	208.246	−205.35	−205.2	−193.7	3.993	29.56	17.96
	g	—	−125.37	−125.7	−128.5	3.511	77.72	13.43
std. state, m = 1	aq	—	—	−208.40	−196.76	—	29.3	—
BaCl₂·H₂O	c	226.261	—	−277.4	−252.32	—	39.9	—
BaCl₂·2H₂O	c	244.277	—	−348.98	−309.86	—	48.5	38.71
Ba(ClO₃)₂	c	272.244	−162.6	−127.0	—	—	47.	—
Ba(ClO₃)₂	c	304.242	—	−184.4	—	—	—	—
Ba(ClO₄)₂	c	336.241	—	−191.2	—	—	—	—
Ba(ClO₄)₂·3H₂O	c	390.287	—	−404.3	−303.7	—	94.	—
BaBr₂	c	297.158	—	−181.0	−176.1	—	35.	—
	g	—	−101.4	−105.	−113.	3.9	79.	14.7
BaBr₂								
std. state, m = 1	aq	—	—	−186.60	−183.72	—	41.7	—
∞ H₂O	aq	—	—	−186.60	—	—	—	—
BaBr₂·H₂O	c	315.173	—	−255.3	—	—	—	—
BaBr₂·2H₂O	c	333.189	—	−326.5	−294.1	—	54.	—
Ba(BrO₃)₂	c	393.154	—	−171.65	−130.1	—	59.	—
Ba(BrO₃)₂								
std. state, m = 1	aq	—	—	−160.56	−125.16	—	79.6	—
Ba(BrO₃)₂·H₂O	c	411.170	−236.99	−243.84	−188.6	9.94	68.9	53.5
BaI₂	c	391.149	—	−143.9	—	—	—	—
std. state, m = 1	aq	—	—	−154.88	−158.68	—	55.5	—
BaI₂·H₂O	c	409.164	—	−219.7	—	—	—	—
BaI₂·2H₂O	c	427.179	—	−290.8	—	—	—	—
BaI₂·7H₂O	c	517.256	—	−639.7	—	—	—	—
Ba(IO₃)₂	c	487.145	−243.31	−245.5	−206.7	8.84	59.6	44.8
std. state, m = 1	aq	—	—	−234.3	−195.2	—	58.9	—
Ba(IO₃)₂·H₂O	c	505.161	—	−316.0	−263.9	—	71.	—
BaS	c	169.404	—	−110.	−109.	—	18.7	11.80
Ba₂S₃	g	338.808	—	−90.	—	—	—	—
BaSO₃	c	217.402	—	−281.9	—	—	—	—
BaSO₄	c	233.402	—	−352.1	−325.6	—	31.6	24.32
std. state, m = 1	aq	—	—	−345.82	−311.99	—	7.1	—
BaSi₂O₃	g	249.466	—	—	—	—	—	40.7
BaSi₂O₄	aq	297.464	—	−415.4	—	—	—	—
Ba(HSO₃)₂	aq	299.480	—	−430.0	—	—	—	—
BaI₂·4SO₂	c	647.400	—	−470.2	—	—	—	—
BaSe	c	216.30	—	−89.	—	—	—	—
BaSeO₃	c	264.298	—	−248.7	−231.4	—	40.	—
BaSeO₄	c	280.298	—	−274.0	−249.7	—	42.	—
BaN₂	c	165.353	—	−41.	—	—	—	—
Ba(N₃)₂·H₂O	c	239.396	—	−73.7	−25.1	—	45.	—
Ba(NO₃)₂	c	229.351	—	−183.6	—	—	—	—

SELECTED VALUES OF CHEMICAL THERMODYNAMIC PROPERTIES (Continued)

Formula and Description	State	Formula weight	ΔHf° 0 K kcal/mol	ΔHf° kcal/mol	ΔGf° kcal/mol	H°₂₉₈ − H°₀	S° cal/deg mol	C°p cal/deg mol
				298.15 K (25°C)				
Ba(NO₃)₂	aq	—	—	−178.5	—	—	—	—
	c	261.350	—	−237.11	−190.42	—	51.1	36.18
std. state, m = 1	aq	—	—	−227.62	−187.24	—	72.3	—
BaHPO₄	c	233.319	—	−433.7	—	—	—	—
Ba(H₂PO₄)₂	c	267.317	—	−421.2	—	—	—	—
Ba(H₂PO₄)₂·H₂O	c	285.332	—	−490.5	—	—	—	—
Ba(H₃PO₄)₂	c	331.315	—	−747.	—	—	—	—
BaHAsO₄·H₂O	c	295.283	—	−412.6	—	—	—	—
Ba(H₂AsO₄)₂·2H₂O	c	455.241	—	−696.9	—	—	—	—
BaC₂	c	161.362	—	−18.	—	—	—	—
BaCO₃								
witherite	c	197.349	—	−290.7	−271.9	—	26.8	20.40
std. state, m = 1	aq	—	—	−290.34	−260.19	—	−11.3	—
BaC₂O₄	c	225.360	—	−327.1	—	—	—	—
BaC₂O₄·2H₂O	c	261.391	—	−471.1	—	—	—	—
Ba(HCO₃)₂								
std. state, m = 1	aq	259.375	—	−459.28	−414.54	—	45.9	—
Ba(C₂H₃O₂)₂	c	255.430	—	−354.8	—	—	—	—
std. state, m = 1	aq	—	—	−360.82	−310.60	—	43.7	—
BaCN₂	c	177.365	—	−63.6	—	—	—	—
Ba(CN)₂	c	189.376	—	−52.2	—	—	—	—
	aq	—	—	−55.0	—	—	—	—
BaO·SiO₂	c	213.424	—	−388.05	−368.13	—	26.2	21.51
glassy	amorp	—	—	−376.	—	—	—	—
BaO·2SiO₂	c	273.509	—	−609.0	−576.2	—	36.6	32.05
2BaO·SiO₂	c	366.764	—	−546.8	−519.8	—	42.1	32.24
2BaO·3SiO₂	c	486.933	—	−1000.2	−947.2	—	61.7	53.68
Ba₂Fe(CN)₆	aq	486.634	—	−145.7	—	—	—	—
Ba₂Fe(CN)₆·6H₂O	c	594.726	—	−567.0	—	—	—	—
Ba₂[Fe(CN)₆]₂								
std. state, m = 1	aq	835.928	—	−116.9	−53.5	—	136.1	—
BaMnO₄								
barium manganate	c	256.276	—	—	−267.5	—	—	—
BaCrO₄	c	253.334	—	−345.6	−321.53	—	379	—
BaMoO₄	c	297.278	—	−370.	−344.1	—	33.	33.6
BaWO₄	c	385.188	—	−407.	—	—	—	—
BaTiO₃	c	233.238	—	−396.7	−375.8	—	25.8	24.49
BaZrO₃	c	276.558	—	−425.3	−405.0	—	29.8	24.31
BERYLLIUM								
Be	c	9.0122	0	0	0	0.466	2.27	3.93
	g	—	76.49	77.5	68.5	1.481	32.543	4.968
Be⁺	g	—	291.474	293.965	—	—	—	—
Be²⁺	g	—	711.426	715.398	—	—	—	—
Std. state, m = 1	aq	—	—	−91.5	−90.75	—	−31.0	—
BeO	c	25.0116	−144.86	−145.7	−138.7	0.669	3.38	6.10
	g	—	—	28.	—	—	—	—
BeO₂²⁻								
std. state, m = 1	aq	41.0110	—	−189.0	−153.0	—	−38.	—
BeH	g	10.0202	75.	75.6	68.3	2.062	42.21	6.95
BeH₂	c	11.0282	—	−4.60	—	—	—	—
Be(OH)₂								
α	c	43.0269	—	−215.7	−194.8	—	12.4	—
β	c	—	—	−216.5	−195.4	—	12.	—
fresh precipitated	amorp	—	—	−214.6	—	—	—	—
Be₂(OH)₃⁺ +								
std. state, m = 1	aq	78.0587	—	—	−430.6	—	—	—
BeF₂								
α, quartz	c	47.0090	−244.85	−245.4	−234.1	2.024	12.75	12.39
β, cristobalite	c	—	—	−244.7	—	—	—	—
glassy	amorp	—	—	−244.3	—	—	—	—
BeCl₂								
α	c	79.9182	−117.40	−117.2	−106.5	2.863	19.76	15.50
β	c	—	−118.57	−118.5	−107.3	2.729	18.12	14.92
BeCl₂·4H₂O	c	151.9796	—	−432.2	—	—	—	—
Be₂Cl₄	g	159.8364	—	−202.	—	—	—	—
BeBr₂	c	168.8302	—	−84.5	—	—	—	—
BeI₂	c	262.8210	—	−46.0	—	—	—	—
BeS	c	41.0762	—	−56.0	—	—	—	—
BeSO₄								
α, tetragonal	c	105.0738	−285.505	−288.05	−261.44	3.125	18.62	20.48
std. state, m = 1	aq	—	—	−308.8	−268.72	—	−26.2	—
BeSO₄·H₂O	c	123.0891	—	−364.2	—	—	—	—
BeSO₄·2H₂O	c	141.1045	−429.814	−435.74	−381.99	5.865	39.01	36.63
BeSO₄·4H₂O								
tetragonal	c	177.1352	−569.682	−579.29	−497.29	8.306	55.68	51.77
BeSeO₄	c	151.970	—	−212.8	—	—	—	—
std. state, m = 1	aq	—	—	−234.7	−196.2	—	−18.1	—
BeSeO₄·2H₂O	c	188.000	—	−360.3	—	—	—	—
BeSeO₄·4H₂O	c	224.031	—	−505.0	—	—	—	—
Be₃N₂								
α, cubic	c	55.0500	—	−140.6	—	—	—	—
β, hexagonal	c	—	—	−136.5	—	—	—	—
Be(NO₃)₂	aq	133.0220	—	−191.0	—	—	—	—
Be₂C	c	30.0356	—	−28.0	—	—	—	—
BeCO₃	c	69.0216	—	−245.	—	—	—	—
Be₂SiO₄	c	110.1080	−510.77	−513.7	−485.8	2.922	15.37	22.84
Be(BO₂)₂	g	94.6318	—	−325.	—	—	—	—
BeO·Al₂O₃								
chrysoberyl	c	126.9728	−546.32	−549.9	−520.7	3.128	15.84	25.19
BeO·3Al₂O₃	c	330.8952	−1335.65	−1344.9	−1271.6	8.153	42.0	63.38
BeMoO₄	c	168.950	—	−328.	—	—	—	—
PuBe₁₃	c	356.21	—	−36.	—	—	—	—
BISMUTH								
Bi	c	208.980	0	0	0	1.536	13.56	6.10
	g	—	49.56	49.5	40.2	1.481	44.669	4.968
Bi⁺	g	—	217.65	219.07	—	—	—	—
Bi²⁺	g	—	602.5	605.4	—	—	—	—
Bi³⁺	g	—	1192.0	1196.4	—	—	—	—
std. state, m = 1	aq	—	—	—	19.8	—	—	—
Bi₂	g	417.960	53.12	52.5	—	2.454	—	8.83

SELECTED VALUES OF CHEMICAL THERMODYNAMIC PROPERTIES (Continued)

Formula and Description	State	Formula weight	0 K ΔHf° kcal/mol	298.15 K (25°C) ΔHf° kcal/mol	ΔGf° kcal/mol	H°₂₉₈ − H°₀	S° cal/deg mol	C°p
BiO⁺								
std. state, m = 1	aq	—	—	-35.0	—	—	—	—
Bi₂O₃	c	465.9582	—	-137.16	-118.0	—	36.2	27.13
BiO·OH	c	241.9868	—	—	-88.0	—	—	—
Bi(OH)₃	c	260.0021	—	-170.0	—	—	—	—
BiCl₃	c	315.339	—	-90.6	-75.3	—	42.3	25.
BiCl₄⁻								
std. state, m = 1	aq	350.792	—	—	-115.1	—	—	—
BiCl₆³⁻								
std. state, m = 1	aq	421.698	—	—	-178.51	—	—	—
BiOCl	c	260.4324	—	-87.7	-77.0	—	28.8	—
Bi(OH)₂Cl	c	278.4477	—	—	-128.71	—	—	—
BiOBr	c	304.8884	—	—	-71.0	—	—	—
BiI₃	c	589.6932	—	—	-41.9	—	—	—
BiI₄⁻								
std. state, m = 1	aq	716.5976	—	—	-49.9	—	—	—
BiS	g	241.044	—	43.	29.	—	68.	—
Bi₂S₃	c	514.152	—	-34.2	-33.6	—	47.9	29.2
Bi₂(SO₄)₃	c	706.1448	—	-608.1	—	—	—	—
Bi₂Te₃	c	800.760	-18.43	-18.5	-18.4	7.387	62.36	28.8
BiAsO₄	c	347.8992	—	—	-148.	—	—	—
BORON								
B								
β	c	10.811	0	0	0	0.290	1.40	2.65
	amorp	—	—	0.9	—	0.315	1.56 + X*	2.86
	g	—	133.28	134.5	124.0	1.511	36.65	4.971
B⁺	g	—	324.64	327.34	—	—	—	—
B²⁺	g	—	904.74	908.92	—	—	—	—
B³⁺	g	—	1779.43	1785.09	—	—	—	—
B⁵⁺	g	—	15606.2	15614.9	—	—	—	—
BO₂⁻								
std. state, m = 1	aq	—	—	-184.60	-162.27	—	-8.9	—
B₂O₃	c	69.6202	-302.731	-304.20	-285.30	2.223	12.90	15.04
	amorp	—	—	-299.84	-282.6	—	18.6	14.6
	g	—	-201.4	-201.67	-198.85	3.426	66.85	15.98
B₄O₇⁻⁻								
std. state, m = 1	aq	155.2398	—	—	-622.6	—	—	—
BH	g	11.8190	106.7	107.46	100.29	2.065	41.05	6.97
BH₃	g	13.8349	—	24.	—	—	—	—
BH₄⁻								
std. state, m = 1	aq	14.8429	—	11.51	27.31	—	26.4	—
B₂H₆	g	27.6698	12.29	8.5	20.7	2.857	55.45	13.60
	—	—	6.0	23.0	—	39.4	—	
B₅H₁₀	g	53.3237	—	15.8	—	—	—	—
H₃BO₃	c	61.8331	-258.312	-261.55	-231.60	—	21.23	19.45
	g	—	—	-237.6	—	—	—	—
un-ionized; std. state, m = 1	aq	—	—	-256.29	-231.56	—	38.8	—
B(OH)₄⁻								
std. state, m = 1	aq	78.8405	—	-321.23	-275.65	—	24.5	—
H₃B₃O₆								
std. state, m = 1, (undissoc.)	aq	157.2557	—	—	-650.1	—	—	—
BF₃	g	67.8062	-271.082	-271.75	-267.77	2.784	60.71	12.06
BF₄⁻								
std. state, m = 1	aq	86.8046	—	-376.4	-355.4	—	43.	—
B₂F₄	g	97.6156	-343.48	-344.2	-337.1	4.08	75.8	18.90
BCl₃	c	117.170	-105.40	—	—	—	—	—
BCl₃	liq	117.170	—	-102.1	-92.6	6.88	49.3	25.5
	g	—	-96.28	-96.50	-92.91	3.362	69.31	14.99
BClF₂	g	84.2608	—	-212.8	-209.4	—	65.	—
BCl₂F	g	100.7154	—	-154.2	-150.9	—	68.	—
BBr₃	liq	250.538	—	-57.3	-57.0	—	54.9	—
	g	—	-43.83	-49.15	-55.56	3.755	77.47	16.20
BBrF₂	g	128.7168	—	—	—	3.054	68.42	13.49
BFBr₂	g	189.6274	—	—	—	3.386	74.06	14.89
BCl₂Br	g	161.626	—	—	—	3.477	74.16	15.39
BClBr₂	g	206.082	—	—	—	3.609	76.90	15.78
BI₃	g	391.5242	18.	17.00	4.96	4.024	83.43	16.92
BS	g	42.875	81.	81.74	69.02	2.085	51.65	7.18
BN	c	24.8177	-60.10	-60.8	-54.6	0.628	3.54	4.71
NH₄BO₂								
std. state, m = 1	aq	60.8484	—	-216.27	-181.24	—	18.2	—
NH₄BO₃	aq	76.8478	—	-195.4	—	—	—	—
B₄C	g	33.6332	181.	—	—	—	—	—
B₄C	c	55.2552	-16.93	-17.	-17.	1.343	6.48	12.62
B(CH₃)₃	liq	55.9162	—	-34.2	-7.7	—	57.1	—
	g	—	-23.38	-29.7	-8.6	3.831	75.2	21.15
B(C₂H₅)₃	liq	97.9974	-42.54	-46.5	2.2	13.02	80.47	57.65
BH(OCH₃)₂								
dimethoxyborane	liq	73.8879	—	-144.7	-113.2	—	57.	—
B(OCH₃)₃								
trimethoxyborane	liq	103.9144	—	-223.2	-178.0	11.45	67.8	45.89
B(OC₂H₅)₃								
triethoxyborane	liq	145.9958	—	-250.8	—	—	—	—
BSi₃	g	66.983	175.	—	—	—	—	—
BSiC	g	50.9082	164.	—	—	—	—	—
BROMINE								
Br	g	79.909	28.189	26.741	19.701	1.481	41.805	4.968
Br⁺	g	—	301.38	301.41	—	—	—	—
Br²⁺	g	—	799.2	800.7	—	—	—	—
Br³⁺	g	—	1627.0	1630.0	—	—	—	—
Br⁻	g	—	-53.0	-55.9	—	—	—	—
std. state, m = 1	aq	—	—	-29.05	-24.85	—	19.7	-33.9
Br₂	c	159.818	0	—	—	—	—	—
	liq	—	—	0	0	5.859	36.384	18.090
	g	—	10.923	7.387	0.751	2.323	58.641	8.61
std. state, m = 1	aq	—	—	-0.62	0.94	—	31.2	—
Br₃⁻								
std. state, m = 1	aq	239.727	—	-31.17	-25.59	—	51.5	—
Br₅⁻								
std. state, m = 1	aq	399.545	—	-34.0	-24.8	—	75.7	—

* X = undetermined residual entropy.

SELECTED VALUES OF CHEMICAL THERMODYNAMIC PROPERTIES (Continued)

Formula and Description	State	Formula weight	ΔHf₀ kcal/mol	ΔHf° kcal/mol	ΔGf° kcal/mol	H₂₉₈ − H₀	S° cal/deg mol	Cp
BrO⁻								
std. state, m = 1	aq	95.9084	—	−22.5	−8.0	—	10.	—
BrO₂	c	111.9078	—	11.6	—	—	—	—
BrO₃⁻								
std. state, m = 1	aq	127.9072	—	−16.03	4.43	—	38.65	—
BrO₄⁻								
std. state, m = 1	aq	—	—	3.1	28.2	—	47.7	—
HBr	g	80.9170	−6.826	−8.70	−12.77	2.067	47.463	6.965
std. state, m = 1	aq	—	—	−29.05	−24.85	—	19.7	−33.9
HBrO₃								
std. state, m = 1	aq	128.9152	—	−16.03	4.43	—	38.65	—
BrF₃	liq	136.9042	—	−71.9	−57.5	—	42.6	29.78
	g	—	−58.41	−61.09	−54.84	3.416	69.89	15.92
BrCl	g	115.362	5.28	3.50	−0.23	2.245	57.36	8.36
CADMIUM								
Cd								
γ	c	112.40	0	0	0	1.491	12.37	6.21
α	c	—	—	−0.14	−0.14	—	12.37	—
	g	—	26.78	26.77	18.51	1.481	40.066	4.968
in Hg; two-phase amalgam	—	—	—	−5.078	−2.328	—	3.145	—
Cd⁺	g	—	234.18	235.65	—	—	—	—
Cd²⁺	g	—	624.09	627.04	—	—	—	—
std. state, m = 1	aq	—	—	−18.14	−18.542	—	−17.5	—
CdO	c	128.399	—	−61.7	−54.6	—	13.1	10.38
Cd(OH)₂								
precipitated	c	146.415	—	−134.0	−113.2	—	23.	—
std. state, m = 1	aq	—	—	−128.08	−93.73	—	−22.6	—
undissoc.;	aq	—	—	—	−105.8	—	—	—
CdF₂	c	150.397	—	−167.4	−154.8	—	18.5	—
std. state, m = 1	aq	—	—	−177.14	−151.82	—	−24.1	—
CdCl₂	c	183.306	−93.677	−93.57	−82.21	3.791	27.55	17.85
std. state, m = 1	aq	—	—	−98.04	−81.286	—	9.5	—
undissoc.; std. state, m = 1	aq	—	—	−96.8	−85.88	—	29.1	—
CdCl₂·H₂O	c	201.321	—	−164.54	−140.310	—	40.1	—
CdCl₂·5/2H₂O	c	228.344	—	−270.54	−225.644	—	54.3	—
Cd(ClO₄)₂								
std. state, m = 1	aq	311.301	—	−79.96	−22.66	—	69.5	—
Cd(ClO₄)₂·6H₂O	c	419.393	—	−490.6	—	—	—	—
CdCl₂·2HCl·7H₂O	c	382.335	—	−654.8	—	—	—	—
CdBr₂	c	272.218	−72.455	−75.57	−70.82	4.235	32.8	18.32
std. state, m = 1	aq	—	—	−76.24	−68.24	—	21.9	—
CdBr₂·4H₂O	c	344.279	—	−356.73	−298.287	—	75.6	—
CdI₂	c	366.209	−48.52	−48.6	−48.13	4.565	38.5	19.11
std. state m = 1	aq	—	—	−44.52	−43.20	—	35.7	—
CdS	c	144.464	—	−38.7	−37.4	—	15.5	—
CdSO₄	c	208.462	−220.720	−223.06	−196.65	4.354	29.407	23.80
std. state, m = 1	aq	—	—	−235.46	−196.51	—	−12.7	—
CdSO₄·H₂O	c	226.477	−292.087	−296.26	−255.46	5.582	36.814	32.16
CdSO₄·8/3H₂O	c	256.502	−406.960	−413.33	−350.224	8.497	54.883	50.97
CdSO₄·2½H₂SO₄	c	453.655	—	−769.6	—	—	—	—
CdSe	c	191.36	—	—	—	—	—	—
CdSeO₃	c	239.358	—	−137.5	−119.0	—	34.0	—
std. state, m = 1	aq	—	—	−139.8	−106.9	—	−14.4	—
CdSeO₄	c	255.358	—	−151.3	−127.1	—	39.3	—
std. state, m = 1	aq	—	—	−161.3	−124.0	—	−4.6	—
CdSeO₄·H₂O	c	273.373	—	−225.2	—	—	—	—
CdTe	c	240.00	—	−22.1	−22.0	—	24.	—
Cd(NO₃)₂	c	236.410	—	−109.06	—	—	—	—
std. state, m = 1	aq	—	—	−117.26	−71.76	—	52.5	—
Cd(NO₃)₂·2H₂O	c	272.440	—	−252.30	—	—	—	—
Cd(NO₃)₂·4H₂O	c	308.471	—	−394.11	—	—	—	—
Cd(NH₃)²⁺								
std. state, m = 1	aq	129.431	—	—	−28.4	—	—	—
Cd(NH₃)₂²⁺								
std. state, m = 1	aq	146.461	—	−63.6	−38.0	—	34.6	—
Cd(NH₃)₄²⁺								
std. state, m = 1	aq	180.522	—	−107.6	−54.1	—	80.4	—
CdCl₂·2NH₃	c	217.367	—	−152.0	−106.1	—	51.	—
CdCl₂·4NH₃	c	251.428	—	−195.4	−116.2	—	79.	—
CdCl₂·6NH₃	c	285.490	—	−237.9	−126.2	—	109.2	—
CdBr₂·NH₃	c	289.249	—	−103.3	—	—	—	—
CdBr₂·2NH₃	c	306.279	—	−131.5	—	—	—	—
CdBr₂·6NH₃	c	374.402	—	−218.4	—	—	—	—
CdI₂·2NH₃	c	400.270	—	−104.0	—	—	—	—
CdI₂·6NH₃	c	468.392	—	−173.0	—	—	—	—
Cd₃(PO₄)₂	c	527.143	—	—	−587.1	—	—	—
CdCO₃	c	172.409	—	−179.4	−160.0	—	22.1	—
CdC₂O₄	c	200.420	—	−218.1	—	—	—	—
std. state, m = 1	aq	—	—	−215.3	−179.6	—	−6.6	—
undissoc.; std. state, m = 1	aq	200.420	—	—	−185.1	—	—	—
CdC₂O₄·3H₂O	c	254.466	—	—	−360.4	—	—	—
CdCN⁺								
std. state, m = 1	aq	138.418	—	—	15.4	—	—	—
Cd(CN)₂	c	164.436	—	38.8	—	—	—	—
std. state, m = 1	aq	—	—	53.9	63.9	—	27.5	—
undissoc.;	aq	—	—	—	49.7	—	—	—
Cd(CN)₃⁻								
std. state, m = 1	aq	190.454	—	—	84.8	—	—	—
Cd(CN)₄²⁻								
std. state, m = 1	aq	216.471	—	102.3	121.3	—	77.	—
Cd(CNS)₂	c	228.564	—	12.43	—	—	—	—
std. state, m = 1	aq	—	—	18.40	25.76	—	51.4	—
undissoc.; std. state, m = 1	aq	—	—	—	23.2	—	—	—
CdSiO₃	c	188.484	—	−284.20	−264.0	—	23.3	21.17
Cd(BO₂)₂	c	198.020	—	—	−354.87	—	—	—
CALCIUM								
Ca	c	40.08	0	0	0	1.364	9.90	6.05
	g	—	42.48	42.6	34.5	1.481	36.992	4.968
Ca⁺	g	—	183.45	185.05	—	—	—	—
Ca²⁺	g	—	457.21	460.29	—	—	—	—

SELECTED VALUES OF CHEMICAL THERMODYNAMIC PROPERTIES (Continued)

			0 K	298.15 K (25°C)				
Formula and Description	State	Formula weight	ΔHf° kcal/mol	ΔHf°	ΔGf° kcal/mol	H°₂₉₈ – H°₀	S°	C°p
							cal/deg mol	
std. state, m = 1	aq	—	—	−129.74	−132.30	—	−12.7	—
CaO	c	56.079	—	−151.79	−144.37	—	9.50	10.23
	g	—	—	+11.	—	—	—	—
CaO₂	c	72.079	—	−156.0	—	—	—	—
CaH	g	41.088	55.	54.7	47.9	2.076	48.19	7.11
CaH₂	c	42.096	—	−44.5	−35.2	—	10.	—
CaOH⁺								
std. state, m = 1	aq	—	—	—	−171.7	—	—	—
Ca(OH)₂	c	74.095	—	−235.68	−214.76	—	19.93	20.91
	g	—	—	−130.	—	—	—	—
std. state, m = 1	aq	—	—	−239.68	−207.49	—	−17.8	—
CaF	g	59.078	−64.76	−65.0	−71.2	2.181	54.8	8.03
CaF₂	c	78.077	—	−291.5	−279.0	—	16.46	16.02
	g	—	−186.35	−186.8	−188.9	3.025	65.55	12.25
std. state, m = 1	aq	—	—	−288.74	−265.58	—	−19.3	—
CaCl	g	75.533	−23.23	−23.4	−29.7	2.292	57.70	8.58
CaCl₂	c	110.986	—	−190.2	−178.8	—	25.0	17.35
	g	—	−112.76	−112.7	−114.54	3.613	69.35	14.18
std. state, m = 1	aq	—	—	−209.64	−195.04	—	14.3	—
CaCl₂· H₂O	c	129.001	—	−265.1	—	—	—	—
CaCl₂· 2H₂O	c	147.017	—	−335.3	—	—	—	—
CaCl₂· 4H₂O	c	183.047	—	−480.3	—	—	—	—
CaCl₂· 6H₂O	c	219.078	—	−623.3	—	—	—	—
CaOCl₂	c	126.985	—	−178.4	—	—	—	—
CaOCl₂· H₂O	c	145.001	—	−249.1	—	—	—	—
Ca(OCl)₂	aq	142.985	—	−180.3	—	—	—	—
Ca(ClO₃)₂	c	174.984	—	−162.1	—	—	—	—
Ca(ClO₄)₂	c	238.981	—	−176.09	—	—	—	—
std. state, m = 1	aq	—	—	−191.56	−136.42	—	74.3	—
8 H₂O	aq	—	—	−188.36	—	—	—	—
10 H₂O	aq	—	—	−189.16	—	—	—	—
Ca(ClO₄)₂· 4H₂O	c	311.043	—	−465.8	−352.97	—	103.6	—
CaBr₂	c	199.898	—	−163.2	−158.6	—	31.	—
	g	—	—	−95.2	—	—	—	—
std. state, m = 1	aq	—	—	−187.84	−182.00	—	26.7	—
CaBr₂· 6H₂O	c	307.990	—	−599.0	−514.6	—	98.	—
Ca(BrO₃)₂	c	295.894	—	−163.9	—	—	—	—
CaI₂	c	293.889	—	−127.5	−126.4	—	34.	—
	g	—	—	−65.	—	—	—	—
std. state, m = 1	aq	—	—	−156.12	−156.96	—	40.5	—
CaI₂· 8H₂O	c	438.012	—	−700.2	—	—	—	—
Ca(IO₃)₂	c	389.885	—	−239.6	−200.6	—	55.	—
Ca(IO₃)₂· H₂O	c	407.901	—	−309.1	—	—	—	—
Ca(IO₃)₂· 6H₂O	c	497.977	—	−664.6	−542.0	—	108.	—
CaS	c	72.144	—	−115.3	−114.1	—	13.5	11.33
	g	—	—	+32.	—	—	—	—
CaSO₃	c	120.142	—	—	—	—	24.23	21.92
CaSO₃· 2H₂O	c	156.173	—	−418.9	−371.7	—	44.	42.7
CaSO₄								
insol., anhydrite	c	136.142	—	−342.76	−315.93	—	25.5	23.82
sol., α	c	—	—	−340.64	−313.93	—	25.9	23.95
sol., β	c	—	—	−339.58	−312.87	—	25.9	23.67
std. state, m = 1	aq	—	—	−347.06	−310.27	—	−7.9	—
CaSO₄· 1/2H₂O	c	145.149	—	−376.85	−343.41	—	31.2	28.54
macro; α								
micro; β	c	—	—	−376.35	−343.18	—	32.1	29.69
CaSO₄· 2H₂O								
selenite	c	172.172	—	−483.42	−429.60	—	46.4	44.46
CaSe	c	119.04	—	−88.0	−86.8	—	16.	—
CaSeO₄	c	183.038	—	−265.25	—	—	—	—
CaSeO₄· 2H₂O	c	219.068	—	−407.9	−355.4	—	53.	—
Ca(N₃)₂	c	124.120	—	+3.5	—	—	—	—
Ca₃N₂	c	148.253	—	−103.	—	—	—	—
Ca(NO₃)₂	c	132.091	—	−177.2	—	—	—	—
in 800 H₂O	aq	—	—	−179.5	—	—	—	—
Ca(NO₃)₂· 4H₂O	c	204.152	—	−450.7	—	—	—	—
Ca(NO₃)₂	c	164.090	—	−224.28	−177.63	—	46.2	35.70
std. state, m = 1	aq	—	—	−228.86	−185.52	—	57.3	—
∞ H₂O	aq	—	—	−228.86	—	—	—	—
Ca(NO₃)₂· 2H₂O	c	200.120	—	−368.25	−293.82	—	64.4	—
Ca(NO₃)₂· 3H₂O	c	218.136	—	−439.3	−351.8	—	76.3	—
Ca(NO₃)₂· 4H₂O	c	236.151	—	−509.64	−409.53	—	89.7	—
Ca(NH₂)₂								
Ca₃P₂	c	182.188	—	−121.	—	—	—	—
Ca₂(PO₄)₂								
β	c	198.024	—	—	—	5.715	35.05	34.68
glassy	amorp	—	—	−587.0	—	—	—	—
Ca₂P₂O₇								
β	c	254.103	−792.88	−798.0	−748.6	7.430	45.23	44.89
Ca₃(PO₄)₂								
β, low temp. form	c	310.183	—	−984.9	−928.5	—	56.4	54.45
α, high temp form	c	—	—	−982.3	−926.3	—	57.58	55.35
std. state, m = 1	aq	—	—	−999.8	−883.9	—	−144.	—
CaHPO₄	c	136.059	−430.299	−433.65	−401.83	4.455	26.62	26.30
std. state, m = 1	aq	—	—	−438.57	−392.64	—	−20.7	—
CaHPO₄· 2H₂O	c	172.090	−568.032	−574.47	−515.00	7.490	45.28	47.10
Ca(H₂PO₄)₂	c	170.057	—	−418.9	—	—	—	—
std. state, m = 1	aq	—	—	−423.1	—	—	—	—
Ca(H₂PO₄)₂	c	234.055	—	−742.04	—	—	—	—
std. state, m = 1	aq	—	—	−749.38	−672.64	—	30.5	—
Ca(H₂PO₄)₂· H₂O	c	252.070	−805.547	−814.93	−730.98	9.950	62.1	61.86
Ca₃(AsO₄)₂	c	398.078	—	−788.4	−732.1	—	54.	—
hydrated precipitate	—	—	—	−799.	—	—	—	—
CaHAsO₄	aq	180.007	—	−345.6	—	—	—	—
Ca(H₂AsO₄)₂	aq	321.950	—	−563.6	—	—	—	—
CaC₂	c	64.102	−15.14	−14.3	−15.5	2.711	16.72	14.99
CaCO₃								
calcite	c	100.089	—	−288.46	−269.80	—	22.2	19.57
aragonite	c	—	—	−288.51	−269.55	—	21.2	19.42
std. state, m = 1	aq	—	—	−291.58	−258.47	—	−26.3	—

ELECTED VALUES OF CHEMICAL THERMODYNAMIC PROPERTIES (Continued)

Formula and Description	State	Formula weight	$\Delta H f_0^\circ$ kcal/mol	$\Delta H f^\circ$ kcal/mol	$\Delta G f^\circ$ kcal/mol	$H_{298}^\circ - H_0^\circ$	S° cal/deg mol	C_p cal/deg mol
CaC$_2$O$_4$	c	128.100	—	−325.2	—	—	—	—
std. state, m = 1	aq	—	—	−326.9	−293.37	—	−1.8	—
CaC$_2$O$_4\cdot$H$_2$O	c	146.115	—	−400.30	−361.85	—	37.4	36.52
CaCN$_2$	c	80.105	—	−83.8	—	—	—	—
Ca(CN)$_2$	c	92.116	—	−44.1	—	—	—	—
	aq	—	—	−56.9	—	—	—	—
CaO·SiO$_2$								
wollastonite	c	116.164	—	−390.76	−370.39	—	19.58	20.38
pseudowollastonite	c	—	—	−389.2	−369.2	—	20.88	20.67
glassy	amorp	—	—	−382.65	—	—	—	—
2CaO·SiO$_2$								
β	c	172.244	−548.95	−551.5	−524.1	5.098	30.53	30.78
γ	c	—	−551.25	−554.0	−526.1	4.898	28.87	30.27
3CaO·SiO$_2$	c	228.323	—	−700.1	−665.4	—	40.3	41.08
3CaO·2SiO$_2$								
rankinite	c	288.408	−942.23	−946.7	−899.0	8.424	50.38	51.24
CaO·Al$_2$O$_3$	c	158.041	−552.87	−556.0	−527.9	4.569	27.30	28.87
CaO·2Al$_2$O$_3$	c	260.002	−944.98	−950.7	−901.2	7.286	42.50	48.00
2CaO·Al$_2$O$_3$	c	214.120	—	−707	—	—	—	—
3CaO·Al$_2$O$_3$	c	270.199	−853.21	−857.5	−815.4	8.216	49.2	50.16
CaO·Al$_2$O$_3$·2SiO$_2$								
anorthite, triclinic	c	278.210	—	−1009.2	−955.5	—	48.4	50.46
anorthite, hexagonal	c	—	—	−1004.3	−949.8	—	45.8	49.76
glassy	amorp	—	—	−991.8	—	—	—	—
CaO·Fe$_2$O$_3$	c	215.772	−361.787	−363.37	−337.67	6.076	34.74	36.71
2CaO·Fe$_2$O$_3$	c	271.851	−508.804	−511.30	−478.44	7.564	45.12	46.19
CaCrO$_4$	aq	156.074	—	−340.9	—	—	—	—
CaMoO$_3$	c	184.018	—	−296.	—	—	—	—
CaMoO$_4$	c	200.018	—	−368.4	−342.9	—	29.3	27.32
	g	—	−197.	—	—	—	—	—
std. state, m = 1	aq	—	—	−368.2	−332.2	—	−6.2	—
CaWO$_4$	c	287.928	−391.272	−393.20	−367.71	4.775	30.21	27.28
std. state, m = 1	aq	—	—	−386.8	—	—	—	—
CaZrO$_3$	c	179.298	—	−422.3	−401.8	—	23.92	23.88
CaHfO$_3$	c	266.568	—	−433.0	—	—	—	—
CaMgC$_2$O$_6$								
dolomite	c	184.411	−552.93	−556.0	−517.1	6.210	37.09	37.65
2CaO·5MgO·8SiO$_2$·H$_2$O								
tremolite	c	812.410	—	−2954.	−2780.	23.34	131.2	156.7
CaUO$_4$	c	342.107	—	−478.4	—	—	—	—
CARBON								
C								
graphite, Acheson spectroscopic	c	12.0112	0	0	0	0.251	1.372	2.038
diamond	c	—	0.5797	0.4533	0.6930	0.125*	0.568	1.4615
	g	—	169.98	171.291	160.442	1.562	37.7597	4.9805
C$^+$	g	—	429.628	432.420	—	—	—	—
C^{2+}	g	—	991.900	996.173	—	—	—	—
C^{3+}	g	—	2095.98	2101.73	—	—	—	—
CO	g	28.0106	−27.199	−26.416	−32.780	2.0716	47.219	6.959
std. state, m = 1	aq	—	—	−28.91	−28.66	—	25.0	—
in CH$_3$COOH	—	—	—	−26.416	−28.15	—	31.7	—
CO$^+$	g	—	295.9	298.16	—	—	—	—
CO^{2+}	g	—	942.	945.5	—	—	—	—
CO$_2$	g	44.0100	−93.963	−94.051	−94.254	2.2378	51.06	8.87
undissoc.; std. state, m = 1	aq	—	—	−98.90	−92.26	—	28.1	—
CO$_2^+$	g	—	223.8	225.23	—	—	—	—
CO$_3^{2-}$								
std. state, m = 1	aq	60.0094	—	−161.84	−126.17	—	−13.6	—
CH	g	13.0191	141.6	142.4	—	—	—	—
CH$^+$	g	—	398.1	400.4	—	—	—	—
CH$_2$	g	14.0271	93.9	93.7	—	—	—	—
CH$_2^+$	g	—	333.6	334.9	—	—	—	—
CH$_3$	g	15.0351	34.0	33.2	—	—	—	—
CH$_3^+$	g	—	261.0	261.7	—	—	—	—
CH$_4$	g	16.0430	−15.970	−17.88	−12.13	2.388	44.492	8.439
std. state, m = 1	aq	—	—	−21.28	−8.22	—	20.0	—
CH$_4^+$	g	—	277.1	276.7	—	—	—	—
HCO	g	29.0185	−4.2	−4.12	−7.76	2.386	53.68	8.26
HCOO$^-$								
std. state, m = 1	aq	45.0180	—	−101.71	−83.9	—	22.	−21.0
HCO$_3^-$								
std. state, m = 1	aq	61.0174	—	−165.39	−140.26	—	21.8	—
HCHO	g	30.0265	−27.1	−28.	−27.	2.394	52.26	8.46
unhydrolyzed	aq	—	—	−35.9	—	—	—	—
HCOOH	liq	—	—	−101.51	−86.38	—	30.82	23.67
	g	—	—	−90.48	—	—	—	—
un-ionized; std. state, m = 1	aq	—	—	−101.68	−89.0	—	39.	—
ionized; std. state, m = 1	aq	—	—	−101.71	−83.9	—	22.	−21.0
in ∞ H$_2$O	aq	—	—	−101.71	—	—	—	—
H$_2$CO$_3$								
std. state, m = 1, (undissoc.)	aq	62.0253	—	−167.22	−148.94	—	44.8	—
CH$_3$OH	liq	32.0424	—	−57.04	−39.76	—	30.3	19.5
	g	—	−45.355	−47.96	−38.72	2.731	57.29	10.49
std. state, m = 1	aq	—	—	−58.779	−41.92	—	31.8	—
CF$_2$	g	69.0064	−113.	−114.	—	—	—	—
CF$_2^+$	g	—	119.9	120.4	—	—	—	—
CF$_4$	g	88.0048	−219.6	−221.	−210.	3.043	62.50	14.60
COF$_2$	g	66.0074	−150.95	−151.7	−148.0	2.642	61.78	11.19
CH$_3$F	g	34.0335	—	—	—	2.422	53.25	8.96
CH$_2$F$_2$	g	52.0239	−104.97	−106.8	−100.2	2.555	58.94	10.25
CHF$_3$	g	70.0143	−162.84	−164.5	−156.3	2.764	62.04	12.20
CCl$_2$	g	118.3702	14.	14.	—	—	—	—
CCl$_4$	liq	153.8232	—	−32.37	−15.60	—	51.72	31.49
	g	—	−24.08	−24.6	−14.49	4.117	74.03	19.91
COCl$_2$	g	98.9166	−51.89	−52.3	−48.9	3.067	67.74	13.78
CH$_3$Cl	g	50.4881	−17.426	−19.32	−13.72	2.489	56.04	9.74
std. state, m = 1	aq	—	—	−24.3	−12.3	—	34.6	—

* Relative to C, diamond.

SELECTED VALUES OF CHEMICAL THERMODYNAMIC PROPERTIES (Continued)

Formula and Description	State	Formula weight	ΔHf° (0 K) kcal/mol	ΔHf° kcal/mol	ΔGf° kcal/mol	$H_{298}^{\circ} - H_0^{\circ}$	S° cal/deg mol	C_p cal/deg mol
CH_2Cl_2	liq	84.9331	—	-29.03	-16.09	—	42.5	23.9
	g		-20.462	-22.10	-15.75	2.830	64.56	12.18
CF_3Cl	g	104.4594	-164.8	-166.	-156.	3.293	68.16	15.98
CF_2Cl_2	g	120.9140	-113.0	-114.	-105.	3.543	71.86	17.27
$CFCl_3$	liq	137.3686	—	-72.02	-56.61	—	53.86	29.05
	g		-65.2	-66.	-57.	3.843	74.05	18.66
COFCl	g	86.4620	—	—	—	2.845	66.11	12.52
CH_2ClF	g	68.4785	—	—	—	2.689	63.17	11.24
$CHClF_2$	g	86.4689	—	—	—	2.955	67.11	13.35
$CHCl_2F$	g	102.9235	—	—	—	3.170	70.02	14.56
CBr_4								
monoclinic	c	331.6472	—	4.5	11.4	—	50.8	34.5
	g		26.10	19.	16.	4.873	85.55	21.79
$COBr_2$	liq	187.8286	—	-30.4	—	—	—	—
	g		-19.19	-23.0	-26.5	3.340	73.85	14.78
CH_3Br	g	94.9441	-4.72	-8.4	-6.2	2.536	58.86	10.14
in C_2H_5OH	—		—	—	-13.08	—	—	—
CH_2Br_2	g	173.8451	—	—	—	3.020	70.06	13.07
$CHBr_3$	liq	252.7461	—	-6.8	-1.2	—	52.8	31.
	g		10.24	4.	2.	3.811	79.07	17.02
CF_3Br	g	148.9154	-150.72	-153.6	-147.3	3.457	71.14	16.57
CCl_3Br	g	198.2792	-8.81	-11.0	-5.1	4.285	79.55	20.38
CH_3I	liq	141.9395	—	-3.7	3.2	—	39.0	30
	g		5.38	3.1	3.5	2.585	60.71	10.54
CHI_3	c	393.7323	—	33.7	—	—	—	—
	g		—	—	—	4.106	85.1	17.92
CF_3I	g	195.9108	—	—	—	3.579	73.44	16.94
CH_2ClI	g	176.3845	—	—	—	3.002	70.7	13.02
CH_2IBr	g	220.8405	—	—	—	3.102	73.5	13.46
CS_2	liq	76.1392	—	21.44	15.60	—	36.17	18.1
	g		27.86	28.05	16.05	2.547	56.82	10.85
	aq		—	21.3	—	—	—	—
COS	g	60.0746	-33.991	-33.96	-40.47	2.373	55.32	9.92
CH_3SH	liq	48.1070	—	-11.08	-1.85	—	40.44	21.64
	g		-2.885	-5.34	-2.23	2.898	60.96	12.01
CN	g	26.0178	108.	109.	102.	2.07	48.4	6.97
CNO^-								
std. state, m = 1	aq	42.0172	—	-34.9	-23.3	—	25.5	—
HCN	liq	27.0258	—	26.02	29.86	—	26.97	16.88
	g		32.39	32.3	29.8	2.208	48.20	8.57
std. state, m = 1	aq		—	36.0	41.2	—	22.5	—
$HCN^·$								
CH_3NH_2								
methylamine	liq	31.0577	—	-11.3	8.5	—	35.90	—
	g		—	-5.49	7.67	—	58.15	12.7
std. state, m = 1	aq		—	-16.77	4.94	—	29.5	—
CH_2N_2								
diazomethane	g		—	—	—	2.887	58.02	12.55
NH_2CN								
cyanamide	c		—	14.1	—	—	—	—
NH_4CN	c	44.0564	—	0.10	—	—	—	32.
std. state, m = 1	aq		—	4.3	22.2	—	49.6	—
	aq		—	7.7	—	—	—	—
$C=NH(NH_2)_2$								
guanidine	c	59.0711	—	-18.1	—	—	—	—
HNCO								
isocyanic acid	g	43.0252	—	—	—	2.615	56.85	10.72
HCNO								
cyanic acid, std. state, m = 1, ionized	aq		—	-34.9	-23.3	—	25.5	—
un-ionized, std. state, m = 1	aq		—	-36.90	-28.0	—	34.6	—
CH_3NO_2								
nitromethane	liq	61.0406	—	-27.03	-3.47	—	41.05	25.33
	g		-14.546	-17.86	-1.65	3.083	65.69	13.70
CH_3ONO								
methyl nitrite	g	61.0406	—	-16.5	—	—	51.9	—
CH_3NO_3	liq	77.0400	-38.82	-38.0	-10.4	8.26	51.9	37.6
	g		—	-29.8	-9.4	—	76.1	—
NH_4HCO_3	c	79.0559	—	-203.0	-159.2	—	28.9	—
std. state, m = 1	aq		—	-197.06	-159.23	—	48.9	—
NH_4CNO	c	60.0558	—	-72.75	—	—	—	—
std. state, m = 1	aq		—	-66.6	-42.3	—	52.6	—
$CO(NH_2)_2$								
urea	c	60.0558	—	-79.71	-47.19	—	25.00	22.26
NH_2COONH_4								
ammonium carbamate	c	78.0712	—	-154.17	-107.09	—	31.9	—
$(NH_4)_2CO_3$								
std. state, m = 1	aq	96.0865	—	-225.18	-164.11	—	40.6	—
CNBr	c	105.9268	—	33.58	—	—	—	—
	g		46.07	44.5	39.5	2.648	59.32	11.22
CNI	c	152.9222	—	39.71	44.22	—	23.0	—
	g		54.04	53.9	47.0	2.724	61.35	11.54
std. state, m = 1	aq		—	42.5	44.95	—	29.9	—
CNS^-								
thiocyanate ion std. state, m = 1	aq	58.0818	—	18.27	22.15	—	34.5	-9.6
HCNS								
thiocyanic acid, undissoc. std. state, m = 1	aq		—	—	23.31	—	—	—
ionized; std. state, m = 1	aq		—	18.27	22.15	—	34.5	-9.6
NH_4CNS								
ammonium thiocyanate	c	76.1204	—	-18.8	—	—	—	—
std. state, m = 1	aq		—	-13.40	3.18	—	61.6	9.5
in 200 H_2O	aq		—	-13.4	—	—	—	—
$CS(NH_2)_2$								
thiourea	c	76.1204	—	-21.1	—	—	—	—
$C_2O_4^{2-}$								
std. state, m = 1	aq	88.0199	—	-197.2	-161.1	—	10.9	—
C_2H_4	g	28.0542	14.515	12.49	16.28	2.525	52.45	10.41
std. state, m = 1	aq		—	8.69	19.43	—	29.2	—

SELECTED VALUES OF CHEMICAL THERMODYNAMIC PROPERTIES (Continued)

Formula and Description	State	Formula weight	ΔHf₀ kcal/mol	ΔHf° kcal/mol	ΔGf° kcal/mol	H°₂₉₈ − H°₀	S° cal/deg mol	C°ₚ cal/deg mol
			0 K	**298.15 K (25°C)**				
C₂H₅								
ethyl radical	g	29.0622	28.	25.	31.	—	59.2	—
C₂H₅⁺	g	—	222.	220.	—	—	—	—
C₂H₆	g	30.0701	−16.523	−20.24	−7.86	2.856	54.85	12.58
std. state, m = 1	aq	—	—	−24.40	−4.09	—	28.3	—
C₂H₅⁻	g	—	252.2	250.0	—	—	—	—
HC₂O₄⁻								
std. state, m = 1	aq	89.0279	—	−195.6	−166.93	—	35.7	—
CH₂CO								
ketene	g	42.0376	−13.86	−14.6	−14.8	2.819	59.16	12.37
(COOH)₂								
oxalic acid	c	90.0358	—	−197.7	—	—	—	28.
std. state, m = 1	aq	—	—	−197.2	−161.1	—	10.9	—
CH₂COO⁻								
std. state, m = 1	aq	59.0450	—	−116.16	−88.29	—	20.7	−1.5
C₂H₃O₃⁻								
glycolate ion, std. state m = 1	aq	75.0444	—	−155.9	—	—	—	—
C₂H₄O								
ethylene oxide	liq	44.0536	—	−18.60	−2.83	—	36.77	21.02
	g	—	−9.589	−12.58	−3.12	2.596	57.94	11.45
CH₃CHO								
acetaldehyde	liq	—	—	−45.96	−30.64	—	38.3	—
1/3 (CH₃CHO)₃								
paraldehyde	liq	—	—	−54.73	—	—	—	—
1/4 (CH₃CHO)₄								
metaldehyde	c	—	—	−56.2	—	—	—	—
HCOOCH₃								
methyl formate	liq	60.0530	—	−90.60	—	—	—	29.
	g	—	—	−83.7	—	—	—	—
CH₃COOH								
acetic acid	liq	60.0530	—	−115.8	−93.2	—	38.2	29.7
	g	—	−99.972	−103.31	−89.4	3.286	67.5	15.9
ionized; std. state, m = 1	aq	—	—	−116.16	−88.29	—	20.7	−1.5
un-ionized; std. state, m = 1	aq	—	—	−116.10	−94.8	—	42.7	—
CH₂OHCOOH								
hydroxyacetic acid (glycolic acid)	c	76.0524	—	−158.7	—	—	—	—
CH(OH)₂COOH								
dihydroxyacetic acid (glyoxylic acid)	c	92.0518	—	−199.7	—	—	—	—
CH₃CH₂O⁻								
std. state, m = 1	aq	45.0616	—	—	−24.5	—	—	—
C₂H₅OH								
ethanol	liq	46.0695	—	−66.37	−41.80	—	38.4	26.64
	g	—	−51.969	−56.19	−40.29	3.390	67.54	15.64
std. state, m = 1	aq	—	—	−68.9	−43.44	—	35.5	—
C₂H₅OH								
in ∞ H₂O	aq	—	—	−68.9	—	—	—	—
(CH₂OH)₂								
ethylene glycol	liq	—	—	−108.70	−77.25	—	39.9	35.8
C₂F₄								
tetrafluoroethylene	g	100.0159	−154.68	−155.5	−147.2	3.903	71.69	19.23
C₂F₆								
hexafluoroethane	g	138.0127	−308.0	−310.	−290.	4.87	79.4	25.5
CH≡CF	g	44.0287	—	—	—	2.739	55.34	12.52
CH₂CH₂F								
ethyl fluoride	g	48.0606	—	—	—	3.06	63.2	14.0
CH₂=CF₂	g	—	−76.95	−78.6	−73.0	2.980	63.6	14.36
CH₃CHF₂	g	66.0510	−110.98	−114.3	−100.6	3.34	67.5	16.2
CH₃CF₃								
1,1,1-trifluoroethane	g	84.0414	−172.93	−176.0	−159.5	3.631	66.87	18.69
CF₃CH₂F	g	102.0318	—	—	—	4.079	75.85	20.82
C₂Cl₄								
tetrachloroethylene	liq	165.8343	—	−12.5	1.1	—	63.8	33.7
	g	—	−2.70	−2.9	5.4	4.686	81.5	22.69
C₂Cl₆								
I, cubic	c	236.7403	—	−46.0	—	—	—	—
II, monoclinic	c	—	—	−47.9	—	—	—	—
III, triclinic	c	—	—	−48.5	—	—	—	—
CH₂=CHCl								
vinyl chloride	liq	62.4992	—	3.5	—	—	—	—
	g	—	10.31	8.5	12.4	2.825	63.07	12.84
1/n(CH₂=CHCl)ₙ								
polyvinyl chloride	c	—	—	−22.5	—	—	—	14.2
C₂H₅Cl								
ethyl chloride	liq	64.5152	—	−32.63	−14.20	—	45.60	24.94
	g	—	−23.331	−26.81	−14.45	3.179	65.94	15.01
CHCl=CHCl								
cis-1,2-dichloroethylene	liq	—	—	−6.6	5.27	—	47.42	27.
trans-1,2-dichloroethylene	liq	—	—	−5.53	6.52	—	46.81	27.
CH₂ClCH₂Cl								
1,2-dichloroethane	liq	—	—	−39.49	−19.03	—	49.84	30.9
	g	—	−28.357	−31.02	−17.67	4.08	73.68	18.8
CHCl=CCl₂								
trichloroethylene	liq	131.3893	—	−10.1	2.9	—	54.6	28.8
	g	—	−1.032	−1.86	4.31	3.975	77.6	19.18
CH₃COCl								
acetyl chloride	liq	78.4986	—	−65.44	−49.73	—	48.0	28.
	g	—	−56.054	−58.20	−49.20	3.53	70.5	16.2
CH₂ClCH₂OH								
ethylene chlorohydrin	liq	80.5146	—	−70.6	—	—	—	—
CHCl₂COOH								
dichloracetic acid	liq	128.9430	—	−119.0	—	—	—	44.
ionized	aq	—	—	−122.4	—	—	—	—
un-ionized	aq	—	—	−120.4	—	—	—	—
CCl₃CHO								
chloral (trichloroacetaldehyde)	liq	147.3887	—	−56.45	—	—	—	36.
CCl₃COOH								
trichloroacetic acid	c	163.3881	—	−120.7	—	—	—	—
ionized	aq	—	—	−123.4	—	—	—	—

SELECTED VALUES OF CHEMICAL THERMODYNAMIC PROPERTIES (Continued)

Formula and Description	State	Formula weight	ΔHf° 0 K kcal/mol	ΔHf° kcal/mol	ΔGf° kcal/mol	H°₂₉₈ − H°₀	S° cal/deg mol	C°ₚ cal/deg mol
CCl₃CH(OH)₂								
chloral hydrate	c	165.4040	—	−137.7	—	—	—	34.
	g	—	—	−107.2	—	—	—	—
in 150 CHCl₃	—	—	—	−131.2	—	—	—	—
CF₂=CFCl	g	116.4705	−132.04	−132.7	−125.2	4.096	76.96	20.06
CF₃CCl₃	g	187.3765	—	—	—	5.61	88.6	28.8
CF₂ClCFCl₂	liq	—	—	−188.37	—	—	—	41.5
	g	—	—	−181.5	—	—	—	—
CCl₃CF₂Cl	g	203.8311	−115.75	−117.1	−97.3	5.65	91.5	29.5
CF₃=CHCl	g	98.4801	−74.25	−75.4	−69.1	3.570	72.39	17.23
CH₃CF₂Cl	liq	100.4960	—	—	—	—	—	31.4
CBr₃CBr₃								
hexabromoethane	g	503.4763	—	—	—	7.108	105.6	33.30
CH₂=CHBr								
vinyl bromide	g	106.9552	22.26	18.7	19.3	2.905	65.90	13.27
C₂H₅Br								
ethyl bromide	liq	108.9712	—	−21.99	−6.64	—	47.5	24.1
	g	—	−10.188	−15.42	−6.34	3.259	68.50	15.42
in 2000 CH₃OH	—	—	−21.71	—	—	—	—	
CHBr=CHBr								
cis 1,2-dibromoethylene	g	185.8562	—	—	—	3.491	74.38	16.44
trans, 1,2-dibromoethylene	g	—	—	—	—	3.682	74.90	16.79
CH₂CHBr₂	g	187.8722	—	—	—	3.94	78.3	19.3
CH₂BrCH₂Br								
ethylene bromide	liq	—	—	−19.4	−5.0	—	53.37	32.51
	g	—	—	−9.16	−2.47	—	79.1	20.7
CH₃COBr								
acetyl bromide	liq	122.9546	—	−53.39	—	—	—	—
CF₂ = CFBr	g	160.9265	—	—	—	4.238	80.0	20.51
CF₂CF₂Br	g	198.9233	—	—	—	4.98	85.7	26.1
CF₂ = CBr₂	g	221.8371	—	—	—	4.537	83.5	21.58
CF₂BrCF₂Br	g	259.8339	—	−186.5	—	—	—	—
CF₂ = CHBr	g	142.9361	—	—	—	3.682	75.1	17.63
CF₂CH₂Br	g	162.9424	—	—	—	4.299	80.59	21.67
CHF₂CF₂Br	g	180.9329	—	−197.0	—	—	—	—
CH₂ClCH₂Br	liq	143.4162	—	—	—	—	—	31.1
CHClBrCHClBr	g	256.7622	—	−8.8	—	—	—	—
CF₂ = CBrCl	g	177.3811	—	—	—	4.385	82.0	21.16
CF₂BrCHCl₂	g	213.8421	—	−107.9	—	—	—	—
CI≡CI	g	277.8311	—	—	—	3.901	74.80	16.81
C₂I₄	c	531.6399	—	73.				
CH₂ = CHI								
vinyl iodide	g	153.9506	—	—	—	3.027	68.1	13.84
C₂H₅I								
ethyl iodide	liq	155.9666	—	−9.6	3.5	—	50.6	27.5
CH₂ICH₂I								
ethylene iodide	c	281.8630	—	0.1	13.8	—	47.	—
	g	—	—	15.9	18.8	—	83.2	19.2
C₂H₅SH								
ethanethiol	liq	—	—	−17.53	−1.28	—	49.48	28.17
	g	—	—	−10.95	−1.05	3.617	70.77	17.37
(CH₃)₂SO								
dimethyl sulfoxide	liq	78.1335	—	−48.6	−23.7	—	45.0	35.2
	g	—	−31.427	−35.96	−19.48	4.132	73.20	21.26
(CH₃)₂SO₂								
dimethyl sulfone	c	94.1329	—	−107.8	−72.3	—	34.	—
	g	—	−83.3	−88.7	−65.2	4.3	74.2	23.9
C₂H₅HSO₄								
ethyl sulfuric acid	aq	126.1317	—	−209.3	—	—	—	—
N≡C−C≡N								
cyanogen	g	52.0357	73.386	73.84	71.07	3.028	57.79	13.58
CH₃CN								
acetonitrile	liq	—	—	12.8	23.7	—	35.76	21.86
	g	—	22.58	20.9	25.0	2.892	58.67	12.48
C₂H₅NH₂								
ethylamine	liq	45.0848	—	−17.7	—	—	—	31.
	g	—	—	−11.27	—	—	—	16.7
(CH₃)₂NH								
dimethylamine	liq	—	—	−10.5	16.7	—	43.58	32.9
	g	—	—	−4.41	16.35	—	65.24	16.9
std. state, m = 1	aq	—	—	−16.88	13.85	—	31.8	—
NH₂CH₂CH₂NH₂								
ethylenediamine	liq	—	—	−5.82	—	—	—	50.
CH₃CONH₂								
acetamide	c	—	—	−76.0	—	—	—	16.
C₂H₅NO₂								
nitroethane	liq	75.0676	—	−33.5	—	—	—	33.
	g	—	—	−23.56	—	—	—	—
CH₃CH₂ONO								
ethyl nitrite	liq	—	—	−30.8	—	—	—	—
NH₂CH₂COOH								
glycine	c	—	−121.415	−126.22	−88.09	3.867	24.74	23.71
ionized; std. state, m = 1	aq	—	—	−112.280	−75.278	—	28.54	—
un-ionized, std. state, m = 1	aq	—	—	−122.846	−88.618	—	37.84	—
CH₃CH₂ONO₂								
ethyl nitrate	c	—	−45.023	—	—	—	—	—
	liq	—	—	−45.49	−10.29	9.242	59.08	40.7
CH₃COONH₄								
ammonium acetate	c	77.0836	—	−147.26	—	—	—	—
std. state, m = 1	aq	—	—	−147.83	−107.26	—	47.8	17.6
CH₂OHCOONH₄								
ammonium glycolate	c	93.0830	—	−190.6	—	—	—	—
(CH = NOH)₂								
glyoxime	c	—	—	−21.2	—	—	—	—
(CH₃)₂NH.NO₃								
dimethylammonium nitrate	c	108.0977	—	−83.7	—	—	—	—
std. state, m = 1	aq	—	—	−78.30	−27.41	—	76.2	—
(NH₄)₂C₂O₄								
ammonium oxalate	c	—	—	−268.4	—	—	—	54.
std. state, m = 1	aq	—	—	−260.5	−199.0	—	65.1	—
(NH₄)₂C₂O₄.H₂O	c	142.1124	—	−340.7	—	—	—	—

SELECTED VALUES OF CHEMICAL THERMODYNAMIC PROPERTIES (Continued)

Formula and Description	State	Formula weight	ΔHf°_0 kcal/mol (0 K)	ΔHf° kcal/mol	ΔGf° kcal/mol	$H^\circ_{298} - H^\circ_0$	S° cal/deg mol	C°_p cal/deg mol
CH_3NCS								
methyl isothiocyanate	c	73.1169	—	19.0	—	—	—	—
	g	—	33.46	31.3	34.5	3.464	69.29	15.65
CH_3SCN								
methyl thiocyanate	liq	—	—	28.4	—	—	—	—
	g	—	—	38.3	—	—	—	—
CH_3COOCH_3	liq	74.0801	—	-106.42	—	—	—	—
$(CH_3)_3N$								
trimethylamine	liq	59.1119	—	-11.0	24.1	—	49.82	32.31
	g	—	—	-5.81	23.65	—	68.6	—
std. state, m = 1	aq	—	—	-18.17	22.22	—	31.9	—
$(C_2H_5)_2O$								
diethyl ether	liq	74.1237	—	-66.82	—	—	—	—
	g	—	—	-60.26	—	—	—	—
$(C_2H_5)_2S$								
diethyl sulfide	liq	90.1883	—	-28.43	2.81	—	64.36	40.97
	g	—	-13.15	-19.86	4.34	5.467	87.96	27.97
$(C_2H_5)_2NH$								
diethylamine	liq	73.1390	—	-24.7	—	—	—	—
	g	—	—	-17.07	—	—	—	—
CERIUM								
Ce								
γ	c	140.12	0	0	0	1.74	16.6	6.44
	g	—	101.2	101.	92	1.594	45.807	5.515
Ce^+	g	—	227.3	228.5	—	—	—	—
Ce^{2+}	g	—	478.	480.	—	—	—	—
Ce^{3+}	g	—	942.7	947.4	—	—	—	—
std. state, m = 1	aq	—	—	-166.4	-160.6	—	-49	—
CeO_2	c	172.119	-256.69	-258.80	-244.9	2.478	14.89	14.73
Ce_2O_3	c	328.238	-427.64	-429.3	-407.8	5.13	36.0	27.4
CeH_2	c	142.136	-47.83	-49.	-39.	1.776	13.3	9.78
CeF_3	c	197.115	—	—	—	4.237	27.5	22.3
$CeF_3 \cdot H_2O$	c	215.131	—	-472.4	—	—	—	—
$CeCl_3$	c	246.479	—	-251.8	-233.7	—	36.	20.9
	g	—	—	-174.	—	—	—	—
std. state, m = 1	aq	—	—	-286.2	-254.7	—	-9.	—
$CeCl_3 \cdot 7H_2O$	c	372.587	—	-757.5	—	—	—	—
$CeOCl$	c	191.572	—	-239.	—	—	—	—
$CeClO_4^{2+}$								
std. state, m = 1	aq	239.571	—	-209.1	-165.2	—	-37.	—
$CeBr^{2+}$								
std. state, m = 1	aq	220.029	—	—	-186.3	—	—	—
CeI_3	c	520.833	—	-155.3	—	—	—	—
$Ce(I_3)_3$	c	664.828	—	-332.	—	—	—	—
$Ce(IO_3)_3 \cdot 2H_2O$	c	700.858	—	—	-378.3	—	—	—
CeS_3	c	204.248	—	-146.3	—	—	—	—
Ce_2S	g	312.304	73.	71.6	60.0	3.25	81.	12.9
Ce_2S_3	c	376.432	—	-284.	—	—	—	—
$CeSO_4^+$								
std. state	aq	236.182	—	-380.2	-343.3	—	-17.	—
$Ce(SO_4)_2^-$								
std. state	aq	332.243	—	-595.9	-523.6	—	2.	—
$Ce_2(SO_4)_3$	c	568.425	—	-945.1	—	—	—	—
$Ce_2(SO_4)_3 \cdot 5H_2O$	c	658.502	—	—	—	—	—	132.
$Ce_2(SO_4)_3 \cdot 8H_2O$	c	712.548	—	—	-1320.6	—	—	—
$Ce(NO_3)_3$	c	326.135	—	-293.0	—	—	—	—
$Ce(NO_3)_3 \cdot 3H_2O$	c	380.181	—	-516.	—	—	—	—
$Ce(NO_3)_3 \cdot 4H_2O$	c	398.196	—	-588.9	—	—	—	—
$Ce(NO_3)_3 \cdot 6H_2O$	c	434.227	—	-729.14	—	—	—	—
CeC_2	c	164.142	—	-15.	15.2	—	20.	—
	g	—	136.	136.2	122.9	2.47	64.	10.5
CeC_4	g	188.165	167.1	168.	152.	3.68	73.	17.3
$CeCrO_3$	c	240.114	—	-368.	-347.	—	25.	—
CESIUM								
Cs	c	132.9054	0	0	0	1.843	20.37	7.69
	g	132.9054	18.542	18.180	11.748	1.481	41.942	4.968
Cs^+	g	132.9054	108.337	109.456	—	—	—	—
Cs^{2+}	g	132.9054	686.63	689.23	—	—	—	—
Cs^+	a	132.9054	—	-61.73	-69.79	—	31.80	-2.5
CsO_2	c	164.9042	—	-68.4	—	—	—	—
Cs_2O	c	281.8102	-82.142	-82.64	-73.65	4.225	35.10	18.16
	g	281.8102	—	-37.	—	—	—	—
CsH	c	133.9134	—	-12.950	—	—	—	—
	g	133.9134	28.4	27.7	23.1	2.114	51.40	7.54
CsOH	c	149.9128	—	-99.72	—	—	—	—
	g	149.9128	-57.9	-59.	-59.1	2.828	60.88	11.88
	a	149.9128	—	-116.70	-107.38	—	29.23	—
$CsOH \cdot H_2O$	c	167.9282	—	-180.22	—	—	—	—
CsF	c	151.9038	-132.205	-132.3	-125.6	2.802	22.18	12.21
	g	151.9038	-85.21	-85.8	-89.8	2.306	58.11	8.57
CsCl	c	168.3584	-105.926	-105.89	-99.08	2.976	24.18	12.54
	g	168.3584	-56.89	-57.41	-61.62	2.42	61.15	8.83
CsClO	a	184.3578	—	-87.3	-78.6	—	42.	—
$CsClO_2$	a	200.3572	—	-77.6	-65.7	—	56.0	—
$CsClO_3$	c	216.3566	—	-98.4	-73.6	—	37.3	—
	a	216.3566	—	-86.58	-71.71	—	70.6	—
$CsClO_4$	c	232.3560	-104.136	-105.90	-75.13	5.325	41.84	25.88
	a	232.3560	—	-92.64	-71.85	—	75.3	—
CsBr	c	212.8144	-95.358	-96.99	-93.55	3.140	27.02	12.65
	g	212.8144	-47.69	-50.0	-57.6	2.46	63.89	8.86
CsBrO	a	228.8138	—	-84.2	-77.8	—	42.	—
$CsBrO_3$	c	260.8126	—	-89.82	-68.11	—	39.1	—
	a	260.8126	—	-77.76	-65.36	—	70.45	—
CsI	c	259.8098	-82.652	-82.84	-81.40	3.232	29.41	12.62
	g	259.8098	-35.40	-36.3	-45.7	2.52	65.77	8.95
CsIO	a	275.8092	—	-87.4	-79.0	—	30.5	—
$CsIO_2$	c	307.8080	—	-103.7	—	—	—	—
$CsIO_3$	a	307.8080	—	-114.6	-100.4	—	60.1	—
$CsIO_4$	c	323.8074	—	-91.0	—	—	—	—
Cs_2S	c	297.8748	—	-86.0	—	—	—	—
	a	297.8748	—	-115.6	-119.1	—	60.1	—

SELECTED VALUES OF CHEMICAL THERMODYNAMIC PROPERTIES (Continued)

Formula and Description	State	Formula weight	ΔHf° kcal/mol	ΔHf° kcal/mol	ΔGf° kcal/mol	H°₂₉₈ − H°₀	S° cal/deg mol	Cp° cal/deg mol
		0 K		**298.15 K (25°C)**				
Cs₂SO₃	c	345.8730	—	−271.2	—	—	—	—
	a	345.8730	—	−275.4	−255.9	—	56.6	—
Cs₂SO₄	c	361.8724	−342.63	−344.89	−316.36	6.63	50.65	32.24
	g	361.8724	—	−263.6	—	—	—	—
	a	361.8724	—	−340.78	−317.55	—	68.4	—
CsHSO₃, from HSO₃⁻	a	213.9756	—	−211.40	−195.94	—	65.2	—
CsHSO₄	c	229.9750	—	−276.8	—	—	—	—
from HSO₄⁻	a	229.9750	—	−273.81	−250.48	—	63.3	—
Cs₂Se	a	344.7708	—	—	−108.7	—	—	—
Cs₂SeO₃	a	392.7690	—	−245.2	−228.0	—	67.	—
Cs₂SeO₄	c	408.7684	—	−272.34	—	—	—	—
	a	408.7684	—	−266.7	−245.1	—	76.5	—
CsN₃	c	174.9255	—	−4.7	—	—	—	—
CsNO₂	a	178.9109	—	−86.7	−77.5	—	61.2	—
CsNO₃	c	194.9103	—	−120.93	−97.18	—	37.1	—
	g	194.9103	—	−89.4	—	—	—	—
	a	194.9103	—	−111.29	−96.40	—	66.8	−23.7
CsPO₃	c	211.8774	—	−296.7	—	—	—	—
Cs₃PO₄	a	493.6876	—	−490.5	−452.9	—	42.	—
Cs₄P₂O₇	a	705.5650	—	−789.7	−737.9	—	99.2	—
CsH₂PO₄	c	229.8928	—	−374.0	—	—	—	—
CsAsO₂	a	239.8258	—	−164.27	−153.45	—	41.5	—
Cs₃AsO₄	a	537.6354	—	−397.46	−364.37	—	56.5	—
Cs₂C₂O₄ oxalate	c	353.8308	—	—	—	7.45	56.92	—
	a	353.8308	—	−320.7	−300.7	—	74.5	—
CsHCO₃	c	193.9228	—	−230.9	—	—	—	—
from HCO₃⁻	a	193.9228	—	−227.12	−210.05	—	53.6	—
CsHC₂O₄ from HC₂O₄⁻	a	221.9334	—	−257.3	−236.72	—	67.5	—
CH₃COOCs acetate	a	191.9506	—	−177.89	−158.08	—	52.5	—
CsCN	c	158.9233	—	—	—	4.330	33.40	15.70
	a	158.9233	—	−25.7	−28.6	—	54.3	—
CsCNS thiocyanate	a	190.9873	—	−43.46	−47.64	—	66.3	—
CsMnO₄	a	251.8410	—	−191.1	−176.7	—	77.5	—
	a	664.7288	—	−305.7	−280.5	—	125.0	—
Cs₂CrO₄	c	381.8044	—	−341.57	—	—	—	—
Cs₂Cr₂O₇	c	481.7986	−497.023	−499.24	−456.09	10.671	78.89	55.34
	a	481.7986	—	−479.7	−450.6	—	126.2	—
Cs₂WO₄	a	513.6584	—	−380.6	—	—	—	—
CsVO₃	a	231.8456	—	−274.0	−257.1	—	44.	—
Cs₃VO₄	a	513.6558	—	—	−424.3	—	—	—
Cs₂UO₄	c	567.8374	−459.01	−461.0	−431.7	7.365	52.50	36.51
CHLORINE								
Cl	g	35.453	28.68	29.082	25.262	1.499	39.457	5.220
Cl⁺	g	—	328.86	330.74	—	—	—	—
Cl²⁺	g	—	877.81	881.17	—	—	—	—
Cl³⁺	g	—	1798.26	1803.10	—	—	—	—
Cl⁻	g	—	−57.7	−58.8	—	—	—	—
std. state, m = 1	aq	—	—	−39.952	−31.372	—	13.5	−32.6
Cl₂	g	70.906	0	0	0	2.193	53.288	8.104
std. state, m = 1	aq	—	—	−5.6	1.65	—	29.	—
Cl₃⁻ std. state, m = 1	aq	106.359	—	—	−28.8	—	—	—
ClO	g	51.4524	24.36	24.34	23.45	2.114	54.14	7.52
ClO⁻ std. state, m = 1	aq	—	—	−25.6	−8.8	—	10.	—
ClO₂	g	67.4518	25.09	24.5	28.8	2.580	61.36	10.03
std. state, m = 1	aq	—	—	17.9	28.7	—	39.4	—
ClO₂⁻ std. state, m = 1	aq	—	—	−15.9	4.1	—	24.2	—
ClO₃	g	83.4512	—	37.	—	—	—	—
ClO₃⁻ std. state, m = 1	aq	—	—	−23.7	−0.8	—	38.8	—
ClO₄⁻ std. state, m = 1	aq	99.4506	—	−30.91	−2.06	—	43.5	—
Cl₂O	g	86.9054	19.71	19.2	23.4	2.719	63.60	10.85
HCl	g	36.4610	−22.020	−22.062	−22.777	2.066	44.646	6.96
std state, m = 1	aq	—	—	−39.952	−31.372	—	13.5	−32.6
HCl ∞ H₂O	aq	—	—	−39.952	—	—	—	—
HClO	g	52.4604	—	—	—	2.440	56.54	8.88
undissoc.; std. state, m = 1	aq	—	—	−28.9	−19.1	—	34.	—
HClO₂ undissoc.; std. state, m = 1	aq	68.4598	—	−12.4	1.4	—	45.0	—
HClO₃ std. state, m = 1	aq	84.4592	—	24.85	−1.92	—	38.8	—
HClO₄	liq	100.4586	—	−9.70	—	—	—	—
std. state, m = 1	aq	—	—	−30.91	−2.06	—	43.5	—
HClO₄ ∞ H₂O	aq	—	—	−30.91	—	—	—	—
ClF	g	54.4514	−13.0	−13.02	−13.37	2.127	52.05	7.66
ClF₃	liq	92.4482	—	−45.3	—	—	—	—
CHROMIUM								
Cr	c	51.996	0	0	0	0.970	5.68	5.58
	g	—	94.29	94.8	84.1	1.481	41.68	4.97
Cr⁺	g	—	250.30	252.29	—	—	—	—
Cr²⁺	g	—	630.7	634.2	—	—	—	—
std. state, m = 1	aq	—	—	−34.3	—	—	—	—
Cr³⁺	g	—	1345.	1350.	—	—	—	—
CrO₂	c	99.9942	—	−140.9	—	—	—	—
	g	—	—	−92.2	—	—	—	—
CrO₄²⁻ std. state, m = 1	aq	115.9936	—	−210.60	−173.96	—	12.00	—
Cr₂O₃	c	151.9902	—	−272.4	−252.9	—	19.4	28.38
Cr₂O₇²⁻ std. state, m = 1	aq	215.9878	—	−356.2	−311.0	—	62.6	—

SELECTED VALUES OF CHEMICAL THERMODYNAMIC PROPERTIES (Continued)

			0 K	298.15 K (25°C)				
Formula and Description	State	Formula weight	ΔHf° kcal/mol	ΔHf° kcal/mol	ΔGf° kcal/mol	H°298 − H°0	S° cal/deg mol	C°p cal/deg mol
Cr₂O₄	c	219.9856	—	−366.	—	—	—	—
HCrO₄								
std. state, m = 1	aq	117.0016	—	−209.9	−182.8	—	44.0	—
Cr(OH)₃								
precipitated	c	103.0181	—	−254.3	—	—	—	—
CrF₂	c	89.9928	—	−186.	—	—	—	—
	g	—	—	−99.	—	—	—	—
CrF₃	c	108.9912	−276.22	−277.	−260.	3.357	22.44	18.82
CrCl₂	c	122.902	−94.93	−94.5	−85.1	3.593	27.56	17.01
	g	—	—	−30.7	—	—	—	—
	aq	—	—	−114.2	—	—	—	—
CrCl₂·2H₂O								
light green	c	158.9327	—	−237.1	—	—	—	—
CrCl₂·3H₂O								
pale blue	c	176.9480	—	−308.9	—	—	—	—
CrCl₂·4H₂O								
dark blue	c	194.9634	—	−384.4	—	—	—	—
CrCl₃	c	158.355	−132.96	−133.0	−116.2	4.22	29.4	21.94
CrO₂Cl₂	liq	154.9008	—	−138.5	−122.1	—	53.0	—
	g	—	−127.68	−128.6	−119.9	4.32	78.8	20.2
CrBr₂	c	211.814	—	−72.2	—	—	—	—
	g	—	—	−17.	—	—	—	—
(CrBr₂)₂	g	423.628	—	−84.	—	—	—	—
CrI₂	c	305.8048	—	−37.5	—	—	—	—
	g	—	—	24.	—	—	—	—
	aq	—	—	−60.1	—	—	—	—
CrI₃	c	432.7092	—	−49.0	—	—	—	—
CrICl₂	c	249.8064	—	−100.	—	—	—	—
CrIBr₂	c	338.7184	—	−79.	—	—	—	—
Cr₂(SO₄)₃	c	392.1768	—	—	—	—	—	67.5
CrN	c	66.0027	—	−29.8	—	—	—	11.0
Cr₂N	c	117.9987	—	−30.5	—	—	—	—
NH₄HCrO₄								
std. state, m = 1	aq	135.0402	—	−241.6	−201.8	—	71.1	—
(NH₄)₂CrO₄	c	152.0708	—	−279.0	—	—	—	—
std. state, m = 1	aq	—	—	−273.94	−211.90	—	66.2	—
in 300 H₂O	aq	—	—	−273.5	—	—	—	—
(NH₄)₂Cr₂O₇	c	252.0650	—	−431.8	—	—	—	—
std. state, m = 1	aq	—	—	−419.5	−348.9	—	116.8	—
Cr₃C₂	c	180.0103	−19.51	−19.3	−19.5	3.621	20.42	23.53
Cr₇C₃	c	400.0054	—	−38.7	−39.9	—	48.0	49.92
Cr₂₃C₆	c	1267.9749	—	−87.2	−89.3	—	145.8	149.2
Cr(CO)₆	c	220.0593	—	−257.4	—	—	—	—
	g	—	—	−240.4	—	—	—	—
PbCrO₄	c	323.184	—	−222.5	—	—	—	—
Tl₂CrO₄	c	524.734	—	−225.8	−205.9	—	67.5	—
Ag₂CrO₄	c	331.7336	—	−174.89	−153.40	—	52.0	34.00
std. state, m = 1	aq	—	—	−160.13	−137.09	—	46.8	—
FeCr₂O₄	c	223.8366	—	−345.3	−321.2	—	34.9	31.94
COBALT								
Co								
α, hexagonal	c	58.9332	0	0	0	1.139	7.18	5.93
β, f. c. cubic	c	—	—	0.11	0.06	—	7.34	—
	g	—	101.119	101.5	90.9	1.520	42.879	5.502
Co⁺	g	—	282.486	284.348	—	—	—	—
Co⁺⁺	g	—	675.82	679.17	—	—	—	—
std. state, m = 1	aq	—	—	−13.9	−13.0	—	−27.	—
Co⁺³	g	—	1448.35	1453.18	—	—	—	—
std. state, m = 1	aq	—	—	22.	32.	—	−73.	—
Co₃CoO	c	74.9326	—	−56.87	−51.20	—	12.66	13.20
Co₃O₄	c	240.7972	—	−213.	−185.	—	24.5	29.5
Co(OH)₂								
blue, precipitated	c	92.9479	—	—	−107.6	—	—	—
pink, precipitated	c	—	—	−129.0	−108.6	—	19.	—
pink, precipitated, aged	c	—	—	—	−109.5	—	—	—
std. state, m = 1	aq	—	—	−123.8	−88.2	—	−32.	—
undissoc.; std. state, m = 1	aq	—	—	—	−100.8	—	—	—
Co(OH)₃								
precipitated	c	109.9553	—	−171.3	—	—	—	—
CoF₂	c	115.9284	—	−193.8	—	—	—	—
CoCl₂	c	129.8392	−74.74	−74.7	−64.5	3.375	26.09	18.76
std. state, m = 1	aq	—	—	−93.8	−75.7	—	0.	—
CoCl₂								
in ∞ H₂O	aq	—	—	−93.8	—	—	—	—
CoCl₂·H₂O	c	147.8545	—	−147.	—	—	—	—
CoCl₂·2H₂O	c	165.8699	—	−220.6	−182.8	—	45.	—
CoCl₂·6H₂O	c	237.9312	—	−505.6	−412.4	—	82.	—
Co(ClO₄)₂								
std. state, m = 1	aq	257.8344	—	−75.7	−17.1	—	60.	—
in ∞ H₂O	aq	—	—	−75.7	—	—	—	—
CoBr₂	c	218.7512	—	−52.8	—	—	—	19.0
std. state, m = 1	aq	—	—	−72.0	−62.7	—	12.	—
CoBr₂·6H₂O	c	326.8432	—	−482.8	—	—	—	—
CoI₂	c	312.7420	—	−21.2	—	—	—	—
std. state, m = 1	aq	—	—	−40.3	−37.7	—	26.	—
Co(IO₃)₂								
std. state, m = 1	aq	408.7384	—	−119.7	−74.2	—	30.	—
Co(IO₃)₂·2H₂O	c	444.7691	—	−258.6	−190.2	—	64.	—
CoS	c	90.9972	—	−19.8	—	—	—	—
Co₃S₃								
precipitated	c	214.0584	—	−35.2	—	—	—	—
CoSO₄	c	154.9948	—	−212.3	−187.0	—	28.2	—
std. state, m = 1	aq	—	—	−231.2	−191.0	—	−22.	—
CoSO₄·6H₂O	c	263.0868	−630.257	−641.4	−534.35	13.525	87.86	84.46
CoSO₄·7H₂O	c	281.1022	−699.547	−712.22	−591.26	15.097	97.05	93.33
CoSe	c	137.893	—	−14.6	—	—	—	—
CoTe₂	c	314.133	—	−31.	—	—	—	—
Co(NO₃)₂	c	182.9430	—	−100.5	—	—	—	—
std. state, m = 1	aq	—	—	−113.0	−66.2	—	43.	—
[Co(NH₃)₆]³⁺								
std. state, m = 1	aq	75.9638	—	−34.7	−22.1	—	3.	—

SELECTED VALUES OF CHEMICAL THERMODYNAMIC PROPERTIES (Continued)

			0 K	298.15 K (25°C)				
Formula and Description	State	Formula weight	ΔHf₀ kcal/mol	ΔHf°	ΔGf° kcal/mol	H₂₉₈ − H₀	S° cal/deg mol	Cp°
[Co(NH₃)₂]²⁺								
std. state, m = 1	aq	92.9944	—	—	−30.5	—	—	—
[Co(NH₃)₃]²⁺								
std. state, m = 1	aq	110.0250	—	—	−38.1	—	—	—
[Co(NH₃)₄]²⁺								
std. state, m = 1	aq	127.0556	—	—	−45.3	—	—	—
[Co(NH₃)₆]²⁺								
std. state, m = 1	aq	161.1169	—	−139.8	−38.9	—	40.0	—
[Co(NH₃)₆]N₃²⁺								
std. state, m = 1	aq	203.1370	—	—	42.9	—	60	—
[Co(NH₃)₆](NO₃)₃	c	347.1316	—	−306.4	−125.5	—	107.	—
std. state, m = 1	aq	—	—	−288.5	−117.4	—	140.	—
CoBr₂·NH₃	c	235.7818	—	−85.1	—	—	—	—
CoBr₂·2NH₃								
rose	c	252.8124	—	−116.0	—	—	—	—
[Co(NH₃)₆]Br²⁺								
std. state, m = 1	aq	241.0259	—	−171.7	−60.5	—	39	—
[Co(NH₃)₆]Br₂	c	320.9349	—	−216.4	—	—	—	—
[Co(NH₃)₆]Br₃	c	400.8439	−217.52	−239.7	−119.8	12.18	77.7	78.1
std. state, m = 1	aq	—	—	−227.0	−112.2	—	94.	—
Co₂P	c	148.8402	—	−45	—	—	—	—
Co₃(PO₄)₂	c	366.7424	—	—	−573.3	—	—	—
CoHPO₄	c	154.9126	—	—	−282.5	—	—	—
CoAs	c	133.8548	—	−9.7	—	—	—	—
CoAs₂	c	208.7764	—	−14.7	—	—	—	—
Co₂As	c	192.7880	—	−9.5	—	—	—	—
Co₅As₂	c	342.6312	—	−23.3	—	—	—	—
Co₂As₃	c	326.6428	—	−19.4	—	—	—	—
Co₅As₃	c	444.5092	—	−19.0	—	—	—	—
Co₃(AsO₄)₂	c	454.6380	—	—	−387.4	—	—	—
CoSb	c	180.683	—	−10.	—	—	—	—
CoSb₂	c	302.433	—	−13.	—	—	—	—
CoSb₃	c	424.183	—	−16.	—	—	—	—
Co₃C	c	129.8776	—	−10.	—	—	—	—
CoCo₃	c	118.9426	—	−170.4	—	—	—	—
CoC₂O₄	c	146.9531	—	−203.5	—	—	—	—
std. state, m = 1	aq	—	—	−211.1	−174.1	—	−16.	—
Co(C₂O₄)₂²⁻								
std. state, m = 1	aq	234.9730	—	−408.5	−344.8	—	26.	—
CoSi	c	87.0192	−23.87	−24.0	−23.6	1.78	10.3	10.6
CoSi₂	c	115.1052	—	−24.6	—	—	—	—
CoSi₃	c	143.1912	—	−25.6	—	—	—	—
Co₂SiO₄	c	209.9500	—	−353.	—	—	—	—
COPPER								
Cu	c	63.54	0	0	0	1.196	7.923	5.840
	g	—	80.58	80.86	71.37	1.481	39.74	4.968
Cu⁺	g	—	258.752	260.513	—	—	—	—
std. state, m = 1	aq	—	—	17.13	11.95	—	9.7	—
Cu²⁺	g	—	726.69	729.93	—	—	—	—
std. state, m = 1	aq	—	—	15.48	15.66	—	−23.8	—
Cu³⁺	g	—	1576.1	1580.8	—	—	—	—
Cu₂	g	127.08	115.7	115.72	103.24	2.370	57.71	8.75
CuO	c	79.539	—	−37.6	−31.0	—	10.19	10.11
Cu₂O	c	143.079	—	−40.3	−34.9	—	22.26	15.21
CuH	c	64.548	—	5.1	—	—	—	—
Cu(OH)₂	c	97.555	—	−107.5	—	—	—	—
std. state, m = 1	aq	—	—	−94.46	−59.53	—	−28.9	—
CuF⁺								
std. state, m = 1	aq	82.538	—	−62.4	−52.7	—	−16.	—
CuF₂	c	101.537	—	−129.7	—	—	—	—
CuF₂·2H₂O	c	137.567	—	—	−234.6	—	—	—
CuCl	c	98.993	—	−32.8	−28.65	—	20.6	11.6
CuCl⁺								
std. state, m = 1	aq	—	—	—	−16.3	—	—	—
CuCl₂	c	134.446	−52.79	−52.6	−42.0	3.581	25.83	17.18
CuCl₂·2H₂O	c	170.477	—	−196.3	−156.8	—	40.	—
CuCl₂								
std. state, m = 1	aq	134.446	—	—	−57.4	—	—	—
Cu(ClO₄)₂								
std. state, m = 1	aq	262.441	—	−46.34	11.54	—	63.2	—
Cu(ClO₄)₂·6H₂O	c	370.533	—	−460.9	—	—	—	—
Cu₃OCl₃	c	213.985	—	−90.	—	—	—	—
CuBr	c	143.449	−23.76	−25.0	−24.1	2.893	22.97	13.08
CuBr⁺								
std. state, m = 1	aq	—	—	—	−11.9	—	—	—
CuBr₂	c	223.358	—	−33.9	—	—	—	—
CuBr₂·4H₂O	c	295.419	—	−317.0	—	—	—	—
CuBr₂·3Cu(OH)₂	c	516.022	—	−378.1	−306.2	—	68.	—
Cu(BrO₃)₂·3Cu(OH)₂	c	612.019	—	—	−244.2	—	—	—
CuI	c	190.444	—	−16.2	−16.6	—	23.1	12.92
Cu(IO₃)₂								
std. state, m = 1	aq	413.345	—	−90.3	−45.5	—	32.8	—
Cu(IO₃)₂·H₂O	c	431.361	—	−165.4	−112.0	—	59.1	—
Cu(IO₃)₂·3Cu(OH)₂	c	706.009	—	—	−160.0	—	—	—
CuS	c	95.604	—	−12.7	−12.8	—	15.9	11.43
Cu₂S								
α	c	159.144	—	−19.0	−20.6	—	28.9	18.24
CuSO₄	c	159.602	—	−184.36	−158.2	—	26.	23.9
std. state, m = 1	aq	—	—	−201.84	−162.31	—	−19.0	—
CuSO₄								
in ∞ H₂O	aq	—	—	−201.84	—	—	—	—
CuSO₄·H₂O	c	177.617	—	−259.52	−219.46	—	34.9	32.
CuSO₄·3H₂O	c	213.648	—	−402.56	−334.65	—	52.9	49.
CuSO₄·5H₂O	c	249.678	—	−544.85	−449.344	—	71.8	67.
Cu₂SO₄	c	223.142	—	−179.6	—	—	—	—
CuO·CuSO₄	c	239.141	—	−223.8	—	—	—	—
CuSO₄·2Cu(OH)₂								
antlerite	c	354.711	—	—	−345.8	—	—	—
CuSO₄·3Cu(OH)₂								
brochantite	c	452.266	—	—	−434.5	—	—	—

SELECTED VALUES OF CHEMICAL THERMODYNAMIC PROPERTIES (Continued)

Formula and Description	State	Formula weight	ΔHf° kcal/mol (0 K)	ΔHf°	ΔGf° kcal/mol	H°₂₉₈ − H°₀	S° cal/deg mol	C°p
CuSO₄·3Cu(OH)₂·H₂O								
langite	c	470.281	—	−594.	−488.6	—	80.	—
CuSe	c	142.50	—	−9.45	—	—	—	—
CuSe₂	c	221.46	—	−10.3	—	—	—	—
Cu₂Se	c	206.04	—	−14.2	—	—	—	—
Cu₂Te	c	254.68	—	5.	—	—	—	—
CuN₃	c	105.560	—	66.7	82.4	—	24.	—
Cu(N₃)₂	c	147.580	—	143.0	—	—	—	—
Cu₃N	c	204.627	—	17.8	—	—	—	22.
Cu(NO₃)₂	c	187.550	—	−72.4	—	—	—	—
std. state, m = 1	aq	—	—	−83.64	−37.56	—	46.2	—
Cu(NO₃)₂								
in ∞ H₂O	aq	—	—	−83.64	—	—	—	—
Cu(NO₃)₂·3H₂O	c	241.596	—	−290.9	—	—	—	—
Cu(NO₃)₂·6H₂O	c	295.642	—	−504.5	—	—	—	—
Cu(NH₃)¹⁺								
std. state, m = 1	aq	80.571	—	−9.3	3.72	—	2.9	—
Cu(NH₃)₂¹⁺								
std. state, m = 1	aq	97.601	—	−34.0	−7.28	—	26.6	—
Cu(NH₃)₃²⁺								
std. state, m = 1	aq	114.632	—	−58.7	−17.48	—	47.7	—
Cu(NH₃)₄²⁺								
std. state, m = 1	aq	131.662	—	−83.3	−26.60	—	65.4	—
Cu(NH₃)₅²⁺								
std. state, m = 1	aq	148.693	—	—	−32.13	—	—	—
CuSO₄·NH₃	c	176.632	—	−218.0	—	—	—	—
CuSO₄·2NH₃	c	193.663	—	−248.2	—	—	—	—
CuSO₄·5NH₃	c	244.755	—	−328.1	—	—	—	—
CuP₂	c	125.488	—	−29.	—	—	—	—
Cu₃(PO₄)₂	c	380.563	—	—	−490.3	—	—	—
Cu₃As	c	265.542	—	−2.8	—	—	—	—
Cu₃(AsO₄)₂	c	468.458	—	—	−310.9	—	—	—
std. state, m = 1	aq	—	—	−378.10	−263.02	—	−149.2	—
CuC₂O₄	c	151.560	—	—	−158.2	—	—	—
std. state, m = 1	aq	—	—	−181.7	−145.5	—	−12.9	—
undissoc.; std. state, m = 1	aq	—	—	—	−153.1	—	—	—
CuCO₃·Cu(OH)₂								
malachite	c	221.104	—	−251.3	−213.6	—	44.5	—
2CuCO₃·Cu(OH)₂								
azurite	c	344.653	—	−390.1	—	—	—	—
CuCN	c	89.558	—	23.0	26.6	—	20.2	—
Cu(CN)₂⁻								
std. state, m = 1	aq	115.576	—	—	61.6	—	—	—
Cu(CN)₃²⁻								
std. state, m = 1	aq	141.594	—	—	96.5	—	—	—
Cu(CN)₄³⁻								
std. state, m = 1	aq	167.611	—	—	135.4	—	—	—
CuONC								
cuprous fulminate	c	105.557	—	26.3	—	—	—	—
CuCNS	c	121.622	—	—	16.7	—	—	—
std. state, m = 1	aq	—	—	35.40	34.10	—	44.2	—
(Cu(CNS)₂								
std. state, m = 1	aq	179.704	—	52.02	59.96	—	45.2	—
CuAl	c	90.522	—	−9.8	—	—	—	—
CuAl₂	c	117.503	—	−9.75	—	—	—	—
Cu₃Al	c	154.062	—	−16.5	—	—	—	—
Cu₃Al	c	217.602	—	−16.8	—	—	—	—
Cu₉Al₄	c	244.583	—	−26.2	—	—	—	—
DYSPROSIUM								
Dy	c	162.50	0	0	0	2.116	17.87	6.73
	g	—	70.04	69.4	60.8	1.48	46.97	4.97
Dy⁺	g	—	206.8	207.8	—	—	—	—
Dy²⁺	g	—	476.	477.0	—	—	—	—
Dy³⁺								
std. state, m = 1	aq	—	—	−167.	−159.	—	−55.2	5.
DyO	g	178.499	−19.	—	—	—	—	—
Dy₂O₃	c	372.998	−443.02	−445.3	−423.4	5.04	35.8	27.79
DyF	g	181.498	—	−43.	—	—	—	—
DyF³⁺								
std. state, m = 1	aq	—	—	—	−231.7	—	—	—
DyCl₃								
β	c	268.859	—	−239.	—	—	—	—
γ	c	—	—	−236.	—	—	—	—
std. state, m = 1	aq	—	—	−286.	−253.	—	−14.8	−93.
DyCl₃								
in ∞ H₂O	aq	—	—	−286.	—	—	—	—
DyCl₃·6H₂O	c	376.951	−676.83	−686.	−586.	14.42	96.00	82.7
DyI₃	c	543.213	—	−145.	—	—	—	—
Dy(IO₃)₃	c	687.208	—	−329.	—	—	—	—
DyC₂	g	186.522	206.	206.1	193.2	2.47	64.	10.5
ERBIUM								
Er	c	167.26	0	0	0	1.765	17.49	6.72
	g	—	76.08	75.8	67.1	1.48	46.72	4.97
Er⁺	g	—	216.8	218.0	—	—	—	—
Er²⁺	g	—	492.	495.	—	—	—	—
Er³⁺								
std. state, m = 1	aq	—	—	−168.6	−159.9	—	−58.4	5.
ErO	g	183.259	−13.	—	—	—	—	—
Er₂O₃	c	382.518	−451.69	−453.6	−432.3	4.78	37.2	25.93
ErH₂	c	169.276	—	−49.0	—	—	—	—
Er³H₃	c	171.288	—	−49.3	—	—	—	—
ErH₃	c	170.284	—	−58.	—	—	—	—
ErF	g	186.258	—	−45.	—	—	—	—
ErF₂	g	205.257	—	−164.	—	—	—	—
ErF₃	c	224.255	—	−409.	—	—	—	—
	g	—	—	−294.	—	—	—	—
ErCl₃	c	273.619	—	−238.7	—	—	—	24.
ErCl₃	g	—	—	−167.	—	—	—	—
ErCl₃·6H₂O	c	381.711	−677.92	−687.0	−586.6	14.39	95.3	82.0
Er(BrO₃)₃·9H₂O)	c	713.120	—	−844.9	—	—	—	—

SELECTED VALUES OF CHEMICAL THERMODYNAMIC PROPERTIES (Continued)

			0 K	298.15 K (25°C)				
Formula and Description	State	Formula weight	$\Delta H f_0^\circ$ kcal/mol	$\Delta H f^\circ$ kcal/mol	$\Delta G f^\circ$ kcal/mol	$H_{298}^\circ - H_0^\circ$	S° cal/deg mol	C_p°
ErI₃	c	547.973	—	−146.5	—	—	—	—
Er(IO₃)₃	c	691.968	—	−330.	—	—	—	—
ErC₂	g	191.282	138.	138.2	125.4	2.47	63.	10.5
EUROPIUM								
Eu	c	151.96	0	0	0	1.914	18.60	6.61
	g	—	42.232	41.9	34.0	1.481	45.097	4.968
Eu⁺	g	—	172.93	174.08	—	—	—	—
Eu³⁺	g	—	432.	435.	—	—	—	—
std. state, m = 1	aq	—	—	−126.	−129.1	—	−2.	—
Eu³⁺								
std. state, m = 1	aq	—	—	−144.6	−137.2	—	−53.	2.
EuO	c	167.959	—	−141.5	−133.1	—	15.	—
Eu₂O₃								
cubic	c	351.918	—	−397.4	—	—	—	29.6
monoclinic	c	—	—	−394.7	−372.1	—	35.	29.2
Eu₃O₄	c	519.878	—	−543.	−512.	—	49.	—
Eu(OH)₃	c	202.982	—	—	−285.5	—	—	—
EuF	g	170.958	—	−70.	—	—	—	—
EuF₃	c	208.955	—	—	—	—	—	—
EuCl₂	c	222.866	—	−197.	—	—	—	—
	g	—	—	−110.	—	—	—	—
EuCl₃	c	258.319	—	−223.7	—	—	—	—
std. state, m = 1	aq	—	—	−264.4	−231.3	—	−13.	−96.
EuCl₃· 6H₂O	c	366.411	—	−665.6	−565.5	—	97.3	87.7
Eu(BrO₃)₃· 9H₂O	c	697.820	—	−823.4	—	—	—	—
Eu(IO₃)₃	c	676.668	—	−308.4	—	—	—	—
EuS	g	184.024	27.	—	—	—	—	—
Eu₂(SO₄)₃· 8H₂O	c	736.228	—	—	—	—	160.6	146.0
EuC₂	c	175.982	—	−15.	−16.	—	24.	—
FLUORINE								
F	g	18.9984	18.38	18.88	14.80	1.558	37.917	5.436
F⁺	g	—	420.16	422.14	—	—	—	—
F²⁺	g	—	1226.98	1230.44	—	—	—	—
F³⁺	g	—	2672.0	2676.9	—	—	—	—
F⁴⁺	g	—	4684.2	4690.6	—	—	—	—
F⁻	g	—	−63.7	−64.7	—	—	—	—
std. state, m = 1	aq	—	—	−79.50	−66.64	—	−3.3	−25.5
F₂	g	37.9968	0	0	0	2.108	48.44	7.48
F₂⁺	g	—	365.1	366.6	—	—	—	—
FO	g	34.9978	41.	41.	—	—	—	—
F₂O	g	53.9962	−4.7	−5.2	−1.1	2.604	59.11	10.35
HF	liq	20.0064	—	−71.65	—	—	18.02 + x*	12.35
	g	—	−64.789	−64.8	−65.3	2.055	41.508	6.963
ionized; std. state, m = 1	aq	—	—	−79.50	−66.64	—	−3.3	−25.5
HF								
∞ H₂O	aq	—	—	−79.50	—	—	—	—
HF₂⁻								
std. state, m = 1	aq	39.0048	—	−155.34	−138.18	—	22.1	—
XeF₄	c	207.294	—	−62.5	—	—	—	—
FRANCIUM								
Fr	cs	223.0000	0	0	0	—	22.8	—
FrF	c	241.9984	—	—	—	2.80	26.0	12.6
FrCl	c	258.4530	—	—	—	3.10	27.0	12.80
FrBr	c	302.9040	—	—	—	3.30	31.0	12.90
FrI	c	349.9045	—	—	—	3.40	33.0	12.90
GADOLINIUM								
Gd	c	157.25	0	0	0	2.178	16.27	8.85
	g	—	95.353	95.0	86.0	1.825	46.416	6.584
Gd⁺	g	—	237.0	238.1	—	—	—	—
Gd²⁺	g	—	517.	519.	—	—	—	—
Gd³⁺								
std. state, m = 1	aq	—	—	−164.	−158.	—	−49.2	0.
GdO	g	173.249	−17.	—	—	—	—	—
Gd₂O₃								
monoclinic	c	362.498	—	−434.9	—	—	—	25.5
cubic	c	—	—	—	—	4.45	36.0	25.22
GdH₂	c	159.266	—	−45.5	—	—	—	—
GdF	g	176.248	—	−41.	—	—	—	—
GdF₂	g	195.247	—	−169.	—	—	—	—
GdF₃	g	214.245	—	−310.	—	—	—	—
GdCl₃	c	263.609	—	−241.	—	—	—	21.
std. state, m = 1	aq	—	—	−284.	−253.	—	−8.8	−98.
GdCl₃· 6H₂O	c	371.701	−675.66	−685.	−586.	14.43	97.56	83.0
Gd(BrO₃)₃· 9H₂O	c	703.110	—	−842.9	—	—	—	—
GdI₃	c	537.963	—	−142.	—	—	—	—
Gd(IO₃)₃	c	681.958	—	−327.	—	—	—	—
GdS	g	189.314	38.	—	—	—	—	—
Gd₂(SO₄)₃· 8H₂O	c	746.808	—	−1513.	−1322.	—	155.8	140.5
Gd(NO₃)₃· 6H₂O	c	451.357	—	—	—	19.16	133.2	106.2
GdPO₄· H₂O	c	270.237	—	—	−490.	—	—	—
GdC₂	c	181.272	—	−25.	—	—	—	—
	g	—	128.2	128.2	—	2.5	—	—
GALLIUM								
Ga	c	69.72	—	—	—	1.331	9.77	6.18
	liq	—	—	1.33	—	—	—	6.06
	g	—	66.	66.2	57.1	1.566	40.38	6.06
Ga⁺	g	—	204.32	206.00	—	—	—	—
Ga²⁺	g	—	677.38	680.54	—	—	—	—
std. state, m = 1	aq	—	—	—	−21.	—	—	—
Ga³⁺	g	—	1385.	1390.	—	—	—	—
std. state, m = 1	aq	—	—	−50.6	−38.0	—	−79.	—
Ga⁴⁺	g	—	2865.	2871.	—	—	—	—
GaO	g	85.719	67.	66.8	60.6	2.127	55.2	7.66
Ga₂O	c	155.439	—	−85.	—	—	—	—
Ga₂O₃								
β, rhombic	c	187.438	—	−260.3	−238.6	—	20.31	22.00
GaH	g	70.728	53.	52.7	46.3	2.07	46.69	7.00
Ga(OH)₃	c	120.742	—	−230.5	−198.7	—	24.	—
GaF	g	88.718	−60.	−60.2	—	2.167	—	7.95
GaCl	g	105.173	−19.	−19.1	−25.4	2.29	57.4	8.50

* x = Undetermined residual entropy

SELECTED VALUES OF CHEMICAL THERMODYNAMIC PROPERTIES (Continued)

Formula and Description	State	Formula weight	ΔHf° 0 K kcal/mol	ΔHf° 298.15 K (25°C) kcal/mol	ΔGf° kcal/mol	H°₂₉₈ − H°₀	S° cal/deg mol	C°p cal/deg mol
GaCl₃	c	176.079	—	−125.4	−108.7	—	34.	—
	g	—	—	−107.0	—	—	—	—
GaBr	g	149.629	−10.	−11.9	−21.5	2.37	60.2	8.70
GaBr₃	c	309.447	—	−92.4	−86.0	—	43.	—
GaI	g	196.624	7.4	6.9	—	2.41	—	8.76
GaI₃	c	450.433	—	−57.1	—	—	—	—
	g	—	—	−34.0	—	—	—	—
Ga₂(SO₄)₃	c	427.625	—	—	—	—	—	62.4
GaN	c	83.727	—	−26.4	—	—	—	—
GaP	c	100.694	—	−21.	—	—	—	—
GaPO₄	c	164.691	—	—	−310.1	—	—	—
GaAs	c	144.642	—	−17.	−16.2	—	15.34	11.05
GaSb	c	191.47	—	−10.0	−9.3	—	18.18	11.60
Ga₂C₂	g	163.462	—	134.	—	—	—	—
GERMANIUM								
Ge	c	72.59	0	0	0	1.105	7.43	5.580
	g	—	89.34	90.0	80.3	1.768	40.103	7.345
Ge⁺	g	—	271.18	273.32	—	—	—	—
Ge²⁺	g	—	638.63	642.25	—	—	—	—
Ge³⁺	g	—	1427.85	1432.95	—	—	—	—
GeO								
brown	c	88.589	—	−62.6	−56.7	—	12.	—
yellow	c	—	—	—	−49.5	—	—	—
	g	—	−11.	−11.04	−17.49	2.102	53.58	7.39
GeO₂								
hexagonal	c	104.589	—	−131.7	−118.8	—	13.21	12.45
	amorp	—	—	−128.4	—	—	—	—
Ge₂O₃	g	177.179	—	−112.	—	—	—	—
Ge₃O₃	g	265.768	—	−212.	—	—	—	—
GeH₄	g	76.622	24.29	21.7	27.1	2.567	51.87	10.76
Ge₂H₆	liq	151.228	—	32.82	—	—	—	—
	g	—	—	38.8	—	—	—	—
Ge₃H₈	liq	225.834	—	46.3	—	—	—	—
	g	—	—	54.2	—	—	—	—
H₂GeO₃	aq	122.604	—	−195.73	—	2.185	—	8.30
GeF	g	91.588	−8.	−7.97	—	2.185	—	8.30
GeF₂	g	110.587	−121.	—	—	—	—	—
GeF₄	g	148.584	—	—	—	4.163	72.36	19.56
GeCl	g	108.043	37.	37.09	29.7	2.290	59.	8.81
GeCl₄	liq	214.402	—	−127.1	−110.6	—	58.7	—
GeH₃Cl	g	111.067	—	—	—	2.865	63.00	13.08
GeHCl₃	g	—	—	—	—	4.192	79.06	19.40
GeBr	g	152.499	58.	56.32	—	2.355	—	8.87
GeBr₂	g	232.408	—	−15.0	−25.5	—	79.1	—
GeBr₄	liq	392.226	—	−83.1	−79.2	—	67.1	—
	g	—	−64.61	−71.7	−76.0	5.736	94.66	24.34
GeI₂	c	326.399	—	−21.	−20.	—	32.	—
	g	—	—	11.2	−1.0	—	76.	—
GeI₄	c	580.208	—	−33.9	−34.5	—	64.8	—
	g	—	−12.31	−13.6	−25.4	6.12	102.49	24.89
GeS	c	104.654	—	−16.5	−17.1	—	17.	—
	g	—	22.0	22.	10.	2.185	56.	8.05
GeS₂	c	136.718	—	−45.3	—	—	—	—
GeSe	c	151.55	—	−22.0	—	—	—	—
GeTe	c	200.19	—	−6.	—	—	—	—
Ge₃N₄	c	273.797	—	−15.1	—	—	—	—
GeP	c	103.564	—	−5.	−4.	—	15.	—
GeC	g	84.601	150.	151.	—	—	—	—
GeC₂	g	96.612	142.	143.	—	—	—	—
Ge₂C	g	157.191	130.	131.	—	—	—	—
GeSi	g	100.676	126.	127.	—	—	—	—
Ge₂Si	g	173.266	123.	124.	—	—	—	—
Ge₃Si	g	245.856	121.	122.	—	—	—	—
GOLD								
Au	c	196.967	—	—	—	1.436	11.33	6.075
	g	—	87.46	87.5	78.0	1.481	43.115	4.968
Au⁺	g	—	300.20	301.73	—	—	—	—
Au²⁺	g	—	773.0	776.0	—	—	—	—
AuO₂³⁻								
std. state, m = 1	aq	244.9652	—	—	−12.4	—	—	—
AuH	g	197.9750	70.9	70.5	63.5	2.068	50.441	6.968
HAuO₃²⁻								
std. state, m = 1	aq	245.9732	—	—	−34.0	—	—	—
H₂AuO₃⁻								
std. state, m = 1	aq	246.9811	—	—	−52.2	—	—	—
Au(OH)₃								
precipitated	c	247.9891	—	−101.5	−75.77	—	45.3	—
AuF₃	c	253.9622	—	−86.9	—	—	—	—
AuCl	c	232.420	—	−8.3	—	—	—	—
AuCl₂⁻								
std. state, m = 1	aq	267.873	—	—	−36.13	—	—	—
AuCl₃	c	303.326	—	−28.1	—	—	—	—
AuCl₃·2H₂O	c	339.357	—	−170.9	—	—	—	—
AuCl₄⁻								
std. state, m = 1	aq	338.779	—	−77.0	−56.22	—	63.8	—
HAuCl₄								
std. state, m = 1	aq	339.787	—	−77.0	−56.22	—	63.8	—
HAuCl₄·3H₂O	c	393.833	—	−284.9	—	—	—	—
HAuCl₄·4H₂O	c	411.848	—	−355.9	—	—	—	—
AuBr	c	276.876	—	−3.34	—	—	—	—
AuBr₂⁻								
std. state, m = 1	aq	356.785	—	−30.7	−27.49	—	52.5	—
AuBr₃	c	436.695	—	−12.73	—	—	—	—
AuBr₄⁻								
std. state, m = 1	aq	516.603	—	−45.8	−40.0	—	80.3	—
HAuBr₄·5H₂O	c	607.688	—	−398.7	—	—	—	—
AuI	c	323.8714	—	0.	—	—	—	—
AuSb₂	c	440.467	—	−2.6	—	—	—	—
Au(CN)₂⁻								
std. state, m = 1	aq	249.003	—	57.9	68.3	—	41.	—

SELECTED VALUES OF CHEMICAL THERMODYNAMIC PROPERTIES (Continued)

Formula and Description	State	Formula weight	0 K ΔHf° kcal/mol	298.15 K (25°C) ΔHf° kcal/mol	ΔGf° kcal/mol	H°₂₉₈ – H°₀ kcal/mol	S° cal/deg mol	C°ₚ cal/deg mol
Au(SCN)₂⁻								
std. state, m = 1	aq	313.131	—	—	60.2	—	—	—
Au(SCN)₄⁻								
std. state, m = 1	aq	429.294	—	—	134.2	—	—	—
Au(SCN)₃²⁻								
std. state, m = 1	aq	487.376	—	—	156.4	—	—	—
Au(SCN)₄²⁻								
std. state, m = 1	aq	545.458	—	—	178.5	—	—	—
AuSn	c	315.657	—	-7.28	-7.15	—	23.22	12.06
AuSn₂	c	434.347	—	-10.14	-9.07	—	32.4	—
AuSn₄	c	671.727	—	-9.25	-9.04	—	59.9	—
AuPb₂	c	611.347	—	-1.5	—	—	—	—
AuIn	c	311.787	—	-10.8	—	—	—	—
AuIn₂	c	426.607	—	-18.0	—	—	—	—
AuCd	c	309.367	—	-9.28	-9.34	—	23.9	—
AuCu	c	260.507	—	-4.46	-4.40	—	19.1	11.9
AuCu₃	c	387.587	—	-6.84	-6.97	—	35.7	24.0
HAFNIUM								
Hf								
α, hexagonal	c	178.49	0	0	0	1.397	10.41	6.15
	g	—	147.92	148.0	137.8	1.481	44.642	4.972
Hf⁺	g	—	309.	311.	—	—	—	—
Hf²⁺	g	—	653.	656.	—	—	—	—
Hf³⁺	g	—	1190.	1195.	—	—	—	—
HfO	g	194.489	—	12.	—	—	—	—
HfO₂	c	210.489	—	-273.6	-260.1	—	14.18	14.40
HfF₄								
monoclinic	c	254.484	—	-461.4	-437.5	—	27.	—
HfCl₄	c	320.302	—	-236.70	-215.42	—	45.6	28.80
HfN	c	192.497	—	-88.3	—	—	—	—
HfC	c	190.501	—	-60.1	—	—	—	—
HfB	c	189.301	—	-47.	—	—	—	—
HfB₂	c	200.112	-80.09	-80.3	-79.4	1.77	10.2	11.89
HELIUM								
He	g	4.0026	0	0	0	1.481	30.1244	4.9679
std. state, m = 1	aq	—	—	-0.4	4.6	—	13.3	—
He⁺	g	—	566.978	568.459	—	—	—	—
HOLMIUM								
Ho	c	164.930	0	0	0	1.91	18.0	6.49
	g	—	72.33	71.9	63.3	1.48	46.72	4.97
Ho⁺	g	—	211.1	212.2	—	—	—	—
Ho²⁺	g	—	483.	486.	—	—	—	—
Ho³⁺								
std. state, m = 1	aq	—	—	-168.5	-161.0	—	-54.2	4.
HoO	g	180.929	-22.	—	—	—	—	—
Ho₂O₃	c	377.858	-447.59	-449.5	-428.1	5.02	37.8	27.48
HoH₂	c	166.946	—	-51.7	—	—	—	—
HoF	g	183.928	—	-43.	—	—	—	—
HoF₂	g	202.927	—	-163.	—	—	—	—
HoF₃	c	221.925	—	-408.	—	—	—	—
	g	—	—	-294.	—	—	—	—
HoCl₃	c	271.289	—	-240.3	—	—	—	21
HoCl₃	g	—	—	-168.	—	—	—	—
∞ H₂O	aq	—	—	-288.4	—	—	—	—
HoCl₃·6H₂O	c	379.381	-678.80	-687.9	-588.0	14.49	97.08	83.0
HoOCl	c	216.382	—	-239.2	—	—	—	—
HoI₃	c	545.643	—	-149.0	—	—	—	—
HoS	g	196.994	43.	—	—	—	—	—
HoAs	c	239.852	—	-72	—	—	—	—
HoC₂	c	188.952	—	-26.	-26.7	—	23.	—
	g	—	135.	135.1	—	2.47	—	10.5
HoC₄	g	212.975	165.	—	—	—	—	—
Ho₂C₃	c	365.893	—	-56.	—	—	—	—
HoAu	g	361.897	100.	99.06	—	2.41	—	8.77
HYDROGEN								
H	g	1.0080	51.626	52.095	48.581	1.481	27.391	4.9679
¹H	g	1.0078	51.626	52.095	48.581	1.481	27.391	4.9679
²H	g	2.0141	52.524	52.981	49.360	1.481	29.455	4.9679
H⁺	g	1.0080	365.211	367.161	—	—	—	—
std. state, m = 1	aq	—	—	0	0	—	0	0
H⁻	g	—	34.40	33.39	—	—	—	—
H₂	g	2.0159	0	0	0	2.0238	31.208	6.889
¹H₂	g	2.0156	0	0	0	2.0238	31.208	6.889
²H₂	g	4.0282	0	0	0	2.0481	34.620	6.978
¹H²H	g	3.0219	0.079	0.076	-0.350	2.0328	34.343	6.978
H₂								
std. state, m = 1	aq	2.0159	—	-1.0	4.2	—	13.8	—
H₂⁺	g	—	355.74	357.23	—	—	—	—
OH	g	17.0074	9.25	9.31	8.18	2.1070	43.890	7.143
OH⁻	g	—	-32.3	-33.67	—	—	—	—
std. state, m = 1	aq	—	—	-54.970	-37.594	—	-2.57	-35.5
HO₂	g	33.0068	6.	5.	—	—	—	—
H₂O	liq	18.0153	—	-68.315	-56.687	—	16.71	17.995
²H₂O	liq	20.0276	—	-70.411	-58.195	—	18.15	20.16
¹H²HO	liq	19.0213	—	-69.285	-57.817	—	18.95	—
H₂O	g	18.0153	-57.102	-57.796	-54.634	2.3667	45.104	8.025
¹H₂O	g	18.0150	-57.102	-57.796	-54.634	2.3667	45.103	8.025
²H₂O	g	20.0276	-58.855	-59.560	-56.059	2.3801	47.378	8.19
¹H²HO	g	19.0213	-57.927	-58.628	-55.719	2.3721	47.658	8.08
H₂O₂	liq	34.0147	—	-44.88	-28.78	-26.2	—	21.3
	g	—	-31.08	-32.58	-25.24	2.594	55.6	10.3
undissoc.; std. state, m = 1	aq	—	—	-45.69	-32.05	—	34.4	—
H₂O₂								
∞ H₂O	aq	—	—	-45.69	—	—	—	—
INDIUM								
In	c	114.82	0	0	0	1.578	13.82	6.39
	g	—	58.25	58.15	49.89	1.48	41.51	4.98
In⁺	g	—	191.68	193.06	—	—	—	—
std. state, m = 1	aq	—	—	—	-2.9	—	—	—

SELECTED VALUES OF CHEMICAL THERMODYNAMIC PROPERTIES (Continued)

Formula and Description	State	Formula weight	0 K ΔHf$\ddagger$ kcal/mol	298.15 K (25°C) ΔHf° kcal/mol	ΔGf° kcal/mol	H°$_{198}$ – H°$\ddagger$	S° cal/deg mol	C°$_p$
In^{2+}	g	—	626.82	629.68	—	—	—	—
std. state, m = 1	aq	—	—	—	-12.1	—	—	—
In^{3+}	g	—	1273.2	1277.56	—	—	—	—
std. state, m = 1	aq	—	—	-25.	-23.4	—	-36.	—
InO	g	130.819	92.	93.	87.1	2.14	56.5	7.78
In$_2$O$_3$	c	277.638	—	-221.27	-198.55	—	24.9	22.
InH	g	115.828	52.	51.5	45.49	2.075	49.60	7.07
InOH	g	131.827	-18.	-19.	—	—	—	—
InF	g	133.818	-48.2	-48.61	—	2.198	—	—
InCl$_3$	c	221.179	—	-128.4	—	—	—	—
In$_2$Cl$_3$	g	335.999	—	-103.6	—	—	—	—
InBr	c	194.729	—	-41.9	-40.4	—	27.	—
	g	—	-11.5	-13.6	-22.54	2.406	61.99	8.76
InBr$_3$	c	354.547	—	-102.5	—	—	—	—
	g	—	—	-67.4	—	—	—	—
InI	c	241.724	—	-27.8	-28.8	—	31.	—
	g	—	2.52	1.8	-9.0	2.437	63.87	8.80
InI$_3$	c	495.533	—	-57.	—	—	—	—
	g	—	—	-28.8	—	—	—	—
InS	c	146.884	—	-33.0	-31.5	—	16.	—
	g	—	—	57.	—	—	—	—
In$_2$S	g	261.704	15.	2.9	—	76.	—	—
In$_2$S$_3$	c	325.832	—	-102.	-98.6	—	39.1	28.20
In$_2$(SO$_4$)$_3$	c	517.825	—	-666.	-583.	—	65.	67.
InN	c	128.827	—	-4.2	—	—	—	—
InP	c	145.794	—	-21.2	-18.4	—	14.3	10.86
InAs	c	189.742	—	-14.0	-12.8	—	18.1	11.42
InSb	c	236.57	—	-7.3	-6.1	—	20.6	11.82
InSb$_3$	g	358.32	—	75.	—	—	—	—
IODINE								
I	g	126.9044	25.631	25.535	16.798	1.481	43.184	4.968
I$^+$	g	—	266.77	268.16	—	—	—	—
I^{2+}	g	—	707.22	710.09	—	—	—	—
I$^-$	g	—	-45.4	-47.0	—	—	—	—
std. state, m = 1	aq	—	—	-13.19	-12.33	—	26.6	-34.0
I$_2$	c	253.8088	0	0	0	3.154	27.757	13.011
	g	—	15.659	14.923	4.627	2.418	62.28	8.82
std. state, m = 1	aq	—	—	5.4	3.92	—	32.8	—
I$_3^-$								
std. state, m = 1	aq	380.7132	—	-12.3	-12.3	—	57.2	—
IO$^-$								
std. state, m = 1	aq	—	—	-25.7	-9.2	—	-1.3	—
IO$_3^-$								
std. state, m = 1	aq	174.9026	—	-52.9	-30.6	—	28.3	—
IO$_4^-$								
std. state, m = 1	aq	190.9020	—	-36.2	-14.0	—	-53.	—
I$_2$O$_5$	c	333.8058	—	-37.78	—	—	—	—
HI	g	127.9124	6.850	6.33	0.41	2.069	49.351	6.969
std. state, m = 1	aq	—	—	-13.19	-12.33	—	26.6	-34.0
HI								
in ∞ H$_2$O	aq	—	—	-13.19	—	—	—	—
HIO								
undissoc.; std. state, m = 1	aq	143.9118	—	-33.0	-23.7	—	22.8	—
HIO$_3$	c	175.9106	—	-55.0	—	—	—	—
undissoc.; std. state, m = 1	aq	—	—	-50.5	-31.7	—	39.9	—
IF	g	145.9028	-22.40	-22.86	-28.32	2.174	56.42	7.99
IF$_5$	liq	221.8964	—	-206.7	—	—	—	—
ICl								
α	c	162.3574	—	-8.4	—	—	—	—
	liq	—	—	-5.71	-3.25	—	32.3	—
	g	—	4.64	4.25	-1.30	2.282	59.140	8.50
std. state, m = 1	aq	—	—	-4.1	—	—	—	—
ICl$_3$	c	233.2634	—	-21.4	-5.34	—	40.0	—
IBr	c	206.8134	—	-2.5	—	—	—	—
	g	—	11.90	9.76	0.89	2.367	61.822	8.71
std. state, m = 1	aq	—	—	—	-1.0	—	—	—
IRIDIUM								
Ir	c	192.2	0	0	0	1.260	8.48	6.00
	g	—	158.78	159.0	147.7	1.481	46.240	4.968
Ir$^+$	g	—	373.	375.	—	—	—	—
IrO$_2$	c	224.20	—	-65.5	—	—	—	13.7
IrO$_3$	g	240.20	—	1.9	—	—	—	—
IrF$_6$	c	306.19	—	-138.54	-110.34	—	59.2	—
IrCl	c	227.65	—	-19.5	—	—	—	—
IrCl$_3$	c	298.56	—	-58.7	—	—	—	—
IrS$_2$	c	256.33	—	-33.	—	—	—	—
Ir$_2$S$_3$	c	480.59	—	-56.	—	—	—	—
IRON								
Fe								
α	c	55.847	0	0	0	1.073	6.52	6.00
	g	—	98.94	99.5	88.6	1.6374	43.112	6.137
Fe$^+$	g	—	281.12	283.16	—	—	—	—
Fe^{2+}	g	—	654.2	657.8	—	—	—	—
std. state, m = 1	aq	—	—	-21.3	-18.85	—	-32.9	—
Fe^{3+}	g	—	1360.8	1365.9	—	—	—	—
std. state, m = 1	aq	—	—	-11.6	-1.1	—	-75.5	—
Fe$_{0.947}$O								
wustite	c	68.8865	-63.85	-63.64	-58.89	2.26	13.74	11.50
FeO	c	71.8464	—	-65.0	—	—	—	—
Fe$_2$O$_3$								
hematite	c	159.6922	-195.46	-197.0	-177.4	3.719	20.89	24.82
Fe$_3$O$_4$								
magnetite	c	231.5386	-265.80	-267.3	-242.7	5.87	35.0	34.28
FeO(OH)								
goethite	c	88.8538	—	-133.6	—	—	—	—
Fe(OH)$_2$								
precipitated	c	89.8617	—	-136.0	-116.3	—	21.	—
Fe(OH)$_3$								
precipitated	c	106.8691	—	-196.7	-166.5	—	25.5	—
undissoc.; std. state, m = 1	aq	—	—	—	-157.6	—	—	—

SELECTED VALUES OF CHEMICAL THERMODYNAMIC PROPERTIES (Continued)

Formula and Description	State	Formula weight	ΔHf₀ kcal/mol (0 K)	ΔHf° kcal/mol	ΔGf° kcal/mol	$H°_{298} - H°_0$	S° cal/deg mol	C°ₚ
FeF₃	c	93.8438	—	—	—	3.049	20.79	16.28
FeF₃⁻								
std. state, m = 1	aq	—	—	-166.4	-150.2	—	-15.	—
FeF₃	aq	112.8422	—	-242.9	—	—	—	—
FeCl₂	c	126.753	-82.313	-81.69	-72.26	3.889	28.19	18.32
std. state, m = 1	aq	—	—	-101.2	-81.59	—	-5.9	—
FeCl₂·2H₂O	c	162.7837	—	-227.8	—	—	—	—
FeCl₂·4H₂O	c	198.8143	—	-370.3	—	—	—	—
FeCl₃	c	162.206	-95.828	-95.48	-79.84	4.710	34.0	23.10
std. state, m = 1	aq	—	—	-131.5	-95.2	—	-35.0	—
FeCl₃·6H₂O	c	270.2980	—	-531.5	—	—	—	—
FeOCl	c	107.2994	—	-90.1	—	—	—	—
Fe(ClO₄)₃								
std. state, m = 1	aq	254.7482	—	-83.1	-22.97	—	54.1	—
FeBr₂	c	215.665	—	-59.7	—	—	—	—
std. state, m = 1	aq	—	—	-79.4	-68.55	—	6.5	—
FeBr₃	c	295.574	—	-64.1	—	—	—	—
std. state, m = 1	aq	—	—	-98.8	-75.7	—	-16.4	—
FeI₂	c	309.6558	—	-27.0	—	—	—	—
std. state, m = 1	aq	—	—	-47.7	-43.51	—	20.3	—
FeI₃	g	436.5602	—	17.	—	—	—	—
std. state, m = 1	aq	—	—	-51.2	-38.1	—	4.3	—
Fe₁.₀₀₀S								
Iron-rich pyrrhotite, σ	c	87.911	-24.01	-23.9	-24.0	2.235	14.41	12.08
FeS₂								
pyrite	c	119.975	-41.72	-42.6	-39.9	2.302	12.65	14.86
markasite	c	—	—	-37.0	—	—	—	—
FeSO₄	c	151.9086	—	-221.9	-196.2	—	25.7	24.04
std. state, m = 1	aq	—	—	-238.6	-196.82	—	-28.1	—
FeSO₄·H₂O	c	169.9239	—	-297.25	—	—	—	—
FeSO₄·4H₂O	c	223.9700	—	-508.9	—	—	—	—
FeSO₄·7H₂O	c	278.0160	—	-720.50	-599.97	—	97.8	94.28
FeSe	c	134.807	—	-18.0	—	—	—	—
FeTe	c	183.447	—	-15.0	—	—	—	—
Fe₄N	c	237.3947	—	-2.5	0.9	—	37.	—
Fe(NO₃)₃	aq	241.8617	—	-161.3	—	—	—	—
std. state, m = 1	aq	—	—	-160.3	-80.9	—	29.5	—
FeP	c	86.8208	—	-30.	—	—	—	—
FeP₁	c	117.7946	—	-46.	—	—	—	—
Fe₂P	c	142.6678	—	-39.	—	—	—	—
Fe₃P	c	198.5148	—	-39.	—	—	—	—
FePO₄	c	150.8184	—	-310.1	—	—	—	—
FePO₄·2H₂O								
strengite	c	186.8491	-445.28	-451.3	-396.2	6.607	40.93	43.15
Fe₃C								
σ, cementite	c	179.5522	—	6.0	4.8	—	25.0	25.3
FeCO₃								
siderite	c	115.8564	—	-177.00	-159.35	—	22.2	19.63
Fe(CO)₅	liq	195.8998	—	-185.0	-168.6	—	80.8	57.5
	g	—	—	-175.4	-166.65	—	106.4	—
HFe(CN)₆³⁻								
std. state, m = 1	aq	212.9621	—	108.9	160.40	—	42.	—
H₂Fe(CN)₆²⁻								
std.state, m = 1	aq	213.9700	—	108.9	157.37	—	52.	—
H₃Fe(CN)₆⁻								
std. state, m = 1	aq	—	—	108.9	—	—	—	—
Fe(SCN)²⁺								
thiocyanate; std. state, m = 1	aq	113.9288	—	5.6	17.0	—	-31.	—
FeSi	c	83.933	-17.67	-17.6	-17.6	1.91	11.0	11.4
FeSi₂								
β-lebeanite	c	112.019	-19.19	-19.4	-18.7	2.40	13.3	15.79
FeSi₂.₃₃								
α-lebeanite	c	121.287	-14.03	-14.	-14.	2.89	16.6	17.62
Fe₃Si	c	195.627	-22.55	-22.4	-22.6	4.14	24.8	23.50
Fe₂SiO₄								
fayalite	c	203.7776	—	-353.7	-329.6	—	34.7	31.76
FeAl₂O₄	c	173.8076	—	-470.	-442.	—	25.4	29.53
KRYPTON								
Kr	g	83.80	0	0	0	1.481	39.1905	4.9679
std. state, m = 1	aq	—	—	-3.7	3.6	—	14.7	—
Kr⁺	g	—	—	322.84	324.32	—	—	—
Kr²⁺	g	—	—	889.47	892.43	—	—	—
Kr³⁺	g	—	—	1741.6	1746.0	—	—	—
LANTHANUM								
La								
σ	c	138.91	0	0	0	1.593	13.6	6.48
	g	—	103.084	103.0	94.07	1.509	43.563	5.438
La⁺	g	—	—	231.690	233.087	—	—	—
La²⁺	g	—	—	487.	490.	—	—	—
La³⁺	g	—	—	929.2	933.2	—	—	—
std. state, m = 1	aq	—	—	-169.0	-163.4	—	-52.0	-3.
LaO	g	154.909	-28.5	-29.01	-34.72	2.121	57.27	7.59
La₂O	g	293.819	-2.	-3.2	—	3.00	—	12.0
La₂O₂	g	309.819	-145.	-146.6	—	3.67	—	16.2
La₂O₃	c	325.818	-427.19	-428.7	-407.7	4.742	30.43	26.00
LaH₂	c	140.926	—	-48.3	—	—	—	—
La(OH)₃	c	189.932	—	-337.0	—	—	—	—
LaF	g	157.908	—	—	—	2.173	56.9	7.98
LaF₃	g	195.905	—	—	—	4.14	78.7	17.4
LaCl₃	c	245.269	—	-256.0	—	—	—	26.0
std. state, m = 1	aq	—	—	-288.9	-257.5	—	-12.	-101.
LaCl₃								
in ∞ H₂O	aq	—	—	-288.9	—	—	—	—
LaCl₃·7H₂O	c	371.376	—	-759.7	-648.5	17.12	110.6	103.0
La(BrO₃)₃·9H₂O	c	684.770	—	-846.0	—	—	—	—
LaI₃	c	519.623	—	-159.4	—	—	—	—
La(IO₃)₃	c	663.618	—	-334.	-270.4	—	62.	—
LaS	c	170.974	-108.9	-109.	-107.9	2.6	17.5	14.
La₂S₃	c	374.012	—	-289	—	—	—	—
La₂(SO₄)₃	c	566.005	—	-942.0	—	—	—	67.
La₂(SO₄)₃·9H₂O	c	728.143	—	-1589.	—	—	—	152.

SELECTED VALUES OF CHEMICAL THERMODYNAMIC PROPERTIES (Continued)

Formula and Description	State	Formula weight	0 K ΔHf° kcal/mol	298.15 K (25°C) ΔHf° kcal/mol	ΔGf° kcal/mol	H°₂₉₈ – H°₀	S° cal/deg mol	C°ₚ cal/deg mol
LaN	c	152.917	—	−72.5	—	—	—	—
La(NO₃)₃	c	324.925	—	−299.8	—	—	—	—
La(NO₃)₃								
in ∞ H₂O	aq	—	—	−317.7	—	—	—	—
La(NO₃)₃·3H₂O	c	378.971	—	−520.0	—	—	—	—
La(NO₃)₃·4H₂O	c	396.986	—	−592.3	—	—	—	—
La(NO₃)₃·6H₂O	c	433.017	—	−732.23	—	—	—	—
LaC₂	c	162.932	—	−17.	−17.3	—	17.	—
La₂(CO₃)₃	c	457.848	—	—	−750.9	—	—	—
LaB₆	c	203.776	—	−31.	—	—	—	—
LaAu	g	335.877	111.4	110.8	98.2	2.46	67.	8.8
LEAD								
Pb	c	207.19	0	0	0	1.644	15.49	6.32
	g	—	46.76	46.6	38.7	1.481	41.889	4.968
Pb⁺	g	—	217.795	219.116	—	—	—	—
Pb²⁺	g	—	564.44	567.25	—	—	—	—
std. state, m = 1	aq	—	—	−0.4	−5.83	—	2.5	—
Pb³⁺	g	—	1300.93	1305.22	—	—	—	—
Pb⁴⁺	g	—	2276.9	2282.7	—	—	—	—
PbO								
yellow	c	223.189	−51.766	−51.466	−44.91	2.207	16.42	10.94
red	c	—	—	−52.34	−45.16	—	15.9	10.95
PbO₂	c	239.189	—	−66.3	−51.95	—	16.4	15.45
Pb₂O₃	c	462.378	—	—	—	—	36.3	25.74
Pb₃O₄	c	685.568	—	−171.7	−143.7	—	50.5	35.1
Pb(OH)₂	c	241.205	—	—	−108.1	—	—	—
precipitated	c	—	—	−123.3	—	—	—	—
PbF₂	c	245.187	—	−158.7	−147.5	—	26.4	—
std. state, m = 1	aq	—	—	−159.4	−139.11	—	−4.1	—
PbF₄	c	283.184	—	−225.1	—	—	—	—
PbCl₂	c	278.096	—	−85.90	−75.98	—	32.5	—
ionized; std. state, m = 1	aq	—	—	−80.3	−68.57	—	29.5	—
PbCl₄	liq	349.002	—	−78.7	—	—	—	—
PbBr₂	c	367.008	—	−66.6	−62.60	—	38.6	19.15
ionized; std. state, m = 1	aq	—	—	−58.5	−55.53	—	41.9	—
PbI₂	c	460.999	−41.81	−41.94	−41.50	4.666	41.79	18.49
ionized; std. state, m = 1	aq	—	—	−26.8	−30.49	—	55.7	—
PbS	c	239.254	—	−24.0	−23.6	—	21.8	11.83
PbSO₃	c	287.252	—	−160.1	—	—	—	—
PbSO₄	c	303.252	−217.82	−219.87	−194.36	4.795	35.51	24.667
PbS₂O₃	c	319.316	—	−161.1	—	—	—	—
PbSe	c	286.15	—	−24.6	−24.3	—	24.5	12.0
PbSeO₄	c	350.148	—	−145.6	−120.7	—	40.1	—
PbTeO₃·0.667H₂O (amorp)	—	394.804	—	−185.5	—	—	—	—
Pb₃(PO₄)₂	c	811.513	—	—	—	—	84.4	61.25
PbCO₃	c	267.199	—	−167.1	−149.5	—	31.3	20.89
PbC₂O₄								
std. state, m = 1	c	295.210	—	−197.6	−166.9	—	34.9	25.2
Pb(CH₃)₄	liq	267.330	—	23.4	—	—	—	—
Pb(C₂H₅)₄	liq	323.439	—	12.6	—	—	—	—
	g	—	—	26.19	—	—	—	—
PSiO₃	c	283.274	—	−273.83	−253.86	—	26.2	21.52
Pb₂SiO₄	c	506.464	—	−325.8	−299.4	—	44.6	32.78
LITHIUM								
Li	c	6.941	0	0	0	1.106	6.96	5.92
	g	6.941	37.715	38.09	30.28	1.481	33.14	4.968
Li⁺	g	6.941	162.045	162.42	—	—	—	—
Li²⁺	g	6.941	—	1936.7	—	—	—	—
Li³⁺	g	6.941	—	4760.5	—	—	—	—
Li⁺	a	6.941	—	−66.56	−70.10	—	3.2	16.4
Li								
in D:99Hg		6.941	—	−19.60	—	—	—	—
Ll								
Hg:x		6.941	—	—	−19.5	—	—	—
LiO	g	22.9404	18.10	18.1	12.5	2.14	50.40	7.75
LiO₂	g	38.9398	—	—	2.60	—	58.27	10.34
Li₂O	c	29.8814	−141.393	−142.91	−134.13	1.732	8.98	12.93
	g	29.8814	−38.18	−38.4	−43.4	3.03	55.30	11.91
Li₂O₂	c	45.8808	—	−151.6	—	—	—	—
	a	45.8808	—	−158.8	—	—	—	—
LiH	c	7.9490	−20.425	−21.64	−16.34	0.903	4.782	6.66
LiD	c	8.9551	−20.684	−21.73	−16.18	1.085	5.640	8.20
LiOH	c	23.9484	−114.517	−115.90	−104.92	1.772	10.23	11.87
LiOD	c	24.9545	—	−116.8	—	—	—	—
LiF	c	25.9394	−146.657	−147.22	−140.47	1.547	8.52	9.94
Li₂F₂	g	51.8788	−223.67	−224.8	−224.7	3.19	61.79	15.09
Li₃F₃	g	77.8182	−360.30	−361.9	−357.	4.88	76.	24.5
LiHF₂	c	45.9458	−223.66	−225.22	−209.11	2.826	17.0	16.77
LiCl	c	42.394	−97.68	−97.66	−91.87	2.224	14.18	11.47
	g	42.394	−46.663	−46.7	−51.8	2.165	50.840	7.94
	a	42.394	—	−106.51	−101.48	—	16.7	−16.2
LiCl·H₂O	c	60.4094	—	−170.31	−151.01	—	24.58	—
LiCl·2H₂O	c	78.4248	—	−242.03	—	—	—	—
LiCl·3H₂O	c	96.4402	—	−313.4	—	—	—	—
Li₂Cl₂	g	84.788	−141.20	−141.9	−142.5	3.70	69.0	17.26
LiClO₃	c	90.3922	—	−88.2	—	—	—	—
LiClO₄	c	106.3916	—	−91.06	—	—	—	—
	a	106.3916	—	−97.472	−72.2	—	46.7	−1.8
LiClO₄·H₂O	c	124.4070	—	−166.6	−121.8	—	37.1	—
LiClO₄·3H₂O	c	160.4378	—	−310.22	−239.30	—	60.9	—
LiBr	c	86.850	—	−83.942	−81.74	—	17.75	—
	a	86.850	—	−95.612	−94.95	—	22.9	−17.5
LiBr·H₂O	c	104.8654	—	−158.36	−142.05	—	26.2	—
LiBr·2H₂O	c	122.8808	—	−230.1	−200.9	—	38.8	—
LiBrO₃	c	134.8482	—	−82.93	—	—	—	—
LiI	c	133.8454	−64.663	−64.63	−64.60	2.716	20.74	12.20
	a	133.8454	—	−79.75	−82.4	—	29.8	−17.6
LiI·H₂O	c	151.8608	—	−141.09	−127.0	—	29.4	—
LiI·2H₂O	c	169.8762	—	−212.81	−186.5	—	44.	—
LiI·3H₂O	c	187.8916	—	−284.93	—	—	—	—
LiIO₃	c	181.8436	—	−120.31	—	—	—	—

SELECTED VALUES OF CHEMICAL THERMODYNAMIC PROPERTIES (Continued)

Substance			0 K		298.15 K (25°C)			
Formula and Description	State	Formula weight	ΔHf_0° kcal/mol	ΔHf° kcal/mol	ΔGf° kcal/mol	$H_{298}^\circ - H_0^\circ$	S° cal/deg mol	C_p° cal/deg mol
	a	181.8436	—	−119.46	−100.70	—	31.4	−13.2
Li₂S	c	45.946	—	−105.5	—	—	—	—
Li₂S₂	c	78.010	—	−104.7	—	—	—	—
Li₂SO₃	c	93.9442	—	−281.3	—	—	—	—
Li₂SO₄	c	109.9436	−340.367	−343.33	−315.91	4.452	27.5	28.10
	a	109.9436	—	−350.44	−318.18	—	11.3	−37.2
LiSO₄·H₂O	c	127.9590	−410.02	−414.8	−374.2	5.697	39.1	36.11
Li₂SO₄·D₂O	c	129.9712	—	−416.9	−375.6	—	40.	—
LiHS	c	40.0130	—	−60.1	—	—	—	—
LiNO₂	c	52.9465	—	−89.0	−72.2	—	23.	—
	a	52.9465	—	−91.56	−77.8	—	32.7	−6.9
LiNO₃	c	68.9459	—	−115.47	−91.1	—	21.5	—
	a	68.9459	—	−116.12	−96.7	—	38.3	−4.3
LiNO₃·3H₂O	c	122.9921	—	−328.5	−263.8	—	53.4	—
Li₃PO₄	c	115.7944	—	−500.9	—	—	—	—
Li₂C₂	c	37.9044	—	−14.2	—	—	—	—
Li₂CO₃	c	73.8914	−288.652	−290.6	−270.58	3.627	21.60	23.69
Li₂SiO₃	c	89.9662	−391.26	−393.9	−372.2	3.453	19.08	23.68
Li₂SiF₆	c	155.9584	—	−704.4	—	—	—	—
LiBO₂	c	49.7508	−245.37	−246.7	−233.3	2.144	12.3	14.3
LiBH₄	c	21.7840	−43.20	−45.6	−29.9	3.049	18.13	19.73
	a	21.7840	—	−55.05	−42.8	—	29.7	—
LiAlSi₂O₆								
α spodumene	c	186.0909	—	−730.1	−688.71	—	30.90	38.0
β spodumene	c	186.0909	—	−723.4	−683.79	—	36.90	38.9
Li₂CrO₄	c	129.8756	—	−331.9	—	—	—	—
	a	129.8756	—	−343.7	−314.2	—	18.5	—
Li₂MoO₄	c	173.8196	—	−363.36	−336.9	—	30.	—
Li₂WO₄	c	261.7296	—	−335.	—	—	—	—
Li₂TiO₃	c	109.7802	−396.780	−399.3	−377.6	3.953	21.93	26.54
Li₂ZrO₃	c	153.1002	—	−420.7	—	—	—	—
Li₂ZrO₃	c	182.9816	—	−567.	—	—	—	—
Li₂PuF₆	c	418.8012	—	−1033.	—	—	—	—
LUTETIUM								
Lu	c	174.97	0	0	0	1.524	12.18	6.42
	g	—	102.242	102.2	96.7	1.482	44.142	4.986
Lu⁺	g	—	227.36	228.80	—	—	—	—
Lu⁺⁺	g	—	547.	549.9	—	—	—	—
Lu⁺⁺								
std. state, m = 1	aq	—	—	−159.	−150.	—	−63.	6.
Lu₂O₃	c	397.938	−446.93	−448.9	−427.6	4.192	26.28	24.32
LuCl₂	c	281.329	—	−226.0	—	—	—	—
std. state, m = 1	aq	—	—	−279.	−244.	—	−23.	−92.
LuCl₃·6H₂O	c	389.421	−667.41	−676.6	−576.3	13.96	89.9	82.0
LuOCl	c	226.422	—	−227.	—	—	—	—
Lu(BrO₃)₃·9H₂O	c	720.830	—	−833.8	—	—	—	—
LuI₃	c	555.683	—	−131.	—	—	—	—
Lu(IO₃)₃	c	699.678	—	320.	—	—	—	—
LuS	g	207.034	48.	—	—	—	—	—
MAGNESIUM								
Mg	c	24.312	0	0	0	1.195	7.81	5.95
	g	—	35.014	35.30	27.04	1.481	35.502	4.968
Mg⁺	g	—	211.333	213.100	—	—	—	—
Mg⁺⁺	g	—	558.052	561.299	—	—	—	—
std. state, m = 1	aq	—	—	−111.58	−108.7	—	−33.0	—
Mg⁺⁺	g	—	2406.08	2410.81	—	—	—	—
Mg⁺⁺	g	—	4927.14	4933.35	—	—	—	—
MgO								
macrocrystal (periclase)	c	40.3114	−142.813	−143.81	−136.10	1.235	6.44	8.88
microcrystal	c	—	−141.954	−142.92	−135.27	1.266	6.67	9.00
MgH₂	c	26.3279	−16.05	−18.0	−8.6	1.270	7.43	8.45
Mg(OH)₂	c	58.3267	−218.402	−220.97	−199.23	2.725	15.10	18.41
precipitate	amorp	—	—	−220.0	—	—	—	—
std. state, m = 1	aq	—	—	−221.52	−183.9	—	−38.1	—
MgF	g	43.3104	−52.89	−53.0	−59.2	2.143	52.79	7.78
MgF₂	c	62.3088	−267.57	−268.5	−255.8	2.370	13.68	14.72
MgCl₂	c	95.218	−153.180	−153.28	−141.45	3.288	21.42	17.06
MgCl₂								
in ∞ H₂O	aq	—	—	−191.48	—	—	—	—
MgCl₂·H₂O	c	113.2333	—	−231.03	−205.98	—	32.8	27.55
MgCl₂·2H₂O	c	131.2487	—	−305.86	−267.24	—	43.0	38.05
MgCl₂·4H₂O	c	167.2793	—	−453.87	−388.03	—	63.1	57.70
MgCl₂·6H₂O	c	203.3100	—	−597.28	−505.49	—	87.5	75.30
Mg(ClO₃)₂	c	223.2132	—	−135.97	—	—	—	—
std. state, m = 1	aq	—	—	−173.40	−112.8	—	54.0	—
Mg(ClO₄)₂								
in ∞ H₂O	aq	—	—	−173.40	—	—	—	—
Mg(ClO₄)₂·2H₂O	c	259.2439	—	−291.3	—	—	—	—
Mg(ClO₄)₂·4H₂O	c	295.2746	—	−439.1	—	—	—	—
Mg(ClO₄)₂·6H₂O	c	331.3052	—	−584.5	−445.3	—	124.5	—
MgBr₂	c	184.130	—	−125.3	−120.4	—	28.0	—
	g	—	—	−74.0	—	—	—	—
std. state, m = 1	aq	—	—	−169.68	−158.4	—	6.4	—
MgBr₂								
in ∞ H₂O	aq	—	—	169.68	—	—	—	—
MgI₂	c	278.1208	—	−87.0	−85.6	—	31.0	—
	g	—	—	−41.	—	—	—	—
std. state, m = 1	aq	—	—	−137.96	−133.4	—	20.2	—
in ∞ H₂O	aq	—	—	−137.96	—	—	—	—
MgS	c	56.376	−82.44	−82.7	−81.7	1.992	12.03	10.89
MgSO₃	c	104.3742	—	−241.0	—	—	—	—
MgSO₄	c	120.3736	—	−307.1	−279.8	—	21.9	23.06
undissociated; std. state,	aq	—	—	−324.1	−289.74	—	−1.7	—
m = 1	aq	—	—	−328.90	−286.7	—	−28.2	—
MgSO₄								
in ∞ H₂O	aq	—	—	−328.90	—	—	—	—
MgSO₄·2H₂O	c	156.4043	—	−453.2	—	—	—	—
MgSO₄·4H₂O	c	192.4350	—	−596.7	—	—	—	—
MgSO₄·6H₂O	c	228.4656	—	−737.8	−629.1	—	83.2	83.20
MgSO₄·7H₂O	c	246.4810	—	−809.92	−686.4	—	89.	—

SELECTED VALUES OF CHEMICAL THERMODYNAMIC PROPERTIES (Continued)

Formula and Description	State	Formula weight	0 K ΔHf° kcal/mol	298.15 K (25°C) ΔHf° kcal/mol	ΔGf° kcal/mol	H°₂₉₈ − H°₀	S° cal/deg mol	C°ₚ cal/deg mol
Mg(NO₃)₂	c	148.3218	—	−188.97	−140.9	—	39.2	33.92
std. state, m = 1	aq	—	—	−210.70	−161.9	—	37.0	—
Mg(NO₃)₂								
in ∞ H₂O	aq	—	—	−210.70	—	—	—	—
Mg(NO₃)₂·2H₂O	c	184.3525	—	−336.8	—	—	—	—
Mg(NO₃)₂·6H₂O	c	256.4138	—	−624.59	−497.3	—	108.	—
Mg₂P₂O₇	c	222.5674	—	—	—	6.467	37.02	42.53
Mg₃(PO₄)₂	c	262.8788	−896.98	−903.6	−845.8	7.825	45.22	51.02
MgC₂	c	48.3343	—	+20.	—	—	—	—
Mg₂C₃	c	84.6574	—	+17	—	—	—	—
MgCO₃								
magnesite	c	84.3214	—	−261.9	−241.9	—	15.7	18.05
MgCO₃·3H₂O								
nesquehonite	c	138.3674	—	—	−412.6	—	—	—
MgCO₃·5H₂O								
lansfordite	c	174.3980	—	—	−525.7	—	—	—
MgC₂O₄	c	112.3319	—	−303.3	—	—	—	—
std. state, m = 1	aq	—	—	−308.8	−269.8	—	−22.1	—
Mg₂Si	c	76.710	—	−18.6	−18.0	—	18.	17.6
MgSiO₃								
clinoenstatite	c	100.3962	−368.039	−370.22	−349.46	2.895	16.19	19.45
Mg₂SiO₄								
forsterite	c	140.7076	−516.42	−519.6	−491.2	4.129	22.74	28.32
Mg₃Si₂O₅(OH)₄								
chrysotile	c	277.1345	—	−1043.4	−965.1	—	52.9	65.41
antigorite	c	—	—	—	—	—	53.2	65.47
Mg₃Si₄O₁₀(OH)₂								
talc	c	379.2887	−1405.57	−1415.5	−1324.8	11.20	62.3	76.9
Mg₂Al₄Si₅O₁₈								
cordierite	c	584.9692	—	−2177.	−2055.	—	97.3	108.1
MgCrO₄	c	140.3056	—	−321.1	—	—	—	—
MgCr₂O₄	c	192.3016	—	−426.3	−398.9	—	25.34	30.30
MgMoO₄	c	184.250	—	−334.81	−309.69	—	28.4	26.57
MgWO₄	c	272.160	−363.88	−366.3	−339.6	4.115	24.18	26.14
MgV₂O₆								
metavanadate	c	222.1924	—	−526.19	−487.43	—	38.4	39.47
Mg₂V₂O₇								
pyrovanadate	c	262.5038	—	−677.80	−632.24	—	47.9	48.63
MgTiO₃								
metatitanate	c	120.210	−373.69	−375.9	−354.8	3.240	17.82	21.96
Mg₂TiO₄								
orthotitanate	c	160.522	−514.32	−517.5	−489.2	4.502	26.13	30.75
MgUO₄	c	326.339	—	−443.9	−418.2	—	31.5	30.6
MgU₃O₁₀	c	898.393	—	—	—	—	80.9	73.
MANGANESE								
Mn								
α	c	54.9380	0	0	0	1.194	7.65	6.29
β	c	—	—	—	—	1.234	8.22	6.34
γ	c	—	.343	.37	.34	1.221	7.75	6.59
	g	—	66.77	67.1	57.0	1.481	41.49	4.97
Mn⁺	g	—	238.20	240.0	—	—	—	—
Mn⁺⁺	g	—	598.87	602.1	—	—	—	—
std. state, m = 1	aq	—	—	−52.76	−54.5	—	−17.6	12.
MnO	c	70.9374	—	−92.07	−86.74	—	14.27	10.86
MnO₂	c	86.9368	—	−124.29	−111.18	—	12.68	12.94
precipitated	amorp	—	—	−120.1	—	—	—	—
Mn₂O₃	c	157.8742	—	−229.2	−210.6	—	26.4	25.73
Mn₃O₄	c	228.8116	—	−331.7	−306.7	—	37.2	33.38
MnH	g	55.9460	—	—	—	2.077	51.03	7.035
Mn(OH)₂								
precipitated	amorp	88.9527	—	−166.2	−147.0	—	23.7	—
MnF	g	73.9364	−5.	−5.2	—	—	—	—
MnF₂	c	92.9348	—	—	—	3.11	22.05	15.96
MnCl₂	c	125.8440	−115.245	−115.03	−105.29	3.602	28.26	17.43
std. state, m = 1	aq	—	—	−132.66	−117.3	—	9.3	−53.
MnCl₂·H₂O	c	143.8593	—	−188.8	−166.4	—	41.6	—
MnCl₂·2H₂O	c	161.8747	—	−261.0	−225.2	—	52.3	—
MnCl₂·4H₂O	c	197.9054	—	−403.3	−340.3	—	72.5	—
MnBr	g	134.8470	20.8	19.1	—	—	—	—
MnBr₂	c	214.7560	—	−92.0	—	—	—	—
std. state, m = 1	aq	—	—	−110.9	—	—	—	—
MnBr₂·H₂O	c	232.7713	—	−168.5	—	—	—	—
MnBr₂·4H₂O	c	286.8174	—	−380.1	—	—	—	—
MnI	g	181.8424	25.7	25.5	—	—	—	—
MnI₂								
std. state, m = 1	aq	308.7468	—	−79.1	—	—	—	—
MnI₂·2H₂O	c	344.7748	—	−201.4	—	—	—	—
MnI₂·4H₂O	c	380.8082	—	−343.9	—	—	—	—
Mn(IO₃)₂	c	404.7432	—	−160.	−124.4	—	63.	—
MnS								
green	c	87.0020	—	−51.2	−52.2	—	18.7	11.94
precipitated, pink	amorp	—	—	−51.1	—	—	—	—
MnSO₄	c	150.9996	—	−254.60	−228.83	—	26.8	24.02
std. state, m = 1	aq	—	—	−270.1	−232.5	—	−12.8	−58.
MnSO₄								
in ∞ H₂O	aq	—	—	−270.1	—	—	—	—
MnSO₄·H₂O								
α	c	169.0149	—	−329.0	—	—	—	—
β	c	—	—	−322.2	—	—	—	—
MnSO₄·4H₂O	c	223.0610	—	−539.7	—	—	—	—
MnSO₄·5H₂O	c	241.0763	—	−610.2	—	—	—	78.
MnSO₄·7H₂O	c	277.1070	—	−750.3	—	—	—	—
Mn(N₃)₂								
manganese azide	c	138.9782	—	92.2	—	—	—	—
Mn₄N₃	c	302.7034	—	−48.8	—	—	—	—
Mn(NO₃)₂	c	178.9478	—	−137.73	—	—	—	—
std. state, m = 1	aq	—	—	−151.9	−107.8	—	52.	−29.
MnP	c	85.9118	—	−27.	—	—	—	—
MnP₃	c	147.8594	—	−51.	—	—	—	—
Mn₃(PO₄)₂	c	354.7568	—	−744.9	—	—	—	—
MnHPO₄	c	150.9174	—	—	−332.5	—	—	—

SELECTED VALUES OF CHEMICAL THERMODYNAMIC PROPERTIES (Continued)

Formula and Description	State	Formula weight	0 K ΔHf°₀ kcal/mol	298.15 K (25°C) ΔHf° kcal/mol	ΔGf° kcal/mol	H°₁₉₈ − H°₀	S° cal/deg mol	Cp° cal/deg mol
Mn₃C	c	176.8252	—	1.1	1.3	—	23.6	22.33
Mn₇C₃	c	420.5995	—	−10.	—	—	—	—
MnCO₃								
natural	c	114.9474	—	−213.7	−195.2	—	20.5	19.48
precipitated	c	—	—	−211.1	—	—	—	—
MnC₂O₄	c	142.9579	—	−245.9	—	—	—	—
undissoc.; std. state, m = 1	aq	—	—	−248.5	−221.0	—	16.1	—
MnC₂O₄·2H₂O	c	178.9886	—	−389.2	−338.2	—	48.	—
MnC₂O₄·3H₂O	c	197.0039	—	−459.1	—	—	—	—
MnSiO₃	c	131.0222	—	−315.7	−296.5	—	21.3	20.66
glassy	amorp	—	—	−307.2	—	—	—	—
Mn₂SiO₄	c	201.9596	—	−413.6	−390.1	—	39.0	31.04
MERCURY								
Hg	c	200.59	0	—	—	—	—	—
	liq	—	—	0	0	2.233	18.17	6.688
	g	—	15.407	14.655	7.613	1.481	41.79	4.968
std. state, m = 1	aq	—	—	9.0	9.4	—	17.	—
Hg⁺	g	—	256.10	256.82	—	—	—	—
Hg²⁺	g	—	688.63	690.83	—	—	—	—
std. state, m = 1	aq	—	—	40.9	39.30	—	−7.7	—
Hg₂²⁺	g	—	1478.	1480.	—	—	—	—
Hg₂²⁺								
std. state, m = 1	aq	—	—	41.2	36.70	—	20.2	—
HgO								
red, orthorhombic	c	216.589	—	−21.71	−13.995	—	16.80	10.53
yellow	c	—	—	−21.62	−13.964	—	17.0	—
hexagonal	c	—	—	−21.4	−13.92	—	17.6	—
HgH	g	201.598	58.36	57.20	51.63	2.078	52.46	7.16
Hg(OH)₂								
undissoc.; std. state, m = 1	aq	234.605	—	−84.9	−65.7	—	34.	—
Hg₂F₂	c	439.177	—	—	−104.1	—	—	—
HgCl⁺								
std. state, m = 1	aq	—	—	−4.5	−1.3	—	18.	—
HgCl₂	c	271.496	—	−53.6	−42.7	—	34.9	—
undissoc.; std. state, m = 1	aq	—	—	−51.7	−41.4	—	37.	—
Hg₂Cl₂	c	472.086	—	−63.39	−50.377	—	46.0	—
HgBr₂	c	360.408	—	−40.8	−36.6	—	41.	—
undissoc.; std. state, m = 1	aq	—	—	−38.4	−34.2	—	41.	—
Hg₂Br₂	c	560.998	—	−49.45	−43.278	—	52.	—
HgI	g	327.494	32.9	31.64	21.14	2.546	67.26	8.99
HgI₂								
red	c	454.399	—	−25.2	−24.3	—	43.	—
yellow	c	—	—	−24.6	—	—	—	—
Hg₂I₂	c	654.989	—	−29.00	−26.53	—	55.8	—
HgS								
red	c	232.654	—	−13.9	−12.1	—	19.7	11.57
black	c	—	—	−12.8	−11.4	—	21.1	—
Hg₂SO₄	c	497.242	—	−177.61	−149.589	—	47.96	31.54
Hg₂(N₃)₂	c	485.220	—	142.0	178.4	—	49.	—
HgC₂O₄	c	288.610	—	−162.1	—	—	—	—
Hg₂CO₃	c	461.189	—	−132.3	−111.9	—	43.	—
Hg₂C₂O₄	c	489.200	—	—	−141.8	—	—	—
HgCH₃	g	215.625	—	40.	—	—	—	—
Hg(CH₃)₂	liq	230.660	—	14.3	33.5	—	50.	—
Hg(C₂H₅)₂	liq	258.715	—	7.2	—	—	—	—
	g	—	—	18.0	—	—	—	—
Hg(CN)₂	c	252.626	—	63.0	—	—	—	—
undissoc.; std. state, m = 1	aq	—	—	66.5	74.6	—	39.5	—
Hg(CN)₃⁻								
std. state, m = 1	aq	278.644	—	94.9	110.7	—	53.8	—
Hg(CN)₄²⁻								
std. state, m = 1	aq	304.662	—	125.8	147.8	—	73.	—
Hg(ONC)₂								
mercuric fulminate	c	284.624	—	64.	—	—	—	—
MOLYBDENUM								
Mo	c	95.94	0	0	0	1.098	6.85	5.75
	g	—	156.92	157.3	146.4	1.4812	43.461	4.968
Mo⁺	g	—	320.6	322.5	—	—	—	—
Mo²⁺	g	—	693.2	696.5	—	—	—	—
Mo³⁺	g	—	1319.0	1323.8	—	—	—	—
Mo⁴⁺	g	—	2388.9	2395.2	—	—	—	—
Mo⁵⁺	g	—	3799.4	3807.2	—	—	—	—
MoO	g	111.939	—	101.	—	—	—	—
MoO₂	c	127.939	—	−140.76	−127.40	—	11.06	13.38
	g	—	—	3.	—	—	—	—
MoO₃	c	143.939	—	−178.08	−159.66	—	18.58	17.92
	g	—	—	−78.	—	—	—	—
	aq	—	—	−172.5	—	—	—	—
MoO₄²⁻	aq	159.938	—	−158.0	—	—	—	—
MoO₄²⁻								
std. state, m = 1	aq	—	—	−238.5	−199.9	—	6.5	—
MoO₃	aq	175.937	—	−139.6	—	—	—	—
H₂MoO₄								
white	c	161.954	—	−250.0	—	—	—	—
H₂MoO₄·H₂O								
yellow	c	179.969	—	−325.	—	—	—	—
MoF₆	liq	209.930	−381.733	−378.95	−352.08	10.205	62.06	40.58
	g	—	−370.608	−372.29	−351.88	5.74	83.75	28.82
MoCl₂	c	166.846	—	−67.4	—	—	—	—
MoCl₃	c	202.299	—	−92.5	—	—	—	—
MoCl₄	c	237.752	—	114.8	—	—	—	—
MoCl₅	c	273.205	—	−126.0	—	—	—	—
MoO₂Cl₂	c	198.845	—	−171.4	—	—	—	—
MoO₂Cl₂·H₂O	c	216.860	—	−245.4	—	—	—	—
MoOCl₄	c	253.751	—	−153.0	—	—	—	—
MoBr₂	c	255.758	—	−62.4	—	—	—	18.3
MoBr₄	c	415.576	—	−76.8	—	—	—	—
MoO₂Br₂	c	287.757	—	−150.4	—	—	—	—
MoS₂	c	160.068	−55.52	−56.2	−54.0	2.528	14.96	15.19
Mo₂S₃	c	288.072	—	−87.	—	—	—	—

SELECTED VALUES OF CHEMICAL THERMODYNAMIC PROPERTIES (Continued)

Formula and Description	State	Formula weight	ΔHf° 0 K kcal/mol	ΔHf° kcal/mol	ΔGf° kcal/mol	H°₂₉₈ − H°₀	S° cal/deg mol	Cp° cal/deg mol
Mo₂N	c	205.887	—	−19.50	—	—	—	—
MoC	c	107.951	—	−2.4	—	—	—	—
Mo₂C	c	203.891	—	−10.9	—	—	—	—
Mo(CO)₆	c	264.003	—	−234.9	−209.8	—	77.9	57.90
FeMoO₄	c	215.785	—	−257.	−233.	—	30.9	28.31
NEODYMIUM								
Nd	c	144.24	0	0	0	1.73	17.1	6.56
	g	—	78.53	78.3	69.9	1.498	45.243	5.280
Nd⁺	g	—	205.1	206.3	—	—	—	—
Nd²⁺	g	—	452.	455.	—	—	—	—
Nd³⁺								
std. state, m = 1	aq	—	—	−166.4	−160.5	—	−49.4	−5.
NdO	g	160.239	−30.2	—	—	—	—	—
Nd₂O₃								
hexagonal	c	336.478	−430.50	−432.1	−411.3	5.00	37.9	26.60
NdH₂	c	146.256	—	−46.	—	—	—	—
NdF	g	163.238	—	−38.	—	—	—	—
NdF₂	g	182.237	—	−165.	—	—	—	—
NdF₃	c	201.235	—	−396.	—	—	—	—
NdCl₂	c	215.146	—	−163.	—	—	—	—
NdCl₃	c	250.599	—	−248.8	—	—	—	27.
NdCl₃								
∞ H₂O	aq	—	—	−286.3	—	—	—	—
NdCl₃·6H₂O	c	358.691	−678.72	−687.0	−588.1	15.14	99.7	86.25
NdI₃	c	524.953	—	−152.8	—	—	—	—
Nd(IO₃)₃	c	668.948	—	−332.	—	—	—	—
NdS	g	176.304	33.	—	—	—	—	—
Nd₂S₃	c	384.672	—	−284.	−280.2	6.160	44.28	29.28
Nd₂(SO₄)₃·8H₂O	c	720.788	—	−1513.1	—	—	160.9	144.9
Nd₂Se₃	c	525.36	—	—	—	7.13	53.6	31.1
Nd₂(SeO₃)₃·8H₂O								
amorp.	c	813.477	—	−1230.3	—	—	—	—
Nd₂(SeO₄)₃·5H₂O	c	807.430	—	−1100.4	—	—	—	—
Nd(NO₃)₃	c	330.255	—	−294.2	—	—	—	—
Nd(NO₃)₃·3H₂O	c	384.301	—	−515.	—	—	—	—
Nd(NO₃)₃·4H₂O	c	402.316	—	−588.6	—	—	—	—
Nd(NO₃)₃·6H₂O	c	438.347	—	−728.39	—	—	—	—
NdC₂	g	168.262	130.5	130.75	117.9	2.48	6.3	10.6
Nd₂(CO₃)₃	c	468.508	—	—	−744.5	—	—	—
NEON								
Ne	g	20.183	0	0	0	1.481	34.9471	4.9679
std. state, m = 1	aq	—	—	−1.1	4.6	—	15.8	—
Ne⁺	g	—	497.29	498.77	—	—	—	—
Ne²⁺	g	1444.7	1447.6	—	—	—	—	—
Ne³⁺	g	—	2915.	2919.	—	—	—	—
Ne⁴⁺	g	—	5156	5162	—	—	—	—
Ne⁵⁺	g	—	8072.	8079.	—	—	—	—
NICKEL								
Ni	c	58.71	0	0	0	1.144	7.14	6.23
	g	—	102.213	102.7	91.9	1.631	43.519	5.583
Ni⁺	g	—	278.275	280.243	—	—	—	—
Ni²⁺	g	—	696.87	700.32	—	—	—	—
std. state, m = 1	aq	—	—	−12.9	−10.9	—	−30.8	—
Ni³⁺	g	—	1508.0	1512.9	—	—	—	—
NiO	c	74.709	−56.7	−57.3	−50.6	1.6	9.08	10.59
Ni₂O₃	c	165.418	—	−117.0	—	—	—	—
Ni(OH)₂	c	92.725	—	−126.6	−106.9	—	21.	—
std. state, m = 1	aq	—	—	−122.8	−86.1	—	−35.9	—
Ni(OH)₂								
precipitated	c	109.732	—	−160.	—	—	—	—
NiF₂	c	96.707	−155.18	−155.7	−144.4	2.729	17.59	15.31
NiCl₂	c	129.616	−73.077	−72.976	−61.918	3.438	23.34	17.13
NiCl₂								
in ∞H₂O	aq	—	—	−92.8	—	—	—	—
NiCl₂·2H₂O	c	165.647	—	−220.4	−181.7	—	42.	—
NiCl₂·4H₂O	c	201.677	—	−362.5	−295.2	—	58.	—
NiCl₂·6H₂O	c	237.708	—	−502.67	−409.54	—	82.3	—
Ni(ClO₃)₂								
std. state, m = 1	aq	257.611	—	−74.7	−15.0	—	56.2	—
Ni(ClO₄)₂·6H₂O	c	365.703	—	−486.6	—	—	—	—
NiBr₂	c	218.528	—	−50.7	—	—	—	—
std. state, m = 1	aq	—	—	−71.0	−60.6	—	8.6	—
NiBr₂·3H₂O	c	272.574	—	−274.0	—	—	—	—
NiI₂	c	312.519	—	−18.7	—	—	—	—
std. state, m = 1	aq	—	—	−39.3	−35.6	—	22.4	—
Ni(IO₃)₂	c	408.515	—	−116.9	−78.0	—	51.	—
NiS	c	90.774	—	−19.6	−19.0	—	12.66	11.26
precipitated	c	—	—	−18.5	—	—	—	—
Ni₃S₂	c	240.258	—	−48.5	−47.1	—	32.0	28.12
NiSO₄	c	154.772	—	−208.63	−181.6	—	22.	33.
std. state, m = 1	aq	—	—	−230.2	−188.9	—	−26.0	—
NiSO₄								
∞ H₂O	aq	—	—	−230.2	—	—	—	—
NiSO₄·4H₂O	c	226.833	—	−502.9	—	—	—	—
NiSO₄·6H₂O								
α, tetragonal, green	c	262.864	−628.887	−641.21	−531.78	12.391	79.94	78.36
β, monoclinic, blue	c	—	—	−638.7	—	—	—	—
NiSO₄·7H₂O	c	280.879	−697.670	−711.36	−588.49	14.085	90.57	87.14
Ni₃N	c	190.137	0.2	—	—	—	—	—
Ni(NO₃)₂	c	182.720	—	−99.2	—	—	—	—
std. state, m = 1	aq	—	—	−112.0	−64.2	—	39.2	—
Ni(NO₃)₂·3H₂O	c	236.766	—	−317.0	—	—	—	—
Ni(NO₃)₂·6H₂O	c	290.812	—	−528.6	—	—	—	111.
Ni₃C	c	188.141	16.1	—	—	—	—	—
NiCO₃	c	118.719	—	—	−146.4	—	—	—
Ni(CO)₄	liq	170.752	—	−151.3	−140.6	—	74.9	48.9
	g	—	−144.877	−144.10	−140.36	7.074	98.1	34.70
Ni(HCO₃)₂	c	148.746	—	−208.4	—	—	—	—
Ni(CN)₂								
precipitated	c	110.746	—	30.5	—	—	—	—
std. state, m = 1	aq	—	—	59.1	71.5	—	14.2	—

SELECTED VALUES OF CHEMICAL THERMODYNAMIC PROPERTIES (Continued)

Formula and Description	State	Formula weight	$\Delta H f^\ddagger$ kcal/mol (0 K)	$\Delta H f^\circ$ kcal/mol	$\Delta G f^\circ$ kcal/mol	$H^\circ_{298} - H^\circ_0$	S° cal/deg mol	C°_p cal/deg mol
Ni(CNS)₂	c	174.874	—	22.8	—	—	—	—
NIOBIUM								
Nb	c	92.906	0	0	0	1.255	8.70	5.88
	g	—	172.758	173.5	162.8	1.997	44.490	7.208
Nb⁺	g	—	330.32	332.54	—	—	—	—
Nb²⁺	g	—	660.55	664.25	—	—	—	—
Nb³⁺	g	—	1238.0	1243.1	—	—	—	—
Nb⁴⁺	g	—	2121.3	2127.9	—	—	—	—
NbO	c	108.9054	—	−97.0	−90.5	—	11.5	9.86
	g	—	51.2	51.	44.	2.099	57.09	7.36
NbO₂	c	124.9048	−189.19	−190.3	−177.0	2.222	13.03	13.74
	g	—	—	−51.3	−52.3	—	61.0	—
Nb₂O₅								
(high temp. form)	c	265.809	−451.63	−454.0	−422.1	5.325	32.80	31.57
NbF₅	c	187.898	−432.68	−433.5	−406.1	5.707	38.3	32.2
NbCl₅	c	270.171	—	−190.6	−163.3	—	50.3	35.4
NbOCl₃	c	179.8114	—	−185.1	—	—	—	—
NbOCl₂	c	215.2644	—	−210.2	~187.	—	34.	—
NbBr₅	c	492.451	—	−132.9	—	—	—	—
NbOBr₃	c	348.6174	—	−179.3	—	—	—	—
NbI₅	c	727.428	—	−64.2	—	—	—	—
NbN	c	106.9127	−55.35	−56.2	−49.2	1.439	8.25	9.32
Nb₂N	c	199.8187	—	−59.9	—	—	—	—
NbC	c	104.9172	−33.12	−33.2	−32.7	1.422	8.46	8.81
Nb₂C	c	197.8232	—	−45.4	−44.4	—	15.3	14.48
NITROGEN								
N	g	14.0067	112.534	112.979	108.886	1.481	36.613	4.968
N⁺	g	—	447.663	449.589	—	—	—	—
N²⁺	g	—	1130.55	1133.96	—	—	—	—
N³⁺	g	—	2224.52	2229.41	—	—	—	—
N⁴⁺	g	—	4011.04	4017.41	—	—	—	—
N⁵⁺	g	—	6268.41	6276.26	—	—	—	—
N₂	g	28.0134	0	0	0	2.072	45.77	6.961
N₃	g	—	45.	43.2	—	—	—	—
std. state, m = 1	aq	—	—	65.76	83.2	—	25.8	—
NO	g	30.0061	21.45	21.57	20.69	—	50.347	7.133
NO₂	g	46.0055	8.60	7.93	12.26	2.438	57.35	8.89
NO₂⁻								
std. state, m = 1	aq	—	—	−25.0	−7.7	—	29.4	−23.3
NO₃⁻								
nitrate; std. state, m = 1	aq	62.0049	—	−49.56	−26.61	—	35.0	−20.7
N₂O	g	44.0128	20.435	19.61	24.90	2.284	52.52	9.19
N₂O₃	liq	76.0116	—	12.02	—	—	—	—
	g	—	21.628	20.01	33.32	3.566	74.61	15.68
N₂O₄	liq	92.0110	—	−4.66	23.29	—	50.0	34.1
	g	—	4.49	2.19	23.38	3.918	72.70	18.47
N₂O₅	c	108.0104	—	−10.3	27.2	—	42.6	34.2
	g	—	5.7	2.7	27.5	4.237	85.0	20.2
NH₃	g	17.0306	−9.34	−11.02	−3.94	2.388	45.97	8.38
undissoc.; std. state, m = 1	aq	—	—	−19.19	−6.35	—	26.6	—
NH₄⁺								
std. state, m = 1	aq	18.0386	—	−31.67	−18.97	—	27.1	19.1
N₂H₄	liq	32.0453	—	12.10	35.67	—	28.97	23.63
	g	—	26.18	22.80	38.07	2.743	56.97	11.85
undissoc.; std. state, m = 1	aq	—	—	8.20	30.6	—	33.	—
HN₃	liq	43.0281	—	63.1	78.2	—	33.6	—
	g	—	71.82	70.3	78.4	2.599	57.09	10.44
undissoc.; std. state, m = 1	aq	—	—	62.16	76.9	—	34.9	—
HNO₂								
cis	g	47.0135	−17.12	−18.64	−10.27	2.608	59.43	10.70
trans	g	—	−17.68	−19.15	−10.82	2.652	59.54	11.01
cis-trans mixture, equil.	g	—	—	−19.0	−11.0	—	60.7	10.9
undissoc.; std. state, m = 1	aq	—	—	−28.5	−12.1	—	32.4	—
HNO₃	liq	63.0129	—	−41.61	−19.31	—	37.19	26.26
	g	—	−29.94	−32.28	−17.87	2.815	63.64	12.75
std. state, m = 1	aq	—	—	−49.56	−26.61	—	35.0	−20.7
HNO₃								
∞ H₂O	aq	—	—	−49.56	—	—	—	—
NH₄OH	liq	35.0460	—	−86.33	−60.74	—	39.57	37.02
undissoc.; std. state, m = 1	aq	—	—	−87.505	−63.04	—	43.3	—
ionized; std. state, m = 1	aq	—	—	−86.64	−56.56	—	24.5	−16.4
NH₄OH								
∞ H₂O	aq	—	—	−86.64	—	—	—	—
NH₄NO₂	c	64.0441	—	−61.3	—	—	—	—
std. state, m = 1	aq	—	—	−56.7	−26.7	—	56.5	−4.2
NH₄NO₃	c	80.0435	—	−87.37	−43.98	—	36.11	33.3
std. state, m = 1	aq	—	—	−81.23	−45.58	—	62.1	−1.6
NH₄NO₃								
∞ H₂O	aq	—	—	−81.23	—	—	—	—
(NH₄)₂O	liq	52.0766	—	−102.94	−63.84	—	63.94	59.08
NF₃	g	71.0019	−28.43	−29.8	−19.9	2.827	62.29	12.7
NH₂F	c	37.0370	−107.41	−110.89	−83.36	2.655	17.20	15.60
std. state, m = 1	aq	—	—	−111.17	−85.61	—	23.8	−6.4
NCl₃	liq	120.3657	—	55.	—	—	—	—
NOCl	g	65.4591	12.81	12.36	15.79	2.716	62.52	10.68
NH₄Cl	c	53.4916	—	−75.15	−48.51	—	22.6	20.1
std. state, m = 1	aq	—	—	−71.62	−50.34	—	40.6	−13.5
NH₄Cl								
∞ H₂O	aq	—	—	−71.62	—	—	—	—
NH₂OH·HCl	c	69.4910	—	−75.9	—	—	—	22.2
NH₄ClO								
std. state, m = 1	aq	69.4910	—	−57.3	−27.8	—	37.	—
NH₄ClO₂								
std. state, m = 1	aq	85.4904	—	−47.6	−14.9	—	51.3	—
std. state, m = 1	aq	101.4898	—	−56.52	−20.89	—	65.9	—
NH₄ClO₄	c	117.4892	—	−70.58	−21.25	—	44.5	—
std. state, m = 1	aq	—	—	−62.58	−21.03	—	70.6	—
NH₄Br	c	94.9477	—	−64.73	−41.9	—	27.	23.
std. state, m = 1	aq	—	—	−60.72	−43.82	—	46.8	−14.8
NH₄Br								
∞ H₂O	aq	—	—	−60.72	—	—	—	—

SELECTED VALUES OF CHEMICAL THERMODYNAMIC PROPERTIES (Continued)

Formula and Description	State	Formula weight	ΔHf‡ 0 K kcal/mol	ΔHf° kcal/mol	ΔGf° kcal/mol	H°₂₉₈ − H°‡	S° cal/deg mol	C°ₚ cal/deg mol
NH₄BrO								
std. state, m = 1	aq	113.9470	—	−54.2	−27.0	—	37.	—
NH₄BrO₃								
std. state, m = 1	aq	145.9458	—	−47.70	−14.54	—	65.75	—
NH₄Br₃Cl								
std. state, m = −72.4	aq	213.3096	—	−72.4	−49.7	—	72.2	—
NH₄I	c	144.9430	—	−48.14	−26.9	—	28.	—
std. state, m = 1	aq	—	—	−44.86	−31.30	—	53.7	−14.9
NH₄I								
∞ H₂O	aq	—	—	−44.86	—	—	—	—
NH₄IO								
std. state, m = 1	aq	160.9424	—	−57.4	−28.2	—	25.8	—
NH₄IO₃	c	192.9412	—	−92.2	—	—	—	—
std. state, m = 1	aq	—	—	−84.6	−49.6	—	55.4	—
NH₄IO₄	aq	208.9406	—	−67.9	—	—	—	—
NH₄HS	c	51.1106	—	−37.5	−12.1	—	23.3	—
std. state, m = 1 (NH₄⁺ + HS⁻)	aq	—	—	−35.9	−16.09	—	42.1	—
(NH₄)₂S								
std. state, m = 1	aq	68.1412	—	−55.4	−17.4	—	50.7	—
(NH₄)₂S₂								
std. state, m = 1	aq	100.2052	—	−56.1	−18.9	—	61.0	—
(NH₄)₂S₃								
std. state, m = 1	aq	132.2692	—	−57.1	−20.3	—	70.0	—
H₃NSO₃H								
sulfamic acid	c	97.0928	—	−161.3	—	—	—	—
	aq	—	—	−156.3	—	—	—	—
NH₄HSO₃	c	99.1088	—	−183.7	—	—	—	—
NH₄HSO₄	c	115.1082	—	−245.45	—	—	—	—
SO₂(NH₂)₂								
sulfamide	c	96.1081	—	−129.3	—	—	—	—
(NH₄)₂SO₃	c	116.1394	—	−211.6	—	—	—	—
std. state, m = 1	aq	—	—	−215.2	−154.2	—	47.2	—
(NH₄)₂SO₄	c	132.1388	−282.23	−215.56	—	52.6	44.81	—
std. state, m = 1, (2NH₄⁺ + SO₄⁼)	aq	—	—	−280.66	−215.91	—	59.0	−31.8
OSMIUM								
Os	c	190.2	0	0	0	0	7.8	5.9
	g	—	—	189.	178.	1.481	46.000	4.968
Os⁺	g	—	—	391.91	—	—	—	—
Os²⁺	g	—	—	785.	—	—	—	—
OsO₂	g	238.20	—	−67.8	—	—	—	—
OsO₄								
yellow	c	254.20	—	−94.2	−72.9	—	34.4	—
white	c	—	—	−92.2	−72.6	—	40.1	—
Os(OH)₄	amorp	258.23	—	—	−161.0	—	—	—
OsCl₃	c	296.56	—	−45.5	—	—	—	—
OsCl₄	c	332.01	—	−60.9	—	—	—	—
OsS₂	c	254.33	—	−34.9	—	—	—	—
O	g	15.9994	58.983	59.553	55.389	1.607	38.467	5.237
O⁺	g	—	373.019	375.070	—	—	—	—
O²⁺	g	—	1183.73	1187.26	—	—	—	—
O³⁺	g	—	2450.87	2455.88	—	—	—	—
O⁴⁺	g	—	4236.1	4242.6	—	—	—	—
O⁵⁺	g	—	6862.8	6870.8	—	—	—	—
O₂	g	31.9988	0	0	0	2.0746	49.003	7.016
std. state, m = 1	aq	—	—	−2.8	3.9	—	26.5	—
O₃	g	47.9982	34.74	34.1	39.0	2.4736	57.08	9.37
	aq	—	—	30.1	—	—	—	—
PALLADIUM								
Pd	c	106.4	0	0	0	1.299	8.98	6.21
	g	—	90.2	90.4	81.2	1.481	39.90	4.968
Pd⁺	g	—	282.436	284.117	—	—	—	—
Pd²⁺	g	—	730.5	733.6	—	—	—	—
std. state, m = 1	aq	—	—	35.6	42.2	—	−44.	—
Pd³⁺	g	—	1490.	1494.	—	—	—	—
PdO	c	122.40	—	−20.4	—	—	—	7.5
Pd₂H	c	213.81	—	−4.7	—	—	—	—
Pd(OH)₂								
precipitated	c	140.41	—	−94.4	—	—	—	—
Pd(OH)₄								
precipitated	c	174.43	—	−171.1	—	—	—	—
PdCl₂	c	177.31	—	−47.5	—	—	—	—
PdBr₂	c	266.22	—	−24.9	—	—	—	—
PdI₂	c	360.21	—	−15.1	−17.1	—	43.	—
PdS	c	138.46	—	−18.	−16.	—	11.	—
PdS₂	c	170.53	—	−19.4	−17.8	—	19.	—
Pd₄S	c	457.66	—	−16.	−16.	—	43.	—
Pd(CN)₂	c	158.44	—	49.1	—	—	—	—
Pd(CNS)₂	c	222.56	—	—	56.0	—	—	—
PHOSPHORUS								
P								
α, white	c	30.9738	0	0	0	1.281*	9.82	5.698
red, triclinic	c	—	−3.78	−4.2	−2.9	0.862	5.45	5.07
black	c	—	—	−9.4	—	—	—	—
red	amorp	—	—	−1.8	—	—	—	—
	g	—	75.	75.20	66.51	1.481	38.978	4.968
in CS₂	—	—	—	0.5	—	—	—	—
P⁺	g	—	328.20	329.88	—	—	—	—
P²⁺	g	—	781.51	784.67	—	—	—	—
P³⁺	g	—	1477.11	1481.75	—	—	—	—
P⁴⁺	g	—	2661.67	2667.79	—	—	—	—
P⁵⁺	g	—	4161.2	4168.8	—	—	—	—
P₂	g	61.9476	34.94	34.5	24.8	2.126	52.108	7.66
P₄	g	123.8952	15.83	14.08	5.85	3.378	66.89	16.05
PO	g	46.9732	−6.7	−6.8	−12.4	2.245	53.22	7.59
P₄O₆	c	219.8916	—	−392.0	—	—	—	—
P₄O₁₀								
hexagonal	c	283.8892	−705.82	−713.2	−644.8	8.117	54.70	50.60
	amorp	—	—	−727.	—	—	—	—
PH₃	g	33.9977	3.20	1.3	3.2	2.420	50.22	8.87

SELECTED VALUES OF CHEMICAL THERMODYNAMIC PROPERTIES (Continued)

Formula and Description	State	Formula weight	ΔHf₀ kcal/mol	ΔHf° kcal/mol	ΔGf° kcal/mol	H₂₉₈ − H₀	S° cal/deg mol	Cₚ cal/deg mol
std. state, m = 1	aq	—	—	−2.27	6.05	—	28.7	—
HPO₃	c	79.9800	—	−226.7	—	—	—	—
H₃PO₃	c	81.9959	—	−230.5	—	—	—	—
H₃PO₄	c	97.9953	−301.29	−305.7	−267.5	4.059	26.41	25.35
ionized, std. state, m = 1	aq	—	—	−305.3	−243.5	—	−53.	—
PF	g	49.9722	—	—	—	2.117	53.74	7.56
PF₃	g	87.9690	−218.25	−219.6	−214.5	3.092	65.28	14.03
PF₅	g	125.9658	—	−381.4	—	—	—	—
PCl₃	liq	137.3328	—	−76.4	−65.1	—	51.9	—
	g	—	−67.85	−68.6	−64.0	3.817	74.49	17.17
PCl₅	c	208.2388	—	−106.0	—	—	—	—
POCl₃	c	153.3322	−145.81	—	—	—	—	—
PBr₃	liq	270.7008	—	−44.1	−42.0	—	57.4	—
	g	—	−27.47	−33.3	−38.9	4.240	83.17	18.16
PBr₅	c	430.5188	—	−64.5	—	—	—	—
POBr₃	c	286.7002	—	−109.6	—	—	—	—
PI₃	c	411.6870	—	−10.9	—	—	—	—
	g	—	—	—	—	4.542	89.45	18.73
PH₄I	c	161.9101	—	−16.7	0.2	—	29.4	26.2
P₄S₃	c	158.1396	—	−19.2	—	—	—	—
PN	g	44.9805	26.5	26.26	20.97	2.080	50.45	7.10
P₃N₅	c	162.9549	—	−71.4	—	—	—	36.
NH₄H₂PO₄	c	115.0259	—	−345.38	−289.33	—	36.32	34.00
(NH₄)₂HPO₄	aq	116.0571	—	−294.9	—	—	—	—
(NH₄)₂HPO₄	c	132.0565	—	−374.50	—	—	—	45.
(NH₄)₃PO₄	c	149.0871	—	−399.6	—	—	—	—
std. state, m = 1	aq	—	—	−400.3	−300.4	—	28.	—
PLATINUM								
Pt	c	195.09	0	0	0	1.372	9.95	6.18
	g	—	134.90	135.1	124.4	1.572	45.960	6.102
Pt⁺	g	—	341.6	343.3	—	—	—	—
std. state, m = 1	aq	—	—	—	60.9	—	—	—
PtO₂	g	227.089	—	41.0	40.1	62.	—	—
Pt₃O₄	c	649.268	—	−39.	—	—	—	—
Pt(OH)₂	c	229.105	—	−84.1	—	—	—	—
PtCl	c	230.543	—	−13.5	—	—	—	—
PtCl₂	c	265.996	—	−29.5	—	—	—	—
std. state, m = 1	aq	—	—	—	—	—	—	—
PtCl₃	c	301.449	—	−43.5	—	—	—	—
PtCl₄	c	336.902	—	−55.4	—	—	—	—
HPtCl₆·2H₂O	c	409.394	—	−242.0	—	—	—	—
H₂PtCl₆·6H₂O	c	517.916	—	566.7	—	—	—	—
PtBr	c	274.999	—	−9.2	—	—	—	—
PtBr₂	c	354.908	—	−19.6	—	—	—	—
PtBr₃	c	434.817	—	−28.9	—	—	—	—
PtBr₄	c	514.726	—	−37.4	—	—	—	—
PtI₄	c	702.708	—	−17.4	—	—	—	—
PtS	c	227.154	−19.020	−19.5	−18.2	1.946	13.16	10.37
PtS₂	c	259.218	−25.333	−26.0	−23.8	2.813	17.85	15.75
POLONIUM								
Po	c	210.	0	0	0	—	—	—
Po²⁺								
std. state, m = 1	aq	—	—	—	17.	—	—	—
Po⁴⁺								
std. state, m = 1	aq	—	—	—	70.	—	—	—
Po(OH)₄	c	278.0	—	—	−130.	—	—	—
PoS	c	242.1	—	—	−1.	—	—	—
POTASSIUM								
K	c	39.1020	0	0	0	1.695	15.34	7.07
	g	39.1020	21.544	21.33	14.49	1.481	38.295	4.968
K⁺	g	39.1020	—	122.92	—	—	—	—
K²⁺	g	39.1020	—	853.70	—	—	—	—
K³⁺	g	39.1020	—	1909.5	—	—	—	—
K⁴⁺	g	39.1020	—	3315.6	—	—	—	—
K⁺	a	39.1020	—	−60.32	−67.70	—	24.5	5.2
K								
in 88.81 Hg		39.1020	—	—	−24.41	—	—	—
KO₂	c	71.1008	—	−68.10	−57.23	—	27.9	18.53
KO₃	c	87.1002	—	−62.2	—	—	—	—
K₂O	c	94.2034	—	−86.4	—	—	—	—
	g	94.2034	—	−15.	—	—	—	—
K₂O₂	c	110.2028	—	−118.1	−101.6	—	24.4	—
	g	110.2028	—	−38.	—	—	—	—
KH	c	40.1100	—	−13.80	—	—	—	—
KD	c	41.1160	—	−13.21	—	—	—	—
KOH	c	56.1094	−100.681	−101.521	−90.61	−2.904	18.85	15.51
	a	56.1094	—	−115.29	−105.29	—	21.9	−30.3
in ∞ H₂O	a	56.1094	—	−115.29	—	—	—	—
KF	c	58.1004	−135.223	−135.58	−128.53	2.392	15.91	11.72
KF								
in ∞ H₂O	a	58.1004	—	−139.82	—	—	—	—
KF·2H₂O	c	94.1312	—	−278.112	−244.17	—	37.1	—
KHF₂	cø	78.1068	−220.560	−221.72	−205.48	3.655	24.92	18.39
KCl	c	74.5550	−104.310	−104.385	−97.79	2.717	19.74	12.26
	a	74.5550	—	−100.27	−99.07	—	38.0	−27.4
KClO	a	90.5544	—	−85.9	−76.5	—	35.	—
KClO₂	a	106.5538	—	−76.2	−63.6	—	48.7	—
KClO₃	c	122.5532	—	−95.06	−70.82	—	34.2	23.96
	a	122.5532	—	−85.17	−69.62	—	63.3	—
KClO₄	c	138.5526	−101.525	−103.43	−72.46	5.036	36.1	26.86
	a	138.5526	—	−91.23	−69.76	—	68.0	—
KBr	c	119.0110	−92.414	−94.120	−90.98	2.919	22.92	12.50
	g	119.0110	−40.832	−43.04	−50.89	2.416	59.85	8.824
	a	119.0110	—	−89.37	−92.55	—	44.2	−28.7
KBr₃	a	278.8290	—	−91.49	−93.29	—	76.0	—
KBrO	a	135.0104	—	−82.8	−75.7	—	34.	—
KBrO₃	c	167.0092	−83.957	−86.10	−64.82	5.593	35.65	28.72
	a	167.0092	—	−76.35	−63.27	—	63.15	—
KBrO₄	c	183.0086	−65.619	−68.80	−41.70	5.593	40.65	28.72
	a	183.0086	—	−57.2	−39.5	—	72.2	—
KI	c	166.0065	−78.137	−78.370	−77.651	3.039	25.41	12.65

SELECTED VALUES OF CHEMICAL THERMODYNAMIC PROPERTIES (Continued)

Formula and Description	State	Formula weight	ΔHf°_0 kcal/mol	ΔHf° kcal/mol	ΔGf° kcal/mol	$H^{\circ}_{298} - H^{\circ}_0$	S° cal/deg mol	C°_p cal/deg mol
			0 K	298.15 K (25°C)				
KI_3	c	419.8152	—	-78.4	—	—	—	—
	a	419.8152	—	-72.6	-80.0	—	81.7	—
KIO	a	182.0059	—	-86.0	-76.9	—	23.2	—
KIO_3	c	214.0047	-118.54	-119.83	-100.00	5.09	36.20	25.45
	a	214.0047	—	-113.2	-98.3	—	52.8	—
KIO_4	c	230.0041	—	-111.67	-86.38	—	42.	—
	a	230.0041	—	-96.5	-81.7	—	77.	—
K_2S	c	110.2680	—	-91.0	-87.0	—	25.0	—
	a	110.2680	—	-112.7	-114.9	—	45.5	—
K_2SO_3	c	158.2662	—	-269.0	—	—	—	—
K_2SO_3	a	158.2662	—	-272.5	-251.7	—	42.	—
K_2SO_4	c	174.2656	-341.126	-343.64	-315.83	6.079	41.96	31.42
	a	174.2656	—	-337.96	-313.37	—	53.8	-60.
$K_2S_2O_3$	c	190.3302	—	-280.5	—	—	—	—
	a	190.3302	—	-276.5	-260.3	—	65.	—
$K_2S_2O_7$	c	254.3278	—	-474.8	-428.2	—	61.	—
	a	254.3278	—	-455.5	—	—	—	—
$KHSO_4$	c	136.1716	—	-277.4	-246.5	—	33.0	—
	a	136.1716	—	272.40	-248.39	—	56.0	-15.
KNO_2 rhombic	c	85.1075	-88.45	-88.39	-73.28	4.871	36.35	25.67
	a	85.1075	—	-85.3	-75.4	—	53.9	—
KNO_3	c	101.1069	-116.86	-118.22	-94.39	4.488	31.80	23.04
KNO_3	a	101.1069	—	-109.88	-94.31	—	59.5	-15.5
KPO_3	c	118.0740	—	—	—	3.886	25.93	21.56
	a	118.0740	—	-293.8	—	—	—	—
K_3PO_4	c	212.2774	—	-466.1	—	—	—	—
	a	212.2774	—	-486.3	-446.6	—	21.0	—
K_3AsO_4	a	256.2252	—	-393.23	-358.10	—	34.6	—
KH_2AsO_4	c	180.0372	-278.60	-282.2	-247.6	5.490	37.05	30.29
K_2CO_3	c	138.2134	-273.76	-275.1	-254.2	5.417	37.17	27.35
$K_2C_2O_4$ oxalate	c	166.2240	—	-321.9	—	—	—	—
HCOOK formate	c	84.1200	—	-162.46	—	—	—	—
$KHCO_3$	c	100.1194	—	-230.2	-206.4	—	27.6	—
CH_3COOK acetate	c	98.1472	—	-172.8	—	—	—	—
	a	98.1472	—	-176.48	-155.99	—	45.2	3.7
KCN	c	65.1199	-28.174	-27.0	-24.35	4.157	30.71	15.84
	g	65.1199	21.494	21.7	15.34	3.188	62.57	12.51
KCNO cyanate	c	81.1193	—	-100.06	—	—	—	—
	a	81.1193	—	-95.2	-91.0	—	50.0	—
$KCNS$	c	97.1839	-47.980	-47.84	-42.62	4.176	29.70	21.16
	a	97.1839	—	-42.05	-45.55	—	59.0	-4.4
K_2SiO_3	c	154.2882	—	—	—	5.230	34.9	28.3
K_2SiF_6	c	220.2804	—	-706.5	-668.9	—	54.0	—
$KAl(SO_4)_2$	c	258.2067	—	-590.4	-535.4	—	48.9	46.12
$KAl(SO_4)_2 \cdot 3H_2O$	c	312.2529	—	-808.1	-711.0	—	75.0	—
$KAl(SO_4)_2 \cdot 12H_2O$	c	474.3915	—	-1448.8	-1228.9	—	164.3	155.6
$KAl_3Si_3O_{10}(OH)_2$ muscovite	c	398.3133	—	-1430.3	-1340.5	—	73.2	—
$K_4Fe(CN)_6$	c	329.2604	—	-59.7	-31.0	—	101.83	—
	a	329.2604	—	-46.7	-28.8	—	138.1	—
$K_3Fe(CN)_6$	c	368.3624	-141.30	-142.0	-108.3	14.871	100.1	79.40
$K_3Fe(CN)_6$	a	368.3624	—	-132.4	-104.71	—	120.7	—
$KMnO_4$	c	158.0376	—	-200.1	-176.3	—	41.04	28.10
	a	158.0376	—	—	—	—	—	-14.4
K_2CrO_4	c	194.1976	-333.796	-335.5	-309.7	6.805	47.83	34.89
	a	194.1976	—	-331.24	-309.36	—	61.0	—
$K_2Cr_2O_7$	c	294.1918	—	-492.7	-449.8	—	69	52.4
	a	294.1918	—	-476.8	-446.4	—	111.6	—
NaK	l	62.0918	—	1.5	—	—	—	—
Na_2K	l	85.0816	—	2.0	—	—	—	—
NaK_2	l	101.1938	—	2.5	—	—	—	—
PRAESEODYMIUM								
Pr	c	140.907	0	0	0	1.74	17.5	6.50
	g	—	85.25	85.0	76.7	1.487	45.339	5.105
Pr^+	g	—	210.3	211.5	—	—	—	—
Pr^{+2}	g	—	453.6	456.3	—	—	—	—
Pr^{+3}	g	—	952.3	956.5	—	—	—	—
std. state, m = 1	aq	—	—	-168.4	-162.3	—	-50.	-7.
Pr^{+4}	g	—	1851	1856.	—	—	—	—
Pr^{+5}	g	—	3176.	3183.	—	—	—	—
PrO	g	156.906	-38.	—	—	—	—	—
PrO_2	c	172.906	—	-226.9	—	—	—	28.06
Pr_2O_3 hexagonal	c	329.812	—	-432.5	—	—	—	28.06
cubic	c	—	—	-432.5	—	—	—	—
PrH_2	c	142.923	-45.47	-47.4	-36.9	1.83	13.6	9.8
$PrOH^{+2}$ std. state, m = 1	aq	157.914	—	—	-206.	—	—	—
$Pr(OH)_2^+$ std. state, m = 1	aq	174.922	—	—	-257.	—	—	—
$Pr(OH)_3$	c	191.929	—	—	-307.1	—	—	—
$PrCl_3$	c	247.266	—	-252.6	—	—	—	24.
	g	—	—	-187.	—	—	—	—
std. state, m = 1	aq	—	—	-288.3	-256.4	—	-10.	-105.
PrOCl	c	192.359	—	-242.	—	—	—	—
PrI_3	c	521.620	—	-156.4	—	—	—	—
$Pr(IO_3)_3$	c	665.615	—	-333.8	—	—	—	—
$PrSO_4^+$ std. state, m = 1	aq	236.969	—	-382.3	-345.1	—	-17.	—
$Pr(SO_4)_2^-$ std. state, m = 1	aq	333.030	—	-598.4	-525.6	—	0.	—
$Pr(NO_3)_3$	c	326.922	—	-293.8	—	—	—	—
in HNO_3(aq)	aq	—	—	-316.7	—	—	—	—

SELECTED VALUES OF CHEMICAL THERMODYNAMIC PROPERTIES (Continued)

Formula and Description	State	Formula weight	0 K ΔHf° kcal/mol	298.15 K (25°C) ΔHf° kcal/mol	ΔGf° kcal/mol	H°₂₉₈ – H°₀	S° cal/deg mol	C°ₚ
Pr(NO₃)₃·6H₂O	c	435.014	—	−731.05	—	—	—	—
PrC	c	152.918	—	−13.0	—	—	—	—
PrC₂	g	164.929	131.	131.3	118.7	2.48	62.6	10.6
Pr₂(CO₃)₃	c	461.842	—	−768.	—	—	—	—
PROMETHIUM								
147ₚₘ	g	146.915	—	—	—	1.545	44.692	5.797
PROTACTINIUM								
Pa	c	231.0359	0	0	0	—	12.4	—
	g	231.0359	—	145.	134.6	1.518	47.31	5.48
Pa⁴⁺	a	231.0359	—	−148.0	—	—	—	—
in HCl + 3.43 H₂O:A		231.0359	—	−159.6	—	—	—	—
Pa⁴⁺								
in HCl + 3.43 H₂O:A		231.0359	—	−161.8	—	—	—	—
Pa⁴⁺								
in HCl + 8.16 H₂O:A		231.0359	—	−144.8	—	—	—	—
in HCl + 54.4 H₂O:A		231.0359	—	−147.7	—	—	—	—
PaO₂	c	263.0347	—	—	—	—	17.8	—
PaCl₄	c	372.8479	—	−249.3	−227.7	—	46.0	—
in HCl + 3.43 H₂O:Au		372.8479	—	−290.1	—	—	—	—
in HCl + 8.16 H₂O:Au		372.8479	—	−291.6	—	—	—	—
in HCl + 54.4 H₂O:Au		372.8479	—	−304.9	—	—	—	—
PaCl₅	c	408.3009	—	−273.6	−247.2	—	57.	—
	g	408.3009	—	−251.	−236.	—	94.	—
in HCl + 3.43 H₂O:Au		408.3009	—	−322.2	—	—	—	—
PaBr₄	c	550.6719	—	−197.0	−188.3	—	56.0	—
in HCl + 8.16 H₂O:Au		550.6719	—	−248.70	—	—	—	—
in HCl + 54.4 H₂O:Au		550.6719	—	−262.44	—	—	—	—
PaBr₅	c	630.5809	—	−206.	−196.	—	69.	—
	g	630.5809	—	−180.	−182.	—	111.	—
in HCl + 3.43 H₂O:Au		630.5809	—	−264.8	—	—	—	—
PaOBr₃	c	406.8533	—	−239.	—	—	—	—
PaI₄	c	738.6535	—	−123.2	—	—	—	—
in HCl + 54.4 H₂O:Au		738.6535	—	−199.1	—	—	—	—
RADIUM								
Ra	c	226.025	0	0	0	—	17.	—
	g	—	—	38.	31.	1.481	42.15	4.97
Ra⁺	g	—	—	161.22	—	—	—	—
Ra⁺⁺	g	—	—	396.70	—	—	—	—
std. state, m = 1	aq	—	—	−126.1	−134.2	—	13.	—
RaO	c	242.0244	—	−125.	—	—	—	—
RaCl₂	c	296.931	—	—	—	—	32.	—
std. state, m = 1	aq	—	—	−206.0	−196.9	—	40.	—
RaCl₂·2H₂O	c	332.9617	—	−350.	−311.4	—	51.	—
Ra(IO₃)₂	c	575.8302	—	−245.4	−207.6	—	65.	—
RaSO₄	c	322.0866	—	−351.6	−326.4	—	33.	—
std. state, m = 1	aq	—	—	−343.4	−312.2	—	18.	—
Ra(NO₃)₂	c	350.0348	—	−237.	−190.3	—	53.	—
std. state, m = 1	aq	—	—	−225.2	−187.4	—	83.	—
RADON								
Rn	g	222.	0	0	0	1.481	42.09	4.968
Rn⁺	g	—	247.86	249.34	—	—	—	—
RHENIUM								
Re	c	186.2	0	0	0	1.296	8.81	6.09
	g	—	183.8	184.0	173.2	1.481	45.131	4.968
Re⁺	g	—	365.5	367.1	—	—	—	—
std. state, m = 1	aq	—	—	—	−8.	—	—	—
ReO₂	c	218.20	—	—	−88.	—	—	—
ReO₃	c	234.20	—	−144.6	—	—	—	—
HReO₄	c	251.21	—	−182.2	−156.9	—	37.8	—
std. state, m = 1	aq	—	—	−188.2	−166.0	—	48.1	−3.2
HReO₄								
∞ H₂O	aq	—	—	−188.2	—	—	—	—
ReCl₃	c	292.56	−62.73	−63.	−45.	4.319	29.6	22.08
ReCl₄	c	363.47	—	−89.	—	—	—	—
H₂ReCl₆	c	330.03	—	−152.	—	—	—	—
ReBr₃	c	425.93	—	−40.	—	—	—	—
ReS₂	c	250.33	—	−43.	—	—	—	—
RHODIUM								
Rh	c	102.905	0	0	0	1.174	7.53	5.97
	g	—	132.79	133.1	122.1	1.483	44.383	5.022
Rh⁺	g	—	304.90	306.69	—	—	—	—
Rh⁺⁺	g	—	721.8	725.0	—	—	—	—
Rh⁺⁺	g	—	1438.	1443.	—	—	—	—
RhO	g	118.9044	—	92.	—	—	—	—
RhO⁺	g	—	—	308.	—	—	—	—
RhO₂	g	134.9038	—	44.	—	—	—	—
Rh₂O₃	c	253.8082	—	−82.	—	—	—	24.8
RhCl₃	g	173.811	—	30.3	—	—	—	—
RhCl₃	c	209.264	—	−71.5	—	—	—	—
RhCl₆³⁻	aq	315.623	—	−202.8	—	—	—	—
RhCl₃·3(C₂H₅)₂S	c	479.8289	—	−166.	—	—	—	—
RUBIDIUM								
Rb	c	85.4678	0	0	0	1.790	18.35	7.424
	g	85.4678	19.639	19.330	12.690	1.481	40.626	4.968
Rb⁺	g	85.4678	115.965	117.137	—	—	—	—
Rb⁺⁺	g	85.4678	745.10	747.76	—	—	—	—
Rb⁺⁺	g	85.4678	1660.0	1664.1	—	—	—	—
Rb⁺	a	85.4678	—	−60.03	−67.87	—	29.04	—
Rb								
in 185 Hg		85.4678	—	—	−25.25	—	—	—
RbO₂	c	117.4666	—	−66.6	—	—	—	—
Rb₂O	c	186.9350	—	−81.	—	—	—	—
Rb₂O₂	c	202.9344	—	−112.8	—	—	—	—
RbH	c	86.4758	—	−12.5	—	—	—	—
RbOH	c	102.4752	—	−99.95	—	—	—	—
RbOH·H₂O	c	120.4906	—	−178.98	—	—	—	—
RbOH·2H₂O	c	138.5060	—	−251.73	—	—	—	—
RbF	c	104.4662	—	−133.3	—	—	—	—
	a	104.4662	—	−139.53	−134.510	—	25.70	—
RbHF₂	c	124.4726	−219.52	−220.5	−204.5	3.932	28.70	18.97

SELECTED VALUES OF CHEMICAL THERMODYNAMIC PROPERTIES (Continued)

Formula and Description	State	Formula weight	ΔHf° (0 K) kcal/mol	ΔHf° kcal/mol	ΔGf° kcal/mol	H°₂₉₈ − H°₀	S° cal/deg mol	Cp° cal/deg mol
RbCl	c	120.9208	−104.080	−104.05	−97.47	2.917	22.92	12.52
RbCl₂	a	191.8268	—	—	−96.7	—	—	—
RbClO	a	136.9202	—	−85.6	−76.7	—	39.	—
RbClO₂	a	152.9196	—	−75.9	−63.8	—	53.2	—
RbClO₃	c	168.9190	—	−96.3	−71.8	—	36.3	24.66
	a	168.9190	—	−84.88	−69.77	—	67.80	—
RbClO₄	c	184.9184	—	−104.50	−73.54	—	39.2	—
	a	184.9184	—	−90.94	−69.93	—	72.5	—
RbBr	c	165.3768	−92.714	−94.31	−91.25	3.124	26.28	12.63
	a	165.3768	—	−89.08	−92.72	—	48.74	—
RbBr₂	c	325.1948	—	−100.0	—	—	—	—
	a	325.1948	—	−91.20	−93.46	—	80.5	—
RbBr₅	a	485.0128	—	−94.0	−92.7	—	104.7	—
RbBrO₃	c	213.3750	—	−87.78	−66.47	—	38.5	—
RbBrO₃	a	213.3750	—	−76.06	−63.44	—	67.69	—
RbBrO₄	a	229.3744	—	−56.9	−39.7	—	76.7	—
RbBrCl₂	c	236.2828	—	−116.2	—	—	—	—
RbI	c	212.3722	−79.603	−79.78	−78.60	3.190	28.30	12.71
	a	212.3722	—	−73.22	−80.20	—	55.6	—
RbI₂	c	466.1810	—	−82.8	−81.0	—	53.9	—
	a	466.1810	—	−72.3	−80.2	—	86.2	—
RbIO₃	c	260.3704	—	—	−101.90	—	—	—
	a	260.3704	—	−112.90	−98.50	—	57.30	—
Rb₂S	c	202.9996	—	−86.2	—	—	—	—
	a	202.9996	—	−112.2	−115.2	—	54.6	—
Rb₂S₂	a	235.0636	—	−112.9	−116.7	—	64.9	—
Rb₂SO₃	a	250.9978	—	−272.0	−252.0	—	51.	—
Rb₂SO₄	c	266.9972	−340.795	−343.12	−314.76	6.458	47.19	32.04
Rb₂SO₄	a	266.9972	—	−337.38	−313.71	—	62.9	—
RbNO₃	c	147.4727	—	−118.32	−94.61	—	35.2	24.4
	a	147.4727	—	−109.59	−94.48	—	64.0	—
Rb₃PO₄	a	351.3748	—	−485.4	−447.1	—	34.	—
Rb₄P₂O₇	a	515.8146	—	−782.9	−730.2	—	88.	—
RbAsO₃	a	192.3882	—	−162.57	−151.53	—	38.7	—
Rb₃AsO₄	a	395.3226	—	−392.36	−358.61	—	48.2	—
Rb₂CO₃	c	230.9450	−270.41	−271.5	−251.2	5.851	43.34	28.11
	a	230.9450	—	−281.90	−261.91	—	44.5	—
RbHCO₃	c	146.4852	—	−230.2	−206.4	—	29.0	—
	a	146.4852	—	−225.42	−208.13	—	50.84	—
RbCN	c	111.4857	—	—	—	4.159	33.67	16.20
	a	111.4857	—	−24.0	−26.7	—	51.5	—
RbCNO cyanate	a	127.4851	—	−94.9	−91.2	—	54.5	—
RbCNS thiocyanate	a	143.5497	—	−41.76	−45.72	—	63.5	—
RbBO₂	c	128.2776	−231.0	−232.0	−218.2	3.181	22.54	17.7
RbBH₄	a	100.3108	—	−48.52	−40.56	—	55.4	—
RbMnO₄	a	204.4034	—	−189.4	−174.8	—	74.7	—
Rb₂MnO₄	a	289.8712	—	−276.	−255.4	—	72.	—
Rb₂CrO₄	c	286.9292	—	−338.0	—	—	—	—
	a	286.9292	—	−330.66	−309.70	—	70.08	—
Rb₂Cr₂O₇	a	386.9234	—	−476.3	−446.7	—	120.7	—
RUTHENIUM								
Ru	c	101.07	0	0	0	1.100	6.82	5.75
	g	—	153.210	153.6	142.4	1.490	44.550	5.144
Ru⁺	g	—	323.07	324.94	—	—	—	—
Ru²⁺	g	—	709.6	713.0	—	—	—	—
Ru³⁺	g	—	1366.0	1370.9	—	—	—	—
RuO₄	c	133.069	—	−72.9	—	—	—	—
RuCl₃ black	c	207.429	—	−49.	—	—	—	—
RuCl₄	g	242.882	—	−12.4	—	—	—	—
RuBr₃	c	340.797	—	−33.	—	—	—	—
RuI₃	c	481.783	—	−15.7	—	—	—	—
RuS₂	c	165.198	—	−47.	—	—	—	—
SAMARIUM								
Sm	c	150.35	0	0	0	1.81	16.63	7.06
	g	—	49.26	49.4	41.3	1.953	43.722	7.255
Sm⁺	g	—	179.1	180.7	—	—	—	—
Sm²⁺ g std. state, m = 1	g	—	434.	438.	—	—	—	—
	aq	—	—	—	−118.9	—	—	—
Sm³⁺ std. state, m = 1	aq	—	—	−165.3	−159.3	—	−50.6	−5.
SmO	g	166.349	−31.	—	—	—	—	—
Sm₂O₃ monoclinic	c	348.698	−433.89	−435.7	−414.6	5.02	36.1	27.37
cubic	c	—	—	—	—	5.0	—	26.86
SmF	g	169.348	—	−63.	—	—	—	—
SmF₃	c	207.345	—	−425	—	—	—	—
SmF₃·H₂O	c	216.353	—	−436.2	—	—	—	—
SmCl₃	c	256.709	—	−245.2	—	—	—	—
∞ H₂O	aq	—	—	−285.2	—	—	—	—
SmCl₃·6H₂O	c	364.801	—	−686.0	−587.1	—	99.	86.4
SmI₃	c	531.063	—	−148.2	—	—	—	—
Sm(IO₃)₃	c	675.058	—	−330.	—	—	—	—
Sm₂(SO₃)₃	c	540.887	—	—	−697.4	—	—	—
Sm₂(SO₄)₃	c	588.885	—	−931.9	—	—	—	—
Sm(NO₃)₃	c	336.365	—	−289.7	—	—	—	—
SmC₂	c	174.372	—	−17.	−18.1	—	23.	—
Sm₂(CO₃)₃	c	480.728	—	—	−741.4	—	—	—
SCANDIUM								
Sc	c	44.956	0	0	0	1.247	8.28	6.10
	g	—	89.87	90.3	80.32	1.674	41.75	5.28
Sc⁺	g	—	240.69	—	—	—	—	—
Sc²⁺	g	—	535.87	—	—	—	—	—
Sc³⁺	g	—	1106.86	—	—	—	—	—
std. state, m = 1	aq	—	—	−146.8	−140.2	—	−61.	—
Sc⁴⁺	g	—	2801.1	—	—	—	—	—

SELECTED VALUES OF CHEMICAL THERMODYNAMIC PROPERTIES (Continued)

Formula and Description	State	Formula weight	ΔHf₀ kcal/mol	ΔHf° kcal/mol	ΔGf° kcal/mol	H°₂₉₈ − H°₀	S° cal/deg mol	C°ₚ cal/deg mol
Sc²⁺	g	—	4914.9	—	—	—	—	—
ScO	g	60.955	-13.5	-13.68	-19.90	2.100	53.65	7.38
Sc₂O	g	105.911	-6.	-6.9	—	2.7	—	11.2
Sc₂O₃	c	137.910	-453.95	-456.22	-434.85	3.34	18.4	22.52
Sc(OH)₃	c	95.978	—	-325.9	-294.8	—	24.	—
ScF	g	63.9544	-33.	-33.2	-39.3	2.138	53.11	7.74
ScF₂	g	82.9528	-153.	-153.5	-156.6	2.84	67.0	11.5
ScF₃	c	101.9512	—	-389.4	-371.8	—	22.	—
ScCl₂	g	115.862	—	—	—	3.1	72.5	12.6
ScCl₃	c	151.315	—	-221.1	—	—	—	—
in 5500 H₂O	aq	—	—	-268.8	—	—	—	—
ScCl₃· 6H₂O	c	259.411	—	-666.6	—	—	—	—
Sc(OH)₂Cl	c	114.424	—	-303.	-276.3	—	26.	—
ScBr₃	c	284.683	—	-177.6	—	—	—	—
ScI₃	g	298.765	—	—	—	3.4	81.6	13.3
ScS	g	77.020	41.9	41.8	29.7	2.2	56.3	8.0
Sc₂(SO₄)₃	c	378.097	—	—	—	—	—	62.0
ScC₂	g	68.978	143.	143.6	—	—	—	—
Sc(HCO₃)₃	c	180.009	—	—	—	—	—	54.
Sc₂(C₂O₄)₃	c	353.971	—	—	—	—	—	124.
SELENIUM								
Se								
hexagonal, black	c	78.96	0	0	0	1.319	10.144	6.062
monoclinic, red	c	—	—	1.6	—	—	—	—
	g	—	54.11	54.27	44.71	1.4815	42.21	4.978
glassy	amorp	—	—	1.2	—	—	—	—
Se²⁻								
std. state, m = 1	aq	—	—	—	30.9	—	—	—
Se₂	g	157.92	35.26	34.9	23.0	2.275	60.2	8.46
Se₆	g	473.76	—	39.2	—	—	—	—
SeO	g	94.959	13.0	12.75	6.41	2.108	55.9	7.47
SeO₂	c	110.959	—	-53.86	—	—	—	—
	aq	—	—	-52.97	—	—	—	—
SeO₃	c	126.958	—	-39.9	—	—	—	—
Se₂O₅	c	237.917	—	-97.6	—	—	—	—
H₂SeO₃	c	128.974	—	-125.35	—	—	—	—
undissoc.; std. state, m = 1	aq	—	—	-121.29	-101.87	—	49.7	—
	aq	—	—	-121.24	—	—	—	—
H₂SeO₄	c	144.974	—	-126.7	—	—	—	—
H₂SeO₄· H₂O	c	162.989	—	-200.9	—	—	—	—
	liq	—	—	-196.1	—	—	—	—
SeF₆	g	192.950	-264.1	-267.	-243.	4.740	74.99	26.4
SeCl₂	g	149.866	—	-7.6	—	—	—	—
SeCl₄	c	220.772	—	-43.8	—	—	—	—
SILICON								
Si	c	28.086	0	0	0	0.769	4.50	4.78
	amorp	—	—	1.0	—	—	—	—
	g	—	107.86	108.9	98.3	1.805	40.12	5.318
Si⁺	g	—	295.83	298.35	—	—	—	—
Si²⁺	g	—	672.71	676.71	—	—	—	—
Si³⁺	g	—	1444.50	1449.98	—	—	—	—
Si⁴⁺	g	—	2485.50	2492.46	—	—	—	—
Si⁵⁺	g	—	6331.3	6339.8	—	—	—	—
Si⁶⁺	g	—	11062.6	11072.6	—	—	—	—
Si₂	g	56.172	141.32	142.	128.	2.22	54.92	8.22
Si₃	g	84.258	146.4	147.	—	2.9	—	12.9
SiO	g	44.0854	-24.08	-23.8	-30.2	2.082	50.55	7.15
SiO₂								
α, quartz	c	60.0848	—	-217.72	-204.75	1.657	10.00	10.62
α, cristobalite	c	—	—	-217.37	-204.46	1.671	10.20	10.56
SiO₂								
α, tridymite	c	—	—	-217.27	-204.42	1.693	10.4	10.66
	amorp	—	—	-215.94	-203.33	—	11.2	10.6
	g	—	—	-77.	—	—	—	—
	aq	—	—	-214.4	—	—	—	—
SiH	g	29.0940	86.	86.28	—	2.069	—	—
SiH₄	g	32.1179	10.30	8.2	13.6	2.517	48.88	10.24
Si₂H₆	g	62.2198	23.04	19.2	30.4	3.768	65.14	19.31
Si₃H₈	liq	92.3218	—	22.1	—	—	—	—
	g	—	—	28.9	—	—	—	—
H₄SiO₃	c	78.1001	—	-284.1	-261.1	—	32.	—
undissoc.; std. state, m = 1	aq	—	-282.7	-258.0	—	26.	—	—
H₄SiO₄	c	96.1155	—	-354.0	-318.6	—	46.	—
SiF	g	47.0844	1.	1.7	-5.8	2.260	53.94	7.80
SiF₂	g	66.0828	-147.75	-148.	-150.	2.630	60.38	10.49
SiF₄	g	104.0796	-384.66	-385.98	-375.88	3.663	67.49	17.60
std. state, m = 1	aq	—	—	—	-384.2	—	—	—
SiF₆²⁻								
std. state, m = 1	aq	142.0764	—	-571.0	-525.7	—	29.2	—
H₂SiF₆	aq	144.0923	—	—	—	—	—	—
SiCl	g	63.539	45.	45.39	—	2.267	—	8.81
SiCl₂	g	98.992	-39.61	-39.59	-42.35	2.98	67.0	12.16
SiCl₄	liq	169.898	—	-164.2	-148.16	—	57.3	34.73
	g	—	-156.508	-157.03	-147.47	4.633	79.02	21.57
SiBr	g	107.995	51.30	50.	—	2.40	—	9.23
SiBr₄	liq	347.722	—	-109.3	-106.1	—	66.4	—
	g	—	—	-99.3	-103.2	5.317	90.29	23.21
SiI₄	c	535.7036	—	-45.3	—	—	—	—
SiH₃I	g	158.0143	—	—	—	2.886	64.73	13.00
SiS	g	60.150	26.6	26.88	14.56	2.135	53.43	7.71
SiS₂	c	92.214	—	-49.5	—	—	—	—
Si₃N₄								
α	c	140.2848	—	-177.7	-153.6	—	24.2	—
SiC								
β, cubic	c	40.0972	-15.36	-15.6	-15.0	0.781	3.97	6.42
α, hexagonal	c	—	—	-15.0	-14.4	—	3.94	6.38
	g	—	175.6	177.	—	2.4	—	—
SiC₂	g	52.1083	145.7	147.	—	2.5	—	—
Si₂C	g	68.1832	131.1	132.	—	2.7	—	—
Si₃C₃	g	80.1943	167.	—	—	—	—	—

SELECTED VALUES OF CHEMICAL THERMODYNAMIC PROPERTIES (Continued)

Formula and Description	State	Formula weight	ΔHf‡ kcal/mol (0 K)	ΔHf° kcal/mol	ΔGf° kcal/mol	H°₂₉₈ − H°₀	S° cal/deg mol	C‡ cal/deg mol
Si₂C₂	g	92.2054	176.	—	—	—	—	—
Si₃C	g	96.2692	161.	—	—	—	—	—
SILVER								
Ag	c	107.870	0	0	0	1.373	10.17	6.059
	g	—	67.90	68.01	58.72	1.481	41.321	4.9679
Ag⁺	g	—	242.00	243.59	—	—	—	—
std. state, m = 1	aq	—	—	25.234	18.433	—	17.37	5.2
Ag₂O	c	231.7394	−7.034	−7.42	−2.68	3.397	29.0	15.74
Ag₂O₂	c	247.7388	—	−5.8	6.6	—	28.	21.
Ag₂O₃	c	263.7382	—	8.1	29.0	—	24.	—
AgF	c	126.8684	—	−48.9	—	—	—	—
std. state, m = 1	aq	—	—	−54.27	−48.21	—	14.1	−20.3
in ∞ H₂O	aq	—	—	−54.27	—	—	—	—
AgF₂	c	145.8668	—	−86.	—	—	—	—
AgCl	c	143.323	—	−30.370	−26.244	—	23.0	12.14
std. state, m = 1	aq	—	—	−14.718	−12.939	—	30.9	−27.4
AgClO₃	c	175.3218	—	2.10	18.1	—	32.16	20.87
AgClO₃	c	191.3212	—	−7.24	15.4	—	34	—
std. state, m = 1	aq	—	—	0.38	16.51	—	56.2	—
AgClO₄	c	207.3206	—	−7.44	—	—	—	—
std. state, m = 1	aq	—	−5.68	16.37	—	60.9	—	—
AgBr	c	187.779	—	−23.99	−23.16	—	25.6	12.52
std. state, m = 1	aq	—	—	−3.82	−6.42	—	37.1	−28.7
AgBrO₃	c	235.7772	—	−2.5	−17.04	—	36.3	—
std. state, m = 1	aq	—	—	9.20	22.86	—	56.02	—
AgI	c	234.7744	—	−14.78	−15.82	—	27.6	13.58
std. state, m = 1	aq	—	—	12.04	6.10	—	44.0	−28.8
AgI₂⁻								
std. state, m = 1	aq	361.6788	—	—	−20.8	—	—	—
AgI₂²⁻								
std. state, m = 1	aq	488.5832	—	−43.5	−36.8	—	60.5	—
AgI₄³⁻								
std. state, m = 1	aq	615.4876	—	—	−50.1	—	—	—
AgIO₃	c	282.7726	—	−40.9	−22.4	—	35.7	24.60
std. state, m = 1	aq	—	—	−27.7	−12.2	—	45.7	—
Ag₂S								
α, orthorhombic	c	247.804	−8.126	−7.79	−9.72	4.136	34.42	18.29
β	c	—	—	−7.03	−9.43	—	36.0	—
Ag₂SO₃	c	295.8022	—	−117.3	−98.3	—	37.8	—
std. state, m = 1	aq	—	—	−101.4	−79.4	—	27.8	—
Ag₂SO₄	c	311.8016	—	−171.10	−147.82	—	47.9	31.40
std. state, m = 1	aq	—	—	−166.85	−141.10	—	39.6	−60.
Ag₂Se	c	294.70	−9.42	−9.	−10.6	4.48	36.02	19.54
AgN₃	c	149.8901	—	73.8	89.9	—	24.9	—
std. state, m = 1	aq	—	—	90.99	101.6	—	43.2	—
Ag₃N	c	337.6167	—	47.6	—	—	—	—
AgNO₂	c	153.8755	—	−10.77	4.56	—	30.64	19.17
std. state, m = 1	aq	—	—	10.7	9.5	—	46.8	−18.1
AgNO₃	c	169.8749	—	−29.73	−8.00	—	33.68	22.24
nitrate; std. state, m = 1	aq	—	—	−24.33	−8.18	—	52.4	−15.5
Ag(NH₃)₂⁺								
std. state, m = 1	aq	141.9312	—	−26.60	−4.12	—	58.6	—
Ag(NH₃)₂NO₃	c	203.9361	—	−85.0	—	—	—	—
std. state, m = 1	aq	—	—	−76.16	−30.73	—	93.6	—
Ag(NH₃)₂Cl								
std. state, m = 1	aq	177.3842	—	−66.55	−35.49	—	72.1	—
undissoc.; std. state, m = 1	aq	—	—	—	−35.01	—	—	—
Ag(NH₃)₂Br								
std. state, m = 1	aq	221.8402	—	−55.65	−28.97	—	78.3	—
Ag₃PO₄	c	418.5814	—	—	−210.	—	—	—
AgCN	c	133.8878	34.354	34.9	37.5	3.206	25.62	15.95
std. state, m = 1	aq	—	—	61.2	59.6	—	39.9	—
Ag(CN)₂⁻								
std. state, m = 1	aq	159.9057	—	64.6	73.0	—	46.	—
AgONC								
silver fulminate	c	149.8872	—	43.	—	—	—	—
AgSCN	c	165.9518	—	21.0	24.23	—	31.3	15.
std. state, m = 1	aq	—	—	43.50	40.58	—	51.9	−4.4
SODIUM								
Na	c	22.9898	0	0	0	1.54	12.24	6.75
	g	22.9898	25.709	25.65	18.354	1.481	36.712	4.968
Na⁺	g	22.9898	—	145.55	—	—	—	—
Na²⁺	g	22.9898	—	1237.48	—	—	—	—
Na³⁺	g	22.9898	—	2891.0	—	—	—	—
Na⁴⁺	g	22.9898	—	5173.5	—	—	—	—
Na⁵⁺	g	22.9898	—	8366.	—	—	—	—
Na⁺	a	22.9898	—	−57.39	−62.593	—	14.1	11.1
NaO	g	38.9892	25.4	25.	19.7	2.22	54.6	8.3
Na₂O	c	54.9886	−62.96	−62.2	−52.2	4.37	27.7	17.24
Na₂O₂	c	61.9790	−97.85	−99.00	−89.74	2.964	17.94	16.52
Na₂O₃	c	77.9784	−120.70	−122.10	−107.00	3.75	22.70	21.33
NaOH	c	39.9972	−100.641	−101.723	−90.709	2.507	15.405	14.23
	g	39.9972	−48.6	−49.5	−50.2	2.72	54.57	11.56
in ∞ H₂O	a	39.9972	—	−112.36	—	—	—	—
NaF	c	41.9882	−136.542	−137.105	−129.902	2.031	12.30	11.20
	a	41.9882	—	−136.89	−129.23	—	10.8	−14.4
NaCl	c	58.4428	−98.168	−98.268	−91.815	2.536	17.24	12.07
	g	58.4428	−41.9	−42.22	−47.00	2.298	54.90	8.55
	a	58.4428	—	−97.34	−93.965	—	27.6	−21.5
in ∞ H₂O	a	58.4428	—	−97.34	—	—	—	—
NaClO	a	74.4422	—	−83.0	−71.4	—	24.	—
in 400 H₂O		74.4422	—	−82.8	—	—	—	—
NaClO₂	c	90.4416	—	−73.38	—	—	—	—
	a	90.4416	—	−73.3	−58.5	—	38.3	—
NaClO₃	c	106.4410	—	−87.422	−62.697	—	29.5	—
	a	106.4410	—	−82.24	−64.51	—	52.9	—
NaClO₄	c	122.4404	—	−91.61	−60.93	—	34.0	—
	a	122.4404	—	−88.30	−64.65	—	57.6	—

SELECTED VALUES OF CHEMICAL THERMODYNAMIC PROPERTIES (Continued)

Formula and Description	State	Formula weight	0 K ΔHf₀ kcal/mol	298.15 K (25°C) ΔHf° kcal/mol	ΔGf° kcal/mol	H°₂₉₈ − H°₀	S° cal/deg mol	C°ₚ cal/deg mol
$NaBr$	c	102.8988	−84.596	−86.296	−83.409	2.770	20.75	12.28
	g	102.8988	−32.08	−34.2	−42.31	2.346	57.62	8.68
	a	102.8988	—	−86.440	−87.440	—	33.8	−22.8
$NaBrO$	a	118.8982	—	−79.9	−70.6	—	24.	—
$NaBrO_2$	c	150.8970	—	−79.85	−58.04	—	30.8	—
	a	150.8970	—	−73.42	−58.16	—	52.8	—
$NaBrO_4$	a	166.8964	—	−54.30	−34.40	—	61.80	—
NaI	c	149.8942	−68.593	−68.78	−68.37	2.93	23.55	12.45
	a	149.8942	—	−70.58	−74.92	—	40.7	−22.9
NaI_2	c	403.7030	—	−56.2	—	—	—	—
NaI_3	a	407.7030	—	−69.7	−74.9	—	71.3	—
$NaIO_3$	c	197.8924	—	−115.150	—	—	—	22.0
	a	197.8924	—	−110.30	−93.20	—	42.4	—
$NaIO_3 \cdot H_2O$	c	215.9078	—	−186.30	−151.56	—	38.8	—
$NaIO_3 \cdot 5H_2O$	c	287.9694	—	−466.60	—	—	—	—
$NaIO_4$	c	213.8918	—	−102.60	−77.22	—	39.0	—
	a	213.8918	—	−93.60	−76.60	—	67.	—
in 2,000 H_2O		213.8918	—	−93.808	—	—	—	—
$NaIO_3 \cdot 3H_2O$	c	267.9380	—	—	−247.8	—	—	—
NaH_4IO_6 from $H_5IO_6\cdot$	au	249.9226	—	−237.8	—	—	—	—
$Na_2H_3IO_6$ from $H_5IO_6\prime$	au	271.9044	—	−294.4	—	—	—	—
$NaIO_2F_2$	c	219.8898	—	−202.50	—	—	—	—
$NaICl_2$	c	220.8002	—	−96.0	—	—	—	—
	a	220.8002	—	—	−101.1	—	—	—
$NaICl_4$	c	291.7062	—	−112.0	—	—	—	—
$NaIBr_2$	c	309.7122	—	−83.0	—	—	—	—
	a	309.7122	—	—	−92.0	—	—	—
$NaBrI_2$	a	356.7076	—	−88.0	−88.9	—	61.3	—
Na_2S	c	78.0436	—	−87.2	−83.6	—	20.0	—
	a	78.0436	—	−106.9	−104.7	—	24.7	—
Na_2S_2	c	110.1076	—	−94.9	−90.5	—	25.	—
Na_2S_3	a	110.1076	—	−107.6	−106.2	—	35.0	—
Na_2SO_3	c	126.0418	−261.21	−263.1	−242.0	5.36	34.88	28.74
	a	126.0418	—	−266.70	−241.50	—	21.0	—
Na_2SO_4	c	142.0412	−328.789	−331.52	−303.59	5.551	35.75	30.64
	a	142.0412	—	−332.10	−303.16	—	33.0	−48.
$Na_2SO_4 \cdot 10H_2O$	c	322.1952	—	−1034.24	−871.75	—	141.5	—
$Na_2S_2O_3$	c	158.1058	—	−268.4	−245.7	—	37.	—
	a	158.1058	—	−270.65	−250.0	—	44.0	—
$Na_2S_2O_3 \cdot 5H_2O$	c	248.1828	—	−623.31	−533.0	—	89.0	—
$Na_2S_2O_4$	c	174.1052	—	−294.5	—	—	—	—
$NaHSO_4$	c	120.0594	—	−269.0	−237.3	—	27.0	—
from HSO_4	a	120.0594	—	−269.47	−243.28	—	45.6	−9.
Na_2SeO_4	c	188.9372	—	255.5	—	—	—	—
	a	188.9372	—	258.0	−230.7	—	41.1	—
NaN_3	c	65.0099	6.316	5.19	22.41	3.522	23.15	18.31
	a	65.0099	—	8.37	20.6	—	39.9	—
$NaNO_2$	c	68.9953	—	−85.72	−68.02	—	24.8	—
	a	68.9953	—	−82.4	−70.3	—	43.5	−12.2
$NaNO_3$	c	84.9947	−110.248	−111.82	−87.73	4.115	27.85	22.20
	a	84.9947	—	−106.95	−89.20	—	49.1	−9.6
$NaNH_2$	c	39.0125	−27.84	−29.6	−15.3	2.842	18.38	15.81
$NaNH_4$	c	40.0205	—	−16.27	−2.94	—	37.2	—
Na_3PO_4	c	163.9408	−454.79	−458.27	−427.55	6.566	41.54	36.68
	a	163.9408	—	−477.5	−431.3	—	−11.	—
$(NaPO_3)_x$	c	305.8854	−865.9	−873.	−808.	10.720	68.47	62.00
$Na_4P_2O_7$	c	265.9026	−756.2	−762.	−709.7	10.180	64.60	57.63
$Na_5P_3O_{10}$ form I, quenched	c	367.8644	−1043.57	−1051.4	−978.5	14.070	91.25	78.16
form II	c2	367.8644	−1045.76	−1054.0	−980.0	13.67	87.37	77.72
in 5,200 H_2O		367.8644	—	−1068.1	—	—	—	—
NaH_2PO_2	c	87.9784	—	−200.5	—	—	—	—
NaH_4PO_4	c	103.9778	—	−288.0	—	—	—	—
NaH_2PO_4	c	119.9772	−363.06	−367.3	−331.3	4.75	30.47	27.93
$NaH_2PO_4 \cdot H_2O$	c	137.9926	—	−438.1	—	—	—	—
Na_2HPO_3	c	125.9596	—	−336.8	—	—	—	—
Na_2HPO_4	c	141.9590	−413.92	−417.8	−384.4	5.646	35.97	32.34
$Na_2HPO_4 \cdot 2H_2O$	c	177.9898	—	−560.7	−499.2	—	52.9	—
$Na_2HPO_4 \cdot 7H_2O$	c	268.0668	—	−913.4	−784.0	—	103.87	—
$Na_2HPO_4 \cdot 12H_2O$	c	358.1438	—	−1266.2	−1068.0	—	151.49	—
$Na_3H_2P_2O_7$	c	221.9390	−654.06	−660.8	−602.9	8.184	52.63	47.36
$NaAsO_2$	c	129.9102	—	−157.87	—	—	—	—
	a	129.9102	—	−159.93	−146.25	—	23.8	—
Na_3AsO_4	c	207.8886	—	−368.	—	—	—	—
	a	207.8886	—	−384.44	−342.78	—	3.4	—
Na_3BiO_3	c	341.9470	—	−291.	—	—	—	—
Na_2CO_3	c	105.9890	−268.76	−270.24	−249.64	4.959	32.26	26.84
	a	105.9890	—	−276.62	−251.36	—	14.6	—
$Na_2CO_3 \cdot H_2O$	c	124.0044	−338.87	−342.08	−307.22	6.296	40.18	34.80
$Na_2CO_3 \cdot 7H_2O$	c	232.0968	—	−764.81	−648.8	—	102.	—
$Na_2CO_3 \cdot 10H_2O$	c	286.1430	−959.62	−975.46	−819.36	21.21	134.8	131.53
$Na_2C_2O_4$	c	133.9996	—	−315.0	—	—	—	34.
$HCOONa$	c	68.0078	−158.19	−159.3	−143.4	3.767	24.80	19.76
$NaHCO_3$	c	84.0072	−225.15	−227.25	−203.4	3.81	24.3	20.94
$NaOCH_3$	c	54.0244	−85.41	−87.9	−70.46	3.374	26.43	16.60
$NaC_2H_3O_2$	c	82.0350	—	−169.41	−145.14	—	29.4	19.1
	a	82.0350	—	−173.55	−150.88	—	34.8	9.6
$NaC_2H_3O_2 \cdot 3H_2O$	c	136.0812	—	−383.2	−317.6	—	58.	—
$NaOC_2H_5$	c	68.0516	—	−98.90	—	—	—	—
	a	68.0516	—	−113.	—	—	—	—
CCl_3COONa sodium trichloroacetate	c	185.3700	—	−178.9	—	—	—	—
$NaCN$ (C,I,cubic)	c	49.0077	−22.56	−20.91	−18.27	4.480	27.63	16.82
(c,II,orthorhombic)	c2	49.0077	—	−21.69	—	—	—	—

SELECTED VALUES OF CHEMICAL THERMODYNAMIC PROPERTIES (Continued)

			0 K			298.15 K (25°C)		
Formula and Description	State	Formula weight	ΔHf‡ kcal/mol	ΔHf° kcal/mol	ΔGf° kcal/mol	H°₉₈ – H°₀	S° cal/deg mol	C‡ cal/deg mol
NaCN·1/2H₂O	c	58.0154	—	−56.35	—	—	—	—
NaCN·2H₂O	c	85.0385	—	−162.47	—	—	—	—
NaCNO	c	65.0071	—	−96.89	−85.6	—	23.1	20.7
	a	65.0071	—	−92.3	−85.9	—	39.6	—
NaCNS	c	81.0717	—	−40.75	—	—	—	—
	a	81.0717	—	−39.12	−40.44	—	48.6	1.5
Na₂SiO₃,								
sodium metasilicate	c	122.0638	—	−371.63	−349.19	—	27.21	—
	gl	122.0638	—	−368.1	—	—	—	—
Na₂SiO₃·5H₂O	c	212.1408	—	−728.6	—	—	—	—
Na₂SiO₃·9H₂O	c	284.2024	—	−1010.7	—	—	—	—
Na₂Si₂O₅	c	182.1486	—	−589.8	−555.0	—	39.21	—
sodium disilicate, stable up to 951K (formerly β)								
	c2	182.1486	—	−589.22	—	—	—	—
stable 951K to m.pt.(1147K)(formerly α)								
unstable	c3	182.1486	—	−587.28	—	—	—	—
	gl	122.0638	—	−368.1	—	—	—	—
Na₄SiO₄	c	184.0428	—	—	—	—	46.76	—
NaBO₂	c	65.7996	−232.38	−233.5	−220.06	2.780	17.57	15.76
NaBO₂·2H₂O	c	101.8304	—	−378.	−337.0	—	37.	—
NaBO₂·4H₂O	c	137.8612	—	−520.	−451.3	—	55.	—
NaBO₃·4H₂O	c	153.8606	—	−505.3	—	—	—	—
Na₂B₄O₇	c	201.2194	−782.36	−786.6	−740.0	7.262	45.30	44.64
	am	201.2194	—	−781.8	−735.4	—	46.1	—
	a	201.2194	—	—	−747.8	—	—	—
Na₂B₄O₇·4H₂O	c	273.2810	—	−1077.3	—	—	—	—
Na₂B₄O₇·5H₂O	c	291.2964	—	−1147.8	—	—	—	—
Na₂B₄O₇·10H₂O								
borax	c	381.3734	—	−1503.0	−1318.5	—	140.	147.
NaBH₄	c	37.8328	−43.100	−45.08	−29.62	3.890	24.21	20.74
	a	37.8328	—	−45.88	−35.28	—	40.5	—
NaBF₄	c	109.7944	−440.04	−440.9	−418.30	5.191	34.73	28.74
NaBF₄	a	109.7944	—	−433.8	−418.0	—	58.	—
NaAlO₂	c	81.9701	—	−271.30	−256.06	—	16.90	17.52
	a	81.9701	—	−277.0	−259.4	—	9.	—
NaAl(SO₄)₂·12H₂O alum	c	458.2793	—	−1434.69	—	—	—	—
NaAlSiO₄								
nepheline, nephelite	c	142.0549	—	−500.2	−472.8	—	29.7	
NaAlSi₂O₆								
jadeite	c	202.1397	—	−724.4	−681.7	—	31.9	—
dehydrated analcite	c2	202.1397	—	−713.5	−673.8	—	41.9	39.30
NaHg	c	223.5798	—	−11.3	−9.76	—	25.	—
NaHg₂	c	424.1698	—	−18.2	−16.2	—	42.	—
NaHg₄	c	825.3498	—	−21.2	−17.7	—	73.	—
Na₃Hg	c	269.5594	—	−11.0	−10.84	—	54.	—
Na₃Hg₂	c	470.1494	—	−21.9	−20.7	—	69.	—
Na₃Hg₄′	c	516.1290	—	—	−21.7	—	—	—
Na₅Hg₆	c	1765.6486	—	−84.4	−76.0	—	203.	—
Na₃Fe(CN)₆	a	280.9238	—	−37.9	−13.5	—	106.9	—
Na₄Fe(CN)₆	a	303.9136	—	−120.7	−84.28	—	79.1	—
NaMnO₄	a	141.9254	—	−186.8	−169.5	—	59.8	—
Na₂MnO₄	c	164.9152	—	−276.3	—	—	—	—
	a	164.9152	—	−271.	−244.9	—	42.	—
Na₂CrO₄	c	161.9732	−318.2	−320.8	−295.17	6.323	42.21	33.97
	a	161.9732	—	−325.38	−299.15	—	40.2	—
Na₂Cr₂O₇	c	261.9674	—	−472.9	—	—	—	—
	a	216.9674	—	−471.0	−436.2	—	90.8	—
Na₂MoO₄	c	205.9172	−348.63	−350.89	−323.71	6.070	38.17	33.87
Na₂MoO₄·2H₂O	c	241.9480	—	−492.1	−437.46	—	57.5	—
Na₂WO₄	c	293.8272	−367.83	−370.2	−342.86	6.05	38.6	33.41
Na₂W₂O₇	c	525.6754	−571.44	−574.8	−529.84	9.36	60.8	51.36
NaVO₃	c	121.9300	−272.31	−273.85	−254.33	4.217	27.17	23.32
Na₃VO₄	c	183.9090	−417.36	−420.14	−391.45	7.093	45.4	39.40
Na₄V₂O₇	c	305.8390	−693.73	−697.62	−650.46	11.75	76.1	64.47
NaNbO₃	c	163.8940	—	−314.5	−294.7	—	28.	—
Na₂TiO₃	c	141.8778	−377.88	−380.3	−357.6	4.917	29.08	30.03
Na₂Ti₃O₇	c	221.7766	—	—	—	6.910	41.56	41.68
NaUO₃	c	309.0170	—	−360.	—	—	—	—
Na₂UO₄								
α form	c	348.0062	−450.02	−452.5	−424.90	6.268	39.68	35.05
β form	c2	348.0062	—	−450.2	—	—	—	—
Na₄UO₅	c	370.9960	−481.15	−484.0	−454.4	7.435	47.37	41.35
STRONTIUM								
Sr	c	87.62	0	0	0	—	12.5	6.3
	g	—	—	39.3	31.3	1.481	39.32	4.968
Sr⁺	g	—	—	172.11	—	—	—	—
Sr⁺⁺	g	—	—	427.96	—	—	—	—
std. state, m = 1	aq	—	—	−130.45	−133.71	—	−7.8	—
Sr⁺⁺	g	—	—	1435.	—	—	—	—
Sr⁺⁺	g	—	—	2751.	—	—	—	—
SrO	c	103.619	—	−141.5	−134.3	—	13.0	10.76
	g	—	—	−2.	—	—	—	—
SrO₂	c	119.619	—	−151.4	—	—	—	—
SrO₂·8H₂O	c	263.742	—	−722.6	—	—	—	—
Sr₂O	c	191.239	—	−154.7	—	—	—	—
SrH	g	88.628	—	+52.	45.	2.080	50.80	7.17
SrH₂	c	89.636	—	−43.1	—	—	—	—
SrOH	g	104.627	—	−41.	—	—	—	—
Sr(OH)₂	c	121.635	—	−229.2	—	—	—	—
	g	—	—	−135.	—	—	—	—
in 800 H₂O	aq	—	—	−240.1	—	—	—	—
SrF	g	106.618	—	−69.0	−75.1	2.219	57.31	8.27
SrF₂	c	125.617	—	−290.7	−278.4	3.125	19.63	16.73
	g	—	—	−182.7	−185.3	3.192	69.54	12.66
SrCl	g	123.073	—	−30.2	−36.5	2.338	60.2	8.73
SrCl₂								
α	c	158.526	—	−198.1	−186.7	3.880	27.45	18.07
	g	—	—	−116.1	−118.6	3.454	74.26	13.33
SrCl₂·H₂O	c	176.541	—	−271.7	−247.7	—	41.	28.7
SrCl₂·2H₂O	c	194.557	—	−343.7	−306.4	—	52.	38.3

SELECTED VALUES OF CHEMICAL THERMODYNAMIC PROPERTIES (Continued)

Formula and Description	State	Formula weight	0 K ΔHf° kcal/mol	ΔHf° kcal/mol	ΔGf° kcal/mol	H°₂₉₈ − H°₀	S° cal/deg mol	C°ₚ
SrCl₂·6H₂O	c	266.618	—	−627.1	−535.67	—	93.4	—
Sr(ClO₄)₂	c	286.521	—	−182.31	—	—	—	—
std. state, m = 1	aq	—	—	−192.27	−137.83	—	79.2	—
∞ H₂O	aq	—	—	−192.27	—	—	—	—
SrBr₂	c	247.438	—	−171.5	−166.6	4.261	32.29	18.01
	g	—	—	−98.	−106.	3.8	77.3	14.5
std. state, m = 1	aq	—	—	−188.55	−183.41	—	31.6	—
∞ H₂O	aq	—	—	−188.55	—	—	—	—
SrBr₂	c	247.438	—	−171.5	−166.6	4.261	32.29	18.01
	g	—	—	−98.	−106	3.8	77.3	14.5
std. state, m = 1	aq	—	—	−188.55	−183.41	—	31.6	—
Sr(IO₃)₂	c	437.425	—	−243.6	−204.4	—	56.	—
Sr(IO₃)₂·H₂O	c	455.441	—	−313.2	−260.4	—	66.	—
Sr(IO₃)₂·6H₂O	c	545.517	—	−666.8	−543.7	—	109.	—
SrS	c	119.684	—	−112.9	−111.8	—	16.3	11.64
SrSO₃	c	167.682	—	−281.3	—	—	—	—
SrSO₄	c	183.682	—	−347.3	−320.5	—	28.	—
precipitate	c	—	—	−346.5	—	—	—	—
std. state, m = 1	aq	—	—	−347.77	−311.68	—	−3.0	—
SrSeO₃	c	214.578	—	−250.4	—	—	—	—
SrSeO₄	c	230.578	—	−273.1	—	—	—	—
Sr(N₃)₂	c	171.660	—	+2.1	—	—	—	—
Sr₃N₂	c	290.873	—	−93.5	—	—	—	—
Sr(NO₂)₂	c	179.631	—	−182.3	—	—	—	—
in 800 H₂O	aq	—	—	−180.5	—	—	—	—
Sr(NO₃)₂	c	211.630	—	−233.80	−186.46	6.854	46.50	35.83
std. state, m = 1	aq	—	—	−229.57	−186.93	—	62.2	—
Sr(NO₃)₂ in ∞ H₂O	aq	—	—	−229.57	—	—	—	—
Sr₃P₂	c	324.808	—	−152.	—	—	—	—
Sr₃(PO₄)₂	c	452.803	—	−985.4	—	—	—	—
SrHPO₄	c	183.599	—	−435.4	−403.6	—	29.	—
Sr(H₂PO₄)₂	c	281.595	—	−749.2	—	—	—	—
Sr₃(AsO₄)₂	c	540.698	—	−792.8	−736.2	—	61.	—
hydrated precipitate	—	—	—	−803.	—	—	—	—
SrHAsO₄	aq	227.547	—	−345.7	—	—	—	—
Sr(H₂AsO₄)₂	aq	369.490	—	−563.2	—	—	—	—
SrC₂	c	111.642	—	−18.	—	—	—	—
SrCO₃								
strontianite	c	147.629	—	−291.6	−272.5	—	23.2	19.46
std. state, m = 1	aq	—	—	−292.29	−259.88	—	−21.4	—
SrC₂O₄	c	175.640	—	−327.6	—	—	—	—
std. state, m = 1	aq	—	—	−327.6	−294.8	—	+2.1	—
SrCN₂	c	127.645	—	−72.5	—	—	—	—
Sr(CN)₂	aq	139.656	—	−57.0	—	—	—	—
Sr(CN)₂·4H₂O	c	211.717	—	−333.7	—	—	—	—
SrSiO₃	c	163.704	—	−390.5	−370.4	—	23.1	21.16
Sr₂SiO₄	c	267.324	—	−550.8	−523.7	—	36.6	32.09
SrMoO₃	c	231.558	—	−306.	—	—	—	—
SrMoO₄	c	247.558	—	−370.	—	—	—	—
SrWO₄	c	335.468	—	−391.9	−366.	—	33.	—
SrTiO₃	c	183.518	—	−399.71	−379.64	—	26.0	23.51
Sr₂TiO₄	c	287.138	—	−546.7	−520.7	—	38.0	34.34
SrZrO₃	c	226.838	—	−422.4	−402.2	—	27.5	24.71
SULFUR								
S								
rhombic	c	32.064	0	0	0	1.054	7.60	5.41
monoclinic	c	—	—	0.08	—	—	—	—
	g	—	66.1	66.636	56.951	1.591	40.084	5.658
S⁺	g	—	304.95	306.97	—	—	—	—
S²⁺	g	—	844.8	848.2	—	—	—	—
S³⁺	g	—	1653.0	1657.9	—	—	—	—
S⁴⁺	g	—	2743.6	2750.0	—	—	—	—
S⁵⁺	g	—	4415.	4422.	—	—	—	—
S⁶⁺	g	—	6445.	6654.	—	—	—	—
S₂	g	64.128	30.647	30.68	18.96	2.141	54.51	7.76
S₂¹⁻								
std. state, m = 1	aq	—	—	7.2	19.0	—	6.8	—
S₃	g	96.192	—	31.7	—	—	—	—
S₄	g	128.256	—	32.7	—	—	—	—
S₅	g	160.320	—	29.6	—	—	—	—
S₆	g	192.384	—	24.5	—	—	—	—
SO	g	48.0634	1.5	1.496	−4.741	2.087	53.02	7.21
SO₂	liq	64.0628	—	−76.6	—	—	—	—
	g	—	−70.336	−70.944	−71.748	2.521	59.30	9.53
undissoc.; std. state, m = 1	aq	—	—	−77.194	−71.871	—	38.7	—
SO₂ in 10,000 H₂O	aq	—	—	−80.584	—	—	—	—
SO₃								
l, β	c	80.0622	—	−108.63	−88.19	—	12.5	—
	liq	—	—	−105.41	−88.04	—	22.85	—
	g	—	−93.21	−94.58	−88.69	2.796	61.34	12.11
SO₃²⁻								
std. state, m = 1	aq	—	—	−151.9	−116.3	—	−7.	—
SO₄²⁻								
std. state, m = 1	aq	96.0616	—	−217.32	−177.97	—	4.8	−70.
S₂O₃²⁻	aq	112.1262	—	−155.9	—	—	—	—
S₂O₄²⁻								
std. state, m = 1	aq	128.1256	—	−180.1	−143.5	—	??	—
S₂O₅²⁻	aq	176.1238	—	−334.9	—	—	—	—
S₂O₆²⁻								
std. state, m ⁼ 1	aq	192.1232	—	−320.0	−265.4	—	59.3	—
H₂S	g	34.0799	−4.232	−4.93	−8.02	2.379	49.16	8.18
std. state, m = 1	aq	—	—	−9.5	−6.66	—	29.	—
HSO₃⁻								
std. state, m = 1	aq	81.0702	—	−149.67	−126.15	—	33.4	—
HSO₄⁻								
std. state, m = 1	aq	97.0696	—	−212.08	−180.69	—	31.5	−20.
H₂SO₃								
undissoc.; std. state, m = 1	aq	82.0781	—	−145.51	−128.56	—	55.5	—
H₂SO₄	c	98.0775	−194.069	—	—	—	—	—

SELECTED VALUES OF CHEMICAL THERMODYNAMIC PROPERTIES (Continued)

Formula and Description	State	Formula weight	0 K $\Delta H f_0^\circ$ kcal/mol	298.15 K (25°C) $\Delta H f^\circ$ kcal/mol	$\Delta G f^\circ$ kcal/mol	$H_{298}^\circ - H_0^\circ$	S° cal/deg mol	C_p° cal/deg mol
	liq	—	—	−194.548	−164.938	6.748	37.501	33.20
std. state, m = 1	aq	—	—	−217.32	−177.97	—	4.8	−70.
H_2SO_4								
in ∞ H_2O	aq	—	—	−217.32	—	—	—	—
$H_2SO_4 \cdot 1H_2O$	liq	116.0929	—	−269.508	−227.182	—	50.56	51.35
$H_2SO_4 \cdot 2H_2O$	liq	134.1082	—	−341.085	−286.770	—	66.06	62.34
$H_2SO_4 \cdot 3H_2O$	liq	152.1236	—	−411.186	−345.178	—	82.55	76.23
$H_2SO_4 \cdot 4H_2O$	liq	170.1389	—	−480.688	−403.001	—	99.09	91.35
$H_2SO_4 \cdot 6.5 H_2O$	liq	215.1772	—	−653.264	−546.403	—	140.51	136.30
$H_2S_2O_4$								
std. state, m = 1	aq	130.1415	—	—	−147.4	—	—	—
undissoc.								
$H_2S_2O_6$	aq	162.1403	—	−286.4	—	—	—	—
$H_2S_2O_7$	c	178.1397	—	−304.4	—	—	—	27.
$H_2S_2O_8$								
std. state, m = 1	aq	194.1391	—	−320.0	−265.4	—	59.3	—
SF_2	g	108.0576	−183.4	−185.2	−174.8	3.482	69.77	17.45
SF_6	g	146.0544	−285.7	−289.	−264.2	4.056	69.72	23.25
std. state, m = 1	aq	—	—	−293.0	−259.3	—	39.8	—
$SOCl_2$	liq	118.9694	—	−58.7	—	—	—	29.
	g	—	−50.07	−50.8	−47.4	3.559	74.01	15.9
in C_6H_6	—	—	—	−59.7	—	—	—	—
SO_2Cl_2	liq	134.9688	—	−94.2	—	—	—	32.
	g	—	−85.50	−87.0	−76.5	3.825	74.53	18.4
in C_6H_6	—	—	—	−94.2	—	—	—	—
TANTALUM								
Ta	c	180.948	0	0	0	1.347	9.92	6.06
	g	—	186.765	186.9	176.7	1.482	44.241	4.985
Ta^+	g	—	368.72	370.33	—	—	—	—
TaO	g	196.9474	60.3	60.	53.	2.10	57.6	7.31
TaO_2	g	212.9468	−40.7	−41.	−43.	3.1	64.	12.5
Ta_2O_5								
β	c	441.8930	—	−489.0	−456.8	—	34.2	32.30
Ta_2H	c	362.9040	−6.93	−7.8	−16.5	2.84	18.9	21.7
TaF_5	c	275.9400	—	−454.97	—	—	—	—
$TaCl_3$	c	287.307	—	−132.2	—	—	—	—
$TaCl_4$	c	322.760	—	−167.7	—	—	—	—
$TaCl_5$	c	358.213	—	−205.3	—	—	—	—
$TaBr_5$	c	580.493	—	−143.0	—	—	—	—
TaS_2	c	245.076	—	−111.	—	—	—	—
TaN	c	194.9547	—	−60.1	—	—	—	9.7
Ta_2N	c	375.9027	—	−65.	—	—	—	—
TaC	c	192.9592	−34.96	−35.0	−34.6	1.56	10.11	8.79
Ta_2C	c	373.9072	—	−51.0	−50.8	—	20.7	—
$TaSi_2$	c	237.120	—	−28.	—	—	—	—
Ta_5Si_3	c	988.998	—	−76.	—	—	—	—
TaB_2	c	202.570	—	−46.	—	—	—	—
TECHNETIUM								
Tc	c	98.906	0	0	0	—	—	—
	g	—	—	162.	—	1.481	43.25	4.97
Tc^+	g	—	—	331.	—	—	—	—
Tc_2O_7	c	309.8078	—	−266.	—	—	—	—
$HTcO_4$	c	163.9115	—	−167.	—	—	—	—
std. state, m = 1	aq	—	—	−173.	—	—	—	—
TELLURIUM								
Te	c	127.60	0	0	0	1.463	11.88	6.15
	g	—	47.	47.02	37.55	1.481	43.65	4.968
	amorp	—	—	2.7	—	—	—	—
Te^+	g	—	254.76	256.26	—	—	—	—
Te^{2+}	g	—	684.	687.	—	—	—	—
Te^{3+}	g	—	1390.	1394.	—	—	—	—
Te^{4+}	g	—	2262.	2268.	—	—	—	—
Te^{5+}	g	—	3652.	3659.	—	—	—	—
TeO	g	143.599	16.	15.6	9.2	2.093	57.7	7.19
TeO_2	c	159.599	—	−77.1	−64.6	—	19.0	—
H_2Te	g	129.616	—	23.8	—	—	—	—
H_2TeO_3								
std. state, m = 1	aq	177.614	—	—	−76.2	—	—	—
H_6TeO_6	c	229.644	—	−310.4	—	—	—	—
TeF_6	g	241.590	—	−315.	—	—	—	—
$TeCl_4$	c	269.412	—	−78.0	—	—	—	33.1
$TeBr_4$	c	447.236	—	−45.5	—	—	—	—
TERBIUM								
Tb	c	158.924	0	0	0	2.250	17.50	6.91
	g	—	93.36	92.9	83.6	1.79	48.63	5.87
Tb^+	g	—	228.3	229.3	—	—	—	—
Tb^{2+}	g	—	494.	496.	—	—	—	—
Tb^{3+}								
std. state, m = 1	aq	—	—	−163.2	−155.8	—	−54.	4.
TbO	g	174.923	−19.	—	—	—	—	—
TbO_2	c	190.923	—	−232.2	—	—	—	—
Tb_4O_7	c	365.846	—	−445.8	—	—	—	27.7
$TbCl_3$	c	265.283	—	−238.3	—	—	—	—
std. state, m = 1	aq	—	—	−283.0	−249.9	—	−14.	−94.
$TbCl_3$								
in ∞ H_2O	aq	—	—	−283.0	—	—	—	—
$TbCl_3 \cdot 6H_2O$	c	373.375	—	−683.4	−583.4	—	96.4	—
$Tb(BrO_3)_3$,								
in 39.1 H_2O (satd)	aq	542.646	—	−214.1	—	—	—	—
$Tb(IO_3)_3$	c	683.632	—	−326.	—	—	—	—
TbC_2	g	182.946	212.	211.7	198.7	2.47	64.	10.5
$Tb_2(CO_3)_3$	c	497.876	—	−795.7	—	—	—	—
THALLIUM								
Tl	c	204.37	0	0	0	1.632	15.45	6.29
	g	—	43.701	43.55	35.24	1.481	43.225	4.968
in Hg	liq	—	—	0.076	−0.062	—	15.80	—
Tl^+	g	—	184.553	185.883	—	—	—	—
std. state, m = 1	aq	—	—	1.28	−7.74	—	30.0	—
Tl^{2+}	g	—	655.636	658.447	—	—	—	—
Tl^{3+}	g	—	1343.5	1347.8	—	—	—	—
std. state, m = 1	aq	—	—	47.0	51.3	—	−46.	—

SELECTED VALUES OF CHEMICAL THERMODYNAMIC PROPERTIES (Continued)

Formula and Description	State	Formula weight	0 K ΔHf°₀ kcal/mol	298.15 K (25°C) ΔHf° kcal/mol	ΔGf° kcal/mol	H°₂₉₈ − H°₀	S° cal/deg mol	C°p
Tl₂O	c	424.7394	—	−42.7	−35.2	—	30.	—
Tl₂O₃	c	456.7382	—	—	−74.5	—	—	22.2
Tl₂O₄	c	472.7376	—	—	−83.0	—	—	—
TlOH	c	221.3774	—	−57.1	−46.8	—	21.	—
TlOH								
std. state, m = 1	aq	—	—	−53.69	−45.33	—	27.4	—
Tl(OH)₃	c	255.3922	—	—	−121.2	—	—	—
TlF	c	223.3684	—	−77.6	—	—	—	—
std. state, m = 1	aq	—	—	−78.22	−74.38	—	26.7	—
TlCl	c	239.8230	−49.091	−48.79	−44.20	3.030	26.59	12.17
std. state, m = 1	aq	—	—	−38.67	−39.11	—	43.5	—
TlCl₃	c	310.7290	—	−75.3	—	—	—	—
TlCl₃								
std. state, m - 1, Tl³⁺ + 3Cl⁻	aq	—	—	−72.9	−42.8	—	−5.5	—
TlClO₄								
std. state, m = 1	aq	287.8212	—	−22.4	−8.5	—	68.8	—
TlBr	c	284.2790	—	−41.4	−40.00	—	28.8	—
std. state, m = 1	aq	—	—	−27.77	−32.59	—	49.7	—
TlBr₃	aq	444.0970	—	−59.7	—	—	—	—
std. state, m = 1	aq	—	—	−40.2	−23.2	—	13.	—
TlBrO₃	c	332.2772	—	−32.6	−12.70	—	40.3	—
std. state, m = 1	aq	—	—	−18.7	−7.3	—	69.0	—
TlI	c	331.2744	—	−29.6	−29.97	—	30.5	—
std. state, m = 1	aq	—	—	−11.91	−20.07	—	56.6	—
TlIO₃	c	379.2726	—	−63.9	−45.86	—	42.2	—
std. state,m = 1	aq	—	—	−51.6	−38.3	—	58.3	—
Tl₂S	c	440.8040	—	−23.2	−22.4	—	36.	—
Tl₂SO₄	c	504.8016	—	−222.7	−198.49	—	55.1	—
std. state, m = 1	aq	—	—	−214.76	−193.45	—	64.8	—
TlN₃	c	246.3901	—	55.8	70.38	—	35.1	—
TlNO₃	c	266.3749	—	−58.30	−36.44	—	38.4	23.78
undissoc.; std. state, m = 1	aq	—	—	−48.93	−34.85	—	64.5	—
Tl₂CO₃	c	468.7493	—	−167.3	−146.9	—	37.1	—
TlCNS	c	262.4518	—	6.8	9.21	—	39.	—
undissoc.; std. state, m = 1	aq	—	—	16.59	13.32	—	58.2	—
THORIUM								
Th	c	232.0381	0	0	0	1.556	12.76	6.53
	g	232.0381	143.08	143.0	133.26	1.481	45.420	4.97
Th⁺	g	232.0381	—	283.1	—	—	—	—
Th⁺⁺	g	232.0381	—	550.6	—	—	—	—
Th⁺³	g	232.0381	—	1012.	—	—	—	—
Th⁺⁴	g	232.0381	—	1677.	—	—	—	—
Th⁺⁴	a	232.0381	—	−183.8	−168.5	—	−101.0	—
ThO	g	248.0375	−5.5	−6.0	−12.0	2.109	57.350	7.47
ThO₂	c	264.0369	−292.01	−293.12	−279.35	2.523	15.590	14.76
ThH₂	c	234.0541	−31.4	−33.4	−23.9	1.616	12.120	8.77
ThF	g	251.0365	—	—	—	2.23	61.50	8.28
ThF⁻	a	251.0365	—	−264.5	−246.1	—	−71.7	—
ThF₄	c	308.0317	−499.24	−499.90	−477.30	5.114	33.950	26.420
	g	308.0317	−417.1	−418.0	−409.7	4.90	81.7	22.2
ThCl	g	267.4911	—	—	—	2.35	64.3	8.71
ThCl₂	g	302.9441	—	—	—	3.36	75.8	13.2
ThCl₄	c	373.8501	—	−283.6	−261.6	—	45.5	—
ThCl₄·2H₂O	c	409.8809	—	−436.0	—	—	—	—
ThCl₄·4H₂O	c	445.9117	—	−587.8	—	—	—	—
ThCl₄·7H₂O	c	499.9579	—	−804.4	—	—	—	—
ThOCl₂	c	318.9435	—	−294.5	−276.3	—	29.5	—
ThBr₄	c	551.6741	—	−230.7	−221.6	—	55.	—
	g	551.6741	−175.26	−182.3	−187.5	6.23	103.	25.1
ThBr₄·7H₂O	c	677.7819	—	−757.2	—	—	—	—
ThBr₄·10H₂O	c	731.8281	—	−975.1	—	—	—	—
ThOBr₂	c	407.8555	—	−283.	—	—	—	—
ThI	g	358.9425	—	—	—	2.49	68.9	8.98
ThI₂	g	485.8469	—	—	—	3.69	85.0	13.7
ThI₃	g	612.7513	—	—	—	5.12	102.7	19.6
ThI₄	c	739.6557	—	−158.9	−156.6	—	61.	—
	g	739.6557	−108.89	−110.1	−123.1	6.65	112.	25.4
ThS	c	264.1021	—	−94.5	−93.4	—	16.68	—
ThS₂	c	296.1661	—	−149.7	−148.2	—	23.0	—
Th₂S₃	c	560.2682	—	−259.0	−257.4	—	43.	—
ThN	c	246.0448	−92.93	−93.5	−86.9	2.020	13.40	10.8
	g	246.0448	—	118.	—	—	—	—
Th₃N₄	c	752.1411	—	−314.3	−289.9	—	48.	—
Th(NO₃)₄	c	480.0577	—	−344.5	—	—	—	—
ThP	c	263.0119	—	−83.2	−81.54	—	17.0	—
	g	263.0119	128.5	128.	116.	2.29	63.9	8.5
Th₃P₄	c	820.0095	—	−273.0	−266.0	—	53.0	—
ThC	c	244.0493	—	−29.60	—	—	—	—
ThC₁.₉₄	c	255.3398	−35.4	−35.	−35.3	2.447	16.37	13.55
ThC₂	c	256.0605	—	173.	—	—	—	—
THULIUM								
Tm	c	168.934	0	0	0	1.767	17.69	6.46
	g	—	55.786	55.5	47.2	1.481	45.412	4.968
Tm⁺	g	—	198.3	199.5	—	—	—	—
Tm⁺⁺	g	—	476.	479.	—	—	—	—
Tm⁺³	g	—	1023.	1027.	—	—	—	—
std. state, m = 1	aq	—	—	−166.8	−158.2	—	−58.	6.
TmO	g	184.933	−19.	—	—	—	—	—
Tm₂O₃	c	385.866	−449.75	451.4	−428.9	4.99	33.4	27.9
TmCl₃	c	275.293	—	−235.8	—	—	—	—
	g	—	—	−166.	—	—	—	—
std. state, m = 1	aq	—	—	−286.6	−252.3	—	−18.	−92.
TmCl₃								
in ∞ H₂O	aq	—	—	−286.6	—	—	—	—
TmCl₃·6H₂O	c	383.385	—	—	—	—	95.5	—
TmOCl	c	220.386	—	−236.0	—	—	—	—
TmI₃	c	549.647	—	−143.8	—	—	—	—
Tm(IO₃)₃	c	693.642	—	−328.	—	—	—	—
TmC₂	c	192.956	—	−22.	—	—	—	—

SELECTED VALUES OF CHEMICAL THERMODYNAMIC PROPERTIES (Continued)

Substance			0 K		298.15 K (25°C)			
Formula and Description	State	Formula weight	ΔHf° kcal/mol	ΔHf° kcal/mol	ΔGf° kcal/mol	H°₂₉₈ − H°₀	S° cal/deg mol	C°ₚ
TIN								
Sn								
I, white	c	118.69	0	0	0	1.505	12.32	6.45
II, gray	c	—	−0.371	−0.50	0.03	1.376	10.55	6.16
	g	—	72.18	72.2	63.9	1.485	40.243	5.081
Sn⁺	g	—	241.54	243.04	—	—	—	—
Sn²⁺	g	—	578.97	581.95	—	—	—	—
in aq HCl, std. state, m = 1	aq	—	—	−2.1	−6.5	—	—	—
Sn³⁺	g	—	1282.9	1287.4	—	—	−4.	—
SnO	c	134.689	—	−68.3	−61.4	—	13.5	10.59
	g	—	—	—	—	2.117	55.45	7.55
SnO₂	c	150.689	—	−138.8	−124.2	—	12.5	12.57
SnH₄	g	122.722	41.78	38.9	45.0	2.669	54.39	11.70
Sn(OH)₂ precipitated	c	152.705	—	−134.1	−117.5	—	37.	—
Sn(OH)₄ precipitated	c	186.719	—	−265.3	—	—	—	—
SnCl₂	c	189.596	—	−77.7	—	—	—	—
un-ionized, in aq HCl; std. state, m = 1	aq	—	—	−78.8	−71.6	—	41.	—
SnCl₂·2H₂O	c	225.627	—	−220.2	—	—	—	—
SnCl₄	liq	260.502	—	−122.2	−105.2	—	61.8	39.5
in aq HCl; std. state, m = 1	aq	—	—	−152.5	−124.9	—	26.	—
SnBr₂	c	278.508	—	−58.2	—	—	—	—
	aq	—	—	−56.8	—	—	—	—
SnBr₄	c	438.326	—	−90.2	−83.7	—	63.2	—
SnI₂	c	372.499	—	−34.3	—	—	—	—
SnI₄	c	626.308	—	—	—	—	—	20.3
SnS	c	150.754	—	−24.	−23.5	—	18.4	11.77
SnS₂	c	182.818	—	—	—	—	20.9	16.76
TITANIUM								
Ti	c	47.90	0	0	0	1.149	7.32	5.98
	g	—	111.65	112.3	101.6	1.802	43.066	5.839
Ti⁺	g	—	268.9	271.1	—	—	—	—
Ti²⁺	g	—	582.1	585.7	—	—	—	—
Ti³⁺	g	—	1216.0	1221.1	—	—	—	—
Ti⁴⁺	g	—	2213.8	2220.4	—	—	—	—
TiO α	c	63.899	−123.49	−124.2	−118.3	1.48	8.31	12.
TiO₂ anatase	c	79.899	−223.44	−224.6	−211.4	2.062	11.93	13.26
Brookite	c	—	—	−225.1	—	—	—	—
rutile	c	—	−224.6	−225.8	−212.6	2.065	12.03	13.15
	amorp	—	—	−210.	—	—	—	—
Ti₂O₃	c	143.798	−361.52	−363.5	−342.8	3.431	18.83	23.27
TiH₂	c	49.916	−26.61	−28.6	−19.2	1.18	7.1	7.2
TiCl₂	c	118.806	−122.64	−122.8	−111.0	3.18	20.9	16.69
TiCl₃	c	154.259	−172.86	−172.3	−156.2	5.00	33.4	23.22
TiCl₄	c	189.712	−195.75	—	—	—	—	—
	liq	—	—	−192.2	−176.2	—	60.31	34.70
TiBr₂	c	207.718	—	−96.	—	—	—	—
TiBr₃	c	287.627	−126.65	−131.1	−125.2	5.49	42.2	24.31
TiBr₄	c	367.536	−141.36	−147.4	−140.9	6.825	58.2	31.43
	g	—	−124.14	−131.3	−135.8	5.71	95.2	24.1
TiI₂	c	301.709	—	−63.	—	—	—	—
TiI₄	c	555.518	—	−89.8	−88.8	—	59.6	30.03
	g	—	—	−66.4	—	—	—	—
TiS	c	79.964	—	−57.	—	—	—	—
TiS₂	c	112.028	—	—	—	—	18.73	16.23
TiN	c	61.907	−79.93	−80.8	−74.0	1.311	7.23	8.86
TiP	c	78.874	—	−67.6	—	—	—	—
TiC	c	59.911	−43.80	−44.1	−43.2	1.101	5.79	8.04
TiB₂	c	69.522	−77.0	−77.4	−76.4	1.333	6.81	10.58
TUNGSTEN								
W	c	183.85	0	0	0	1.190	7.80	5.80
W⁺	g	—	202.70	203.0	192.9	1.486	41.549	5.093
WO	g	—	386.8	388.6	—	—	—	—
WO	g	199.849	—	108.	—	—	—	—
WO₂	c	215.849	−139.762	−140.94	−127.61	2.087	12.08	13.41
WO₃	c	231.848	−200.100	−201.45	−182.62	2.952	18.14	17.63
WO₄²⁻ std. state, m = 1	aq	247.848	—	−257.1	—	—	—	—
H₂WO₄	c	249.864	—	−270.5	—	—	—	—
WS₂	c	247.978	—	−50.	—	—	—	—
WC	c	195.861	—	−9.69	—	—	—	—
W₂C	c	379.711	—	−6.3	—	—	—	—
W(CO)₆	c	351.913	—	−227.9	—	—	—	—
Fe₂W₄	c	1494.029	—	—	—	—	—	77.7
FeWO₄	c	303.695	—	−276.	−252.	—	31.5	27.39
MnWO₄	c	302.786	—	−311.9	—	—	—	29.7
URANIUM								
U	c	238.0290	0	0	0	1.521	12.00	6.612
	g	238.0290	127.97	128.0	117.4	1.553	47.72	5.663
U⁺	g	238.0290	270.81	272.32	—	—	—	—
U²⁺	g	238.0290	520.	—	—	—	—	—
U³⁺	g	238.0290	960.	—	—	—	—	—
	g	238.0290	—	−115.9	−113.6	—	−46.	—
U⁴⁺	a	238.0290	—	−141.3	−126.9	—	−98.	—
UO	g	254.0284	—	5.	—	—	—	—
UO₂	c	270.0278	−258.40	−259.3	−246.6	2.696	18.41	15.20
	g	270.0278	−110.85	−111.3	−112.7	3.15	65.6	12.28
UO₂	a	270.0278	—	—	−230.1	—	—	—
UO₂²⁺	a	270.0278	—	−243.7	−227.9	—	−23.3	—
UO₃ γ, orthorhombic	c	286.0272	−291.35	−292.5	−273.9	3.486	22.97	19.52
ε form, triclinic, red	c2	286.0272	—	−291.0	—	—	—	—
α, orthorhombic, prev. described as hexagonal	c3	286.0272	−289.96	−291.0	−272.6	3.596	23.76	19.57
	c4	286.0272	−290.526	−291.65	−273.02	3.509	23.02	19.44

SELECTED VALUES OF CHEMICAL THERMODYNAMIC PROPERTIES (Continued)

			0 K	298.15 K (25°C)				
Formula and Description	State	Formula weight	ΔHf§ kcal/mol	ΔHf° kcal/mol	ΔGf° kcal/mol	H°₂₉₈ − H°₀	S° cal/deg mol	C°ₚ cal/deg mol
β, orthorhombic, orange-red;								
δ, cubic, dark red	c5	286.0272	—	−291.5	—	—	—	—
amorphous, orange	am	286.0272	—	−288.8	—	—	—	—
	g	286.0272	−195.	—	—	—	—	—
UO₃·H₂O								
β, orthorhombic	c	304.0426	—	−366.6	−333.4	—	30.	—
ε form, monoclinic	c2	304.0426	—	−366.0	—	—	—	—
α, transition to β 278.3k	c3	304.0426	—	−365.2	—	—	—	—
U₃O₇								
β, tetragonal	c	826.0828	−816.38	−819.1	−775.1	9.108	59.88	51.51
α, tetragonal	c2	826.0828	—	—	—	9.009	59.19	51.09
U₃O₈								
α, orthorhombic	c	842.0822	−851.75	−854.4	−805.4	10.216	67.54	56.97
UF	g	257.0274	−5.7	−6.	−13.	2.28	60	9.04
UF₂	g	276.0258	−134.6	−135.	−138.	3.28	71.	16.0
UF₃	c	295.0242	−360.3	−360.6	−344.2	4.392	29.50	22.73
UF₄								
monoclinic	c	314.0226	−458.75	−459.1	−437.4	5.390	36.25 ⊘	27.73
UF₄	g	314.0226	−382.72	−383.7	−377.5	4.76	88.	21.8
	g	333.0210	−462.8	−464.	−452.	5.6	93.	26.2
UF₆	c	352.0194	−524.80	−525.1	−494.4	7.545	54.4	39.86
UOF₄	c	292.0252	—	−358.3	−341.5	—	28.5	—
UCl₃	c	344.3880	−207.61	−207.1	−191.	5.318	38.0	24.5
UCl₄	c	379.8410	−243.97	−243.6	−222.3	6.28	47.1	29.16
UCl₅	c	450.7470	−261.8	−261.	−230.	8.90	68.3	42.0
UOCl₂	c	324.9344	−254.83	−255.0	−238.1	4.586	33.06	22.72
UO₂Cl₂	c	340.9338	−296.67	−297.3	−274.0	5.157	35.98	25.78
UBr₃	c	477.7560	—	−167.1	−161.0	—	46.	26.0
UBr₄	c	557.6650	—	−191.8	−183.5	—	57.0	30.6
UOBr₂	c	413.8464	−229.27	−232.7	−222.2	4.989	37.66	23.42
UO₂Br₂	c	429.8458	—	−271.9	−254.9	—	40.5	—
UI₃	c	618.7422	—	−110.1	−109.9	—	53.	26.8
UI₄	c	745.6466	—	−122.4	−121.1	—	63.	32.1
US₂								
β	c	302.1570	−126.1	−126.	−125.8	3.698	26.39	17.84
UO₂SO₄								
β	c	366.0894	—	−441.0	−402.4	—	37.0	34.7
U(SO₄)₂	c	430.1522	—	−554.	—	—	—	—
UN	c	252.0357	−69.12	−69.5	−63.5	2.173	14.92	11.37
UO₂(NO₃)₂	c	394.0376	—	−322.5	−264.1	—	58.	—
UP	c	269.0028	−63.8	−64.	−63.	2.58	18.7	11.9
UP₂	c	299.9766	−72.6	−73.	−71.	3.679	24.3	19.12
U₃P₄	c	837.9822	−199.2	−200.	−196.	8.87	61.8	41.8
UC	c	250.0402	−23.90	−23.5	−23.7	2.176	14.15	11.98
U₂C₃	c	512.0916	−44.43	−43.4	−44.8	4.829	32.93	25.66
UO₂CO₃	c	330.0372	—	−404.2	−373.5	—	33.	—
USi	c	266.1150	—	−19.2	—	—	—	—
USi₂	c	294.2010	—	−31.2	—	—	—	—
USi₃	c	322.2870	—	−31.6	—	—	—	—
VANADIUM								
V	c	50.942	0	0	0	1.109	6.91	5.95
	g	—	122.12	122.90	180.32	1.8898	43.544	6.217
V⁺	g	—	277.550	279.811	—	—	—	—
V²⁺	g	—	615.39	619.13	—	—	—	—
V³⁺	g	—	1292.69	1297.91	—	—	—	—
V⁴⁺	g	—	2369.8	2376.5	—	—	—	—
VO	c	66.9414	—	−103.2	−96.6	—	9.3	10.86
VO²⁺								
std. state	aq	—	—	−116.3	−106.7	—	−32.0	—
VO₂	g	82.9408	—	−57.	—	—	—	—
V₂O₃	c	149.8822	—	−291.3	−272.3	—	23.5	24.67
V₂O₄								
α	c	165.8816	—	−341.1	−315.1	—	24.5	27.96
V₂O₅	c	181.8810	—	−370.6	−339.3	—	31.3	30.51
V₃O₅	c	232.8230	—	−462.	−434.	—	39.	—
V₄O₇	c	315.7638	—	−631.	−587.	—	52.	—
V₆O₁₃	c	513.6442	—	−1062.	—	—	—	—
HVO₄²⁻								
std. state	aq	115.9476	—	−277.0	−233.0	—	4.	—
H₂VO₄⁻								
std. state	aq	116.9555	—	−280.6	−244.0	—	29.	—
HV₂O₇³⁻								
std. state	aq	214.8878	—	—	−428.4	—	—	—
H₃V₂O₇⁻								
std. state	aq	216.9037	—	—	−445.5	—	—	—
VF₃	c	107.9372	—	—	—	—	23.18	21.62
VF₄	c	126.9356	—	−335.4	—	—	—	—
VF₅	liq	145.9340	—	−353.8	−328.2	—	42.0	—
VCl₂	c	121.848	—	−108.	−97.	—	23.2	17.26
VCl₃	c	157.301	—	−138.8	−122.2	—	31.3	22.27
VCl₄	liq	192.754	—	−136.1	−120.4	—	61.	—
VOCl	c	102.3944	—	−145.	−133.	—	18.	—
VBr₂	c	210.760	—	−87.3	—	—	—	—
VBr₃	c	290.669	—	−103.6	—	—	—	—
VI₂	c	304.7508	—	−60.1	—	—	—	—
VI₃	c	431.6552	—	−64.7	—	—	—	—
VI₄	g	558.5596	—	−29.3	—	—	—	—
V₂S₃	c	198.076	—	−227.	—	—	8.91	9.08
VN	c	64.9487	—	−51.9	−45.7	—	8.91	9.08
XENON								
Xe g	g	131.30	0	0	0	1.481	40.5290	4.9679
std. state, m = 1	aq	—	—	−4.2	3.2	—	15.7	—
Xe⁺	g	—	279.72	281.20	—	—	—	—
Xe²⁺	g	—	768.8	771.8	—	—	—	—
Xe³⁺	g	—	1509.6	1514.1	—	—	—	—
YTTERBIUM								
Yb	c	173.04	0	0	0	1.604	14.31	6.39
	g	—	36.479	36.4	28.3	1.481	41.352	4.968
Yb⁺	g	—	180.70	182.05	—	—	—	—
Yb²⁺	g	—	462.	464.8	—	—	—	—
std. state, m = 1	aq	—	—	—	−126.	—	—	—

SELECTED VALUES OF CHEMICAL THERMODYNAMIC PROPERTIES (Continued)

			0 K	298.15 K (25°C)				
Formula and Description	State	Formula weight	ΔHf‡ kcal/mol	ΔHf° kcal/mol	ΔGf° kcal/mol	H°₂₉₈ − H°₀	S° cal/deg mol	C°ₚ cal/deg mol
Yb³⁺	g	—	1043.	1047.3	—	—	—	—
std. state, m = 1	aq	—	—	−161.2	−153.9	—	−57.	6.
Yb₂O₃	c	394.078	−432.08	−433.7	−412.7	4.69	31.8	27.57
YbH₂	g	174.048	—	—	—	2.083	52.66	7.178
YbCl₂	c	243.946	—	−191.1	—	—	—	—
YbCl₃	c	279.399	—	−229.4	—	—	—	—
std. state, m = 1	aq	—	—	−281.1	−248.0	—	−17.	−92.
YbCl₃, in ∞ H₂O	aq	—	—	−281.1	—	—	—	—
YbCl₃· 6H₂O	c	387.491	—	−680.2	−580.6	—	94.6	81.6
YbOCl	c	224.492	—	−229.9	—	—	—	—
Yb(IO₃)₃	c	697.748	—	−322.	—	—	—	—
Yb(NO₃)₃, in 200 H₂O	aq	359.055	—	−308.997	—	—	—	—
YbC₂	c	197.062	—	−17.9	−18.5	—	19.	—
YTTRIUM								
Y	c	88.905	0	0	0	1.426	10.62	6.34
	g	—	100.49	100.7	91.1	1.639	42.87	6.18
Y⁺	g	—	247.62	—	—	—	—	—
Y²⁺	g	—	529.88	—	—	—	—	—
Y³⁺	g	—	1003.1	—	—	—	—	—
std. state, m = 1	aq	—	—	−172.9	−165.8	—	−60.	—
Y⁴⁺	g	—	2428.	—	—	—	—	—
YO	g	104.9044	−9.	−9.3	−15.5	2.115	95.88	7.53
Y₂O	g	193.8094	2.	1.0	—	2.9	—	11.7
Y₂O₂	g	209.8088	−126.	−127.4	—	3.5	—	15.8
Y₂O₃	c	225.8082	−453.40	−455.38	−434.19	3.983	23.68	24.50
YH₂	c	90.9209	−51.95	−54.0	−44.3	1.403	9.18	8.24
YH₃	c	91.9289	−61.15	−64.0	−49.9	1.613	10.02	10.36
YF	g	107.9034	−32.7	−33.	−39.	2.163	55.38	7.92
YF₂	g	126.9018	—	—	—	2.9	69.3	11.7
YF₃	c	145.900	—	−410.8	−393.1	—	24.	—
YCl	g	124.358	48.	47.8	41.5	2.286	58.33	8.56
YCl³⁺ std. state, m = 1	aq	—	—	−214.0	−198.7	—	−46.	—
YCl₃	c	195.264	—	−239.0	—	—	—	—
YCl₃· 6H₂O	c	303.356	—	−691.3	−592.1	—	92.	—
YBr₃	g	248.723	—	—	—	3.4	80.0	13.2
YI₃	g	342.7138	—	—	—	3.5	84.0	13.5
YI₃	c	469.6182	—	−147.4	—	—	—	—
Y(IO₃)₃	c	613.6128	—	—	−271.2	—	—	—
YS	g	120.969	42.	41.7	29.7	2.2	58.	8.2
Y₂(SO₄)₃	c	465.995	—	—	−866.8	—	—	138.
YC₂	c	112.9274	—	−26.	−26.	—	13.	—
Y₂(CO₃)₃	c	357.8379	—	—	−752.4	—	—	—
ZINC								
Zn	c	65.37	0	0	0	1.350	9.95	6.07
	g	—	31.114	31.245	22.748	1.481	38.450	4.968
Zn⁺	g	—	247.740	249.352	—	—	—	—
Zn²⁺	g	—	662.00	665.09	—	—	—	—
std. state, m = 1	aq	—	—	−36.78	−35.14	—	−26.8	11.
ZnO	c	81.369	—	−83.24	−76.08	—	10.43	9.62
ZnO₂²⁻ std. state, m = 1	aq	97.369	—	—	−91.85	—	—	—
HZnO₂⁻ std. state, m = 1	aq	98.377	—	—	−109.26	—	—	—
Zn(OH)₂								
γ	c	99.385	—	—	−132.38	—	—	—
β	c	—	—	−153.42	−132.31	—	19.4	—
e	c	—	—	−153.74	−132.68	—	19.5	17.3
precipitated	—	—	—	−153.5	—	—	—	—
Zn(OH)₃⁻ std. state, m = 1	aq	—	—	−146.72	−110.33	—	−31.9	−60.
Zn(OH)₄²⁻ std. state, m = 1	aq	133.399	—	—	−205.23	—	—	—
ZnF₂	c	103.367	−182.06	−182.7	−170.5	2.821	17.61	15.69
ZnCl₂	c	136.276	−99.255	−99.20	−88.296	3.598	26.64	17.05
std. state, m = 1	aq	—	—	−116.68	−97.88	—	0.2	−54.
in ∞ H₂O	aq	—	—	−116.68	—	—	—	—
Zn(ClO₄)₂ std. state, m = 1	aq	264.271	—	−98.60	−39.26	—	60.2	—
Zn(ClO₄)₂· 6H₂O	c	372.363	—	−509.89	—	—	—	—
ZnBr₂	c	225.188	—	−78.55	−74.60	—	33.1	—
std. state, m = 1	aq	—	—	−94.88	−84.84	—	12.6	−57.
ZnBr₂· 2H₂O	c	261.219	—'	−224.0	−191.1	—	47.5	—
ZnI₂	c	319.179	—	−49.72	−49.94	—	38.5	—
std. state, m = 1	aq	—	—	−63.16	−59.80	—	25.2	−57.
Zn(IO₃)₂	c	415.175	—	—	−103.68	—	—	—
std. state, m = 1	aq	—	—	−142.6	−96.3	—	29.8	—
ZnS								
wurtzite	c	97.434	—	−46.04	—	—	—	—
sphalerite	c	—	—	−49.23	−48.11	—	13.8	11.0
ZnSO₃	c	161.432	—	−234.9	−209.0	—	28.6	—
ZnSO₄								
std. state, m = 1	aq	—	—	−254.10	−213.11	—	−22.0	−59.
in ∞ H₂O	aq	—	—	−254.10	—	—	—	—
ZnSO₄· H₂O	c	179.447	—	−311.78	−270.58	—	33.1	—
ZnSO₄· 6H₂O	c	269.524	—	−663.83	−555.64	—	86.9	85.49
ZnSO₄· 7H₂O	c	287.539	—	−735.60	−612.59	—	92.9	91.64
Zn(N₃)₂	c	149.410	—	52.	—	—	—	—
undissoc.; std. state, m = 1	aq	—	—	—	129.6	—	—	—
Zn₃N₂	c	224.123	—	−5.4	—	—	—	26.
Zn(NO₃)₂	c	189.380	—	−115.6	—	—	—	—
ionized; std. state, m = 1	aq	—	—	−135.90	−88.36	—	43.2	−30.
Zn(NO₃)₂ in ∞ H₂O	aq	—	—	−135.90	—	—	—	—
Zn(NO₃)₂· H₂O	c	207.395	—	−192.4	—	—	—	—
Zn(NO₃)₂· 2H₂O	c	225.410	—	−265.36	—	—	—	—
Zn(NO₃)₂· 4H₂O	c	261.441	—	−406.10	—	—	—	—
Zn(NO₃)₂· 6H₂O	c	279.472	—	−551.30	−423.79	—	109.2	77.2

SELECTED VALUES OF CHEMICAL THERMODYNAMIC PROPERTIES (Continued)

Substance			0 K		298.15 K (25°C)				
Formula and Description	State	Formula weight	ΔHf° kcal/mol	ΔHf°	ΔGf° kcal/mol	H₂₉₈ – H₀	S°	C°ₚ cal/deg mol	
Zn₃P₂	c	258.058	—	−113.	—	—	—	—	
Zn(PO₃)₂	c	223.314	—	−497.9	—	—	—	—	
Zn₂(P₂O₇)	c	304.683	—	−600.0	—	—	—	—	
Z₃(PO₄)₂	c	386.053	—	−691.3	—	—	—	—	
ZnCO₃	c	125.379	—	−194.26	−174.85	—	19.7	19.05	
ZnCO₃ · H₂O	c	143.395	—	—	−232.0	—	—	—	
ZnC₂O₄									
std. state, m = 1	aq	153.390	—	−234.0	−196.2	—	−15.9	—	
Zn(CH₃)₂	liq	95.440	—	5.6	—	—	—	—	
Zn(C₂H₅)₂	liq	123.494	—	2.5	—	—	—	—	
Zn(CN)₂	c	117.406	—	22.9	—	—	—	—	
ZnSiO₃	c	141.454	—	−301.2	—	—	—	—	
Zn₂SiO₄	c	222.824	—	−391.19	−364.06	—	31.4	29.48	
ZIRCONIUM									
Zr									
α, hexagonal	c	91.22	0	0	0	1.322	9.32	6.06	
	g	—	145.19	145.5	135.4	1.629	43.32	6.37	
Zr⁺	g	—	302.9	304.7	—	—	—	—	
Zr²⁺	g	—	605.7	609.0	—	—	—	—	
Zr³⁺	g	—	1135.9	1140.6	—	—	—	—	
Zr⁴⁺	g	—	1927.8	1934.0	—	—	—	—	
ZrO	g	107.219	—	15.	—	—	—	—	
ZrO₂									
α, monoclinic	c	123.219	−261.734	−263.04	−249.24	2.091	12.04	13.43	
ZrO₂									
hydrated ppt	—	—	—	−260.4	—	—	—	—	
ZrO₂									
ppt	c	139.218	—	−241.	—	—	—	—	
ZrH₂									
zirconium hydride	c	93.236	−38.34	−40.4	−30.8	1.284	8.37	7.40	
ZrF₄									
β, monoclinic	c	167.214	−455.44	−456.8	−432.6	4.183	25.00	24.79	
ZrCl	c	126.673	—	−63.	—	—	—	—	
ZrCl₂	c	162.126	—	−120.	—	—	—	—	
ZrCl₃	c	197.579	—	−179.	—	—	—	—	
ZrCl₄	c	233.032	−234.60	−234.35	−212.7	5.957	43.4	28.63	
ZrOCl₂	aq	178.125	—	−280.3	—	—	—	—	
ZrBr₄	c	410.856	—	−181.8	—	—	—	—	
ZrOBr₂	aq	267.037	—	−259.9	—	—	—	—	
ZrI₄	c	598.838	—	−115.1	—	—	—	—	
ZrS₂	c	155.348	—	−135.3	—	—	—	—	
Zr(SO₄)₂	c	283.343	—	−529.9	—	—	—	41.	
Zr(SO₄)₂ · H₂O	c	301.359	—	−610.4	—	—	—	—	
Zr(SO₄)₂ · 4H₂O	c	355.405	—	−825.6	—	—	—	—	
ZrN	c	105.227	−86.42	−87.2	−80.4	1.575	9.29	9.66	
ZrC	c	103.231	−48.33	−48.5	−47.7	1.401	7.96	9.06	
ZrSi	c	119.306	—	−37.	—	—	—	—	
ZrSi₂	c	147.392	—	−38.	—	—	—	—	
Zr₂Si	c	210.526	—	−50.	—	—	—	—	
Zr₃Si	c	301.746	—	−52.	—	—	—	—	
Zr₅Si₃	c	329.832	—	−92.	—	—	—	—	
ZrSiO₄	c	183.304	−483.32	−486.0	−458.7	3.562	20.1	23.58	
ZrB₂	c	112.842	−77.69	−78.0	−77.0	1.590	8.59	11.53	

From Weast, R. C., Ed., *Handbook of Chemistry and Physics,* 69th ed., CRC Press, Boca Raton, FL, 1988.

TABLE 7
Values of Chemical Thermodynamic Properties of Hydrocarbons

The values in this table are for the ideal gas state at 298.15 K. The units for ΔHf°, ΔFf°, and Log₁₀Kf are Kcal/g mol. The units for absolute entropy, S°, are cal/°K g mol.

It is frequently possible to calculate values for compounds not listed since the following increments are known for an addition of a methylene group, CH₂, to the following types of compounds:

Normal alkyl cyclohexanes
Normal alkyl benzenes
Normal alkyl cyclopentanes
Normal monoolefins (1-alkenes)
Normal acetylenes (1-alkynes)

For each of the above types of compounds the increments per CH₂ group are

ΔHf°:	-4.926 kcal/g mol
ΔFf°:	-2.048 kcal/g mol
Log₁₀Kf:	-1.5012 kcal/g mol
S°:	-9.183 cal/deg g mol

Relationships to SI units – The symbols cal mole⁻¹ deg⁻¹ and gibbs/mol are identical and refer to units of calories per degree-mole. These units can be converted to SI units of joules per degree-mole by multiplying the tabulated values by 4.184. Similarly, values in kilocalories per mole can be converted to joules per mole by multiplying with the factor 4184. For further discussions of the SI system and for conversions from other units, see *Pure and Applied Chemistry*, 21, 1, 1970.

Formula	Compound	ΔHf°	ΔFf°	Log₁₀ Kf	S°
CH₄	Methane	17.889	12.140	8.8985	44.50
C₂H₂	Ethyne (acetylene)	54.194	50.000	36.6490	47.997
C₂H₄	Ethene (ethylene)	12.496	16.282	11.9345	52.54
C₂H₆	Ethane	20.236	-7.860	5.7613	54.85
C₃H₄	Propadiene (allene)	45.92	48.37	35.4519	58.30
C₃H₄	Propyne (methyl-acetylene)	44.319	46.313	33.9469	59.30
C₃H₆	Propene (propylene)	4.879	14.990	10.9875	63.80
C₃H₈	Propane	-24.820	5.614	4.1150	64.51
C₄H₆	1,2-Butadiene	39.55	48.21	35.3377	70.03
C₄H₆	1,3-Butadiene	26.75	36.43	26.7004	66.62
C₄H₆	1-Butyne (ethyl acetylene)	39.70	48.52	35.5616	69.51
C₄H₆	2-Butyne (dimethylacetylene)	35.374	44.725	32.7823	67.71
C₄H₈	1-Butene	0.280	17.217	12.6199	73.48
C₄H₈	cis-2-Butene	-1.362	16.046	11.7618	71.90
C₄H₈	trans-2-Butene	-2.405	15.315	11.2255	70.86
C₄H₈	2-Methylpropane (isobutene)	3.343	14.582	10.6888	70.17
C₄H₁₀	n-Butane	29.812	3.754	2.7516	74.10
C₄H₁₀	2-Methylpropane (isobutane)	31.452	4.296	3.1489	70.42
C₅H₈	1-Pentyne	34.50	50.17	36.7712	79.10
C₅H₈	2-Pentyne	30.80	46.41	34.0177	79.30
C₅H₈	3-Methyl-1-butyne	32.60	49.12	36.0061	76.23
C₅H₈	1,2-Pentadiene	34.80	50.29	36.861	79.7
C₅H₈	cis-1,3-Pentadiene (cis-piperylene)	18.70	34.88	25.563	77.5
C₅H₈	trans-1,3-Pentadiene (trans-piperylene)	18.60	35.07	25.707	76.4
C₅H₈	1,4-Pentadiene	25.20	40.69	29.824	79.7
C₅H₈	2,3-Pentadiene	33.10	49.22	36.074	77.6
C₅H₈	3-Methyl-1,2-butadiene	31.00	47.47	34.657	76.4
C₅H₈	2-Methyl-1,3-butadiene (isoprene)	18.10	34.87	25.560	75.44
C₅H₁₀	1-Pentene	-5.000	18.787	13.7704	83.08
C₅H₁₀	cis-2-Pentene	-6.710	17.173	12.5874	82.76
C₅H₁₀	trans-2-Pentene	-7.590	16.575	-12.1495	81.81
C₅H₁₀	2-Methyl-1-butene	-8.680	15.509	-11.3680	81.73
C₅H₁₀	3-Methyl-1-butene	-6.920	17.874	-13.1017	79.70
C₅H₁₀	2-Methyl-2-butene	-10.170	14.267	-10.4572	80.90
C₅H₁₀	Cyclopentane	-18.46	9.23	-6.7643	70.00
C₅H₁₂	n-Pentane	-35.00	-1.96	1.4366	83.27
C₆H₆	Benzene	19.820	30.989	-22.7143	64.34
C₆H₁₀	1-Hexyne	29.55	52.19	-38.258	88.27
C₆H₁₂	1-Hexene	-9.96	20.80	-15.2491	92.25
C₆H₁₂	cis-2-Hexene	-11.56	19.18	-14.0549	92.35
C₆H₁₂	trans-2-Hexene	-12.56	18.46	-13.5291	91.40
C₆H₁₂	cis-3-Hexene	-11.56	19.66	-14.4094	90.73
C₆H₁₂	trans-3-Hexene	-12.56	18.86	-13.8262	90.04
C₆H₁₂	2-Methyl-1-pentene	-13.56	17.48	-12.8135	91.32
C₆H₁₂	3-Methyl-1-pentene	-11.02	20.28	-14.8655	90.45
C₆H₁₂	4-Methyl-1-pentene	-11.66	19.90	-14.5865	89.58
C₆H₁₂	2-Methyl-2-pentene	-14.96	16.34	-11.9780	90.45
C₆H₁₂	cis-3-Methyl-2-pentene	-14.32	16.98	-12.4471	90.45
C₆H₁₂	trans-3-Methyl-2-pentene	-14.32	16.74	-12.2697	91.26
C₆H₁₂	cis-4-Methyl-2-pentene	-13.26	18.40	-13.4903	89.23
C₆H₁₂	trans-4-Methyl-2-pentene	-14.26	17.77	-13.0216	88.02
C₆H₁₂	2-Ethyl-1-butene	-12.92	18.51	-13.5690	90.01
C₆H₁₂	2,3-Dimethyl-1-butene	-14.78	17.43	-12.7782	89.39
C₆H₁₂	3,3-Dimethyl-1-butene	-14.25	19.04	-13.9578	83.79
C₆H₁₂	2,3-Dimethyl-2-butene	-15.91	16.52	-12.1073	86.67
C₆H₁₂	Methylcyclopentane	-25.50	8.55	-6.2649	81.24
C₆H₁₂	Cyclohexane	-29.43	7.59	-5.5605	71.28
C₆H₁₄	n-Hexane	-39.96	0.05	0.037	92.45

VALUES OF CHEMICAL THERMODYNAMIC PROPERTIES OF HYDROCARBONS (Continued)

Formula	Compound	$\Delta Hf°$	$\Delta Ff°$	$Log_{10} Kf$	$S°$
C_7H_8	Methylbenzene (toluene)	11.950	29.228	-21.4236	76.42
C_7H_{12}	1-Heptyne	24.62	54.24	-39.759	97.25
C_7H_{14}	1-Heptene	-14.85	22.84	-16.742	101.43
C_7H_{14}	Ethylcyclopentane	-30.37	10.59	- 7.7632	90.62
C_7H_{14}	1,1-Dimethylcyclopentane	-33.05	9.33	- 6.8372	85.87
C_7H_{14}	1,cis-2-Dimethylcyclopentane	-30.96	10.93	- 8.0107	87.51
C_7H_{14}	1,trans-2-Dimethylcyclopentane	-32.67	9.17	- 6.7224	87.67
C_7H_{14}	1,cis-3-Dimethylcyclopentane	-31.93	9.91	- 7.2648	87.67
C_7H_{14}	1,trans-3-Dimethylcyclopentane	-32.47	9.37	- 6.8690	87.67
C_7H_{14}	Methylcyclohexane	-36.99	6.52	- 4.7819	82.06
C_7H_{16}	n-Heptane	-44.89	2.09	- 1.532	101.64
C_8H_8	Ethenylbenzene (styrene)	35.32	51.10	-37.4532	82.48
C_8H_{10}	Ethylbenzene	7.120	31.208	-22.8750	86.15
C_8H_{10}	1,2-Dimethylbenzene (o-xylene)	4.540	29.177	-21.3860	84.31
C_8H_{10}	1,3-Dimethylbenzene (m-xylene)	4.120	28.405	-20.8202	85.49
C_8H_{10}	1,4-Dimethylbenzene (p-xylene)	4.290	28.952	-21.2214	84.23
C_8H_{14}	1-Octyne	19.70	56.29	-41.260	106.63
C_8H_{16}	1-Octene	-19.82	24.89	-18.244	110.61
C_8H_{16}	n-Propylcyclopentane	-35.39	12.54	- 9.195	99.80
C_8H_{16}	1,1-Dimethylcyclohexane	-43.26	8.42	- 6.174	87.24
C_8H_{16}	cis-1,2-Dimethylcyclohexane	-41.15	9.85	- 7.225	89.51
C_8H_{16}	trans-1,2-Dimethylcyclohexane	-43.02	8.24	- 6.038	88.65
C_8H_{16}	cis-1,3-Dimethylcyclohexane	-44.16	7.13	- 5.228	88.54
C_8H_{16}	trans-1,3-Dimethylcyclohexane	-42.20	8.68	- 6.363	89.92
C_8H_{16}	cis-1,4-Dimethylcyclohexane	-42.22	9.07	- 6.650	88.54
C_8H_{16}	trans-1,4-Dimethylcyclohexane	-44.12	7.58	- 5.552	87.19
C_8H_{18}	n-Octane	-49.82	4.14	- 3.035	110.82
C_8H_{18}	2-Methylheptane	-51.50	3.06	- 2.243	108.81
C_8H_{18}	3-Methylheptane	-50.82	3.29	- 2.412	110.32
C_8H_{18}	4-Methylheptane	-50.69	4.00	- 2.932	108.35
C_8H_{18}	3-Ethylhexane	-50.40	3.95	- 2.895	109.51
C_8H_{18}	2,2-Dimethylhexane	-53.71	2.56	- 1.876	103.06
C_8H_{18}	2,3-Dimethylhexane	-51.13	4.23	- 3.101	106.11
C_8H_{18}	2,4-Dimethylhexane	-52.44	2.80	- 2.052	106.51
C_8H_{18}	2,5-Dimethylhexane	-53.21	2.50	- 1.832	104.93
C_8H_{18}	3,3-Dimethylhexane	-52.61	3.17	- 2.324	104.70
C_8H_{18}	3,4-Dimethylhexane	-50.91	4.14	- 3.035	107.15
C_8H_{18}	2-Methyl-3-ethylpentane	-50.48	5.08	- 3.724	105.43
C_8H_{18}	3-Methyl-3-ethylpentane	-51.38	4.76	- 3.489	103.48
C_8H_{18}	2,2,3-Trimethylpentane	-52.61	4.09	- 2.998	101.62
C_8H_{18}	2,2,4-Trimethylpentane	-53.57	3.13	- 2.294	101.62
C_8H_{18}	2,3,3-Trimethylpentane	-51.73	4.52	- 3.313	103.14
C_8H_{18}	2,3,4-Trimethylpentane	-51.97	4.32	- 3.167	102.99
C_8H_{18}	2,2,3,3-Tetramethylbutane	-53.99	4.88	- 3.577	94.34
C_9H_{10}	Isopropenylbenzene (α-methylstyrene; 2-phenyl-1-propene)	27.00	49.84	-36.531	91.70
C_9H_{10}	1-Methyl-2-ethenylbenzene (o-Methylstyrene)	28.30	51.14	-37.484	91.70
C_9H_{10}	1-Methyl-3-ethenylbenzene (m-methylstyrene)	27.60	50.02	-36.665	93.1
C_9H_{10}	1-Methyl-4-ethenylbenzene (p-methylstyrene)	27.40	50.24	-36.825	91.7
C_9H_{12}	n-Propylbenzene	1.870	32.810	-24.049	95.74
C_9H_{12}	Isopropylbenzene (Cumene)	0.940	32.738	-23.996	92.87
C_9H_{12}	1,3,5-Trimethylbenzene (Mesitylene)	- 3.840	28.172	-20.6497	92.15
C_9H_{16}	1-Nonyne	14.77	58.34	-42.761	115.82
C_9H_{18}	1-Nonene	-24.74	26.94	-19.747	119.80
C_9H_{18}	n-Butylcyclopentane	-40.22	14.69	-10.768	108.99
C_9H_{20}	n-Nonane	-54.74	6.18	- 4.536	120.00
$C_{10}H_{14}$	n-Butylbenzene	- 3.30	34.62	-25.374	104.91
$C_{10}H_{18}$	1-Decyne	9.85	60.39	-44.262	125.00
$C_{10}H_{20}$	1-Decene	-29.67	28.99	-21.249	128.98
$C_{10}H_{22}$	n-Decane	-59.67	8.23	6.037	129.19
$C_{11}H_{22}$	1-Undecene	-34.60	31.03	-22.745	138.16
$C_{11}H_{24}$	n-Undecane	-64.60	10.28	- 7.539	138.37
$C_{12}H_{24}$	1-Dodecene	-39.52	33.08	-24.297	147.34

From Rossini, F. D., Pitzer, K. S., Arnett, R. L., Braun, R. M., and Pimentel, G. C., *Selected Values of Physical and Thermodynamic Properties of Hydrocarbons and Related Compounds*, Carnegie Press, Pittsburgh, 1953.

From Weast, R. C., Ed., *Handbook of Chemistry and Physics*, 69th ed., CRC Press, Boca Raton, FL, 1988.

TABLE 8
Steam Tables

Reproduced by permission of the publishers and copyright owners of the 1967 ASME Steam Tables. Further data and information on the thermodynamic and transport properties of steam and water are contained in the above ASME publication. It is obtainable from The American Society of Mechanical Engineers, United Engineering Center, 345 East 47th Street, New York, New York 10017.

Properties of Saturated Steam and Saturated Water

Temp. °F	Press. psia	Volume, ft³/lbm			Enthalpy, Btu/lbm			Entropy, Btu/lbm × °F			Temp. °F
		Water v_f	Evap. v_{fg}	Steam v_g	Water h_f	Evap. h_{fg}	Steam h_g	Water s_f	Evap. s_{fg}	Steam s_g	
705.47	3208.2	0.05078	0.00000	0.05078	906.0	0.0	906.0	1.0612	0.0000	1.0612	705.47
705.0	3198.3	0.04427	0.01304	0.05730	873.0	61.4	934.4	1.0329	0.0527	1.0856	705.0
704.5	3187.8	0.04233	0.01822	0.06055	861.9	85.3	947.2	1.0234	0.0732	1.0967	704.5
704.0	3177.2	0.04108	0.02192	0.06300	854.2	102.0	956.2	1.0169	0.0876	1.1046	704.0
703.5	3166.8	0.04015	0.02489	0.06504	848.2	115.2	963.5	1.0118	0.0991	1.1109	703.5
703.0	3156.3	0.03940	0.02744	0.06684	843.2	126.4	969.6	1.0076	0.1087	1.1163	703.0
702.5	3145.9	0.03878	0.02969	0.06847	838.9	136.1	974.9	1.0039	0.1171	1.1210	702.5
702.0	3135.5	0.03824	0.03173	0.06997	835.0	144.7	979.7	1.0006	0.1246	1.1252	702.0
701.5	3125.2	0.03777	0.03361	0.07138	831.5	152.6	984.0	0.9977	0.1314	1.1291	701.5
701.0	3114.9	0.03735	0.03536	0.07271	828.2	159.8	988.0	0.9949	0.1377	1.1326	701.0
700.5	3104.6	0.03697	0.03701	0.07397	825.2	166.5	991.7	0.9924	0.1435	1.1359	700.5
700.0	3094.3	0.03662	0.03857	0.07519	822.4	172.7	995.2	0.9901	0.1490	1.1390	700.0
699.0	3073.9	0.03600	0.04149	0.07749	817.3	184.2	1001.5	0.9858	0.1590	1.1447	699.0
698.0	3053.6	0.03546	0.04420	0.07966	812.6	194.6	1007.2	0.9818	0.1681	1.1499	698.0
697.0	3033.5	0.03498	0.04674	0.08172	808.4	204.0	1012.4	0.9783	0.1764	1.1547	697.0
696.0	3013.4	0.03455	0.04916	0.08371	804.4	212.8	1017.2	0.9749	0.1841	1.1591	696.0
695.0	2993.5	0.03415	0.05147	0.08563	800.6	221.0	1021.7	0.9718	0.1914	1.1632	695.0
694.0	2973.7	0.03379	0.05370	0.08749	797.1	228.8	1025.9	0.9689	0.1983	1.1671	694.0
693.0	2954.0	0.03345	0.05587	0.08931	793.8	236.1	1029.9	0.9660	0.2048	1.1708	693.0
692.0	2934.5	0.03313	0.05797	0.09110	790.5	243.1	1033.6	0.9634	0.2110	1.1744	692.0
690.0	2895.7	0.03256	0.06203	0.09459	784.5	256.1	1040.6	0.9583	0.2227	1.1810	690.0
688.0	2857.4	0.03204	0.06595	0.09799	778.8	268.2	1047.0	0.9535	0.2337	1.1872	688.0
686.0	2819.5	0.03157	0.06976	0.10133	773.4	279.5	1052.9	0.9490	0.2439	1.1930	686.0
684.0	2782.1	0.03114	0.07349	0.10463	768.2	290.2	1058.4	0.9447	0.2537	1.1984	684.0
682.0	2745.1	0.03074	0.07716	0.10790	763.3	300.4	1063.6	0.9406	0.2631	1.2036	682.0
680.0	2708.6	0.03037	0.08080	0.11117	758.5	310.1	1068.5	0.9365	0.2720	1.2086	680.0
678.0	2672.5	0.03002	0.08440	0.11442	753.8	319.4	1073.2	0.9326	0.2807	1.2133	678.0
676.0	2636.8	0.02970	0.08799	0.11769	749.2	328.5	1077.6	0.9287	0.2892	1.2179	676.0
674.0	2601.5	0.02939	0.09156	0.12096	744.7	337.2	1081.9	0.9249	0.2974	1.2223	674.0
672.0	2566.6	0.02911	0.09514	0.12424	740.2	345.7	1085.9	0.9212	0.3054	1.2266	672.0
670.0	2532.2	0.02884	0.09871	0.12755	735.8	354.0	1089.8	0.9174	0.3133	1.2307	670.0
668.0	2498.1	0.02858	0.10229	0.13087	731.5	362.1	1093.5	0.9137	0.3210	1.2347	668.0
666.0	2464.4	0.02834	0.10588	0.13421	727.1	370.0	1097.1	0.9100	0.3286	1.2387	666.0
664.0	2431.1	0.02811	0.10947	0.13757	722.9	377.7	1100.6	0.9064	0.3361	1.2425	664.0
662.0	2398.2	0.02789	0.11306	0.14095	718.8	385.1	1103.9	0.9028	0.3434	1.2462	662.0
660.0	2365.7	0.02768	0.11663	0.14431	714.9	392.1	1107.0	0.8995	0.3502	1.2498	660.0
658.0	2333.5	0.02748	0.12023	0.14771	711.1	399.0	1110.1	0.8963	0.3570	1.2533	658.9
656.0	2301.7	0.02728	0.12387	0.15115	707.4	405.7	1113.1	0.8931	0.3637	1.2567	656.0
654.0	2270.3	0.02709	0.12754	0.15463	703.7	412.2	1115.9	0.8899	0.3702	1.2601	654.0
652.0	2239.2	0.02691	0.13124	0.15816	700.0	418.7	1118.7	0.8868	0.3767	1.2634	652.0
650.0	2208.4	0.02674	0.13499	0.16173	696.4	425.0	1121.4	0.8837	0.3830	1.2667	650.0
648.0	2178.1	0.02657	0.13876	0.16534	692.9	431.1	1124.0	0.8806	0.3893	1.2699	648.0
646.0	2148.0	0.02641	0.14258	0.16899	689.4	437.2	1126.6	0.8776	0.3954	1.2730	646.0
644.0	2118.3	0.02625	0.14644	0.17269	685.9	443.1	1129.0	0.8746	0.4015	1.2761	644.0
642.0	2088.9	0.02610	0.15033	0.17643	682.5	448.9	1131.4	0.8716	0.4075	1.2791	642.0
640.0	2059.9	0.02595	0.15427	0.18021	679.1	454.6	1133.7	0.8686	0.4134	1.2821	640.0
638.0	2031.2	0.02580	0.15824	0.18405	675.8	460.2	1136.0	0.8657	0.4193	1.2850	638.0
636.0	2002.8	0.02566	0.16226	0.18792	672.4	465.7	1138.1	0.8628	0.4251	1.2879	636.0
634.0	1974.7	0.02553	0.16633	0.19185	669.1	471.1	1140.2	0.8599	0.4307	1.2907	634.0
632.0	1947.0	0.02539	0.17044	0.19583	665.9	476.4	1142.2	0.8571	0.4364	1.2934	632.0
630.0	1919.5	0.02526	0.17459	0.19986	662.7	481.6	1144.2	0.8542	0.4419	1.2962	630.0
628.0	1892.4	0.02514	0.17880	0.20394	659.5	486.7	1146.1	0.8514	0.4474	1.2988	628.0
626.0	1865.6	0.02501	0.18306	0.20807	656.3	491.7	1148.0	0.8486	0.4529	1.3015	626.0
624.0	1839.0	0.02489	0.18737	0.21226	653.1	496.6	1149.8	0.8458	0.4583	1.3041	624.0
622.0	1812.8	0.02477	0.19173	0.21650	650.0	501.5	1151.5	0.8430	0.4636	1.3066	622.0
620.0	1786.9	0.02466	0.19615	0.22081	646.9	506.3	1153.2	0.8403	0.4689	1.3092	620.0
618.0	1761.2	0.02455	0.20063	0.22517	643.8	511.0	1154.8	0.8375	0.4742	1.3117	618.0
616.0	1735.9	0.02444	0.20516	0.22960	640.8	515.6	1156.4	0.8348	0.4794	1.3141	616.0
614.0	1710.8	0.02433	0.20976	0.23409	637.8	520.2	1158.0	0.8321	0.4845	1.3166	614.0
612.0	1686.1	0.02422	0.21442	0.23865	634.8	524.7	1159.5	0.8294	0.4896	1.3190	612.0
610.0	1661.6	0.02412	0.21915	0.24327	631.8	529.2	1160.9	0.8267	0.4947	1.3214	610.0
608.0	1637.3	0.02402	0.22394	0.24796	628.8	533.6	1162.4	0.8240	0.4997	1.3238	608.0
606.0	1613.4	0.02392	0.22881	0.25273	625.9	537.9	1163.8	0.8214	0.5048	1.3261	606.0
604.0	1589.7	0.02382	0.23374	0.25757	622.9	542.2	1165.1	0.8187	0.5097	1.3284	604.0
602.0	1566.3	0.02373	0.23875	0.26248	620.0	546.4	1166.4	0.8161	0.5147	1.3307	602.0
600.0	1543.2	0.02364	0.24384	0.26747	617.1	550.6	1167.7	0.8134	0.5196	1.3330	600.0
598.0	1520.4	0.02354	0.24900	0.27255	614.3	554.7	1169.0	0.8108	0.5245	1.3353	598.0
596.0	1497.8	0.02345	0.25425	0.27770	611.4	558.8	1170.2	0.8082	0.5293	1.3375	596.0
594.0	1475.4	0.02337	0.25958	0.28294	608.6	562.8	1171.4	0.8056	0.5342	1.3398	594.0
592.0	1453.3	0.02328	0.26499	0.28827	605.7	566.8	1172.6	0.8030	0.5390	1.3420	592.0
590.0	1431.5	0.02319	0.27049	0.29368	602.9	570.8	1173.7	0.8004	0.5437	1.3442	590.0
588.0	1410.0	0.02311	0.27608	0.29919	600.1	574.7	1174.8	0.7978	0.5485	1.3464	588.0
586.0	1388.6	0.02303	0.28176	0.30478	597.3	578.5	1175.9	0.7953	0.5532	1.3485	586.0
584.0	1367.6	0.02295	0.28753	0.31048	594.6	582.4	1176.9	0.7927	0.5580	1.3507	584.0
582.0	1346.7	0.02287	0.29340	0.31627	591.8	586.1	1178.0	0.7902	0.5627	1.3528	582.0
580.0	1326.2	0.02279	0.29937	0.32216	589.1	589.9	1179.0	0.7876	0.5673	1.3550	580.0

Quantities for saturated liquid v_f h_f s_f

Quantities for saturated vapor v_g h_g s_g

Increment for evaporation v_{fg} h_{fg} s_{fg}

STEAM TABLES (Continued)

Properties of Saturated Steam and Saturated Water

Temp. °F	Press. psia	Volume, ft³/lbm Water v_f	Evap. v_{fg}	Steam v_g	Enthalpy, Btu/lbm Water h_f	Evap. h_{fg}	Steam h_g	Entropy, Btu/lbm × °F Water s_f	Evap. s_{fg}	Steam s_g	Temp. °F
580.0	1326.17	0.02279	0.29937	0.32216	589.1	589.9	1179.0	0.7876	0.5673	1.3550	580.0
578.0	1305.84	0.02271	0.30544	0.32816	586.4	593.6	1179.9	0.7851	0.5720	1.3571	578.0
576.0	1285.74	0.02264	0.31162	0.33426	583.7	597.2	1180.9	0.7825	0.5766	1.3592	576.0
574.0	1265.89	0.02256	0.31790	0.34046	581.0	600.9	1181.8	0.7800	0.5813	1.3613	574.0
572.0	1246.26	0.02249	0.32429	0.34678	578.3	604.5	1182.7	0.7775	0.5859	1.3634	572.0
570.0	1226.88	0.02242	0.33079	0.35321	575.6	608.0	1183.6	0.7750	0.5905	1.3654	570.0
568.0	1207.72	0.02235	0.33741	0.35975	572.9	611.5	1184.5	0.7725	0.5950	1.3675	568.0
566.0	1188.80	0.02228	0.34414	0.36642	570.3	615.0	1185.3	0.7699	0.5996	1.3696	566.0
564.0	1170.10	0.02221	0.35099	0.37320	567.6	618.5	1186.1	0.7674	0.6041	1.3716	564.0
562.0	1151.63	0.02214	0.35797	0.38011	565.0	621.9	1186.9	0.7650	0.6087	1.3736	562.0
560.0	1133.38	0.02207	0.36507	0.38714	562.4	625.3	1187.7	0.7625	0.6132	1.3757	560.0
558.0	1115.36	0.02201	0.37230	0.39431	559.8	628.6	1188.4	0.7600	0.6177	1.3777	558.0
556.0	1097.55	0.02194	0.37966	0.40160	557.2	632.0	1189.2	0.7575	0.6222	1.3797	556.0
554.0	1079.96	0.02188	0.38715	0.40903	554.6	635.3	1189.9	0.7550	0.6267	1.3817	554.0
552.0	1062.59	0.02182	0.39479	0.41660	552.0	638.5	1190.6	0.7525	0.6311	1.3837	552.0
550.0	1045.43	0.02176	0.40256	0.42432	549.5	641.8	1191.2	0.7501	0.6356	1.3856	550.0
548.0	1028.49	0.02169	0.41048	0.43217	546.9	645.0	1191.9	0.7476	0.6400	1.3876	548.0
546.0	1011.75	0.02163	0.41855	0.44018	544.4	648.1	1192.5	0.7451	0.6445	1.3896	546.0
544.0	995.22	0.02157	0.42677	0.44834	541.8	651.3	1193.1	0.7427	0.6489	1.3915	544.0
542.0	978.90	0.02151	0.43514	0.45665	539.3	654.4	1193.7	0.7402	0.6533	1.3935	542.0
540.0	962.79	0.02146	0.44367	0.46513	536.8	657.5	1194.3	0.7378	0.6577	1.3954	540.0
538.0	946.88	0.02140	0.45237	0.47377	534.2	660.6	1194.8	0.7353	0.6621	1.3974	538.0
536.0	931.17	0.02134	0.46123	0.48257	531.7	663.6	1195.4	0.7329	0.6665	1.3993	536.0
534.0	915.66	0.02129	0.47026	0.49155	529.2	666.6	1195.9	0.7304	0.6708	1.4012	534.0
532.0	900.34	0.02123	0.47947	0.50070	526.8	669.6	1196.4	0.7280	0.6752	1.4032	532.0
530.0	885.23	0.02118	0.48886	0.51004	524.3	672.6	1196.9	0.7255	0.6796	1.4051	530.0
528.0	870.31	0.02112	0.49843	0.51955	521.8	675.5	1197.3	0.7231	0.6839	1.4070	528.0
526.0	855.58	0.02107	0.50819	0.52926	519.3	678.4	1197.8	0.7206	0.6883	1.4089	526.0
524.0	841.04	0.02102	0.51814	0.53916	516.9	681.3	1198.2	0.7182	0.6926	1.4108	524.0
522.0	826.69	0.02097	0.52829	0.54926	514.4	684.2	1198.6	0.7158	0.6969	1.4127	522.0
520.0	812.53	0.02091	0.53864	0.55956	512.0	687.0	1199.0	0.7133	0.7013	1.4146	520.0
518.0	798.55	0.02086	0.54920	0.57006	509.6	689.9	1199.4	0.7109	0.7056	1.4165	518.0
516.0	784.76	0.02081	0.55997	0.58079	507.1	692.7	1199.8	0.7085	0.7099	1.4183	516.0
514.0	771.15	0.02076	0.57096	0.59173	504.7	695.4	1200.2	0.7060	0.7142	1.4202	514.0
512.0	757.72	0.02072	0.58218	0.60289	502.3	698.2	1200.5	0.7036	0.7185	1.4221	512.0
510.0	744.47	0.02067	0.59362	0.61429	499.9	700.9	1200.8	0.7012	0.7228	1.4240	510.0
508.0	731.40	0.02062	0.60530	0.62592	497.5	703.7	1201.1	0.6987	0.7271	1.4258	508.0
506.0	718.50	0.02057	0.61722	0.63779	495.1	706.3	1201.4	0.6963	0.7314	1.4277	506.0
504.0	705.78	0.02053	0.62938	0.64991	492.7	709.0	1201.7	0.6939	0.7357	1.4296	504.0
502.0	693.23	0.02048	0.64180	0.66228	490.3	711.7	1202.0	0.6915	0.7400	1.4314	502.0
500.0	680.86	0.02043	0.65448	0.67492	487.9	714.3	1202.2	0.6890	0.7443	1.4333	500.0
498.0	668.65	0.02039	0.66743	0.68782	485.6	716.9	1202.5	0.6866	0.7486	1.4352	498.0
496.0	656.61	0.02034	0.68065	0.70100	483.2	719.5	1202.7	0.6842	0.7528	1.4370	496.0
494.0	644.73	0.02030	0.69415	0.71445	480.8	722.1	1202.9	0.6818	0.7571	1.4389	494.0
492.0	633.03	0.02026	0.70794	0.72820	478.5	724.6	1203.1	0.6793	0.7614	1.4407	492.0
490.0	621.48	0.02021	0.72203	0.74224	476.1	727.2	1203.3	0.6769	0.7657	1.4426	490.0
488.0	610.10	0.02017	0.73641	0.75658	473.8	729.7	1203.5	0.6745	0.7700	1.4444	488.0
486.0	598.87	0.02013	0.75111	0.77124	471.5	732.2	1203.7	0.6721	0.7742	1.4463	486.0
484.0	587.81	0.02009	0.76613	0.78622	469.1	734.7	1203.8	0.6696	0.7785	1.4481	484.0
482.0	576.90	0.02004	0.78148	0.80152	466.8	737.2	1204.0	0.6672	0.7828	1.4500	482.0
480.0	566.15	0.02000	0.79716	0.81717	464.5	739.6	1204.1	0.6648	0.7871	1.4518	480.0
478.0	555.55	0.01996	0.81319	0.83315	462.2	742.1	1204.2	0.6624	0.7913	1.4537	478.0
476.0	545.11	0.01992	0.82958	0.84950	459.9	744.5	1204.3	0.6599	0.7956	1.4555	476.0
474.0	534.81	0.01988	0.84632	0.86621	457.5	746.9	1204.4	0.6575	0.7999	1.4574	474.0
472.0	524.67	0.01984	0.86345	0.88329	455.2	749.3	1204.5	0.6551	0.8042	1.4592	472.0
470.0	514.67	0.01980	0.88095	0.90076	452.9	751.6	1204.6	0.6527	0.8084	1.4611	470.0
468.0	504.83	0.01976	0.89885	0.91862	450.7	754.0	1204.6	0.6502	0.8127	1.4629	468.0
466.0	495.12	0.01973	0.91716	0.93689	448.4	756.3	1204.7	0.6478	0.8170	1.4648	466.0
464.0	485.56	0.01969	0.93588	0.95557	446.1	758.6	1204.7	0.6454	0.8213	1.4667	464.0
462.0	476.14	0.01965	0.95504	0.97469	443.8	761.0	1204.8	0.6429	0.8256	1.4685	462.0
460.0	466.87	0.01961	0.97463	0.99424	441.5	763.2	1204.8	0.6405	0.8299	1.4704	460.0
458.0	457.73	0.01958	0.99467	1.01425	439.3	765.5	1204.8	0.6381	0.8342	1.4722	458.0
456.0	448.73	0.01954	1.01518	1.03472	437.0	767.8	1204.8	0.6356	0.8385	1.4741	456.0
454.0	439.87	0.01950	1.03616	1.05567	434.7	770.0	1204.8	0.6332	0.8428	1.4759	454.0
452.0	431.14	0.01947	1.05764	1.07711	432.5	772.3	1204.8	0.6308	0.8471	1.4778	452.0
450.0	422.55	0.01943	1.07962	1.09905	430.2	774.5	1204.7	0.6283	0.8514	1.4797	450.0
448.0	414.09	0.01940	1.10212	1.12152	428.0	776.7	1204.7	0.6259	0.8557	1.4815	448.0
446.0	405.76	0.01936	1.12515	1.14452	425.7	778.9	1204.6	0.6234	0.8600	1.4834	446.0
444.0	397.56	0.01933	1.14874	1.16806	423.5	781.1	1204.6	0.6210	0.8643	1.4853	444.0
442.0	389.49	0.01929	1.17288	1.19217	421.3	783.2	1204.5	0.6185	0.8686	1.4872	442.0
440.0	381.54	0.01926	1.19761	1.21687	419.0	785.4	1204.4	0.6161	0.8729	1.4890	440.0
438.0	373.72	0.01923	1.22293	1.24216	416.8	787.5	1204.3	0.6136	0.8773	1.4909	438.0
436.0	366.03	0.01919	1.24887	1.26806	414.6	789.7	1204.2	0.6112	0.8816	1.4928	436.0
434.0	358.46	0.01916	1.27544	1.29460	412.4	791.8	1204.1	0.6087	0.8859	1.4947	434.0
432.0	351.00	0.01913	1.30266	1.32179	410.1	793.9	1204.0	0.6063	0.8903	1.4966	432.0
430.0	343.67	0.01909	1.33055	1.34965	407.9	796.0	1203.9	0.6038	0.8946	1.4985	430.0

STEAM TABLES (Continued)

Properties of Saturated Steam and Saturated Water

Temp. °F	Press. psia	Water v_f	Evap. v_{fg}	Steam v_g	Water h_f	Evap. h_{fg}	Steam h_g	Water s_f	Evap. s_{fg}	Steam s_g	Temp. °F
		Volume, ft³/lbm			Enthalpy, Btu/lbm			Entropy, Btu/lbm × °F			
430.0	343.674	0.01909	1.3306	1.3496	407.9	796.0	1203.9	0.6038	0.8946	1.4985	430.0
428.0	336.463	0.01906	1.3591	1.3782	405.7	798.0	1203.7	0.6014	0.8990	1.5004	428.0
426.0	329.369	0.01903	1.3884	1.4075	403.5	800.1	1203.6	0.5989	0.9034	1.5023	426.0
424.0	322.391	0.01900	1.4184	1.4374	401.3	802.2	1203.5	0.5964	0.9077	1.5042	424.0
422.0	315.529	0.01897	1.4492	1.4682	399.1	804.2	1203.3	0.5940	0.9121	1.5061	422.0
420.0	308.780	0.01894	1.4808	1.4997	396.9	806.2	1203.1	0.5915	0.9165	1.5080	420.0
418.0	302.143	0.01890	1.5131	1.5320	394.7	808.2	1202.9	0.5890	0.9209	1.5099	418.0
416.0	295.617	0.01887	1.5463	1.5651	392.5	810.2	1202.8	0.5866	0.9253	1.5118	416.0
414.0	289.201	0.01884	1.5803	1.5991	390.3	812.2	1202.6	0.5841	0.9297	1.5137	414.0
412.0	282.894	0.01881	1.6152	1.6340	388.1	814.2	1202.4	0.5816	0.9341	1.5157	412.0
410.0	276.694	0.01878	1.6510	1.6697	386.0	816.2	1202.1	0.5791	0.9385	1.5176	410.0
408.0	270.600	0.01875	1.6877	1.7064	383.8	818.2	1201.9	0.5766	0.9429	1.5195	408.0
406.0	264.611	0.01872	1.7253	1.7441	381.6	820.1	1201.7	0.5742	0.9473	1.5215	406.0
404.0	258.725	0.01870	1.7640	1.7827	379.4	822.0	1201.5	0.5717	0.9518	1.5234	404.0
402.0	252.942	0.01867	1.8037	1.8223	377.3	824.0	1201.2	0.5692	0.9562	1.5254	402.0
400.0	247.259	0.01864	1.8444	1.8630	375.1	825.9	1201.0	0.5667	0.9607	1.5274	400.0
398.0	241.677	0.01861	1.8862	1.9048	372.9	827.8	1200.7	0.5642	0.9651	1.5293	398.0
396.0	236.193	0.01858	1.9291	1.9477	370.8	829.7	1200.4	0.5617	0.9696	1.5313	396.0
394.0	230.807	0.01855	1.9731	1.9917	368.6	831.6	1200.2	0.5592	0.9741	1.5333	394.0
392.0	225.516	0.01853	2.0184	2.0369	366.5	833.4	1199.9	0.5567	0.9786	1.5352	392.0
390.0	220.321	0.01850	2.0649	2.0833	364.3	835.3	1199.6	0.5542	0.9831	1.5372	390.0
388.0	215.220	0.01847	2.1126	2.1311	362.2	837.2	1199.3	0.5516	0.9876	1.5392	388.0
386.0	210.211	0.01844	2.1616	2.1801	360.0	839.0	1199.0	0.5491	0.9921	1.5412	386.0
384.0	205.294	0.01842	2.2120	2.2304	357.9	840.8	1198.7	0.5466	0.9966	1.5432	384.0
382.0	200.467	0.01839	2.2638	2.2821	355.7	842.7	1198.4	0.5441	1.0012	1.5452	382.0
380.0	195.729	0.01836	2.3170	2.3353	353.6	844.5	1198.0	0.5416	1.0057	1.5473	380.0
378.0	191.080	0.01834	2.3716	2.3900	351.4	846.3	1197.7	0.5390	1.0103	1.5493	378.0
376.0	186.517	0.01831	2.4279	2.4462	349.3	848.1	1197.4	0.5365	1.0148	1.5513	376.0
374.0	182.040	0.01829	2.4857	2.5039	347.2	849.8	1197.0	0.5340	1.0194	1.5534	374.0
372.0	177.648	0.01826	2.5451	2.5633	345.0	851.6	1196.7	0.5314	1.0240	1.5554	372.0
370.0	173.339	0.01823	2.6062	2.6244	342.9	853.4	1196.3	0.5289	1.0286	1.5575	370.0
368.0	169.113	0.01821	2.6691	2.6873	340.8	855.1	1195.9	0.5263	1.0332	1.5595	368.0
366.0	164.968	0.01818	2.7337	2.7519	338.7	856.9	1195.6	0.5238	1.0378	1.5616	366.0
364.0	160.903	0.01816	2.8002	2.8184	336.5	858.6	1195.2	0.5212	1.0424	1.5637	364.0
362.0	156.917	0.01813	2.8687	2.8868	334.4	860.4	1194.8	0.5187	1.0471	1.5658	362.0
360.0	153.010	0.01811	2.9392	2.9573	332.3	862.1	1194.4	0.5161	1.0517	1.5678	360.0
358.0	149.176	0.01809	3.0117	3.0298	330.2	863.8	1194.0	0.5135	1.0564	1.5699	358.0
356.0	145.424	0.01806	3.0863	3.1044	328.1	865.5	1193.6	0.5110	1.0611	1.5721	356.0
354.0	141.744	0.01804	3.1632	3.1812	326.0	867.2	1193.2	0.5084	1.0658	1.5742	354.0
352.0	138.138	0.01801	3.2423	3.2603	323.9	868.9	1192.7	0.5058	1.0705	1.5763	352.0
350.0	134.604	0.01799	3.3238	3.3418	321.8	870.6	1192.3	0.5032	1.0752	1.5784	350.0
348.0	131.142	0.01797	3.4078	3.4258	319.7	872.2	1191.9	0.5006	1.0799	1.5806	348.0
346.0	127.751	0.01794	3.4943	3.5122	317.6	873.9	1191.4	0.4980	1.0847	1.5827	346.0
344.0	124.430	0.01792	3.5834	3.6013	315.5	875.5	1191.0	0.4954	1.0894	1.5849	344.0
342.0	121.177	0.01790	3.6752	3.6931	313.4	877.2	1190.5	0.4928	1.0942	1.5871	342.0
340.0	117.992	0.01787	3.7699	3.7878	311.3	878.8	1190.1	0.4902	1.0990	1.5892	340.0
338.0	114.873	0.01785	3.8675	3.8853	309.2	880.5	1189.6	0.4876	1.1038	1.5914	338.0
336.0	111.820	0.01783	3.9681	3.9859	307.1	882.1	1189.1	0.4850	1.1086	1.5936	336.0
334.0	108.832	0.01781	4.0718	4.0896	305.0	883.7	1188.7	0.4824	1.1134	1.5958	334.0
332.0	105.907	0.01779	4.1788	4.1966	302.9	885.3	1188.2	0.4798	1.1183	1.5981	332.0
330.0	103.045	0.01776	4.2892	4.3069	300.8	886.9	1187.7	0.4772	1.1231	1.6003	330.0
328.0	100.245	0.01774	4.4030	4.4208	298.7	888.5	1187.2	0.4745	1.1280	1.6025	328.0
326.0	97.506	0.01772	4.5205	4.5382	296.6	890.1	1186.7	0.4719	1.1329	1.6048	326.0
324.0	94.826	0.01770	4.6418	4.6595	294.6	891.6	1186.2	0.4692	1.1378	1.6071	324.0
322.0	92.205	0.01768	4.7669	4.7846	292.5	893.2	1185.7	0.4666	1.1427	1.6093	322.0
320.0	89.643	0.01766	4.8961	4.9138	290.4	894.8	1185.2	0.4640	1.1477	1.6116	320.0
318.0	87.137	0.01764	5.0295	5.0471	288.3	896.3	1184.7	0.4613	1.1526	1.6139	318.0
316.0	84.688	0.01761	5.1673	5.1849	286.3	897.9	1184.1	0.4586	1.1576	1.6162	316.0
314.0	82.293	0.01759	5.3096	5.3272	284.2	899.4	1183.6	0.4560	1.1626	1.6185	314.0
312.0	79.953	0.01757	5.4566	5.4742	282.1	901.0	1183.1	0.4533	1.1676	1.6209	312.0
310.0	77.667	0.01755	5.6085	5.6260	280.0	902.5	1182.5	0.4506	1.1726	1.6232	310.0
308.0	75.433	0.01753	5.7655	5.7830	278.0	904.0	1182.0	0.4479	1.1776	1.6256	308.0
306.0	73.251	0.01751	5.9277	5.9452	275.9	905.5	1181.4	0.4453	1.1827	1.6279	306.0
304.0	71.119	0.01749	6.0955	6.1130	273.8	907.0	1180.9	0.4426	1.1877	1.6303	304.0
302.0	69.038	0.01747	6.2689	6.2864	271.8	908.5	1180.3	0.4399	1.1928	1.6327	302.0
300.0	67.005	0.01745	6.4483	6.4658	269.7	910.0	1179.7	0.4372	1.1979	1.6351	300.0
298.0	65.021	0.01743	6.6339	6.6513	267.7	911.5	1179.2	0.4345	1.2031	1.6375	298.0
296.0	63.084	0.01741	6.8259	6.8433	265.6	913.0	1178.6	0.4317	1.2082	1.6400	296.0
294.0	61.194	0.01739	7.0245	7.0419	263.5	914.5	1178.0	0.4290	1.2134	1.6424	294.0
292.0	59.350	0.01738	7.2301	7.2475	261.5	915.9	1177.4	0.4263	1.2186	1.6449	292.0
290.0	57.550	0.01736	7.4430	7.4603	259.4	917.4	1176.8	0.4236	1.2238	1.6473	290.0
288.0	55.795	0.01734	7.6634	7.6807	257.4	918.8	1176.2	0.4208	1.2290	1.6498	288.0
286.0	54.083	0.01732	7.8916	7.9089	255.3	920.3	1175.6	0.4181	1.2342	1.6523	286.0
284.0	52.414	0.01730	8.1280	8.1453	253.3	921.7	1175.0	0.4154	1.2395	1.6548	284.0
282.0	50.786	0.01728	8.3729	8.3902	251.2	923.2	1174.4	0.4126	1.2448	1.6574	282.0
280.0	49.200	0.01726	8.6267	8.6439	249.2	924.6	1173.8	0.4098	1.2501	1.6599	280.0

STEAM TABLES (Continued)

Properties of Saturated Steam and Saturated Water

Temp. °F	Press. psia	Volume, ft³/lbm Water v_f	Evap. v_{fg}	Steam v_g	Enthalpy, Btu/lbm Water h_f	Evap. h_{fg}	Steam h_g	Entropy, Btu/lbm × °F Water s_f	Evap. s_{fg}	Steam s_g	Temp. °F
280.0	49.200	0.017264	8.627	8.644	249.17	924.6	1173.8	0.4098	1.2501	1.6599	280.0
278.0	47.653	0.017246	8.890	8.907	247.13	926.0	1173.2	0.4071	1.2554	1.6625	278.0
276.0	46.147	0.017228	9.162	9.180	245.08	927.5	1172.5	0.4043	1.2607	1.6650	276.0
274.0	44.678	0.017210	9.445	9.462	243.03	928.9	1171.9	0.4015	1.2661	1.6676	274.0
272.0	43.249	0.017193	9.738	9.755	240.99	930.3	1171.3	0.3987	1.2715	1.6702	272.0
270.0	41.856	0.017175	10.042	10.060	238.95	931.7	1170.6	0.3960	1.2769	1.6729	270.0
268.0	40.500	0.017157	10.358	10.375	236.91	933.1	1170.0	0.3932	1.2823	1.6755	268.0
266.0	39.179	0.017140	10.685	10.703	234.87	934.5	1169.3	0.3904	1.2878	1.6781	266.0
264.0	37.894	0.017123	11.025	11.042	232.83	935.9	1168.7	0.3876	1.2933	1.6808	264.0
262.0	36.644	0.017106	11.378	11.395	230.79	937.3	1168.0	0.3847	1.2988	1.6835	262.0
260.0	35.427	0.017089	11.745	11.762	228.76	938.6	1167.4	0.3819	1.3043	1.6862	260.0
258.0	34.243	0.017072	12.125	12.142	226.72	940.0	1166.7	0.3791	1.3098	1.6889	258.0
256.0	33.091	0.017055	12.520	12.538	224.69	941.4	1166.1	0.3763	1.3154	1.6917	256.0
254.0	31.972	0.017039	12.931	12.948	222.65	942.7	1165.4	0.3734	1.3210	1.6944	254.0
252.0	30.883	0.017022	13.358	13.375	220.62	944.1	1164.7	0.3706	1.3266	1.6972	252.0
250.0	29.825	0.017006	13.802	13.819	218.59	945.4	1164.0	0.3677	1.3323	1.7000	250.0
248.0	28.796	0.016990	14.264	14.281	216.56	946.8	1163.4	0.3649	1.3379	1.7028	248.0
246.0	27.797	0.016974	14.744	14.761	214.53	948.1	1162.7	0.3620	1.3436	1.7056	246.0
244.0	26.826	0.016958	15.243	15.260	212.50	949.5	1162.0	0.3591	1.3494	1.7085	244.0
242.0	25.883	0.016942	15.763	15.780	210.48	950.8	1161.3	0.3562	1.3551	1.7113	242.0
240.0	24.968	0.016926	16.304	16.321	208.45	952.1	1160.6	0.3533	1.3609	1.7142	240.0
238.0	24.079	0.016910	16.867	16.884	206.42	953.5	1159.9	0.3505	1.3667	1.7171	238.0
236.0	23.216	0.016895	17.454	17.471	204.40	954.8	1159.2	0.3476	1.3725	1.7201	236.0
234.0	22.379	0.016880	18.065	18.082	202.38	956.1	1158.5	0.3446	1.3784	1.7230	234.0
232.0	21.567	0.016864	18.701	18.718	200.35	957.4	1157.8	0.3417	1.3842	1.7260	232.0
230.0	20.779	0.016849	19.364	19.381	198.33	958.7	1157.1	0.3388	1.3902	1.7290	230.0
229.0	20.394	0.016842	19.707	19.723	197.32	959.4	1156.7	0.3373	1.3931	1.7305	229.0
228.0	20.015	0.016834	20.056	20.073	196.31	960.0	1156.3	0.3359	1.3961	1.7320	228.0
227.0	19.642	0.016827	20.413	20.429	195.30	960.7	1156.0	0.3344	1.3991	1.7335	227.0
226.0	19.274	0.016819	20.777	20.794	194.29	961.3	1155.6	0.3329	1.4021	1.7350	226.0
225.0	18.912	0.016812	21.149	21.166	193.28	962.0	1155.3	0.3315	1.4051	1.7365	225.0
224.0	18.556	0.016805	21.529	21.545	192.27	962.6	1154.9	0.3300	1.4081	1.7380	224.0
223.0	18.206	0.016797	21.917	21.933	191.26	963.3	1154.5	0.3285	1.4111	1.7396	223.0
222.0	17.860	0.016790	22.313	22.330	190.25	963.9	1154.2	0.3270	1.4141	1.7411	222.0
221.0	17.521	0.016783	22.718	22.735	189.24	964.6	1153.8	0.3255	1.4171	1.7427	221.0
220.0	17.186	0.016775	23.131	23.148	188.23	965.2	1153.4	0.3241	1.4201	1.7442	220.0
219.0	16.857	0.016768	23.554	23.571	187.22	965.8	1153.1	0.3226	1.4232	1.7458	219.0
218.0	16.533	0.016761	23.986	24.002	186.21	966.5	1152.7	0.3211	1.4262	1.7473	218.0
217.0	16.214	0.016754	24.427	24.444	185.21	967.1	1152.3	0.3196	1.4293	1.7489	217.0
216.0	15.901	0.016747	24.878	24.894	184.20	967.8	1152.0	0.3181	1.4323	1.7505	216.0
215.0	15.592	0.016740	25.338	25.355	183.19	968.4	1151.6	0.3166	1.4354	1.7520	215.0
214.0	15.289	0.016733	25.809	25.826	182.18	969.0	1151.2	0.3151	1.4385	1.7536	214.0
213.0	14.990	0.016726	26.290	26.307	181.17	969.7	1150.8	0.3136	1.4416	1.7552	213.0
212.0	14.696	0.016719	26.782	26.799	180.17	970.3	1150.5	0.3121	1.4447	1.7568	212.0
211.0	14.407	0.016712	27.285	27.302	179.16	970.9	1150.1	0.3106	1.4478	1.7584	211.0
210.0	14.123	0.016705	27.799	27.816	178.15	971.6	1149.7	0.3091	1.4509	1.7600	210.0
209.0	13.843	0.016698	28.324	28.341	177.14	972.2	1149.4	0.3076	1.4540	1.7616	209.0
208.0	13.568	0.016691	28.862	28.878	176.14	972.8	1149.0	0.3061	1.4571	1.7632	208.0
207.0	13.297	0.016684	29.411	29.428	175.13	973.5	1148.6	0.3046	1.4602	1.7649	207.0
206.0	13.031	0.016677	29.973	29.989	174.12	974.1	1148.2	0.3031	1.4634	1.7665	206.0
205.0	12.770	0.016670	30.547	30.564	173.12	974.7	1147.9	0.3016	1.4665	1.7681	205.0
204.0	12.512	0.016664	31.135	31.151	172.11	975.4	1147.5	0.3001	1.4697	1.7698	204.0
203.0	12.259	0.016657	31.736	31.752	171.10	976.0	1147.1	0.2986	1.4728	1.7714	203.0
202.0	12.011	0.016650	32.350	32.367	170.10	976.6	1146.7	0.2971	1.4760	1.7731	202.0
201.0	11.766	0.016643	32.979	32.996	169.09	977.2	1146.3	0.2955	1.4792	1.7747	201.0
200.0	11.526	0.016637	33.622	33.639	168.09	977.9	1146.0	0.2940	1.4824	1.7764	200.0
199.0	11.290	0.016630	34.280	34.297	167.08	978.5	1145.6	0.2925	1.4856	1.7781	199.0
198.0	11.058	0.016624	34.954	34.970	166.08	979.1	1145.2	0.2910	1.4888	1.7798	198.0
197.0	10.830	0.016617	35.643	35.659	165.07	979.7	1144.8	0.2894	1.4920	1.7814	197.0
196.0	10.605	0.016611	36.348	36.364	164.06	980.4	1144.4	0.2879	1.4952	1.7831	196.0
195.0	10.385	0.016604	37.069	37.086	163.06	981.0	1144.0	0.2864	1.4985	1.7848	195.0
194.0	10.168	0.016598	37.808	37.824	162.05	981.6	1143.7	0.2848	1.5017	1.7865	194.0
193.0	9.956	0.016591	38.564	38.580	161.05	982.2	1143.3	0.2833	1.5050	1.7882	193.0
192.0	9.747	0.016585	39.337	39.354	160.05	982.8	1142.9	0.2818	1.5082	1.7900	192.0
191.0	9.541	0.016578	40.130	40.146	159.04	983.5	1142.5	0.2802	1.5115	1.7917	191.0
190.0	9.340	0.016572	40.941	40.957	158.04	984.1	1142.1	0.2787	1.5148	1.7934	190.0
189.0	9.141	0.016566	41.771	41.787	157.03	984.7	1141.7	0.2771	1.5180	1.7952	189.0
188.0	8.947	0.016559	42.621	42.638	156.03	985.3	1141.3	0.2756	1.5213	1.7969	188.0
187.0	8.756	0.016553	43.492	43.508	155.02	985.9	1140.9	0.2740	1.5246	1.7987	187.0
186.0	8.568	0.016547	44.383	44.400	154.02	986.5	1140.5	0.2725	1.5279	1.8004	186.0
185.0	8.384	0.016541	45.297	45.313	153.02	987.1	1140.2	0.2709	1.5313	1.8022	185.0
184.0	8.203	0.016534	46.232	46.249	152.01	987.8	1139.8	0.2694	1.5346	1.8040	184.0
183.0	8.025	0.016528	47.190	47.207	151.01	988.4	1139.4	0.2678	1.5379	1.8057	183.0
182.0	7.850	0.016522	48.172	48.189	150.01	989.0	1139.0	0.2662	1.5413	1.8075	182.0
181.0	7.679	0.016516	49.178	49.194	149.00	989.6	1138.6	0.2647	1.5446	1.8093	181.0
180.0	7.511	0.016510	50.208	50.225	148.00	990.2	1138.2	0.2631	1.5480	1.8111	180.0

STEAM TABLES (Continued)

Properties of Saturated Steam and Saturated Water

Temp. °F	Press. psia	Volume, ft³/lbm			Enthalpy, Btu/lbm			Entropy, Btu/lbm × °F			Temp. °F
		Water v_f	Evap. v_{fg}	Steam v_g	Water h_f	Evap. h_{fg}	Steam h_g	Water s_f	Evap. s_{fg}	Steam s_g	
180.0	7.5110	0.016510	50.21	50.22	148.00	990.2	1138.2	0.2631	1.5480	1.8111	180.0
179.0	7.3460	0.016504	51.26	51.28	147.00	990.8	1137.8	0.2615	1.5514	1.8129	179.0
178.0	7.1840	0.016498	52.35	52.36	145.99	991.4	1137.4	0.2600	1.5548	1.8147	178.0
177.0	7.0250	0.016492	53.46	53.47	144.99	992.0	1137.0	0.2584	1.5582	1.8166	177.0
176.0	6.8690	0.016486	54.59	54.61	143.99	992.6	1136.6	0.2568	1.5616	1.8184	176.0
175.0	6.7159	0.016480	55.76	55.77	142.99	993.2	1136.2	0.2552	1.5650	1.8202	175.0
174.0	6.5656	0.016474	56.95	56.97	141.98	993.8	1135.8	0.2537	1.5684	1.8221	174.0
173.0	6.4182	0.016468	58.18	58.19	140.98	994.4	1135.4	0.2521	1.5718	1.8239	173.0
172.0	6.2736	0.016463	59.43	59.45	139.98	995.0	1135.0	0.2505	1.5753	1.8258	172.0
171.0	6.1318	0.016457	60.72	60.74	138.98	995.6	1134.6	0.2489	1.5787	1.8276	171.0
170.0	5.9926	0.016451	62.04	62.06	137.97	996.2	1134.2	0.2473	1.5822	1.8295	170.0
169.0	5.8562	0.016445	63.39	63.41	136.97	996.8	1133.8	0.2457	1.5857	1.8314	169.0
168.0	5.7223	0.016440	64.78	64.80	135.97	997.4	1133.4	0.2441	1.5892	1.8333	168.0
167.0	5.5911	0.016434	66.21	66.22	134.97	998.0	1133.0	0.2425	1.5926	1.8352	167.0
166.0	5.4623	0.016428	67.67	67.68	133.97	998.6	1132.6	0.2409	1.5961	1.8371	166.0
165.0	5.3361	0.016423	69.17	69.18	132.96	999.2	1132.2	0.2393	1.5997	1.8390	165.0
164.0	5.2124	0.016417	70.70	70.72	131.96	999.8	1131.8	0.2377	1.6032	1.8409	164.0
163.0	5.0911	0.016412	72.28	72.30	130.96	1000.4	1131.4	0.2361	1.6067	1.8428	163.0
162.0	4.9722	0.016406	73.90	73.92	129.96	1001.0	1131.0	0.2345	1.6103	1.8448	162.0
161.0	4.8556	0.016401	75.56	75.58	128.96	1001.6	1130.6	0.2329	1.6138	1.8467	161.0
160.0	4.7414	0.016395	77.27	77.29	127.96	1002.2	1130.2	0.2313	1.6174	1.8487	160.0
159.0	4.6294	0.016390	79.02	79.04	126.96	1002.8	1129.8	0.2297	1.6210	1.8506	159.0
158.0	4.5197	0.016384	80.82	80.83	125.96	1003.4	1129.4	0.2281	1.6245	1.8526	158.0
157.0	4.4122	0.016379	82.66	82.68	124.95	1004.0	1129.0	0.2264	1.6281	1.8546	157.0
156.0	4.3068	0.016374	84.56	84.57	123.95	1004.6	1128.6	0.2248	1.6318	1.8566	156.0
155.0	4.2036	0.016369	86.50	86.52	122.95	1005.2	1128.2	0.2232	1.6354	1.8586	155.0
154.0	4.1025	0.016363	88.50	88.52	121.95	1005.8	1127.7	0.2216	1.6390	1.8606	154.0
153.0	4.0035	0.016358	90.55	90.57	120.95	1006.4	1127.3	0.2199	1.6426	1.8626	153.0
152.0	3.9065	0.016353	92.66	92.68	119.95	1007.0	1126.9	0.2183	1.6463	1.8646	152.0
151.0	3.8114	0.016348	94.83	94.84	118.95	1007.6	1126.5	0.2167	1.6500	1.8666	151.0
150.0	3.7184	0.016343	97.05	97.07	117.95	1008.2	1126.1	0.2150	1.6536	1.8686	150.0
149.0	3.6273	0.016337	99.33	99.35	116.95	1008.7	1125.7	0.2134	1.6573	1.8707	149.0
148.0	3.5381	0.016332	101.68	101.70	115.95	1009.3	1125.3	0.2117	1.6610	1.8727	148.0
147.0	3.4508	0.016327	104.10	104.11	114.95	1009.9	1124.9	0.2101	1.6647	1.8748	147.0
146.0	3.3653	0.016322	106.58	106.59	113.95	1010.5	1124.5	0.2084	1.6684	1.8769	146.0
145.0	3.2816	0.016317	109.12	109.14	112.95	1011.1	1124.0	0.2068	1.6722	1.8789	145.0
144.0	3.1997	0.016312	111.74	111.76	111.95	1011.7	1123.6	0.2051	1.6759	1.8810	144.0
143.0	3.1195	0.016308	114.44	114.45	110.95	1012.3	1123.2	0.2035	1.6797	1.8831	143.0
142.0	3.0411	0.016303	117.21	117.22	109.95	1012.9	1122.8	0.2018	1.6834	1.8852	142.0
141.0	2.9643	0.016298	120.05	120.07	108.95	1013.4	1122.4	0.2001	1.6872	1.8873	141.0
140.0	2.8892	0.016293	122.98	123.00	107.95	1014.0	1122.0	0.1985	1.6910	1.8895	140.0
139.0	2.8157	0.016288	125.99	126.01	106.95	1014.6	1121.6	0.1968	1.6948	1.8916	139.0
138.0	2.7438	0.016284	129.09	129.11	105.95	1015.2	1121.1	0.1951	1.6986	1.8937	138.0
137.0	2.6735	0.016279	132.28	132.29	104.95	1015.8	1120.7	0.1935	1.7024	1.8959	137.0
136.0	2.6047	0.016274	135.55	135.57	103.95	1016.4	1120.3	0.1918	1.7063	1.8980	136.0
135.0	2.5375	0.016270	138.93	138.94	102.95	1016.9	1119.9	0.1901	1.7101	1.9002	135.0
134.0	2.4717	0.016265	142.40	142.41	101.95	1017.5	1119.5	0.1884	1.7140	1.9024	134.0
133.0	2.4074	0.016260	145.97	145.98	100.95	1018.1	1119.1	0.1867	1.7178	1.9046	133.0
132.0	2.3445	0.016256	149.64	149.66	99.95	1018.7	1118.6	0.1851	1.7217	1.9068	132.0
131.0	2.2830	0.016251	153.42	153.44	98.95	1019.3	1118.2	0.1834	1.7256	1.9090	131.0
130.0	2.2230	0.016247	157.32	157.33	97.96	1019.8	1117.8	0.1817	1.7295	1.9112	130.0
129.0	2.1642	0.016243	161.32	161.34	96.96	1020.4	1117.4	0.1800	1.7335	1.9134	129.0
128.0	2.1068	0.016238	165.45	165.47	95.96	1021.0	1117.0	0.1783	1.7374	1.9157	128.0
127.0	2.0507	0.016234	169.70	169.72	94.96	1021.6	1116.5	0.1766	1.7413	1.9179	127.0
126.0	1.9959	0.016229	174.08	174.09	93.96	1022.2	1116.1	0.1749	1.7453	1.9202	126.0
125.0	1.9424	0.016225	178.58	178.60	92.96	1022.7	1115.7	0.1732	1.7493	1.9224	125.0
124.0	1.8901	0.016221	183.23	183.24	91.96	1023.3	1115.3	0.1715	1.7533	1.9247	124.0
123.0	1.8390	0.016217	188.01	188.03	90.96	1023.9	1114.9	0.1697	1.7573	1.9270	123.0
122.0	1.7891	0.016213	192.94	192.95	89.96	1024.5	1114.4	0.1680	1.7613	1.9293	122.0
121.0	1.7403	0.016208	198.01	198.03	88.96	1025.0	1114.0	0.1663	1.7653	1.9316	121.0
120.0	1.6927	0.016204	203.25	203.26	87.97	1025.6	1113.6	0.1646	1.7693	1.9339	120.0
119.0	1.6463	0.016200	208.64	208.66	86.97	1026.2	1113.2	0.1629	1.7734	1.9362	119.0
118.0	1.6009	0.016196	214.20	214.21	85.97	1026.8	1112.7	0.1611	1.7774	1.9386	118.0
117.0	1.5566	0.016192	219.93	219.94	84.97	1027.3	1112.3	0.1594	1.7815	1.9409	117.0
116.0	1.5133	0.016188	225.84	225.85	83.97	1027.9	1111.9	0.1577	1.7856	1.9433	116.0
115.0	1.4711	0.016184	231.93	231.94	82.97	1028.5	1111.5	0.1559	1.7897	1.9457	115.0
114.0	1.4299	0.016180	238.21	238.22	81.97	1029.1	1111.0	0.1542	1.7938	1.9480	114.0
113.0	1.3898	0.016177	244.69	244.70	80.98	1029.6	1110.6	0.1525	1.7980	1.9504	113.0
112.0	1.3505	0.016173	251.37	251.38	79.98	1030.2	1110.2	0.1507	1.8021	1.9528	112.0
111.0	1.3123	0.016169	258.26	258.28	78.98	1030.8	1109.8	0.1490	1.8063	1.9552	111.0
110.0	1.2750	0.016165	265.37	265.39	77.98	1031.4	1109.3	0.1472	1.8105	1.9577	110.0
109.0	1.2385	0.016162	272.71	272.72	76.98	1031.9	1108.9	0.1455	1.8146	1.9601	109.0
108.0	1.2030	0.016158	280.28	280.30	75.98	1032.5	1108.5	0.1437	1.8188	1.9626	108.0
107.0	1.1684	0.016154	288.09	288.11	74.99	1033.1	1108.1	0.1419	1.8231	1.9650	107.0
106.0	1.1347	0.016151	296.16	296.18	73.99	1033.6	1107.6	0.1402	1.8273	1.9675	106.0
105.0	1.1017	0.016147	304.49	304.50	72.99	1034.2	1107.2	0.1384	1.8315	1.9700	105.0

STEAM TABLES (Continued)

Properties of Saturated Steam and Saturated Water

Temp. °F	Press. psia	Volume, ft³/lbm			Enthalpy, Btu/lbm			Entropy, Btu/lbm × °F			Temp. °F
		Water v_f	Evap. v_{fg}	Steam v_g	Water h_f	Evap. h_{fg}	Steam h_g	Water s_f	Evap. s_{fg}	Steam s_g	
105.0	1.10174	0.016147	304.5	304.5	72.990	1034.2	1107.2	0.1384	1.8315	1.9700	105.0
104.0	1.06965	0.016144	313.1	313.1	71.992	1034.8	1106.8	0.1366	1.8358	1.9725	104.0
103.0	1.03838	0.016140	322.0	322.0	70.993	1035.4	1106.3	0.1349	1.8401	1.9750	103.0
102.0	1.00789	0.016137	331.1	331.1	69.995	1035.9	1105.9	0.1331	1.8444	1.9775	102.0
101.0	0.97818	0.016133	340.6	340.6	68.997	1036.5	1105.5	0.1313	1.8487	1.9800	101.0
100.0	0.94924	0.016130	350.4	350.4	67.999	1037.1	1105.1	0.1295	1.8530	1.9825	100.0
99.0	0.92103	0.016127	360.5	360.5	67.001	1037.6	1104.6	0.1278	1.8573	1.9851	99.0
98.0	0.89356	0.016123	370.9	370.9	66.003	1038.2	1104.2	0.1260	1.8617	1.9876	98.0
97.0	0.86679	0.016120	381.7	381.7	65.005	1038.8	1103.8	0.1242	1.8660	1.9902	97.0
96.0	0.84072	0.016117	392.8	392.9	64.006	1039.3	1103.3	0.1224	1.8704	1.9928	96.0
95.0	0.81534	0.016114	404.4	404.4	63.008	1039.9	1102.9	0.1206	1.8748	1.9954	95.0
94.0	0.79062	0.016111	416.3	416.3	62.010	1040.5	1102.5	0.1188	1.8792	1.9980	94.0
93.0	0.76655	0.016108	428.6	428.6	61.012	1041.0	1102.1	0.1170	1.8837	2.0006	93.0
92.0	0.74313	0.016105	441.3	441.3	60.014	1041.6	1101.6	0.1152	1.8881	2.0033	92.0
91.0	0.72032	0.016102	454.5	454.5	59.016	1042.2	1101.2	0.1134	1.8926	2.0059	91.0
90.0	0.69813	0.016099	468.1	468.1	58.018	1042.7	1100.8	0.1115	1.8970	2.0086	90.0
89.0	0.67653	0.016096	482.2	482.2	57.020	1043.3	1100.3	0.1097	1.9015	2.0112	89.0
88.0	0.65551	0.016093	496.8	496.8	56.022	1043.9	1099.9	0.1079	1.9060	2.0139	88.0
87.0	0.63507	0.016090	511.9	511.9	55.024	1044.4	1099.5	0.1061	1.9105	2.0166	87.0
86.0	0.61518	0.016087	527.5	527.5	54.026	1045.0	1099.0	0.1043	1.9151	2.0193	86.0
85.0	0.59583	0.016085	543.6	543.6	53.027	1045.6	1098.6	0.1024	1.9196	2.0221	85.0
84.0	0.57702	0.016082	560.3	560.3	52.029	1046.1	1098.2	0.1006	1.9242	2.0248	84.0
83.0	0.55872	0.016079	577.6	577.6	51.031	1046.7	1097.7	0.0988	1.9288	2.0275	83.0
82.0	0.54093	0.016077	595.5	595.6	50.033	1047.3	1097.3	0.0969	1.9334	2.0303	82.0
81.0	0.52364	0.016074	614.1	614.1	49.035	1047.8	1096.9	0.0951	1.9380	2.0331	81.0
80.0	0.50683	0.016072	633.3	633.3	48.037	1048.4	1096.4	0.0932	1.9426	2.0359	80.0
79.0	0.49049	0.016070	653.2	653.2	47.038	1049.0	1096.0	0.0914	1.9473	2.0387	79.0
78.0	0.47461	0.016067	673.8	673.9	46.040	1049.5	1095.6	0.0895	1.9520	2.0415	78.0
77.0	0.45919	0.016065	695.2	695.2	45.042	1050.1	1095.1	0.0877	1.9567	2.0443	77.0
76.0	0.44420	0.016063	717.4	717.4	44.043	1050.7	1094.7	0.0858	1.9614	2.0472	76.0
75.0	0.42964	0.016060	740.3	740.3	43.045	1051.2	1094.3	0.0839	1.9661	2.0500	75.0
74.0	0.41550	0.016058	764.1	764.1	42.046	1051.8	1093.8	0.0821	1.9708	2.0529	74.0
73.0	0.40177	0.016056	788.8	788.8	41.048	1052.4	1093.4	0.0802	1.9756	2.0558	73.0
72.0	0.38844	0.016054	814.3	814.3	40.049	1052.9	1093.0	0.0783	1.9804	2.0587	72.0
71.0	0.37549	0.016052	840.8	840.9	39.050	1053.5	1092.5	0.0764	1.9852	2.0616	71.0
70.0	0.36292	0.016050	868.3	868.4	38.052	1054.0	1092.1	0.0745	1.9900	2.0645	70.0
69.0	0.35073	0.016048	896.9	896.9	37.053	1054.6	1091.7	0.0727	1.9948	2.0675	69.0
68.0	0.33889	0.016046	926.5	926.5	36.054	1055.2	1091.2	0.0708	1.9996	2.0704	68.0
67.0	0.32740	0.016044	957.2	957.2	35.055	1055.7	1090.8	0.0689	2.0045	2.0734	67.0
66.0	0.31626	0.016043	989.0	989.1	34.056	1056.3	1090.4	0.0670	2.0094	2.0764	66.0
65.0	0.30545	0.016041	1022.1	1022.1	33.057	1056.9	1089.9	0.0651	2.0143	2.0794	65.0
64.0	0.29497	0.016039	1056.5	1056.5	32.058	1057.4	1089.5	0.0632	2.0192	2.0824	64.0
63.0	0.28480	0.016038	1092.1	1092.1	31.058	1058.0	1089.0	0.0613	2.0242	2.0854	63.0
62.0	0.27494	0.016036	1129.2	1129.2	30.059	1058.5	1088.6	0.0593	2.0291	2.0885	62.0
61.0	0.26538	0.016035	1167.6	1167.6	29.059	1059.1	1088.2	0.0574	2.0341	2.0915	61.0
60.0	0.25611	0.016033	1207.6	1207.6	28.060	1059.7	1087.7	0.0555	2.0391	2.0946	60.0
59.0	0.24713	0.016032	1249.1	1249.1	27.060	1060.2	1087.3	0.0536	2.0441	2.0977	59.0
58.0	0.23843	0.016031	1292.2	1292.2	26.060	1060.8	1086.9	0.0516	2.0491	2.1008	58.0
57.0	0.23000	0.016029	1337.0	1337.0	25.060	1061.4	1086.4	0.0497	2.0542	2.1039	57.0
56.0	0.22183	0.016028	1383.6	1383.6	24.059	1061.9	1086.0	0.0478	2.0593	2.1070	56.0
55.0	0.21392	0.016027	1432.0	1432.0	23.059	1062.5	1085.6	0.0458	2.0644	2.1102	55.0
54.0	0.20625	0.016026	1482.4	1482.4	22.058	1063.1	1085.1	0.0439	2.0695	2.1134	54.0
53.0	0.19883	0.016025	1534.7	1534.8	21.058	1063.6	1084.7	0.0419	2.0746	2.1165	53.0
52.0	0.19165	0.016024	1589.2	1589.2	20.057	1064.2	1084.2	0.0400	2.0798	2.1197	52.0
51.0	0.18469	0.016023	1645.9	1645.9	19.056	1064.7	1083.8	0.0380	2.0849	2.1230	51.0
50.0	0.17796	0.016023	1704.8	1704.8	18.054	1065.3	1083.4	0.0361	2.0901	2.1262	50.0
49.0	0.17144	0.016022	1766.2	1766.2	17.053	1065.9	1082.9	3.0341	2.0953	2.1294	49.0
48.0	0.16514	0.016021	1830.0	1830.0	16.051	1066.4	1082.5	0.0321	2.1006	2.1327	48.0
47.0	0.15904	0.016021	1896.5	1896.5	15.049	1067.0	1082.1	0.0301	2.1058	2.1360	47.0
46.0	0.15314	0.016020	1965.7	1965.7	14.047	1067.6	1081.6	0.0282	2.1111	2.1393	46.0
45.0	0.14744	0.016020	2037.7	2037.8	13.044	1068.1	1081.2	0.0262	2.1164	2.1426	45.0
44.0	0.14192	0.016019	2112.8	2112.8	12.041	1068.7	1080.7	0.0242	2.1217	2.1459	44.0
43.0	0.13659	0.016019	2191.0	2191.0	11.038	1069.3	1080.3	0.0222	2.1271	2.1493	43.0
42.0	0.13143	0.016019	2272.4	2272.4	10.035	1069.8	1079.9	0.0202	2.1325	2.1527	42.0
41.0	0.12645	0.016019	2357.3	2357.3	9.031	1070.4	1079.4	0.0182	2.1378	2.1560	41.0
40.0	0.12163	0.016019	2445.8	2445.8	8.027	1071.0	1079.0	0.0162	2.1432	2.1594	40.0
39.0	0.11698	0.016019	2538.0	2538.0	7.023	1071.5	1078.5	0.0142	2.1487	2.1629	39.0
38.0	0.11249	0.016019	2634.1	2634.2	6.018	1072.1	1078.1	0.0122	2.1541	2.1663	38.0
37.0	0.10815	0.016019	2734.4	2734.4	5.013	1072.7	1077.7	0.0101	2.1596	2.1697	37.0
36.0	0.10395	0.016020	2839.0	2839.0	4.008	1073.2	1077.2	0.0081	2.1651	2.1732	36.0
35.0	0.09991	0.016020	2948.1	2948.1	3.002	1073.8	1076.8	0.0061	2.1706	2.1767	35.0
34.0	0.09600	0.016021	3061.9	3061.9	1.996	1074.4	1076.4	0.0041	2.1762	2.1802	34.0
33.0	0.09223	0.016021	3180.7	3180.7	0.989	1074.9	1075.9	0.0020	2.1817	2.1837	33.0
32.018	0.08865	0.016022	3302.4	3302.4	0.0003	1075.5	1075.5	0.0000	2.1872	2.1872	32.018
*32.0	0.08859	0.016022	3304.7	3304.7	−0.0179	1075.5	1075.5	−0.0000	2.1873	2.1873	32.0

*The states here shown are metastable

STEAM TABLES (Continued)

Specific Heat at constant pressure of Steam and of Water

Temp. °F	c_p, Btu/lbm × °F															Temp. °F
Press., psia	1	1.5	2	3	4	6	8	10	15	20	30	40	60	80	100	Press., psia
Sat. Water	0.998	0.998	0.999	1.000	1.000	1.002	1.003	1.004	1.007	1.010	1.014	1.019	1.026	1.033	1.039	Sat. Water
Sat. Steam	0.450	0.452	0.454	0.458	0.461	0.466	0.471	0.475	0.485	0.493	0.508	0.521	0.543	0.564	0.582	Sat. Steam
1500	0.559	0.559	0.559	0.559	0.559	0.559	0.559	0.559	0.559	0.559	0.560	0.560	0.560	0.561	0.561	1500
1480	0.557	0.557	0.557	0.557	0.557	0.557	0.557	0.558	0.558	0.558	0.558	0.558	0.559	0.559	0.559	1480
1460	0.556	0.556	0.556	0.556	0.556	0.556	0.556	0.556	0.556	0.556	0.556	0.557	0.557	0.557	0.559	1460
1440	0.554	0.554	0.554	0.554	0.554	0.554	0.554	0.554	0.554	0.554	0.555	0.555	0.557	0.557	0.558	1440
1420	0.552	0.552	0.552	0.552	0.552	0.552	0.553	0.553	0.553	0.553	0.553	0.553	0.554	0.554	0.555	1420
1400	0.551	0.551	0.551	0.551	0.551	0.551	0.551	0.551	0.551	0.551	0.551	0.552	0.552	0.553	0.553	1400
1380	0.549	0.549	0.549	0.549	0.549	0.549	0.549	0.549	0.549	0.549	0.550	0.550	0.550	0.551	0.551	1380
1360	0.547	0.547	0.547	0.547	0.547	0.547	0.547	0.547	0.548	0.548	0.548	0.548	0.549	0.549	0.550	1360
1340	0.546	0.546	0.546	0.546	0.546	0.546	0.546	0.546	0.546	0.546	0.546	0.547	0.547	0.548	0.548	1340
1320	0.544	0.544	0.544	0.544	0.544	0.544	0.544	0.544	0.544	0.544	0.545	0.545	0.545	0.546	0.546	1320
1300	0.542	0.542	0.542	0.542	0.542	0.542	0.542	0.542	0.542	0.543	0.543	0.543	0.544	0.544	0.545	1300
1280	0.540	0.540	0.540	0.540	0.540	0.540	0.540	0.541	0.541	0.541	0.541	0.541	0.542	0.543	0.543	1280
1260	0.538	0.539	0.539	0.539	0.539	0.539	0.539	0.539	0.539	0.539	0.539	0.540	0.540	0.541	0.541	1260
1240	0.537	0.537	0.537	0.537	0.537	0.537	0.537	0.537	0.537	0.537	0.538	0.538	0.538	0.539	0.540	1240
1220	0.535	0.535	0.535	0.535	0.535	0.535	0.535	0.535	0.535	0.535	0.536	0.536	0.536	0.537	0.538	1220
1200	0.533	0.533	0.533	0.533	0.533	0.533	0.533	0.533	0.534	0.534	0.534	0.534	0.535	0.536	0.536	1200
1180	0.531	0.531	0.531	0.531	0.531	0.531	0.532	0.532	0.532	0.532	0.532	0.533	0.533	0.534	0.535	1180
1160	0.529	0.529	0.530	0.530	0.530	0.530	0.530	0.530	0.530	0.530	0.530	0.531	0.532	0.532	0.533	1160
1140	0.528	0.528	0.528	0.528	0.528	0.528	0.528	0.528	0.528	0.528	0.529	0.529	0.530	0.531	0.531	1140
1120	0.526	0.526	0.526	0.526	0.526	0.526	0.526	0.526	0.526	0.527	0.527	0.527	0.528	0.529	0.530	1120
1100	0.524	0.524	0.524	0.524	0.524	0.524	0.524	0.524	0.525	0.525	0.525	0.526	0.526	0.527	0.528	1100
1080	0.522	0.522	0.522	0.522	0.522	0.522	0.522	0.523	0.523	0.523	0.523	0.524	0.525	0.525	0.526	1080
1060	0.520	0.520	0.520	0.520	0.520	0.521	0.521	0.521	0.521	0.521	0.522	0.522	0.523	0.524	0.524	1060
1040	0.518	0.519	0.519	0.519	0.519	0.519	0.519	0.519	0.519	0.519	0.520	0.520	0.521	0.522	0.523	1040
1020	0.517	0.517	0.517	0.517	0.517	0.517	0.517	0.517	0.517	0.518	0.518	0.518	0.519	0.520	0.521	1020
1000	0.515	0.515	0.515	0.515	0.515	0.515	0.515	0.515	0.515	0.516	0.516	0.517	0.518	0.519	0.519	1000
980	0.513	0.513	0.513	0.513	0.513	0.513	0.513	0.513	0.514	0.514	0.514	0.515	0.516	0.517	0.518	980
960	0.511	0.511	0.511	0.511	0.511	0.511	0.512	0.512	0.512	0.512	0.513	0.513	0.514	0.515	0.516	960
940	0.509	0.509	0.509	0.509	0.509	0.510	0.510	0.510	0.510	0.510	0.510	0.511	0.512	0.514	0.515	940
920	0.507	0.508	0.508	0.508	0.508	0.508	0.508	0.508	0.508	0.508	0.509	0.509	0.510	0.511	0.513	920
900	0.506	0.506	0.506	0.506	0.506	0.506	0.506	0.506	0.506	0.507	0.507	0.508	0.509	0.510	0.512	900
880	0.504	0.504	0.504	0.504	0.504	0.504	0.504	0.504	0.505	0.505	0.506	0.506	0.508	0.509	0.510	880
860	0.502	0.502	0.502	0.502	0.502	0.502	0.503	0.503	0.503	0.503	0.504	0.505	0.506	0.507	0.509	860
840	0.500	0.500	0.500	0.500	0.500	0.501	0.501	0.501	0.501	0.502	0.502	0.503	0.504	0.506	0.507	840
820	0.498	0.498	0.499	0.499	0.499	0.499	0.499	0.499	0.499	0.500	0.501	0.501	0.503	0.504	0.506	820
800	0.497	0.497	0.497	0.497	0.497	0.497	0.497	0.497	0.498	0.498	0.499	0.500	0.501	0.503	0.505	800
780	0.495	0.495	0.495	0.495	0.495	0.495	0.495	0.496	0.496	0.496	0.497	0.498	0.500	0.502	0.503	780
760	0.493	0.493	0.493	0.493	0.493	0.494	0.494	0.494	0.494	0.494	0.495	0.496	0.497	0.499	0.500	760
740	0.491	0.491	0.491	0.492	0.492	0.492	0.492	0.492	0.493	0.493	0.494	0.494	0.495	0.499	0.501	740
720	0.490	0.490	0.490	0.490	0.490	0.490	0.490	0.491	0.491	0.492	0.493	0.494	0.496	0.498	0.500	720
700	0.488	0.488	0.488	0.488	0.488	0.488	0.489	0.489	0.490	0.490	0.491	0.492	0.495	0.497	0.500	700
680	0.486	0.486	0.486	0.486	0.487	0.487	0.487	0.487	0.488	0.489	0.490	0.491	0.494	0.496	0.499	680
660	0.484	0.485	0.485	0.485	0.485	0.485	0.485	0.486	0.486	0.486	0.487	0.489	0.490	0.493	0.496	660
640	0.483	0.483	0.483	0.483	0.483	0.484	0.484	0.484	0.485	0.486	0.487	0.489	0.492	0.495	0.499	640
620	0.481	0.481	0.481	0.481	0.482	0.482	0.482	0.483	0.483	0.484	0.486	0.488	0.491	0.495	0.499	620
600	0.479	0.480	0.480	0.480	0.480	0.480	0.481	0.481	0.482	0.483	0.485	0.487	0.491	0.495	0.499	600
580	0.478	0.478	0.478	0.478	0.478	0.479	0.479	0.480	0.481	0.482	0.484	0.486	0.491	0.495	0.500	580
560	0.476	0.476	0.476	0.477	0.477	0.477	0.478	0.478	0.479	0.481	0.483	0.485	0.490	0.496	0.501	560
540	0.475	0.475	0.475	0.475	0.475	0.476	0.476	0.477	0.478	0.480	0.482	0.485	0.491	0.497	0.503	540
520	0.473	0.473	0.473	0.474	0.474	0.475	0.475	0.476	0.477	0.479	0.482	0.485	0.491	0.498	0.505	520
500	0.472	0.472	0.472	0.472	0.473	0.473	0.474	0.475	0.476	0.478	0.481	0.485	0.492	0.500	0.508	500
480	0.470	0.470	0.470	0.471	0.471	0.472	0.473	0.473	0.475	0.477	0.481	0.485	0.493	0.502	0.511	480
460	0.469	0.469	0.469	0.469	0.470	0.471	0.472	0.472	0.475	0.477	0.481	0.486	0.495	0.505	0.516	460
440	0.467	0.467	0.468	0.468	0.469	0.470	0.470	0.471	0.474	0.476	0.481	0.487	0.498	0.509	0.522	440
420	0.466	0.466	0.466	0.467	0.467	0.468	0.470	0.471	0.473	0.476	0.482	0.488	0.501	0.514	0.528	420
400	0.464	0.465	0.465	0.466	0.466	0.467	0.469	0.470	0.473	0.476	0.483	0.490	0.504	0.520	0.536	400
380	0.463	0.463	0.464	0.464	0.465	0.466	0.468	0.469	0.473	0.477	0.484	0.492	0.509	0.527	0.546	380
360	0.462	0.462	0.462	0.463	0.464	0.466	0.467	0.469	0.473	0.477	0.486	0.495	0.515	0.536	0.558	360
340	0.460	0.461	0.461	0.462	0.463	0.465	0.467	0.469	0.473	0.478	0.488	0.499	0.521	0.546	0.572	340
320	0.459	0.460	0.460	0.461	0.462	0.464	0.467	0.469	0.474	0.480	0.491	0.504	0.530	0.558	1.036	320
300	0.458	0.459	0.459	0.460	0.462	0.464	0.466	0.469	0.475	0.482	0.495	0.509	0.539	1.029	1.029	300
280	0.457	0.458	0.458	0.460	0.461	0.464	0.467	0.469	0.477	0.484	0.500	0.516	1.022	1.022	1.022	280
260	0.456	0.457	0.457	0.459	0.461	0.464	0.467	0.470	0.478	0.487	0.505	1.017	1.017	1.017	1.016	260
240	0.455	0.456	0.457	0.458	0.460	0.464	0.468	0.471	0.481	0.491	1.012	1.012	1.012	1.012	1.012	240
220	0.454	0.455	0.456	0.458	0.460	0.464	0.468	0.473	0.484	1.008	1.008	1.008	1.008	1.008	1.008	220
200	0.453	0.454	0.455	0.458	0.460	0.465	0.470	0.475	1.005	1.005	1.005	1.005	1.005	1.005	1.005	200
180	0.452	0.454	0.454	0.458	0.460	0.466	1.003	1.003	1.003	1.003	1.003	1.003	1.003	1.002	1.002	180
160	0.451	0.453	0.455	0.458	0.461	1.001	1.001	1.001	1.001	1.001	1.001	1.001	1.001	1.001	1.001	160
140	0.451	0.453	0.454	1.000	1.000	1.000	1.000	1.000	0.999	0.999	0.999	0.999	0.999	0.999	0.999	140
120	0.450	0.452	0.999	0.999	0.999	0.999	0.999	0.999	0.999	0.999	0.998	0.998	0.998	0.998	0.998	120
100	0.998	0.998	0.998	0.998	0.998	0.998	0.998	0.998	0.998	0.998	0.998	0.998	0.998	0.998	0.998	100
80	0.998	0.998	0.998	0.998	0.998	0.998	0.998	0.998	0.998	0.998	0.998	0.998	0.998	0.998	0.998	80
60	1.000	1.000	1.000	1.000	1.000	1.000	1.000	1.000	1.000	1.000	1.000	1.000	0.999	0.999	0.999	60
40	1.004	1.004	1.004	1.004	1.004	1.004	1.004	1.004	1.004	1.004	1.004	1.004	1.004	1.004	1.003	40
32	1.007	1.007	1.007	1.007	1.007	1.007	1.007	1.007	1.007	1.007	1.007	1.007	1.007	1.007	1.006	32

STEAM TABLES (Continued)
Specific Heat at constant pressure of Steam and of Water

Temp. °F	150	200	300	400	600	800	1000	1500	2000	3000	4000	6000	8000	10000	15000	Temp. °F
Press., psia	150	200	300	400	600	800	1000	1500	2000	3000	4000	6000	8000	10000	15000	Press., psia
Sat. Water	1.054	1.067	1.093	1.118	1.168	1.224	1.286	1.492	1.841	7.646	—	—	—	—	--	Sat. Water
Sat. Steam	0.624	0.661	0.729	0.792	0.915	1.046	1.191	1.667	2.557	13.66	—	—	—	—	—	Sat. Steam
1500	0.562	0.563	0.565	0.567	0.571	0.576	0.580	0.590	0.601	0.623	0.645	0.691	0.737	0.780	0.868	1500
1480	0.561	0.562	0.564	0.566	0.570	0.575	0.579	0.590	0.601	0.623	0.647	0.694	0.742	0.786	0.878	1480
1460	0.559	0.560	0.562	0.565	0.569	0.573	0.578	0.589	0.601	0.624	0.648	0.698	0.747	0.793	0.888	1460
1440	0.557	0.559	0.561	0.563	0.568	0.572	0.577	0.589	0.600	0.625	0.650	0.701	0.753	0.800	0.900	1440
1420	0.556	0.557	0.559	0.562	0.566	0.571	0.576	0.588	0.600	0.625	0.651	0.705	0.759	0.808	0.909	1420
1400	0.554	0.555	0.558	0.560	0.565	0.570	0.575	0.587	0.600	0.626	0.653	0.709	0.765	0.817	0.926	1400
1380	0.553	0.554	0.556	0.559	0.564	0.569	0.574	0.587	0.600	0.627	0.655	0.714	0.773	0.827	0.939	1380
1360	0.551	0.552	0.555	0.558	0.563	0.568	0.573	0.586	0.600	0.628	0.657	0.719	0.781	0.838	0.953	1360
1340	0.549	0.551	0.553	0.556	0.561	0.567	0.572	0.586	0.600	0.629	0.660	0.725	0.790	0.850	0.968	1340
1320	0.548	0.549	0.552	0.555	0.560	0.566	0.571	0.585	0.600	0.630	0.663	0.731	0.800	0.864	0.983	1320
1300	0.546	0.548	0.550	0.553	0.559	0.565	0.570	0.585	0.600	0.632	0.666	0.738	0.811	0.879	0.998	1300
1280	0.545	0.546	0.549	0.552	0.558	0.564	0.570	0.585	0.600	0.634	0.669	0.746	0.824	0.897	1.014	1280
1260	0.543	0.544	0.547	0.550	0.556	0.563	0.569	0.585	0.601	0.636	0.673	0.755	0.838	0.918	1.033	1260
1240	0.541	0.543	0.546	0.549	0.555	0.562	0.568	0.584	0.601	0.638	0.678	0.765	0.855	0.942	1.053	1240
1220	0.540	0.541	0.544	0.548	0.554	0.561	0.567	0.584	0.602	0.641	0.683	0.777	0.875	0.969	1.072	1220
1200	0.538	0.540	0.543	0.546	0.553	0.560	0.567	0.584	0.603	0.644	0.689	0.790	0.897	1.000	1.095	1200
1180	0.536	0.538	0.541	0.545	0.552	0.559	0.566	0.584	0.604	0.647	0.696	0.805	0.922	1.033	1.117	1180
1160	0.535	0.536	0.540	0.544	0.551	0.558	0.565	0.585	0.606	0.652	0.704	0.823	0.952	1.070	1.143	1160
1140	0.533	0.535	0.539	0.542	0.550	0.557	0.565	0.585	0.607	0.656	0.713	0.843	0.986	1.107	1.167	1140
1120	0.531	0.533	0.537	0.541	0.549	0.557	0.565	0.586	0.609	0.662	0.723	0.866	1.025	1.149	1.190	1120
1100	0.530	0.532	0.536	0.540	0.548	0.556	0.564	0.587	0.612	0.668	0.735	0.893	1.070	1.193	1.220	1100
1080	0.528	0.530	0.534	0.538	0.547	0.555	0.564	0.588	0.615	0.676	0.749	0.924	1.120	1.242	1.240	1080
1060	0.527	0.529	0.533	0.537	0.546	0.555	0.564	0.590	0.618	0.685	0.765	0.960	1.176	1.295	1.260	1060
1040	0.525	0.527	0.532	0.536	0.545	0.555	0.565	0.592	0.622	0.695	0.783	1.002	1.238	1.351	1.282	1040
1020	0.523	0.526	0.530	0.535	0.545	0.555	0.565	0.594	0.627	0.707	0.804	1.051	1.306	1.399	1.298	1020
1000	0.522	0.524	0.529	0.534	0.544	0.555	0.566	0.597	0.633	0.721	0.829	1.110	1.382	1.471	1.306	1000
980	0.520	0.523	0.528	0.533	0.544	0.555	0.567	0.600	0.640	0.737	0.858	1.180	1.475	1.531	1.312	980
960	0.519	0.521	0.527	0.532	0.543	0.556	0.568	0.605	0.648	0.756	0.893	1.267	1.598	1.595	1.310	960
940	0.517	0.520	0.526	0.531	0.543	0.556	0.570	0.610	0.658	0.778	0.934	1.376	1.708	1.639	1.299	940
920	0.516	0.519	0.525	0.531	0.544	0.558	0.573	0.617	0.669	0.803	0.984	1.520	1.819	1.667	1.281	920
900	0.515	0.518	0.524	0.530	0.544	0.559	0.576	0.624	0.683	0.834	1.048	1.716	1.932	1.660	1.259	900
880	0.513	0.516	0.523	0.530	0.545	0.561	0.580	0.633	0.699	0.872	1.130	1.993	2.000	1.633	1.232	880
860	0.512	0.515	0.523	0.530	0.546	0.564	0.584	0.644	0.718	0.918	1.240	2.316	2.019	1.593	1.212	860
840	0.511	0.514	0.522	0.530	0.548	0.568	0.590	0.657	0.740	0.977	1.395	2.653	1.978	1.547	1.192	840
820	0.510	0.514	0.522	0.531	0.550	0.572	0.597	0.672	0.767	1.054	1.620	2.886	1.888	1.503	1.175	820
800	0.509	0.513	0.522	0.532	0.553	0.577	0.605	0.690	0.800	1.160	1.967	2.872	1.768	1.459	1.157	800
780	0.508	0.513	0.522	0.533	0.557	0.584	0.615	0.712	0.840	1.312	2.550	2.547	1.670	1.416	1.142	780
760	0.507	0.512	0.523	0.535	0.561	0.592	0.628	0.738	0.892	1.542	4.462	2.156	1.576	1.370	1.126	760
740	0.507	0.512	0.524	0.537	0.567	0.602	0.642	0.770	0.960	1.913	8.119	1.886	1.493	1.332	1.114	740
720	0.506	0.512	0.525	0.540	0.574	0.613	0.660	0.811	1.052	2.584	3.458	1.696	1.421	1.290	1.100	720
700	0.506	0.513	0.528	0.544	0.582	0.627	0.681	0.861	1.181	6.145°	2.237	1.557	1.358	1.250	1.089	700
680	0.506	0.514	0.530	0.549	0.592	0.644	0.707	0.927	1.365	2.469	1.789	1.450	1.303	1.217	1.079	680
660	0.507	0.515	0.534	0.555	0.604	0.665	0.738	1.015	1.639	1.851	1.587	1.369	1.256	1.187	1.071	660
640	0.507	0.517	0.538	0.562	0.619	0.690	0.777	1.135	2.219	1.601	1.454	1.303	1.216	1.157	1.063	640
620	0.509	0.519	0.543	0.571	0.637	0.720	0.826	1.308	1.614	1.455	1.362	1.252	1.184	1.136	1.056	620
600	0.510	0.522	0.550	0.582	0.659	0.757	0.888	1.586	1.453	1.358	1.295	1.211	1.157	1.118	1.052	600
580	0.513	0.526	0.558	0.595	0.685	0.804	0.969	1.393	1.351	1.289	1.243	1.178	1.134	1.102	1.046	580
560	0.516	0.531	0.568	0.611	0.717	0.862	1.079	1.309	1.281	1.237	1.202	1.151	1.115	1.087	1.039	560
540	0.519	0.538	0.580	0.630	0.756	0.937	1.272	1.249	1.229	1.196	1.169	1.128	1.098	1.074	1.033	540
520	0.524	0.545	0.594	0.653	0.804	1.035	1.221	1.204	1.189	1.164	1.142	1.109	1.083	1.062	1.024	520
500	0.530	0.554	0.611	0.680	0.865	1.187	1.181	1.169	1.157	1.137	1.120	1.092	1.069	1.051	1.017	500
480	0.537	0.565	0.632	0.714	1.159	1.154	1.150	1.140	1.131	1.115	1.101	1.077	1.057	1.041	1.010	480
460	0.545	0.578	0.657	0.755	1.132	1.128	1.125	1.117	1.110	1.096	1.084	1.064	1.047	1.033	1.004	460
440	0.556	0.594	0.687	1.113	1.110	1.107	1.104	1.098	1.092	1.080	1.070	1.052	1.038	1.025	0.999	440
420	0.568	0.614	0.724	1.094	1.091	1.089	1.087	1.081	1.076	1.067	1.058	1.042	1.029	1.018	0.994	420
400	0.583	0.636	1.079	1.078	1.076	1.074	1.072	1.067	1.063	1.055	1.047	1.034	1.022	1.011	0.990	400
380	0.601	1.066	1.065	1.065	1.063	1.061	1.059	1.056	1.052	1.044	1.038	1.026	1.015	1.006	0.986	380
360	0.622	1.054	1.054	1.053	1.052	1.050	1.049	1.045	1.042	1.036	1.030	1.019	1.009	1.001	0.982	360
340	1.045	1.044	1.044	1.043	1.042	1.040	1.039	1.036	1.033	1.028	1.022	1.013	1.004	0.996	0.979	340
320	1.036	1.036	1.035	1.034	1.033	1.032	1.031	1.028	1.026	1.021	1.016	1.007	0.999	0.992	0.976	320
300	1.028	1.028	1.028	1.027	1.026	1.025	1.024	1.022	1.019	1.015	1.010	1.002	0.995	0.988	0.973	300
280	1.022	1.022	1.021	1.021	1.020	1.019	1.018	1.016	1.014	1.009	1.005	0.998	0.991	0.985	0.971	280
260	1.016	1.016	1.016	1.015	1.014	1.013	1.013	1.011	1.009	1.005	1.001	0.994	0.988	0.982	0.968	260
240	1.012	1.011	1.011	1.011	1.010	1.009	1.008	1.006	1.004	1.001	0.997	0.991	0.985	0.979	0.966	240
220	1.008	1.008	1.007	1.007	1.006	1.005	1.005	1.003	1.001	0.998	0.994	0.988	0.982	0.977	0.964	220
200	1.005	1.004	1.004	1.004	1.003	1.002	1.002	1.000	0.998	0.995	0.992	0.986	0.980	0.975	0.963	200
180	1.002	1.002	1.002	1.001	1.001	1.000	0.999	0.998	0.996	0.993	0.989	0.983	0.978	0.973	0.961	180
160	1.000	1.000	1.000	0.999	0.999	0.998	0.997	0.996	0.994	0.991	0.987	0.981	0.976	0.971	0.959	160
140	0.999	0.999	0.998	0.998	0.997	0.997	0.996	0.994	0.992	0.989	0.986	0.980	0.974	0.969	0.958	140
120	0.998	0.998	0.997	0.997	0.996	0.996	0.995	0.993	0.991	0.988	0.984	0.978	0.972	0.967	0.957	120
100	0.997	0.997	0.997	0.996	0.996	0.995	0.994	0.992	0.990	0.986	0.983	0.976	0.970	0.965	0.955	100
80	0.998	0.997	0.997	0.996	0.995	0.994	0.994	0.991	0.989	0.985	0.981	0.974	0.968	0.962	0.951	80
60	0.999	0.998	0.998	0.997	0.996	0.995	0.994	0.991	0.989	0.984	0.979	0.970	0.963	0.956	0.942	60
40	1.003	1.003	1.002	1.001	1.000	0.998	0.997	0.993	0.989	0.983	0.976	0.965	0.954	0.945	0.920	40
32	1.006	1.006	1.005	1.004	1.002	1.000	0.999	0.994	0.990	0.983	0.975	0.962	0.949	0.937	0.904	32

°Critical point.

STEAM TABLES (Continued)

Thermal Conductivity of Steam and Water

Temp. °F Press., psia	1	2	5	10	20	50	100	200	500	1000	2000	5000	7500
						k, Btu/hr × ft × °F × 10³							
Sat. Water	364.0	373.1	383.8	390.4	395.2	397.4	394.7	386.2	361.7	327.6	271.8	—	—
Sat. Steam	11.6	12.2	13.0	13.8	14.8	16.6	18.4	21.1	27.2	36.5	61.3	—	—
1500	63.7	63.7	63.7	63.7	63.7	63.8	64.0	64.3	65.4	67.1	70.7	82.0	92.2
1450	61.4	61.4	61.5	61.5	61.5	61.6	61.8	62.1	63.2	64.9	68.5	80.1	90.6
1400	59.2	59.2	59.2	59.2	59.3	59.4	59.6	59.9	60.9	62.7	66.3	78.2	89.2
1350	57.0	57.0	57.0	57.0	57.1	57.2	57.3	57.7	58.7	60.5	64.2	76.3	87.9
1300	54.8	54.8	54.8	54.8	54.9	54.9	55.1	55.5	56.5	58.3	62.0	74.6	86.9
1250	52.6	52.6	52.6	52.6	52.6	52.7	52.9	53.2	54.3	56.1	59.9	73.0	86.3
1200	50.4	50.4	50.4	50.4	50.4	50.5	50.7	51.0	52.1	53.9	57.8	71.6	86.2
1150	48.2	48.2	48.2	48.2	48.2	48.3	48.5	48.9	49.9	51.8	55.7	70.5	87.0
1100	46.0	46.0	46.0	46.0	46.1	46.2	46.3	46.7	47.8	49.6	53.7	69.8	89.0
1050	43.9	43.9	43.9	43.9	43.9	44.0	44.2	44.6	45.6	47.5	51.8	69.7	93.4
1000	41.7	41.7	41.8	41.8	41.8	41.9	42.1	42.4	43.5	45.5	50.0	70.7	102.9
950	39.6	39.6	39.7	39.7	39.7	39.8	40.0	40.3	41.4	43.5	48.3	73.5	115.5
900	37.6	37.6	37.6	37.6	37.6	37.7	37.9	38.3	39.4	41.5	46.8	80.2	138.7
850	35.5	35.6	35.6	35.6	35.6	35.7	35.9	36.3	37.4	39.7	45.6	96.7	178.8
800	33.6	33.6	33.6	33.6	33.6	33.7	33.9	34.3	35.5	37.9	44.9	129.6	223.2
750	31.6	31.6	31.6	31.6	31.6	31.8	32.0	32.3	33.6	36.3	45.2	202.5	258.3
700	29.7	29.7	29.7	29.7	29.8	29.9	30.1	30.4	31.8	35.0	47.5®	262.8	295.1
650	27.8	27.8	27.9	27.9	27.9	28.0	28.2	28.6	30.1	34.1	55.7	304.3	326.7
600	26.0	26.0	26.1	26.1	26.1	26.2	26.4	26.9	28.7	34.1	301.9	333.7	349.3
550	24.3	24.3	24.3	24.3	24.4	24.5	24.7	25.2	27.5	36.1	333.7	356.1	368.0
500	22.6	22.6	22.6	22.6	22.7	22.8	23.0	23.6	26.9	350.8	357.4	373.8	383.6
450	21.0	21.0	21.0	21.0	21.0	21.2	21.4	22.3	368.1	370.6	375.3	387.9	396.5
400	19.4	19.4	19.4	19.4	19.5	19.6	20.0	21.3	383.0	384.9	388.5	398.6	406.4
350	17.9	17.9	17.9	17.9	18.0	18.2	18.8	392.0	392.9	394.4	397.4	406.1	413.2
300	16.5	16.5	16.5	16.5	16.6	16.9	396.9	397.2	398.0	399.3	402.0	409.9	416.4
250	15.1	15.1	15.1	15.2	15.3	396.9	397.0	397.3	398.1	399.4	402.1	409.7	415.8
200	13.8	13.8	13.9	14.0	391.6	391.6	391.8	392.1	393.0	394.4	397.2	404.9	410.6
150	12.7	12.7	380.5	380.5	380.6	380.7	380.8	381.1	382.1	383.7	386.7	394.7	400.3
100	363.3	363.3	363.3	363.3	363.3	363.4	363.6	363.9	365.0	366.6	369.8	378.3	384.1
50	339.1	339.1	339.1	339.1	339.2	339.3	339.4	339.8	340.8	342.5	345.7	354.6	361.0
32	328.6	328.6	328.6	328.6	328.6	328.7	328.9	329.2	330.3	331.9	335.1	344.1	350.8

®Critical point.

From Weast, R. C., Ed., *Handbook of Chemistry and Physics,* 69th ed., CRC Press, Boca Raton, FL, 1988.

TABLE 9
Thermodynamic Properties

Ammonia (NH₃)

Temp. (°F)	Abs. press. sat. vap lb/in²	kg/cm²	Enthalpy above −40°F, Btu/lb Liq.	Vap.	Ht. of vaporiz. BTU/lb	Enthalpy above −40°F, cal/g Liq.	Vap.	Ht. of vaporiz. cal/gm	Spec. vol. sat. vap. ft³/lb	m³/kg	Density sat. vap. lb/ft³	kg/m³	Dens. liq. lb/ft³	Entropy from −40°F, Btu·lb⁻¹·°F⁻¹ Liq.	Vap.	Temp. (°C)
−60	5.55	0.390	−21.2	589.6	610.8	−11.8	327.6	339.3	44.73	2.792	0.02235	0.3580	43.91	−0.0517	1.4769	−51.11
−58	5.93	0.417	−19.1	590.4	609.5	−10.6	328.0	338.6	42.05	2.625	0.02378	.3809		−0.0464	1.4713	−50.00
−56	6.33	0.445	−17.0	591.2	608.2	−9.44	328.4	337.9	39.56	2.470	0.02528	.4049		−0.0412	1.4658	−48.89
−54	6.75	0.475	−14.8	592.1	606.9	−8.22	328.9	337.2	37.24	2.325	0.02685	.4301		−0.0360	1.4604	−47.78
−52	7.20	0.506	−12.7	592.9	605.6	−7.06	329.4	336.4	35.09	2.191	0.02850	.4565		−0.0307	1.4551	−46.67
−50	7.67	0.539	−10.6	593.7	604.3	−5.89	329.8	335.7	33.08	2.065	0.03023	0.4842	43.49	−0.0256	1.4497	−45.56
−48	8.16	0.574	−8.5	594.4	602.9	−4.7	330.2	334.9	31.20	1.948	0.03205	.5134		−0.0204	1.4445	−44.44
−46	8.68	0.610	−6.4	595.2	601.6	−3.6	330.7	334.2	29.45	1.839	0.03395	.5438		−0.0153	1.4393	−43.33
−44	9.23	0.649	−4.3	596.0	600.3	−2.4	331.1	333.5	27.82	1.737	0.03595	.5758		−0.0102	1.4342	−42.22
−42	9.81	0.690	−2.1	596.8	598.9	−1.2	331.6	332.7	26.29	1.641	0.03804	.6093		−0.0051	1.4292	−41.11
−40	10.41	0.7319	0.0	597.6	597.6	0.0	332.0	332.0	24.86	1.552	0.04022	0.6442	43.08	0.0000	1.4242	−40.00
−38	11.04	0.7762	2.1	598.3	596.2	1.2	332.4	331.2	23.53	1.469	0.04251	.6809		0.0051	1.4193	−38.89
−36	11.71	0.8233	4.3	599.1	594.8	2.4	332.8	330.4	22.27	1.390	0.04489	.7190		0.0101	1.4144	−37.78
−34	12.41	0.8725	6.4	599.9	593.5	3.6	333.3	329.7	21.10	1.317	0.04739	.7591		0.0151	1.4096	−36.67
−32	13.14	0.9238	8.5	600.6	592.1	4.7	333.7	328.9	20.00	1.249	0.04999	.8007		0.0201	1.4048	−35.56
−30	13.90	0.9773	10.7	601.4	590.7	5.94	334.1	328.2	18.97	1.184	0.05271	0.8443	42.65	0.0250	1.4001	−34.44
−28	14.71	1.034	12.8	602.1	589.3	7.11	334.5	327.4	18.00	1.124	0.05555	.8898		0.0300	1.3955	−33.33
−26	15.55	1.093	14.9	602.8	587.9	8.28	334.9	326.6	17.09	1.067	0.05850	.9371		0.0350	1.3909	−32.22
−24	16.42	1.154	17.1	603.6	586.5	9.50	335.3	325.8	16.24	1.014	0.06158	.9864		0.0399	1.3863	−31.11
−22	17.34	1.219	19.2	604.3	585.1	10.7	335.7	325.1	15.43	0.9633	0.06479	1.038		0.0448	1.3818	−30.00
−20	18.30	1.287	21.4	605.0	583.6	11.9	336.1	324.2	14.68	0.9164	0.06813	1.091	42.22	0.0497	1.3774	−28.89
−18	19.30	1.357	23.5	605.7	582.2	13.1	336.5	323.4	13.97	0.8721	0.07161	1.147		0.0545	1.3729	−27.78
−16	20.34	1.430	25.6	606.4	580.8	14.2	336.9	322.7	13.29	0.8297	0.07522	1.205		0.0594	1.3686	−26.67
−14	21.43	1.507	27.8	607.1	579.3	15.4	337.3	321.8	12.66	0.7903	0.07898	1.265		0.0642	1.3643	−25.56
−12	22.56	1.586	30.0	607.8	577.8	16.7	337.7	321.0	12.06	0.7529	0.08289	1.328		0.0690	1.3600	−24.44
−10	23.74	1.669	32.1	608.5	576.4	17.8	338.1	320.2	11.50	0.7179	0.08695	1.393	41.78	0.0738	1.3558	−23.33
−8	24.97	1.756	34.3	609.2	574.9	19.1	338.4	319.4	10.97	0.6848	0.09117	1.460		0.0786	1.3516	−22.22
−6	26.26	1.846	36.4	609.8	573.4	20.2	338.8	318.6	10.47	0.6536	0.09555	1.531		0.0833	1.3474	−21.11
−4	27.59	1.940	38.6	610.5	571.9	21.4	339.2	317.7	9.991	0.6237	0.1001	1.603		0.0880	1.3433	−20.00
−2	28.98	2.037	40.7	611.1	570.4	22.6	339.5	316.9	9.541	0.5956	0.1048	1.679		0.0928	1.3393	−18.89
0	30.42	2.139	42.9	611.8	568.9	23.8	339.9	316.1	9.116	0.5691	0.1097	1.757	41.34	0.0975	1.3352	−17.78
2	31.92	2.244	45.1	612.4	567.3	25.1	340.2	315.2	8.714	0.5440	0.1148	1.839		0.1022	1.3312	−16.67
4	33.47	2.353	47.2	613.0	565.8	26.2	340.6	314.3	8.333	0.5202	0.1200	1.922		0.1069	1.3273	−15.56
6	35.00	2.467	49.4	613.6	564.2	27.4	340.9	313.4	7.971	0.4976	0.1254	2.009		0.1115	1.3234	−14.44

THERMODYNAMIC PROPERTIES (continued)

Temp. (°F)	Abs. press. sat. vap lb/in²	kg/cm²	Enthalpy above −40°F, Btu/lb Liq.	Vap.	Ht. of vaporiz. Btu/lb	Enthalpy above −40°C, cal/g Liq.	Vap.	Ht. of vaporiz. cal/g	Spec. vol. sat. vap. ft³/lb	m³/kg	Density sat. vap. lb/ft³	kg/m³	Dens. liq. lb/ft³	Entropy from −40°F, Btu·lb⁻¹·°F⁻¹ Liq.	Vap.	Temp. (°C)
8	36.77	2.585	51.6	614.3	562.7	28.7	341.3	312.6	7.629	0.4763	0.1311	2.100		0.1162	1.3195	−13.33
10	38.51	2.708	53.8	614.9	561.1	29.9	341.6	311.7	7.304	0.4560	0.1369	2.193	40.89	0.1208	1.3157	−12.22
12	40.31	2.834	56.0	615.5	559.5	31.1	341.9	310.8	6.996	0.4367	0.1429	2.289		0.1254	1.3118	−11.11
14	42.18	2.966	58.2	616.1	557.9	32.3	342.3	309.9	6.703	0.4185	0.1492	2.390		0.1300	1.3081	−10.00
16	44.12	3.102	60.3	616.6	556.3	33.5	342.6	309.1	6.425	0.4011	0.1556	2.492		0.1346	1.3043	−8.89
18	46.13	3.243	62.5	617.2	554.7	34.7	342.9	308.2	6.161	0.3846	0.1623	2.600		0.1392	1.3006	−7.78
20	48.21	3.390	64.7	617.8	553.1	35.9	343.2	307.3	5.910	0.3690	0.1692	2.710	40.43	0.1437	1.2969	−6.67
22	50.36	3.541	66.9	618.3	551.4	37.2	343.5	306.3	5.671	0.3540	0.1763	2.824		0.1483	1.2933	−5.56
24	52.59	3.697	69.1	618.9	549.8	38.4	343.8	305.4	5.443	0.3398	0.1837	2.943		0.1528	1.2897	−4.44
26	54.90	3.860	71.3	619.4	548.1	39.6	344.1	304.5	5.227	0.3263	0.1913	3.064		0.1573	1.2861	−3.33
28	57.28	4.027	73.5	619.9	546.4	40.8	344.4	303.6	5.021	0.3135	0.1992	3.191		0.1618	1.2825	−2.22
30	59.74	4.200	75.7	620.5	544.8	42.1	344.7	302.7	4.825	0.3012	0.2073	3.321	39.96	0.1663	1.2790	−1.11
32	62.29	4.379	77.9	621.0	543.1	43.3	345.0	301.7	4.637	0.2895	0.2156	3.453		0.1708	1.2755	0.00
34	64.91	4.564	80.1	621.5	541.4	44.5	345.3	300.8	4.459	0.2784	0.2243	3.593		0.1753	1.2721	1.11
36	67.63	4.755	82.3	622.0	539.7	45.7	345.6	299.8	4.289	0.2678	0.2332	3.735		0.1797	1.2686	2.22
38	70.43	4.952	84.6	622.5	537.9	47.0	345.8	298.8	4.126	0.2576	0.2423	3.881		0.1841	1.2652	3.33
40	73.32	5.155	86.8	623.0	536.2	48.2	346.1	297.9	3.971	0.2479	0.2518	4.003	39.49	0.1885	1.2618	4.44
42	76.31	5.365	89.0	623.4	534.4	49.4	346.3	296.9	3.823	0.2387	0.2616	4.190		0.1930	1.2585	5.56
44	79.38	5.581	91.2	623.9	532.7	50.7	346.6	295.9	3.682	0.2299	0.2716	4.350		0.1974	1.2552	6.67
46	82.55	5.804	93.5	624.4	530.9	51.9	346.9	294.9	3.547	0.2214	0.2819	4.515		0.2018	1.2519	7.78
48	85.82	6.034	95.7	624.8	529.1	53.2	347.1	293.9	3.418	0.2134	0.2926	4.687		0.2062	1.2486	8.89
50	89.19	6.271	97.9	625.2	527.3	54.4	347.3	292.9	3.294	0.2056	0.3036	4.863	39.00	0.2105	1.2453	10.00
52	92.66	6.515	100.2	625.7	525.5	55.67	347.6	291.9	3.176	0.1983	0.3149	5.044		0.2149	1.2421	11.11
54	96.23	6.766	102.4	626.1	523.7	56.89	347.8	290.9	3.063	0.1912	0.3265	5.230		0.2192	1.2389	12.22
56	99.91	7.024	104.7	626.5	521.8	58.17	348.1	289.9	2.954	0.1844	0.3385	5.422		0.2236	1.2357	13.33
58	103.7	7.291	106.9	626.9	520.0	59.39	348.3	288.9	2.851	0.1780	0.3508	5.619		0.2279	1.2325	14.44
60	107.6	7.565	109.2	627.3	518.1	60.67	348.5	287.8	2.751	0.1717	0.3635	5.823	38.50	0.2322	1.2294	15.56
62	111.6	7.846	111.5	627.7	516.2	61.94	348.7	286.8	2.656	0.1658	0.3765	6.031		0.2365	1.2262	16.67
64	115.7	8.135	113.7	628.0	514.3	63.17	348.9	285.7	2.565	0.1601	0.3899	6.245		0.2408	1.2231	17.78
66	120.0	8.437	116.0	628.4	512.4	64.44	349.1	284.7	2.477	0.1546	0.4037	6.466		0.2451	1.2201	18.89
68	124.3	8.739	118.3	628.8	510.5	65.72	349.3	283.6	2.393	0.1494	0.4179	6.694		0.2404	1.2170	20.00
70	128.8	9.056	120.5	629.1	508.6	66.94	349.5	282.6	2.312	0.1443	0.4325	6.928	38.00	0.2537	1.2140	21.11
72	133.4	9.379	122.8	629.4	506.6	68.22	349.7	281.4	2.235	0.1395	0.4474	7.166		0.2579	1.2110	22.22
74	138.1	9.709	125.1	629.8	504.7	69.50	349.9	280.4	2.161	0.1349	0.4628	7.413		0.2622	1.2080	23.33
76	143.0	10.05	127.4	630.1	502.7	70.78	350.1	279.3	2.089	0.1304	0.4786	7.666		0.2664	1.2050	24.44
78	147.9	10.40	129.7	630.4	500.7	72.06	350.2	278.2	2.021	0.1262	0.4949	7.927		0.2706	1.2020	25.56
80	153.0	10.76	132.0	630.7	498.7	73.33	350.4	277.1	1.955	0.1220	0.5115	8.193	37.48	0.2749	1.1991	26.67
82	158.3	11.13	134.3	631.0	496.7	74.61	350.6	275.9	1.892	0.1181	0.5287	8.469		0.2791	1.1962	27.78
84	163.7	11.51	136.6	631.3	494.7	75.89	350.7	274.8	1.831	0.1143	0.5462	8.749		0.2833	1.1933	28.89
86	169.2	11.90	138.9	631.5	492.6	77.17	350.8	273.7	1.772	0.1106	0.5643	9.039		0.2875	1.1904	30.00
88	174.8	12.29	141.2	631.8	490.6	78.44	351.0	272.6	1.716	0.1071	0.5828	9.335		0.2917	1.1875	31.11
90	180.6	12.70	143.5	632.0	488.5	79.72	351.1	271.4	1.661	0.1037	0.6019	9.641	36.95	0.2958	1.1846	32.22
92	186.6	13.12	145.8	632.2	486.4	81.00	351.2	270.2	1.609	0.1004	0.6214	9.954		0.3000	1.1818	33.33
94	192.7	13.55	148.2	632.5	484.3	82.33	351.4	269.1	1.559	0.09733	0.6415	10.28		0.3041	1.1789	34.44
96	198.9	13.98	150.5	632.6	482.1	83.61	351.4	267.8	1.510	0.09427	0.6620	10.60		0.3083	1.1761	35.56
98	205.3	14.43	152.9	632.9	480.0	84.94	351.6	266.7	1.464	0.09140	0.6832	10.94		0.3125	1.1733	36.67
100	211.9	14.90	155.2	633.0	477.8	86.22	351.7	265.4	1.419	0.08859	0.7048	11.29	36.40	0.3166	1.1705	37.78
102	218.6	15.37	157.6	633.2	475.6	87.56	351.8	264.2	1.375	0.08584	0.7270	11.65		0.3207	1.1677	38.89
104	225.4	15.85	159.9	633.4	473.5	88.83	351.9	263.1	1.334	0.08328	0.7498	12.01		0.3248	1.1649	40.00
106	232.5	16.35	162.3	633.5	471.2	90.17	351.9	261.8	1.293	0.08072	0.7732	12.39		0.3289	1.1621	41.11
108	239.7	16.85	164.6	633.6	469.0	91.44	352.0	260.6	1.254	0.07829	0.7972	12.77		0.3330	1.1593	42.22
110	247.0	17.37	167.0	633.7	466.7	92.78	352.1	259.3	1.217	0.07598	0.8219	13.17	35.84	0.3372	1.1566	43.33
112	254.5	17.89	169.4	633.8	464.4	94.11	352.1	258.0	1.180	0.07367	0.8471	13.57		0.3413	1.1538	44.44
114	262.2	18.43	171.8	633.9	462.1	95.44	352.2	256.7	1.145	0.07148	0.8730	13.98		0.3453	1.1510	45.56
116	270.1	18.99	174.2	634.0	459.8	96.78	352.2	255.4	1.112	0.06942	0.8996	14.41		0.3495	1.1483	46.67
118	278.2	19.56	176.6	634.0	457.4	98.11	352.2	254.1	1.079	0.06736	0.9269	14.85		0.3535	1.1455	47.78
120	286.4	20.14	179.0	634.0	455.0	99.45	352.2	252.8	1.047	0.06536	0.9549	15.30	35.26	0.3576	1.1427	48.89
122	294.8	20.73	181.4	634.0	452.6	100.8	352.2	251.4	1.017	0.06349	0.9837	15.76		0.3618	1.1400	50.00
124	303.4	21.33	183.9	634.0	450.1	102.2	352.2	250.1	0.987	0.0616	1.0132	16.229		0.3659	1.1372	51.11

Carbon Dioxide (CO₂)

Temp. (°F)	Abs. press sat. vap lb/in²	kg/cm²	Enthalpy above 32°F, Btu/lb Liq.	Vap	Heat of vaporiz. Btu/lb	Enthalpy above 0°C, cal/g Liq.	Vap	Heat of vaporiz. cal/g	Spec. vol. sat. vap. ft³/lb	m³/kg	Density sat. vap. lb/ft³	kg/m³	Dens. liq. lb/ft³	Entropy from 32°F, Btu·lb⁻¹·°F⁻¹ Liq.	Vap.	Temp. (°C)
−20	220.6	15.51	−23.96	102.0	126.0	−13.31	56.67	70.00	0.4166	0.02601	2.401	38.46	64.34	−0.0514	0.2353	−28.89
−18	228.4	16.06	−23.13	102.1	125.2	−12.85	56.72	69.56	0.4018	0.02508	2.489	39.87	64.15	−0.0495	0.2342	−27.78
−16	236.4	16.62	−22.30	102.2	124.5	−12.39	56.78	69.17	0.3876	0.02420	2.580	41.33	63.94	−0.0476	0.2331	−26.67
−14	244.6	17.20	−21.46	102.2	123.7	−11.92	56.78	68.72	0.3739	0.02334	2.674	42.83	63.73	−0.0458	0.2319	−25.56

THERMODYNAMIC PROPERTIES (continued)

Temp. (°F)	Abs. press sat. vap lb/in²	kg/cm²	Enthalpy above 32°F, Btu/lb Liq.	Vap.	Heat of vaporiz. Btu/lb	Enthalpy above 0°C, cal/g Liq.	Vap.	Heat of vaporiz. cal/g	Spec. vol. sat. vap. ft³/lb	m³/kg	Density sat. vap. lb/ft³	kg/m³	Dens. liq. lb/ft³	Entropy from 32°F, Btu·lb⁻¹·°F⁻¹ Liq.	Vap.	Temp. (°C)
-12	253.0	17.79	-20.61	102.3	122.9	-11.45	56.83	68.28	0.3608	0.02252	2.772	44.40	63.49	-0.0439	0.2307	-24.44
-10	261.7	18.40	-19.76	102.3	122.0	-10.98	56.83	67.78	0.3482	0.02174	2.872	46.00	63.25	-0.0420	0.2296	-23.33
-8	270.6	19.03	-18.90	102.3	121.2	-10.50	56.83	67.33	0.3360	0.02098	2.976	47.67	63.01	-0.0401	0.2284	-22.22
-6	279.7	19.66	-18.04	102.3	120.3	-10.02	56.83	66.83	0.3243	0.02025	3.083	49.38	62.76	-0.0382	0.2273	-21.11
-4	289.1	20.33	-17.17	102.3	119.5	-9.539	56.83	66.39	0.3131	0.01955	3.194	51.16	62.50	-0.0362	0.2261	-20.00
-2	298.7	21.00	-16.29	102.3	118.6	-9.050	56.83	65.89	0.3022	0.01887	3.309	53.00	62.23	-0.0343	0.2249	-18.89
0	308.6	21.70	-15.41	102.2	117.7	-8.561	56.78	65.39	0.2918	0.01822	3.427	54.89	61.95	-0.0324	0.2237	-17.78
2	318.7	22.41	-14.51	102.2	116.7	-8.061	56.78	64.83	0.2817	0.01759	3.550	56.86	61.65	-0.0304	0.2225	-16.67
4	329.1	23.14	-13.61	102.1	115.8	-7.561	56.72	64.33	0.2720	0.01698	3.676	58.88	61.36	-0.0285	0.2213	-15.56
6	339.8	23.89	-12.71	102.1	114.8	-7.061	56.72	63.78	0.2627	0.01640	3.807	60.98	61.07	-0.0266	0.2201	-14.44
8	350.7	24.66	-11.79	102.0	113.8	-6.550	56.67	63.22	0.2537	0.01584	3.942	63.14	60.77	-0.0246	0.2189	-13.33
10	361.8	25.44	-10.87	101.9	112.8	-6.039	56.61	62.67	0.2450	0.01529	4.082	65.39	60.48	-0.0226	0.2176	-12.22
12	373.3	26.25	-9.934	101.8	111.7	-5.519	56.56	62.06	0.2366	0.01477	4.227	67.71	60.18	-0.0206	0.2164	-11.11
14	385.0	27.07	-8.992	101.7	110.7	-4.996	56.50	61.50	0.2285	0.01426	4.377	70.11	59.88	-0.0186	0.2151	-10.00
16	397.1	27.92	-8.038	101.5	109.6	-4.466	56.39	60.89	0.2207	0.01378	4.532	72.59	59.58	-0.0166	0.2139	-8.89
18	409.4	28.78	-7.076	101.4	108.5	-3.931	56.33	60.28	0.2131	0.01330	4.692	75.16	59.27	-0.0146	0.2126	-7.78
20	422.0	29.67	-6.102	101.2	107.3	-3.390	56.22	59.61	0.2058	0.01285	4.859	77.83	58.95	-0.0126	0.2113	-6.67
22	434.9	30.58	-5.117	101.0	106.1	-2.843	56.11	58.94	0.1987	0.01240	5.031	80.59	58.64	-0.0105	0.2100	-5.56
24	448.1	31.50	-4.121	100.8	104.9	-2.289	56.00	58.28	0.1919	0.01198	5.211	83.47	58.31	-0.0085	0.2087	-4.44
25	454.8	31.98	-3.618	100.7	104.3	-2.010	55.94	57.94	0.1886	0.01177	5.303	84.94	58.14	-0.0074	0.2080	-3.89
27	468.5	32.94	-2.601	100.5	103.1	-1.445	55.83	57.28	0.1821	0.01137	5.492	87.97	57.81	-0.0053	0.2066	-2.78
29	482.5	33.92	-1.570	100.2	101.8	-0.8722	55.67	56.56	0.1758	0.01097	5.688	91.11	57.47	-0.0032	0.2053	-1.67
31	496.8	34.93	-0.525	99.98	100.5	-0.292	55.54	55.83	0.1697	0.01059	5.892	94.38	57.12	-0.0011	0.2039	-0.56
33	511.4	35.95	0.531	99.69	99.16	0.295	55.38	55.09	0.1639	0.01023	6.103	97.76	56.77	0.0011	0.2025	0.56
35	526.4	37.01	1.604	99.38	97.77	0.8911	55.21	54.32	0.1581	0.009870	6.323	101.3	56.41	0.0033	0.2010	1.67
37	541.7	38.09	2.697	99.05	96.35	1.498	55.03	53.53	0.1526	0.009527	6.553	105.0	56.03	0.0055	0.1996	2.78
39	557.4	39.19	3.806	98.69	94.88	2.114	54.83	52.71	0.1472	0.009189	6.792	108.8	55.65	0.0077	0.1981	3.89
41	573.4	40.31	4.932	98.31	93.37	2.740	54.62	51.87	0.1420	0.008865	7.040	112.8	55.25	0.0099	0.1965	5.00
43	589.8	41.47	6.080	97.90	91.82	3.378	54.39	51.01	0.1370	0.008553	7.300	116.9	54.84	0.0122	0.1950	6.11
45	606.5	42.64	7.251	97.46	90.21	4.028	54.14	50.12	0.1321	0.008247	7.571	121.3	54.41	0.0146	0.1934	7.22
47	623.6	43.84	8.443	96.99	88.55	4.691	53.88	49.19	0.1273	0.007947	7.854	125.8	53.97	0.0169	0.1918	8.33
49	641.1	45.07	9.664	95.50	86.83	5.369	53.61	48.24	0.1227	0.007660	8.151	130.6	53.51	0.0193	0.1901	9.44
51	659.0	46.33	10.91	95.97	85.06	6.061	53.32	47.26	0.1182	0.007379	8.461	135.5	53.04	0.0218	0.1884	10.56
53	677.3	47.62	12.19	95.40	83.21	6.772	53.00	46.23	0.1138	0.007104	8.787	140.8	52.55	0.0243	0.1867	11.67
55	695.9	48.93	13.49	94.78	81.29	7.494	52.66	45.16	0.1095	0.006836	9.132	146.3	52.05	0.0268	0.1849	12.78
57	714.9	50.26	14.84	94.13	79.30	8.244	52.29	44.06	0.1053	0.006574	9.497	152.1	51.53	0.0294	0.1830	13.89
59	734.3	51.63	16.22	93.44	77.22	9.011	51.91	42.90	0.1012	0.006318	9.880	158.3	50.99	0.0321	0.1811	15.00
61	754.2	53.03	17.65	92.69	75.04	9.806	51.49	41.69	0.0972	0.00607	10.29	164.8	50.42	0.0348	0.1790	16.11
63	774.5	54.45	19.13	91.88	72.75	10.63	51.04	40.42	0.0933	0.00582	10.72	171.7	49.80	0.0377	0.1770	17.22
65	795.1	55.90	20.66	91.01	70.35	11.48	50.56	39.08	0.0894	0.00558	11.18	179.1	49.14	0.0406	0.1748	18.33
67	816.2	57.38	22.25	90.07	67.81	12.36	50.04	37.67	0.0856	0.00534	11.67	186.9	48.44	0.0436	0.1725	19.44
69	837.8	58.90	23.92	89.04	65.12	13.29	49.47	36.18	0.0819	0.00511	12.21	195.6	47.69	0.0468	0.1701	20.56
71	859.8	60.45	25.67	87.92	62.25	14.26	48.84	34.58	0.0782	0.00488	12.82	205.4	46.87	0.0501	0.1676	21.67
73	882.2	62.02	27.52	86.69	59.17	15.29	48.16	32.87	0.0745	0.00465	13.43	215.1	45.99	0.0535	0.1647	22.78
75	905.1	63.63	29.50	85.33	55.83	16.39	47.41	31.02	0.0708	0.00442	14.13	226.3	45.05	0.0573	0.1618	23.89
77	928.4	65.27	31.62	83.80	52.17	17.57	46.56	28.98	0.0671	0.00419	14.90	238.7	44.06	0.0613	0.1585	25.00
79	952.2	66.95	33.95	82.06	48.11	18.86	45.59	26.73	0.0633	0.00395	15.81	253.2	43.04	0.0656	0.1550	26.11
81	976.5	68.65	36.54	80.03	43.49	20.30	44.46	24.16	0.0592	0.00370	16.90	270.7	41.95	0.0704	0.1509	27.22
83	1001.0	70.377	39.53	77.60	38.07	21.96	43.11	21.15	0.0548	0.00342	18.25	292.3	40.62	0.0759	0.1461	28.33
85	1027.0	72.205	43.18	74.47	31.29	23.99	41.37	17.38	0.0500	0.00312	20.00	320.4	38.76	0.0826	0.1401	29.44
86	1039.0	73.049	45.45	72.46	27.00	25.25	40.26	15.00	0.0474	0.00296	21.09	337.8	37.41	0.0868	0.1363	30.00
87	1052.0	73.963	48.32	69.84	21.52	26.84	38.80	11.96	0.0446	0.00278	22.42	359.1	35.34	0.0921	0.1314	30.56
88	1065.0	74.877	52.78	65.62	12.84	29.32	36.46	7.133	0.0401	0.00250	24.95	399.6	32.79	0.1002	0.1237	31.11

Sulfur Dioxide (SO₂)

Temp. (°F)	Abs. press. sat. vap. lb/in²	kg/cm²	Enthalpy above -40°F, Btu/lb Liq.	Vap.	Ht. of vaporiz. Btu/lb	Enthalpy above -40°C, cal/g Liq.	Vap.	Ht. of vaporiz. cal/g	Spec. vol. sat. vap. ft³/lb	m³/kg	Density sat. vap. lb/ft³	kg/m³	Dens. liq. lb/ft³	Entropy from 40°F, Btu·lb⁻¹·°F⁻¹ Liq.	Vap.	Temp. (°C)
-40	3.136	0.2205	0.00	178.61	178.61	0.00	99.228	99.228	22.42	1.400	0.04460	0.7144	95.79	0.00000	0.42562	-40.00
-30	4.331	0.3045	2.93	179.90	176.97	1.63	99.945	98.317	16.56	1.034	0.06039	0.9673	94.94	0.00674	0.41864	-34.44
-20	5.883	0.4136	5.98	181.07	175.09	3.32	100.59	97.272	12.42	0.7754	0.08119	1.301	94.10	0.01366	0.41192	-28.89
-10	7.863	0.5528	9.16	182.13	172.97	5.09	101.18	96.095	9.44	0.5893	0.1025	1.642	93.27	0.02075	0.40544	-23.33
0	10.35	0.7277	12.44	183.07	170.63	6.911	101.71	94.795	7.280	0.4545	0.1374	2.201	92.42	0.02795	0.39917	-17.78
2	10.91	0.7670	13.12	183.25	170.13	7.289	101.81	94.517	6.923	0.4322	0.1444	2.313	92.25	0.02941	0.39794	-16.67
4	11.50	0.8085	13.78	183.41	169.63	7.656	101.89	94.239	6.584	0.4110	0.1501	2.404	92.08	0.03084	0.39670	-15.56
5	11.81	0.8303	14.11	183.49	169.38	7.839	101.94	94.100	6.421	0.4009	0.1558	2.496	92.00	0.03155	0.39609	-15.00
6	12.12	0.8521	14.45	183.57	169.12	8.028	101.98	93.956	6.266	0.3912	0.1596	2.556	91.91	0.03228	0.39547	-14.44
8	12.75	0.8964	15.13	183.73	168.60	8.406	102.07	93.667	5.967	0.3725	0.1676	2.685	91.74	0.03373	0.39426	-13.33

THERMODYNAMIC PROPERTIES (continued)

Temp. (°F)	Abs. press. sat. vap. lb/in²	Abs. press. sat. vap. kg/cm²	Enthalpy above −40°F, Btu/lb Liq.	Enthalpy above −40°F, Btu/lb Vap.	Ht. of vaporiz. Btu/lb	Enthalpy above −40°C, cal/g Liq.	Enthalpy above −40°C, cal/g Vap.	Ht. of vaporiz. g-cal/g	Spec. vol. sat. vap. ft³/lb	Spec. vol. sat. vap. m³/kg	Density sat. vap. lb/ft³	Density sat. vap. kg/m³	Dens. liq. lb/ft³	Entropy from −40°F, Btu·lb⁻¹·°F⁻¹ Liq.	Entropy from −40°F, Btu·lb⁻¹·°F⁻¹ Vap.	Temp. (°C)
10	13.42	0.9435	15.80	183.87	168.07	8.778	102.15	93.372	5.682	0.3547	0.1760	2.819	91.58	0.03519	0.39306	−12.22
12	14.12	0.9927	16.48	184.01	167.53	9.156	102.23	93.072	5.417	0.3382	0.1846	2.957	91.41	0.03664	0.39185	−11.11
14	14.84	1.043	17.15	184.14	166.97	9.528	102.30	92.761	5.164	0.3224	0.1936	3.101	91.24	0.03808	0.39065	−10.00
16	15.59	1.096	17.84	184.28	166.44	9.911	102.38	92.467	4.926	0.3075	0.2030	3.252	91.07	0.03953	0.38946	−8.89
18	16.37	1.1509	18.52	184.40	165.88	10.29	102.44	92.156	4.701	0.2935	0.2127	3.407	90.89	0.04098	0.38827	−7.78
20	17.18	1.208	19.20	184.52	165.32	10.67	102.51	91.845	4.487	0.2801	0.2228	3.569	90.71	0.04241	0.38707	−6.67
22	18.03	1.268	19.90	184.64	164.74	11.06	102.58	91.522	4.287	0.2676	0.2332	3.735	90.53	0.04385	0.38589	−5.56
24	18.89	1.328	20.58	184.74	164.16	11.43	102.63	91.200	4.096	0.2557	0.2441	3.910	90.33	0.04528	0.38471	−4.44
26	19.80	1.392	21.26	184.84	163.58	11.81	102.69	90.878	3.915	0.2444	0.2559	4.099	90.15	0.04671	0.38354	−3.33
28	20.73	1.457	21.96	184.94	162.98	12.20	102.74	90.545	3.744	0.2337	0.2671	4.278	89.96	0.04814	0.38237	−2.22
30	21.70	1.526	22.64	185.02	162.38	12.58	102.79	90.211	3.581	0.2236	0.2800	4.485	89.76	0.04956	0.38119	−1.11
32	22.71	1.597	23.33	185.10	161.77	12.96	102.83	89.872	3.437	0.2146	0.2909	4.660	89.58	0.05099	0.38003	0.00
34	23.75	1.670	24.03	185.18	161.15	13.35	102.88	89.528	3.283	0.2050	0.3046	4.879	89.39	0.05242	0.37887	1.11
36	24.82	1.745	24.72	185.25	160.53	13.73	102.92	89.183	3.144	0.1963	0.3181	5.095	89.18	0.05384	0.37772	2.22
38	25.95	1.824	25.42	185.31	159.89	14.12	102.95	88.828	3.013	0.1881	0.3319	5.316	89.00	0.05527	0.37657	3.33
40	27.10	1.905	26.12	185.37	159.25	14.51	102.98	88.472	2.887	0.1802	0.3464	5.549	88.81	0.05668	0.37541	4.44
42	28.29	1.989	26.81	185.42	158.61	14.89	103.01	88.117	2.769	0.1729	0.3611	5.784	88.62	0.05809	0.37427	5.56
44	29.52	2.075	27.51	185.46	157.95	15.28	103.03	87.750	2.656	0.1658	0.3765	6.031	88.43	0.05949	0.37311	6.67
46	30.79	2.165	28.21	185.50	157.29	15.67	103.06	87.383	2.548	0.1591	0.3925	6.287	88.24	0.06090	0.37197	7.78
48	32.10	2.257	28.92	185.54	156.62	16.07	103.08	87.011	2.446	0.1527	0.4088	6.548	88.05	0.06231	0.37083	8.89
50	33.45	2.352	29.61	185.56	155.95	16.45	103.09	86.639	2.348	0.1466	0.4259	6.822	87.87	0.06370	0.36969	10.00
52	34.86	2.451	30.31	185.58	155.27	16.84	103.10	86.261	2.256	0.1408	0.4433	7.101	87.67	0.06509	0.36857	11.11
54	36.31	2.553	31.00	185.59	154.59	17.22	103.11	85.883	2.167	0.1353	0.4615	7.392	87.51	0.06646	0.36743	12.22
56	37.80	2.658	31.72	185.61	153.89	17.62	103.12	85.495	2.083	0.1300	0.4801	7.690	87.31	0.06785	0.36629	13.33
58	39.33	2.765	32.42	185.61	153.19	18.01	103.12	85.106	2.003	0.1250	0.4992	7.996	87.13	0.06923	0.36517	14.44
60	40.93	2.878	33.10	185.59	152.49	18.39	103.11	84.717	1.926	0.1202	0.5194	8.320	86.95	0.07066	0.36405	15.56
62	42.58	2.994	33.79	185.57	151.78	18.77	103.09	84.322	1.853	0.1157	0.5396	8.643	86.77	0.07196	0.36293	16.67
64	44.27	3.112	34.49	185.55	151.06	19.16	103.08	83.922	1.783	0.1113	0.5609	8.984	86.59	0.07333	0.36181	17.78
66	46.00	3.234	35.19	185.53	150.34	19.55	103.07	83.522	1.716	0.1071	0.5827	9.334	86.41	0.07469	0.36070	18.89
68	47.78	3.359	35.88	185.50	149.62	19.93	103.06	83.122	1.652	0.1031	0.6054	9.697	86.22	0.07602	0.35958	20.00
70	49.62	3.489	36.58	185.46	148.88	20.32	103.03	82.711	1.590	0.09926	0.6290	10.08	86.02	0.07736	0.35846	21.11
72	51.54	3.624	37.28	185.42	148.14	20.71	103.01	82.300	1.532	0.09564	0.6527	10.45	85.82	0.07871	0.35734	22.22
74	53.48	3.760	37.97	185.37	147.40	21.09	102.98	81.889	1.476	0.09214	0.6777	10.86	85.62	0.08003	0.35624	23.33
76	55.48	3.901	38.67	185.31	146.64	21.48	102.95	81.467	1.422	0.08877	0.7030	11.26	85.42	0.08135	0.35512	24.44
78	57.56	4.047	39.36	185.24	145.88	21.87	102.91	81.045	1.371	0.08559	0.7295	11.69	85.23	0.08268	0.35401	25.56
80	59.68	4.196	40.05	185.17	145.12	22.25	102.87	80.622	1.321	0.08247	0.7570	12.13	85.03	0.8399	0.35291	26.67
82	61.88	4.351	40.73	185.09	144.36	22.63	102.83	80.200	1.274	0.07953	0.7850	12.57	84.84	0.08525	0.35177	27.78
84	64.14	4.509	41.43	185.01	143.58	23.02	102.78	79.767	1.229	0.07672	0.8140	13.04	84.64	0.08653	0.35065	28.89
86	66.45	4.672	42.12	184.92	142.80	23.40	102.73	79.333	1.185	0.07398	0.8440	13.52	84.44	0.08783	0.34954	30.00
88	68.84	4.840	42.80	184.82	142.02	23.78	102.68	78.900	1.144	0.07142	0.8740	14.00	84.25	0.08910	0.34843	31.11
90	71.25	5.009	43.50	184.72	141.22	24.17	102.62	78.456	1.104	0.06892	0.9058	14.51	84.05	0.09038	0.34731	32.22
92	73.70	5.182	44.19	184.61	140.42	24.55	102.56	78.011	1.065	0.06649	0.9390	15.04	83.86	0.09165	0.34620	33.33
94	76.30	5.364	44.86	184.49	139.62	24.92	102.49	77.545	1.028	0.06418	0.9730	15.59	83.67	0.09293	0.34508	34.44
96	79.03	5.556	45.54	184.37	138.83	25.30	102.43	77.128	0.9931	0.06200	1.007	16.13	83.47	0.09411	0.34397	35.56
98	81.77	5.749	46.22	184.25	138.03	25.68	102.36	76.683	0.9591	0.05987	1.043	16.71	83.27	0.09532	0.34285	36.67
100	84.52	5.942	46.90	184.10	137.20	26.06	102.28	76.222	0.9262	0.05782	1.080	17.30	83.07	0.09657	0.34173	37.78

Butane (CH₃(CH₂)₂CH₃)

Temp. (°F)	Abs. press. sat. vap. lb/in²	Abs. press. sat. vap. kg/cm²	Enthalpy above 32°F, Btu/lb Liq.	Enthalpy above 32°F, Btu/lb Vap.	Ht. of vaporiz. Btu/lb	Enthalpy above 0°C, cal/g Liq.	Enthalpy above 0°C, cal/g Vap.	Ht. of vaporiz. cal/g	Spec. vol. sat. vap. ft³/lb	Spec. vol. sat. vap. m³/kg	Density of sat. vap. lb/ft³	Density of sat. vap. kg/m³	Density of liq. lb/ft³	Density of liq. kg/m³	Temp. (°C)
0	7.3	0.51	−17.2	153.3	170.5	−9.56	85.17	94.72	11.1	0.693	0.0901	1.44	38.59	618.1	−17.78
10	9.2	0.65	−11.7	156.8	168.5	−6.50	87.11	93.61	8.95	0.559	0.112	1.79	38.24	612.5	−12.22
20	11.6	0.816	−6.7	160.3	167.0	−3.7	89.06	92.78	7.23	0.451	0.138	2.21	37.89	606.9	−6.67
30	14.4	1.01	−1.2	164.3	165.5	−0.67	91.28	91.94	5.90	0.368	0.169	2.71	37.54	601.3	−1.11
40	17.7	1.24	4.3	167.8	163.5	2.4	93.22	90.83	4.88	0.305	0.205	3.28	37.19	595.7	4.44
50	21.6	1.52	9.8	171.3	161.5	5.4	95.17	89.72	4.07	0.254	0.246	3.94	36.82	589.8	10.00
60	26.3	1.85	15.8	175.3	159.5	8.78	97.39	88.61	3.40	0.212	0.294	4.71	36.45	583.9	15.56
70	31.6	2.22	21.3	178.8	157.5	11.8	99.33	87.50	2.88	0.180	0.347	5.56	36.06	577.6	21.11
80	37.6	2.64	27.3	182.3	155.0	15.2	101.3	86.11	2.46	0.154	0.407	6.52	35.65	571.0	26.67
90	44.5	3.13	33.8	185.8	152.0	18.8	103.2	84.44	2.10	0.131	0.476	7.62	35.24	564.5	32.22
100	52.2	3.67	39.8	189.3	149.5	22.1	105.2	83.06	1.81	0.113	0.552	8.84	34.84	558.1	37.78
110	60.8	4.27	46.3	193.3	147.0	25.7	107.4	81.67	1.58	0.0986	0.633	10.1	34.41	551.2	43.33
120	70.8	4.98	52.8	196.3	143.5	29.3	109.1	79.72	1.38	0.0862	0.725	11.6	33.96	544.0	48.89
130	81.4	5.72	59.3	199.8	140.5	32.9	111.0	78.06	1.21	0.0755	0.826	13.2	33.49	536.4	54.44
140	92.6	6.51	66.3	203.8	137.5	36.8	113.2	76.39	1.07	0.0668	0.934	15.0	32.98	528.3	60.00

THERMODYNAMIC PROPERTIES (continued)

Isobutane ((CH₃)₃CH)

Temp. (°F)	Abs. press. sat. vap. lb/in²	Abs. press. sat. vap. kg/cm²	Enthalpy above 32°F, Btu/lb Liq.	Enthalpy above 32°F, Btu/lb Vap.	Ht. of vaporiz. Btu/lb	Enthalpy above 0°C, cal/g Liq.	Enthalpy above 0°C, cal/g Vap.	Ht. of vaporiz. cal/g	Spec. vol. sat. vap. ft³/lb	Spec. vol. sat. vap. m³/kg	Density of sat. vap. lb/ft³	Density of sat. vap. kg/m³	Density of liq. lb/ft³	Density of liq. kg/m³	Temp. (°C)
−20	7.50	0.527	−25.5	140.0	165.5	−14.2	77.78	91.94	10.5	0.655	0.0952	1.52	38.35	614.3	−28.89
−10	9.28	0.652	−21.0	142.0	163.0	−11.7	78.89	90.56	8.91	0.556	0.112	1.79	37.95	607.9	−23.33
0	11.6	0.816	−16.5	144.0	160.5	−9.17	80.00	89.17	7.17	0.448	0.139	2.23	37.60	602.3	−17.78
10	14.6	1.03	−11.5	147.0	158.5	−6.39	81.67	88.06	5.75	0.359	0.174	2.79	37.20	595.9	−12.22
20	18.2	1.28	−6.5	149.5	156.0	−3.6	83.06	86.67	4.68	0.292	0.214	3.43	36.80	589.5	−6.67
30	22.3	1.57	−1.0	152.5	153.5	−0.56	84.72	85.28	3.86	0.241	0.259	4.15	36.40	583.1	−1.11
40	26.9	1.89	4.5	155.5	151.0	2.5	86.39	83.89	3.22	0.201	0.311	4.98	36.00	576.6	4.44
50	32.5	2.28	10.5	159.0	148.5	5.83	88.33	82.50	2.71	0.169	0.369	5.91	35.60	570.2	10.00
60	38.7	2.72	16.5	162.5	146.0	9.17	90.28	81.11	2.28	0.142	0.439	7.03	35.20	563.8	15.56
70	45.8	3.22	23.0	166.5	143.5	12.8	92.50	79.72	1.94	0.121	0.515	8.25	34.80	557.4	21.11
80	53.9	3.79	30.0	170.5	140.5	16.7	94.72	78.06	1.66	0.104	0.602	9.64	34.35	550.2	26.67
90	63.3	4.45	37.0	174.5	137.5	20.6	96.94	76.39	1.42	0.0886	0.704	11.3	33.90	543.0	32.22
100	73.7	5.18	44.5	179.0	134.5	24.7	99.44	74.72	1.23	0.0768	0.813	13.0	33.35	535.8	37.78
110	85.1	5.98	52.5	183.5	131.0	29.2	101.9	72.78	1.07	0.0668	0.935	15.0	33.00	528.6	43.33
120	98.0	6.89	60.5	188.0	127.5	33.6	104.4	70.83	0.926	0.0578	1.08	17.3	32.50	520.6	48.89
130	112.0	7.87	69.5	193.5	124.0	38.6	107.5	68.89	0.811	0.0506	1.23	19.7	32.00	512.6	54.44
140	126.8	8.915	78.5	199.0	120.5	43.6	110.6	66.94	0.716	0.0447	1.32	21.1	31.80	509.4	60.00

Propane (C₃H₈)

Temp. (°F)	Abs. press. sat. vap. lb/in²	Abs. press. sat. vap. kg/cm²	Enthalpy abv. 32°F, Btu/lb Liq.	Enthalpy abv. 32°F, Btu/lb Vap.	Ht. of vaporiz. Btu/lb	Enthalpy abv. 0°C, cal/g Liq.	Enthalpy abv. 0°C, cal/g Vap.	Ht. of vaporiz. cal/g	Spec. vol. sat. vap. ft³/lb	Spec. vol. sat. vap. m³/kg	Density of sat. vap. lb/ft³	Density of sat. vap. kg/m³	Density of liq. lb/ft³	Density of liq. kg/m³	Temp. (°C)
−70	7.37	0.518	−55.2	134.3	189.5	−30.7	74.61	105.3	12.9	0.805	0.0775	1.24	37.40	599.1	−56.67
−60	9.72	0.683	−50.2	136.8	187.0	−27.9	76.00	103.9	9.93	0.620	0.111	1.78	37.00	592.7	−51.11
−50	12.6	0.886	−44.7	139.8	184.5	−24.8	77.67	102.5	7.74	0.483	0.129	2.07	36.60	586.3	−45.56
−40	16.2	1.14	−39.7	141.8	181.5	−22.1	78.78	100.8	6.13	0.383	0.163	2.61	36.19	579.7	−40.00
−30	20.3	1.43	−34.2	144.8	179.0	−19.0	80.44	99.44	4.93	0.308	0.203	3.25	35.78	573.1	−34.44
−20	25.4	1.79	−29.2	146.8	176.0	−16.2	81.56	97.78	4.00	0.250	0.250	4.00	35.37	566.6	−28.89
−10	31.4	2.21	−23.7	149.8	173.5	−13.2	83.22	96.39	3.26	0.204	0.307	4.92	34.96	560.0	−23.33
0	38.2	2.69	−18.2	152.3	170.5	−10.1	84.61	94.72	2.71	0.169	0.369	5.91	34.54	553.3	−17.78
10	46.0	3.23	−12.7	155.3	168.0	−7.06	86.28	93.33	2.27	0.142	0.441	7.06	34.12	546.5	−12.22
20	55.5	3.90	−7.2	157.8	165.0	−4.0	87.67	91.67	1.90	0.119	0.526	8.43	33.67	539.3	−6.67
30	66.3	4.66	−1.2	160.8	162.0	−0.67	89.33	90.00	1.60	0.0999	0.625	10.0	33.20	531.8	−1.11
40	78.0	5.48	4.8	163.8	159.0	2.7	91.00	88.33	1.37	0.0855	0.730	11.7	32.73	524.3	4.44
50	91.8	6.45	10.8	166.8	156.0	6.00	92.67	86.67	1.18	0.0737	0.847	13.6	32.24	516.4	10.00
60	107.1	7.530	16.8	169.8	153.0	9.33	94.33	85.00	1.01	0.0631	0.990	15.9	31.75	508.6	15.56
70	124.0	8.718	22.8	172.3	149.5	12.7	95.72	83.06	0.883	0.0551	1.13	18.1	31.24	500.4	21.11
80	142.8	10.04	29.3	175.3	146.0	16.3	97.39	81.11	0.770	0.0481	1.30	20.8	30.70	491.8	26.67
90	164.0	11.53	35.8	178.3	142.5	19.9	99.06	79.17	0.673	0.0420	1.49	23.9	30.15	482.9	32.22
100	187.0	13.15	42.3	180.8	138.5	23.5	100.4	76.94	0.591	0.0369	1.69	27.1	29.58	473.8	37.78
110	213.0	14.98	48.8	182.8	134.0	27.1	101.6	74.44	0.519	0.0324	1.96	31.4	28.85	462.1	43.33
120	240.0	16.87	55.3	184.3	129.0	30.7	102.4	71.67	0.459	0.0287	2.18	34.9	28.30	453.3	48.89

Difluorodichloromethane (CCl₂F₂)("F-12")

Temp. (°F)	Abs. press. sat. vap. lb/in²	Abs. press. sat. vap. kg/cm²	Enthalpy abv. −40°F, Btu/lb Liq.	Enthalpy abv. −40°F, Btu/lb Vap.	Ht. of vaporiz. Btu/lb	Enthalpy abv. −40°C, cal/g Liq.	Enthalpy abv. −40°C, cal/g Vap.	Ht. of vaporiz. cal/g	Spec. vol sat. vap. ft³/lb	Spec. vol sat. vap. m³/kg	Density of vap. lb/ft³	Density of vap. kg/m³	Dens. liq. lb/ft³	Entropy from −40°F, Btu·lb⁻¹·°F Liq.	Entropy from −40°F, Btu·lb⁻¹·°F Vap.	Temp. (°C)
−40	9.32	0.655	0	73.50	73.50	0	40.83	40.83	3.911	0.2442	0.2557	4.096	94.58	0	0.17517	−40.00
−30	12.02	0.845	2.03	74.70	72.67	1.13	41.50	40.37	3.088	0.1928	0.3238	5.187	93.59	0.00471	0.17387	−34.44
−20	15.28	1.074	4.07	75.87	71.80	2.26	42.15	39.89	2.474	0.1544	0.4042	6.474	92.58	0.00940	0.17275	−28.89
−10	19.20	1.350	6.14	77.05	70.91	3.41	42.81	39.39	2.003	0.1250	0.4993	7.998	91.57	0.01403	0.17175	−23.33
0	23.87	1.678	8.25	78.21	69.96	4.58	43.45	38.87	1.637	0.1022	0.6109	9.785	90.52	0.01869	0.17091	−17.78
10	29.35	2.064	10.39	79.36	68.97	5.771	44.09	38.32	1.351	0.08434	0.7402	11.86	89.45	0.02328	0.17015	−12.22
20	35.75	2.513	12.55	80.49	67.94	6.972	44.72	37.74	1.121	0.06998	0.8921	14.29	88.37	0.02783	0.16949	−6.67
30	43.16	3.034	14.76	81.61	66.85	8.200	45.34	37.14	0.939	0.0586	1.065	17.06	87.24	0.03233	0.16887	−1.11
40	51.68	3.633	17.00	82.71	65.71	9.444	45.95	36.51	0.792	0.0494	1.263	20.23	86.10	0.03680	0.16833	4.44
50	61.39	4.316	19.27	83.78	64.51	10.71	46.54	35.84	0.673	0.0420	1.485	23.79	84.94	0.04126	0.16785	10.00
60	72.41	5.091	21.57	84.82	63.25	11.98	47.12	35.14	0.575	0.0359	1.740	27.87	83.78	0.04568	0.16741	15.56
70	84.82	5.963	23.90	85.82	61.92	13.28	47.68	34.40	0.493	0.0308	2.028	32.48	82.60	0.05009	0.16701	21.11
80	98.76	6.944	26.28	86.80	60.52	14.60	48.22	33.62	0.425	0.0265	2.353	37.69	81.39	0.05446	0.16662	26.67
90	114.3	8.036	28.70	87.74	59.04	15.94	48.74	32.80	0.368	0.0230	2.721	43.58	80.11	0.05882	0.16624	32.22
100	131.6	9.252	31.16	88.62	57.46	17.31	49.23	31.92	0.319	0.0199	3.135	50.22	78.80	0.06316	0.16584	37.78
110	150.7	10.60	33.65	89.43	55.78	18.69	49.68	30.99	0.277	0.0173	3.610	57.82	77.46	0.06749	0.16542	43.33
120	171.8	12.08	36.16	90.15	53.99	20.09	50.08	29.99	0.240	0.0150	4.167	66.75	76.02	0.07180	0.16495	48.89

THERMODYNAMIC PROPERTIES (continued)

Carbon Disulfide (CS$_2$)

Temp. (°F)	Abs. press. sat. vap. lb/in²	Abs. press. sat. vap. kg/cm²	Enthalpy abv. 32°F, Btu/lb Liq.	Enthalpy abv. 32°F, Btu/lb Vap.	Ht. of vaporiz. Btu/lb	Enthalpy abv. 0°C, cal/g Liq.	Enthalpy abv. 0°C, cal/g Vap.	Ht. of vaporiz. cal/g	Spec. vol. sat. vap. ft³/lb	Spec. vol. sat. vap. m³/kg	Density sat. vap. lb/ft³	Density sat. vap. kg/m³	Temp. (°C)
0	1.10	0.0773	−8.60	156.90	165.5	−4.78	87.167	91.94	53.76	3.356	0.0186	0.2979	−17.78
10	1.46	0.103	−5.60	158.90	164.5	−3.11	88.278	91.39	43.47	2.714	0.0230	0.3684	−12.22
20	1.89	0.133	−3.00	160.20	163.2	−1.67	89.000	90.67	34.84	2.175	0.0287	0.4597	−6.67
30	2.36	0.166	−0.50	161.70	162.2	−0.28	89.833	90.11	29.49	1.841	0.0339	0.5430	−1.11
40	3.03	0.213	2.05	163.25	161.2	1.14	90.695	89.56	23.52	1.468	0.0425	0.6808	4.44
50	3.90	0.274	4.24	164.24	160.0	2.36	91.245	88.89	20.60	1.286	0.0482	0.7721	10.00
60	4.95	0.348	7.20	166.40	159.2	4.00	92.445	88.44	18.00	1.124	0.0555	0.8890	15.56
70	5.85	0.411	9.80	167.90	158.1	5.44	92.278	87.83	13.20	0.824	0.0758	1.214	21.11
80	7.30	0.513	11.70	168.60	156.9	6.500	93.667	87.17	10.40	0.649	0.0961	1.539	26.67
90	9.15	0.643	13.80	169.40	155.6	7.667	94.111	86.44	8.30	0.518	0.1204	1.920	32.22
100	11.08	0.7790	16.15	170.55	154.4	8.972	94.750	85.78	7.03	0.439	0.1369	2.193	37.78
110	13.50	0.9491	18.30	171.50	153.2	10.17	95.278	85.11	5.80	0.362	0.1724	2.762	43.33
120	16.10	1.132	20.01	172.01	152.0	11.12	95.561	84.44	5.10	0.318	0.1960	3.140	48.89

Carbon Tetrachloride (CCl$_4$)

Temp. (°F)	Abs. press. sat. vap. lb/in²	Abs. press. sat. vap. kg/cm²	Enthalpy abv. 32°F, Btu/lb Liq.	Enthalpy abv. 32°F, Btu/lb Vap.	Ht. of vaporiz. Btu/lb	Enthalpy abv. 0°C, cal/g Liq.	Enthalpy abv. 0°C, cal/g Vap.	Ht. of vaporiz. cal/g	Spec. vol. sat. vap. ft³/lb	Spec. vol. sat. vap. m³/kg	Density sat. vap. lb/ft³	Density sat. vap. kg/m³	Temp. (°C)
20	0.40	0.028	−2.00	92.45	94.45	−1.11	51.36	52.47	69.5	4.34	0.01438	0.2303	−6.67
30	0.60	0.042	−0.25	93.45	93.70	−0.14	51.92	52.06	53.0	3.31	0.01886	0.3021	−1.11
40	0.84	0.059	1.60	94.80	93.20	0.889	52.67	51.78	40.0	2.50	0.02500	0.4005	4.44
60	1.42	0.100	5.95	98.15	92.20	3.31	54.53	51.22	24.0	1.50	0.04166	0.6673	15.56
70	1.85	0.130	8.20	99.53	91.40	4.56	55.29	50.78	19.5	1.22	0.05128	0.8214	21.11
80	2.40	0.169	9.80	99.87	90.07	5.44	55.48	50.04	16.0	0.999	0.06345	1.016	26.67
90	3.12	0.219	11.60	101.62	90.02	6.444	56.46	50.01	13.0	0.812	0.07692	1.232	32.22
100	4.00	0.281	13.40	102.80	89.40	7.444	57.11	49.28	10.0	0.624	0.1000	1.602	37.78
110	4.89	0.344	15.80	104.50	88.70	8.778	58.06	49.28	8.5	0.53	0.1176	1.884	43.33
120	5.95	0.418	18.06	105.90	87.90	10.03	58.83	48.83	7.5	0.47	0.1333	2.135	48.89

Ethyl Ether ((C$_2$H$_5$)$_2$O)

Temp. (°F)	Abs. press. sat. vap. lb/in²	Abs. press. sat. vap. kg/cm²	Enthalpy abv. 32°F, Btu/lb Liq.	Enthalpy abv. 32°F, Btu/lb Vap.	Ht. of vaporiz. Btu/lb	Enthalpy abv. 0°C, cal/g Liq.	Enthalpy abv. 0°C, cal/g Vap.	Ht. of vaporiz. cal/g	Spec. vol. sat. vap. ft³/lb	Spec. vol. sat. vap. m³/kg	Density sat. vap. lb/ft³	Density sat. vap. kg/m³	Temp. (°C)
0	1.3	0.091	−18.00	153.00	171.0	−10.00	85.000	95.00	38.0	2.37	0.0263	0.4213	−17.78
10	1.8	0.13	−12.0	158.43	170.4	−6.67	88.017	94.67	32.5	2.03	0.0332	0.5318	−12.22
20	2.5	0.18	−6.50	163.50	170.0	−3.61	90.833	94.44	27.0	1.69	0.0372	0.5959	−6.67
30	3.4	0.24	−1.50	167.90	169.4	−0.833	93.278	94.11	21.4	1.34	0.0468	0.7496	−1.11
40	4.4	0.31	4.00	172.40	168.4	2.22	95.778	93.56	17.0	1.06	0.0588	0.9419	4.44
50	5.5	0.39	9.57	177.17	167.6	5.32	98.428	93.11	13.2	0.824	0.0757	1.213	10.00
70	8.8	0.62	20.04	185.44	165.4	11.13	103.02	91.89	7.8	0.49	0.1280	2.050	21.11
80	10.9	0.766	26.40	190.60	164.2	14.67	105.89	91.22	6.2	0.39	0.1620	2.595	26.67
90	13.4	0.942	31.50	194.50	163.0	17.50	108.60	90.56	5.1	0.32	0.1960	3.140	32.22
100	16.0	1.12	36.50	197.50	161.5	20.28	109.72	89.72	4.5	0.28	0.2220	3.556	37.78

From Weast, R. C., Ed., *Handbook of Chemistry and Physics,* 69th ed., CRC Press, Boca Raton, FL, 1988.

TABLE 10
Brand Names of Polymeric Materials

Material	Chart classification	Material	Chart classification
Aeroflex	Polyethylene	Mylar	Polyester
Alathon	Polyethylene	Nylon	Nylon
Araldite	Epoxy	Penton	Polyether
Avisco	Urea	Plexiglas	Methyl methacrylate
Bakelite	Phenolic	Plioflex	Vinyl
Beelte	Urea	Polythene	Polyethylene
Dacron	Polyester	Pro-Fax	Polypropylene
Durcon	Epoxy	PVC	Polyvinyl chloride
Durez	Phenolic	Resinox	Phenolic
Dypol	Polyester	Saran	Vinyl
Epon	Epoxy	Styron	Polystyrene
Excon	Polypropylene	Teflon	Fluorocarbon
Kel F	Fluorocarbon	Tygon	Vinyl
Lauxite	Urea	Vibrin	Polyester
Lucite	Methyl methacrylate	Vinylite	Vinyl
Lustrex	Polystryene	Viton	Fluorocarbon
Moplen	Polypropylene		

From Bonner, T. A., Cornett, C. L., Dasai, B. O., Fullenkamp, J. M., Hughes, T. W., Johnson, M. L., Kennedy, E. D., McCormick, R. J., Peters, J. A., and Zanders, D. L., Engineering Handbook for Hazardous Waste Incineration, Rep. MRC-DA 1090, U.S. Environmental Protection Agency, Washington, D.C., 1981.

TABLE 11
Property Comparisons — Natural and Synthetic Rubbers

Property	Natural rubber	Butyl (GR-I)	Buna S (GR-S)	Neoprene	Nitrile (buna N)	Polyacrylic rubber	Silicone rubber
Hardness range (Shore "A")[a]	40—100	40—90	40—100	30—90	45—100	50—90	40—80
Tensile strength, psi[b]	4500	3000	3500	3500	4000	1500	900
Max. elongation	900	900	600	1000	700	200	250
Abrasion resistance[c]	Excellent	Good	Excellent	Very good	Excellent	Fair	Poor
Resistance to compression set at 158°F[c]	Good	Fair	Excellent	Good	Excellent	Good	Excellent
Resistance to compression set up to 2500°F[c]	Poor	Poor	Excellent	Fair	Excellent	Good	Excellent
Aging resistance (normal temp)	Good	Excellent	Excellent	Excellent	Excellent	Excellent	Excellent
Max ambient temp allowable (°F)	160	275	275	225	300	400	580
Resistance to weather and ozone	Fair	Very good	Fair	Excellent	Fair	Excellent	Excellent
Resistance to flexing	Excellent	Excellent	Good	Excellent	Fair	Excellent	Poor
Resistance to diffusion of gases	Fair	Excellent	Fair	Very good	Fair	Excellent	
Resilience	Excellent	Poor at low temperature; good at high temperature	Fair	Very good	Fair	Poor	
Resistance to petroleum oils and greases	Poor	Poor	Poor	Good	Excellent	Very good	Good
Resistance to vegetable oils	Good	Good					
Resistance to nonaromatic fuels and solvents	Poor	Poor	Poor	Fair to good	Very good		Fair
Resistance to aromatic fuels and solvents	Poor	Poor	Poor	Fair	Good		Poor
Resistance to water and antifreezes[c]	Good	Good	Good	Fair	Excellent	Poor	Fair
Resistance to dilute acids	Good	Good	Good	Good	Good		
Resistance to oxidizing agents	Poor	Fair	Poor	Poor	Poor		
Resistance to alkali	Fair	Fair	Fair	Good	Fair		
Bielectric strength[c]	Excellent	Good	Excellent	Fair	Fair		
Flame resistance	Poor	Poor	Poor	Good	Poor		
Processing characteristics	Excellent	Good	Good	Good	Good	Fair	Poor
Low temperature resistance[c]	Very good	Fair	Good	Fair	Good	Poor	Excellent
Tear resistance	Excellent	Excellent	Good	Good	Good	Fair	Poor

[a] 100 durometer reading is bone hard and indicates that ebonite or hard rubber can be made.

[b] Indicates soft-rubber type. Hard-rubber types run higher in value.

[c] These properties available in specific compounds.

From Huibregtse, K. R., Sholz, R. C., Wullschleger, R. E., Moser, J. M., Bollinge, E. R., and Hansen, C. A., Manual for the Control of Hazardous Material Spills, Vol. 1, Spill Assessment and Water Treatment Techniques, Rep. EPA-600/2-77-227, U.S. Environmental Protection Agency, Cincinnati, OH, 1977.

TABLE 12
Properties of Commercially Available Plastics

Material	Acids		Alkalies		Organic solvents	Water absorption, %/24 h	Oxygen and ozone	High vacuum	Ionizing radiation	Temperature resistance	
	Weak	Strong	Weak	Strong						High	Low
Thermoplastics											
Fluorocarbons	Inert	Inert	Inert	Inert	Inert	0.0	Inert	—	P	550	G-275
Methyl methacrylate	R	A-O	R	A	A	0.2	R	decomp.	P	180	—
Nylon	G	A	R	R	R	1.5	SA	—	F	300	G-70
Polyether (chlorinated)	R	A-O	R	R	G	0.01	R	—	—	280	G
Polyethylene (low density)	R	A-O	R	R	G	0.15	A	F	F	140	G-80
Polyethylene (high density)	R	A-O	R	R	G	0.1	A	F	G	160	G-100
Polypropylene	R	A-O	R	R	R	<0.01	A	F	G	300	P
Polystyrene	R	A-O	R	R	A	0.04	SA	P	G	160	P
Rigid polyvinyl chloride	R	R	R	R	A	0.10	R	—	P	150	P
Vinyls (chloride)	R	R	R	R	A	0.45	R	P	P	160	—
Thermosetters											
Epoxy (cast)	R	SA	R	R	G	0.1	SA	—	G	400	L
Phenolics	SA	A	SA	A	SA	0.6	—	—	G	400	L
Polyesters	SA	A	A	A	SA	0.2	A	—	G	350	L
Silicones	SA	SA	SA	SA	A	0.15	R	—	F	550	L
Ureas	A	A	A	A	A	0.6	A	—	P	170	L

Note: R = resistant, A = attacked, SA = slight attack, A-O = attacked by oxidizing acids, G = good, F = fair, P = poor, L = little change.

From Huibregtse, K. R., Sholz, R. C., Wullschleger, R. E., Moser, J. M., Bollinge, E. R., and Hansen, C. A., Manual for the Control of Hazardous Material Spills, Vol. 1, Spill Assessment and Water Treatment Techniques, Rep. EPA-600/2-77-227, U.S. Environmental Protection Agency, Cincinnati, OH, 1977.

TABLE 13
General Characteristics of Silica and Aluminosilicate Refractory Brick

Type	Typical composition	Fusion temperature, °F[a]	Resistant to	Degraded by
Silica	95% SiO_2	3100	HCl, NH_3, acid slags	Basic slags, Al, Na, Mg, F_2, Cl_2, H_2 (>2550°F)
High-duty fire-clay	54% SiO_2 40% Al_2O_3	3125	Most acids, slag conditions	High-lime slags, other bases at high temperature
Super-duty fireclay	52% SiO_2 42% Al_2O_3	3170	HCl, NH_3, SO_2, most acids	Basic slags, Na, Mg, F_2, Cl_2, H_2 (>2550°F)
Acid resistant (type H)	59% SiO_2	3040	Excellent for most acids; bases in moderate concentration	HF, H_3PO_4
High alumina	50—85% Al_2O_3	3200—3400	HCl, NH_3, SO_2	Basic slags, Na, Mg, F_2, Cl_2, H_2 (>2550°F)
Extra-high alumina	90—99% Al_2O_3	3000—3650	HCl, HF, NH_3, SO_2, S_2, HNO_3, H_2SO_4, Cl_2	Na, F_2 (>1800°F)
Mullite	71% Al_2O_3	3290	HCl, SO_2, NH_3	Na, F_2, Cl_2, H_2 (>2550°F)

[a] A safety factor of at least several hundred degrees between refractory fusion temperature and incinerator operating temperature is advisable.

From Bonner, T. A., Cornett, C. L., Desai, B. O., Fullenkamp, J. M., Hughes, T. W., Johnson, M. L., Kennedy, E. D., McCormick, R. J., Peters, J. A., and Zanders, D. L., Engineering Handbook for Hazardous Waste Incineration, Rep. MRC-DA 1090, U.S. Environmental Protection Agency, Washington, D.C., 1981.

TABLE 14
Standard Types of Stainless and Heat Resisting Steels

Chemical Ranges and Limits
Subject to Tolerances for Check Analyses
By permission of American Iron and Steel Institute

Chemical composition (%)

Type number	C	Mn Max.	P Max	S Max.	Si Max.	Cr	Ni	Mo	Zr	Se	Cb-Ta	Ta	Al	N
201[c]	0.15 Max.	5.50/ 7.50	0.060	0.030	1.00	16.00/ 18.00	3.50/ 5.50							0.25 Max.
202[c]	0.15 Max.	7.50/ 10.00	0.060	0.030	1.00	17.00/ 19.00	4.00/ 6.00							0.25 Max.
301[c]	0.15 Max.	2.00	0.045	0.030	1.00	16.00/ 18.00	6.00/ 8.00							
302[c]	0.15 Max.	2.00	0.045	0.030	1.00	17.00/ 19.00	8.00/ 10.00							
302B[c]	0.15 Max.	2.00	0.045	0.030	2.00/ 3.00	17.00/ 19.00	8.00/ 10.00							
303[c]	0.15 Max.	2.00	0.20	0.15 Min.	1.00	17.00/ 19.00	8.00/ 10.00	0.60 Max.[a]	0.60 Max.[a]					
303 Se[c]	0.15 Max.	2.00	0.20	0.06	1.00	17.00/ 19.00	8.00/ 10.00			0.15 Min.				
304[c]	0.08 Max.	2.00	0.045	0.030	1.00	18.00/ 20.00	8.00/ 12.00							
304L[c]	0.03 Max.	2.00	0.045	0.030	1.00	18.00/ 20.00	8.00/ 12.00							
305[c]	0.12 Max.	2.00	0.045	0.030	1.00	17.00/ 19.00	10.00/ 13.00							
308[c]	0.08 Max.	2.00	0.045	0.030	1.00	19.00/ 21.00	10.00/ 12.00							
309[c]	0.20 Max.	2.00	0.045	0.030	1.00	22.00/ 24.00	12.00/ 15.00							
309S[c]	0.08 Max.	2.00	0.045	0.030	1.00	22.00/ 24.00	12.00/ 15.00							
310[c]	0.25 Max.	2.00	0.045	0.030	1.50	24.00/ 26.00	19.00/ 22.00							
310S[c]	0.08 Max.	2.00	0.045	0.030	1.50	24.00/ 26.00	19.00/ 22.00							
314[c]	0.25 Max.	2.00	0.045	0.030	1.50/ 3.00	23.00/ 26.00	19.00/ 22.00							
316[c]	0.08 Max.	2.00	0.045	0.030	1.00	16.00/ 18.00	10.00/ 14.00	2.00/ 3.00						
316L[c]	0.03 Max.	2.00	0.045	0.030	1.00	16.00/ 18.00	10.00/ 14.00	2.00 3.00						
317[c]	0.08 Max.	2.00	0.045	0.030	1.00	18.00/ 20.00	11.00/ 15.00	3.00/ 4.00						
321[c]	0.08 Max.	2.00	0.045	0.030	1.00	17.00/ 19.00	9.00/ 12.00				5 × C Min			
347[c]	0.08 Max.	2.00	0.045	0.030	1.00	17.00/ 19.00	9.00/ 13.00					10 × C Min.		
348[c]	0.08 Max.	2.00	0.045	0.030	1.00	17.00/ 19.00	9.00/ 13.00				10 × C Min	0.10 Max		
403[b]	0.15 Max.	1.00	0.040	0.030	0.50	11.50/ 13.00								
405[d]	0.08 Max.	1.00	0.040	0.030	1.00	11.50/ 14.50							0.10/ 0.30	
410[b]	0.15 Max.	1.00	0.040	0.030	1.00	11.50/ 13.50								
414[b]	0.15 Max.	1.00	0.040	0.030	1.00	11.50/ 13.50	1.25/ 2.50							
416[b]	0.15 Max.	1.25	0.06	0.15 Min.	1.00	12.00/ 14.00		0.60 Max.[a]	0.60 Max.[a]					
416 Se[b]	0.15 Max.	1.25	0.06	0.06	1.00	12.00/ 14.00				0.15 Min.				
420[b]	Over 0.15	1.00	0.040	0.030	1.00	12.00/ 14.00								
430[a]	0.12 Max.	1.00	0.040	0.030	1.00	14.00/ 18.00								
430F[d]	0.12 Max.	1.25	0.06	0.15 Min.	1.00	14.00/ 18.00		0.60 Max.	0.60 Max.[a]					
430F Se[d]	0.12 Max.	1.25	0.06	0.06	1.00	14.00/ 18.00				0.15 Min.				
431[b]	0.20 Max.	1.00	0.040	0.030	1.00	15.00/ 17.00	1.25/ 2.50							
440A[b]	0.60/ 0.75	1.00	0.040	0.030	1.00	16.00/ 18.00		0.75 Max.						

STANDARD TYPES OF STAINLESS AND HEAT RESISTING STEELS (continued)

Chemical composition (%)

Type number	C	Mn Max.	P Max.	S Max.	Si Max.	Cr	Ni	Mo	Zr	Se	Cb-Ta	Ta	Al	N
400B[b]	0.75/ 0.95	1.00	0.040	0.030	1.00	16.00/ 18.00		0.75 Max.						
440C[b]	0.95/ 1.20	1.00	0.040	0.030	1.00	16.00/ 18.00		0.75 Max.						
446[d]	0.20 Max.	1.50	0.040	0.030	1.00	23.00/ 27.00								0.25 Max.
501[b]	Over 0.10	1.00	0.040	0.030	1.00	4.00/ 6.00		0.40/ 0.65						
502[b]	0.10 Max.	1.00	0.040	0.030	1.00	4.00/ 6.00		0.40/ 0.65						

[a] At producer's option; reported only when intentionally added.
[b] Heat treatable
[c] Not heat treatable.
[d] Essentially not heat treatable.

From Weast, R. C., Ed., *Handbook of Chemistry and Physics*, 69th ed., CRC Press, Boca Raton, FL, 1988.

TABLE 15
Properties of High-Temperature Commercial Alloys

Alloy: Haynes Alloy 556[a]
Nominal composition (%):

Fe	Ni	Co	Cr	Mo	W	Ta	N	Si	Mn	Al	C	La	Zr
Bal.	20.0	18.0	22.0	3.0	2.5	0.6	0.2	0.4	1.0	0.2	0.10	0.02	0.02

Mechanical properties at elevated temperatures:

Property	Units	Temperature (°F) 1000	1200	1400	1600	1800	2000
Ultimate tensile strength[b]	ksi	93.4	85.4	68.5	47.6	28.0	14.8
0.2% offset yield strength	ksi	34.9	32.8	32.0	28.6	15.5	8.0
10,000 h rupture stress	ksi		27.5	11.9	4.9	1.9	
10,000 h, 1% creep stress	ksi		18.5	8.5	4.1	1.6	
Thermal conductivity	Btu/h-ft-°F	12.3	13.3	14.4	15.4	16.4	17.5
Specific heat	Btu/lb-°F	0.130	0.133	0.135	0.140	0.147	0.152
Thermal expansion coefficient[c]	10^{-6} in./in.-°F	8.8	9.0	9.2	9.4	9.5	9.6
Dynamic modulus of elasticity	10^6 psi	24.4	23.1	21.8	20.9	20.1	
Poisson's ratio	—						

[a] Product of Haynes International, Inc.
[b] Cold rolled solution annealed sheet, 0.033 to 0.109 in. thick.
[c] Mean value between 70°F and indicated temperature.

TABLE 15 (continued)
Properties of High-Temperature Commercial Alloys

Alloy: Haynes Alloy 214[a]
Nominal composition (%):

Ni	Cr	Fe	Al	Y
Bal.	16.0	3.0	4.5	Present

Mechanical properties at elevated temperatures:

Property	Units	Temperature (°F)					
		1000	1200	1400	1600	1800	2000
Ultimate tensile strength[b]	ksi		120.2	101.7	74.7	14.8	7.5
0.2% Offset yield strength[b]	ksi		85.4	79.6	56.6	7.9	2.7
10,000 h rupture stress[c]	ksi	—	—	13.0	5.0	1.0	—
1000 h, 1% creep stress	ksi	—	—	16.0	5.0	0.8	—
Thermal conductivity	Btu/h-ft-°F	12.8	14.6	16.7	17.9	18.8	19.5
Specific heat	Btu/lb-°F	0.136	0.154	0.166	0.173	0.177	0.179
Thermal expansion coefficient[d]	10^{-6} in./in.-°F	8.2	8.6	9.0	9.6	10.2	11.1
Dynamic modulus of elasticity	10^6 psi	26.3	25.3	23.9	22.3	20.2	19.0

[a]   Product of Haynes International, Inc.
[b]   Sheet, 0.030 to 0.5 in. thick, annealed at 2000°F.
[c]   Extrapolated data.
[d]   Mean value between 86°F and indicated temperature.

Alloy: Haynes Alloy 230[a]
Nominal composition (%):

Ni	Cr	W	Mo	Fe	Co	Mn	Si	Al	C	La	B
Bal.	22	14	2	3[b]	5[b]	0.5	0.4	0.3	0.10	0.02	0.005

Mechanical properties at elevated temperatures:

Property	Units	Temperature (°F)					
		1000	1200	1400	1600	1800	2000
Ultimate tensile strength[c]	ksi	103.1	97.2	84.9	58.3	32.5	17.3
0.2% Offset yield strength[c]	ksi	39.7	39.0	41.2	32.4	17.3	8.2
10,000 h rupture stress[d]	ksi	—	29.0	14.2	6.2	1.6	—
10,000 h, 1% creep stress[d]	ksi	—	17.5	8.0	4.4	1.1	—
Thermal conductivity	Btu/h-ft-°F	11.1	12.3	13.7	14.9	16.3	
Specific heat	Btu/lb-°F	0.112	0.134	0.140	0.145	0.147	
Thermal expansion coefficient[e]	10^{-6} in./in.-°F	7.9	8.1	8.3	8.6	8.9	
Dynamic modulus of elasticity	10^6 psi	26.4	25.3	24.1	23.1	21.9	

[a]   Product of Haynes International, Inc.
[b]   Maximum.
[c]   Cold rolled sheet, 2250°F solution annealed.
[d]   Plate and Bar, 2250°F solution annealed.
[e]   Mean value between 70°F and indicated temperature.

TABLE 15 (continued)
Properties of High-Temperature Commercial Alloys

Alloy: Inconel Alloy 625[a]
Nominal composition (%):

Fe	Ni	Cr	Mo	Cb + Ta	Mn	Si	Al	Ti	Co
5 max	58 min	20—23	8—10	3.15—4.15	0.50 max	0.50 max	0.40 max	0.40 max	1.0 max

Mechanical properties at elevated temperatures

Property	Units	Temperature (°F)					
		1000	1200	1400	1600	1800	2000
Ultimate tensile strength[b]	ksi	133	120	78	41	19	10
10,000 h rupture stress[c]	ksi		44	16	4.8		
10,000 h, 1% creep stress[c]	ksi		42	12	3.3		
Thermal conductivity	Btu/h-ft-°F	10.1	11.0	12.0	13.2	14.6	
Specific heat	Btu/lb-°F	0.128	0.135	0.141	0.148	0.154	0.160
Thermal expansion coefficient[d]	10^{-6} in./in.-°F	7.8	8.2	8.5	8.8		
Dynamic modulus of elasticity[e]	10^6 psi	25.9	24.7	23.3	21.4		
Poisson's ratio	—		0.312	0.314	0.305	0.289	

[a] Product of Inco Alloys International.
[b] Annealed bar.
[c] Solution treated.
[d] Mean value between 70°F and indicated temperature.
[e] ¾-in. hot rolled bar.

Alloy: Incolony 800[a]
Nominal composition (%):

Fe	Ni	Cr	C	Mn	S	Si	Cu	Al	Ti
39.5 min	30—35	19—23	0.10 max	1.50 max	0.015 max	1.0 max	0.75 max	0.15—0.60	0.15—0.60

Mechanical properties at elevated temperatures:

Property	Units	Temperature (°F)					
		1000	1200	1400	1600	1800	2000
Ultimate tensile strength[b]	ksi	73.7	58.7	34.5			
10,000 h rupture stress[c]	ksi	37	15				
10,000 h, 1% creep stress[c]	ksi	34	12				
Thermal conductivity	Btu/h-ft-°F	11.6	12.7	13.8	15.1	17.8	
Specific heat	Btu/lb-°F						
Thermal expansion coefficient[d]	10^{-6} in./in.-°F	9.4	9.6	9.9	10.2		
Dynamic modulus of elasticity	10^6 psi	23.5	22.4	21.1	19.2		
Poisson's ratio	—		0.367	0.377	0.389	0.408	

[a] Product of Inco Alloys International.
[b] Hot rolled bar, annealed at 1800°F.
[c] Annealed.
[d] Mean value between 70°F and indicated temperature.

TABLE 15 (continued)
Properties of High-Temperature Commercial Alloys

Alloy: Incoloy Alloy 800HT[a]
Nominal composition (%):

Ni	Cr	Fe	C	Mn	S	Si	Cu	Al	Ti	Al + Ti
30-35	19-23	39.5 min	0.06-0.10	1.50 max	0.015 max	1.0 max	0.75 max	0.15-0.60	0.15-0.60	0.85-1.20

Mechanical properties at elevated temperatures:

Property	Units	Temperature (°F)					
		1000	1200	1400	1600	1800	2000
Ultimate tensile strength[b]	ksi	68	58	39	21	12	5
10,000 h rupture stress	ksi		17.5	7.3	3.5	1.2	0.47
10,000 h, 1% creep stress	ksi		17	5.8	3.5	1.0	
Thermal conductivity	Btu/h-ft-°F	11.6	12.7	13.8	15.1	17.8	
Specific heat	Btu/lb-°F						
Thermal expansion coefficient[c]	10^{-6} in./in.-°F	9.4	9.6	9.9	10.2		
Dynamic modulus of elasticity	10^6 psi	23.5	22.4	21.1	19.2		
Poisson's ratio	—		0.367	0.377	0.389	0.408	

[a] Product of Inco Alloys International.
[b] Annealed extruded tubing 5 in. OD × 0.5 in. wall.
[c] Mean value between 70°F and indicated temperature.

Alloy: Inconel Alloy 617[a]
Nominal composition (%):

Ni	Cr	Co	Mo	Al	C	Fe	Mn	Si	S	Ti	Cu
52	22	12.5	9.0	1.2	0.07	1.5	0.5	0.5	0.008	0.3	0.2

Mechanical properties at elevated temperatures:

Property	Units	Temperature (°F)					
		1000	1200	1400	1600	1800	2000
Ultimate tensile strength[b]	ksi	88	85	77	49	25	14
10,000 h rupture stress[c]	ksi		39	16	5.5	2.2	0.8
10,000 h, 1% creep stress[c]	ksi		39	12	3.7	1.7	0.8
Thermal conductivity	Btu/h-ft-°F	12.4	13.4	14.4	15.4	16.4	17.4
Specific heat	Btu/lb-°F	0.131	0.137	0.144	0.150	0.157	0.163
Thermal expansion coefficient[d]	10^{-6} in./in.-°F	7.7	8.0	8.4	8.7	9.0	9.2
Dynamic modulus of elasticity	10^6 psi	25.8	24.6	23.3	21.9	20.5	18.8
Poisson's ratio	—	0.30	0.30	0.30	0.30	0.31	0.32

[a] Product of Inco Alloys International.
[b] Solution-annealed hot rolled rod.
[c] Solution annealed.
[d] Mean value between 78°F and indicated temperature.

TABLE 15 (continued)
Properties of High-Temperature Commercial Alloys

Alloy: Inconel Alloy 601[a]
Nominal composition (%):

Ni	Cr	Fe	Al	C	Mn	S	Si	Cu
58—63	21—25	bal	1.0—1.7	0.10 max	1.0 max	0.015 max	0.50 max	1.0 max

Mechanical properties at elevated temperatures:

Property	Units	Temperature (°F)					
		1000	1200	1400	1600	1800	2000
Ultimate tensile strength[b]	ksi	95	72	37	18	8	5
10,000 h rupture stress[c]	ksi	14.2	12.1	6.1	2.6	1.2	
10,000 h, 1% creep stress[c]	ksi	43.0	18.0	4.0	2.0	0.71	0.43
Thermal conductivity	Btu/h-ft-°F	11.6	12.8	13.8	14.8	15.8	16.9
Specific heat	Btu/lb-°F	0.140	0.147	0.155	0.162	0.169	0.176
Thermal expansion coefficient[d]	10^{-6} in./in.-°F	8.50	8.87	9.19	9.51	9.82	10.18
Dynamic modulus of elasticity	10^6 psi	25.4	24.1	22.5	20.5	18.4	16.2
Poisson's ratio	—	0.314	0.333	0.351	0.366	0.390	0.426

[a] Product of Inco Alloys International.
[b] Hot finished rod annealed at 1800°F.
[c] Solution treated at 2100°F.
[d] Mean value between 80°F and indicated temperature.

Alloy: Incoloy Alloy 825[a]
Nominal composition (%):

Ni	Cr	Fe	Mo	Cu	Ti	C	Mn	S	Si	Al
38—46	19.5—23.5	22 min	2.5—3.5	1.50—3.0	0.6—1.2	0.05 max	1.0 max	0.03 max	0.5 max	0.2 max

Mechanical properties at elevated temperatures:

Property	Units	Temperature (°F)					
		1000	1200	1400	1600	1800	2000
Ultimate tensile strength[b]	ksi	85	67	40	20	12	6
10,000 h rupture stress	ksi						
10,000 h, 1% creep stress	ksi						
Thermal conductivity	Btu/h-ft-°F	10.9	11.8	12.9	14.3	16.0	
Specific heat	Btu/lb-°F						
Thermal expansion coefficient[c]	10^{-6} in./in.-°F	8.8	9.1	9.5	9.7		
Dynamic modulus of elasticity	10^6 psi	25	23.8	22.5	20.9	19.0	16.8
Poisson's ratio	—	0.45	0.46	0.47	0.47	0.47	0.51

[a] Product of Inco Alloys International.
[b] Cold drawn rod 0.75 in. diameter annealed at 1725°F.
[c] Mean value between 80°F and indicated temperature.

TABLE 15 (continued)
Properties of High-Temperature Commercial Alloys

Alloy: RA 253 MA[a]

Nominal composition (%):

Ni	Cr	Fe	C	Si	Mn	P	S	N	Ce
10—12	20—22	bal	0.05—0.10	1.4—2.0	0.80 max	0.04 max	0.03 max	0.14—0.20	0.03—0.08

Mechanical properties at elevated temperatures:

Property	Units	Temperature (°F)					
		1000	1200	1400	1600	1800	2000
Ultimate tensile strength	ksi						
10,000 h rupture stress	ksi		14.0	5.2	2.5	1.15	0.68
10,000 h, 1% creep stress	ksi		11.6	5.0	2.3	0.89	
Thermal conductivity	Btu/h-ft-°F		13.0	14.0		16.6	
Specific heat	Btu/lb-°F		0.142	0.149		0.164	
Thermal expansion coefficient[b]	10^{-6} in./in.-°F	9.97	10.14	10.3	10.5	10.8	
Dynamic modulus of elasticity	10^6 psi		21.7	20.2		17.6	
Poisson's ratio	—						

[a]   Product of Rolled Alloys, Inc.
[b]   Mean value between 68°F and indicated temperature.

Alloy: RA 310[a]

Nominal composition (%):

Ni	Cr	Fe	C[b]	Si[c]	Mn	P	S	Mo	Cu
19—22	24—26	bal	0.08 max	0.75 max	2.00 max	0.04 max	0.030 max	0.75 max	0.5

Mechanical properties at elevated temperatures:

Property	Units	Temperature (°F)					
		1000	1200	1400	1600	1800	2000
Ultimate tensile strength[d]	ksi	67.8	54.1	35.1	19.1	10.4	
10,000 h rupture stress[d]	ksi		11	3.9	1.5	0.54	
10,000 h, 1% creep stress	ksi		7.0			0.12	
Thermal conductivity	Btu/h-ft-°F						
Specific heat	Btu/lb-°F	0.132	0.139	0.146	0.153	0.160	0.167
Thermal expansion coefficient[e]	10^{-6} in./in.-°F	9.5	9.8	10.05	10.15	10.30	10.60
Dynamic modulus of elasticity	10^6 psi	23	21.8	20.5	19.2		
Poisson's ratio	—		0.32		0.34		

[a]   Product of Rolled Alloys, Inc.
[b]   All forms except pipe, C 0.15 max.
[c]   All forms except bars, forgings, and rings, Si 0.3—0.8.
[d]   Annealed.
[e]   Mean value between 70°F and indicated temperature.

TABLE 15 (continued)
Properties of High-Temperature Commercial Alloys

Alloy: RA 330[a]
Nominal composition (%):

Ni	Cr	Fe	C	Si[b]	Mn	P	S	Cu
34—37	17—20	bal	0.08 max	1—1.5	2.0 max	0.030 max	0.03 max	1.00 max

Mechanical properties at elevated temperatures:

Property	Units	Temperature (°F)					
		1000	1200	1400	1600	1800	2000
Ultimate tensile strength	ksi	71	56.7	35.9	21.1	10.4	
10,000 h rupture stress	ksi	29	11	4.3	1.7	0.63	
10,000 h, 1% creep stress	ksi	21	7.6	3.6	2.1	0.5	
Thermal conductivity	Btu/h-ft-°F	12.8	13.2	13.7	14.2	14.7	15.1
Specific heat	Btu/lb-°F						0.140
Thermal expansion coefficient[c]	10^{-6} in./in.-°F	9.3			9.8	10.0	
Dynamic modulus of elasticity	10^6 psi	23.8	22.3	21.0	19.5	18.0	
Poisson's ratio	—						

[a] Product of Rolled Alloys, Inc.
[b] All product forms except welded pipe and tube, Si 0.75—1.50.
[c] Mean value between 77°F and indicated temperature.

Alloy: RA 333[a]
Nominal composition (%):

Ni	Cr	Fe	Mo	Co	W	Si	Mn	C	P	S
45	25	18	3.00	3.00	3.00	1.25	1.50	0.05	0.015	0.015

Mechanical properties at elevated temperatures:

Property	Units	Temperature (°F)					
		1000	1200	1400	1600	1800	2000
Ultimate tensile strength[b]	ksi		75	57.2	33.3	16.2	
10,000 h rupture stress	ksi			9.4	3.1	1.05	0.35
10,000 h, 1% creep stress	ksi			6.2	3.1	0.90	0.26
Thermal conductivity	Btu/h-ft-°F	11.3	12.4	13.5	14.6	15.7	
Specific heat	Btu/lb-°F						
Thermal expansion coefficient[c]	10^{-6} in./in.-°F	8.9	9.1	9.3	9.5	9.8	
Dynamic modulus of elasticity	10^6 psi				19.5	18.0	
Poisson's ratio	—						

[a] Product of Rolled Alloys, Inc.
[b] Mill annealed.
[c] Mean value between 70°F and indicated temperature.

Appendix B.2

COMBUSTION PARAMETERS

I. Combustion 506

II. Thermal Decomposition 517

III. Incinerability 518

I. COMBUSTION

TABLE 1
Heat of Combustion for Organic Compounds

The heat of combustion is given in kilogram calores per gram molecular weight of the substance when combustion takes place at atmospheric pressure and at either 20°C or 25°C. If the data are for 20°C there is no asterisk for the numerical value of the heat of combustion. If the numerical value is for 25°C there is an asterisk marking the numerical value of the heat of combustion. The final products of combustion are gaseous carbon dioxide, liquid water and nitrogen gas for C, H, N compounds. For method of computing heats of formation see statement following this table.

Name	Formula	Physical state	Heat of combustion, kg. calories
Acetaldehyde	CH_3CHO	liquid	278.77*
Acetamide	CH_3CONH_2	solid	282.6
Acetanilide	$CH_3CONHC_6H_5$	solid	1010.4
Acetic acid	CH_3CO_2H	liquid	209.02*
Acetic anhydride	$(CH_3CO)_2O$	liquid	431.70*
Acetone	$(CH_3)_2CO$	liquid	427.92*
Acetonitrile	CH_3CN	liquid	302.4
Acetophenone	$C_6H_5COCH_3$	liquid	991.60*
Acetylacetone	$CH_3COCH_2COCH_3$	liquid	615.9
Acetylene	$(CH)_2$	gas	310.61*
Acrolein	CH_2CHO	liquid	389.6
Acrylic acid	CH_2CO_2H	liquid	327.0*
Adipic acid	$(CH_2)_4(CH_2H)_2$	solid	668.29*
Alanine	$CH_3CH(NH_2)CO_2H$	solid	387.7
Aldol, see β-hydroxybutyr-aldehyde			
Alizarin, see Dihydroxyanthraquinone			
Allyl alcohol	CH_2CH_2OH	liquid	442.4
Allylene	CH_3C	gas	465.1
p-Aminoazobenzene	$H_2NC_6H_4N_2C_6H_5$	solid	1574.0
p-Aminophenol	$HOC_6H_4NH_2$	solid	760.0
Amygdalin	$C_{20}H_{27}O_{11}N$	solid	2348.4
Amyl acetate	$C_5H_{11}CO_2C_2H_5$	liquid	1042.5
Amyl aocohol	$(CH_3)_2CHCH_2CH_2OH$	liquid	793.7
Amylene	C_5H_{10}	liquid	803.4
Anethole	$C_{10}H_{12}O$	solid	1324.4
Aniline	$C_6H_5NH_2$	liquid	811.7
p-Anisidine	$CH_3OC_6H_4NH_2$	solid	924.0
Anisole	$C_6H_5OCH_3$	liquid	905.1
Anthracene	$C_{14}H_{10}$	solid	1712.0*
Anthraquinone	$C_{14}H_8O_2$	solid	1544.5
Arabinose	$C_5H_{10}O_5$	solid	559.9
Arabitol	$C_5H_{12}O_5$	solid	661.2
Arachidic acid	$C_{20}H_{40}O_2$	solid	3025.9
Azelaic acid	$(CH_2)_7(CO_2H)_2$	solid	1141.7
Azobenzene	$(C_6H_5N)_2$	solid	1545.9
Azoxybenzene	$(C_6H_5N)_2O$	solid	1534.5
Behenic acid	$C_{22}H_{44}O_2$	solid	3338.4
Benzalacetone	$C_6H_5CHCOCH_3$	solid	1257.4
Benzaldehyde	C_6H_5CHO	liquid	843.2*
Benzamide	$C_6H_5CONH_2$	solid	847.6
Benzanilide	$C_6H_5CONHC_6H_5$	solid	1575.5
Benzene	C_6H_6	liquid	780.96*
Benzenediazonium nitrate	$C_6H_5N_2NO_3$	solid	782.6
Benzidine	$(C_6H_4NH_2)_2$	solid	1560.9
Benzil	$(C_6H_5CO)_2$	solid	1624.6
Benzoic acid*	$C_6H_5CO_2H$	solid	771.24*
Benzoic anhydride	$(C_6H_5CO)_2O$	solid	1555.1
Benzoin	$C_6H_5.CHOH.COC_6H_5$	solid	1671.4
Benzonitrile	C_6H_5CN	liquid	865.5
Benzophenone	$(C_6H_5)_2CO$	solid	1556.5
Benzoyl chloride	C_6H_5COCl	liquid	782.8
Benzoyl peroxide	$(C_6H_5CO)_2O_2$	solid	1551.7
Benzyl alcohol	$C_6H_5CH_2OH$	liquid	894.3
Benzylamine	$C_6H_5CH_2NH_2$	liquid	969.4
Benzyl carbylamine	$C_6H_5CH_2NC$	liquid	1046.5
Benzyl chloride	$C_6H_5CH_2Cl$	liquid	886.4
Benzyl cyanide	$C_6H_5CH_2CN$	liquid	1023.5
Borneol	$C_{10}H_{18}O$	liquid	1469.6
Brucine	$C_{23}H_{26}O_4N_2$	gas	687.68*
n-Butyl alcohol	C_4H_9OH	liquid	639.53*
tert-Butyl alcohol, see Trimethyl carbinol			
n-Butylamine	$C_4H_9NH_2$	liquid	710.6
sec-Butylamine	$(CH_3)(C_2H_5)NH_2$	liquid	713.0
tert-Butylamine	$(CH_3)_3CNH_2$	liquid	716.0
tert-Butylbenzene	$C_6H_5C(CH_3)_3$	liquid	1400.4
n-Butyramide	$C_3H_7CONH_2$	solid	596.0
n-Butyric acid	$C_3H_7CO_2H$	liquid	521.87*
n-Butyronitrile	C_3H_7CN	liquid	613.3
Caffeine	$C_8H_{10}O_2N_4$	solid	1014.2
Camphene	$C_{10}H_{16}$	solid	1468.8
Camphor	$C_{10}H_{16}O$	solid	1411.0
Cane sugar, see Sucrose			
Capric acid	$C_9H_{19}O_2$	solid	1453.07*
Caproic acid	$C_5H_{11}CO_2H$	liquid	834.49*
Carbon disulfide	CS_2	liquid	246.6
Carbon subnitride	$(C.CN)_2$	solid	514.8

* 25°C

HEAT OF COMBUSTION
For Organic Compounds (Continued)

Name	Formula	Physical state	Heat of combustion. kg. calories
Carbon tetrachloride	CCl₄	liquid	37.3
Carbonyl sulfide	COS	gas	130.5
Carvacrol	C₁₀H₁₄O	liquid	1354.5
Cetyl alcohol	C₁₆H₃₄O	solid	2504.4
Cetyl palmitate	C₃₂H₆₄O	solid	4872.8
Chloracetic acid	ClCH₂CO₂H	solid	171.0
o-Chlorobenzoic acid	ClC₆H₄CO₂H	solid	734.5
Chloroform	CHCl₃	liquid	89.2
Chrysene	C₁₈H₁₂	solid	2139.1
Cinnamic acid (trans)	C₆H₅CH:CHCO₂H	solid	1040.2
Cinnamic aldehyde	C₆H₅CH:CHCHO	liquid	1112.3
Cinnamic anhydride	C₁₈H₁₄O₃	solid	2091.3
d-Citrene	C₁₀H₁₆	liquid	1483.0
Citric acid (anhydr)	C₆H₈O₇	solid	468.6*
Codeine	C₁₈H₂₁O₃N.H₂O	solid	2327.6
Coniine	C₈H₁₇N	liquid	1275.5
Creatine (anhydr)	C₄H₉O₂N₃	solid	559.8
Creatinine	C₄H₇ON₃	solid	563.4
o-Cresol	CH₃C₆H₄OH	liquid	882.6
o-Cresol	CH₃C₆H₄OH	solid	882.72*
m-Cresol	CH₃C₆H₄OH	liquid	880.5
p-Cresol	CH₃C₆H₄OH	liquid	882.5
p-Cresol	CH₃C₆H₄OH	solid	883.99*
m-Cresolmethyl ether	CH₃C₆H₄OCH₃	liquid	1057.0
Crotonaldehyde	C₃H₅CHO	liquid	542.1
Cyanoacetic acid	NCCH₂CO₂H	solid	298.8
Cyanogen	(CN)₂	gas	258.3
Cyclobutane	C₄H₈	liquid	650.22*
Cycloheptane	(CH₂)₇	liquid	1087.3
Cycloheptanol	CH₂(CH₂)₅CHOH	liquid	1050.2
Cycloheptene	C₇H₁₂	liquid	1099.09*
Cyclohexane	(CH₂)₆	liquid	936.87*
Cyclohexanol	CH₂(CH₂)₄CHOH	liquid	890.7
Cyclohexene, see Tetrahydrobenzene			
Cyclopentane	(CH₂)₅	liquid	786.55*
Cyclopropane, see Trimethylene			
Cymene	C₆H₄(CH₃)(CH₃COCO₂)-(1,4)	liquid	1402.8
Decahydronaphthalene (cis)	C₁₀H₁₈	liquid	1502.5
Decahydronaphthalene (trons)	C₁₀H₁₈	liquid	1499.5
Decane	C₁₀H₂₂	liquid	1610.2
Dextrose, see Glucose			
Diallyl	(CH₂CHCH₂)₂	vapor	903.4
Diamyl ether	(C₅H₁₁)₂O	liquid	1609.3
Diamylene	C₁₀H₂₀	liquid	1582.2
Dibenzyl	(C₆H₅CH₂)₂	solid	1810.6
Dibenzyl amine	(C₆H₅CH₂)₂NH	solid	1853.0
o-Dichlorobenzene	C₆H₄Cl₂	liquid	671.8
Diethylacetic acid	(C₂H₅)₂CHCO₂H	liquid	830.8
Diethyl amine	(C₂H₅)₂NH	liquid	716.9
Diethylaniline	C₆H₅(C₂H₅)₂	liquid	1451.6
Diethyl carbonate	CO(OC₂H₅)₂	liquid	647.9
Diethyl ether	(C₂H₅)₂O	liquid	657.52*
Diethyl ketone	(C₂H₅)₂CO	liquid	735.6
Diethyl malonate	CH₂(CO₂C₂H₅)₂	liquid	860.4
Diethyl oxalate	(CO₂C₂H₅)₂	liquid	716.0
Diethyl succintae	(CH₂CO₂C₂H₅)	liquid	1007.3
Dihydrobenzene	C₆H₈	liquid	847.8
δ₁-Dihydronaphthalene	C₁₀H₁₀	liquid	1296.3
δ₁-Dihydronaphthalene	C₁₀H₁₀	solid	1298.3
Dihydroxyanthraquinone	C₁₄ H₆O₂(OH)₂-(1,2)	solid	1448.9
Diisoamyl	[(CH₃)₂CHCH₂CH₂]₂	liquid	1615.8
Diisobutylene	[(CH₃)₂CHCH₂]₂	liquid	1252.4
Diisopropyl	[(CH₃)₂CH]₂	vapor	993.9
Diisopropyl ketone	[(CH₃)₂CH]₂CO	liquid	1045.5
Dimethyl amine	(CH₃)₂NH	liquid	416.7
Dimethylaniline	C₆H₅N(CH₃)₂	liquid	1142.7
Dimethyl carbonate	CO(OCH₃)₂	liquid	340.8
Dimethyl ether	(CH₃)₂O	gas	
Dimethylethyl carbinol	C₂H₅(CH₃)₂CHOH	liquid	784.6
Dimethyl fumarate	(CHCO₂CH₃)₂	solid	663.3*
2,5-Dimethylhexane	(CH₃)₂CH.C₂H₄.CH(CH₃)₂	liquid	1303.3
3,4-Dimethylhexane	[(C₂H₅)(CH₃)CH]₂	liquid	1303.7
Dimethyl maleate	(CHCO₂CH₃)₂	liquid	669.4*
Dimethyl oxalate	(CO₂CH₃)₂	liquid	400.2*
2,2-Dimethylpentane	(CH₃)₃C.C₄H₇	liquid	1148.9
2,3-Dimethylpentane	(CH₃)₂CHCH(CH₃)C₂H₅	liquid	1148.9
2,4-Dimethylpentane	(CH₃)₂CHCH₂CH(CH₃)₂	liquid	1148.9
3,3-Dimethylpentane	(CH₃)₂C(C₂H₅)₂	liquid	1147.9
Dimethyl phthalate	C₆H₄(CO₂CH₃)₂	liquid	1119.7
Dimethyl succinate	(CH₂CO₂CH₃)₂	solid	706.3*
m-Dinitrobenzene	C₆H₄(NO₂)₂	solid	696.8

HEAT OF COMBUSTION
For Organic Compounds (Continued)

Name	Formula	Physical state	Heat of combustion. kg. calories
Dinitrophenol	$C_6H_3(OH)(NO_2)_2$—(1, 2,4)	solid	648.0
Dinitrotoluene	$C_6H_3(NO_2)_2$—(1,2,4)	solid	852.8
Diphenyl	$(C_6H_5)_2$	solid	1493.6
Diphenyl amine	$(C_6H_5)_2NH$	solid	1536.2
Diphenyl carbinol	$(C_6H_5)_2CHOH$	solid	1615.4
Diphenylmethane	$(C_6H_5)_2CH_2$	solid	1655.0
Diphenylnitrosamine	$(C_6H_5)_2N.NO$	solid	1532.6
Dipropargyl	$(CH: C.CH_2)_2$	vapor	882.9
Dipropyl ketone	$(C_3H_7)_2CO$	liquid	1050.5
Dulcitol	$C_6H_{14}O_6$	solid	729.1
Durene	$C_6H_2(CH_3)_4$—(1,2,4,5)	solid	1393.6
Eicosane	$C_{20}H_{42}$	solid	3183.1
Erythritol	$C_4H_{10}O_4$	solid	504.1
Ethane	C_2H_6	gas	372.81*
Ethine, see Acetylene			
Ethyl acetate	$CH_3CO_2C_2H_5$	liquid	536.9
Ethyl acetoacetate	$CH_3COCH_2CO_2C_2H_5$	liquid	690.8
Ethyl alcohol	C_2H_6OH	liquid	326.68*
Ethyl amine	$C_2H_5NH_2$	liquid	409.5*
Ethylaniline	$C_6H_5NHC_2H_5$	liquid	1121.5
Ethylbenzene	$C_2H_5C_6H_5$	liquid	1091.2
Ethyl benzoate	$C_6H_5CO_2C_2H_5$	liquid	1098.7
Ethyl bromide	C_2H_5Br	vapor	340.5
Ethyl n-butyrate	$C_3H_7CO_2C_2H_5$	liquid	851.2
Ethyl carbylamine	C_2H_5NC	liquid	477.1
Ethyl chloride	C_2H_5Cl	vapor	316.7
Ethylcycloheptane	$C_2H_5C_7H_{13}$	liquid	1406.8
Ethyl formate	$HCO_2C_2H_5$	liquid	391.7
3-Ethylhexane	$(C_2H_5)_2CH.C_3H_7$	liquid	1302.3
Ethyl iodide	C_2H_5I	liquid	356.0
Ethyl isobutyrate	$(CH_3)_2CHCH_2CO_2C_2H_5$	liquid	845.7
Ethyl isocyanate	C_2H_5NCO	liquid	424.5
Ethyl nitrate	$C_2H_5ONO_2$	vapor	322.4
Ethyl nitrite	C_2H_5ONO	vapor	332.6
3-Ethylpentane	$(C_2H_5)_3CH$	liquid	1149.9
Ethyl propionate	$C_2H_5CO_2C_2H_5$	liquid	690.8
Ethyl salicylate	$HOC_6H_4CO_2C_2H_5$	liquid	1051.2
Ethyl valerate	$C_4H_6CO_2C_2H_5$	liquid	1017.5
Ethylene	CH_2CH_2	gas	337.23*
Ethylene chloride	$(CH_2Cl)_2$	vapor	271.0
Ethylene diamine	$(CH_2NH_2)_2$	liquid	452.6
Ethylene glycol	$(CH_2OH)_2$	liquid	281.9
Ethylene iodide	$(CH_2I)_2$	solid	324.8
Ethylene oxide	CH_2CH_2O	liquid	302.1
Ethylidene chloride	CH_3CHCl_2	liquid	267.1
Eugenol	$C_{10}H_{12}O_2$	liquid	1286.6
Fenchane	$C_{10}H_{18}$	liquid	1502.6
Fluorene	$(C_6H_4)_2: CH_2$	solid	1584.9
Fluorobenzene	C_6H_5F	liquid	747.2
Formaldehyde	CH_2O	gas	136.42*
Formamide	$HCONH_2$	solid	134.9
Formic acid	HCO_2H	liquid	60.86*
β-D-Fructose	$C_6H_{12}O_6$	solid	672.0*
Fumaric acid *(trans)*	$(CHCO_2H)_2$	solid	318.99*
Furfural	C_4H_3OCHO	liquid	559.5
a-D-Galactose	$C_6H_{12}O_6$	solid	670.1*
Gallic acid	$C_6H_2(OH)_3CO_2H$—(1,3,5,6)	solid	633.7
a-D-Glucose	$C_6H_{12}O_6$	solid	669.94*
Glutaric acid	$(CH_2)_3(CO_2H)_2$	solid	514.08*
Glycerol	$(CH_2OH)_2CHOH$	liquid	397.0
Glyceryl tributyrate	$C_{15}H_{26}O_6$	liquid	1941.1
Glycine	$H_2NCH_2CO_2H$	solid	232.67*
Glycogen	$(C_6H_{10}O_5)x$ per kg	solid	4186.8
Glycollic acid	CH_2OHCO_2H	solid	166.1*
Glycylglycine	$C_4H_8O_3N_2$	solid	470.7
n-Heptaldehyde	$CH_3(CH_2)_5CHO$	liquid	1062.2*
n-Heptane	C_7H_{16}	liquid	1149.9
Heptine-1	$CH: C(CH_2)_4CH_3$	liquid	1091.2
n-Heptyl alcohol	$CH_3(CH_2)_5CH_2OH$	liquid	1104.9
Heptyl amine	$C_7H_{15}NH_2$	liquid	1178.9
Heptylic acid	$C_7H_{14}O_2$	liquid	986.1
n-Hexane	C_6H_{14}	liquid	995.01*
Hexachlorbenzene	C_6Cl_6	solid	509.0
Hexachlorethane	C_2Cl_6	solid	110.0
Hexadecane	$C_{16}H_{34}$	solid	2559.1
Hexahydronaphthalene	$C_{10}H_{14}$	liquid	1419.3
Hexamethylbenzene	$C_6(CH_3)_6$	solid	1711.9
Hexamethylenetetramine	$(CH_2)_6N_4$	solid	1006.7
Hexamethylethane	$(CH_3)_2Cl_2$	solid	1301.8
Hexyl amine	$C_6H_{13}NH_2$	liquid	1022.2
Hexylene	C_6H_{12}	liquid	952.6

HEAT OF COMBUSTION
For Organic Compounds (Continued)

Name	Formula	Physical state	Heat of combustion. kg. calories
Hippuric acid	$C_6H_5CONHCH_2CO_2H$	solid	1012.4
Hydantoic acid	$C_3H_6O_3N_2$	solid	308.6
Hydrazobenzene	$(C_6H_5NH)_2$	solid	1597.3
Hydroquinol	$C_6H_4(OH)_2$	solid	681.78*
Hydroquinoldimethyl ether	$(CH_3O)_2C_6H_4$	solid	1014.7
p-Hydroxyazobenzene	$HOC_6H_4N_2C_6H_5$	solid	1502.0
o-Hydroxybenzaldehyde	$C_6H_4(OH)CHO$	liquid	796.4*
m-Hydroxybenzaldehyde	$C_6H_4(OH)CHO$	solid	788.7
p-Hydroxybenzaldehyde	$C_6H_4(OH)CHO$	solid	792.7
m-Hydroxybenzoic acid	$HOC_6H_4CO_2H$	solid	726.1
p-Hydroxybenzoic acid	$HOC_6H_4CO_2H$	solid	725.4
β-Hydroxybutyraldehyde	$CH_3CHOHCH_2CHO$	liquid	546.6
Indigo	$C_{16}H_{10}O_2N_2$	solid	1815.0
Indole	C_8H_7N	solid	1022.2
Inositol	$C_6H_{12}O_6$	solid	662.1
Iodoform	CHI_3	solid	161.9
Isoamyl amine	$(CH_3)_2CHC_2H_4NH_2$	liquid	866.8
Isobutane	$(CH_3)_3CH$	gas	683.4
Isobutyl alcohol	$(CH_3)_2CHCH_2OH$	liquid	638.2
Isobutyl amine	$C_4H_9NII_2$	liquid	713.6
Isobutylene	$(CH_3)_2C:CH_2$	gas	647.2
Isobutyraldehyde	$(CH_3)_2CHCHO$	vapor	596.8
Isobutyramide	$(CH_3)_2CHCONH_2$	solid	595.9
Isobutyric acid	$(CH_3)_2CHCO_2H$	liquid	517.4
Isoeugenol	$C_{10}H_{12}O_2$	liquid	1277.6
Isopentane	C_5H_{12}	gas	843.5(?)
Isopentane	C_5H_{12}	liquid	838.3(?)
Isophthalic acid	$C_6H_4(CO_2H)_2$	solid	768.3
Isopropyl alcohol	$(CH_3)_2CHOH$	liquid	474.8
Isopropylbenzene	$(CH_3)_2CHC_6H_5$	liquid	1247.3
Isopropyltoluene	$C_6H_4(CH_3)(CH_3CHC_3)—(1,3)$	liquid	1409.5
Isopropyltoluene, see Cymene			
Isosafrole	$C_{10}H_{10}O_2$	liquid	1233.9
Lactic acid, DL	$CH_3CHOHCO_2H$	liquid	326.8*
Lactose (anhydr.)	$C_{12}H_{22}O_{11}$	solid	1350.0*
Lauric acid	$C_{12}H_{24}O_2$	solid	1763.25*
Leucine	$C_6H_{13}O_2N$	solid	855.6
d-Limonene	$C_{10}H_{16}$	liquid	1471.2
Maleic acid *(cis)*	$(CHCO_2H)_2$	solid	323.89*
Maleic anhydride	$(CHCO)_2O$	solid	332.10*
l-Malic acid	$(CHOHCH_2):(CO_2H)_2$	solid	317.37*
Malonic acid	$CH_2(CO_2H)_2$	solid	205.82*
Maltose	$C_{12}H_{22}O_{11}$	solid	1349.3*
Mandelic acid	$C_6H_5CHOHCO_2H$	solid	890.3
d-Mannitol	$C_6H_{14}O_6$	solid	727.6
Menthene	$C_{10}H_{18}$	liquid	1523.2
Menthol	$C_{10}H_{20}O$	solid	1508.8
Mesitylene	$(CH_3)_3C_6H_3—(1,3,5)$	liquid	1243.6
Mesityl oxide	$(CH_3)_2C:CHCOCH_3$	liquid	846.7
Mesotartaric acid	$(CHOH)_2(CO_2H)_2$	solid	276.0
Methane	CH_4	gas	212.79*
Methyl acetate	$CH_3CO_2CH_3$	liquid	381.2
Methyl alcohol	CH_3OH	liquid	173.64*
Methyl amine	CH_3NH_2	liquid	253.5*
Methylaniline	$C_6H_5NHCH_3$	liquid	973.5
Methyl benzoate	$C_6H_5CO_2CH_3$	liquid	945.9*
Methyl bromide	CH_3Br	vapor	184.0
Methyl butyl ketone	$CH_3COC_4H_9$	liquid	895.2
Methyl tert-butyl ketone, see Pinacoline			
Methyl butyrate	$C_3H_7COOCH_3$	liquid	692.8
Methyl carbylamine	CH_3NC	liquid	320.1
Methyl chloride	CH_3Cl	gas	164.2
Methyl cinnamate	$C_{10}H_{10}O_2$	solid	1213.0
Methylcyclobutane	$CH_3CHCH_2CH_2CH_2$	liquid	784.2
Methylcycloheptane	$CH_3C_7H_{13}$	liquid	1244.5
Methylcyclohexane	$CH_3C_6H_{11}$	liquid	1091.8
Methylcyclopentane	$CH_3CH.C_3H_6CH_2$	liquid	937.9
Methyldiethyl carbinol	$CH_3(C_2H_5)_2CHOH$	liquid	927.0
Methylene chloride	CH_2Cl_2	vapor	106.8
Methylene iodide	CH_2I_2	liquid	178.4
Methylethyl ether	$CH_3OC_2H_5$	vapor	503.69*
Methylethyl ketone	$CH_3COC_2H_5$	liquid	584.17*
Methyl formate	HCO_2CH_3	liquid	234.1*
2-Methylheptane	$(CH_3)_2CH.C_5H_{11}$	liquid	1306.1
2-Methylhexane	$(CH_3)_2CHC_4H_9$	liquid	1148.9
3-Methylhexane	$(C_2H_5)(CH_3)CHC_2H_7$	liquid	1148.9
Methylhexyl ketone	$CH_3COC_6H_{13}$	liquid	1205.1
Methyl iodide	CH_3I	liquid	194.7
Methyl isobutyrate	$(CH_3)_2CHCO_2CH_3$	liquid	694.2
Methyl isocyanate	CH_3NCO	liquid	269.4
Methylisopropyl ketone	$CH_3COCH(CH_3)_2$	liquid	733.9
Methyl lactate	$CH_3CHOHCO_2CH_3$	liquid	497.2

HEAT OF COMBUSTION
For Organic Compounds (Continued)

Name	Formula	Physical state	Heat of combustion. kg. calories
Methyl propionate	$C_2H_5CO_2CH_3$	vapor	552.3
Methylpropyl ketone	$CH_3COC_3H_7$	liquid	740.78*
Methyl salicylate	$HOC_6H_4CO_2CH_3$	liquid	898.6*
Milk sugar, see Lactose			
Morphine	$C_{17}H_{19}O_3N.H_2O$	solid	2146.3
Mucic acid	$C_6H_{10}O_8$	solid	483.0*
Myristic acid	$C_{14}H_{28}O_2$	solid	2073.91*
Naphthalene	$C_{10}H_8$	solid	1231.8*
α-Naphthoic acid	$C_{10}H_7CO_2H$	solid	1231.8
β-Naphthoic acid	$C_{10}H_7CO_2H$	solid	1227.6
α-Naphthol	$C_{10}H_7OH$	solid	1185.4
β-Naphthol	$C_{10}H_7OH$	solid	1187.2
α-Naphthonitrile	$C_{10}H_7CN$	solid	1326.2
β-Naphthonitrile	$C_{10}H_7CN$	solid	1321.0
α-Naphthoquinone	$C_{10}H_6O_2$	solid	1100.8
β-Naphthoquinone	$C_{10}H_6O_2$	solid	1106.4
α-Naphthyl amine	$C_{10}H_7NH_2$	solid	1263.5
β-Naphthyl amine	$C_{10}H_7NH_2$	solid	1261.0
Narceine	$C_{23}H_{27}O_8N.2H_2O$	solid	2802.9
Narcotine	$C_{22}H_{23}O_7N$	solid	2644.5
Nicotine	$C_{10}H_{14}N_2$	liquid	1427.7
o-Nitraniline	$C_6H_4(NH_2)(NO_2)$	solid	765.8
m-Nitraniline	$C_6H_4(NH_2)(NO_2)$	solid	765.2
p-Nitraniline	$C_6H_4(NH_2)(NO_2)$	solid	761.0
m-Nitrobenzaldehyde	$O_2NC_6H_4CHO$	solid	800.4
Nitrobenzene	$C_6H_5NO_2$	liquid	739.2
m-Nitrobenzoic acid	$O_2NC_6H_4CO_2H$	solid	729.1
Nitroethane	$C_2H_5NO_2$	liquid	322.2
Nitroglycerine, see Trinitroglycerol			
Nitromethane	CH_3NO_2	liquid	169.4
o-Nitrophenol	$HOC_6H_4NO_2$	solid	689.1
m-Nitrophenol	$HOC_6H_4NO_2$	solid	684.4
p-Nitrophenol	$HOC_6H_4NO_2$	solid	688.8
Nitropropane	$C_3H_7NO_2$	liquid	477.9
p-Nitrotoluene	$CH_3C_6H_4NO_2$	liquid	897.0
p-Nitrotoluene	$CH_3C_6H_4NO_2$	solid	888.6
Octahydronaphthalene	$C_{10}H_{16}$	liquid	1461.7
n-Octane	C_8H_{18}	liquid	1302.7
Octyl alcohol	$C_8H_{18}O$	liquid	1262.0
Oleic acid	$C_{18}H_{34}O_2$	liquid	2657.4*
Oxalic acid, a	$(CO_2H)_2$	solid	58.7*
Oxamide	$(CONH_2)_2$	solid	203.2
Palmitic acid	$C_{16}H_{32}O_2$	solid	2384.76*
Papaverine	$C_{20}H_{21}O_4N$	solid	2478.1
Pentamethylbenzene	$C_6H(CH_3)_5$	solid	1554.0
n-Pentane	C_5H_{12}	gas	845.16*
n-Pentane	C_5H_{12}	liquid	838.78*
Phenacetin	$C_{10}H_{13}O_2N$	solid	1285.2
Phenanthraquinone	$C_{14}H_8O_2$	solid	1544.0
Phenanthrene	$C_{14}H_{10}$	solid	1685.6*
Phenetole	$C_6H_5OC_2H_5$	liquid	1060.3
Phenol	C_6H_5OH	solid	729.80*
Phenylacetic acid	$C_6H_5CH_2CO_2H$	solid	930.4*
Phenylacetylene	C_6H_5C	liquid	1024.2
Phenylalanine	$C_9H_{11}O_2N$	solid	1111.3
p-Phenylenediamine	$C_6H_4(NH_2)_2$	solid	843.4
Phenylethylene, see Styrene			
Phenylglycine	$C_2H_5NHCH_2CO_2H$	solid	955.1
Phenylhydrazine	$C_6H_5N_2H_3$	solid	875.4
Phenylhydroxylamine	C_6H_5NHOH	liquid	803.7
Phenyl iodide	C_6H_5I	liquid	770.7
Phloroglucinol	$C_6H_3(OH)_3$	solid	635.7
o-Phthalic acid	$C_6H_4(CO_2H)_2$	solid	770.44*
Phthalic anhydride	$C_6H_4(CO)_2O$	solid	783.4
Phthalimide	$C_8H_5O_2N$	solid	849.5
Pieric acid	$C_6H_2(OH)(NO_2)_3 - (1,2,4,6)$	solid	611.8
Pinacoline	$CH_3COC(CH_3)_3$	solid	891.8
Piperidine	$C_5H_{11}N$	liquid	826.6
Piperonal	$C_8H_6O_3$	solid	870.7
Propane	C_3H_8	gas	530.57*
Propine, see Allylene			
Propionaodehyde	C_2H_5CHO	liquid	434.1*
Propionamide	$C_2H_5CONH_2$	solid	439.9
Propionic acid	$C_2H_5CO_2H$	liquid	365.03*
Propionic anhydride	$(C_2H_5CO)_2O$	liquid	746.6
Propionitrile	C_2H_5CN	liquid	456.4
n-Propyl alcohol	C_3H_7OH	liquid	482.75*
Propyl amine	$C_3H_7NH_2$	liquid	565.3*
n-Propylbenzene	$C_3H_7C_6H_5$	liquid	1246.4
Propyl bromide	C_3H_7Br	vapor	497.3
Propyl carbylamine	C_3H_7NC	liquid	639.6

HEAT OF COMBUSTION
For Organic Compounds (Continued)

Name	Formula	Physical state	Heat of combustion. kg. calories
Propyl chloride	C_3H_7Cl	vapor	478.3
Propylene	CH_3CH_2	gas	490.2
Propylene glycol	$CH_3CHOHCH_2OH$	liquid	431.0
n-Propyl iodide	C_3H_7I	liquid	514.3
n-Propyltoluene	$C_6H_4(CH_3)(C_3H_7)-(1,3)$	liquid	1405.4
Pseudocumene	$C_6H_3(CH_3)_3-(1,2,4)$	liquid	1241.7
Pyridine	C_5H_5N	liquid	665.0*
Pyrocatechol	$C_6H_4(OH)_2$	solid	683.0*
Pyrogallol	$C_6H_3(OH)_3$	solid	638.7
Pyrrole	C_4H_5N	liquid	567.7
Quercitol	$C_6H_{12}O_5$	solid	704.2
Quinoline	C_9H_7N	liquid	1123.5
Quinone	$O:C_6H_4:O$	solid	656.6
Raffinose	$C_{18}H_{32}O_{16}$	solid	2025.5
Retene	$C_{19}H_{18}$	solid	2306.8
Resorcinol	$C_6H_4(OH)_2$	solid	681.30*
Resorcinoldimethyl ether	$(CH_3O)_2C_6H_4$	liquid	1022.6
Rhamnose	$C_6H_{12}O_5$	solid	718.3
Safrole	$C_{10}H_{10}O_2$	liquid	1244.1
Salicylaldehyde, see o-Hydroxybenzaldehyde			
Salicylic acid*	$HOC_6H_4CO_2H-(1,2)$	solid	722.4**
Sarcosine	$CH_3NHCH_2CO_2H$	solid	401.1
Sebacic acid	$(CH_2)_8(CO_2H)_2$	solid	1297.3
Skatole	C_9H_9N	liquid	1170.5
d-Sorbose	$C_6H_{12}O_6$	solid	668.3
Starch	$(C_6H_{10}O_5)x$ per kg	solid	4178.8
Stearic acid	$C_{18}H_{36}O_2$	solid	2696.12*
Strychnine	$C_{21}H_{22}O_2N_2$	solid	2685.7
Styrene	$C_6H_5CH_2$	liquid	1047.1
Suberic acid	$(CH_2)_6(CO_2H)_2$	solid	985.2
Succinic acid	$(CH_2CO_2H)_2$	solid	356.36*
Succinic acid nitrile	$(CH_2CN)_2$	liquid	545.7
Succinic anhydride	$(CH_2CO)_2O$	solid	369.0*
Succinimide	$C_4H_5O_2N$	solid	437.9
Sucrose	$C_{12}H_{22}O_{11}$	solid	1348.2*
Sylvestrene	$C_{10}H_{15}$	liquid	1464.7
l-Tartaric acid	$(CHOH)_2(CO_2H)_2$	solid	274.7*
d,l-Tartaric acid (anhydr)	$(CHOH)_2(CO_2H)_2$	solid	272.6*
Terephthalic acid	$C_6H_4(CO_2H)_2$	solid	770.4
Terpin hydrate	$C_{10}H_{22}O_3$	solid	1451.0
Terpineol	$C_{10}H_{18}O$	solid	1469.5
Tetrahydrobenzene	C_6H_{10}	liquid	891.9
Tetrahydronaphthalene	$C_{10}H_{12}$	liquid	1352.4
Tetramethylmethane	$(CH_3)_4C$	gas	842.6
Tetraphenylmethane	$(C_6H_5)_4C$	solid	3102.4
Tetryl	$C_6H_5N_5O_8$	solid	842.3
Thebaine	$C_{19}H_{21}O_3N$	solid	2441.3
Thiophene	C_4H_4S	liquid	670.5
Thujane	$C_{10}H_{18}$	liquid	1506.4
Thymol	$C_{10}H_{14}O$	liquid	1353.4
Thymol	$C_{10}H_{14}O$	solid	1349.7
Thymoquinone	$C_{10}H_{12}O_2$	solid	1271.3
Toluene	$CH_3C_6H_5$	liquid	934.2
o-Toluic acid	$CH_3C_6H_4CO_2H$	solid	928.9
m-Toluic acid	$CH_3C_6H_4CO_2H$	solid	928.6
p-Toluic acid	$CH_3C_6H_4CO_2H$	solid	926.9
o-Toluidine	$CH_3C_6H_4NH_2$	liquid	964.3
m-Toluidine	$CH_3C_6H_4NH_2$	liquid	965.3
p-Toluidine	$Ch_3C_6H_4NH_2$	solid	958.4
o-Tolunitrile	$CH_3C_6H_4CN$	liquid	1030.3
Toluquinone	$C_7H_6O_3$	solid	803.2
Triaminotriphenyl carbinol	$(C_6H_4NH_2)_3COH$	solid	2483.5
Tribenzyl amine	$(C_6H_5CH_2)_3N$	solid	2762.1
Trichloracetic acid	$Cl_3C.CO_2H$	solid	92.8
Triethyl amine	$(C_2H_5)_3N$	liquid	1036.8
Triethyl carbinol	$(C_2H_5)_3CHOH$	liquid	1080.0
Triisoamyl amine	$[(CH_3)_2CHCH_2CH_2]_3N$	liquid	2459.3
Triisobutyl amine	$[(CH_3)_2CHCH_2]_3N$	liquid	1973.6
Trimethyl amine	$(CH_3)_3N$	liquid	578.6
2,2,3-Trimethylbutane	$(CH_3)_3C.CH(CH_3)_2$	liquid	1147.9
Trimethyl carbinol	$(CH_3)_3COH$	liquid	629.3
Trimethylene	$CH_2CH_2CH_2$	gas	499.89*
Trimethylethylene	$(CH_3)_2CCH_3$	liquid	796.0
Trimethylethylene	$(CH_3)_2CCH_3$	vapor	803.6
2,2,4-Trimethylpentane	$(CH_3)_3C.C.CH_2CH(CH_3)_2$	liquid	1303.9
Trinitrobenzene	$C_6H_3(NO_2)_3-(1,3,5)$	solid	663.7
Trinitroglycerol	$C_3H_5(NO_3)_3$	liquid	368.4
Trinitrotoluene	$C_6H_2(CH_3)(NO_2)_3-(1,2,4,6)$	solid	820.7
Triphenyl amine	$(C_6H_5)_3N$	solid	2267.8
Triphenylbenzene	$C_6H_3(C_6H_5)_3-(1,3,5)$	solid	2936.7
Triphenyl carbinol	$(C_6H_5)_3CHOH$	sed	2340.8

** Recommended as a secondary thermochemical standard.

HEAT OF COMBUSTION
For Organic Compounds (Continued)

Name	Formula	Physical state	Heat of combustion. kg. calories
Triphenylmethane	$(C_6H_5)_3CH$	solid	2388.7
Triphenyl methyl	$(C_6H_5)_3C$	solid	2378.5
Tyrosine	$C_9H_{11}O_3N$	solid	1070.2
Undecyclic acid	$C_{11}H_{22}O_2$	solid	1615.9
Urea	$(NH_2)_2CO$	solid	150.97*
Urethane	$NH_2CO_2C_2H_5$	solid	397.2
Uric acid	$C_5H_4O_3N_4$	solid	460.2
n-Valeric acid	$C_4H_9CO_2H$	liquid	678.12*
Vanillin	$C_6H_3(OH)(OCH_3)CHO$ (1,2,4)	solid	914.1
o-Xylene	$(CH_3)_2C_6H_4$	liquid	1091.7
m-Xylene	$(CH_3)_2C_6H_4$	liquid	1088.4
p-Xylene	$(CH_3)_2C_6H_4$	liquid	1089.1
Xylose	$C_5H_{10}O_5$	solid	559.0*

HEAT OF FORMATION

The thermal change involved in the formation of 1 mol of substance from its elements is the heat of formation, ΔH, of the substance. If all of the substances involved in the reaction are each in their standard states and each substance is at unit activity, the thermal change is the standard heat of formation, ΔH°. By definition, all elements in their standard state have a heat of formation of zero. By further definition, the sign of ΔH is negative if heat is evolved and is positive if heat is absorbed. Thus, for the thermochemical reaction:

$$7C (s) + 3H_2(g) + O_2(g) = C_6H_5COOH (s) \qquad ΔH° = -771.24 \text{ kcal}$$
$$\text{(benzoic acid)}$$

Heats of formation may be calculated from heats of combustion. The heat of formation of compound "A" is equal to the sum of the heats of formation of the products of combustion of compound "A" minus the heat of combustion of compound "A". Some heats of formation of some products of combustion of organic compounds are:

Substance	ΔH, heat of formation (kcal/ g mole)
Free Elements	0
CO	− 26.416
CO_2	− 93.963
½ H_2O (l), from 1 H	− 34.158
H_2O (l)	− 68.317
HF (Dilute aqueous solution)	− 76.531
HCl (Dilute aqueous solution)	− 39.850
HBr (Dilute aqueous solution)	− 28.958
HI (Dilute aqueous solution)	− 13.106
HNO_3 (Dilute aqueous solution)	− 48.484
H_2SO_4 (Dilute aqueous solution)	−213.552
SO_2 (g)	− 70.336

Two examples of calculations are:

Example I

Calculate the heat of formation of methane (CH_4) where

Heat of combustion of CH_4 = −210.8 kcal/g mol
Heat of formation of CO_2 = −93.963 kcal/g mol
Heat of formation of H_2O = −68.317 kcal/g mol

and where the combustion reaction occurs according to the following equation:

$$CH_4(gas) + 2O_2(gas) = CO_2(gas) + 2H_2O \text{ (liquid)}$$

Heat of formation of CH_4 equals:

Heat of formation of CO_2	=	−93.963 kcal
+ Two times heat of formation of H_2O	=	2(−68.317)kcal
− Heat of combustion of CH_4	=	−(−210.8) kcal
		−19.8 kcal/g mol

Example II

Calculate the heat of formation of ethylene (C_2H_4) where

Heat of combustion of C_2H_4 = −337.23 kcal/g mol

and where the combustion reaction is as follows:

$$C_2H_4(gas) + 3O_2(gas) = 2CO_2(gas) + 2H_2O \text{ (liquid)}$$

Two times heat of formation of CO_2	=	2(−93.963)kcal
+ Two times heat of formation of H_2O	=	2(−68.317)kcal
− Heat of combustion of C_2H_4	=	−(−337.23) kcal
		+12.67 kcal/g mol

From Weast, R. C. Ed., *Handbook of Chemistry and Physics*, 69th ed., CRC Press, Boca Raton, FL, 1988.

TABLE 2
Limits of Inflammability of Gases and Vapor in Air and Oxygen

Reprinted from "Combustion Flame and Explosions of Gases", B. Lewis and G. von Elbe, authors, Academic Press (1951), publishers, by special permission.

The limits of inflammability given in the following tables were all determined at atmospheric pressure and room* temperature for upward propagation in a tube or bomb 2 inches or more in diameter. Values are on a percentage-by-volume basis.

LIMITS OF INFLAMMABILITY OF GASES AND VAPORS IN AIR

Compound	Empirical formula	Limits of inflammability Lower	Limits of inflammability Upper	Compound	Empirical formula	Limits of inflammability Lower	Limits of inflammability Upper
Paraffin hydrocarbons				Methyl propyl ketone	$C_5H_{10}O$	1.55	8.15
Methane	CH_4	5.00	15.00	Methylbutyl ketone	$C_6H_{12}O$	1.35	7.60
Ethane	C_2H_6	3.00	12.50	Acids			
Propane	C_3H_8	2.12	9.35	Acetic acid	$C_2H_4O_2$	5.40	—
Butane	C_4H_{10}	1.86	8.41	Hydrocyanic acid	HCN	5.60	40.00
Isobutane	C_4H_{10}	1.80	8.44	Esters			
Pentane	C_5H_{12}	1.40	7.80	Methyl formate	$C_2H_4O_2$	5.05	22.70
Isopentane	C_5H_{12}	1.32	—	Ethyl formate	$C_3H_6O_2$	2.75	16.40
2,2-Dimethylpropane	C_5H_{12}	1.38	7.50	Methyl acetate	$C_3H_6O_2$	3.15	15.60
Hexane	C_6H_{14}	1.18	7.40	Ethyl acetate	$C_4H_8O_2$	2.18	11.40
Heptane	C_7H_{16}	1.10	6.70	Propyl acetate	$C_5H_{10}O_2$	1.77	8.00
2,3-Dimethylpentane	C_7H_{16}	1.12	6.75	Isopropyl acetate	$C_5H_{10}O_2$	1.78	7.80
Octane	C_8H_{18}	0.95	—	Butyl acetate	$C_6H_{12}O_2$	1.39	7.55
Nonane	C_9H_{20}	0.83	—	Amyl acetate	$C_7H_{14}O_2$	1.10	—
Decane	$C_{10}H_{22}$	0.77	5.35	Hydrogen			
Olefins				Hydrogen	H_2	4.00	74.20
Ethylene	C_2H_4	2.75	28.60	Nitrogen compounds			
Propylene	C_3H_6	2.00	11.10	Ammonia	NH_3	15.50	27.00
Butene-1	C_4H_8	1.65	9.95	Cyanogen	C_2N_2	6.60	42.60
Butene-2	C_4H_8	1.75	9.70	Pyridine	C_5H_5N	1.81	12.40
Amylene	C_5H_{10}	1.42	8.70	Ethyl nitrate	$C_2H_5NO_3$	3.80	—
Acetylenes				Ethyl nitrite	$C_2H_5NO_2$	3.01	50.00
Acetylene	C_2H_2	2.50	80.00	Oxides			
Aromatics				Carbon monoxide	CO	12.50	74.20
Benzene	C_6H_6	1.40	7.10	Ethylene oxide	C_2H_4O	3.00	80.00
Toluene	C_7H_8	1.27	6.75	Propylene oxide	C_3H_6O	2.00	22.00
o-Xylene	C_8H_{10}	1.00	6.00	Dioxan	$C_4H_8O_2$	1.97	22.25
Cyclic hydrocarbons				Diethyl peroxide	$C_4H_{10}O_2$	2.34	—
Cyclopropane	C_3H_6	2.40	10.40	Sulfides			
Cyclohexane	C_6H_{12}	1.26	7.75	Carbon disulfide	CS_2	1.25	50.00
Methylcyclohexane	C_7H_{14}	1.15	—	Hydrogen sulfide	H_2S	4.30	45.50
Terpenes				Carbon oxysulfide	COS	11.90	28.50
Turpentine	$C_{10}H_{16}$	0.80	—	Chlorides			
Alcohols				Methyl chloride	CH_3Cl	8.25	18.70
Methyl alcohol	CH_4O	6.72	36.50	Ethyl chloride	C_2H_5Cl	4.00	14.80
Ethyl alcohol	C_2H_6O	3.28	18.95	Propyl chloride	C_3H_7Cl	2.60	11.10
Allyl alcohol	C_3H_6O	2.50	18.00	Butyl chloride	C_4H_9Cl	1.85	10.10
η-Propyl alcohol	C_3H_8O	2.15	13.50	Isobutyl chloride	C_4H_9Cl	2.05	8.75
Isopropyl alcohol	C_3H_8O	2.02	11.80	Allyl chloride	C_3H_5Cl	3.28	11.15
η-Butyl alcohol	$C_4H_{10}O$	1.45	11.25	Amyl chloride	$C_5H_{11}Cl$	1.60	8.63
Isobutyl alcohol	$C_4H_{10}O$	1.68	—	Vinyl chloride	C_2H_3Cl	4.00	21.70
η-Amyl alcohol	$C_5H_{12}O$	1.19	—	Ethylene dichloride	$C_2H_4Cl_2$	6.20	15.90
Isoamyl alcohol	$C_5H_{12}O$	1.20	—	Propylene dichloride	$C_3H_6Cl_2$	3.40	14.50
Aldehydes				Bromides			
Acetaldehyde	C_2H_4O	3.97	57.00	Methyl bromide	CH_3Br	13.50	14.50
Crotonic aldehyde	C_4H_6O	2.12	15.50	Ethyl bromide	C_2H_5Br	6.75	11.25
Furfural	$C_5H_4O_2$	2.10	—	Allyl bromide	C_3H_5Br	4.36	7.25
Paraldehyde	$C_6H_{12}O_3$	1.30	—	Amines			
Ethers				Methyl amine	CH_5N	4.95	20.75
Methylethyl ether	C_3H_8O	2.00	10.00	Ethyl amine	C_2H_7N	3.55	13.95
Diethyl ether	$C_4H_{10}O$	1.85	36.50	Dimethyl amine	C_2H_7N	2.80	14.40
Divinyl ether	C_4H_6O	1.70	27.00	Propyl amine	C_3H_9N	2.01	10.35
Ketones				Diethyl amine	$C_4H_{11}N$	1.77	10.10
Acetone	C_3H_6O	2.55	12.80	Trimethyl amine	C_3H_9N	2.00	11.60
Methylethyl ketone	C_4H_8O	1.81	9.50	Triethyl amine	$C_6H_{15}N$	1.25	7.90

* The upper limits of some vapors were determined at somewhat higher temperatures because of their low vapor pressures.

LIMITS OF INFLAMMABILITY OF GASES AND VAPORS IN OXYGEN

Compound	Formula	Limits of inflammability Lower	Limits of inflammability Upper	Compound	Formula	Limits of inflammability Lower	Limits of inflammability Upper
Hydrogen	H_2	4.65	93.9	Propylene	C_3H_6	2.10	52.8
Deuterium	D_2	5.00	95.0	Cyclopropane	C_3H_6	2.45	63.1
Carbon monoxide	CO	15.50	93.9	Ammonia	NH_3	13.50	79.0
Methane	CH_4	5.40	59.2	Diethyl ether	$C_4H_{10}O$	2.10	82.0
Ethane	C_2H_6	4.10	50.5	Divinyl ether	C_4H_6O	1.85	85.5
Ethylene	C_2H_4	2.90	79.9				

From Weast, R. C., Ed., *Handbook of Chemistry and Physics*, 69th ed., CRC Press, Boca Raton, FL, 1988.

TABLE 3
Heat of Formation of Gaseous Atoms from Elements in their Standard States

Atom	$\Delta H^{\circ}_{f(298)}$/kcal mol⁻¹	$\Delta H^{\circ}_{f(298)}$/kJ mol⁻¹	Ref.	Atom	$\Delta H^{\circ}_{f(298)}$/kcal mol⁻¹	$\Delta H^{\circ}_{f(298)}$/kJ mol⁻¹	Ref.	Atom	$\Delta H^{\circ}_{f(298)}$/kcal mol⁻¹	$\Delta H^{\circ}_{f(298)}$/kJ mol⁻¹	Ref.
Ag	68.1 ± 0.2	284.9 ± 0.8	1	Hf	148 ± 1	619 ± 4	2	Re	185 ± 1.5	774 ± 6.3	2
Al	78.8 ± 1.0	329.7 ± 4.0	1	Hg	14.69 ± 0.03	61.46 ± 0.13	2	Rh	133 ± 1	557 ± 4	2
As	72.3 ± 3	302.5 ± 13	2	I	25.518 ± 0.010	106.765 ± 0.040	3	Ru	155.5 ± 1.5	648.5 ± 6.3	2
Au	88.0 ± 0.5	368.2 ± 2.1	2	In	58 ± 1	243 ± 4	2	S	66.20 ± 0.06	276.98 ± 0.25	1
B	139 ± 3	560 ± 12	1	Ir	160 ± 1	669 ± 4	2	Sb	63.2 ± 0.6	264.4 ± 2.5	2
Ba	42.5 ± 1	177.8 ± 4	2	K	21.42 ± 0.05	89.62 ± 0.21	2	Sc	90.3 ± 1	377.8 ± 4	2
Be	77.5 ± 1.5	324.3 ± 6.3	2	Li	38.6 ± 0.4	161.5 ± 1.7	2	Se	54.3 ± 1	227.2 ± 4	2
Bi	50.1 ± 0.5	209.6 ± 2.1	2	Mg	35.0 ± 0.3	146.4 ± 1.3	2	Sn	72.2 ± 0.5	302.1 ± 2.1	2
Br	26.735	111.857	3	Mn	67.7 ± 1	283.3 ± 4	2	Sr	39.1 ± 0.5	163.6 ± 2.1	2
C	171.29 ± 0.11	716.67 ± 0.44	1	Mo	157.3 ± 0.5	658.1 ± 2.1	2	Ta	186.9 ± 0.6	782.0 ± 2.5 •	2
Ca	42.6 ± 0.4	178.2 ± 1.7	2	N	112.97 ± 0.10	472.68 ± 0.40	1	Te	47.0 ± 0.5	196.7 ± 2.1	2
Cd	26.72 ± 0.15	111.80 ± 0.63	2	Na	25.85 ± 0.15	108.16 ± 0.63	2	Th	137.5 ± 0.5	575.3 ± 2.1	2
Ce	101 ± 3	423 ± 13	2	Nb	172.4 ± 1	721.3 ± 4	2	Ti	112.3 ± 0.5	469.9 ± 2.1	2
Cl	28.989 ± 0.002	121.290 ± 0.008	3	Ni	102.8 ± 0.5	430.1 ± 2.1	2	Tl	43.55 ± 0.1	182.21 ± 0.4	2
Co	102.4 ± 1	428.4 ± 4	2	O	59.553 ± 0.024	249.17 ± 0.10	1	U	126 ± 3	527 ± 13	2
Cr	95 ± 1	398 ± 4	2	Os	188 ± 1.5	787 ± 6.3	2	V	122.9 ± 0.3	514.2 ± 1.3	2
Cs	18.7 ± 0.1	78.2 ± 0.4	2	P	79.4 ± 1.0	332.2 ± 4.2	2	W	203.1 ± 1	849.8 ± 4	2
Cu	80.7 ± 0.3	337.6 ± 1.2	1	Pb	46.62 ± 0.3	195.06 ± 1.3	2	Y	101.5 ± 0.5	424.7 ± 2.1	2
Er	75.8 ± 1	317.1 ± 4	2	Pd	90.0 ± 0.5	376.6 ± 2.1	2	Yb	36.35 ± 0.2	152.09 ± 0.8	2
F	18.98	79.41	3	Pt	135.2 ± 0.3	565.7 ± 1.3	2	Zn	31.17 ± 0.05	130.42 ± 0.20	1
Ge	89.5 ± 0.5	374.5 ± 2.1	2	Pu	87.1 ± 4	364.4 ± 17	2	Zr	145.5 ± 1	608.8 ± 4	2
H	52.102 ± 0.001	217.995 ± 0.005	3	Rb	19.6 ± 0.1	82.0 ± 0.4	2				

From Weast, R. C., Ed., *Handbook of Chemistry and Physics*, 69th ed., CRC Press, Boca Raton, FL, 1988.

REFERENCES

1. CODATA recommended Key values for thermodynamics, 1975, *J. Chem. Thermodyn.*, 8, 603, 1976.
2. Brewer, L. and Rosenblatt, G. M., *Adv. High Temp. Chem.*, 2, 1, 1969.
3. Calculated from D°₀ taken from Huber, K. P. and Herzberg, G., *Molecular Spectra and Molecular Structure Constants of Diatomic Molecules*, Van Nostrand, New York, 1979 and enthalpy functions from *J.A.N.A.F. Thermochemical Tables*, NSRDS-NBS 37, 1971.

TABLE 4
Stoichiometric Oxygen Requirements and Combustion Product Yields

Elemental waste component	Stoichiometric oxygen requirement	Combustion product yield
C	2.67 lb/lb C	3.67 lb CO_2/lb C
H_2	8.0 lb/lb H_2	9.0 lb H_2O/lb H_2
O_2	−1.0 lb/lb O_2	
N_2	—	1.0 lb N_2/lb N_2
H_2O	—	1.0 lb H_2O/lb H_2O
Cl_2	−0.23 lb/lb Cl_2	1.03 lb HCl/lb Cl_2 −0.25 lb H_2O/lb. Cl_2
F_2	−0.42 lb/lb F_2	1.05 lb HF/lb F_2 −0.47 lb H_2O/lb F_2
Br_2	—	1.0 lb Br_2/lb Br_2
I_2	—	1.0 lb I_2/lb I_2
S	1.0 lb/lb S	2.0 lb SO_2/lb S
P	1.29 lb/lb P	2.29 lb P_2O_5/lb P
Air N_2	—	3.31 lb N_2/lb $(O_2)_{\text{stoich}}$

Note: Stoichiometric air requirement = $4.31 \times (O_2)_{\text{stoich}}$

From Bonner, T. A., Cornett, C. L., Desai, B. O., Fullenkamp, J. M., Hughes, T. W., Johnson, M. L., Kennedy, E. D., McCormick, R. J., Peters, J. A., and Zanders, D. L., Engineering Handbook for Hazardous Waste Incineration, Rep. MRC-DA 1090, U.S. Environmental Protection Agency, Washington, D.C., 1981.

TABLE 5
Approximate Net Heating
Values for Common Auxiliary
Fuels

Fuel	Approximate net heating value
Residual fuel oil (e.g., no. 6)	17,500 Btu/lb
Distillate fuel oil (e.g., no. 2)	18,300 Btu/lb
Natural gas	19,700 Btu/lb (1000 Btu/scf)

From Bonner, T. A., Cornett, C. L., De-
sai, B. O., Fullenkamp, J. M., Hughes, T.
W., Johnson, M. L., Kennedy, E. D.,
McCormick, R. J., Peters, J. A., and Zan-
ders, D. L., Engineering Handbook for
Hazardous Waste Incineration, Rep. MRC-
DA 1090, U.S. Environmental Protection
Agency, Washington, D.C., 1981.

Figure 1 may be used to check the internal consistency of the proposed excess air rate and temperature, as long as the amount of carbon, hydrogen, and oxygen in the stream and its net heating value are known. To use the figure, first find the weight fraction of carbon on the scale marked C (on the far left) and the weight fraction of hydrogen on scale H. Connect these two points with a ruler and read the value at its intersection with arbitrary scale 1. Subtract the weight fraction of oxygen in the feed stream from this number. Plot this value on the middle graph, using arbitrary scale 1 as the vertical axis and the excess air scale as the horizontal axis. Interpolate between the set of curves to find a value for arbitrary scale 2, which is then plotted on the vertical axis of the right-hand graph, with the net heating value of the feed plotted on the horizontal axis to determine the combustion gas temperature.

For example, suppose methane (CH_4 — 75% carbon, 25% hydrogen; net heating value of 19,700 Btu/lb) is burned with 50% excess air. Connecting 0.75 on the C scale with 0.25 on the H gives 4.9 at the intersection with arbitrary scale 1. Using this value and 50% excess air gives 7.25 on the middle graph. Plotting 7.25 vertically and 19.7 kBtu/lb horizontally on the right-hand graph shows a temperature of 2700°F. If a temperature of 2000°F is desired and excess air is to be calculated, plotting 19.7 kBtu/lb and 2000°F gives 9.8 on the arbitrary scale 2. Next, using the middle graph, 9.8 on arbitrary scale 2 and 4.9 on arbitrary scale 1 shows an excess air rate of 100%. Figure 1 is accurate so long as the combustion gases consist of air, CO_2, and H_2O; i.e., the wastes consist mainly of carbon, hydrogen, and oxygen.

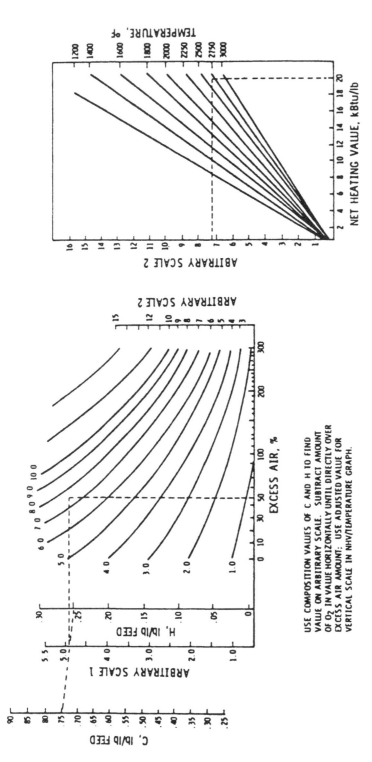

USE COMPOSITION VALUES OF C AND H TO FIND
VALUE ON ARBITRARY SCALE. SUBTRACT AMOUNT
OF O₂ IN VALUE HORIZONTALLY UNTIL DIRECTLY OVER
EXCESS AIR AMOUNT; USE ADJUSTED VALUE FOR
VERTICAL SCALE IN NHV/TEMPERATURE GRAPH.

FIGURE 1. Nomograph for checking the internal consistency of proposed excess air rate and combustion temperature in hazardous waste incinerators. (From Bonner, T. A., Cornett, C. L., Desai, B. O., Fullenkamp, J. M., Hughes, T. W., Johnson, M. L., Kennedy, E. D., McCormick, R. J., Peters, J. A., and Zanders, D. L., Engineering Handbook for Hazardous Waste Incineration, Rep. MRC-DA 1090, U.S. Environmental Protection Agency, Washington, D.C., 1981.)

II. THERMAL DECOMPOSITION

TABLE 6
Summary of Thermal Decomposition Data[a] for 20 Selected Hazardous Organic Compounds

Compound	Empirical formula	T_{onset}[b] (°C)	T_{99}[c] (°C)	$T_{99.99}$[d] (°C)	$A(s^{-1})$[e]	E_a[f] (kcal/mol)
Acetonitrile	C_2H_3N	760	900	~950	4.7×10^7	40
Tetrachloroethylene	C_2Cl_4	660	850	920	2.6×10^6	33
Acrylonitrile	C_3H_3N	650	830	860	1.3×10^6	31
Methane	CH_4	660	830	870	3.5×10^9	48
Hexachlorobenzene	C_6Cl_6	650	820	880	2.5×10^8	41
1,2,3,4-Tetrachlorobenzene	$C_6H_2Cl_4$	660	800	850	1.9×10^6	30
Pyridine	C_6H_5N	620	770	840	1.1×10^5	24
Dichloromethane	CH_2Cl_2	650	770	780	3.0×10^{13}	64
Carbon tetrachloride	CCl_4	600	750	820	2.8×10^5	26
Hexachlorobutadiene	C_4Cl_6	620	750	780	6.3×10^{12}	59
1,2,4-Trichlorobenzene	$C_6H_3Cl_3$	640	750	790	2.2×10^8	39
1,2-Dichlorobenzene	$C_6H_4Cl_2$	630	740	780	3.0×10^8	39
Ethane	C_2H_6	500	735	785	1.3×10^5	24
Benzene	C_6H_6	630	730	760	2.8×10^8	38
Aniline	C_6H_7N	620	730	750	9.3×10^{15}	71
Monochlorobenzene	C_6H_5Cl	540	710	780	8.0×10^4	23
Nitrobenzene	$C_6H_5NO_2$	570	670	700	1.4×10^{15}	64
Hexachloroethane	C_2Cl_6	470	600	640	1.9×10^7	29
Chloroform	$CHCl_3$	410	590	620	2.9×10^{12}	49
1,1,1-Trichloroethane	$C_2H_3Cl_3$	390	570	600	1.9×10^8	32

[a]   Data for 2-s residence time in flowing air.
[b]   Onset of decomposition.
[c]   Interpolated temperature for 99% destruction.
[d]   Extrapolated temperature for 99.99% destruction.
[e]   Arrhenius coefficient.
[f]   Activation energy.

From Dellinger, B., Torres, J. L., Rubey, W. A., Hall, D. L., Graham, J. L., and Carnes, R. A., Determination of the thermal decomposition properties of 20 selected hazardous organic compounds, in Incineration and Treatment of Hazardous Waste — Proc. 10th Annu. Res. Symp., Rep. EPA-600/9-84-022, U.S. Environmental Protection Agency, Cincinnati, OH, 1984.

TABLE 7
Thermal Decomposition Parameters for Selected
Chlorinated Benzenes in Air at 2.0-s Residence Time

Compound	Temperature for onset of thermal decomposition (°C)	Extrapolated temperature for 99.99% DRE (°C)
Benzene	610	760
Monoclorobenzene	520	790
1,2-Dichlorobenzene	615	790
1,2,4-Trichlorobenzene	620	795
1,2,3,5-Tetrachlorobenzene	620	800
Pentachlorobenzene	675	840
Hexachlorobenzene	615	880

From Dellinger, B., Hall, D. L., Rubey, W. A., Torres, J. L., and Carnes, R. A., Factors affecting the gas-phase thermal decomposition of chlorinated aromatic hydrocarbons, in Incineration and Treatment of Hazardous Waste — Proc. 9th Annu. Res. Symp., Rep. EPA-600/9-84-015, U.S. Environmental Protection Agency, Cincinnati, OH, 1984, 71.

III. INCINERABILITY

TABLE 8
Composition of EPA Soup One

Compound	Parts by volume	Mass fraction	Chlorine content (%)
CCl_4	2.75	0.3813	92.19
C_2HCl_3	2.95	0.3466	80.95
C_7H_8	1.00	0.1043	00.00
C_6H_5Cl	1.56	0.1678	31.50
Total		1.0000	68.50

From Carnes, R. A., USEPA combustion research facility permit compliance test burn, in Incineration and Treatment of Hazardous Waste — Proc. 11th Annu. Res. Symp., Rep. EPA-600/9-85-028, U.S. Environment Protection Agency, Cincinnati, OH, 1985, 144.

TABLE 9
Summary of Strengths and Shortcomings of Various Incinerability Ranking Parameters

Parameter	Strengths ($+$) and shortcomings ($-$)
Heat of combustion Enthalpy change from reactants to products	$+$ Data available or can be calculated $+$ Relates to heat release and temperature rise $-$ Some apparent inconsistencies in rankings $-$ Correlation with other methods
Autoignition temperature Temperature at which compound ignites spontaneously in air	$+$ Correlates with thermal decomposition data $+$ Extensive data available $-$ Value varies with experiment $-$ Some compounds do not autoignite
Chemical kinetics Reaction path used as measure of incinerability	$+$ Considers destruction as a rate process $+$ Considers both unimolecular and bimolecular processes $-$ Does not consider physical processes $-$ Limited kinetic data available
Thermal decomposition Experimental, nonflame determination of destruction efficiency	$+$ Simple experimental system $+$ Can get 99.99% destruction efficiency directly $-$ Does not consider flame reactions $-$ Validity for incineration unknown
Multiple linear regression Relates physical and chemical properties to decomposition	$+$ Considers both AI and structure $+$ Several variables included $-$ Many adjustable parameters $-$ Coefficients obtained from thermal decomposition
Other methods	Heat of formation Gibbs free energy Ionization potential Flash point Combustion ignition delay Thermal decomposition in a flame environment

From Miller, D. L., Cundy, V. A., and Matula, A. A., Incinerability characteristics of selected chlorinated hydrocarbons, in Incineration and Treatment of Hazardous Waste — Proc. 9th Annu. Res. Symp., Rep-EPA-600/9-84-015, U.S. Environmental Protection Agency, Cincinnati, OH, 1984, 113.

TABLE 10
Hazardous Wastes Rated as Good, Potential, or Poor Candidates for Incineration by Appropriate Technologies

This table was prepared using available background documents for some of the listed waste, trial burn data, and engineering judgment based on chemical formula(s) of compound(s) present in the waste.

The following criteria were used to structure engineering judgment:

Waste containing	Incineration category
Carbon, hydrogen, and/or oxygen	Good
Carbon, hydrogen, ≤30% by weight chlorine and/or oxygen	Good
Carbon, hydrogen, and/or oxygen >30% by weight chlorine, phosphorous, sulfur, bromine, iodine, or nitrogen	Potential
Unknown percentage of chlorine	Potential
Inorganic compounds	Poor
Compounds containing metals	Poor

Other factors to be considered in evaluating waste for incineration are

1.   Moisture content
2.   Potential pollutants present in incinerator effluents
3.   Inert content
4.   Heating value and auxiliary fuel requirements
5.   Potential health and environmental effects
6.   Physical form
7.   Corrosiveness
8.   Quality
9.   Known carcinogenic content
10.   PCB content

TABLE 10 (continued)
Hazardous Wastes Rated as Good, Potential, or Poor Candidates for Incineration by Appropriate Technologies

EPA hazardous waste number	Hazardous waste	Candidate for incineration			Incinerator type		
		Good	Potential	Poor	Liquid injection	Rotary kiln	Fluidized bed
Generic							
F001	The spent halogenated solvents used in degreasing; tetrachloroethylene, trichloroethylene, methylene chloride, 1,1,1-trichloroethylene, carbon tetrachloride, the chlorinated fluorocarbons, and sludges from the recovery of these solvents in degreasing operations		✓		✓	✓	✓
F002	The spent halogenated solvents tetrachloroethylene, methylene chloride, trichloroethylene, 1,1,1-trichloroethane, chlorobenzene, 1,1,2-trichloro-1,2,2-trifluoroethane, o-dichlorobenzene, trichlorofluoromethane, and the still bottoms from the recovery of these solvents		✓		✓	✓	✓
F003	The spent nonhalogenated solvents of xylene, acetone, ethyl acetate, ethyl benzene, ethyl ether, n-butyl alcohol, cyclohexanone, and the still bottoms from the recovery of these solvents	✓			✓	✓	✓
F004	The spent nonhalogenated solvents, cresols and cresylic acid, nitrobenzene, and the still bottoms from the recovery of these solvents	✓			✓	✓	✓
F005	The spent nonhalogenated solvents methanol, toluene, methyl ethyl ketone, carbon disulfide, isobutanol, pyridine, and the still bottoms from the recovery of these solvents	✓			✓	✓	✓
F006	Wastewater treatment sludges from electroplating operations			✓			
F007	Spent plating bath solutions from electroplating operations			✓			
F008	Plating bath sludges from the bottom of plating baths from electroplating operations			✓			
F009	Spent stripping and cleaning bath solutions from electroplating operations			✓			
F010	Quenching bath sludge from oil baths from metal heat-treating operations			✓			
F011	Spent solutions from salt bath pot cleaning from metal heat-treating operations			✓			
F012	Quenching wastewater treatment sludges from metal heat-treating operations			✓			

TABLE 10 (continued)
Hazardous Wastes Rated as Good, Potential, or Poor Candidates for Incineration by Appropriate Technologies

EPA hazardous waste number	Hazardous waste	Candidate for incineration			Incinerator type		
		Good	Potential	Poor	Liquid injection	Rotary kiln	Fluidized bed
F013	Flotation tailings from selective flotation from mineral metal recovery operations			✓			
F014	Cyanidation wastewater treatment tailing pond sediment from mineral metal recovery operations			✓			
F015	Spent cyanide bath solutions from mineral metal recovery operations			✓			
F016	Dewatered air pollution control scrubber sludges from coke ovens and blast furnaces			✓			
Wood preservation							
K001	Bottom sediment sludge from the treatment of wastewaters from wood preserving processes that use creosote and/or pentachlorophenol		✓			✓	✓
Inorganic pigments							
K002	Wastewater treatment sludge from the production of chrome yellow and orange pigments			✓			
K003	Wastewater treatment sludge from the production of molybdate orange pigments			✓			
K004	Wastewater treatment sludge from the production of zinc yellow pigments			✓			
K005	Wastewater treatment sludge from the production of chrome green pigments			✓			
K006	Wastewater treatment sludge from the production of chrome oxide green pigments (anhydrous and hydrated)			✓			
K007	Wastewater treatment sludge from the production of iron blue pigments			✓			
K008	Oven residue from the production of chrome oxide green pigments			✓			
Organic chemicals							
K009	Distillation bottoms from the production of acetaldehyde from ethylene		✓				
K010	Distillation side cuts from the production of acetaldehyde from ethylene		✓				

Waste No.	Description					
K011	Bottom stream from the wastewater stripper in the production of acrylonitrile	✓				
K012	Still bottoms from the final purification of acrylonitrile in the production of acrylonitrile	✓		✓	✓	✓
K013	Bottom stream from the acetonitrile column in the production of acrylonitrile	✓				✓
K1014	Bottoms from the acetonitrile purification column in the production of acrylonitrile	✓				
K015	Still bottoms from the distillation of benzyl chloride	✓		✓	✓	
K016	Heavy ends or distillation residues from the production of carbon tetrachloride	✓			✓	✓
K017	Heavy ends (still bottoms) from the purification column in the production of epichlorohydrin	✓			✓	✓
K018	Heavy ends from fractionation of ethylchloride production	✓			✓	✓
K019	Heavy ends from the distillation of ethylene dichloride in ethylene dichloride production	✓			✓	✓
K020	Heavy ends from the distillation of vinyl chloride in vinyl chloride monomer production	✓			✓	✓
K021	Aqueous spent antimony catalyst waste from fluoromethane production		✓			
K022	Distillation bottom tars from the production of phenol/acetone from cumene	✓		✓	✓	✓
K023	Distillation light ends from the production of phthalic anhydride from naphthalene	✓			✓	✓
K024	Distillation bottoms from the production of phthalic anhydride from naphthalene	✓			✓	
K025	Distillation bottoms from the production of nitrobenzene by the nitration of benzene	✓				✓
K026	Stripping still tails from the production of methyl ethyl pyridines	✓				
K027	Centrifuge residue from toluene diisocyanate production	✓				
K028	Spent catalyst from the hydrochlorinator reactor in the production of 1,1,1-trichloroethane		✓			
K029	Waste from the product stream stripper in the production of 1,1,1-trichloroethane	✓				
K030	Column bottoms or heavy ends from the combined production of trichloroethylene and perchloroethylene	✓		✓	✓	✓

Pesticides

| K031 | By products salts generated in the production of MSMA and cacodylic acid | | ✓ | | | |

TABLE 10 (continued)
Hazardous Wastes Rated as Good, Potential, or Poor Candidates for Incineration by Appropriate Technologies

EPA hazardous waste number	Hazardous waste	Candidate for incineration			Incinerator type		
		Good	Potential	Poor	Liquid injection	Rotary kiln	Fluidized bed
K032	Wastewater treatment sludge from the production of chlordane		✓				
K033	Wastewater and scrub water from the chlorination of cyclopentadiene in the production of chlordane		✓				
K034	Filter solids from the filtration of hexachlorocyclopentadiene in the production of chlordane			✓			
K035	Wastewater treatment sludges generated in the production of creosote		✓				
K036	Still bottoms from toluene reclamation distillation in the production of disulfoton		✓				
K037	Wastewater treatment sludges from the production of disulfoton		✓				
K038	Wastewater from the washing and stripping of phorate production		✓				
K039	Filter cake from the filtration of diethylphosphorodithoric acid in the production of phorate		✓			✓	✓
K040	Wastewater treatment sludge from the production of phorate			✓			
K041	Wastewater treatment sludge from the production of toxaphene		✓				
K042	Heavy ends or distillation residues from the distillation of tetrachlorobenzene in the production of 2,4,5-T		✓				
K043	2,6-Dichlorophenol waste from the production of 2,4-D		✓				
Explosives							
K044	Wastewater treatment sludges from the manufacturing and processing of explosives		✓			✓	✓
K045	Spent carbon from the treatment of wastewater containing explosives	✓				✓	✓
K046	Wastewater treatment sludges from the manufacturing, formulation, and loading of lead-based initiating compounds			✓			
K047	Pink/red water from TNT operations	✓			✓	✓	✓
Petroleum refining							
K048	Dissolved air flotation (DAF) float from the petroleum refining industry	✓				✓	✓
K049	Slop oil emulsion solids from the petroleum refining industry	✓				✓	✓
K050	Heat exchanger bundle cleaning sludge from the petroleum refining industry		✓			✓	✓

Code	Description					
K051	API separator sludge from the petroleum refining industry			√		
K052	Tank bottoms (leaded) from the petroleum refining industry			√	√	√
Leather tanning finishing						
K053	Chrome (blue) trimmings generated by the following subcategories of the leather tanning and finishing industry; hair pulp/chrome tan/retan/wet finish; hair save/chrome tan/retan/wet finish; retan/wet finish; no beamhouse; through-the-blue; shearling		√			
K054	Chrome (blue) shavings generated by the following subcategories of the leather tanning and finishing industry: hair pulp/chrome tan/retan/wet finish; hair save/chrome tan/retan/wet finish; retan/wet finish; no beamhouse; through-the blue; shearling		√			
K055	Buffing dust generated by the following subcategories of the leather tanning and finishing industry: hair pulp/chrome tan/retan/wet finish; hair save/chrome tan/retan/wet finish; retan/wet finish; no beamhouse; through-the-blue		√			
K056	Sewer screenings generated by the following subcategories of the leather tanning and finishing industry: hair pulp/chrome tan/retan/wet finish; hair save/chrome tan/retan/wet finish; retan/wet finish; no beamhouse; through-the-blue; shearling		√			
K057	Wastewater treatment sludges generated by the following subcategories of the leather tanning and finishing industry; hair pulp/chrome tan/retan/wet finish, hair save/chrome tan/retan/wet finish; retan/wet finish; no beamhouse; through-the-blue and shearling		√			
K058	Wastewater treatment sludges generated by the following subcategories of the leather tanning and finishing industry; hair pulp/chrome tan/retan/wet finish; hair save/chrome tan/retan/wet finish; through-the-blue	√				
K059	Wastewater treatment sludges generated by the following subcategory of the leather tanning and finishing industry; hair save/nonchrome tan/retan/wet finish		√			
Iron and steel						
K060	Ammonia still lime sludge from coking operations		√			
K061	Emission control dust/sludge from the electric furnace production of steel		√			
K062	Spent pickle liquor from steel finishing operations		√			
K063	Sludge from lime treatment of spent pickle liquor from steel finishing operations		√			

TABLE 10 (continued)
Hazardous Wastes Rated as Good, Potential, or Poor Candidates for Incineration by Appropriate Technologies

EPA hazardous waste number	Hazardous waste	Candidate for incineration			Incinerator type		
		Good	Potential	Poor	Liquid injection	Rotary kiln	Fluidized bed
Primary copper							
K064	Acid plant blowdown slurry/sludge resulting from the thickening of blowdown slurry from primary copper production			✓			
Primary lead							
K065	Surface impoundment solids contained in and dredged from surface impoundments at primary lead smelting facilities			✓			
Primary zinc							
K066	Sludge from treatment of process wastewater and/or acid plant blow-down from primary zinc production			✓			
K067	Electrolytic anode slimes/sludges from primary zinc production			✓			
K068	Cadmium plant leach residue (iron oxide) from primary zinc production			✓			
Secondary lead							
K069	Emission control dust/sludge from secondary lead smelting			✓			
	Discarded commercial chemical products, off-specification species, containers, and spill residues thereof						
P001	3-(α-Acetonylbenzyl)-4-hydroxycoumarin and salts	✓			✓	✓	✓
P002	1-Acetyl-2-thiourea		✓		✓	✓	✓
P003	Acrolein	✓			✓	✓	✓
P004	Aldrin		✓			✓	✓
P005	Allyl alcohol	✓			✓	✓	✓
P006	Aluminum phosphide			✓			
P007	5-(Aminomethyl)-3-isoxazolol		✓			✓	✓
P008	4-Aminopyridine		✓			✓	✓
P009	Ammonium picrate		✓			✓	✓
P010	Arsenic acid			✓			
P011	Arsenic pentoxide			✓			
P012	Arsenic trioxide			✓			
P013	Barium cyanide			✓			
P014	Benzenethiol		✓		✓	✓	✓

P015	Beryllium dust	
P016	Bis(chloromethyl) ether	
P017	Bromoacetone	
P018	Brucine	
P019	2-Butanone peroxide	
P020	2-sec-butyl-4,6-dinitrophenol	
P021	Calcium cyanide	
P022	Carbon disulfide	
P023	Chloroacetaldehyde	
P024	p-Chloroaniline	
P025	1-(p-Chlorobenzoyl)-5-methoxy-2-methylindole-3-acetic acid	
P026	1-(o-Chlorophenyl) thiourea	
P027	3-Chloropropionitrile	
P028	α-Chlorotoluene	
P029	Copper cyanide	
P030	Cyanides	
P031	Cyanogen	
P032	Cyanogen bromide	
P033	Cyanogen chloride	
P034	2-Cyclohexyl-4,6-dinitrophenol	
P035	2,4-Dichlorophenoxyacetic acid (2,4-D)	
P036	Dichlorophenylarsine	
P037	Dieldrin	
P038	Diethylarsine	
P039	O,O-Diethyl-S-[2-(ethylthio)ethyl] ester of phosphorothioic acid	
P040	O,O-Diethyl-O-(2-pyrazinyl) phosphorothioate	
P041	O,O-Diethyl phosphoric acid, O-p-nitrophenyl ester	
P042	3,4-Dihydroxy-α-(methylamino)-methyl benzyl alcohol	
P043	Di-isopropylfluorophosphate	
P044	Dimethoate	
P045	3,3-Dimethyl-1-(methylthio)-2-butanone-O-[(methylamino)carbonyl] oxime	
P046	Dimethoate	
P047	4,6-Dinitro-o-cresol and salts	
P048	2,4-Dintrophenol	
P049	2,4-Dithiobiuret	
P050	Endosulfan	

TABLE 10 (continued)
Hazardous Wastes Rated as Good, Potential, or Poor Candidates for Incineration by Appropriate Technologies

EPA hazardous waste number	Hazardous waste	Candidate for incineration			Incinerator type		
		Good	Potential	Poor	Liquid injection	Rotary kiln	Fluidized bed
P051	Endrin		✓			✓	✓
P052	Ethylcyanide	✓			✓	✓	✓
P053	Ethylenediamine	✓			✓	✓	✓
P054	Ethyleneimine	✓			✓	✓	
P055	Ferric cyanide			✓			
P056	Fluorine			✓			
P057	2-Fluoroacetamide		✓			✓	✓
P058	Fluoroacetic acid, sodium salt			✓			
P059	Heptachlor		✓			✓	✓
P060	1,2,3,4,10,10-Hexachloro-1,4,4a,5,8,8a-hexahydro-1,4:5,8-endo, endo-dimethanonaphthalene		✓		✓	✓	✓
P061	Hexachloropropene	✓				✓	✓
P062	Hexaethyl tetraphosphate		✓			✓	✓
P063	Hydrocyanic acid	✓				✓	✓
P064	Isocyanic acid, methyl ester	✓				✓	✓
P065	Mercury fulminate			✓			✓
P066	Methomyl						
P067	2-Methylaziridine	✓			✓	✓	✓
P068	Methyl hydrazine	✓			✓	✓	✓
P069	2-Methyllactonitrile		✓		✓	✓	✓
P070	2-Methyl-2-(methylthio)propionaldehyde-*o*-(methylcarbonyl) oxime		✓			✓	✓
P071	Methyl parathion		✓			✓	✓
P072	1-Naphthyl-2-thiourea					✓	✓
P073	Nickel carbonyl			✓			
P074	Nickel cyanide			✓			
P075	Nicotine and salts		✓			✓	✓
P076	Nitric oxide			✓			
P077	*p*-Nitroaniline		✓			✓	
P078	Nitrogen dioxide			✓		✓	✓
P079	Nitrogen peroxide			✓		✓	✓

P080	Nitrogen tetroxide
P081	Nitroglycerine
P082	N-Nitrosodimethylamine
P083	N-Nitrosodiphenylamine
P084	N-Nitrosomethylvinylamine
P085	Octamethylpyrophosphoramide
P086	Oleyl alcohol condensed with 2 mol ethylene oxide
P087	Osmium tetroxide
P088	7-Oxabicyclo[2.2.1]heptane-2,3-dicarboxylic acid
P089	Parathion
P090	Pentachlorophenol
P091	Phenyl dichloroarsine
P092	Phenylmercury acetate
P093	N-Phenylthiourea
P094	Phorate
P095	Phosgene
P096	Phosphine
P097	Phosphorothioic acid, *O*,*O*-dimethyl ester, *O*-ester with *N*,*N*-dimethyl benzene sulfonamide
P098	Potassium cyanide
P099	Potassium silver cyanide
P100	1,2-Propanediol
P101	Propiontrile
P102	2-Propyn-1-ol
P103	Selenourea
P104	Silver cyanide
P105	Sodium azide
P106	Sodium cyanide
P107	Strontium sulfide
P108	Strychnine and salts
P109	Tetraethyldithiopyrophosphate
P110	Tetraethyl lead
P111	Tetraethylpyrophosphate
P112	Tetranitromethane
P113	Thallic oxide
P114	Thallium selenite
P115	Thallium (1) sulfate
P116	Thiosemicarbazide

TABLE 10 (continued)
Hazardous Wastes Rated as Good, Potential, or Poor Candidates for Incineration by Appropriate Technologies

EPA hazardous waste number	Hazardous waste	Candidate for incineration			Incinerator type		
		Good	Potential	Poor	Liquid injection	Rotary kiln	Fluidized bed
P117	Thiuram		✓			✓	✓
P118	Trichloromethanethiol		✓			✓	✓
P119	Vanadic acid, ammonium salt			✓			
P120	Vanadium pentoxide			✓			
P121	Zinc cyanide			✓			
P122	Zinc phosphide			✓			
U001	Acetaldehyde	✓			✓	✓	✓
U002	Acetone	✓			✓	✓	✓
U003	Acetonitrile		✓		✓	✓	✓
U004	Acetophenone		✓			✓	✓
U005	2-Acetylaminoflourene		✓			✓	✓
U006	Acetyl chloride		✓		✓	✓	✓
U007	Acrylamide				✓	✓	✓
U008	Acrylic acid	✓			✓	✓	✓
U009	Acrylonitrile	✓			✓	✓	✓
U010	6-Amino-1,1a,2,8,8a,8b-hexahydro-8-(hydroxymethyl)8-methoxy-5-methylcarbamate azirino(2',3':3,4) pyrrolo(1,2-a)indole-4, 7-dione (ester)		✓			✓	✓
U011	Amitrole		✓			✓	✓
U012	Aniline	✓			✓	✓	✓
U013	Asbestos			✓		✓	✓
U014	Auramine		✓			✓	✓
U015	Azaserine			✓		✓	
U016	Benz[c]acridine		✓			✓	✓
U017	Benzal chloride		✓			✓	✓
U018	Benz[a]anthracene		✓		✓	✓	
U019	Benzene	✓			✓	✓	✓
U020	Benzenesulfonyl chloride	✓			✓	✓	✓
U021	Benzidine	✓			✓	✓	✓
U022	Benzo[a]pyrene		✓			✓	✓

Code	Name	1	2	3	4	5
U023	Benzotrichloride		✓	✓	✓	
U024	Bis(2-chloroethoxy)methane		✓	✓	✓	✓
U025	Bis(2-chloroethyl)ether		✓	✓	✓	✓
U026	N,N-Bis(2-chloroethyl)2-naphthylamine	✓			✓	✓
U027	Bis(2-chloroisopropyl) ether		✓	✓	✓	✓
U029	Bromoethane		✓		✓	✓
U030	4-Bromophenyl phenyl ether		✓	✓	✓	✓
U031	n-Butyl alcohol	✓		✓	✓	✓
U032	Calcium chromate			✓		
U033	Carbonyl fluoride		✓		✓	✓
U034	Chloral		✓	✓	✓	✓
U035	Chlorambucil		✓		✓	✓
U036	Chlordane		✓	✓	✓	
U037	Chlorobenzene	✓		✓	✓	
U038	Chlorobenzilate	✓		✓	✓	
U039	p-Chloro-m-cresol	✓			✓	
U040	Chlorodibromomethane		✓		✓	✓
U041	1-Chloro-2,3-epoxypropane		✓	✓	✓	✓
U042	Chloroethyl vinyl ether		✓	✓	✓	
U043	Chloroethene		✓		✓	
U044	Chloroform		✓	✓	✓	
U045	Chloromethane		✓		✓	
U046	Chloromethyl methyl ether		✓	✓	✓	
U047	2-Chloronaphthalene	✓			✓	✓
U048	2-Chlorophenol	✓		✓	✓	
U049	4-Chloro-o-toluidine hydrochloride		✓	✓	✓	✓
U050	Chrysene	✓			✓	✓
U051	Cresote	✓		✓	✓	✓
U052	Cresols	✓			✓	✓
U053	Crotonaldehyde	✓		✓	✓	✓
U054	Cresylic acid	✓		✓	✓	✓
U055	Cumene	✓		✓	✓	✓
U056	Cyclohexane	✓		✓	✓	✓
U057	Cyclohexanone	✓		✓	✓	✓
U058	Cyclophosphamide		✓		✓	✓
U059	Daunomycin		✓		✓	✓
U060	DDD		✓		✓	

TABLE 10 (continued)
Hazardous Wastes Rated as Good, Potential, or Poor Candidates for Incineration by Appropriate Technologies

EPA hazardous waste number	Hazardous waste	Candidate for incineration			Incinerator type		
		Good	Potential	Poor	Liquid injection	Rotary kiln	Fluidized bed
U061	DDT			✓		✓	
U062	Diallate		✓			✓	✓
U063	Dibenz[a,h]anthracene	✓				✓	✓
U064	Dibenzo[a,i]pyrene	✓				✓	✓
U065	Dibromochloromethane		✓		✓	✓	✓
U066	1,2-Dibromo-3-chloropropane		✓		✓	✓	✓
U067	1,2-Dibromoethane		✓			✓	✓
U068	Dibromomethane	✓					
U069	Di-n-butyl phthalate		✓				
U070	1,2-Dichlorobenzene		✓		✓	✓	✓
U071	1,3-Dichlorobenzene		✓		✓	✓	✓
U072	1,4-Dichlorobenzene		✓		✓	✓	✓
U073	3,3'-Dichlorobenzidine		✓		✓	✓	✓
U074	1,4-Dichloro-2-butene		✓		✓	✓	✓
U075	Dichlorodifluoromethane		✓			✓	✓
U076	1,1-Dichloroethane		✓		✓	✓	✓
U077	1,2-Dichloroethane		✓		✓	✓	✓
U078	1,1-Dichloroethylene		✓		✓	✓	✓
U079	1,2-trans-dichloroethylene		✓		✓	✓	✓
U080	Dichloromethane		✓		✓		
U081	2,4-Dichlorophenol		✓			✓	✓
U082	2,6-Dichlorophenol		✓			✓	✓
U083	1,2-Dichloropropane		✓		✓	✓	✓
U084	1,3-Dichloropropane		✓		✓	✓	✓
U085	Diepoxybutane	✓					
U086	1,2-Diethylhydrazine	✓					
U087	O,O-Diethyl-S-methyl ester of phosphorodithioic acid		✓		✓		
U088	Diethyl phthalate	✓			✓	✓	✓
U089	Diethylstilbestrol	✓					✓
U090	Dihydrosafrole	✓				✓	✓

U091	3,3'-Dimethoxybenzidine	✓	✓	
U092	Dimethylamine	✓	✓	
U093	p-Dimethylaminoazobenzene	✓	✓	✓
U094	7,12-Dimethylbenz[a]anthracene	✓	✓	✓
U095	3,3'-Dimethylbenzidine	✓	✓	✓
U096	αα-Dimethylbenzylhydroperoxide	✓	✓	
U097	Dimethylcarbamoyl chloride	✓	✓	✓
U098	1,1-Dimethylhydrazine	✓	✓	✓
U099	1,2-Dimethylhydrazine	✓	✓	✓
U100	Dimethylnitrosoamine	✓	✓	
U101	2,4-Dimethylphenol	✓	✓	✓
U102	Dimethyl phthalate	✓	✓	✓
U103	Dimethyl sulfate	✓	✓	✓
U104	2,4-Dinitrophenol	✓	✓	
U105	2,4-Dinitrotoluene	✓	✓	✓
U106	2,6-Dinitrotoluene	✓	✓	✓
U107	Di-n-octyl phthalate	✓	✓	✓
U108	1,4-Dioxane	✓	✓	
U109	1,2-Diphenylhydrazine	✓	✓	✓
U110	Dipropylamine	✓	✓	✓
U111	Di-n-propylnitrosamine	✓	✓	
U112	Ethyl acetate	✓	✓	✓
U113	Ethyl acrylate	✓	✓	✓
U114	Ethylenebisdithiocarbamate	✓	✓	
U115	Ethylene oxide	✓	✓	✓
U116	Ethylene thiourea	✓	✓	✓
U117	Ethyl ether	✓	✓	✓
U118	Ethylmethacrylate	✓	✓	✓
U119	Ethyl methanesulfonate	✓	✓	✓
U120	Fluoranthene	✓	✓	✓
U121	Fluorotrichloromethane	✓	✓	
U122	Formaldehyde	✓	✓	✓
U123	Formic acid	✓	✓	✓
U124	Furan	✓	✓	✓
U125	Furfural	✓	✓	✓
U126	Glycidylaldehyde	✓	✓	✓
U127	Hexachlorobenzene	✓	✓	✓

TABLE 10 (continued)
Hazardous Wastes Rated as Good, Potential, or Poor Candidates for Incineration by Appropriate Technologies

EPA hazardous waste number	Hazardous waste	Candidate for incineration			Incinerator type		
		Good	Potential	Poor	Liquid injection	Rotary kiln	Fluidized bed
U128	Hexachlorobutadiene		✓		✓	✓	✓
U129	Hexachlorocyclophexane		✓			✓	✓
U130	Hexachlorocyclopentadiene		✓		✓	✓	✓
U131	Hexachloroethane		✓			✓	✓
U132	Hexachlorophene		✓		✓	✓	✓
U133	Hydrazine			✓			
U134	Hydrofluoric acid			✓			
U135	Hydrogen sulfide		✓			✓	✓
U136	Hydroxydimethyl arsine oxide			✓			
U137	Indeno(1,2,3-cd)pyrene	✓			✓	✓	✓
U138	Iodomethane		✓		✓		✓
U139	Iron dextran	✓					
U140	Isobutyl alcohol	✓			✓	✓	✓
U141	Isosafrole	✓			✓	✓	✓
U142	Kepone	✓				✓	✓
U143	Lasiocarpine		✓				
U144	Lead acetate			✓			
U145	Lead phosphate			✓			
U146	Lead subacetate			✓			
U147	Maleic anhydride	✓				✓	✓
U148	Maleic hydrazide		✓			✓	✓
U149	Malononitrile		✓			✓	✓
U150	Melphalan			✓			
U151	Mercury			✓			
U152	Methacrylonitrile		✓		✓	✓	✓
U153	Methanethiol		✓		✓	✓	✓
U154	Methanol	✓			✓	✓	✓
U155	Methapyrilene		✓			✓	✓
U156	Methyl chlorocarbonate		✓		✓	✓	✓
U157	3-Methylcholanthrene	✓				✓	✓

U158	4,4'-Methylene-bis-(2-chloroaniline)
U159	Methyl ethyl ketone
U160	Methyl ethyl ketone peroxide
U161	Methyl isobutyl ketone
U162	Methyl methacrylate
U163	*N*-Methyl-*N*'-nitro-*N*-nitrosoquanidine
U164	Methylthiouracil
U165	Naphthalene
U166	1,4-Naphthoquinone
U167	1-Naphthylamine
U168	2-Naphthylamine
U169	Nitrobenzene
U170	4-Nitrophenol
U171	2-Nitropropane
U172	N-Nitrosodi-*n*-butylamine
U173	N-Nitrosodiethanolamine
U174	N-Nitrosodiethylamine
U175	N-Nitrosodi-*n*-propylamine
U176	N-Nitroso-*n*-ethylurea
U177	N-Nitroso-*n*-methylurea
U178	N-Nitroso-*n*-methylurethane
U179	N-Nitrosopiperidine
U180	N-Nitrosopyrrolidine
U181	5-Nitro-*o*-toluidine
U182	Paraldehyde
U183	Pentachlorobenzene
U184	Pentachloroethane
U185	Pentachloronitrobenzene
U186	1,3-Pentadiene
U187	Phenacetin
U188	Phenol
U189	Phosphorous sulfide
U190	Phthalic anhydride
U191	2-Picoline
U192	Pronamide
U193	1,3-Propane sultone
U194	*n*-Propylamine

TABLE 10 (continued)
Hazardous Wastes Rated as Good, Potential, or Poor Candidates for Incineration by Appropriate Technologies

EPA hazardous waste number	Hazardous waste	Candidate for incineration			Incinerator type		
		Good	Potential	Poor	Liquid injection	Rotary kiln	Fluidized bed
U196	Pyridine		✓		✓	✓	✓
U197	Quinones	✓				✓	✓
U200	Reserpine		✓			✓	✓
U201	Resorcinol	✓				✓	✓
U202	Saccharin		✓			✓	✓
U203	Safrole	✓			✓	✓	✓
U204	Selenious acid			✓			
U205	Selenium sulfide			✓			
U206	Streptozotocin						
U207	1,2,4,5-Tetrachlorobenzene		✓		✓	✓	✓
U208	1,1,1,2-Tetrachloroethane		✓		✓	✓	✓
U209	1,1,2,2-Tetrachloroethane		✓		✓	✓	✓
U210	Tetrachloroethane		✓			✓	✓
U211	Tetrachloromethane		✓			✓	✓
U212	2,3,4,6-Tetrachlorophenol		✓			✓	✓
U213	Tetrahydrofuran	✓			✓	✓	✓
U214	Thallium acetate			✓			
U215	Thallium carbonate			✓			
U216	Thallium chloride			✓			
U217	Thallium nitrate			✓			
U218	Thioacetamide		✓			✓	✓
U219	Thiourea		✓			✓	✓
U220	Toluene	✓				✓	✓
U221	Toluenediamine		✓			✓	✓
U222	o-Toluidine hydrochloride		✓			✓	✓
U223	Toluene diisocyanate		✓		✓	✓	✓
U224	Toxaphene		✓			✓	✓
U225	Tribromomethane		✓		✓	✓	✓
U226	1,1,1-Trichloroethane		✓		✓	✓	✓
U227	1,1,2-Trichloroethane		✓		✓	✓	✓

Code		Columns

U228   Trichloroethane

U229   Trichlorofluoromethane

U230   2,4,5-Trichlorophenol

U231   2,4,6-Trichlorophenol

U232   2,4,5-Trichlorophenoxyacetic acid

U233   2,4,5-Trichlorophenoxypropionic acid α,α,α-trichlorotoluene

U234   Trinitrobenzene

U235   Tris(2,3-dibromopropyl) phosphate

U236   Trypan blue

U237   Uracil mustard

U238   Urethane

U239   Xylene

Other hazardous wastes

SIC code number

2865   Vacuum still bottoms from the production of maleic anhydride

2865   Distillation residues from fractionating tower for recovery of benzene and chlorobenzenes

2865   Vacuum distillation residues from purification of 1-chloro-4-nitrobenzene

2865   Still bottoms or heavy ends from methanol recovery in methyl methacrylate production

2869   Heavy ends and distillation from production of carbaryl

2869   Residues from the production of hexachlorophenol, trichlorophenol, and 2,4,5-T

2869   Heavy ends from distillation of ethylene dichloride in vinyl chloride production

2869   Solid waste discharge from ion exchange column in production of acrylonitrile

2869   Bottom stream from quench column in acrylonitrile production of acrylonitrile

2869   Still bottoms from aniline production

2869   Tars from manufacture of bicycloheptadiene and cyclopentadiene

2869   Still bottom from production of furfural

2869   Unrecovered triester from production of disulfoton

2869   Waste polyvinyl chloride (PVC) from the manufacture of coated fabrics

TABLE 10 (continued)
Hazardous Wastes Rated as Good, Potential, or Poor Candidates for Incineration by Appropriate Technologies

EPA hazardous waste number	Hazardous waste	Candidate for incineration			Incinerator type		
		Good	Potential	Poor	Liquid injection	Rotary kiln	Fluidized bed
2869	Still bottoms from the production of pentachiloronitrobenzene		√			√	√
2869	Process clean out sludges from production of 1,1,1-trichloroethane		√			√	√
2869	Heavy ends and light ends from the production of methyl acrylate	√				√	√
2822	Polyvinyl chloride sludge from the manufacture of polyvinyl chloride					√	√
2869	Still bottoms from the purification of fluoromethanes in the production of fluoromethanes		√			√	√
2869	Heavy ends and light ends from the production of ethyl acrylate	√				√	√
2869	Heavy ends from the production of glycerine from allyl chloride	√				√	√
2869	Heavy ends from the distillation of acetic anhydride in the production of acetic anhydride	√				√	√
2869	Light ends from the distillation of acetaldehyde in the production of acetic anhydride	√			√	√	√
	Reactor cleanup wastes from the chlorination, dehydrochlorination, or oxychlorination of aliphatic hydrocarbons		√		√	√	√
	Fractionation bottoms from the separation of chlorinated aliphatic hydrocarbons		√		√	√	√
	Distillation bottoms from the separation of chlorinated aliphatic hydrocarbons		√		√	√	√
	Reactor cleanup wastes from the chlorination or oxychlorination of cyclic aliphatic hydrocarbons		√		√	√	√
	Fractionation bottoms from the separation of chlorinated cyclic aliphatic hydrocarbons		√		√	√	√
	Distillation bottoms from the separation of chlorinated cyclic aliphatic hydrocarbons		√		√	√	√
	Batch residues from the batch production of chlorinated polymers		√		√	√	√
	Solution residues from the production of chlorinated polymers	√			√	√	√
	Reactor cleanup wastes from the chlorination of aromatic hydrocarbon		√		√	√	√
	Fractionation bottoms from the separation of chlorinated aromatic hydrocarbons		√		√	√	√

SIC Code	Waste	✓
3333	Distillation bottoms from the separation of chlorinated aromatic hydrocarbons	✓
3339	Zinc production: oxide furnace residue and acid plant sludge	✓
3339	Ferromanganese emissions control: baghouse dusts and scrubwater solids	✓
3339	Ferrochrome silicon furnace emission control dust or sludge	✓
3339	Ferrochrome emissions control: furnace baghouse dust or sludge	✓
3341	Primary antimony-pyrometallurgical blast furnace slag	✓
3341	Secondary lead, scrubber sludge from SO_2 emission control, soft lead production	✓
3341	Secondary lead-white metal production furnace dust	✓
3341	Secondary copper-pyrometallurgical, blast furnace slag	✓
3341	Secondary copper-electrolytic refining wastewater treatment sludge	✓
3341	Secondary aluminum dross smelting-high salt slag plant residue	✓
3341	Zinc-cadmium metal reclamation, cadmium plant residue	✓
3691	Lead acid storage battery production wastewater treatment sludges	✓
3691	Lead acid storage battery production cleanup wastes from cathode and anode paste production	✓
3691	Nickel cadmium battery production wastewater treatment sludges	✓
3691	Cadmium silver oxide battery production wastewater treatment sludges	✓
3691	Mercury cadmium battery production wastewater treatment sludges	✓
3692	Magnesium carbon battery production chronic acid wastewater treatment sludges	✓
2816	Ash from incinerated still bottoms (paint and pigment production)	✓
2819	Arsenic bearing wastewater treatment sludges from production of boric acid	✓
2834	Arsenic or organo-arsenic containing wastewater treatment sludges from production of veterinary pharmaceuticals	✓
2851	Wastewater treatment sludges from paint production	✓
2851	Air pollution control sludges from paint production	✓
2869	By-product salts in production of MSMA	✓
2869	By-product salts in production of carcodylic acid	✓
2869	Lead slag from lead alkyl production	✓
3312	Steel finishing: alkaline cleaning waste	✓
	Waste pickle liquor	
	Cyanide-bearing wastes from electrolytic coating	
	Chromate and dichromate wastes from chemical treatment	
	Descaling acid	

TABLE 10 (continued)
Hazardous Wastes Rated as Good, Potential, or Poor Candidates for Incineration by Appropriate Technologies

EPA hazardous waste number	Hazardous waste	Candidate for incineration			Incinerator type		
		Good	Potential	Poor	Liquid injection	Rotary kiln	Fluidized bed
3322	Lead/phenolic sand-casting waste from malleable iron foundries			✓	✓		
3331	Primary copper smelting and refining electric furnace slag, converter dust, acid plant sludge, and reverberatory dust (T)			✓	✓		
3332	Primary lead blast furnace dust			✓	✓		
3339	Primary antimony-electrolytic sludge			✓	✓		
3339	Primary tungsten-digestion residue			✓	✓		
1094	Waste rock and overburden from uranium mining			✓	✓		
1099	Chlorinator residues and clarifier sludge from zirconium extraction			✓	✓		
1475	Overburden and slimes from phosphate surface mining			✓	✓		
2874	Waste gypsum from phosphoric acid production			✓	✓		
2819—2874	Slag and fluid bed prills from elemental phosphorus production			✓	✓		
2812	Sodium calcium sludge from production of chlorine by Down Cell process			✓	✓		
2812	Mercury-bearing brine purification muds from mercury cell process in chlorine production			✓	✓		
2816	Mercury-bearing wastewater treatment sludges from the production of mercuric sulfide pigment			✓	✓		
2816	Chromium-bearing wastewater treatment sludges from the production of TiO_2 pigment by the chloride process			✓	✓		
2816	Arsenic-bearing sludges from purification process in the production of antimony oxide			✓	✓		
2816	Antimony-bearing wastewater treatment sludge from production of antimony oxide			✓	✓		
3312	Ironmaking: ferromaganese blast furnace dust Ferromanganese blast furnace sludge Electric furnace dust and sludge			✓	✓		

a   Use this table for indicative guidance only. For decision making, read the material presented in the text.

Note: Table 10 should be used with caution. The information is indicative rather than conclusive. Conclusive decisions can be made only after studying the actual physical, chemical, and thermodynamic characteristics of the material(s) along with trial burn data (if available), and comparing expected behavior with the known behavior of a similar material (similar composition or physical, chemical, and thermodynamic characteristics) undergoing thermal destruction. It may be possible to blend different wastes or wastes and fuel oils to change poor or potential candidates into good candidates for incineration. Such blending may change the characteristics of a waste, making it incinerable in a different incinerator type than is identified in the table. It is also possible that some wastes identified in the table as good or potential candidates may turn out to be poor candidates if mixed with or contaminated by poor incineration candidates such as metals (arsenic, chromium, etc.) Therefore, factors such as blending and waste contamination should be considered on a case-by-case basis in making decisions. As mentioned earlier, Table 10 should be used with caution for indicative guidance rather than conclusive decisions.

From Bonner, T. A., Cornett, C. L., Desai, B. O., Fullenkamp, J. M., Hughes, T. W., Johnson, M. L., Kennedy, E. D., McCormick, R. J., Peters, J. A., and Zanders, D. L., Engineering Handbook for Hazardous Waste Incineration, Rep. MRC-DA 1090, U.S. Environmental Protection Agency, Washington, D.C., 1981.

TABLE 11
Waste Stream Constituents Which May Be Subjected to Ultimate Disposal by Controlled Incineration

Organic

Acetaldehyde
Acetic acid
Acetic anhydride
Acetone
Acetone cyanohydrin: oxides of nitrogen are removed from the effluent gas by scrubbers and/or thermal devices
Acetonitrile: oxides of nitrogen are removed from the effluent gas by scrubbers and/or thermal devices
Acetyl chloride
Acetylene
Acridine: oxides of nitrogen are removed from the effluent gas by scrubber, catalytic, or thermal device
Acrolein: 1500°F, 0.5-s minimum for primary combustion; 2000°F, 1.0 s for secondary combustion, combustion products CO_2 and H_2O
Acrylic acid
Acrylonitrile: NO_x removed from effluent gas by scrubbers and/or thermal devices
Adipic acid
Allyl alcohol
Allyl chloride: 1800°F, 2-s minimum
Aminoethylethanolamine: incinerator equipped with a scrubber of thermal unit to reduce NO_x emissions
Amyl acetate
Amyl alcohol
Aniline: oxides of nitrogen removed from the effluent gas by scrubber, catalytic, or thermal device
Anthracene
Benzene
Benzene sulfonic acid: incineration followed by scrubbing to remove the SO_2 gas
Benzoic acid
Benzyl chloride: 1500°F, 0.5-s minimum for primary combustion; 2200°F, 1.0-s for secondary combustion; elemental chlorine formation may be alleviated through injection of steam or methane into the combustion process
Butadiene
Butane
Butanols
1-Butene
Butyl acrylate
n-Butylamine: incinerator equipped with a scrubber or thermal unit to reduce NO_x emissions
Butylenes
Butyl phenol
Butyraldehyde
Camphor
Carbolic acid (phenol)
Carbon disulfide: sulfur dioxide scrubber necessary when combusting significant quantities of carbon disulfide
Carbon monoxide
Carbon tetrachloride: preferably after mixing with another combustible fuel; exercise care to assure complete combustion to prevent formation of phosgene; acid scrubber necessary to remove halo acids produced
Chloral hydrate: same as carbon tetrachloride
Chlorobenzene: same as carbon tetrachloride
Chloroform: same as carbon tetrachloride
Creosote
Cresol
Crotonaldehyde
Cumene
Cyanoacetic acid: oxides of nitrogen removed from the effluent gas by scrubbers and/or thermal devices
Cyclohexane
Cyclohexanol
Cyclohexanone
Cyclohexylamine: incinerator equipped with a scrubber or thermal unit to reduce NO_x emissions

TABLE 11 (continued)
Waste Stream Constituents Which May Be Subjected to Ultimate Disposal by
Controlled Incineration

Decyl alcohol

Di-*n*-butyl phthalate

Dichlorobenzene: incineration, preferably after mixing with another combustible fuel; exercise care to assure complete combustion to prevent formation of phosgene; Acid scrubber necessary to remove the halo acids produced

Dichlorodifluoromethane (freon): same as dichlorobenzene

Dichloroethyl ether: same as dichlorobenzene

Dichloromethane (methylene chloride); same as dichlorobenzene

1,2-Dichloropropane: same as dichlorobenzene

Dichlorotetrafluoroethane: same as dichlorobenzene

Dicyclopentadiene

Diethanolamine: incinerator equipped with scrubber or thermal unit to reduce NO_x emissions

Diethylamine: same as diethanolamine

Diethylene glycol

Diethyl ether: concentrated waste containing no peroxides — discharge liquid at a controlled rate near a pilot flame; concentrated waste containing peroxides — perforation of a container of the waste from a safe distance followed by open burning

Diethyl phthalate

Diethylstilbestrol

Diisobutylene

Diisobutyl ketone

Diisopropanolamine: incinerator equipped with scrubber or thermal unit to reduce NO_x emissions

Dimethylamine: same as diisopropanolamine

Dimethyl sulfate: incineration (1800°F), 1.5-s minimum) of dilute, neutralized dimethyl sulfate waste recommended; incinerator must be equipped with efficient scrubbing devices for oxides of sulfur

2,4-Dinitroaniline: controlled incineration whereby oxides of nitrogen are removed from the effluent gas by scrubber, catalytic, or thermal device

Dinitrobenzol: incineration (1800°F, 2.0-s minimum) followed by removal of the oxides of nitrogen formed using scrubbers and/or catalytic or thermal devices; dilute wastes should be concentrated before incineration

Dinitrocresol: incineration (1100°F minimum) with adequate scrubbing and ash disposal facilities

Dinitrophenol: incinerated (1800°F, 2.0-s minimum) with adequate scrubbing equipment for the removal of NO_x

Dinitrotoluene: pretreatment involves contact of the dinitrotoluene contaminated waste with $NaHCO_3$ and solid combustibles followed by incineration in an alkaline scrubber equipped incinerator unit

Dioxane: concentrated waste containing no peroxides — discharge liquid at a controlled rate near a pilot flame; concentrated waste containing peroxides — perforation of a container of the waste from a safe distance followed by open burning

Dipropylene glycol

Dodecylbenzene

Epichlorohydrin: incineration, preferably after mixing with another combustible fuel; Exercise care to assure complete combustion to prevent formation of phosgene; acid scrubber necessary to remove halo acids produced

Ethane

Ethanol

Ethanolamine: controlled incineration; incinerator equipped with scrubber of thermal unit to reduce NO_x emissions

Ethyl acetate

Ethyl acrylate

Ethylamine: controlled incineration; incinerator equipped with scrubber or thermal unit to reduce NO_x emissions

Ethylbenzene

Ethyl chloride: incineration, preferably after mixing with another combustible fuel; exercise care to assure complete combustion to prevent formation of phosgene; acid scrubber necessary to remove the halo acids produced

Ethylene

TABLE 11 (continued)
Waste Stream Constituents Which May Be Subjected to Ultimate Disposal by
Controlled Incineration

Ethylene cyanohydrin: controlled incineration (oxides of nitrogen removed from the effluent gas by scrubbers and/or thermal devices)

Ethylenediamine: same as ethylene cyanohydrin

Ethylene dibromide: controlled incineration with adequate scrubbing and ash disposal facilities

Ethylene dichloride: incineration, preferably after mixing with another combustible fuel; exercise care to assure complete combustion to prevent formation of phosgene; acid scrubber necessary to remove the halo acids produced

Ethylene glycol

Ethylene glycol monoethyl ether: concentrated waste containing no peroxides — discharge liquid at a controlled rate near a pilot flame; concentrated waste containing peroxides — perforation of a container of the waste from a safe distance followed by open burning

Ethyl mercaptan: incineration (2000°F) followed by scrubbing with a caustic solution

Fatty acids

Formaldehyde

Formic acid

Furfural

Glycerin

n-Heptane

Hexamethylenediamine: incinerator equipped with scrubber or thermal unit to reduce NO_x emissions

Hexane

Hydroquinone: incineration (1800°F, 2.0-s minimum), then scrub to remove harmful combustion products

Isobutyl acetate

Isopentane

Isophorone

Isoprene

Isopropanol

Isopropyl acetate

Isopropylamine: controlled incineration (incinerator equipped with scrubber or thermal unit to reduce NO_x emissions)

Isopropyl ether: concentrated waste containing no peroxides — discharge liquid at a controlled rate near a pilot flame; concentrated waste containing perioxides — perforation of a container of the waste from a safe distance followed by open burning

Maleic anhydride: controlled incineration; exercise care so that complete oxidation to nontoxic products occurs

Mercury compounds (organic): incineration followed by recovery/removal of mercury from the gas stream

Mesityl oxide

Methanol

Methyl acetate

Methyl acrylate

Methylamine: controlled incineration (incinerator equipped with scrubber or thermal unit to reduce NO_x emissions)

Methyl amyl alcohol

n-Methylaniline: controlled incineration whereby oxides of nitrogen are removed from the effluent gas by scrubber or by catalytic or thermal device

Methyl bromide: controlled incineration with adequate scrubbing and ash disposal facilities

Methyl chloride: same as methyl bromide

Methyl chloroformate: incineration, preferably after mixing with another combustible fuel; exercise care to assure complete combustion to prevent formation of phosgene; acid scrubber necessary to remove the halo acids produced

Methyl ethyl ketone

Methyl formate

Methyl isobutyl ketone

Methyl mercaptan: incineration followed by effective scrubbing of the effluent gas

Methyl methacrylate monomer

Morpholine: controlled incineration (incinerator equipped with scrubber or thermal unit to reduce NO_x emissions)

TABLE 11 (continued)
Waste Stream Constituents Which May Be Subjected to Ultimate Disposal by
Controlled Incineration

Naphtha

Naphthalene

β-Naphthylamine: controlled incineration whereby oxides of nitrogen are removed from the effluent gas by scrubber catalyst, or thermal device

Nitroaniline: incineration (1800°F, 2.0-s minimum) with scrubbing for NO_x abatement

Nitrobenzene: same as nitroaniline

Nitrocellulose: incinerator equipped with scrubber for NO_x abatement

Nitrochlorobenzene: incineration (1500°F, 0.5 s for primary combustion; 2200°F, 1.0 s for secondary combustion); formation of elemental chlorine can be prevented through injection of steam or methane into combustion process; NO_x may be abated through use of thermal or catalytic devices

Nitroethane: incineration; large quantities of material may require NO_x removal by catalytic or scrubbing processes

Nitromethane: same as nitroethane

Nitrophenol: controlled incineration; exercise care to maintain complete combustion at all times; incineration of large quantities may require scrubbers to control emission of NO_x

Nitropropane: same as nitroethane

4-Nitrotoluene: same as nitrophenol

Nonyl phenol

Octyl alcohol

Oleic acid

Oxalic acid: pretreatment involves chemical reaction with limestone or calcium oxide forming calcium oxalate; may then be incinerated utilizing particulate collection equipment to collect calcium oxide for recycling

Paraformaldehyde

Pentachlorophenol: incineration (600 to 900°C) coupled with adequate scrubbing and ash disposal facilities

n-Pentane

Perchloroethylene: incineration, preferably after mixing with another combustible fuel; exercise care to assure complete combustion to prevent formation of phosgene; acid scrubber necessary to remove the halo acids produced

Phenylhydrazine hydrochloride: controlled incineration whereby oxides of nitrogen are removed from the effluent gas by scrubber or by catalytic or thermal device

Phthalic anhydride

Polychlorinated biphenyls (PCBs): incineration (3000°F) with scrubbing to remove any chlorine-containing products

Polypropylene glycol methyl ether: concentrated waste containing no peroxides — discharge liquid at a controlled rate near a pilot flame; concentrated waste containing peroxides — perforation of a container of the waste from a safe distance followed by open burning

Polyvinyl chloride: incineration, preferably after mixing with another combustible fuel; exercise care to assure complete combustion to prevent formation of phosgene; acid scrubber necessary to remove the halo acids produced

Propane

Propionaldehyde

Propionic acid

Propyl acetate

Propyl alcohol

Propylamine: controlled incineration (incinerator equipped with scrubber or thermal unit to reduce NO_x emissions)

Propylene

Propylene oxide: concentrated waste containing no peroxides — discharge liquid at a controlled rate near a pilot flame; concentrated waste containing peroxides — perforation of a container of the waste from a safe distance followed by open burning

Pyridine: controlled incineration whereby oxides of nitrogen are removed from the effluent gas by scrubber or by catalytic or thermal devices

Quinone: controlled incineration (1800°F, 2.0-s minimum)

Salicylic acid

Sorbitol

Styrene

Tetrachloroethane: incineration, preferably after mixing with another combustible fuel; exercise care to assure complete combustion to prevent formation of phosgene; acid scrubber necessary to remove the halo acids produced

Tetraethyllead: controlled incineration with scrubbing for collection of lead oxides which may be recycled or landfilled

Tetrahydrofuran: concentrated waste containing peroxides; perforation of a container of the waste from a safe distance followed by open burning

Tetrapropylene

Toluene

Toluene diisocyanate: controlled incineration (oxides of nitrogen removed from the effluent gas by scrubbers and/or thermal devices)

Toluidine: same as toluene diisocyanate

Trichlorobenzene: incineration, preferably after mixing with another combustible fuel; exercise care to assure complete combustion to prevent formation of phosgene; acid scrubber necessary to remove the halo acids produced

Trichloroethane: same as trichlorobenzene

Trichloroethylene: same as trichlorobenzene

Trichlorofluoromethane: same as trichlorobenzene

Triethanolamine: controlled incineration (incinerator equipped with scrubber or thermal unit to reduce NO_x emissions)

Triethylamine: same as triethanolamine

Triethylene glycol

Triethylene tetramine: same as triethanolamine

Turpentine

Urea: same as triethanolamine

Vinyl acetate

Vinyl chloride: incineration, preferably after mixing with another combustible fuel; exercise care to assure complete combustion to prevent formation of phosgene; acid scrubber necessary to remove the halo acids produced

Xylene

Inorganic

Boron hydrides: with aqueous scrubbing of exhaust gases to remove B_2O_3 particulates

Fluorine: pretreatment involves reaction with a charcoal bed; product of reaction is carbon tetrafluoride, which is usually vented; residual fluorine can be combusted by means of a fluorine-hydrocarbon air burner followed by a caustic scrubber and stack

Hydrazine: controlled incineration with facilities for effluent scrubbing to abate any ammonia formed in the combustion process

Hydrazine/hydrazine azide: blends should be diluted with water and sprayed into an incinerator equipped with a scrubber

Mercuric chloride: incineration followed by recovery/removal of mercury from the gas stream

Mercuric nitrate: same as mercuric chloride

Mercuric sulfate: same as mercuric chloride

Phosphorus (white or yellow): controlled incineration followed by alkaline scrubbing and particulate removal equipment

Sodium azide: disposal via reaction with sulfuric acid solution and sodium nitrate in a hard rubber vessel; nitrogen dioxide generated by this reaction, and gas run through scrubber before release to atmosphere; controlled incineration also acceptable (after mixing with other combustible wastes) with adequate scrubbing and ash disposal facilities

Sodium formate: pretreatment involves conversion to formic acid followed by controlled incineration

Sodium oxalate: pretreatment involves conversion to oxalic acid followed by controlled incineration

Sodium-potassium alloy: controlled incineration with subsequent effluent scrubbing

Appendix B.3

EQUIPMENT PERFORMANCE

I.   Plant Equipment 548

II.   Instrumentation 558

III. Additional Information 569

I. PLANT EQUIPMENT

CALCULATION OF PERFORMANCE OF CENTRIFUGAL BLOWERS*

In the following formulas these symbols are used:

P   = pressure (psi or inches of mercury)
V   = volume in cubic feet per minute (ft³/min)
N   = speed in revolutions per minute (rpm)
D   = density in pounds per cubic foot (lb/ft³)
H   = height of air or gas column (ft)
SG = specific gravity (ratio of density of gas to density of air)

Standard air — Air at 68°F (absolute temperature 528°R) and 29.92 inHg (barometric pressure at sea level.) The density of such air is 0.075 lb/ft³ and the specific volume is 13.29 ft³/lb. The specific gravity is 1.0.

1. VARIATION OF BLOWER SPEED

Changing the speed of a centrifugal blower influences volume, pressure, and horsepower input.

1-A. The volume changes in *direct* ratio to the speed. *Example:* A blower is operating at 3500 rpm and delivering 1000 ft³/min. If the speed is reduced to 3000 rpm, what is the new volume?

V_1 = original volume (1000 ft³/min) N_1 = original speed (3500 rpm)
V_2 = new volume N_2 = new speed (3000 rpm)

$$V_2 = V_1 \times (N_2/N_1) = 1000 \times (3000/3500) = 857 \text{ ft}^3/\text{min}$$

1-B. The pressure changes as the *square* of the speed ratio. *Example:* A blower is operating at a speed of 3500 rpm and delivering air at 5.0 psi. If the speed is reduced to 3000 rpm, what is the new pressure?

P_1 = original pressure (5 psi) N_1 = original speed (3500 rpm)
P_2 = new pressure N_2 = new speed (3000 rpm)

$$P_2 = P_1 \times (N_2/N_1)^2 = 5 \times (3000/3500)^2 = 3.68 \text{ psi}$$

1-C. The power changes as the *cube* of the speed ratio. *Example:* A blower is operating at a speed of 3500 rpm and requiring 50 horsepower. If the speed is reduced to 3000 rpm, what is the new required power?

HP_1 = original power (50 hp) N_1 = original speed (3500 rpm)
HP_2 = new power N_2 = new speed (3000 rpm)

$$HP_2 = HP_1 \times (N_2/N_1)^3 = 50 \times (3000/3500)^3 = 31.5 \text{ hp}$$

* From Multistage Centrifugal Air or Gas Blowers and Exhausters, File #101-2-2 8/87 5M, Centrifugal Air Systems Division, Lamson Corporation, 1987. With permission.

2. RELATION OF INLET DENSITY TO OUTLET PRESSURE

The outlet pressure of a blower depends on the condition of the air or gas at the inlet. The inlet condition is influenced by:

1.   Specific gravity
2.   Altitude of blower
3.   Temperature of inlet air or gas

2-A. Pressure varies in direct proportion to the density. *Example:* A 3-psi (standard air) blower is to be used to handle gas having a specific gravity of 0.5. What pressure does the blower create when handling the gas?

P_a = air pressure (3 psi)
P_g = gas pressure
SG = specific gravity (0.5)

$$P_g = P_a \times SG = 3 \times 0.5 = 1.5 \text{ psi}$$

If one is required to handle a gas having a specific gravity of 0.5 at 1.5 psi of pressure, one can determine the standard air pressure blower as follows:

$$P_a = P_g / SG = 1.5/0.5 = 3 \text{ psi}$$

2-B. Pressure (barometric) varies in direct proportion to altitude. The following table gives the barometric pressure for various altitudes:

Absolute Pressure at Altitudes above Sea Level (Based on U.S. Standard Atmosphere)

Altitude (ft)	Pressure InHg	psia
0	29.92	14.70
500	29.38	14.43
600	29.28	14.38
700	29.18	14.33
800	29.07	14.28
900	28.97	14.23
1000	28.86	14.18
1500	28.33	13.90
2000	27.82	13.67
2500	27.31	13.41
3000	26.81	13.19
3500	26.32	12.92
4000	25.84	12.70
4500	25.36	12.45
5000	24.89	12.23
5500	24.23	12.00
6000	23.98	11.77
6500	23.53	11.56
7000	23.09	11.34
7500	22.65	11.12
8000	22.22	10.90
8500	21.80	10.70

Absolute Pressure at Altitudes above Sea Level (Based on U.S. Standard Atmosphere) (continued)

Altitude	Pressure	
(ft)	InHg	psia
9000	21.38	10.50
9500	20.98	10.90
10,000	20.58	10.10

Example: A blower is to operate at an elevation of 6000 ft and is to deliver 3 psi of pressure. What standard air blower pressure is required?

$$\text{Pressure} = 3 \times 29.92/23.98 = 3.75 \text{ psi}$$

If it is desired to determine what pressure a 3-psi standard air blower will deliver at 6000 ft:

$$\text{Pressure} = 3 \times 23.98/29.92 = 2.4 \text{ psi}$$

2-C. The air density varies in inverse proportion to the absolute temperature. *Example:* A blower is to handle 200°F air at 3 psi pressure. What standard air pressure blower is required?

P_1 = pressure hot air (3 psi)
P_2 = pressure standard air
AT_1 = absolute temperature hot air (200 + 460 = 660°R)
AT_2 = absolute temperature standard air (68 = 460 = 528°R)

$$P_2 = P_1 \times AT_1/AT_2 = 3 \times 660/528 = 3.75$$

Example: A blower is capable of delivering 3 psi of pressure with standard air. What pressure will it develop handling 200°F air?

$$P_1 = P_2 \times AT_2/AT_1 = 3 \times 528/660 = 2.4 \text{ psi}$$

3. RELATION OF INLET DENSITY TO POWER

The power varies in direct proportion to the specific gravity. *Example:* A standard air blower requires a 10-hp motor. What power is required when this blower is to handle a gas whose specific gravity is 0.5?

HP_1 = original power (10) SG = specific gravity of gas
HP_2 = new power

$$HP_2 = HP_1 \times SG = 10 \times 0.5 = 5 \text{ hp}$$

4. RELATION OF DENSITY TO INLET VOLUME

When a blower is to operate at a high altitude, it is frequently specified that the blower be capable of handling a given volume of "standard air". It is then necessary to determine the equivalent volume of air at the higher altitude. *Example:* A blower is to operate at 6000 ft altitude and is to handle 1000 ft^3/min of standard air. What is the volume of air the blower must handle at 6000 ft elevation?

V_1 = volume of standard air (1000 ft³/min)
V_2 = volume of air at altitude
Hg_1 = barometric pressure at sea level (29.92)
Hg_2 = barometric pressure at 6000 ft

$$V_2 = V_1 \times Hg_1/Hg_2 = 1000 \times 29.92/23.98 = 1225 \text{ ft}^3/\text{min}$$

Several of the above adjustments may be required on any given installation. Therefore, it may be necessary to use various combinations of these formulas.

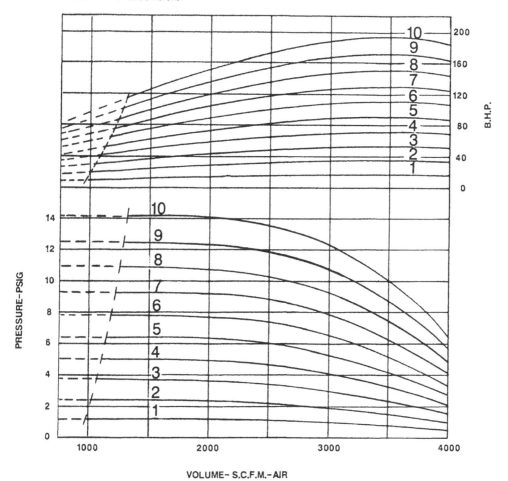

FIGURE 1. Typical performance curve for a centrifugal blower. (Courtesy of Lamson Corp.).

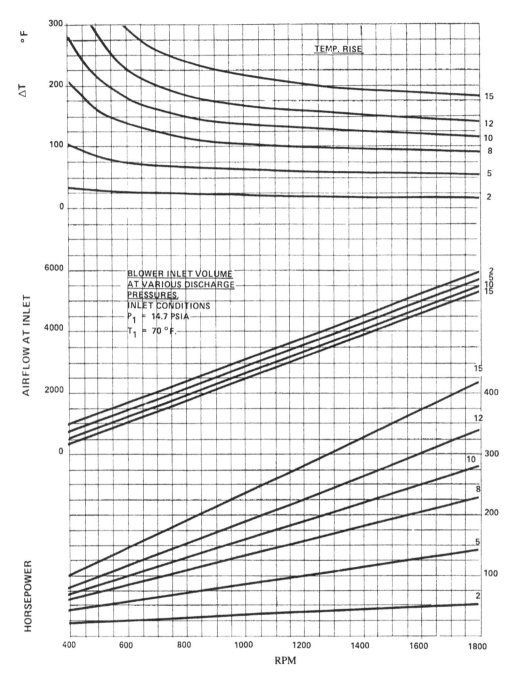

FIGURE 2.   Typical performance curve for a positive displacement blower. (Courtesy of Tuthill Corp., M-D Pneumatics Division, Springfield, MO.)

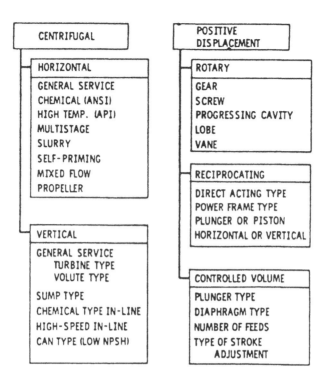

FIGURE 3. Pump classification chart. (From Bonner, T. A., Cornett, C. L., Desai, B. O., Fullenkamp, J. M., Hughes, T. W., Johnson, M. L., Kennedy, E. D., McCormick, R. J., Peters, J. A., and Zanders, D. L., Engineering Handbook for Hazardous Waste Incineration, Rep. MRC-DA 1090, U.S. Environmental Protection Agency, Washington, D.C., 1981.

TABLE 1
Kinematic Viscosity and Solids Handling Limitations of Various Atomization Techniques

Atomization type	Maximum kinematic viscosity (SSU)	Maximum solids mesh size	Maximum solids concentration
Rotary cup	175—300	35—100	20%
Single-fluid pressure	150		Essentially zero
Internal low pressure air (≤30 psi)	100		Essentially zero
External low pressure air	200—1500	200 (depends on nozzle ID)	30% (depends on nozzle ID)
External high pressure air	150—5000	100—200 (depends on nozzle ID)	70%
External high pressure steam	150—5000	100—200 (depends on nozzle ID)	70%

From Bonner, T. A., Cornett, C. L., Desai, B. O., Fullenkamp, J. M., Hughes, T. W., Johnson, M. L., Kennedy, E. D., McCormick, R. J., Peters, J. A., and Zanders, D. L., Engineering Handbook for Hazardous Waste Incineration, Rep. MRC-DA 1090, U.S. Environmental Protection Agency, Washingtion, D.C., 1981.

TABLE 2
Removal of Particulates and Acid Gases from Flue Gas Streams — Summary of Available Methods

Type of Separator	Applicable Particle Size (μm)	Applicable Dust Loading (gr/ACF)	Typical Operating Conditions	Remarks
Centrifugal collector (Cydone)	10—1000	50—1000	$\Delta P = 3$—10″	(1) High temp is o.k (2) Inertial force
Gravity settler	100~5000(+)	>100	Gas vel. < terminal vel.	(1) Settling chamber Gas 1—3 t/sec
Electrostatic precipitator	0.01—50	<5	N/A	(1) High temp is o.k (2) No acid gas removal
Conventional Baghouse	1—100	5—50	$\Delta P = 6$″ water	(1) Temp < 450°F (2) No acid gas removal
HEAF, CHEAF, Paper/Fiber filter	0.1—100	0.1—10	Wet/Dry type	(1) Temp > 250°F (2) Apply to small gas flow
HEPA	>1.0 μm	>0.01	Min 99.97% of 0.3 μm	(1) Low temp (2) No acid removal
Wet scrubber	0.3—100	0.05—5	$\Delta P = 20 - 40°$ W.G	(1) High temp (2) Acid removal

Note: (1) Inertial force = Cyclone, wet scrubber; (2) Gravity = Gravity settler; (3) Flow line interception = Baghouse, Filters; (4) Electric charge = Electrostatic precipitator.

From Shu, A., personal communication.

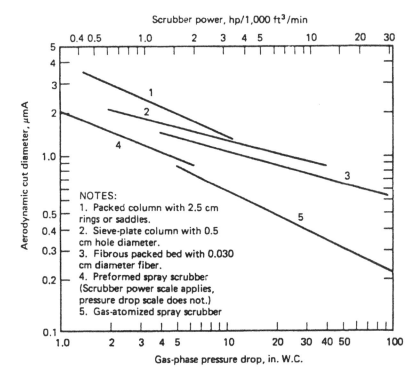

FIGURE 4. 50% Cut-power relations for several representative scrubber concepts. (From Calvert, p. 106.)

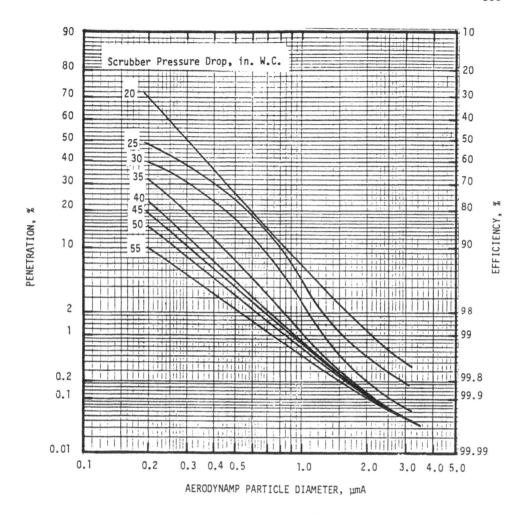

CALVERT Collision Scrubber™ Performance

FIGURE 5.   Performance curves for proprietary wet scrubber system. (From Calvert.)

TABLE 3
Flow of Water in Steel Pipe

Pressure Drop per 100 feet and Velocity in Schedule 40 Pipe for Water at 60 F.

Pipe-size labels (shown floating in the table) appear in the cell where each new size begins. The eight Velocity/Drop column-pairs host successively larger pipe sizes going down the table (the staircase layout of the original). Velocity = Feet per Second; Pressure Drop = Lbs per Sq. In.

Gallons per Minute	Cubic Ft per Second	Vel	Drop	Vel	Drop	Vel	Drop	Vel	Drop	Vel	Drop	Vel	Drop	Vel	Drop	Vel	Drop
		1/8″		**1/4″**		**3/8″**		**1/2″**		**3/4″**		**1″**		**1 1/4″**		**1 1/2″**	
.2	0.000446	1.13	1.86	0.616	0.359												
.3	0.000668	1.69	4.22	0.924	0.903	0.504	0.159	0.317	0.061								
.4	0.000891	2.26	6.98	1.23	1.61	0.672	0.345	0.422	0.086								
.5	0.00111	2.82	10.5	1.54	2.39	0.840	0.539	0.528	0.167	0.301	0.033						
.6	0.00134	3.39	14.7	1.85	3.29	1.01	0.751	0.633	0.240	0.361	0.041						
.8	0.00178	4.52	25.0	2.46	5.44	1.34	1.25	0.844	0.408	0.481	0.102						
1	0.00223	5.65	37.2	3.08	8.28	1.68	1.85	1.06	0.600	0.602	0.155	0.371	0.048				
2	0.00446	11.29	134.4	6.16	30.1	3.36	6.58	2.11	2.10	1.20	0.526	0.743	0.164	0.429	0.044		
3	0.00668	**2″**		9.25	64.1	5.04	13.9	3.17	4.33	1.81	1.09	1.114	0.336	0.644	0.090	0.473	0.043
4	0.00891			12.33	111.2	6.72	23.9	4.22	7.42	2.41	1.83	1.49	0.565	0.858	0.150	0.630	0.071
5	0.01114					8.40	36.7	5.28	11.2	3.01	2.75	1.86	0.835	1.073	0.223	0.788	0.104
6	0.01337	0.574	0.044	**2 1/2″**		10.08	51.9	6.33	15.8	3.61	3.84	2.23	1.17	1.29	0.309	0.946	0.145
8	0.01782	0.705	0.073			13.44	91.1	8.45	27.7	4.81	6.60	2.97	1.99	1.72	0.518	1.26	0.241
10	0.02228	0.956	0.108	0.670	0.046			10.56	42.4	6.02	9.99	3.71	2.99	2.15	0.774	1.58	0.361
15	0.03342	1.43	0.224	1.01	0.094	**3″**				9.03	21.6	5.57	6.36	3.22	1.63	2.37	0.755
20	0.04456	1.91	0.375	1.34	0.158	0.868	0.056	**3 1/2″**		12.03	37.8	7.43	10.9	4.29	2.78	3.16	1.28
25	0.05570	2.39	0.561	1.68	0.234	1.09	0.083	0.812	0.041	**4″**		9.28	16.7	5.37	4.22	3.94	1.93
30	0.06684	2.87	0.786	2.01	0.327	1.30	0.114	0.974	0.056			11.14	23.8	6.44	5.92	4.73	2.72
35	0.07798	3.35	1.05	2.35	0.436	1.52	0.151	1.14	0.071	0.882	0.041	12.99	32.2	7.51	7.90	5.52	3.64
40	0.08912	3.83	1.35	2.68	0.556	1.74	0.192	1.30	0.095	1.01	0.052	14.85	41.5	8.59	10.24	6.30	4.65
45	0.1003	4.30	1.67	3.02	0.668	1.95	0.239	1.46	0.117	1.13	0.064			9.67	12.80	7.09	5.85
50	0.1114	4.78	2.03	3.35	0.839	2.17	0.288	1.62	0.142	1.26	0.076	**5″**		10.74	15.66	7.88	7.15
60	0.1337	5.74	2.87	4.02	1.18	2.60	0.406	1.95	0.204	1.51	0.107			12.89	22.2	9.47	10.21
70	0.1560	6.70	3.84	4.69	1.59	3.04	0.540	2.27	0.261	1.76	0.143	1.12	0.047			11.05	13.71
80	0.1782	7.65	4.97	5.36	2.03	3.47	0.687	2.60	0.334	2.02	0.180	1.28	0.060			12.62	17.59
90	0.2005	8.60	6.20	6.03	2.53	3.91	0.861	2.92	0.416	2.27	0.224	1.44	0.074	**6″**		14.20	22.0
100	0.2228	9.56	7.59	6.70	3.09	4.34	1.05	3.25	0.509	2.52	0.272	1.60	0.090	1.11	0.036	15.78	26.9
125	0.2785	11.97	11.76	8.38	4.71	5.43	1.61	4.06	0.769	3.15	0.415	2.01	0.135	1.39	0.055	19.72	41.4
150	0.3342	14.36	16.70	10.05	6.69	6.51	2.24	4.87	1.08	3.78	0.580	2.41	0.190	1.67	0.077	**8″**	
175	0.3899	16.75	22.3	11.73	8.97	7.60	3.00	5.68	1.44	4.41	0.774	2.81	0.253	1.94	0.102		
200	0.4456	19.14	28.8	13.42	11.68	8.68	3.87	6.49	1.85	5.04	0.985	3.21	0.323	2.22	0.130		
225	0.5013			15.09	14.63	9.77	4.83	7.30	2.32	5.67	1.23	3.61	0.401	2.50	0.162	1.44	0.043
250	0.557					10.85	5.93	8.12	2.84	6.30	1.46	4.01	0.495	2.78	0.195	1.60	0.051
275	0.6127					11.94	7.14	8.93	3.40	6.93	1.79	4.41	0.583	3.05	0.234	1.76	0.061
300	0.6684					13.00	8.36	9.74	4.02	7.56	2.11	4.81	0.683	3.33	0.275	1.92	0.072
325	0.7241					14.12	9.89	10.53	4.09	8.19	2.47	5.21	0.797	3.61	0.320	2.08	0.083
350	0.7798							11.36	5.41	8.82	2.84	5.62	0.919	3.89	0.367	2.24	0.095
375	0.8355							12.17	6.18	9.45	3.25	6.02	1.05	4.16	0.416	2.40	0.108
400	0.8912							12.98	7.03	10.08	3.68	6.42	1.19	4.44	0.471	2.56	0.121
425	0.9469							13.80	7.89	10.71	4.12	6.82	1.33	4.72	0.529	2.73	0.136
450	1.003							14.61	8.80	11.34	4.60	7.22	1.48	5.00	0.590	2.89	0.151
475	1.059	1.93	0.054							11.97	5.12	7.62	1.64	5.27	0.653	3.04	0.166
500	1.114	2.03	0.059							12.60	5.65	8.02	1.81	5.55	0.720	3.21	0.182
550	1.225	2.24	0.071							13.85	6.79	8.82	2.17	6.11	0.861	3.53	0.219
600	1.337	2.44	0.083							15.12	8.04	9.63	2.55	6.66	1.02	3.85	0.258
650	1.448	2.64	0.097	**12″**								10.43	2.98	7.22	1.18	4.17	0.301
700	1.560	2.85	0.112	2.01	0.047	**14″**						11.23	3.43	7.78	1.35	4.49	0.343
750	1.671	3.05	0.127	2.15	0.054							12.03	3.92	8.33	1.55	4.81	0.392
800	1.782	3.25	0.143	2.29	0.061							12.83	4.43	8.88	1.75	5.13	0.443
850	1.894	3.46	0.160	2.44	0.068	2.02	0.042					13.64	5.00	9.44	1.96	5.45	0.497
900	2.005	3.66	0.179	2.58	0.075	2.13	0.047					14.44	5.58	9.99	2.18	5.77	0.554
950	2.117	3.86	0.198	2.72	0.083	2.25	0.052	**16″**				15.24	6.21	10.55	2.42	6.09	0.613
1000	2.228	4.07	0.218	2.87	0.091	2.37	0.057					16.04	6.84	11.10	2.68	6.41	0.675
1100	2.451	4.48	0.260	3.15	0.110	2.61	0.068					17.65	8.23	12.22	3.22	7.05	0.807
1200	2.674	4.88	0.306	3.43	0.128	2.85	0.080	2.18	0.042					13.33	3.81	7.70	0.948
1300	2.896	5.29	0.355	3.73	0.150	3.08	0.093	2.36	0.048					14.43	4.45	8.33	1.11
1400	3.119	5.70	0.409	4.01	0.171	3.32	0.107	2.54	0.055	**18″**				15.55	5.13	8.98	1.28
1500	3.342	6.10	0.466	4.30	0.195	3.56	0.122	2.72	0.063					16.66	5.85	9.62	1.46
1600	3.565	6.51	0.527	4.59	0.219	3.79	0.138	2.90	0.071					17.77	6.61	10.26	1.65
1800	4.010	7.32	0.663	5.16	0.276	4.27	0.172	3.27	0.088	2.58	0.050			19.99	8.37	11.54	2.08
2000	4.456	8.14	0.808	5.73	0.339	4.74	0.209	3.63	0.107	2.87	0.060	**20″**		22.21	10.3	12.82	2.55
2500	5.570	10.17	1.24	7.17	0.515	5.93	0.321	4.54	0.163	3.59	0.091			**24″**		16.03	3.94
3000	6.684	12.20	1.76	8.60	0.731	7.11	0.451	5.45	0.232	4.30	0.129	3.46	0.075			19.24	5.59
3500	7.798	14.24	2.38	10.03	0.982	8.30	0.607	6.35	0.311	5.02	0.173	4.04	0.101			22.44	7.56
4000	8.912	16.27	3.08	11.47	1.27	9.48	0.787	7.26	0.401	5.74	0.222	4.62	0.129	3.19	0.052	25.65	9.80
4500	10.03	18.31	3.87	12.90	1.60	10.67	0.990	8.17	0.503	6.46	0.280	5.20	0.162	3.59	0.065	28.87	12.2
5000	11.14	20.35	4.71	14.33	1.95	11.85	1.21	9.08	0.617	7.17	0.340	5.77	0.199	3.99	0.079		
6000	13.37	24.41	6.74	17.20	2.77	14.23	1.71	10.89	0.877	8.61	0.483	6.93	0.280	4.79	0.111		
7000	15.60	28.49	9.11	20.07	3.74	16.60	2.31	12.71	1.18	10.04	0.652	8.08	0.376	5.59	0.150		
8000	17.82			22.93	4.84	18.96	2.99	14.47	1.51	11.47	0.839	9.23	0.488	6.38	0.192		
9000	20.05			25.79	6.09	21.34	3.76	16.34	1.90	12.91	1.05	10.39	0.608	7.18	0.242		
10000	22.28			28.66	7.46	23.71	4.61	18.15	2.34	14.34	1.28	11.54	0.739	7.98	0.294		
12000	26.74			34.40	10.7	28.45	6.59	21.79	3.33	17.21	1.83	13.85	1.06	9.58	0.416		
14000	31.19					33.19	8.89	25.42	4.49	20.08	2.45	16.16	1.43	11.17	0.562		
16000	35.65							29.05	5.83	22.95	3.18	18.47	1.85	12.77	0.723		
18000	40.10							32.68	7.31	25.82	4.03	20.77	2.32	14.36	0.907		
20000	44.56							36.31	9.03	28.69	4.93	23.08	2.86	15.96	1.12		

For pipe lengths other than 100 feet, the pressure drop is proportional to the length. Thus, for 50 feet of pipe, the pressure drop is approximately one-half the value given in the table . . . for 300 feet, three times the given value, etc.

Velocity is a function of the cross sectional flow area; thus, it is constant for a given flow rate and is independent of pipe length.

TABLE 4
Flow of Air in Steel Pipe

For lengths of pipe other than 100 feet, the pressure drop is proportional to the length. Thus, for 50 feet of pipe, the pressure drop is approximately one-half the value given in the table . . . for 300 feet, three times the given value, etc.

The pressure drop is also inversely proportional to the absolute pressure and directly proportional to the absolute temperature.

Therefore, to determine the pressure drop for inlet or average pressures other than 100 psi and at temperatures other than 60 F, multiply the values given in the table by the ratio:

$$\left(\frac{100 + 14.7}{P + 14.7}\right)\left(\frac{460 + t}{520}\right)$$

where:

"P" is the inlet or average gauge pressure in pounds per square inch, and,

"t" is the temperature in degrees Fahrenheit under consideration.

The cubic feet per minute of compressed air at any pressure is inversely proportional to the absolute pressure and directly proportional to the absolute temperature.

To determine the cubic feet per minute of compressed air at any temperature and pressure other than standard conditions, multiply the value of cubic feet per minute of free air by the ratio:

$$\left(\frac{14.7}{14.7 + P}\right)\left(\frac{460 + t}{520}\right)$$

Calculations for Pipe Other than Schedule 40

To determine the velocity of water, or the pressure drop of water or air, through pipe other than Schedule 40, use the following formulas:

$$v_a = v_{40}\left(\frac{d_{40}}{d_a}\right)^2$$

$$\Delta P_a = \Delta P_{40}\left(\frac{d_{40}}{d_a}\right)^5$$

Subscript "a" refers to the Schedule of pipe through which velocity or pressure drop is desired.

Subscript "40" refers to the velocity or pressure drop through Schedule 40 pipe, as given in the tables on these facing pages.

Free Air q'_m Cubic Feet Per Minute at 60 F and 14.7 psia	Compressed Air Cubic Feet Per Minute at 60 F and 100 psig	Pressure Drop of Air In Pounds per Square Inch Per 100 Feet of Schedule 40 Pipe For Air at 100 Pounds per Square Inch Gauge Pressure and 60 F Temperature								
		1/8"	1/4"	3/8"	1/2"					
1	0.128	0.361	0.083	0.018						
2	0.256	1.31	0.285	0.064	0.020					
						3/4"				
3	0.384	3.06	0.605	0.133	0.042					
4	0.513	4.83	1.04	0.226	0.071					
5	0.641	7.45	1.58	0.343	0.106	0.027				
							1"			
6	0.769	10.6	2.23	0.408	0.148	0.037				
8	1.025	18.6	3.89	0.848	0.255	0.062	0.019			
								1¼"	1½"	
10	1.282	28.7	5.96	1.26	0.356	0.094	0.029			
15	1.922	...	13.0	2.73	0.834	0.201	0.062			
20	2.563	...	22.8	4.76	1.43	0.345	0.102	0.026		
25	3.204	...	35.6	7.34	2.21	0.526	0.156	0.039	0.019	
30	3.845	...	...	10.5	3.15	0.748	0.219	0.055	0.026	
35	4.486	...	...	14.2	4.24	1.00	0.293	0.073	0.035	
40	5.126	...	...	18.4	5.49	1.30	0.379	0.095	0.044	
										2"
45	5.767	...	...	23.1	6.90	1.62	0.474	0.116	0.055	
50	6.408			28.5	8.49	1.99	0.578	0.149	0.067	0.019
		2½"								
60	7.690			40.7	12.2	2.85	0.819	0.200	0.094	0.027
70	8.971			...	16.5	3.83	1.10	0.270	0.126	0.036
80	10.25	0.019		...	21.4	4.96	1.43	0.350	0.162	0.046
90	11.53	0.023		...	27.0	6.25	1.80	0.437	0.203	0.058
			3"							
100	12.82	0.029			33.2	7.69	2.21	0.534	0.247	0.070
125	16.02	0.044			...	11.9	3.39	0.825	0.380	0.107
150	19.22	0.062	0.021		...	17.0	4.87	1.17	0.537	0.151
175	22.43	0.083	0.028		...	23.1	6.60	1.58	0.727	0.205
				3½"						
200	25.63	0.107	0.036		...	30.0	8.54	2.05	0.937	0.264
225	28.84	0.134	0.045	0.022		37.9	10.8	2.59	1.19	0.331
250	32.04	0.164	0.055	0.027		...	13.3	3.18	1.45	0.404
275	35.24	0.191	0.066	0.032		...	16.0	3.83	1.75	0.484
300	38.45	0.232	0.078	0.037		...	19.0	4.56	2.07	0.573
325	41.65	0.270	0.090	0.043		...	22.3	5.32	2.42	0.673
					4"					
350	44.87	0.313	0.104	0.050		...	25.8	6.17	2.80	0.776
375	48.06	0.356	0.119	0.057	0.030	...	29.6	7.05	3.20	0.887
400	51.26	0.402	0.134	0.064	0.034	...	33.6	8.02	3.64	1.00
425	54.47	0.452	0.151	0.072	0.038	...	37.9	9.01	4.09	1.13
450	57.67	0.507	0.168	0.081	0.042	...	...	10.2	4.59	1.26
475	60.88	0.562	0.187	0.089	0.047		...	11.3	5.09	1.40
500	64.08	0.623	0.206	0.099	0.052		...	12.5	5.61	1.55
550	70.49	0.749	0.248	0.118	0.062		...	15.1	6.79	1.87
600	76.90	0.887	0.293	0.139	0.073		...	18.0	8.04	2.21
						5"				
650	83.30	1.04	0.342	0.163	0.086		...	21.1	9.43	2.60
700	89.71	1.19	0.395	0.188	0.099	0.032		24.3	10.9	3.00
750	96.12	1.36	0.451	0.214	0.113	0.036		27.9	12.6	3.44
800	102.5	1.55	0.513	0.244	0.127	0.041		31.8	14.2	3.90
850	108.9	1.74	0.576	0.274	0.144	0.046		35.9	16.0	4.40
							6"			
900	115.3	1.95	0.642	0.305	0.160	0.05[illegible]		40.2	18.0	4.91
950	121.8	2.18	0.715	0.340	0.178	0.057	0.023	...	20.0	5.47
1 000	128.2	2.40	0.788	0.375	0.197	0.063	0.025	...	22.1	6.06
1 100	141.0	2.89	0.948	0.451	0.236	0.075	0.030	...	26.7	7.29
1 200	153.8	3.44	1.13	0.533	0.279	0.089	0.035	...	31.8	8.63
1 300	166.6	4.01	1.32	0.626	0.327	0.103	0.041	...	37.3	10.1
1 400	179.4	4.65	1.52	0.718	0.377	0.119	0.047			11.8
1 500	192.2	5.31	1.74	0.824	0.431	0.136	0.054			13.5
								8"		
1 600	205.1	6.04	1.97	0.932	0.490	0.154	0.061			15.3
1 800	230.7	7.65	2.50	1.18	0.616	0.193	0.075			19.3
2 000	256.3	9.44	3.06	1.45	0.757	0.237	0.094	0.023		23.9
									10"	
2 500	320.4	14.7	4.76	2.25	1.17	0.366	0.143	0.035		37.3
3 000	384.5	21.1	6.82	3.20	1.67	0.524	0.204	0.051	0.016	
3 500	448.6	28.8	9.23	4.33	2.26	0.709	0.276	0.068	0.022	
4 000	512.6	37.6	12.1	5.66	2.94	0.919	0.358	0.088	0.028	
										12"
4 500	576.7	47.6	15.3	7.16	3.69	1.16	0.450	0.111	0.035	
5 000	640.8	...	18.8	8.85	4.56	1.42	0.552	0.136	0.043	0.018
6 000	769.0	...	27.1	12.7	6.57	2.03	0.794	0.195	0.061	0.025
7 000	897.1	...	36.9	17.2	8.94	2.76	1.07	0.262	0.082	0.034
8 000	1025	...	...	22.5	11.7	3.59	1.39	0.339	0.107	0.044
9 000	1153	...	...	28.5	14.9	4.54	1.76	0.427	0.134	0.055
10 000	1282	...	...	35.2	18.4	5.60	2.16	0.526	0.164	0.067
11 000	1410	...	...	...	22.2	6.78	2.62	0.633	0.197	0.081
12 000	1538	...	...	...	26.4	8.07	3.09	0.753	0.234	0.096
13 000	1666	...	...	...	31.0	9.47	3.63	0.884	0.273	0.112
14 000	1794	...	...	...	36.0	11.0	4.21	1.02	0.316	0.129
15 000	1922	...	...	...	...	12.6	4.84	1.17	0.364	0.148
16 000	2051	...	...	...	...	14.3	5.50	1.33	0.411	0.167
18 000	2307	...	...	...	...	18.2	6.96	1.68	0.520	0.213
20 000	2563	...	...	...	...	22.4	8.60	2.01	0.642	0.260
22 000	2820	...	...	...	...	27.1	10.4	2.50	0.771	0.314
24 000	3076	...	...	...	...	32.3	12.4	2.97	0.918	0.371
26 000	3332	...	...	...	...	37.9	14.5	3.49	1.12	0.435
28 000	3588	...	...	...	...	...	16.9	4.04	1.25	0.505
30 000	3845	...	...	...	...	...	19.3	4.64	1.42	0.520

From Flow of Fluids through Valves, Fittings, and Pipe, Technical Paper 410, Crane Co., 1978. With permission.

II. INSTRUMENTATION

TABLE 5
Standard Thermocouple Types

Type J	Iron vs. constantan (modified 1913 calibration)
Type K	Originally Chromel-P vs. Alumel
Type R	Platinum 13% rhodium vs. platinum
Type S	Platinum 10% rhodium vs. platinum
Type T	Copper vs. constantan
Type E	Originally Chromel-P vs. constantan
Type B	Platinum 30% rhodium vs. platinum 6% rhodium

Note: Standard thermocouple letter designations are defined in ANSI Standard C96.1.

From Bonner, T. A., Cornett, C. L., Desai, B. O., Fullenkamp, J. M., Hughes, T. W., Johnson, M. L., Kennedy, E. D., McCormick, R. J., Peters, J. A., and Zanders, D. L., Engineering Handbook for Hazardous Waste Incineration, Rep. MRC-DA 1090, U.S. Environmental Protection Agency, Washington, D.C. 1981.

TABLE 6
Standard Calibration Tables for Thermocouples

The following tables which represent the Temperature-E.M.F. functions of various thermocouples should be used with appropriate correction curves if precise results are desired. These curves must be determined for each individual couple by plotting ΔE, the difference between the observed and the standard E.M.F., against the standard E.M.F. at three or more fixed temperature points. The value ΔE as shown by such a correction curve is then subtracted algebraically from the observed E.M.F. to give the true E.M.F. reading.

In the following tables the fixed or "cold junction" is at $0°C$.; when the cold junction is not maintained at $0°C$. the readings of the E.M.F. must be corrected as follows: $Et = E_{(t-u)} + Etc$ where $E_{(t-u)}$ is the observed reading, Etc is the E.M.F. for the temperature corresponding to the cold junction temperature as read from the standard table and Et is the E.M.F. produced by the jot junction corrected to the value which would be obtained with the cold junction at $0°C$. The temperature corresponding to Et is then obtained by reference to the standard table.

Since the E.M.F.-temperature function is not linear the cold junction should be maintained at a temperature very close to that at whch the thermocouple was calibrated. Otherwise considerable error will result despite the above correction.

PLATINUM VERSUS PLATINUM-10-PERCENT RHODIUM THERMOCOUPLES

(Electromotive Force in Absolute Millivolts. Temperatures in Degrees C (Int. 1948). Reference Junctions at 0°C.)

°C	0	10	20	30	40	50	60	70	80	90
					Millivolts					
0 0		0.06	0.11	0.17	0.24	0.30	0.36	0.43	0.50	0.57
100	0.64	0.72	0.79	0.87	0.95	1.03	1.11	1.19	1.27	1.35
200	1.44	1.52	1.61	1.69	1.78	1.87	1.96	2.05	2.14	2.23
300	2.32	2.41	2.50	2.59	2.69	2.78	2.87	2.97	3.06	3.16
400	3.25	3.35	3.44	3.54	3.64	3.73	3.83	3.93	4.02	4.12
500	4.22	4.32	4.42	4.52	4.62	4.72	4.82	4.92	5.02	5.12
600	5.22	5.33	5.43	5.53	5.64	5.74	5.84	5.95	6.05	6.16
700	6.26	6.37	6.47	6.58	6.68	6.79	6.90	7.01	7.11	7.22
800	7.33	7.44	7.55	7.66	7.77	7.88	7.99	8.10	8.21	8.32
900	8.43	8.55	8.66	8.77	8.88	9.00	9.11	9.23	9.34	9.46
1000	9.57	9.6	9.80	9.92	10.04	10.15	10.27	10.39	10.51	10.62
1100	10.74	10.86	10.98	11.10	11.22	11.34	11.46	11.58	11.70	11.82
1200	11.94	12.06	12.18	12.30	12.42	12.54	12.66	12.78	12.90	13.02
1300	13.14	13.26	13.38	13.50	13.62	13.74	13.86	13.98	14.10	14.22
1400	14.34	14.46	14.58	14.70	14.82	14.94	15.05	15.17	15.29	15.41
1500	15.53	15.65	15.77	15.89	16.01	16.12	16.24	16.36	16.48	16.60
1600	16.72	16.83	16.95	17.07	17.19	17.31	17.42	17.54	17.66	17.77
1700	17.89	18.01	18.12	18.24	18.36	18.47	18.59	—	—	—

(Electromotive Force in Absolute Millivolts. Temperatures in Degrees F.* Reference Junctions at 32°F.)

°F	0	10	20	30	40	50	60	70	80	90
					Millivolts					
0	—	—	—	0.02	0.06	0.09	0.12	0.15	0.19	
100	0.22	0.26	0.29	0.33	0.36	0.40	0.44	0.48	0.52	0.56
200	0.60	0.64	0.68	0.72	0.76	0.80	0.84	0.89	0.93	0.97
300	1.02	1.06	1.11	1.15	1.20	1.24	1.29	1.33	1.38	1.43
400	1.47	1.52	1.57	1.62	1.66	1.71	1.76	1.81	1.86	1.91
500	1.96	2.01	2.06	2.11	2.16	2.21	2.26	2.31	2.36	2.41
600	2.46	2.51	2.56	2.61	2.66	2.72	2.77	2.82	2.87	2.92
700	2.98	3.03	3.08	3.14	3.19	3.24	3.29	3.35	3.40	3.45
800	3.51	3.56	3.61	3.67	3.72	3.78	3.83	3.88	3.94	3.99
900	4.05	4.10	4.16	4.21	4.26	4.32	4.37	4.43	4.49	4.54
1000	4.60	4.65	4.71	4.76	4.82	4.87	4.93	4.99	5.04	5.10
1100	5.16	5.21	5.27	5.33	5.38	5.44	5.50	5.56	5.61	5.67
1200	5.73	5.78	5.84	5.90	5.96	6.02	6.07	6.13	6.19	6.25
1300	6.31	6.37	6.42	6.48	6.54	6.60	6.66	6.72	6.78	6.84
1400	6.90	6.96	7.02	7.08	7.14	7.20	7.26	7.32	7.38	7.44
1500	7.50	7.56	7.62	7.68	7.74	7.80	7.86	7.93	7.99	8.05
1600	8.11	8.17	8.23	8.30	8.36	8.42	8.48	8.55	8.61	8.67
1700	8.73	8.80	8.86	8.92	8.98	9.05	9.11	9.17	9.24	9.30
1800	9.37	9.43	9.49	9.56	9.62	9.69	9.75	9.82	9.88	9.94
1900	10.01	10.07	10.14	1.20	10.27	10.33	10.40	10.47	10.53	10.60
2000	10.66	10.73	10.79	10.86	10.93	10.99	11.06	11.12	11.19	11.26
2100	11.32	11.39	11.46	11.52	11.59	11.66	11.72	11.79	11.86	11.92
2200	11.99	12.06	12.12	12.19	12.26	12.32	12.39	12.46	12.52	12.59
2300	12.66	12.72	12.79	12.86	12.92	12.99	13.06	13.12	13.19	13.26
2400	13.33	13.39	13.46	13.53	13.59	13.66	13.73	13.79	13.86	13.92
2500	13.99	14.06	14.12	14.19	14.26	14.32	14.39	14.46	14.52	14.59
2600	14.66	14.72	14.79	14.86	14.92	14.99	15.05	15.12	15.19	15.25
2700	15.32	15.39	15.45	15.52	15.58	15.65	15.72	15.78	15.85	15.91
2800	15.98	16.05	16.11	16.18	16.24	16.31	16.37	16.44	16.51	16.57
2900	16.64	16.70	16.77	16.83	16.90	16.97	17.03	17.10	17.16	17.23
3000	17.29	17.36	17.42	17.49	17.55	17.62	17.68	17.75	17.81	17.88
3100	17.94	18.01	18.07	18.14	18.20	18.27	18.33	18.40	18.46	18.53
3200	18.59	18.66	—	—	—	—	—	—	—	—

PLATINUM VERSUS PLATINUM-13-PERCENT RHODIUM THERMOCOUPLES

(Electromotive Force in Absolute Millivolts. Temperatures in Degrees C (Int. 1948) Reference Junctions at 0°C.)

°C	0	10	20	30	40	50	60	70	80	90
					Millivolts					
0	0.00	0.06	0.11	0.17	0.23	0.30	0.36	0.43	0.50	0.57
100	0.65	0.72	0.80	0.88	0.96	1.04	1.12	1.21	1.29	1.38
200	1.47	1.55	1.64	1.73	1.83	1.92	2.01	2.11	2.20	2.30
300	2.40	2.49	2.59	2.69	2.79	2.89	2.99	3.09	3.19	3.30
400	3.40	3.50	3.61	3.71	3.82	3.92	4.03	4.13	4.24	4.35
500	4.46	4.56	4.67	4.78	4.89	5.00	5.12	5.23	5.34	5.45
600	5.56	5.68	5.79	5.91	6.02	6.14	6.25	6.37	6.49	6.60
700	6.72	6.84	6.96	7.08	7.20	7.32	7.44	7.56	7.68	7.80
800	7.92	8.05	8.17	8.29	8.42	8.54	8.67	8.80	8.92	9.05
900	9.18	9.30	9.43	9.56	9.69	9.82	9.95	10.08	10.21	10.34
1000	10.47	10.60	10.74	10.87	11.00	11.14	11.27	11.41	11.54	11.68
1100	11.82	11.94	12.09	12.23	12.37	12.50	12.64	12.78	12.92	13.06
1200	13.19	13.33	13.47	13.61	13.75	13.89	14.03	14.17	14.30	14.44
1300	14.58	14.72	14.86	15.00	15.14	15.28	15.42	15.55	15.69	15.83
1400	15.97	16.11	16.25	16.39	16.52	16.66	16.80	16.94	17.08	17.22
1500	17.36	17.49	17.63	17.77	17.91	18.04	18.18	18.32	18.45	18.59
1600	18.73	18.86	19.00	19.14	19.27	19.41	19.55	19.68	19.82	19.95
1700	20.09									

(Electromotive Force in Absolute Millivolts. Temperatures in Degrees F° Reference Junctions at 32°F.)

°F	0	10	20	30	40	50	60	70	80	90
					Millivolts					
0	—	—	—	0.02	0.06	0.09	0.12	0.15	0.19	
100	0.22	0.26	0.29	0.33	0.36	0.40	0.44	0.48	0.52	0.56
200	0.60	0.64	0.68	0.72	0.76	0.81	0.85	0.89	0.94	0.98
300	1.03	1.08	1.12	1.17	1.21	1.26	1.31	1.36	1.41	1.46
400	1.50	1.55	1.60	1.65	1.70	1.75	1.81	1.86	1.91	1.96
500	2.01	2.07	2.12	2.17	2.22	2.28	2.33	2.38	2.44	2.49
600	2.55	2.60	2.66	2.71	2.77	2.82	2.88	2.94	2.99	3.05
700	3.10	3.16	3.22	3.27	3.33	3.39	3.45	3.50	3.56	3.62
800	3.68	3.74	3.79	3.85	3.91	3.97	4.03	4.09	4.14	4.21
900	4.26	4.32	4.38	4.44	4.50	4.56	4.62	4.69	4.75	4.81
1000	4.87	4.93	4.99	5.05	5.12	5.18	5.24	5.30	5.36	5.43
1100	5.49	5.55	5.61	5.68	5.74	5.81	5.87	5.93	6.00	6.06
1200	6.13	6.19	6.25	6.32	6.38	6.45	6.51	6.58	6.64	6.71
1300	6.77	6.84	6.90	6.97	7.04	7.10	7.17	7.24	7.30	7.37
1400	7.44	7.50	7.57	7.64	7.71	7.77	7.84	7.91	7.98	8.05
1500	8.12	8.18	8.25	8.32	8.39	8.46	8.53	8.60	8.67	8.74
1600	8.81	8.88	8.95	9.02	9.09	9.16	9.23	9.30	9.37	9.45
1700	9.52	9.59	9.66	9.73	9.80	9.87	9.95	10.02	10.09	10.16
1800	10.24	10.31	10.38	10.46	10.53	10.60	10.68	10.75	10.82	10.90
1900	10.97	11.05	11.12	11.20	11.27	11.35	11.42	11.50	11.58	11.65
2000	11.73	11.80	11.88	11.95	12.03	12.11	12.18	12.26	12.34	12.41
2100	12.49	12.56	12.64	12.72	12.80	12.87	12.95	13.03	13.10	13.18
2200	13.26	13.33	13.41	13.49	13.56	13.64	13.72	13.80	13.87	13.95
2300	14.03	14.10	14.18	14.26	14.34	14.41	14.49	14.57	14.64	14.72
2400	14.80	14.88	15.95	15.03	15.11	15.18	15.26	15.34	15.42	15.49
2500	15.57	15.65	15.72	15.80	15.88	15.95	16.03	16.11	16.19	16.26
2600	16.34	16.42	16.49	16.57	16.65	16.73	16.80	16.88	16.96	17.03
2700	17.11	17.19	17.26	17.34	17.42	17.49	17.57	17.65	17.72	17.80
2800	17.88	17.95	18.03	18.10	18.18	18.26	18.33	18.41	18.48	18.56
2900	18.64	18.71	18.79	18.86	18.94	19.02	19.09	19.17	19.24	19.32
3000	19.39	19.47	19.55	19.62	19.70	19.77	19.85	19.92	20.00	20.08
3100	20.15	—	—	—	—	—	—	—	—	—

* Based on the International Temperature Scale of 1948.

From Weast, R. C., Ed., *Handbook of Chemistry and Physics,* 69th ed., CRC Press, Boca Raton, FL, 1988.

CHROMEL-ALUMEL THERMOCOUPLES

(Electromotive Force in Absolute Millivolts. Temperatures in Degrees C (Int. 1948) Reference Junctions at 0°C)

°C	0	1	2	3	4	5	6	7	8	9
	Millivolts									
−190	−5.60	−5.62	−5.63	−5.65	−5.67	−5.68	−5.70	−5.71	−5.73	−5.74
−180	−5.43	−5.45	−5.46	−5.48	−5.50	−5.52	−5.53	−5.55	−5.57	−5.58
−170	−5.24	−5.26	−5.28	−5.30	−5.32	−5.34	−5.35	−5.37	−5.39	−5.41
−160	−5.03	−5.05	−5.08	−5.10	−5.12	−5.14	−5.16	−5.18	−5.20	−5.22
−150	−4.81	−4.84	−4.86	−4.88	−4.90	−4.92	−4.95	−4.97	−4.99	−5.01
−140	−4.58	−4.60	−4.62	−4.65	−4.67	−4.70	−4.72	−4.74	−4.77	−4.79
−130	−4.32	−4.35	−4.37	−4.40	−4.42	−4.45	−4.48	−4.50	−4.52	−4.55
−120	−4.06	−4.08	−4.11	−4.14	−4.16	−4.19	−4.22	−4.24	−4.27	−4.30
−110	−3.78	−3.81	−3.84	−3.86	−3.89	−3.92	−3.95	−3.98	−4.00	−4.03
−100	−3.49	−3.52	−3.55	−3.58	−3.61	−3.64	−3.66	−3.69	−3.72	−3.75
−90	−3.19	−3.22	−3.25	−3.28	−3.31	−3.34	−3.37	−3.40	−3.43	−3.46
−80	−2.87	−2.90	−2.93	−2.96	−3.00	−3.03	−3.06	−3.09	−3.12	−3.16
−70	−2.54	−2.57	−2.61	−2.64	−2.67	−2.71	−2.74	−2.77	−2.80	−2.84
−60	−2.29	−2.24	−2.27	−2.30	−2.34	−2.37	−2.41	−2.33	−2.47	−2.51
−50	−1.86	−1.89	−1.93	−1.96	−2.00	−2.03	−2.07	−2.10	−2.13	−2.17
−40	−1.50	−1.54	−1.57	−1.61	−1.64	−1.68	−1.72	−1.75	−1.79	−1.82
−30	−1.14	−1.17	−1.21	−1.25	−1.28	−1.32	−1.36	−1.39	−1.43	−1.37
−20	−0.77	−0.80	−0.84	−0.88	−0.92	−0.95	−0.99	−1.03	−1.06	−1.10
−10	−0.39	−0.42	−0.46	−0.50	−0.54	−0.58	−0.62	−0.66	−0.69	−0.73
(−)0	−0.00	−0.04	−0.08	−0.12	−0.16	−0.19	−0.23	−0.27	−0.31	−0.35
(+)0	0.00	0.04	0.08	0.12	0.16	0.20	0.24	0.28	0.32	0.36
10	0.40	0.44	0.48	0.52	0.56	0.60	0.64	0.68	0.72	0.76
20	0.80	0.84	0.88	0.92	0.96	1.00	1.04	1.08	1.12	1.16
30	1.20	1.24	1.28	1.32	1.36	1.40	1.44	1.49	1.53	1.57
40	1.61	1.65	1.69	1.73	1.77	1.81	1.85	1.90	1.94	1.98
50	2.02	2.06	2.10	2.14	2.18	2.23	2.27	2.31	2.35	2.39
60	2.43	2.47	2.51	2.56	2.60	2.64	2.68	2.72	2.76	2.80
70	2.85	2.89	2.93	2.97	3.01	3.05	3.10	3.14	3.18	3.22
80	3.26	3.30	3.35	3.39	3.43	3.47	3.51	3.56	3.60	3.63
90	3.68	3.72	3.76	3.81	3.85	3.89	3.93	3.97	4.01	4.06
100	4.10	4.14	4.18	4.22	4.26	4.31	4.35	4.39	4.43	4.47
110	4.51	4.55	4.60	4.64	4.68	4.72	4.76	4.80	4.84	4.88
120	4.92	4.96	5.01	5.05	5.09	5.13	5.17	5.21	5.25	5.29
130	5.33	5.37	5.41	5.45	5.49	5.53	5.57	5.61	5.65	5.69
140	5.73	5.77	5.81	5.85	5.89	5.93	5.97	6.01	6.05	6.09
150	6.13	6.17	6.21	6.25	6.29	6.33	6.37	6.41	6.45	6.49
160	6.53	6.57	6.61	6.65	6.69	6.73	6.77	6.81	6.85	6.89
170	6.93	6.97	7.01	7.05	7.09	7.13	7.17	7.21	7.25	7.29
180	7.33	7.37	7.41	7.45	7.49	7.53	7.57	7.61	7.65	7.69
190	7.73	7.77	7.81	7.85	7.89	7.93	7.97	8.01	8.05	8.09
200	8.13	8.17	8.21	8.25	8.29	8.33	8.37	8.41	−8.46	8.50
210	8.54	8.58	8.62	8.66	8.70	8.74	8.78	8.82	8.86	8.90
220	8.94	8.98	9.02	9.06	9.10	9.14	9.18	9.22	9.26	9.30
230	9.34	9.38	9.42	9.46	9.50	9.54	9.59	9.63	9.67	9.71
240	9.75	9.79	9.83	9.87	9.91	9.95	9.99	10.03	10.07	10.11
250	10.16	10.20	10.24	10.28	10.32	10.36	10.40	10.44	10.48	10.52
260	10.57	10.61	10.65	10.69	10.73	10.77	10.81	10.85	10.89	10.93
270	10.98	11.02	11.06	11.10	11.14	11.18	11.22	11.26	11.30	11.34
280	11.39	11.43	11.47	11.51	11.55	11.59	11.63	11.67	11.72	11.76
290	11.80	11.84	11.88	11.92	11.96	12.01	12.05	12.09	12.13	12.17
300	12.21	12.25	12.29	12.34	12.38	12.42	12.46	12.50	12.54	12.58
310	12.63	12.67	12.71	12.75	12.79	12.83	12.88	12.92	12.96	13.00
320	13.04	13.08	13.12	13.17	13.21	13.25	13.29	13.33	13.37	13.42
330	13.46	13.50	13.54	13.58	13.62	13.67	13.71	13.75	13.79	13.83
340	13.88	13.92	13.96	14.00	14.04	14.09	14.13	14.17	14.21	14.25
350	14.29	14.34	14.38	14.42	14.46	14.50	14.55	14.59	14.63	14.67
360	14.71	14.76	14.80	14.84	14.88	14.92	14.97	15.01	15.05	15.09
370	15.13	15.18	15.22	15.26	15.30	15.34	15.39	15.43	15.47	15.51
380	15.55	15.60	15.64	15.68	15.72	15.76	15.81	15.85	15.89	15.93
390	15.98	16.02	16.06	16.10	16.14	16.19	16.23	16.27	16.31	16.36
400	16.40	16.44	16.48	16.52	16.57	16.61	16.65	16.69	16.74	16.78
410	16.82	16.86	16.91	16.95	16.99	17.03	17.07	17.12	17.16	17.20
420	17.24	17.29	17.33	17.37	17.41	17.46	17.50	17.54	17.58	17.62
430	17.67	17.71	17.75	17.79	17.84	17.88	17.92	17.96	18.01	18.05
440	18.09	18.13	18.17	18.22	18.26	18.30	18.34	18.39	18.43	18.47
450	18.51	18.56	18.60	18.64	18.68	18.73	18.77	18.81	18.85	18.90
460	18.94	18.98	19.02	19.07	19.11	19.15	19.19	19.24	19.28	19.32
470	19.36	19.41	19.45	19.49	19.54	19.58	19.62	19.66	19.71	19.75
480	19.79	19.84	19.88	19.92	19.96	20.01	20.03	20.09	20.13	20.18
490	20.22	20.26	20.31	20.35	20.39	20.43	20.48	20.53	20.56	20.60
500	20.65	20.69	20.73	20.77	20.82	20.86	20.90	20.94	20.99	21.03
510	21.07	21.11	21.16	21.20	21.24	21.28	21.32	21.37	21.41	21.45
520	21.50	21.54	21.58	21.63	21.67	21.71	21.75	21.80	21.84	21.88
530	21.92	21.97	22.01	22.05	22.09	22.14	22.18	22.22	22.26	22.31
540	22.35	22.39	22.43	22.48	22.52	22.56	22.61	22.65	22.69	22.73
550	22.78	22.82	22.86	22.90	22.95	22.99	23.03	23.07	23.12	23.16
560	23.20	23.25	23.29	23.33	23.38	23.42	23.46	23.50	23.54	23.59
570	23.63	23.67	23.72	23.76	23.80	23.84	23.89	23.93	23.97	24.01

°C	0	1	2	3	4	5	6	7	8	9
	Millivolts									
580	24.06	24.10	24.14	24.18	24.23	24.27	24.31	24.36	24.40	24.44
590	24.49	24.53	24.57	24.61	25.65	24.70	24.74	24.78	24.83	24.87
600	24.91	24.95	25.00	25.04	25.08	25.12	25.17	25.21	25.25	25.29
610	25.34	25.38	25.42	25.47	25.51	25.55	25.59	25.64	25.68	25.72
620	25.76	25.81	25.85	25.89	25.93	25.98	26.02	26.06	26.10	26.15
630	26.19	26.23	26.27	26.32	26.36	26.40	26.44	26.48	26.53	26.57
640	26.61	26.65	26.70	26.74	26.78	26.82	26.86	26.91	26.95	26.99
650	27.03	27.07	27.12	27.16	27.20	27.24	27.28	27.33	27.37	27.41
660	27.45	27.49	27.54	27.58	27.62	27.66	27.71	27.75	27.79	27.83
670	27.87	27.92	27.96	28.00	28.04	28.08	28.13	28.17	28.21	28.25
680	28.29	28.34	28.38	28.42	28.46	28.50	28.55	28.59	28.63	28.67
690	28.72	28.76	28.80	28.84	28.88	28.93	28.97	29.01	29.05	29.10
700	29.14	29.18	29.22	29.26	29.30	29.35	29.39	29.43	29.47	29.52
710	29.56	29.60	29.64	29.68	29.72	29.77	29.81	29.85	29.89	29.93
720	29.97	30.02	30.06	30.10	30.14	30.18	30.23	30.27	30.31	30.35
730	30.39	30.44	30.48	30.52	30.56	30.60	30.65	30.69	30.73	30.77
740	30.81	30.85	30.90	30.94	30.98	31.02	31.06	31.10	31.15	31.19
750	31.23	31.27	31.31	31.35	31.40	31.44	31.48	31.52	31.56	31.60
760	31.65	31.69	31.73	31.77	31.81	31.85	31.90	31.94	31.98	32.02
770	32.06	32.10	32.15	32.19	32.23	32.27	32.31	32.35	32.39	32.43
780	32.48	32.52	32.56	32.60	32.64	32.68	32.72	32.76	32.81	32.85
790	32.89	32.93	32.97	33.01	33.05	33.09	33.13	33.18	33.22	33.26
800	33.30	33.34	33.38	33.42	33.46	33.50	33.54	33.59	33.63	33.67
810	33.71	33.75	33.79	33.83	33.87	33.91	33.95	33.99	34.04	34.08
820	34.12	34.16	34.20	34.24	34.28	34.32	34.36	34.40	34.44	34.48
830	34.53	34.57	34.61	34.65	34.69	34.73	34.77	34.81	34.85	34.89
840	34.93	34.97	35.02	35.06	35.10	35.14	35.18	35.22	35.26	35.30
850	35.34	35.38	35.42	35.46	35.50	35.54	35.58	35.63	35.67	35.71
860	35.75	35.79	35.83	35.87	35.91	35.95	35.99	36.03	36.07	36.11
870	36.15	36.19	36.23	36.27	36.31	36.35	36.39	36.43	36.47	36.51
880	36.55	36.59	36.63	36.67	36.72	36.76	36.80	36.84	36.88	36.92
890	36.96	37.00	37.04	37.08	37.12	37.16	37.20	37.24	37.28	37.32
900	37.36	37.40	37.44	37.48	37.52	37.56	37.60	37.64	37.68	37.72
910	37.76	37.80	37.84	37.88	37.92	37.96	38.00	38.04	38.08	38.12
920	38.16	38.20	38.24	38.28	38.32	38.36	38.40	38.44	38.48	38.52
930	38.56	38.60	38.64	38.68	38.72	38.76	38.80	38.84	38.88	38.92
940	38.95	38.99	39.03	39.07	39.11	39.15	39.19	39.23	39.27	39.31
950	39.35	39.39	39.43	39.47	39.51	39.55	39.59	39.63	39.67	39.71
960	39.75	39.79	39.83	39.86	39.90	39.94	39.98	40.02	40.06	40.10
970	40.14	40.18	40.22	40.26	40.30	40.34	40.38	40.41	40.45	40.49
980	40.53	40.57	40.61	40.65	40.69	40.73	40.77	40.81	40.85	40.89
990	30.92	40.96	41.00	41.04	41.08	41.12	41.16	41.20	41.24	41.28
1000	41.31	41.35	41.39	41.43	41.47	41.51	41.55	41.59	41.63	41.67
1010	41.70	41.74	41.78	41.82	41.86	41.90	41.94	41.98	42.02	42.05
1020	42.09	42.13	42.17	42.21	42.25	42.29	42.33	42.36	42.40	42.44
1030	42.48	42.52	42.56	42.60	42.63	42.67	42.71	42.75	42.79	42.83
1040	42.87	42.90	42.94	42.98	43.02	43.06	43.10	43.14	43.17	43.21
1050	43.25	43.29	43.33	43.37	43.41	43.44	43.48	43.52	43.56	43.60
1060	43.63	43.67	43.71	43.75	43.79	43.83	43.87	43.90	43.94	43.98
1070	44.02	44.06	44.10	44.13	44.17	44.21	44.25	44.29	44.33	44.36
1080	44.40	44.44	44.48	44.52	44.55	44.59	44.63	44.67	44.71	44.74
1090	44.78	44.82	44.86	44.90	44.93	44.97	45.01	45.05	45.09	45.12
1100	45.16	45.20	45.24	45.27	45.31	45.35	45.39	45.43	45.46	45.50
1110	45.54	45.58	45.62	45.65	45.69	45.73	45.77	45.80	45.84	45.88
1120	45.92	45.96	45.99	46.03	46.07	46.11	46.14	46.18	46.22	46.26
1130	46.29	46.33	46.37	46.41	46.44	46.48	46.52	46.56	46.59	46.63
1140	46.67	46.70	46.74	46.78	46.82	46.85	46.89	46.93	46.97	47.00
1150	47.04	47.08	47.12	47.15	47.19	47.23	47.26	47.30	47.34	47.38
1160	47.41	47.45	47.49	47.52	47.56	47.60	47.63	47.67	47.71	47.75
1170	47.78	47.82	47.86	47.89	47.93	47.97	48.00	48.04	48.08	48.12
1180	48.15	48.19	48.23	48.26	48.30	48.34	48.37	48.41	48.45	48.48
1190	48.52	48.56	48.59	48.63	48.67	48.70	48.74	48.78	48.81	48.85
1200	48.89	48.92	48.96	49.00	49.03	49.07	49.11	49.14	49.18	49.22
1210	49.25	49.29	49.32	49.36	49.40	49.43	49.47	49.51	49.54	49.58
1220	49.62	49.65	49.69	49.72	49.76	49.80	49.83	49.87	49.90	49.94
1230	49.98	50.01	50.05	50.08	50.12	50.16	50.19	50.23	50.26	50.30
1240	50.34	50.37	50.41	50.44	50.48	50.52	50.55	50.59	50.62	50.66
1250	50.69	50.73	50.77	50.80	50.84	50.87	50.91	50.94	50.98	51.02
1260	51.05	51.09	51.12	51.16	51.19	51.23	51.27	51.30	51.34	51.37
1270	51.41	51.44	51.48	51.51	51.55	51.58	51.62	51.66	51.69	51.73
1280	51.76	51.80	51.83	51.87	51.90	51.94	51.97	52.01	52.04	52.08
1290	52.11	52.15	52.18	52.22	52.25	52.29	52.32	52.36	52.39	52.43
1300	52.46	52.50	52.53	52.57	52.60	52.64	52.67	52.71	52.74	52.78
1310	52.81	52.85	52.88	52.92	52.95	52.99	53.02	53.06	53.09	53.13
1320	53.16	53.20	53.23	53.27	53.30	53.34	53.37	53.41	53.44	53.47
1330	53.51	53.54	53.58	53.61	53.65	53.68	53.72	53.75	53.79	53.82
1340	53.85	53.89	53.92	53.96	53.99	54.03	54.06	54.10	54.13	54.16
1350	54.20	54.23	54.27	54.30	54.34	54.37	54.40	54.44	54.47	54.51
1360	54.54	54.57	54.61	54.64	54.68	54.71	54.74	54.78	54.81	54.85
1370	54.88	54.91	—	—	—	—	—	—	—	—

* Based on the International Temperature Scale of 1948.

CHROMEL-ALUMEL THERMOCOUPLES (Continued)

(Electromotive Force in Absolute Millivolts. Temperatures in Degrees F* Reference Junctions at 32°F.)

°F	0	1	2	3	4	5	6	7	8	9
	Millivolts									
−300	−5.51	−5.52	−5.53	−5.54	−5.54	−5.55	−5.56	−5.57	−5.58	−5.59
−290	−5.41	−5.42	−5.43	−5.44	−5.45	−5.46	−5.47	−5.48	−5.49	−5.50
−280	−5.30	−5.31	−5.32	−5.34	−5.35	−5.36	−5.37	−5.38	−5.39	−5.40
−270	−5.20	−5.21	−5.22	−5.23	−5.24	−5.25	−5.26	−5.27	−5.28	−5.29
−260	−5.08	−5.09	−5.10	−5.12	−5.13	−5.14	−5.15	−5.16	−5.17	−5.18
−250	−4.96	−4.97	−4.99	−5.00	−5.01	−5.02	−5.03	−5.04	−5.06	−5.07
−240	−4.84	−4.85	−4.86	−4.88	−4.89	−4.90	−4.91	−4.92	−4.94	−4.95
−230	−4.71	−4.72	−4.74	−4.75	−4.76	−4.77	−4.79	−4.80	−4.81	−4.82
−220	−4.58	−4.59	−4.60	−4.62	−4.63	−4.64	−4.66	−4.67	−4.68	−4.70
−210	−4.44	−4.45	−4.46	−4.48	−4.49	−4.51	−4.52	−4.53	−4.55	−4.56
−200	−4.29	−4.31	−4.32	−4.34	−4.35	−4.36	−4.38	−4.39	−4.41	−4.42
−190	−4.15	−4.16	−4.18	−4.19	−4.21	−4.22	−4.24	−4.25	−4.26	−4.28
−180	−4.00	−4.01	−4.03	−4.04	−4.06	−4.07	−4.09	−4.10	−4.12	−4.13
−170	−3.84	−3.86	−3.88	−3.89	−3.91	−3.92	−3.94	−3.95	−3.97	−3.98
−160	−3.69	−3.70	−3.72	−3.73	−3.75	−3.76	−3.78	−3.80	−3.81	−3.83
−150	−3.52	−3.54	−3.56	−3.57	−3.59	−3.60	−3.62	−3.64	−3.65	−3.67
−140	−3.36	−3.38	−3.39	−3.41	−3.42	−3.44	−3.46	−3.47	−3.49	−3.51
−130	−3.19	−3.20	−3.22	−3.24	−3.25	−3.27	−3.29	−3.31	−3.32	−3.34
−120	−3.01	−3.03	−3.05	−3.06	−3.08	−3.10	−3.12	−3.13	−3.15	−3.17
−110	−2.83	−2.85	−2.87	−2.89	−2.90	−2.92	−2.94	−2.96	−2.98	−2.99
−100	−2.65	−2.67	−2.69	−2.71	−2.72	−2.74	−2.76	−2.78	−2.80	−2.82
−90	−2.47	−2.49	−2.50	−2.52	−2.54	−2.56	−2.58	−2.60	−2.62	−2.63
−80	−2.28	−2.30	−2.32	−2.34	−2.36	−2.37	−2.39	−2.41	−2.43	−2.45
−70	−2.09	−2.11	−2.13	−2.15	−2.17	−2.18	−2.20	−2.22	−2.24	−2.26
−60	−1.90	−1.92	−1.94	−1.96	−1.97	−1.99	−2.01	−2.03	−2.05	−2.07
−50	−1.70	−1.72	−1.74	−1.76	−1.78	−1.80	−1.82	−1.84	−1.86	−1.88
−40	−1.50	−1.52	−1.54	−1.56	−1.58	−1.60	−1.62	−1.64	−1.66	−1.68
−30	−1.30	−1.32	−1.34	−1.36	−1.38	−1.40	−1.42	−1.44	−1.46	−1.48
−20	−1.10	−1.12	−1.14	−1.16	−1.18	−1.20	−1.22	−1.24	−1.26	−1.28
−10	−0.89	−0.91	−0.93	−0.95	−0.97	−0.99	−1.01	−1.03	−1.06	−1.08
(−)0	−0.68	−0.70	−0.72	−0.75	−0.77	−0.79	−0.81	−0.83	−0.85	−0.87
(+)0	−0.68	−0.66	−0.64	−0.62	−0.60	−0.58	−0.56	−0.54	−0.52	−0.49
10	−0.47	−0.45	−0.43	−0.41	−0.39	−0.37	−0.34	−0.32	−0.30	−0.28
20	−0.26	−0.24	−0.22	−0.19	−0.17	−0.15	−0.13	−0.11	−0.09	−0.07
30	−0.04	−0.02	0.00	0.02	0.04	0.07	0.09	0.11	0.13	0.15
40	0.18	0.20	0.22	0.24	0.26	0.29	0.31	0.33	0.35	0.37
50	0.40	0.42	0.44	0.46	0.48	0.51	0.53	0.55	0.57	0.60
60	0.62	0.64	0.65	0.68	0.71	0.73	0.75	0.77	0.80	0.82
70	0.84	0.86	0.88	0.91	0.93	0.95	0.97	1.00	1.02	1.04
80	1.06	1.09	1.11	1.13	1.15	1.18	1.20	1.22	1.24	1.27
90	1.29	1.31	1.33	1.36	1.38	1.40	1.43	1.45	1.47	1.49
100	1.52	1.54	1.56	1.58	1.61	1.63	1.65	1.68	1.70	1.72
110	1.74	1.77	1.79	1.81	1.84	1.86	1.88	1.90	1.93	1.95
120	1.97	2.00	2.02	2.04	2.06	2.09	2.11	2.13	2.16	2.18
130	2.20	2.23	2.25	2.27	2.29	2.32	2.34	2.36	2.39	2.41
140	2.43	2.46	2.48	2.50	2.52	2.55	2.57	2.59	2.62	2.64
150	2.66	2.69	2.71	2.73	2.75	2.78	2.80	2.82	2.85	2.87
160	2.89	2.92	2.94	2.96	2.98	3.01	3.03	3.05	3.08	3.10
170	3.12	3.15	3.17	3.19	3.22	3.24	3.26	3.29	3.31	3.33
180	3.36	3.38	3.40	3.43	3.45	3.47	3.49	2.52	3.54	3.56
190	3.59	3.61	3.63	3.66	3.68	3.70	3.73	3.75	3.77	3.80
200	3.82	3.84	3.87	3.89	3.91	3.94	3.96	3.98	4.01	4.03
210	4.05	4.08	4.10	4.12	4.15	4.17	4.19	4.21	4.24	4.26
220	4.28	4.31	4.33	4.35	4.38	4.40	4.42	4.44	4.47	4.49
230	4.51	4.54	4.56	4.58	4.61	4.63	4.65	4.67	4.70	4.72
240	4.74	4.77	4.79	4.81	4.83	4.86	4.88	4.90	4.92	4.95
250	4.97	4.99	5.02	5.04	5.06	5.08	5.11	5.13	5.15	5.17
260	5.20	5.22	5.24	5.26	5.29	5.31	5.33	5.35	5.38	5.40
270	5.42	5.44	5.47	5.49	5.51	5.53	5.56	5.58	5.60	5.62
280	5.65	5.67	5.69	5.71	5.73	5.76	5.78	5.80	5.82	5.85
290	5.87	5.89	5.91	5.93	5.96	5.98	6.00	6.02	6.05	6.07
300	6.09	6.11	6.13	6.16	6.18	6.20	6.22	6.25	6.27	6.29
310	6.31	6.33	6.36	6.38	6.40	6.42	6.45	6.47	6.49	6.51
320	6.53	6.56	6.58	6.60	6.62	6.65	6.67	6.69	6.71	6.73
330	6.76	6.78	6.80	6.82	6.84	6.87	6.89	6.91	6.93	6.96
340	6.98	7.00	7.02	7.04	7.07	7.09	7.11	7.13	7.15	7.18
350	7.20	7.22	7.24	7.26	7.29	7.31	7.33	7.35	7.38	7.40
360	7.42	7.44	7.46	7.49	7.51	7.53	7.55	7.58	7.60	7.62
370	7.64	7.66	7.69	7.71	7.73	7.75	7.78	7.80	7.82	7.84
380	7.87	7.89	7.91	7.93	7.95	7.98	8.00	8.02	8.04	8.07
390	8.09	8.11	8.13	8.16	8.18	8.20	8.22	8.24	8.27	8.29
400	8.31	8.33	8.36	8.38	8.40	8.42	8.45	8.47	8.49	8.51
410	8.54	8.56	8.58	8.60	8.62	8.65	8.67	8.69	8.71	8.74
420	8.76	8.78	8.80	8.82	8.85	8.87	8.89	8.91	8.94	8.96
430	8.98	9.00	9.03	9.05	9.07	9.09	9.12	9.14	9.16	9.18
440	9.21	9.23	9.25	9.27	9.30	9.32	9.34	9.36	9.39	9.41
450	9.43	9.45	9.48	9.50	9.52	9.54	9.57	9.59	9.61	9.63
460	9.66	9.68	9.70	9.73	9.75	9.77	9.79	9.82	9.84	9.86
470	9.88	9.91	9.93	9.95	9.97	10.00	10.02	10.04	10.06	10.09
480	10.11	10.13	10.16	10.18	10.20	10.22	10.25	10.27	10.29	10.31
490	10.34	10.36	10.38	10.40	10.43	10.45	10.47	10.50	10.52	10.54
500	10.57	10.59	10.61	10.63	10.66	10.68	10.70	10.72	10.75	10.77

°F	0	1	2	3	4	5	6	7	8	9
	Millivolts									
510	10.79	10.82	10.84	10.86	10.88	10.91	10.93	10.95	10.98	11.00
520	11.02	11.04	11.07	11.09	11.11	11.13	11.16	11.18	11.20	11.23
530	11.25	11.27	11.29	11.32	11.34	11.36	11.39	11.41	11.43	11.45
540	11.48	11.50	11.52	11.55	11.57	11.59	11.61	11.64	11.66	11.68
550	11.71	11.73	11.75	11.78	11.80	11.82	11.84	11.87	11.89	11.91
560	11.94	11.98	11.98	12.01	12.03	12.05	12.07	12.10	12.12	12.14
570	12.17	12.19	12.21	12.24	12.26	12.28	12.30	12.33	12.35	12.37
580	12.40	12.42	12.44	12.47	12.49	12.51	12.53	12.56	12.58	12.60
590	12.63	12.65	12.67	12.70	12.72	12.74	12.76	12.79	12.81	12.83
600	12.86	12.88	12.90	12.93	12.95	12.97	13.00	13.02	13.04	13.06
610	13.09	13.11	13.13	13.16	13.18	13.20	13.23	13.25	13.27	13.30
620	13.32	13.34	13.36	13.39	13.41	13.44	13.46	13.48	13.50	13.53
630	13.55	13.57	13.60	13.62	13.63	13.67	13.69	13.71	13.74	13.76
640	13.78	13.81	13.83	13.85	13.88	13.90	13.92	13.95	13.97	13.99
650	14.02	14.04	14.06	14.09	14.11	14.13	14.15	14.18	14.20	14.22
660	14.25	14.27	14.29	14.32	14.34	14.36	14.39	14.41	14.43	14.46
670	14.48	14.50	14.53	14.55	14.57	14.60	14.62	14.64	14.67	14.69
680	14.71	14.74	14.76	14.78	14.81	14.83	14.85	14.88	14.90	14.92
690	14.95	14.97	14.99	15.02	15.04	15.06	15.09	15.11	15.13	15.16
700	15.18	15.20	15.23	15.25	15.27	15.30	15.32	15.34	15.37	15.39
710	15.41	15.44	15.46	15.48	15.51	15.53	15.55	15.58	15.60	15.62
720	15.65	15.67	15.69	15.72	15.74	15.76	15.79	15.81	15.83	15.86
730	15.88	15.90	15.93	15.95	15.98	16.00	16.02	16.05	16.07	16.09
740	16.12	16.14	16.16	16.19	16.21	16.23	16.26	16.28	16.30	16.33
750	16.35	16.37	16.40	16.42	16.45	16.47	16.49	16.52	16.54	16.56
760	16.59	16.61	16.63	16.66	16.68	16.70	16.73	16.75	16.77	16.80
770	16.82	16.84	16.87	16.89	16.92	16.94	16.96	16.99	17.01	17.03
780	17.06	17.08	17.10	17.13	17.15	17.17	17.20	17.22	17.24	17.27
790	17.29	17.31	17.34	17.36	17.39	17.41	17.43	17.46	17.48	17.50
800	17.53	17.55	17.57	17.60	17.62	17.64	17.67	17.69	17.71	17.74
810	17.76	17.78	17.81	17.83	17.86	17.88	17.90	17.93	17.95	17.97
820	18.00	18.02	18.04	18.07	18.09	18.11	18.14	18.16	18.18	18.21
830	18.23	18.25	18.28	18.30	18.33	18.35	18.37	18.40	18.42	18.44
840	18.47	18.49	18.51	18.54	18.56	18.58	18.61	18.63	18.65	18.68
850	18.70	18.73	18.75	18.77	18.80	18.82	18.84	18.87	18.89	18.91
860	18.94	18.96	18.90	19.01	19.03	19.06	19.08	19.10	19.13	19.15
870	19.18	19.20	19.22	19.25	19.27	19.29	19.32	19.34	19.36	19.39
880	19.41	19.44	19.46	19.48	19.51	19.53	19.55	19.58	19.60	19.63
890	19.65	19.67	19.70	19.72	19.75	19.77	19.79	19.82	19.84	19.86
900	19.89	19.91	19.94	19.96	19.98	20.01	20.03	20.05	20.08	20.10
910	20.13	20.15	20.17	20.20	20.22	20.24	20.27	20.29	20.32	20.24
920	20.36	20.39	20.41	20.43	20.46	20.48	20.50	20.53	20.55	20.58
930	20.60	20.62	20.65	20.67	20.69	20.72	20.74	20.76	20.79	20.81
940	20.84	20.86	20.88	20.91	20.93	20.95	20.98	21.00	21.03	21.05
950	21.07	21.10	21.12	21.14	21.17	21.19	21.21	21.24	21.26	21.28
960	21.31	21.33	21.36	21.38	21.40	21.43	21.45	21.47	21.50	21.52
970	21.54	21.57	21.59	21.62	21.64	21.66	21.69	21.71	21.73	21.76
980	21.78	21.81	21.83	21.85	21.88	21.90	21.92	21.95	21.97	21.99
990	22.02	22.04	22.07	22.09	22.11	22.14	22.16	22.18	22.21	22.23
1000	22.26	22.28	22.30	22.33	22.35	22.37	22.40	22.42	22.44	22.47
1010	22.49	22.52	22.54	22.56	22.59	22.61	22.63	22.66	22.68	22.71
1020	22.73	22.75	22.78	22.80	22.82	22.85	22.87	22.90	22.92	22.94
1030	22.97	22.99	23.01	23.04	23.06	23.08	23.11	23.13	23.16	23.18
1040	23.20	23.23	23.25	23.27	23.30	23.32	23.35	23.37	23.39	23.42
1050	23.44	23.46	23.49	23.51	23.54	23.56	23.58	23.61	23.63	23.65
1060	23.68	23.70	23.72	23.75	23.77	23.80	23.82	23.84	23.87	23.89
1070	23.91	23.94	23.96	23.99	24.01	24.03	24.06	24.08	24.10	24.13
1080	24.15	24.18	24.20	24.22	24.25	24.27	24.29	24.32	24.34	24.36
1090	24.39	24.41	24.44	24.46	24.49	24.51	24.53	24.55	24.58	24.60
1100	24.63	24.65	24.67	24.70	24.72	24.74	24.77	24.79	24.82	24.84
1110	24.86	24.89	24.91	24.93	24.96	24.98	25.01	25.03	25.05	25.08
1120	25.10	25.12	25.15	25.17	25.20	25.22	25.24	25.27	25.29	25.31
1130	25.34	25.36	25.38	25.41	25.43	25.46	25.48	25.50	25.53	25.55
1140	25.57	25.60	25.62	25.65	25.67	25.69	25.72	25.74	25.76	25.79
1150	25.81	25.83	25.86	25.88	25.91	25.93	25.95	25.98	26.00	26.02
1160	26.05	26.07	26.09	26.12	26.14	26.16	26.19	26.21	26.24	26.26
1170	26.28	26.31	26.33	26.35	26.38	26.40	26.42	26.45	26.47	26.49
1180	26.52	26.54	26.56	26.59	26.61	26.63	26.66	26.68	26.70	26.73
1190	26.75	26.77	26.80	26.82	26.85	26.87	26.89	26.91	26.94	26.96
1200	26.98	27.01	27.03	27.06	27.08	27.10	27.12	27.15	27.17	27.20
1210	27.22	27.24	27.27	27.29	27.31	27.34	27.36	27.38	27.40	27.43
1220	27.45	27.48	27.50	27.52	27.55	27.57	27.59	27.62	27.64	27.66
1230	27.69	27.71	27.73	27.76	27.78	27.80	27.83	27.85	27.87	27.90
1240	27.92	27.94	27.97	27.99	28.01	28.04	28.06	28.08	28.11	28.13
1250	28.15	28.18	28.20	28.22	28.25	28.27	28.29	28.32	28.34	28.37
1260	28.39	28.41	28.44	28.46	28.48	28.50	28.53	28.55	28.58	28.60
1270	28.62	28.65	28.67	28.69	28.72	28.74	28.76	28.79	28.81	28.83
1280	28.86	28.88	28.90	28.93	28.95	28.97	29.00	29.02	29.04	29.07
1290	29.09	29.11	29.14	29.16	29.18	29.21	29.23	29.25	29.28	29.30
1300	29.32	29.35	29.37	29.39	29.42	29.44	29.46	29.49	29.51	29.53
1310	29.56	29.58	29.60	29.63	29.65	29.67	29.70	29.72	29.74	29.77
1320	29.79	29.81	29.84	29.86	29.88	29.91	29.93	29.95	29.97	30.00

* Based on on the International Temperature Scale of 1948.

CHROMEL-ALUMEL THERMOCOUPLES (Continued)

Millivolts

°F	0	1	2	3	4	5	6	7	8	9
1330	30.02	30.05	30.07	30.09	30.11	30.14	30.16	30.18	30.21	30.23
1340	30.25	30.28	30.30	30.32	30.35	30.37	30.39	30.42	30.44	30.46
1350	30.49	30.51	30.53	30.56	30.58	30.60	30.63	30.65	30.67	30.70
1360	30.72	30.74	30.77	30.79	30.81	30.83	30.86	30.88	30.90	30.93
1370	30.95	30.97	31.00	31.02	31.04	31.07	31.09	31.11	31.14	31.16
1380	31.18	31.21	31.23	31.25	31.28	31.30	31.32	31.34	31.37	31.39
1390	31.42	31.44	31.46	31.48	31.51	31.53	31.55	31.58	31.60	31.62
1400	31.65	31.67	31.69	31.72	31.74	31.76	31.78	31.81	31.83	31.85
1410	31.88	31.90	31.92	31.95	31.97	31.99	32.02	32.04	32.06	32.08
1420	32.11	32.13	32.15	32.18	32.20	32.22	32.25	32.27	32.29	32.31
1430	32.34	32.36	32.38	32.41	32.43	32.45	32.48	32.50	32.52	32.54
1440	32.57	32.59	32.61	32.64	32.66	32.68	32.70	32.73	32.75	32.77
1450	32.80	32.82	32.84	32.86	32.89	32.91	32.93	32.96	32.98	33.00
1460	33.02	33.05	33.07	33.09	33.12	33.14	33.16	33.18	33.21	33.23
1470	33.25	33.28	33.30	33.32	33.34	33.37	33.39	33.41	33.43	33.46
1480	33.48	33.50	33.53	33.55	33.57	33.59	33.62	33.64	33.66	33.69
1490	33.71	33.73	33.75	33.78	33.80	33.82	33.84	33.87	33.89	33.91
1500	33.93	33.96	33.98	34.00	34.03	34.05	34.07	34.09	34.12	34.14
1510	34.16	34.18	34.21	34.23	34.25	34.28	34.30	34.32	34.34	34.37
1520	34.39	34.41	34.43	34.46	34.48	34.50	34.53	34.55	34.57	34.59
1530	34.62	34.64	34.66	34.68	34.71	34.73	34.75	34.77	34.80	34.82
1540	34.84	34.87	34.89	34.91	34.93	34.96	34.98	35.00	35.02	35.05
1550	35.07	35.09	35.11	35.14	35.16	35.18	35.21	35.23	35.25	35.27
1560	35.29	35.32	35.34	35.36	35.39	35.41	35.43	35.45	35.48	35.50
1570	35.52	35.54	35.57	35.59	35.61	35.63	35.66	35.68	35.70	35.72
1580	35.75	35.77	35.79	35.81	35.84	35.86	35.88	35.90	35.93	35.95
1590	35.97	35.99	36.02	36.04	36.06	36.08	36.11	36.13	36.15	36.17
1600	36.19	36.19	36.22	36.24	36.29	36.31	36.33	36.35	36.37	36.40
1610	36.42	36.44	36.46	36.49	36.51	36.53	36.55	36.58	36.60	36.62
1620	36.64	36.67	36.69	36.71	36.73	36.76	36.78	36.80	36.82	36.84
1630	36.87	36.89	36.91	36.93	36.96	36.98	37.00	37.02	37.05	37.07
1640	37.09	37.11	37.14	37.16	37.18	37.20	37.23	37.25	37.27	37.29
1650	37.31	37.34	37.36	37.38	37.40	37.43	37.45	37.47	37.49	37.52
1660	37.54	37.56	37.58	37.60	37.63	37.65	37.67	37.69	37.72	37.74
1670	37.76	37.78	37.81	37.83	37.85	37.87	37.89	37.92	37.94	37.96
1680	37.98	38.01	38.03	38.05	38.07	38.09	38.12	38.14	38.16	38.18
1690	38.20	38.23	38.25	38.27	38.29	38.32	38.34	38.36	38.38	38.40
1700	38.43	38.45	38.47	38.49	38.51	38.54	38.56	38.58	38.60	38.62
1710	38.65	38.67	38.69	38.71	38.73	38.76	38.78	38.80	38.82	38.84
1720	38.87	38.89	38.91	38.93	38.95	38.98	39.00	39.02	39.04	39.06
1730	39.09	39.11	39.13	39.15	39.17	39.20	39.22	39.24	39.26	39.28
1740	39.31	39.33	39.35	39.37	39.39	39.42	39.44	39.46	39.48	39.50
1750	39.53	39.55	39.57	39.59	39.61	39.64	39.66	39.68	39.70	39.72
1760	39.75	39.77	39.79	39.81	39.83	39.86	39.88	39.69	39.92	39.94
1770	39.96	39.99	40.01	40.03	40.05	40.07	40.10	40.12	40.14	40.16
1780	40.18	40.20	40.23	40.25	40.27	40.29	40.31	40.34	40.36	40.38
1790	40.40	40.42	40.44	40.47	40.49	40.51	40.53	40.55	40.58	40.60
1800	40.62	40.64	40.66	40.68	40.71	40.73	40.75	40.77	40.79	40.82
1810	40.84	40.86	40.88	40.90	40.92	40.95	40.97	40.99	41.01	41.03
1820	41.05	41.08	41.10	41.12	41.14	41.16	41.18	41.21	41.23	41.25
1830	41.27	41.29	41.31	41.34	41.36	41.38	41.40	41.42	41.45	41.47
1840	41.49	41.51	41.53	41.55	41.57	41.60	41.62	41.64	41.66	41.68
1850	41.70	41.73	41.75	41.77	41.79	41.81	41.83	41.85	41.88	41.90
1860	41.92	41.94	41.96	41.99	42.01	42.03	42.05	42.07	42.09	42.11
1870	42.14	42.16	42.18	42.20	42.22	42.24	42.26	42.29	42.31	42.33
1880	42.35	42.37	42.39	42.42	42.44	42.46	42.48	42.50	42.52	42.55
1890	42.57	42.59	42.61	42.63	42.65	42.67	42.69	42.72	42.74	42.76
1900	42.78	42.80	42.82	42.84	42.87	42.89	42.91	42.93	42.95	42.97
1910	42.99	43.01	43.04	43.06	43.08	43.10	43.12	43.14	43.17	43.19

Millivolts

°F	0	1	2	3	4	5	6	7	8	9
1920	43.21	43.23	43.25	43.27	43.29	43.21	43.34	43.36	43.38	43.40
1930	43.42	43.44	43.47	43.49	43.51	43.53	43.55	43.57	43.59	43.61
1940	43.63	43.66	43.68	43.70	43.72	43.75	43.76	43.78	43.81	43.83
1950	43.85	43.87	43.89	43.91	43.93	43.95	43.98	44.00	44.02	44.04
1960	44.06	44.08	44.10	44.13	44.15	44.17	44.19	44.21	44.23	44.25
1970	44.27	44.30	44.32	44.34	44.36	44.38	44.40	44.42	44.44	44.47
1980	44.49	44.51	44.53	44.55	44.57	44.59	44.61	44.63	44.66	44.68
1990	44.70	44.72	44.74	44.76	44.78	44.80	44.82	44.85	44.87	44.89
2000	44.91	44.93	44.95	44.97	44.99	45.01	45.03	45.06	45.08	45.10
2010	45.12	45.14	45.16	45.18	45.20	45.22	45.24	45.27	45.29	45.31
2020	45.33	45.35	45.87	45.39	45.41	45.43	45.45	45.48	45.50	45.52
2030	45.54	45.56	45.58	45.60	45.62	45.64	45.66	45.69	45.71	45.73
2040	45.75	45.77	45.79	45.81	45.83	45.85	45.87	45.90	45.92	45.94
2050	45.96	45.98	46.00	46.02	46.04	46.06	46.08	46.11	46.13	46.15
2060	46.17	46.19	46.21	46.23	46.25	46.27	46.29	46.31	46.33	46.36
2070	46.38	46.40	46.42	46.44	46.46	46.48	46.50	46.52	46.54	46.56
2080	46.58	46.60	46.63	46.65	46.67	46.69	46.71	46.73	46.75	46.77
2090	46.79	46.81	46.83	46.85	46.87	46.90	46.92	46.94	46.96	46.98
2100	47.00	47.02	47.04	47.06	47.08	47.10	47.12	47.14	47.17	47.19
2110	47.21	47.23	47.25	47.27	47.29	47.31	47.33	47.35	47.37	47.39
2120	47.41	47.43	47.45	47.47	47.49	47.52	47.54	47.56	47.58	47.60
2130	47.62	47.64	47.66	47.68	47.70	47.72	47.74	47.76	47.78	47.80
2140	47.82	47.84	47.86	47.89	47.91	47.93	47.95	47.97	47.99	48.01
2150	48.03	48.05	48.07	48.09	48.11	48.13	48.15	48.17	48.19	48.21
2160	48.23	48.25	48.27	48.29	48.32	48.34	48.36	48.38	48.40	48.42
2170	48.44	48.46	48.48	48.50	48.52	48.54	48.56	48.58	48.60	48.62
2180	48.64	48.66	48.68	48.70	48.72	48.74	48.76	48.79	48.81	48.83
2190	48.85	48.87	48.89	48.91	48.93	48.95	48.97	48.99	49.01	49.03
2200	49.05	49.07	49.09	49.11	49.13	49.15	49.17	49.19	49.21	49.23
2210	49.25	49.27	49.29	49.31	49.33	49.35	49.37	49.39	49.41	49.43
2220	49.45	49.47	49.49	49.51	49.53	49.55	49.57	49.59	49.61	49.63
2230	49.65	49.67	49.69	49.71	49.73	49.76	49.78	49.80	49.82	49.84
2240	49.86	49.88	49.90	49.92	49.94	49.96	49.98	50.00	50.02	50.04
2250	50.06	50.08	50.10	50.12	50.14	50.16	50.18	50.20	50.22	50.24
2260	50.26	50.28	50.30	50.32	50.34	50.36	50.38	50.40	50.42	50.44
2270	50.46	50.48	50.50	50.52	50.54	50.56	50.57	50.59	50.61	50.63
2280	50.65	50.67	50.69	50.71	50.73	50.75	50.77	50.79	50.81	50.83
2290	50.85	50.87	50.89	50.91	50.93	50.95	50.97	50.99	51.01	51.03
2300	51.05	51.07	51.09	51.11	51.13	51.15	51.17	51.19	51.21	51.23
2310	51.25	51.27	51.29	51.31	51.33	51.35	51.37	51.39	51.41	51.43
2320	51.45	51.47	51.48	51.50	51.52	51.54	51.56	51.58	51.60	51.62
2330	51.64	51.66	51.68	51.70	51.72	51.74	51.76	51.78	51.80	51.82
2340	51.84	51.86	51.88	51.90	51.92	51.94	51.96	51.98	52.00	52.01
2350	52.03	52.05	52.07	52.09	52.11	52.13	52.15	52.17	52.19	52.21
2360	52.23	52.25	52.27	52.29	52.31	52.33	52.35	52.37	52.39	52.41
2370	52.42	52.44	52.46	52.48	52.50	52.52	52.54	52.56	52.58	52.60
2380	52.62	52.64	52.66	52.68	52.70	52.72	52.74	52.76	52.77	52.79
2390	52.81	52.83	52.85	52.87	52.89	52.91	52.93	52.95	52.97	52.99
2400	53.01	53.03	53.05	53.07	53.08	53.10	53.12	53.14	53.16	53.18
2410	53.20	53.22	53.24	53.26	53.28	53.30	53.32	53.34	53.35	53.37
2420	53.39	53.41	53.43	53.45	53.47	53.49	53.51	53.53	53.55	53.57
2430	53.59	53.60	53.62	53.64	53.66	53.68	53.70	53.72	53.74	53.76
2440	53.78	53.80	53.82	53.83	53.85	53.87	53.89	53.91	53.93	53.95
2450	53.97	53.99	54.01	54.03	54.04	54.06	54.08	54.10	54.12	54.14
2460	54.16	54.18	54.20	54.22	54.24	54.25	54.27	54.29	54.31	54.33
2470	54.35	54.37	54.39	54.41	54.43	54.44	54.46	54.48	54.50	54.52
2480	54.54	54.56	54.58	54.60	54.62	54.63	54.65	54.67	54.69	54.71
2490	54.73	54.75	54.77	54.79	54.81	54.82	54.84	54.86	54.88	54.90
2500	54.92	—	—	—	—	—	—	—	—	—

IRON-CONSTANTAN THERMOCOUPLES (MODIFIED 1913)

(Electromotive Force in Absolute Millivolts. Temperatures in Degrees C (Int. 1948). Reference Junctions at 0°C.)

Millivolts

°C	0	1	2	3	4	5	6	7	8	9
−190	−7.66	−7.69	−7.71	−7.73	−7.76	−7.78				
−180	−7.40	−7.43	−7.46	−7.49	−7.51	−7.54	−7.56	−7.59	−7.61	−7.64
−170	−7.12	−7.15	−7.18	−7.21	−7.24	−7.27	−7.30	−7.32	−7.35	−7.38
−160	−6.82	−6.85	−6.88	−6.91	−6.94	−6.97	−7.00	−7.03	−7.06	−7.09
−150	−6.50	−6.53	−5.56	−6.60	−6.63	−6.66	−6.69	−6.72	−6.76	−6.79
−140	−6.16	−6.19	−6.22	−6.26	−6.29	−6.33	−6.36	−6.40	−6.43	−6.46
−130	−5.80	−5.84	−5.87	−5.91	−5.94	−5.98	−6.01	−6.05	−6.08	−6.12
−120	−5.42	−5.46	−5.50	−5.54	−5.58	−5.61	−5.65	−5.69	−5.72	−5.76
−110	−5.03	−5.07	−5.11	−5.15	−5.19	−5.23	−5.27	−5.31	−5.35	−5.38
−100	−4.63	−4.67	−4.71	−4.75	−4.79	−4.83	−4.87	−4.91	−4.95	−4.99
−90	−4.21	−4.25	−4.30	−4.34	−4.38	−4.42	−4.46	−4.50	−4.55	−4.59
−80	−3.78	−3.82	−3.87	−3.91	−3.96	−4.00	−4.04	−4.08	−4.13	−4.17
−70	−3.34	−3.38	−3.43	−3.47	−3.52	−3.56	−3.60	−3.65	−3.69	−3.74
−60	−2.89	−2.94	−2.98	−3.03	−3.07	3.12	−3.16	−3.21	−3.25	−3.30
−50	−2.43	−2.48	−2.52	−2.57	−2.62	−2.66	−2.71	−2.75	−2.80	−2.84

Millivolts

°C	0	1	2	3	4	5	6	7	8	9
−40	−1.96	−2.01	−2.06	−2.10	−2.15	−2.20	−2.24	−2.29	−2.34	−2.38
−30	−1.48	−1.53	−1.58	−1.63	−1.67	−1.72	−1.77	−1.82	−1.87	−1.91
−20	−1.00	−1.04	−1.09	−1.14	−1.19	−1.24	−1.29	−1.34	−1.39	−1.43
−10	−0.50	−0.55	−0.60	−0.65	−0.70	−0.75	−0.80	−0.85	−0.90	−0.95
(−)0	0.00	−0.05	−0.10	−0.15	−0.20	−0.25	−0.30	−0.35	−0.40	−0.45
(+)0	0.00	0.05	0.10	0.15	0.20	0.25	0.30	0.35	0.40	0.45
10	0.50	0.56	0.61	0.66	0.71	0.76	0.81	0.86	0.91	0.97
20	1.02	1.07	1.12	1.17	1.22	1.28	1.33	1.38	1.43	1.48
30	1.54	1.59	1.64	1.69	1.74	1.80	1.85	1.90	1.95	2.00
40	2.06	2.11	2.16	2.22	2.27	2.32	2.37	2.42	2.48	2.53
50	2.58	2.64	2.69	2.74	2.80	2.85	2.90	2.96	3.01	3.06
60	3.11	3.17	3.22	3.27	3.33	3.38	3.43	3.49	3.54	3.60
70	3.65	3.70	3.76	3.81	3.86	3.92	3.97	4.02	4.08	4.13
80	4.19	4.24	4.29	4.35	4.40	4.46	4.51	4.56	4.62	4.67
90	4.73	4.78	4.83	4.89	4.94	5.00	5.05	5.10	5.16	5.21

IRON-CONSTANTAN THERMOCOUPLES (MODIFIED 1913) (Continued)

°C	0	1	2	3	4	5	6	7	8	9
	Millivolts									
100	5.27	5.32	5.38	5.43	5.48	5.54	5.59	5.65	5.70	5.76
110	5.81	5.86	5.92	5.97	6.03	6.08	6.14	6.19	6.25	6.30
120	6.36	6.41	6.47	6.52	6.58	6.63	6.68	6.74	6.79	6.85
130	6.90	6.96	7.01	7.07	7.12	7.18	7.23	7.29	7.34	7.40
140	7.45	7.51	7.56	7.62	7.67	7.73	7.78	7.84	7.89	7.95
150	8.00	8.06	8.12	8.17	8.23	8.28	8.34	8.39	8.45	8.50
160	8.56	8.61	8.67	8.72	8.78	8.84	8.89	8.95	9.00	9.06
170	9.11	9.17	9.22	9.28	9.33	9.39	9.44	9.50	9.56	9.61
180	9.67	9.72	9.78	9.83	9.89	9.95	10.00	10.06	10.11	10.17
190	10.22	10.28	10.34	10.39	10.45	10.50	10.56	10.61	10.67	10.72
200	10.78	10.84	10.89	10.95	11.00	11.06	11.12	11.17	11.23	11.28
210	11.34	11.39	11.45	11.50	11.56	11.62	11.67	11.73	11.78	11.84
220	11.89	11.95	12.00	12.06	12.12	12.17	12.23	12.28	12.34	12.39
230	12.45	12.50	12.56	12.62	12.67	12.73	12.78	12.84	12.89	12.95
240	13.01	13.06	13.12	13.17	13.23	13.28	13.34	13.40	13.45	13.51
250	13.56	13.62	13.67	13.73	13.78	13.84	13.89	13.95	14.00	14.06
260	14.12	14.17	14.23	14.28	14.34	14.39	14.45	14.50	14.56	14.61
270	14.67	14.72	14.78	14.83	14.89	14.94	15.00	15.06	15.11	15.17
280	15.22	15.28	15.33	15.39	15.44	15.50	15.55	15.61	15.66	15.72
290	15.77	15.83	15.88	15.94	16.00	16.05	16.11	16.16	16.22	16.27
300	16.33	16.38	16.44	16.49	16.55	16.60	16.66	16.71	16.77	16.82
310	16.88	16.93	16.99	17.04	17.10	17.15	17.21	17.26	17.32	17.37
320	17.43	17.48	17.54	17.60	17.65	17.71	17.76	17.82	17.87	17.93
330	17.98	18.04	18.09	18.15	18.20	18.26	18.32	18.37	18.43	18.48
340	18.54	18.59	18.65	18.70	18.76	18.81	18.87	18.92	18.98	19.03
350	19.09	19.14	19.20	19.26	19.31	19.37	19.42	19.48	19.53	19.59
360	19.64	19.70	19.75	19.81	19.86	19.92	19.97	20.03	20.08	20.14
370	20.20	20.25	20.31	20.36	20.42	20.47	20.53	20.58	20.64	20.69
380	20.75	20.80	20.86	20.91	20.97	21.02	21.08	21.13	21.19	21.24
390	21.30	21.35	21.41	21.46	21.52	21.57	21.63	21.68	21.74	21.79
400	21.85	21.90	21.96	22.02	22.07	22.13	22.18	22.24	22.29	22.35
410	22.40	22.46	22.51	22.57	22.62	22.68	22.73	22.79	22.84	22.90
420	22.95	23.01	23.06	23.12	23.17	23.23	23.28	23.34	23.39	23.45
430	23.50	23.56	23.61	23.67	23.72	23.78	23.83	23.89	23.94	24.00

°C	0	1	2	3	4	5	6	7	8	9
	Millivolts									
440	24.06	24.11	24.17	24.22	24.28	24.33	24.39	24.44	24.50	24.55
450	24.61	24.66	24.72	24.77	24.83	24.88	24.94	25.00	25.05	25.11
460	25.16	25.22	25.27	25.33	25.38	25.44	25.49	25.55	25.60	25.66
470	25.72	25.77	25.83	25.88	25.94	25.99	26.05	26.10	26.16	26.22
480	26.27	26.33	26.38	26.44	26.49	26.55	26.61	26.66	26.72	26.77
490	26.83	26.89	26.94	27.00	27.05	27.11	27.17	27.22	27.28	27.33
500	27.39	27.45	27.50	27.56	27.61	27.67	27.73	27.78	27.84	27.90
510	27.95	28.01	28.07	28.12	28.18	28.23	28.29	28.35	28.40	28.46
520	28.52	28.57	28.63	28.69	28.74	28.80	28.86	28.91	28.97	29.02
530	29.08	29.14	29.20	29.25	29.31	29.37	29.42	29.48	29.54	29.59
540	29.65	29.71	29.76	29.82	29.88	29.94	29.99	30.05	30.11	30.16
550	30.22	30.28	30.34	30.39	30.45	30.51	30.57	30.62	30.68	30.74
560	30.80	30.85	30.91	30.97	31.02	31.08	31.14	31.20	31.26	31.31
570	31.37	31.43	31.49	31.54	31.60	31.66	31.72	31.78	31.83	31.89
580	31.95	32.01	32.06	32.12	32.18	32.24	32.30	32.36	32.41	32.47
590	32.53	32.59	32.65	32.71	32.76	32.82	32.88	32.94	33.00	33.06
600	33.11	33.17	33.23	33.29	33.35	33.41	33.46	33.52	33.58	33.64
610	33.70	33.76	33.82	33.88	33.94	33.99	34.05	34.11	34.17	34.23
620	34.29	34.35	34.41	34.47	34.53	34.58	34.64	34.70	34.76	34.82
630	34.88	34.94	35.00	35.06	35.12	35.18	35.24	35.30	35.36	35.42
640	35.48	35.54	35.60	36.66	35.72	35.78	35.84	35.90	35.96	36.02
650	36.08	36.14	36.20	36.26	36.32	36.38	36.44	36.50	36.56	36.62
660	36.69	36.75	36.81	36.87	36.93	36.99	37.05	37.11	37.18	37.24
670	37.30	37.36	37.42	37.48	37.54	37.60	37.66	37.73	37.79	37.85
680	37.91	37.97	38.04	38.10	38.16	38.22	38.28	38.34	38.41	38.47
690	38.53	38.59	38.66	38.72	38.78	38.84	38.90	38.97	39.03	39.09
700	39.15	39.22	39.28	39.34	39.40	39.47	39.53	39.59	39.65	39.72
710	39.78	39.84	39.91	39.97	40.03	40.10	40.16	40.22	40.28	40.35
720	40.41	40.48	40.54	40.60	40.66	40.73	40.79	40.86	40.92	40.98
730	41.05	41.11	41.17	41.24	41.30	41.36	41.43	41.49	41.56	41.62
740	41.68	41.75	41.81	41.87	41.94	42.00	42.07	42.13	42.19	42.26
750	42.32	42.38	42.45	42.51	42.58	42.64	42.70	42.77	42.83	42.90
760	42.96	—	—	—	—	—	—	—	—	—

(Electromotive Force in Absolute Millivolts. Temperatures in Degrees F* Reference Junctions at 32°F.)

°F	0	1	2	3	4	5	6	7	8	9
	Millivolts									
-310	-7.66	-7.68	-7.69	-7.70	-7.71	-7.73	-7.74	-7.75	-7.76	-7.78
-300	-7.52	-7.54	-7.55	-7.57	-7.58	-7.59	-7.61	-7.62	-7.64	-7.65
-290	-7.38	-7.39	-7.40	-7.42	-7.44	-7.45	-7.46	-7.48	-7.49	-7.51
-280	-7.22	-7.24	-7.25	-7.27	-7.28	-7.30	-7.31	-7.33	-7.34	-7.36
-270	-7.06	-7.07	-7.09	-7.11	-7.12	-7.14	-7.15	-7.17	-7.19	-7.20
-260	-6.89	-6.90	-6.92	-6.94	-6.96	-6.97	-6.99	-7.01	-7.02	-7.04
-250	-6.71	-6.73	-6.75	-6.77	-6.78	-6.80	-6.82	-6.84	-6.85	-6.87
-240	-6.53	-6.55	-6.57	-6.59	-6.61	-6.62	-6.64	-6.66	-6.68	-6.70
-230	-6.35	-6.37	-6.38	-6.40	-6.42	-6.44	-6.46	-6.48	-6.50	-6.52
-220	-6.16	-6.18	-6.19	-6.21	-6.23	-6.25	-6.27	-6.29	-6.31	-6.33
-210	-5.96	-5.98	-6.00	-6.02	-6.04	-6.06	-6.08	-6.10	-6.12	-6.14
-200	-5.76	-5.78	-5.80	-5.82	-5.84	-5.86	-5.88	-5.90	-5.92	-5.94
-190	-5.55	-5.57	-5.59	-5.61	-5.63	-5.65	-5.67	-5.70	-5.72	-5.74
-180	-5.34	-5.36	-5.38	-5.40	-5.42	-5.44	-5.46	-5.49	-5.51	-5.53
-170	-5.12	-5.14	-5.16	-5.19	-5.21	-5.23	-5.25	-5.27	-5.30	-5.32
-160	-4.90	-4.92	-4.94	-4.97	-4.99	-5.01	-5.03	-5.06	-5.08	-5.10
-150	-4.68	-4.70	-4.72	-4.74	-4.76	-4.79	-4.81	-4.83	-4.86	-4.88
-140	-4.44	-4.47	-4.49	-4.51	-4.54	-4.56	-4.58	-4.61	-4.63	-4.65
-130	-4.21	-4.23	-4.26	-4.28	-4.30	-4.33	-4.35	-4.38	-4.40	-4.42
-120	-3.97	-4.00	-4.02	-4.04	-4.07	-4.09	-4.12	-4.14	-4.16	-4.19
-110	-3.73	3.76	-3.78	-3.81	-3.83	-3.85	-3.88	-3.90	-3.93	-3.95
-100	-3.49	-3.51	-3.54	-3.56	-3.59	-3.61	-3.64	-3.66	-3.68	-3.71
-90	-3.24	-3.27	-3.29	-3.32	-3.34	-3.36	-3.39	-3.41	-3.44	-3.46
-80	-2.99	-3.02	-3.04	-3.07	-3.09	-3.12	-3.14	-3.17	-3.19	-3.22
-70	-2.74	-2.76	-2.79	-2.81	-2.84	-2.86	-2.89	-2.92	-2.94	-2.97
-60	-2.48	-2.51	-2.53	-2.56	-2.58	-2.61	-2.64	-2.66	-2.69	-2.71
-50	-2.22	-2.25	-2.27	-2.30	-2.33	-2.35	-2.38	-2.40	-2.43	-2.46
-40	-1.96	-1.99	-2.01	-2.04	-2.06	-2.09	-2.12	-2.14	-2.17	-2.20
-30	-1.70	-1.72	-1.75	-1.78	-1.80	-1.83	-1.86	-1.88	-1.91	-1.94
-20	-1.43	-1.46	-1.48	-1.51	-1.54	-1.56	-1.59	-1.62	-1.64	-1.67
-10	-1.16	-1.19	-1.21	-1.24	-1.27	-1.29	-1.32	-1.35	-1.38	-1.40
(-)0	-0.89	-0.91	-0.94	-0.97	-1.00	-1.02	-1.05	-1.08	-1.10	-1.13
+ (0)	-0.89	-0.86	-0.83	-0.80	-0.78	-0.75	-0.72	-0.70	-0.67	-0.64
10	-0.61	-0.58	-0.56	-0.53	-0.50	-0.48	-0.45	-0.42	-0.39	-0.36
20	-0.34	-0.31	-0.28	-0.25	-0.22	-0.20	-0.17	-0.14	-0.11	-0.09
30	-0.06	-0.03	0.00	0.03	0.05	0.08	0.11	0.14	0.17	0.19
40	0.22	0.25	0.28	0.31	0.34	0.36	0.39	0.42	0.45	0.48
50	0.50	0.53	0.56	0.59	0.62	0.65	0.67	0.70	0.73	0.76
60	0.79	0.82	0.84	0.87	0.90	0.93	0.96	0.99	1.02	1.04
70	1.07	1.10	1.13	1.16	1.19	1.22	1.25	1.28	1.30	1.33
80	1.36	1.39	1.42	1.45	1.48	1.51	1.54	1.56	1.59	1.62

°F	0	1	2	3	4	5	6	7	8	9
	Millivolts									
90	1.65	1.68	1.71	1.74	1.77	1.80	1.83	1.85	1.88	1.91
100	1.94	1.97	2.00	2.03	2.06	2.09	2.12	2.14	2.17	2.20
110	2.23	2.26	2.29	2.32	2.35	2.38	2.41	2.44	2.47	2.50
120	2.52	2.55	2.58	2.61	2.64	2.67	2.70	2.73	2.76	2.79
130	2.82	2.85	2.88	2.91	2.94	2.97	3.00	3.03	3.06	3.08
140	3.11	3.14	3.17	3.20	3.23	3.26	3.29	3.32	3.35	3.38
150	3.41	3.44	3.47	3.50	3.53	3.56	3.59	3.62	3.65	3.68
160	3.71	3.74	3.77	3.80	3.83	3.86	3.89	3.92	3.95	3.98
170	4.01	4.04	4.07	4.10	4.13	4.16	4.19	4.22	4.25	4.28
180	4.31	4.34	4.37	4.40	4.43	4.46	4.49	4.52	4.55	4.58
190	4.61	4.64	4.67	4.70	4.73	4.76	4.79	4.82	4.85	4.88
200	4.91	4.94	4.97	5.00	5.03	5.06	5.09	5.12	5.15	5.18
210	5.21	5.24	5.27	5.30	5.33	5.36	5.39	5.42	5.45	5.48
220	5.51	5.54	5.57	5.60	5.63	5.66	5.69	5.72	5.75	5.78
230	5.81	5.84	5.87	5.90	5.93	5.96	5.99	6.02	6.05	6.08
240	6.11	6.14	6.17	6.20	6.24	6.27	6.30	6.33	6.36	6.39
250	6.42	6.45	6.48	6.51	6.54	6.57	6.60	6.63	6.66	6.69
260	6.72	6.75	6.78	6.81	6.84	6.87	6.90	6.93	6.96	7.00
270	7.03	7.06	7.09	7.12	7.15	7.18	7.21	7.24	7.27	7.30
280	7.33	7.36	7.39	7.42	7.45	7.48	7.51	7.54	7.58	7.61
290	7.64	7.67	7.70	7.73	7.76	7.79	7.82	7.85	7.88	7.91
300	7.94	7.97	8.00	8.04	8.07	8.10	8.13	8.16	8.19	8.22
310	8.25	8.28	8.31	8.34	8.37	8.40	8.44	8.47	8.50	8.53
320	8.56	8.59	8.62	8.65	8.68	8.71	8.74	8.77	8.80	8.84
330	8.87	8.90	8.93	8.96	8.99	9.02	9.05	9.08	9.11	9.14
340	9.17	9.20	9.24	9.27	9.30	9.33	9.36	9.39	9.42	9.45
350	9.48	9.51	9.54	9.58	9.61	9.64	9.67	9.70	9.73	9.76
360	9.79	9.82	9.85	9.88	9.92	9.95	9.98	10.01	10.04	10.07
370	10.10	10.13	10.16	10.19	10.22	10.25	10.28	10.32	10.35	10.38
380	10.41	10.44	10.47	10.50	10.53	10.56	10.60	10.63	10.66	10.69
390	10.72	10.75	10.78	10.81	10.84	10.87	10.90	10.94	10.97	11.00
400	11.03	11.06	11.00	11.12	11.15	11.18	11.21	11.24	11.28	11.31
410	11.34	11.37	11.40	11.43	11.46	11.49	11.52	11.55	11.58	11.62
420	11.65	11.68	11.71	11.74	11.77	11.80	11.83	11.86	11.89	11.92
430	11.96	11.99	12.02	12.05	12.08	12.11	12.14	12.17	12.20	12.23
440	12.26	12.30	12.33	12.36	12.39	12.42	12.45	12.48	12.51	12.54
450	12.57	12.60	12.64	12.67	12.70	12.73	12.76	12.79	12.82	12.85
460	12.88	12.91	12.94	12.98	13.01	13.04	13.07	13.10	13.13	13.16
470	13.19	13.22	13.25	13.28	13.31	13.34	13.38	13.41	13.44	13.47
480	13.50	13.53	13.56	13.59	13.62	13.65	13.68	13.72	13.75	13.78
490	13.81	13.84	13.87	13.90	13.93	13.96	13.99	14.02	14.05	14.08

* Based on the International Temperature Scale of 1948.

IRON-CONSTANTAN THERMOCOUPLES (MODIFIED 1913) (Continued)

°F	0	1	2	3	4	5	6	7	8	9
					Millivolts					
500	14.12	14.15	14.18	14.21	14.24	14.27	14.30	14.33	14.36	14.39
510	14.42	14.45	14.48	14.52	14.55	14.58	14.61	14.64	14.67	14.70
520	14.73	14.76	14.79	14.82	14.85	14.88	14.91	14.94	14.98	15.01
530	15.04	15.07	15.10	15.13	15.16	15.19	15.22	15.25	15.28	15.31
540	15.34	15.37	15.40	15.44	15.47	15.50	15.53	15.56	15.59	15.62
550	15.65	15.68	15.71	15.74	15.77	15.80	15.84	15.87	15.90	15.93
560	15.96	15.99	16.02	16.05	16.08	16.11	16.14	16.17	16.20	16.23
570	16.26	16.30	16.33	16.36	16.39	16.42	16.45	16.48	16.51	16.54
580	16.57	16.60	16.63	16.66	16.69	16.72	16.75	16.78	16.82	16.85
590	16.88	16.91	16.94	16.97	17.00	17.03	17.06	17.09	17.12	17.15
600	17.18	17.21	17.24	17.28	17.31	17.34	17.37	17.40	17.43	17.46
610	17.49	17.52	17.55	17.58	17.61	17.64	17.68	17.71	17.74	17.77
620	17.80	17.83	17.86	17.89	17.92	17.95	17.98	18.01	18.04	18.08
630	18.11	18.14	18.17	18.20	18.23	18.26	18.29	18.32	18.35	18.38
640	18.41	18.44	18.47	18.50	18.54	18.57	18.60	18.63	18.66	18.69
650	18.72	18.75	18.78	18.81	18.84	18.87	18.90	18.94	18.97	19.00
660	19.03	19.06	19.09	19.12	19.15	19.18	19.21	19.24	19.27	19.30
670	19.34	19.37	19.40	19.43	19.46	19.49	19.52	19.55	19.58	19.61
680	19.64	19.67	19.70	19.74	19.77	19.80	19.83	19.86	19.89	19.92
690	19.95	19.98	20.01	20.04	20.07	20.10	20.13	20.16	20.20	20.23
700	20.26	20.29	20.32	20.35	20.38	20.41	20.44	20.47	20.50	20.53
710	20.56	20.59	20.62	20.66	20.69	20.72	20.75	20.78	20.81	20.84
720	20.87	20.90	20.93	20.96	20.99	21.02	21.05	21.08	21.11	21.14
730	21.18	21.21	21.24	21.27	21.30	21.33	21.36	21.39	21.42	21.45
740	21.48	21.51	21.54	21.57	21.60	21.64	21.67	21.70	21.73	21.76
750	21.79	21.82	21.85	21.88	21.91	21.94	21.97	22.00	22.03	22.06
760	22.10	22.13	22.16	22.19	22.22	22.25	22.28	22.31	22.34	22.37
770	22.40	22.43	22.46	22.49	22.52	22.55	22.58	22.62	22.65	22.68
780	22.71	22.74	22.77	22.80	22.83	22.86	22.89	22.92	22.95	22.98
790	23.01	23.04	23.08	23.11	23.14	23.17	23.20	23.23	23.26	23.29
800	23.32	23.35	23.38	23.41	23.44	23.47	23.50	23.53	23.56	23.60
810	23.63	23.66	23.69	23.72	23.75	23.78	23.81	23.84	23.87	23.90
820	23.93	23.96	23.99	24.02	24.06	24.09	24.12	24.15	24.18	24.21
830	24.24	24.27	24.30	24.33	24.36	24.39	24.42	24.45	24.48	24.52
840	24.55	24.58	24.61	24.64	24.67	24.70	24.73	24.76	24.79	24.82
850	24.85	24.88	24.91	24.94	24.98	25.01	25.04	25.07	25.10	25.13
860	25.16	25.19	25.22	25.25	25.28	25.32	25.35	25.38	25.41	25.44
870	25.47	25.50	25.53	25.56	25.59	25.62	25.65	25.68	25.72	25.75
880	25.78	25.81	25.84	25.87	25.90	25.93	25.96	25.99	26.02	26.06
890	26.09	26.12	26.15	26.18	26.21	26.24	26.27	26.30	26.33	26.36
900	26.40	26.43	26.46	26.49	26.52	26.55	26.58	26.61	26.64	26.67
910	26.70	26.74	26.77	26.80	26.83	26.86	26.89	26.92	26.95	26.98
920	27.02	27.05	27.08	27.11	27.14	27.17	27.20	27.23	27.26	27.30
930	27.33	27.36	27.39	27.42	27.45	27.48	27.51	27.54	27.58	27.61
940	27.64	27.67	27.70	27.73	27.76	27.80	27.83	27.86	27.89	27.92
950	27.95	27.98	28.02	28.05	28.08	28.11	28.14	28.17	28.20	28.23
960	28.26	28.30	28.33	28.36	28.39	28.42	28.45	28.48	28.52	28.55
970	28.58	28.61	28.64	28.67	28.70	28.74	28.77	28.80	28.83	28.86
980	28.89	28.92	28.96	28.99	29.02	29.05	29.08	29.11	29.14	29.18
990	29.21	29.24	29.27	29.30	29.33	29.37	29.40	29.43	29.46	29.49
1000	29.52	29.56	29.59	29.62	29.65	29.68	29.71	29.75	29.78	29.81
1010	29.84	29.87	29.90	29.94	29.97	30.00	30.03	30.06	30.10	30.13
1020	30.16	30.19	30.22	30.25	30.28	30.32	30.35	30.38	30.41	30.44
1030	30.48	30.51	30.54	30.57	30.60	30.64	30.67	30.70	30.73	30.76
1040	30.80	30.83	30.86	30.89	30.92	30.96	30.99	31.02	31.05	31.08
1050	31.12	31.15	31.18	31.21	31.24	31.28	31.31	31.34	31.37	31.40
1060	31.44	31.47	31.50	31.53	31.56	31.60	31.63	31.66	31.69	31.72
1070	31.76	31.79	31.82	31.85	31.88	31.92	31.95	31.98	32.01	32.05
1080	32.08	32.11	32.14	32.18	32.21	32.24	32.27	32.30	32.34	32.37
1090	32.40	32.43	32.47	32.50	32.53	32.56	32.60	32.63	32.66	32.69
1100	32.72	32.76	32.79	32.82	32.86	32.89	32.92	32.95	32.98	33.02
1110	33.05	33.08	33.11	33.15	33.18	33.21	33.24	33.28	33.31	33.34
1120	33.37	33.41	33.44	33.47	33.50	33.54	33.57	33.60	33.64	33.67
1130	33.70	33.73	33.76	33.80	33.83	33.86	33.90	33.93	33.96	33.99
1140	34.03	34.06	34.09	34.12	34.16	34.19	34.22	34.26	34.29	34.32
1150	34.36	34.39	34.42	34.45	34.49	34.52	34.55	34.58	34.62	34.65
1160	34.68	34.72	34.75	34.78	34.82	34.85	34.88	34.92	34.95	34.98
1170	35.01	35.05	35.08	35.11	35.15	35.18	35.21	35.25	35.28	35.31
1180	35.35	35.38	35.41	35.45	35.48	35.51	35.54	35.58	35.61	35.64
1190	35.68	35.71	35.74	35.78	35.81	35.84	35.88	35.91	35.94	35.98
1200	36.01	36.05	36.08	36.11	36.15	36.18	36.21	36.25	36.28	36.31
1210	36.35	36.38	36.42	36.45	36.48	36.52	36.55	36.58	36.62	36.65
1220	36.69	36.72	36.75	36.79	36.82	36.86	36.89	36.92	36.96	36.99
1230	37.02	37.06	37.09	37.13	37.16	37.20	37.23	37.26	37.30	37.33
1240	37.36	37.40	37.43	37.47	37.50	37.54	37.57	37.60	37.64	37.67
1250	37.71	37.74	37.78	37.81	37.84	37.88	37.91	37.95	37.98	38.02
1260	38.05	38.08	38.12	38.15	38.19	38.22	38.26	38.29	38.32	38.36
1270	38.39	38.43	38.46	38.50	38.53	38.57	38.60	38.64	38.67	38.70
1280	38.74	38.77	38.81	38.84	38.88	38.91	38.95	38.98	39.02	39.05
1290	39.08	39.12	39.15	39.19	39.22	39.26	39.29	39.33	39.36	39.40
1300	39.43	39.47	39.50	39.54	39.57	39.61	39.64	39.68	39.71	39.75
1310	39.78	39.82	39.85	39.89	39.92	39.96	39.99	40.03	40.06	40.10
1320	40.13	40.17	40.20	40.24	40.27	40.31	40.34	40.38	40.41	40.45
1330	40.48	40.52	40.55	40.59	40.62	40.66	40.69	40.73	40.76	40.80
1340	40.83	40.87	40.90	40.94	40.98	41.01	41.05	41.08	41.12	41.15
1350	41.19	41.22	41.26	41.29	41.33	41.36	41.40	41.43	41.47	41.50
1360	41.54	41.58	41.61	41.65	41.68	41.72	41.75	41.79	41.82	41.86
1370	41.90	41.93	41.97	42.00	42.04	42.07	42.11	42.14	42.18	42.22
1380	42.25	42.29	42.32	42.36	42.39	42.43	42.46	42.50	42.53	42.57
1390	42.61	42.64	42.68	42.71	42.75	42.78	42.82	42.85	42.89	42.92
1400	42.96	—	—	—	—	—	—	—	—	—

TEMPERATURE-E. M. F. VALUES FOR COPPER-CONSTANTAN

E. M. F. values are in millivolts; reference junctions at 0°C.;
temperatures are in degrees C.
Roeser and Wensel, National Bureau of Standards

°C	0	10	20	30	40	50	60	70	80	90
					Millivolts					
−200	−5.54									
−100	−3.35	−3.62	−3.89	−4.14	−4.38	−4.60	−4.82	−5.02	−5.20	−5.38
0	0	−0.38	−0.75	−1.11	−1.47	−1.81	−2.14	−2.46	−2.77	−3.06
0	0	0.39	0.79	1.19	1.61	2.03	2.47	2.91	3.36	3.81
100	4.28	4.75	5.23	5.71	6.20	6.70	7.21	7.72	8.23	8.76
200	9.29	9.82	10.36	10.91	11.46	12.01	12.57	13.14	13.71	14.28
300	14.86	15.44	16.03	16.62	17.22	17.82	18.42	19.03	19.64	20.25
400	20.87									

E. M. F. values are in millivolts; reference junctions at 32°F.;
temperatures are in degrees F.
Roeser and Wensel, National Bureau of Standards

°F	0	10	20	30	40	50	60	70	80	90
					Millivolts					
−300	−5.28									
−200	−4.11	−4.25	−4.38	−4.50	−4.63	−4.75	−4.86	−4.97	−5.08	−5.18
−100	−2.56	−2.73	−2.90	−3.06	−3.22	−3.38	−3.53	−3.68	−3.83	−3.97
0	−0.67	−0.87	−1.07	−1.27	−1.47	−1.66	−1.84	−2.03	−2.21	−2.39
0	−0.67	−0.46	−0.26	−0.04	+0.17	0.39	0.61	0.83	1.06	1.29
100	1.52	1.75	1.99	2.23	2.47	2.71	2.96	3.21	3.46	3.71
200	3.97	4.22	4.48	4.75	5.01	5.28	5.55	5.82	6.09	6.37
300	6.64	6.92	7.21	7.49	7.77	8.06	8.35	8.64	8.93	9.23
400	9.52	9.82	10.12	10.42	10.72	11.03	11.33	11.64	11.95	12.26
500	12.57	12.89	13.20	13.52	13.83	14.15	14.47	14.79	15.12	15.44
600	15.77	16.10	16.42	16.75	17.08	17.42	17.75	18.08	18.42	18.75
700	19.09	19.43	19.77	20.11	20.45	20.80				

REFERENCE TABLE FOR Pt TO Pt—10 PER CENT Rh THERMOCOUPLE

Emfs are expressed in microvolts and temperatures in °C. Cold junctions at 0°C. Roeser and Wensel, National Bureau of Standards

E(μv)	0	1,000	2,000	3,000	4,000	5,000	6,000	7,000	8,000	9,000	10,000	11,000	12,000	13,000	14,000	15,000	16,000	17,000
0	0	146.9	265.0	373.7	477.7	578.1	675.3	769.5	861.0	950.2	1,037.2	1,122.3	1,206.4	1,290.0	1,373.8	1,458.0	1,542.6	1,627.8
	17.7	12.5	11.2	10.5	10.2	9.9	9.5	9.3	9.0	8.8	8.6	8.5	8.3	8.3	8.4	8.5	8.5	8.6
100	17.7	159.4	276.2	384.2	487.9	588.0	684.8	778.8	870.0	959.0	1,045.8	1,130.8	1,214.7	1,298.3	1,382.2	1,466.4	1,551.1	1,636.4
	16.7	12.3	11.1	10.5	10.2	9.8	9.5	9.2	9.0	8.8	8.6	8.4	8.4	8.4	8.4	8.4	8.5	8.5
200	34.4	171.7	287.3	394.7	498.1	597.8	694.3	788.0	879.0	967.8	1,054.4	1,139.2	1,223.1	1,306.7	1,390.6	1,474.8	1,559.6	1,644.9
	15.8	12.1	11.0	10.5	10.1	9.8	9.5	9.2	9.0	8.7	8.5	8.4	8.3	8.4	8.4	8.5	8.5	8.6
300	50.2	183.8	298.3	405.2	508.2	607.6	703.8	797.2	888.0	976.5	1,062.9	1,147.6	1,231.4	1,315.1	1,399.0	1,483.3	1,568.1	1,653.5
	15.2	12.0	11.0	10.5	10.1	9.8	9.5	9.2	9.0	8.8	8.6	8.4	8.4	8.4	8.4	8.5	8.5	8.6
400	65.4	195.8	309.3	415.7	518.3	617.4	713.3	806.4	897.0	985.3	1,071.5	1,156.0	1,239.8	1,323.5	1,407.4	1,491.8	1,576.6	1,662.1
	14.6	11.8	10.9	10.4	10.1	9.7	9.4	9.2	8.9	8.7	8.5	8.4	8.4	8.3	8.4	8.4	8.5	8.6
500	80.0	207.6	320.2	426.1	528.4	627.1	722.7	815.6	905.9	994.0	1,080.0	1,164.4	1,248.2	1,331.8	1,415.8	1,500.2	1,585.1	1,670.7
	14.1	11.7	10.8	10.4	10.0	9.7	9.4	9.1	8.9	8.7	8.5	8.4	8.3	8.4	8.4	8.5	8.6	8.6
600	94.1	219.3	331.0	436.5	538.4	636.8	732.1	824.7	914.8	1,002.7	1,088.5	1,172.8	1,256.5	1,340.2	1,424.2	1,508.7	1,593.7	1,679.3
	13.7	11.6	10.7	10.3	10.0	9.7	9.4	9.1	8.9	8.6	8.5	8.4	8.4	8.6	8.5	8.4	8.6	8.6
700	107.8	230.9	341.7	446.8	548.4	646.5	741.5	833.8	923.7	1,011.3	1,097.0	1,181.2	1,264.9	1,348.6	1,432.7	1,517.2	1,602.2	1,687.9
	13.3	11.5	10.7	10.3	9.9	9.6	9.4	9.1	8.9	8.7	8.4	8.4	8.4	8.4	8.4	8.4	8.5	8.6
800	121.1	242.4	352.4	457.1	558.3	656.1	750.9	842.9	932.6	1,020.0	1,105.4	1,189.6	1,273.2	1,357.0	1,441.1	1,525.6	1,610.7	1,696.5
	13.0	11.3	10.7	10.3	9.9	9.6	9.3	9.1	8.8	8.6	8.5	8.4	8.4	8.4	8.4	8.5	8.6	8.6
900	134.1	253.7	363.0	467.4	568.2	665.7	760.2	852.0	941.4	1,028.6	1,113.9	1,198.0	1,281.6	1,365.4	1,449.5	1,534.1	1,619.3	1,705.1
	12.8	11.3	10.6	10.3	9.9	9.6	9.3	9.0	8.8	8.6	8.4	8.4	8.4	8.4	8.5	8.5	8.5	8.6
1,000	146.9	265.0	373.7	477.8	578.1	675.3	769.5	861.0	950.2	1,037.2	1,122.3	1,206.4	1,290.0	1,373.8	1,458.0	1,542.6	1,627.8	1,713.7

Emfs are expressed in microvolts and temperatures in °C. Cold junctions at 0°C. Roeser and Wensel, National Bureau of Standards

E(μv)	0	1,000	2,000	3,000	4,000	5,000	6,000	7,000	8,000	9,000	10,000	11,000	12,000	13,000	14,000	15,000	16,000	17,000
0	32.0	296.4	509.0	704.7	891.9	1,072.6	1,247.5	1,417.1	1,581.8	1,742.4	1,899.0	2,052.1	2,203.5	2,354.0	2,504.8	2,656.4	2,808.7	2,962.0
	31.9	22.5	20.1	19.0	18.4	17.7	17.1	16.7	16.2	15.8	15.5	15.2	15.0	15.0	15.2	15.1	15.4	15.4
100	63.9	318.9	529.1	723.7	910.3	1,090.3	1,264.6	1,433.8	1,598.0	1,758.2	1,914.5	2,067.3	2,218.5	2,369.0	2,520.0	2,671.5	2,824.0	2,977.4
	30.0	22.1	20.0	18.9	18.3	17.7	17.1	16.6	16.2	15.8	15.4	15.2	15.0	15.1	15.1	15.2	15.3	15.4
200	93.9	341.0	549.1	742.6	928.6	1,108.0	1,281.7	1,450.4	1,614.2	1,774.0	1,929.9	2,082.5	2,233.5	2,384.1	2,535.1	2,686.7	2,839.3	2,992.8
	28.5	21.8	19.9	18.8	18.2	17.7	17.1	16.6	16.2	15.7	15.4	15.2	15.0	15.1	15.1	15.2	15.3	15.5
300	122.4	362.8	569.0	761.4	946.8	1,125.7	1,298.8	1,467.0	1,630.4	1,789.7	1,945.3	2,097.7	2,248.5	2,399.2	2,550.2	2,701.9	2,854.6	3,008.3
	27.3	21.6	19.8	18.8	18.1	17.6	17.1	16.5	16.1	15.8	15.4	15.1	15.1	15.1	15.3	15.3	15.3	15.5
400	149.7	384.4	588.8	780.2	964.9	1,143.3	1,315.9	1,483.5	1,646.5	1,805.5	1,960.7	2,112.8	2,263.6	2,414.3	2,565.3	2,717.2	2,869.9	3,023.8
	26.3	21.3	19.6	18.8	18.1	17.5	17.0	16.5	16.1	15.7	15.3	15.1	15.0	15.0	15.1	15.2	15.3	15.5
500	176.0	405.7	608.4	799.0	983.0	1,160.8	1,332.9	1,500.0	1,662.6	1,821.2	1,976.0	2,127.9	2,278.7	2,429.3	2,580.4	2,732.4	2,885.2	3,039.3
	25.4	21.0	19.4	18.7	18.1	17.5	16.9	16.5	16.0	15.6	15.3	15.1	15.0	15.1	15.2	15.3	15.4	15.4
600	201.4	426.7	627.8	817.7	1,001.1	1,178.3	1,349.8	1,516.5	1,678.6	1,836.8	1,991.3	2,143.0	2,293.7	2,444.4	2,595.6	2,747.7	2,900.6	3,054.7
	24.6	20.9	19.3	18.6	18.0	17.4	16.9	16.4	16.0	15.6	15.2	15.1	15.1	15.1	15.2	15.3	15.4	15.5
700	226.0	447.6	647.1	836.3	1,019.1	1,195.7	1,366.7	1,532.9	1,694.6	1,852.4	2,006.5	2,158.2	2,308.8	2,459.5	2,610.8	2,763.0	2,916.0	3,070.2
	24.0	20.7	19.3	18.5	17.9	17.3	16.9	16.3	16.0	15.6	15.2	15.1	15.0	15.1	15.2	15.3	15.3	15.5
800	250.0	468.3	666.4	854.8	1,037.0	1,213.0	1,383.6	1,549.2	1,710.6	1,868.0	2,021.7	2,173.3	2,323.8	2,474.6	2,626.0	2,778.2	2,931.3	3,085.7
	23.4	20.4	19.2	18.6	17.8	17.3	16.8	16.3	15.9	15.5	15.2	15.1	15.1	15.1	15.2	15.2	15.4	15.5
900	273.4	488.7	685.6	873.4	1,054.8	1,230.3	1,400.4	1,565.5	1,726.5	1,883.5	2,036.9	2,188.4	2,338.9	2,489.7	2,641.2	2,793.4	2,946.7	3,101.2
	23.0	20.3	19.1	18.5	17.8	17.2	16.7	16.3	15.9	15.5	15.2	15.1	15.1	15.1	15.2	15.4	15.3	15.5
1,000	296.4	509.0	704.7	891.9	1,072.6	1,247.5	1,417.1	1,581.8	1,742.4	1,899.0	2,052.1	2,203.5	2,354.0	2,504.8	2,656.4	2,808.7	2,962.0	3,116.7

REFERENCE TABLE FOR Pt TO Pt—13 PER CENT Rh THERMOCOUPLE

Emfs are expressed in microvolts and temperatures in °C. Cold junctions at 0°C. Roeser and Wensel, National Bureau of Standards

(Each cell shows temperature value / increment to next row)

E(µv)	0	1,000	2,000	3,000	4,000	5,000	6,000	7,000	8,000	9,000	10,000	11,000	12,000	13,000	14,000	15,000	16,000	17,000	18,000	19,000
0	0 / 17.9	145.3 / 12.2	258.8 / 10.6	361.0 / 9.9	457.4 / 9.5	549.8 / 9.0	638.3 / 8.7	723.5 / 8.4	806.0 / 8.1	886.1 / 7.9	964.1 / 7.6	1,040.0 / 7.4	1,113.9 / 7.3	1,186.9 / 7.2	1,259.3 / 7.3	1,331.8 / 7.2	1,404.3 / 7.3	1,476.9 / 7.3	1,550.0 / 7.4	1,623.6 / 7.4
100	17.9 / 16.7	157.5 / 12.0	269.4 / 10.5	370.9 / 9.8	466.9 / 9.4	558.8 / 9.0	647.0 / 8.6	731.9 / 8.4	814.1 / 8.1	894.0 / 7.8	971.7 / 7.7	1,047.4 / 7.5	1,121.2 / 7.4	1,194.1 / 7.3	1,266.6 / 7.2	1,339.0 / 7.2	1,411.6 / 7.3	1,484.2 / 7.3	1,557.4 / 7.4	1,631.0 / 7.4
200	34.6 / 15.8	169.5 / 11.7	279.9 / 10.4	380.7 / 9.8	476.3 / 9.4	567.8 / 9.0	655.6 / 8.6	740.3 / 8.3	822.2 / 8.1	901.8 / 7.9	979.4 / 7.6	1,054.9 / 7.4	1,128.6 / 7.3	1,201.4 / 7.3	1,273.8 / 7.3	1,346.2 / 7.3	1,418.9 / 7.2	1,491.5 / 7.3	1,564.8 / 7.3	1,638.4 / 7.4
300	50.4 / 15.1	181.2 / 11.5	290.3 / 10.3	390.5 / 9.7	485.7 / 9.3	576.8 / 8.9	664.2 / 8.6	748.6 / 8.3	830.3 / 8.0	909.7 / 7.8	987.0 / 7.7	1,062.3 / 7.4	1,135.9 / 7.3	1,208.7 / 7.2	1,281.1 / 7.2	1,353.5 / 7.3	1,426.1 / 7.2	1,498.8 / 7.3	1,572.1 / 7.4	1,645.8 / 7.4
400	65.5 / 14.5	192.7 / 11.4	300.6 / 10.2	400.2 / 9.7	495.0 / 9.3	585.7 / 8.9	672.8 / 8.5	756.9 / 8.2	838.3 / 8.0	917.5 / 7.8	994.7 / 7.6	1,069.7 / 7.4	1,143.2 / 7.3	1,215.9 / 7.3	1,288.3 / 7.2	1,360.8 / 7.2	1,433.3 / 7.2	1,506.1 / 7.3	1,579.5 / 7.3	1,653.2 / 7.4
500	80.0 / 13.9	204.1 / 11.2	310.8 / 10.2	409.9 / 9.6	504.3 / 9.2	594.6 / 8.8	681.3 / 8.5	765.1 / 8.2	846.3 / 8.0	925.3 / 7.8	1,002.3 / 7.6	1,077.1 / 7.3	1,150.5 / 7.3	1,223.2 / 7.2	1,295.5 / 7.2	1,368.0 / 7.3	1,440.5 / 7.3	1,513.4 / 7.3	1,586.8 / 7.4	1,660.6 / 7.4
600	93.9 / 13.4	215.3 / 11.1	321.0 / 10.1	419.5 / 9.5	513.5 / 9.2	603.4 / 8.8	689.8 / 8.5	773.3 / 8.2	854.3 / 8.0	933.1 / 7.8	1,009.9 / 7.6	1,084.4 / 7.4	1,157.8 / 7.3	1,230.4 / 7.2	1,302.7 / 7.2	1,375.3 / 7.3	1,447.8 / 7.2	1,520.7 / 7.3	1,594.2 / 7.3	1,668.0 / 7.4
700	107.3 / 13.0	226.4 / 10.9	331.1 / 10.0	429.0 / 9.5	522.7 / 9.1	612.2 / 8.7	698.3 / 8.4	781.5 / 8.2	862.3 / 8.0	940.9 / 7.8	1,017.5 / 7.5	1,091.8 / 7.4	1,165.1 / 7.2	1,237.6 / 7.2	1,309.9 / 7.3	1,382.6 / 7.2	1,455.0 / 7.3	1,528.0 / 7.3	1,601.5 / 7.4	1,675.4 / 7.4
800	120.3 / 12.6	237.3 / 10.8	341.1 / 10.0	438.5 / 9.5	531.8 / 9.0	620.9 / 8.7	706.7 / 8.4	789.7 / 8.1	870.3 / 7.9	948.7 / 7.7	1,025.0 / 7.5	1,099.2 / 7.3	1,172.3 / 7.3	1,244.8 / 7.3	1,317.2 / 7.3	1,389.8 / 7.3	1,462.3 / 7.3	1,535.3 / 7.4	1,608.9 / 7.3	1,682.8 / 7.4
900	132.9 / 12.4	248.1 / 10.7	351.1 / 9.9	448.0 / 9.4	540.8 / 9.0	629.6 / 8.7	715.1 / 8.4	797.8 / 8.2	878.2 / 7.9	956.4 / 7.7	1,032.5 / 7.5	1,106.5 / 7.4	1,179.6 / 7.3	1,252.1 / 7.2	1,324.5 / 7.3	1,397.1 / 7.2	1,469.6 / 7.3	1,542.7 / 7.3	1,616.2 / 7.4	1,690.2 / 7.4
1,000	145.3	258.8	361.0	457.4	549.8	638.3	723.5	806.0	886.1	964.1	1,040.0	1,113.9	1,186.9	1,259.3	1,331.8	1,404.3	1,476.9	1,550.0	1,623.6	1,697.6

Emfs are expressed in microvolts and temperatures in °F. Cold junctions at 32°F. Roeser and Wensel, National Bureau of Standards

E(µv)	0	1,000	2,000	3,000	4,000	5,000	6,000	7,000	8,000	9,000	10,000	11,000	12,000	13,000	14,000	15,000	16,000	17,000	18,000	19,000
0	32.0 / 32.2	293.5 / 22.0	497.8 / 19.1	681.8 / 17.8	855.3 / 17.0	1,021.6 / 16.2	1,180.9 / 15.6	1,334.3 / 15.1	1,482.8 / 14.6	1,627.0 / 14.2	1,767.4 / 13.8	1,904.0 / 13.4	2,037.0 / 13.2	2,168.4 / 13.0	2,298.7 / 13.1	2,429.2 / 13.0	2,559.7 / 13.1	2,690.4 / 13.2	2,822.0 / 13.3	2,954.5 / 13.3
100	64.2 / 30.1	315.5 / 21.6	516.9 / 18.9	699.6 / 17.7	872.3 / 17.0	1,037.8 / 16.2	1,196.5 / 15.6	1,349.4 / 15.1	1,497.4 / 14.6	1,641.2 / 14.1	1,781.2 / 13.7	1,917.4 / 13.4	2,050.2 / 13.2	2,181.4 / 13.1	2,311.8 / 13.0	2,442.2 / 13.0	2,572.8 / 13.1	2,703.6 / 13.1	2,835.3 / 13.3	2,967.8 / 13.3
200	94.3 / 28.4	337.1 / 21.1	535.8 / 18.7	717.3 / 17.6	889.3 / 16.9	1,054.0 / 16.2	1,212.1 / 15.5	1,364.5 / 15.0	1,512.0 / 14.5	1,655.3 / 14.1	1,794.9 / 13.7	1,930.8 / 13.3	2,063.4 / 13.2	2,194.5 / 13.1	2,324.8 / 13.1	2,455.2 / 13.1	2,585.9 / 13.0	2,716.7 / 13.1	2,848.6 / 13.2	2,981.1 / 13.3
300	122.7 / 27.2	358.2 / 20.7	554.5 / 18.6	734.9 / 17.5	906.2 / 16.8	1,070.2 / 16.1	1,227.6 / 15.4	1,379.5 / 14.9	1,526.5 / 14.4	1,669.4 / 14.1	1,808.6 / 13.8	1,944.1 / 13.3	2,076.6 / 13.2	2,207.6 / 13.0	2,337.9 / 13.0	2,468.3 / 13.1	2,598.9 / 13.0	2,729.8 / 13.2	2,861.8 / 13.3	2,994.4 / 13.4
400	149.9 / 26.1	378.9 / 20.4	573.1 / 18.4	752.4 / 17.4	923.0 / 16.7	1,086.3 / 16.0	1,243.0 / 15.3	1,394.4 / 14.8	1,540.9 / 14.4	1,683.5 / 14.0	1,822.4 / 13.7	1,957.4 / 13.3	2,089.8 / 13.1	2,220.6 / 13.1	2,350.9 / 13.0	2,481.4 / 13.0	2,611.9 / 13.0	2,743.0 / 13.1	2,875.1 / 13.2	3,007.8 / 13.3
500	176.0 / 25.0	399.3 / 20.2	591.5 / 18.3	769.8 / 17.3	939.7 / 16.6	1,102.3 / 15.9	1,258.3 / 15.3	1,409.2 / 14.8	1,555.3 / 14.4	1,697.5 / 14.1	1,836.1 / 13.7	1,970.7 / 13.3	2,102.9 / 13.1	2,233.7 / 13.0	2,363.9 / 13.0	2,494.4 / 13.1	2,624.9 / 13.1	2,756.1 / 13.2	2,888.3 / 13.3	3,021.1 / 13.3
600	201.0 / 24.1	419.5 / 20.0	609.8 / 18.2	787.1 / 17.2	956.3 / 16.4	1,118.2 / 15.8	1,273.6 / 15.3	1,424.0 / 14.8	1,569.7 / 14.4	1,711.6 / 14.0	1,849.8 / 13.6	1,984.0 / 13.3	2,116.0 / 13.1	2,246.7 / 13.0	2,376.9 / 13.0	2,507.5 / 13.1	2,638.0 / 13.0	2,769.3 / 13.1	2,901.6 / 13.2	3,034.4 / 13.3
700	225.1 / 23.4	439.5 / 19.7	628.0 / 18.0	804.3 / 17.1	972.7 / 16.4	1,134.0 / 15.7	1,288.9 / 15.2	1,438.8 / 14.7	1,584.1 / 14.4	1,725.6 / 14.0	1,863.4 / 13.6	1,997.3 / 13.3	2,129.1 / 13.1	2,259.7 / 13.0	2,389.9 / 13.1	2,520.6 / 13.0	2,651.0 / 13.1	2,782.4 / 13.2	2,914.8 / 13.2	3,047.7 / 13.3
800	248.5 / 22.7	459.2 / 19.4	646.0 / 18.0	821.4 / 17.0	989.1 / 16.3	1,149.7 / 15.6	1,304.1 / 15.2	1,453.5 / 14.6	1,598.5 / 14.3	1,739.6 / 13.9	1,877.0 / 13.5	2,010.6 / 13.2	2,142.2 / 13.1	2,272.7 / 13.0	2,403.0 / 13.1	2,533.6 / 13.1	2,664.1 / 13.1	2,795.6 / 13.2	2,928.0 / 13.2	3,061.0 / 13.4
900	271.2 / 22.3	478.6 / 19.2	664.0 / 17.8	838.4 / 16.9	1,005.4 / 16.2	1,165.3 / 15.6	1,319.2 / 15.1	1,468.1 / 14.7	1,612.8 / 14.2	1,753.5 / 13.9	1,890.5 / 13.5	2,023.8 / 13.2	2,155.3 / 13.1	2,285.7 / 13.0	2,416.1 / 13.1	2,546.7 / 13.0	2,677.2 / 13.2	2,808.8 / 13.2	2,941.2 / 13.3	3,074.4 / 13.3
1,000	293.5	497.8	681.8	855.3	1,021.6	1,180.9	1,334.3	1,482.8	1,627.0	1,767.4	1,904.0	2,037.0	2,168.4	2,298.7	2,429.2	2,559.7	2,690.4	2,822.0	2,954.5	3,087.7

TABLE 7
Devices for Liquid Flow Measurement

Flow measurement device	Advantages	Disadvantages	Flow range, gpm (applicable pipe diameter)
Venturi meter	Low permanent pressure drop. Applicable to streams with appreciable solids content. Accurate.	Flow disrupted and plumbing modifications required for installation. Expensive.	0—750 (1—18 in.)
Orifice meter	Inexpensive.	Flow disrupted and plumbing modifications required for installation. Large permanent pressure drop. Solids may deposit behind device. Moderately accurate.	0—750 (0.5—30 in.)
Flow tube	Applicable to streams with appreciable solids content.	Flow disrupted and plumbing modifications required for installation. Intermediate permanent pressure drop. Moderately expensive. Moderately accurate.	0—750 (1—18 in.)
Pitot tube	Low permanent pressure drop. Inexpensive method for pipes of large diameter.	Flow disrupted and plumbing modifications required for installation. Solids may cause plugging. High flow velocities may cause instability. Moderately accurate.	250—50,000
Magnetic meter	Minimum permanent pressure drop. Applicable to streams with appreciable solids content. Accurate.	Flow disrupted and plumbing modifications required for installation. Expensive. Electrodes may be fouled by waste waters containing oil and grease. Susceptible to electromagnetic interference from nearby equipment.	250—20,000 (0.1—100 in.)
Acoustic meter	Installation without flow disruption. Relatively accurate. No head loss or pressure drop. Applicable to streams with appreciable solids content. Portable.	Expensive. Moderately accurate.	250—20,000 (pipes of all diameters)

From Bonner, T. A., Cornett, C. L., Desai, B. O., Fullenkamp, J. M., Hughes, T. W., Johnson, M. L., Kennedy, E. D., McCormick, R. J., Peters, J. A., and Zanders, D. L., Engineering Handbook for Hazardous Waste Incineration, Rep. MRC-DA 1900, U.S. Environmental Protection Agency, Washington, D.C., 1981.

TABLE 8
Infrared Band Centers of Some
Common Gases

Gas	Location of band centers (μm)	Wave number (cm^{-1})
NO	5.0—5.5	1800—2000
NO$_2$	5.5—20	500—1800
SO$_2$	8—14	700—1250
H$_2$O	3.1	1,000—1400
	5.0—5.5	1800—2000
	7.1—10	3200
CO	2.3	2200
	4.6	4300
CO$_2$	2.7	850—1250
	5.2	1900
	8—12	3700
NH$_3$	10.5	950
CH$_4$	3.3	1300
	7.7	3000
Aldehydes	3.4—3.9	2550—2950

From Bonner T. A., Cornett, C. L., Desai, B. O., Fullenkamp, J. M., Hughes, T. W., Johnson, M. L., Kennedy, E. D., McCormick, R. J., Peters, J. A., and Zanders, D. L., Engineering Handbook for Hazardous Waste Incineration, Rep. MRC-DA 1090, U.S. Environmental Protection Agency, Washington, D.C., 1981.

III. ADDITIONAL INFORMATION

TABLE 9
Standard Test Sieves (Wire Cloth)

Sieve Designation		Nominal Sieve Opening in	Permissible Variation of Average Opening from the Standard Sieve Designation	Maximum Opening Size for Not More than 5 percent of Openings	Maximum Individual Opening	Nominal Wire Diameter. mm[a]
Standard	Alternative					
(1)	(2)	(3)	(4)	(5)	(6)	(7)
125 mm	5 in.	5	± 3.7 mm	130.0 mm	130.9 mm	8.0
106 mm	4.24 in.	4.24	± 3.2 mm	110.2 mm	111.1 mm	6.40
100 mm	4 in.	4	± 3.0 mm	104.0 mm	104.8 mm	6.30
90 mm	3½ in.	3.5	± 2.7 mm	93.6 mm	94.4 mm	6.08
75 mm	3 in.	3	± 2.2 mm	78.1 mm	78.7 mm	5.80
63 mm	2½ in.	2.5	± 1.9 mm	65.6 mm	66.2 mm	5.50
53 mm	2.12 in.	2.12	± 1.6 mm	55.2 mm	55.7 mm	5.15
50 mm	2 in.	2	± 1.5 mm	52.1 mm	52.6 mm	5.05
45 mm	1¾ in.	1.75	± 1.4 mm	46.9 mm	47.4 mm	4.85
37.5 mm	1½ in.	1.5	± 1.1 mm	39.1 mm	39.5 mm	4.59
31.5 mm	1¼ in.	1.25	± 1.0 mm	32.9 mm	33.2 mm	4.23
26.5 mm	1.06 in.	1.06	± 0.8 mm	27.7 mm	28.0 mm	3.90
25.0 mm	1 in.	1	± 0.8 mm	26.1 mm	26.4 mm	3.80
22.4 mm	⅞ in.	0.875	± 0.7 mm	23.4 mm	23.7 mm	3.50
19.0 mm	¾ in.	0.750	± 0.6 mm	19.9 mm	20.1 mm	3.30
16.0 mm	⅝ in.	0.625	± 0.5 mm	16.7 mm	17.0 mm	3.00
13.2 mm	0.530 in.	0.530	± 0.41 mm	13.83 mm	14.05 mm	2.75
12.5 mm	½ in.	0.500	± 0.39 mm	13.10 mm	13.31 mm	2.67
11.2 mm	⁷⁄₁₆ in.	0.438	± 0.35 mm	11.75 mm	11.94 mm	2.45
9.5 mm	⅜ in.	0.375	± 0.30 mm	9.97 mm	10.16 mm	2.27
8.0 mm	⁵⁄₁₆ in.	0.312	± 0.25 mm	8.41 mm	8.58 mm	2.07
6.7 mm	0.265 in.	0.265	± 0.21 mm	7.05 mm	7.20 mm	1.87
6.3 mm	¼ in.	0.250	± 0.20 mm	6.64 mm	6.78 mm	1.82
5.6 mm	No. 3½	0.223	± 0.18 mm	5.90 mm	6.04 mm	1.68
4.75 mm	No. 4	0.187	± 0.15 mm	5.02 mm	5.14 mm	1.54
4.00 mm	No. 5	0.157	± 0.13 mm	4.23 mm	4.35 mm	1.37
3.35 mm	No. 6	0.132	± 0.11 mm	3.55 mm	3.66 mm	1.23
2.80 mm	No. 7	0.111	± 0.095 mm	2.975 mm	3.070 mm	1.10
2.36 mm	No. 8	0.0937	± 0.080 mm	2.515 mm	2.600 mm	1.00
2.00 mm	No. 10	0.0787	± 0.070 mm	2.135 mm	2.215 mm	0.900
1.70 mm	No. 12	0.0661	± 0.060 mm	1.820 mm	1.890 mm	0.810
1.40 mm	No. 14	0.0555	± 0.050 mm	1.505 mm	1.565 mm	0.725
1.18 mm	No. 16	0.0469	± 0.045 mm	1.270 mm	1.330 mm	0.650
1.00 mm	No. 18	0.0394	± 0.040 mm	1.080 mm	1.135 mm	0.580
850 μm	No. 20	0.0331	± 35 μm	925 μm	970 μm	0.510
710 μm	No. 25	0.0278	± 30 μm	775 μm	815 μm	0.450
600 μm	No. 30	0.0234	± 25 μm	660 um	695 μm	0.390
500 μm	No. 35	0.0197	± 20 μm	550 μm	585 μm	0.340
425 μm	No. 40	0.0165	± 19 μm	471 μm	502 μm	0.290
355 μm	No. 45	0.0139	± 16 μm	396 μm	425 μm	0.247
300 μm	No. 50	0.0117	± 14 μm	337 μm	363 μm	0.215
250 μm	No. 60	0.0098	± 12 μm	283 μm	306 μm	0.180
212 μm	No. 70	0.0083	± 10 μm	242 μm	263 μm	0.152
180 μm	No. 80	0.0070	± 9 μm	207 μm	227 μm	0.131
150 μm	No. 100	0.0059	± 8 μm	174 μm	192 μm	0.110
125 μm	No. 120	0.0049	± 7 μm	147 μm	163 μm	0.091
106 μm	No. 140	0.0041	± 6 μm	126 μm	141 μm	0.076
90 μm	No. 170	0.0035	± 5 μm	108 μm	122 μm	0.064
75 μm	No. 200	0.0029	± 5 μm	91 μm	103 μm	0.053
63 μm	No. 230	0.0025	± 4 μm	77 μm	89 μm	0.044
53 μm	No. 270	0.0021	± 4 μm	66 μm	76 μm	0.037
45 μm	No. 325	0.0017	± 3 μm	57 μm	66 μm	0.030
38 μm	No. 400	0.0015	± 3 μm	48 μm	57 μm	0.025

[a] The average diameter of the warp and of the shoot wires, taken separately, of the cloth of any sieve shall not deviate from the nominal values by more than the following:

Sieves coarser than 600 μm	5 percent
Sieves 600 to 125 μm	7½ percent
Sieves finer than 125 μm	10 percent

From Weast, R. C., Ed., *Handbook of Chemistry and Physics*, 69th ed., CRC Press, Boca Raton, FL, 1988.

TABLE 10
Pipe Data: Carbon and Alloy Steel — Stainless Steel

Nom. Pipe Size Inches	Outside Diam. Inches	Iron Pipe Size	Sched. No.	Stainless Steel Sched. No.	Wall Thickness (t) Inches	Inside Diameter (d) Inches	Area of Metal Square Inches	Transverse Internal Area (a) Square Inches	(A) Square Feet	Moment of Inertia (I) Inches^4	Weight Pipe Pounds per foot	Weight Water Pounds per foot of pipe	External Surface Sq. Ft. per foot of pipe	Section Modulus $\left(2\frac{I}{O.D.}\right)$
1/8	0.405	...	...	10S	.049	.307	.0548	.0740	.00051	.00088	.19	.032	.106	.00437
		STD	40	40S	.068	.269	.0720	.0568	.00040	.00106	.24	.025	.106	.00523
		XS	80	80S	.095	.215	.0925	.0364	.00025	.00122	.31	.016	.106	.00602
1/4	0.540	...	...	10S	.065	.410	.0970	.1320	.00091	.00279	.33	.057	.141	.01032
		STD	40	40S	.088	.364	.1250	.1041	.00072	.00331	.42	.045	.141	.01227
		XS	80	80S	.119	.302	.1574	.0716	.00050	.00377	.54	.031	.141	.01395
3/8	0.675	...	...	10S	.065	.545	.1246	.2333	.00162	.00586	.42	.101	.178	.01736
		STD	40	40S	.091	.493	.1670	.1910	.00133	.00729	.57	.083	.178	.02160
		XS	80	80S	.126	.423	.2173	.1405	.00098	.00862	.74	.061	.178	.02554
1/2	0.840	...	...	5S	.065	.710	.1583	.3959	.00275	.01197	.54	.172	.220	.02849
		...	...	10S	.083	.674	.1974	.3568	.00248	.01431	.67	.155	.220	.03407
		STD	40	40S	.109	.622	.2503	.3040	.00211	.01709	.85	.132	.220	.04069
		XS	80	80S	.147	.546	.3200	.2340	.00163	.02008	1.09	.102	.220	.04780
		...	160	...	.187	.466	.3836	.1706	.00118	.02212	1.31	.074	.220	.05267
		XXS	...	...	.294	.252	.5043	.050	.00035	.02424	1.71	.022	.220	.05772
3/4	1.050	...	...	5S	.065	.920	.2011	.6648	.00462	.02450	.69	.288	.275	.04667
		...	...	10S	.083	.884	.2521	.6138	.00426	.02969	.86	.266	.275	.05655
		STD	40	40S	.113	.824	.3326	.5330	.00371	.03704	1.13	.231	.275	.07055
		XS	80	80S	.154	.742	.4335	.4330	.00300	.04479	1.47	.188	.275	.08531
		...	160	...	.219	.612	.5698	.2961	.00206	.05269	1.94	.128	.275	.10036
		XXS	...	...	.308	.434	.7180	.148	.00103	.05792	2.44	.064	.275	.11032
1	1.315	...	...	5S	.065	1.185	.2553	1.1029	.00766	.04999	.87	.478	.344	.07603
		...	...	10S	.109	1.097	.4130	.9452	.00656	.07569	1.40	.409	.344	.11512
		STD	40	40S	.133	1.049	.4939	.8640	.00600	.08734	1.68	.375	.344	.1328
		XS	80	80S	.179	.957	.6388	.7190	.00499	.1056	2.17	.312	.344	.1606
		...	160	...	.250	.815	.8365	.5217	.00362	.1251	2.84	.230	.344	.1903
		XXS	...	...	.358	.599	1.0760	.282	.00196	.1405	3.66	.122	.344	.2136
1¼	1.660	...	...	5S	.065	1.530	.3257	1.839	.01277	.1038	1.11	.797	.435	.1250
		...	...	10S	.109	1.442	.4717	1.633	.01134	.1605	1.81	.708	.435	.1934
		STD	40	40S	.140	1.380	.6685	1.495	.01040	.1947	2.27	.649	.435	.2346
		XS	80	80S	.191	1.278	.8815	1.283	.00891	.2418	3.00	.555	.435	.2913
		...	160	...	.250	1.160	1.1070	1.057	.00734	.2839	3.76	.458	.435	.3421
		XXS	...	...	.382	.896	1.534	.630	.00438	.3411	5.21	.273	.435	.4110
1½	1.900	...	...	5S	.065	1.770	.3747	2.461	.01709	.1579	1.28	1.066	.497	.1662
		...	...	10S	.109	1.682	.6133	2.222	.01543	.2468	2.09	.963	.497	.2598
		STD	40	40S	.145	1.610	.7995	2.036	.01414	.3099	2.72	.882	.497	.3262
		XS	80	80S	.200	1.500	1.068	1.767	.01225	.3912	3.63	.765	.497	.4118
		...	160	...	.281	1.338	1.429	1.406	.00976	.4824	4.86	.608	.497	.5078
		XXS	...	...	.400	1.100	1.885	.950	.00660	.5678	6.41	.42	.497	.5977
2	2.375	...	...	5S	.065	2.245	.4717	3.958	.02749	.3149	1.61	1.72	.622	.2652
		...	...	10S	.109	2.157	.7760	3.654	.02538	.4992	2.64	1.58	.622	.4204
		STD	40	40S	.154	2.067	1.075	3.355	.02330	.6657	3.65	1.45	.622	.5606
		XS	80	80S	.218	1.939	1.477	2.953	.02050	.8679	5.02	1.28	.622	.7309
		...	160	...	.344	1.687	2.190	2.241	.01556	1.162	7.46	.97	.622	.979
		XXS	...	...	.436	1.503	2.656	1.774	.01232	1.311	9.03	.77	.622	1.104
2½	2.875	...	...	5S	.083	2.709	.7280	5.764	.04002	.7100	2.48	2.50	.753	.4939
		...	...	10S	.120	2.635	1.039	5.453	.03787	.9873	3.53	2.36	.753	.6868
		STD	40	40S	.203	2.469	1.704	4.788	.03322	1.530	5.79	2.07	.753	1.064
		XS	80	80S	.276	2.323	2.254	4.238	.02942	1.924	7.66	1.87	.753	1.339
		...	160	...	.375	2.125	2.945	3.546	.02463	2.353	10.01	1.54	.753	1.638
		XXS	...	...	.552	1.771	4.028	2.464	.01710	2.871	13.69	1.07	.753	1.997
3	3.500	...	...	5S	.083	3.334	.8910	8.730	.06063	1.301	3.03	3.78	.916	.7435
		...	...	10S	.120	3.260	1.274	8.347	.05796	1.822	4.33	3.62	.916	1.041
		STD	40	40S	.216	3.068	2.228	7.393	.05130	3.017	7.58	3.20	.916	1.724
		XS	80	80S	.300	2.900	3.016	6.605	.04587	3.894	10.25	2.86	.916	2.225
		...	160	...	.438	2.624	4.205	5.408	.03755	5.032	14.32	2.35	.916	2.876
		XXS	...	...	.600	2.300	5.466	4.155	.02885	5.993	18.58	1.80	.916	3.424

Identification, wall thickness and weights are extracted from ANSI B36.10 and B36.19. The notations STD, XS, and XXS indicate Standard, Extra Strong, and Double Extra Strong pipe respectively.

Transverse internal area values listed in "square feet" also represent volume in cubic feet per foot of pipe length.

TABLE 10 (continued)
Pipe Data: Carbon and Alloy Steel — Stainless Steel

Nominal Pipe Size Inches	Outside Diam. Inches	Identification Steel — Iron Pipe Size	Identification Steel — Sched. No.	Identification Stainless Steel Sched. No.	Wall Thickness (t) Inches	Inside Diameter (d) Inches	Area of Metal Square Inches	Transverse Internal Area (a) Square Inches	Transverse Internal Area (A) Square Feet	Moment of Inertia (I) Inches⁴	Weight Pipe Pounds per foot	Weight Water Pounds per foot of pipe	External Surface Sq. Ft. per foot of pipe	Section Modulus $\left(2\frac{I}{O.D.}\right)$
3½	4.000	...	...	5S	.083	3.834	1.021	11.545	.08017	1.960	3.48	5.00	1.047	.9799
		...	...	10S	.120	3.760	1.463	11.104	.07711	2.755	4.97	4.81	1.047	1.378
		STD	40	40S	.226	3.548	2.680	9.886	.06870	4.788	9.11	4.29	1.047	2.394
		XS	80	80S	.318	3.364	3.678	8.888	.06170	6.280	12.50	3.84	1.047	3.140
4	4.500	...	...	5S	.083	4.334	1.152	14.75	.10245	2.810	3.92	6.39	1.178	1.249
		...	...	10S	.120	4.260	1.651	14.25	.09898	3.963	5.61	6.18	1.178	1.761
		STD	40	40S	.237	4.026	3.174	12.73	.08840	7.233	10.79	5.50	1.178	3.214
		XS	80	80S	.337	3.826	4.407	11.50	.07986	9.610	14.98	4.98	1.178	4.271
		...	120	...	.438	3.624	5.595	10.31	.0716	11.65	19.00	4.47	1.178	5.178
		...	160	...	.531	3.438	6.621	9.28	.0645	13.27	22.51	4.02	1.178	5.898
		XXS	...	...	.674	3.152	8.101	7.80	.0542	15.28	27.54	3.38	1.178	6.791
5	5.563	...	...	5S	.109	5.345	1.868	22.44	.1558	6.947	6.36	9.72	1.456	2.498
		...	...	10S	.134	5.295	2.285	22.02	.1529	8.425	7.77	9.54	1.456	3.029
		STD	40	40S	.258	5.047	4.300	20.01	.1390	15.16	14.62	8.67	1.456	5.451
		XS	80	80S	.375	4.813	6.112	18.19	.1263	20.67	20.78	7.88	1.456	7.431
		...	120	...	.500	4.563	7.953	16.35	.1136	25.73	27.04	7.09	1.456	9.250
		...	160	...	.625	4.313	9.696	14.61	.1015	30.03	32.96	6.33	1.456	10.796
		XXS	...	...	.750	4.063	11.340	12.97	.0901	33.63	38.55	5.61	1.456	12.090
6	6.625	...	...	5S	.109	6.407	2.231	32.24	.2239	11.85	7.60	13.97	1.734	3.576
		...	...	10S	.134	6.357	2.733	31.74	.2204	14.40	9.29	13.75	1.734	4.346
		STD	40	40S	.280	6.065	5.581	28.89	.2006	28.14	18.97	12.51	1.734	8.496
		XS	80	80S	.432	5.761	8.405	26.07	.1810	40.49	28.57	11.29	1.734	12.22
		...	120	...	.562	5.501	10.70	23.77	.1650	49.61	36.39	10.30	1.734	14.98
		...	160	...	.719	5.187	13.32	21.15	.1469	58.97	45.35	9.16	1.734	17.81
		XXS	...	...	.864	4.897	15.64	18.84	.1308	66.33	53.16	8.16	1.734	20.02
8	8.625	...	...	5S	.109	8.407	2.916	55.51	.3855	26.44	9.93	24.06	2.258	6.131
		...	...	10S	.148	8.329	3.941	54.48	.3784	35.41	13.40	23.61	2.258	8.212
		...	20	...	.250	8.125	6.57	51.85	.3601	57.72	22.36	22.47	2.258	13.39
		...	30	...	.277	8.071	7.26	51.16	.3553	63.35	24.70	22.17	2.258	14.69
		STD	40	40S	.322	7.981	8.40	50.03	.3474	72.49	28.55	21.70	2.258	16.81
		...	60	...	.406	7.813	10.48	47.94	.3329	88.73	35.64	20.77	2.258	20.58
		XS	80	80S	.500	7.625	12.76	45.66	.3171	105.7	43.39	19.78	2.258	24.51
		...	100	...	.594	7.437	14.96	43.46	.3018	121.3	50.95	18.83	2.258	28.14
		...	120	...	.719	7.187	17.84	40.59	.2819	140.5	60.71	17.59	2.258	32.58
		...	140	...	.812	7.001	19.93	38.50	.2673	153.7	67.76	16.68	2.258	35.65
		XXS	...	...	.875	6.875	21.30	37.12	.2578	162.0	72.42	16.10	2.258	37.56
		...	160	...	.906	6.813	21.97	36.46	.2532	165.9	74.69	15.80	2.258	38.48
10	10.750	...	...	5S	.134	10.482	4.36	86.29	.5992	63.0	15.19	37.39	2.814	11.71
		...	...	10S	.165	10.420	5.49	85.28	.5922	76.9	18.65	36.95	2.814	14.30
		...	20	...	.250	10.250	8.24	82.52	.5731	113.7	28.04	35.76	2.814	21.15
		...	30	...	.307	10.136	10.07	80.69	.5603	137.4	34.24	34.96	2.814	25.57
		STD	40	40S	.365	10.020	11.90	78.86	.5475	160.7	40.48	34.20	2.814	29.90
		XS	60	80S	.500	9.750	16.10	74.66	.5185	212.0	54.74	32.35	2.814	39.43
		...	80	...	.594	9.562	18.92	71.84	.4989	244.8	64.43	31.13	2.814	45.54
		...	100	...	.719	9.312	22.63	68.13	.4732	286.1	77.03	29.53	2.814	53.22
		...	120	...	.844	9.062	26.24	64.53	.4481	324.2	89.29	27.96	2.814	60.32
		XXS	140	...	1.000	8.750	30.63	60.13	.4176	367.8	104.13	26.06	2.814	68.43
		...	160	...	1.125	8.500	34.02	56.75	.3941	399.3	115.64	24.59	2.814	74.29
12	12.75	...	...	5S	.156	12.438	6.17	121.50	.8438	122.4	20.98	52.65	3.338	19.2
		...	...	10S	.180	12.390	7.11	120.57	.8373	140.4	24.17	52.25	3.338	22.0
		...	20	...	.250	12.250	9.82	117.86	.8185	191.8	33.38	51.07	3.338	30.2
		...	30	...	.330	12.090	12.87	114.80	.7972	248.4	43.77	49.74	3.338	39.0
		STD	...	40S	.375	12.000	14.58	113.10	.7854	279.3	49.56	49.00	3.338	43.8
		...	40	...	.406	11.938	15.77	111.93	.7773	300.3	53.52	48.50	3.338	47.1
		XS	...	80S	.500	11.750	19.24	108.43	.7528	361.5	65.42	46.92	3.338	56.7
		...	60	...	.562	11.626	21.52	106.16	.7372	400.4	73.15	46.00	3.338	62.8
		...	80	...	.688	11.374	26.03	101.64	.7058	475.1	88.63	44.04	3.338	74.6
		...	100	...	.844	11.062	31.53	96.14	.6677	561.6	107.32	41.66	3.338	88.1
		XXS	120	...	1.000	10.750	36.91	90.76	.6303	641.6	125.49	39.33	3.338	100.7
		...	140	...	1.125	10.500	41.08	86.59	.6013	700.5	139.67	37.52	3.338	109.9
		...	160	...	1.312	10.126	47.14	80.53	.5592	781.1	160.27	34.89	3.338	122.6

Identification, wall thickness and weights are extracted from ANSI B36.10 and B36.19. The notations STD, XS, and XXS indicate Standard, Extra Strong, and Double Extra Strong pipe respectively.

Transverse internal area values listed in "square feet" also represent volume in cubic feet per foot of pipe length.

TABLE 10 (continued)
Pipe Data: Carbon and Alloy Steel — Stainless Steel

Nominal Pipe Size Inches	Outside Diam. Inches	Identification — Steel Iron Pipe Size	Identification — Steel Sched. No.	Identification — Stainless Steel Sched. No.	Wall Thickness (t) Inches	Inside Diameter (d) Inches	Area of Metal Square Inches	Transverse Internal Area (a) Square Inches	Transverse Internal Area (A) Square Feet	Moment of Inertia (I) Inches⁴	Weight Pipe Pounds per foot	Weight Water Pounds per foot of pipe	External Surface Sq. Ft. per foot of pipe	Section Modulus $\left(2\frac{I}{O.D.}\right)$
		...	...	5S	.156	13.688	6.78	147.15	1.0219	162.6	23.07	63.77	3.665	23.2
		...	...	10S	.188	13.624	8.16	145.78	1.0124	194.6	27.73	63.17	3.665	27.8
		...	10	...	.250	13.500	10.80	143.14	.9940	255.3	36.71	62.03	3.665	36.6
		...	20	...	.312	13.376	13.42	140.52	.9758	314.4	45.61	60.89	3.665	45.0
		STD	30	...	.375	13.250	16.05	137.88	.9575	372.8	54.57	59.75	3.665	53.2
		...	40	...	.438	13.124	18.66	135.28	.9394	429.1	63.44	58.64	3.665	61.3
14	14.00	XS	...	...	.500	13.000	21.21	132.73	.9217	483.8	72.09	57.46	3.665	69.1
		...	60	...	.594	12.812	24.98	128.96	.8956	562.3	85.05	55.86	3.665	80.3
		...	80	...	.750	12.500	31.22	122.72	.8522	678.3	106.13	53.18	3.665	98.2
		...	100	...	.938	12.124	38.45	115.49	.8020	824.4	130.85	50.04	3.665	117.8
		...	120	...	1.094	11.812	44.32	109.62	.7612	929.6	150.79	47.45	3.665	132.8
		...	140	...	1.250	11.500	50.07	103.87	.7213	1027.0	170.28	45.01	3.665	146.8
		...	160	...	1.406	11.188	55.63	98.31	.6827	1117.0	189.11	42.60	3.665	159.6
		...	...	5S	.165	15.670	8.21	192.85	1.3393	257.3	27.90	83.57	4.189	32.2
		...	...	10S	.188	15.624	9.34	191.72	1.3314	291.9	31.75	83.08	4.189	36.5
		...	10	...	.250	15.500	12.37	188.69	1.3103	383.7	42.05	81.74	4.189	48.0
		...	20	...	.312	15.376	15.38	185.69	1.2895	473.2	52.27	80.50	4.189	59.2
		STD	30	...	.375	15.250	18.41	182.65	1.2684	562.1	62.58	79.12	4.189	70.3
16	16.00	XS	40	...	.500	15.000	24.35	176.72	1.2272	731.9	82.77	76.58	4.189	91.5
		...	60	...	.656	14.688	31.62	169.44	1.1766	932.4	107.50	73.42	4.189	116.6
		...	80	...	.844	14.312	40.14	160.92	1.1175	1155.8	136.61	69.73	4.189	144.5
		...	100	...	1.031	13.938	48.48	152.58	1.0596	1364.5	164.82	66.12	4.189	170.5
		...	120	...	1.219	13.562	56.56	144.50	1.0035	1555.8	192.43	62.62	4.189	194.5
		...	140	...	1.438	13.124	65.78	135.28	.9394	1760.3	223.64	58.64	4.189	220.0
		...	160	...	1.594	12.812	72.10	128.96	.8956	1893.5	245.25	55.83	4.189	236.7
		...	...	5S	.165	17.670	9.25	245.22	1.7029	367.6	31.43	106.26	4.712	40.8
		...	...	10S	.188	17.624	10.52	243.95	1.6941	417.3	35.76	105.71	4.712	46.4
		...	10	...	.250	17.500	13.94	240.53	1.6703	549.1	47.39	104.21	4.712	61.1
		...	20	...	.312	17.376	17.34	237.13	1.6467	678.2	58.94	102.77	4.712	75.5
		STD	...	...	.375	17.250	20.76	233.71	1.6230	806.7	70.59	101.18	4.712	89.6
		...	30	...	.438	17.124	24.17	230.30	1.5990	930.3	82.15	99.84	4.712	103.4
18	18.00	XS	...	...	.500	17.000	27.49	226.98	1.5763	1053.2	93.45	98.27	4.712	117.0
		...	40	...	.562	16.876	30.79	223.63	1.5533	1171.5	104.67	96.93	4.712	130.1
		...	60	...	.750	16.500	40.64	213.83	1.4849	1514.7	138.17	92.57	4.712	168.3
		...	80	...	.938	16.124	50.23	204.24	1.4183	1833.0	170.92	88.50	4.712	203.8
		...	100	...	1.156	15.688	61.17	193.30	1.3423	2180.0	207.96	83.76	4.712	242.3
		...	120	...	1.375	15.250	71.81	182.66	1.2684	2498.1	244.14	79.07	4.712	277.6
		...	140	...	1.562	14.876	80.66	173.80	1.2070	2749.0	274.22	75.32	4.712	305.5
		...	160	...	1.781	14.438	90.75	163.72	1.1369	3020.0	308.50	70.88	4.712	335.6
		...	...	5S	.188	19.624	11.70	302.46	2.1004	574.2	39.78	131.06	5.236	57.4
		...	...	10S	.218	19.564	13.55	300.61	2.0876	662.8	46.06	130.27	5.236	66.3
		...	10	...	.250	19.500	15.51	298.65	2.0740	765.4	52.73	129.42	5.236	75.6
		STD	20	...	.375	19.250	23.12	290.04	2.0142	1113.0	78.60	125.67	5.236	111.3
		XS	30	...	.500	19.000	30.63	283.53	1.9690	1457.0	104.13	122.87	5.236	145.7
		...	40	...	.594	18.812	36.15	278.00	1.9305	1703.0	123.11	120.46	5.236	170.4
20	20.00	...	60	...	.812	18.376	48.95	265.21	1.8417	2257.0	166.40	114.92	5.236	225.7
		...	80	...	1.031	17.938	61.44	252.72	1.7550	2772.0	208.87	109.51	5.236	277.1
		...	100	...	1.281	17.438	75.33	238.83	1.6585	3315.2	256.10	103.39	5.236	331.5
		...	120	...	1.500	17.000	87.18	226.98	1.5762	3754.0	296.37	98.35	5.236	375.5
		...	140	...	1.750	16.500	100.33	213.82	1.4849	4216.0	341.09	92.66	5.236	421.7
		...	160	...	1.969	16.062	111.49	202.67	1.4074	4585.5	379.17	87.74	5.236	458.5
		...	...	5S	.188	21.624	12.88	367.25	2.5503	766.2	43.80	159.14	5.760	69.7
		...	...	10S	.218	21.564	14.92	365.21	2.5362	884.8	50.71	158.26	5.760	80.4
		...	10	...	.250	21.500	17.08	363.05	2.5212	1010.3	58.07	157.32	5.760	91.8
		STD	20	...	.375	21.250	25.48	354.66	2.4629	1489.7	86.61	153.68	5.760	135.4
		XS	30	...	.500	21.000	33.77	346.36	2.4053	1952.5	114.81	150.09	5.760	117.5
22	22.00	...	60	...	.875	20.250	58.07	322.06	2.2365	3244.9	197.41	139.56	5.760	295.0
		...	80	...	1.125	19.75	73.78	306.35	2.1275	4030.4	250.81	132.76	5.760	366.4
		...	100	...	1.375	19.25	89.09	291.04	2.0211	4758.5	302.88	126.12	5.760	432.6
		...	120	...	1.625	18.75	104.02	276.12	1.9175	5432.0	353.61	119.65	5.760	493.8
		...	140	...	1.875	18.25	118.55	261.59	1.8166	6053.7	403.00	113.36	5.760	550.3
		...	160	...	2.125	17.75	132.68	247.45	1.7184	6626.4	451.06	107.23	5.760	602.4

Identification, wall thickness and weights are extracted from ANSI B36.10 and B36.19. The notations STD, XS, and XXS indicate Standard, Extra Strong, and Double Extra Strong pipe respectively.

Transverse internal area values listed in "square feet" also represent volume in cubic feet per foot of pipe length.

TABLE 10 (continued)
Pipe Data: Carbon and Alloy Steel — Stainless Steel

Nominal Pipe Size Inches	Outside Diam. Inches	Identification Steel Iron Pipe Size	Identification Steel Sched. No.	Identification Stainless Steel Sched. No.	Wall Thickness (t) Inches	Inside Diameter (d) Inches	Area of Metal Square Inches	Transverse Internal Area (a) Square Inches	Transverse Internal Area (A) Square Feet	Moment of Inertia (I) Inches⁴	Weight Pipe Pounds per foot	Weight Water Pounds per foot of pipe	External Surface Sq. Ft. per foot of pipe	Section Modulus $\left(2\frac{I}{O.D.}\right)$
24	24.00	...	...	5S	.218	23.564	16.29	436.10	3.0285	1151.6	55.37	188.98	6.283	96.0
		...	10	10S	.250	23.500	18.65	433.74	3.0121	1315.4	63.41	187.95	6.283	109.6
		STD	20	...	.375	23.250	27.83	424.56	2.9483	1942.0	94.62	183.95	6.283	161.9
		XS	...	...	.500	23.000	36.91	415.48	2.8853	2549.5	125.49	179.87	6.283	212.5
		...	30	...	.562	22.876	41.39	411.00	2.8542	2843.0	140.68	178.09	6.283	237.0
		...	40	...	.688	22.624	50.31	402.07	2.7921	3421.3	171.29	174.23	6.283	285.1
		...	60	...	.969	22.062	70.04	382.35	2.6552	4652.8	238.35	165.52	6.283	387.7
		...	80	...	1.219	21.562	87.17	365.22	2.5362	5672.0	296.58	158.26	6.283	472.8
		...	100	...	1.531	20.938	108.07	344.32	2.3911	6849.9	367.39	149.06	6.283	570.8
		...	120	...	1.812	20.376	126.31	326.08	2.2645	7825.0	429.39	141.17	6.283	652.1
		...	140	...	2.062	19.876	142.11	310.28	2.1547	8625.0	483.12	134.45	6.283	718.9
		...	160	...	2.344	19.312	159.41	292.98	2.0346	9455.9	542.13	126.84	6.283	787.9
26	26.00	...	10	...	.312	25.376	25.18	505.75	3.5122	2077.2	85.60	219.16	6.806	159.8
		STD	...	...	.375	25.250	30.19	500.74	3.4774	2478.4	102.63	216.99	6.806	190.6
		XS	20	...	.500	25.000	40.06	490.87	3.4088	3257.0	136.17	212.71	6.806	250.5
28	28.00	...	10	...	.312	27.376	27.14	588.61	4.0876	2601.0	92.26	255.07	7.330	185.8
		STD	...	...	.375	27.250	32.54	583.21	4.0501	3105.1	110.64	252.73	7.330	221.8
		XS	20	...	.500	27.000	43.20	572.56	3.9761	4084.8	146.85	248.11	7.330	291.8
		...	30	...	.625	26.750	53.75	562.00	3.9028	5037.7	182.73	243.53	7.330	359.8
30	30.00	...	...	5S	.250	29.500	23.37	683.49	4.7465	2585.2	79.43	296.18	7.854	172.3
		...	10	10S	.312	29.376	29.10	677.76	4.7067	3206.3	98.93	293.70	7.854	213.8
		STD	...	...	.375	29.250	34.90	671.96	4.6664	3829.4	118.65	291.18	7.854	255.3
		XS	20	...	.500	29.000	46.34	660.52	4.5869	5042.2	157.53	286.22	7.854	336.1
		...	30	...	.625	28.750	57.68	649.18	4.5082	6224.0	196.08	281.31	7.854	414.9
32	32.00	...	10	...	.312	31.376	31.06	773.19	5.3694	3898.9	105.59	335.05	8.378	243.7
		STD	...	...	.375	31.250	37.26	766.99	5.3263	4658.5	126.66	332.36	8.378	291.2
		XS	20	...	.500	31.000	49.48	754.77	5.2414	6138.6	168.21	327.06	8.378	383.7
		...	30	...	.625	30.750	61.60	742.64	5.1572	7583.4	209.43	321.81	8.378	474.0
		...	40	...	.688	30.624	68.07	736.57	5.1151	8298.3	230.08	319.18	8.378	518.6
34	34.00	...	10	...	.344	33.312	36.37	871.55	6.0524	5150.5	123.65	377.67	8.901	303.0
		STD	...	...	.375	33.250	39.61	868.31	6.0299	5599.3	134.67	376.27	8.901	329.4
		XS	20	...	.500	33.000	52.62	855.30	5.9396	7383.5	178.89	370.63	8.901	434.3
		...	30	...	.625	32.750	65.53	842.39	5.8499	9127.6	222.78	365.03	8.901	536.9
		...	40	...	.688	32.624	72.00	835.92	5.8050	9991.6	244.77	362.23	8.901	587.7
36	36.00	...	10	...	.312	35.376	34.98	982.90	6.8257	5569.5	118.92	425.92	9.425	309.4
		STD	...	...	.375	35.250	41.97	975.91	6.7771	6658.9	142.68	422.89	9.425	369.9
		XS	20	...	.500	35.000	55.76	962.11	6.6813	8786.2	189.57	416.91	9.425	481.1
		...	30	...	.625	34.750	69.46	948.42	6.5862	10868.4	236.13	417.22	9.425	603.8
		...	40	...	.750	34.500	83.06	934.82	6.4918	12906.1	282.35	405.09	9.425	717.0

Identification, wall thickness and weights are extracted from ANSI B36.10 and B36.19. The notations STD, XS, and XXS indicate Standard, Extra Strong, and Double Extra Strong pipe respectively.

Transverse internal area values listed in "square feet" also represent volume in cubic feet per foot of pipe length.

From Flow of Fluids through Valves, Fittings and Pipe, Technical Paper 410, Crane Co., 1978.

TABLE 11
Conversion Factors

To convert from	To	Multiply by
Acre	Square kilometer (km²)	0.00404047
	Square meter (m²)	4,046.86
	Square mile (mi²)	0.0015625
Btu (British thermal unit)	Calorie (cal)	251.99576
	Joule (J)	1,054.35
	Kilocalorie (kcal)	0.251996
Btu/minute (Btu/min)	Joule/second (J/s)	17.5725
Btu/pound (Btu/lb)	Calorie/gram (cal/g)	0.555555
Btu/pound·°F[Btu/(lb·°F)]	Calorie/(gram · °C) [cal/(g · °C)]	1.0
Btu/second (Btu/s)	Kilocalorie/hour (kcal/h)	970.185
	Kilocalorie/minute	15.1197
Calorie (cal)	Btu	0.0039683207
	Kilocalorie (kcal)	0.001
	Joule (J)	4.184
Calorie/gram (cal/g)	Btu/pound (Btu/lb)	1.8
Calorie/hour (cal/h)	Btu/hour (Btu/h)	0.0039683207
	Erg/second (erg/s)	11,622.222
Centigrade (°C)	Fahrenheit (°F)	°F = (1.8 × °C) + 32
	Kelvin (°K)	°K = °C + 273.17
Centimeter (cm)	Inch (in.)	0.39370079
Centipoise (cP)	Gram/(centimeter · second) [g/(cm · s)]	0.01
Centistokes (cSt)	Saybolt seconds (SSU)	See Table 12
Cubic centimeter (cm³)	Cubic foot (ft³)	3.5314667×10^{-5}
	Cubic inch (in.³)	0.061023744
	Cubic yard (yd³)	1.3079506×10^{-6}
Cubic foot (ft³)	Cubic centimeter (cm³)	28,316.847
	Cubic meter (m³)	0.028316847
	Gallon (U.S. liquid)	7.4805195
	Liter (l)	28.316847
Cubic meter (m³)	Cubic foot (ft³)	35.314667
	Cubic yard (yd³)	1.3079506
	Liter (l)	1,000
Dyne/square centimeter (dyne/cm²)	Atmosphere (atm)	9.86923×10^{-7}
	Bar	1×10^{6}
	Centimeter of mercury @ 0°C (cmHg @ 0°C)	7.50062×10^{-5}
	Centimeter of water @ 4°C (cmH₂O @ 4°C)	0.00109745
	Inch of mercury @ 32°F (inHg @ 32°F)	2.95300×10^{-5}
	Inch of water @ 4°C (inH₂O @ 4°C)	0.000401474
	Pascal (Pa)	0.1
	Pound/square inch (lb/in.²)	1.450377×10^{-5}
Fahrenheit (°F)	Centigrade (°C)	°C = 0.5556 (°F − 32°)
	Rankine (°R)	°R = °F + 459.7°
Foot (ft)	Centimeter (cm)	30.48
	Inch (in.)	12
	Meter (m)	0.3048
	Millimeter (mm)	304.8
Gallon (U.K. liquid) (gal)	Gallon (U.S. liquid) (gal)	1.20095
	Liter (l)	0.00668932
Gallon (U.S. liquid) (gal)	Cubic centimeter (cm³)	3,785.4118
	Cubic foot (ft³)	0.133680555
	Cubic inch (in.³)	231
	Cubic meter (m³)	0.0037854118

TABLE 11 (continued)
Conversion Factors

To convert from	To	Multiply by
	Liter (l)	3.7854118
Grains/standard cubic foot (gr/scf)	Milligrams/standard cubic meter	2288.3
Gram (g)	Kilogram (kg)	0.001
	Pound (lb)	0.0022046226
Gram/(centimeter · second)	Poise (P)	1
Gram/cubic centimeter (g/cm³)	Grain/milliliter (gr/ml)	15.43279
	Gram/milliliter (g/ml)	1
	Pound/cubic foot (lb/ft³)	62.427961
	Pound/cubic inch (lb/in.³)	0.036127292
	Pound/gallon (U.S. liquid) (lb/gal)	8.3454044
Gram/cubic meter (g/m³)	Grain/cubic foot (gr/ft³)	0.43699572
Gram/liter (g/l)	Part/million (ppm)	1,000
	Pound/cubic foot (lb/ft³)	0.06242621
Gram/milliliter (g/ml)	Gram/cubic centimeter (g/cm³)	1
	Pound/cubic foot (lb/ft³)	62.4261
	Pound/gallon (U.S.) (lb/gal)	8.345171
Inch of water @ 4°C (in H$_2$O @ 4°C)	Atmosphere (atm)	0.0024582
	Inch of mercury @ 32°F (inHg @ 32°F)	0.0735529
	Kilopascal (kPa)	249,082
	Pascal (Pa)	249.082
	Pound/square inch (psi)	0.03612628
Joule (J)	Btu	0.000948451
Joule/second (J/s)	Btu/minute (Btu/min)	0.0569071
	Btu/hour (Btu/h)	3.414426
Kilocalorie (kcal)	Btu	3.9683207
	Erg	4.184 × 10¹
	Joule (J)	4,184
Kilogram (kg)	Pound (avoirdupois) [lb (avdp)]	2.2046226
	Ton (short, 2000-lb mass)	0.0011023113
Liter (l)	Cubic foot (ft³)	0.035314667
	Quart (U.S. liquid) (qt)	1.0566882
Meter (m)	Foot (ft)	3.2808399
	Inch (in.)	39.370079
	Mile (statute) (mi)	0.00062137119
	Millimicrons (mμ)	1 × 10⁹
	Yard (yd)	1.0936133
Pascal (Pa)	Atmosphere (standard) (atm)	9.869233 × 10⁻⁶
	Dyne/square centimeter (dyne/cm²)	10
	Inch of water @ 39.2°F (inH$_2$O @ 39.2°F)	0.004014742
	Inch of water @ 60°F (inH$_2$O @ 60°F)	0.004018647
	Pound-force/square inch (lb-force/in.²) (psi)	0.0001450377
Pascal · seconds (pa · s)	Poise	10.00
Poise (P)	Centipose (cP)	100.00
	Dyne · second/square centimeter	1
	Gram/(centimeter · second) [g/(cm · s)]	1
	Pound/(second · foot) [lb/s · ft]	0.0672
Pound (lb)	Gram (g)	953.59237
Pound/(foot · second) [lb/(ft · s)]	Poise (P)	14.88
Pound/cubic foot (lb/ft³)	Gram/cubic centimeter (g/cm³)	0.016018463
	Kilogram/cubic meter (kg/m³)	16.018463

TABLE 11 (continued)
Conversion Factors

To convert from	To	Multiply by
Pound/cubic inch (lb/in.³)	Gram/cubic centimeter (g/cm³)	27.679905
	Gram/liter (g/l)	27.68068
	Kilogram/cubic meter (kg/m³)	27,679.905
Pound/gallon (U.K. liquid) (lb/gal)	Pound/cubic foot (lb/ft³)	6.228839
Pound/gallon (U.S. liquid) (lb/gal)	Gram/cubic centimeter (g/cm³)	0.11982643
	Pound/cubic foot (lb/ft³)	7.4805195
Pound/square inch (psi)	Atmosphere (atm)	0.0680460
Saybolt seconds (SSU)	Centistokes (cSt)	see Table 12
Square foot (ft²)	Acre	2.295684×10^5
	Square centimeter (cm²)	929.0304
	Square inch (in.²)	144
	Square meter (m²)	0.09290304
Square kilometer (km²)	Acre	247.10538
	Square meter (m²)	1,000,000
	Square mile (mi²)	0.38610216
Square meter (m²)	Acre	0.00024710538
	Square foot (ft²)	10.763910
	Square kilometer (km²)	0.000001
Stoke (St)	Centistoke (cSt)	1×10^2
	Saybolt seconds (SSU)	See Table 12
	Square centimeter/second (cm²/s)	1
	Square foot/hour (ft²/h)	3.875
	Square foot/second (ft²/s)	0.001076
Ton (metric)	Kilogram (kg)	1000
	Ton (short, 2000-lb mass)	1.1023113

From Bonner, T. A., Cornett, C. L., Desai, B. O., Fullenkamp, J. M., Hughes, T. W., Johnson, M. L., Kennedy, E. D., McCormick, R. J., Peters, J. A., and Zanders, D. L., Engineering Handbook for Hazardous Waste Incineration, Rep. MRC-DA 1900, U.S. Environmental Protection Agency, Washington, D.C., 1981.

TABLE 12
Viscosity Conversion

Kinematic

To convert from	To	Multiply by	To convert from	To	Multiply by
cm²/s (stokes)	Centistokes	10^2	ft²/s	cm²/s (stokes)	9.29×10^2
	ft²/h	3.875		cm²/s $\times 10^2$ (centistokes)	9.29×10^4
	ft²/s	1.076×10^{-3}		ft²/h	3.60×10^3
	in²/s	1.550×10^{-1}		in²/h	1.44×10^2
	m²/h	3.600×10^{-1}		m²/h	3.345×10^2
cm²/s $\times 10^2$ (centistokes)	cm²/s (stokes)	1×10^{-2}	in²/s	cm²/s (stokes)	6.452
	ft²/h	3.875×10^{-2}		cm²/s $\times 10^2$ (centistokes)	6.452×10^2
	ft²/s	1.076×10^{-5}		ft²/h	2.50×10
	in²/s	1.550×10^{-3}		ft²/s	6.944×10^{-3}
	m²/h	3.600×10^{-3}		m²/h	2.323
ft²/s	cm²/s (stokes)	2.581×10^{-1}	m²/h	cm²/s (stokes)	2.778
	cm/s $\times 10^2$ (centistokes)	2.581×10		cm²/s $\times 10^2$ (centistokes)	2.778×10^2
	ft²/h	2.778×10^{-4}		ft²/h	1.076×10
	in²/s	4.000×10^{-2}		ft²/s	2.990×10^{-3}
	m²/h	9.290×10^{-2}		in²/s	4.306×10^{-1}

Absolute

Absolute viscosity = kinematic viscosity × density; lb = mass pounds, lb_f = force pounds

To convert from	To	Multiply by	To convert from	To	Multiply by
g/(cm)(s) [poise]	g/(cm)(s)(10^2) [centipoise]	10^2	g/(cm)(s)(10^2) [centipoise]	g/(cm)(s) [poise]	10^{-2}
	kg/(m)(h)	3.6×10^2		kg/(m)(h)	3.6
	lb/(ft)(s)	6.72×10^{-2}		lb/(ft)(s)	6.72×10^{-4}
	lb/(ft)(h)	2.419×10^2		lb/(ft)(h)	2.419
	lb/(in)(s)	5.6×10^{-3}		lb/(in)(s)	5.60×10^{-5}
	$(g_f)(s)/cm^2$	1.02×10^{-3}		$(lb_f)(s)/cm^2$	1.02×10^{-5}
	$(lb_f)(s)/in^2$ [Reyn]	1.45×10^{-5}		$(lb_f)(s)/in^2$ [Reyn]	1.45×10^{-7}
	$(lb_f)(s)/ft^2$	2.089×10^{-3}		lb/(ft)(s)	3.217×10
g/(cm)(s)(10^2) [centipoise]	$(lb_f)(s)/ft^2$	2.089×10^{-5}		lb/(ft)(h)	1.158×10^5
kg/(m)(h)	g/(cm)(s)	2.778×10^{-3}		lb/(in)(s)	2.681
	g/(cm)(s)(10^2) [centipoise]	2.778×10^{-1}		$(g_f)(s)/cm^2$	4.882×10^{-1}
	lb/(ft)(s)	1.867×10^{-4}		$(lb_f)(s)/in^2$ [Reyn]	6.944×10^{-3}
	lb/(ft)(h)	6.720×10^{-1}	lb/(ft)(s)	g/(cm)(s) [poise]	1.488×10^1
	lb/(in)(s)	1.555×10^{-5}		g/(cm)(s)(10^2) [centipoise]	1.488×10^2
	$(lb_f)(s)/in^2$ [Reyn]	4.029×10^{-8}		kg/(m)(h)	5.357×10^3
	$(lb_f)(s)/ft^2$	5.801×10^{-6}		lb/(ft)(h)	3.60×10^3
lb/(in)(s)	$(lb_f)(s)/ft^2$	3.73×10^{-1}		lb/(in)(s)	8.333×10^{-2}
$(g_f)(s)/cm^2$	g/(cm)(s)	9.807×10^2		$(g_f)(s)/cm^3$	1.518×10^{-2}
	g/(cm)(s)(10^2) [centipoise]	9.807×10^4		$(lb_f)(s)/in^2$ [Reyn]	2.158×10^{-4}
	kg/(m)(h)	3.530×10^5		$(lb_f)(s)/ft^2$	3.108×10^{-2}
	lb/(ft)(s)	6.590×10	lb/(ft)(h)	g/(cm)(s) [poise]	4.134×10^{-3}
	lb/(ft)(h)	2.372×10^5		g/(cm)(s)(10^2) [centipoise]	4.134×10^{-1}
	lb/(in)(s)	5.492		kg/(m)(h)	1.488
	$(lb_f)(s)/in^2$ [Reyn]	1.422×10^{-2}		lb/(ft)(s)	2.778×10^{-4}
	$(lb_f)(s)/ft^2$	2.048		lb/(in)(s)	2.315×10^{-5}
$(lb_f)(s)/in^2$ [Reyn]	g/(cm)(s) [poise]	6.895×10^4		$(g_f)(s)/cm^2$	4.215×10^{-6}
	g/(cm)(s)(10^2) [centipoise]	6.895×10^6		$(lb_f)(s)/in^2$ [Reyn]	5.996×10^{-8}
	kg/(m)(h)	2.482×10^7		$(lb_f)(s)/ft^2$	8.634×10^{-6}
	lb/(ft)(s)	4.633×10^3	lb/(in)(s)	g/(cm)(s) [poise]	1.786×10^2
	lb/(ft)(h)	1.668×10^7		g/(cm)(s)(10^2) [centipoise]	1.786×10^4
	lb/(in)(s)	3.861×10^2		kg/(m)(h)	6.429×10^4
	$(g_f)(s)/cm^2$	7.031×10		lb/(ft)(s)	1.2×10
	$(lb_f)(s)/ft^2$	1.440×10^2		lb/(ft)(h)	4.32×10^4
$(lb_f)(s)/ft^2$	g/(cm)(s) [poise]	4.788×10^2		$(g_f)(s)/cm^2$	1.821×10^{-1}
	g/(cm)(s)(10^2) [centipoise]	4.788×10^4		$(lb_f)(s)/in^2$ [Reyn]	2.590×10^{-3}
	kg/(m)(h)	1.724×10^5			

From Weast, R. C., **Ed.**, *Handbook of Chemistry and Physics*, 69th ed., CRC Press, Boca Raton, FL, 1988.

INDEX

A

Absorbers, in air pollution control, 19—20
Acenaphthene, wet air oxidation and, 287
Acid gas neutralization, in fluid bed incinerators, 19
Acid gas removal
  in air pollution control, 19, 47
  from flue gas streams, 554—555
  in general orientation of incinerator subsystems, 13
  infrared systems and, 274—275
Acrolein, wet air oxidation and, 287
Acrylonitrile, wet air oxidation and, 287
Administrative appeals, in permitting process, 130—131
Advanced electrical reactor, 257, 277—280, 291
Advisory groups, for public in hazardous waste siting, 207
Advocates, in siting process, 186
Afterburner, design of, 305
Afterburner heat, for rotary kiln, 302—304
Agency review, of permit application, 125—128
Aggregate kilns, light-weight, 32
Agricultural lands, zoning and protection of, 83
Air
  flow of in steel pipe, 557
  limits of inflammability of gases and vapor in, 513
Airlock feeders, in waste feeding to combustion chamber, 12
Air pollution control
  devices for, 20
  in general orientation of incinerator subsystems, 13
  incorporation of into waste incineration system, 12, 19—20
  infrared systems and, 272
  history of laws for, 71
  local agencies for, 83
  local districts for, 83
  residue quality and, 29, 37—38
  summary of federal legislation for, 73
  systems for, 216, 246—252
Air Pollution Control Act
  PL 84-159, 6/14/55, 69 Stat. 3221, 73
  PL 87-761, 10/9/62, 76 Stat. 760, 73
Air Pollution Control Act Extension, PL 86-365, 9/22/59, 73 Stat. 646, 73
Air pollution, overall risks of long-term, 44—47
Air Quality Act of 1967, PL 90-148, 11/21/67, 81 Stat. 485, 73
Air Quality Control Regions (AQCRs), establishment of, 77
Aisle space, required, in TSD facility, 100
Alaska, ranking of in use of public participation in siting hazardous waste facilities, 193
Alumina particles, in fluidized bed incinerators, 18
Aluminosilicate refractory brick, general characteristics of, 495

Ammonia, thermodynamic properties of, 485—486
Antimony
  emission of, 33
  infrared systems and, 276
  waste feed solid matrix properties and, 274
AQCRs, see Air Quality Control Regions
Archaeological guidelines, in siting and operation of incinerators, 83
Arizona, ranking of in use of public participation in siting hazardous waste facilities, 193
Arkansas, ranking of in use of public participation in siting hazardous waste facilities, 193
Arsanilic acid, molten salt combustion and, 271
Arsenic
  emission of, 32
  infrared systems and, 275—276
  waste feed solid matrix properties and, 274
Ash handling
  in general orientation of incinerator subsystems, 13
  incorporation of into waste incineration system, 12, 20—21
  system for, 216
  waste feed solid matrix properties and, 274
Ash pollution, control residue quality and, 29, 37—38
Ash quantity
  for circulating bed combustion, 307
  for rotary kiln, 307
Ash residue, infrared systems and, 275
Asphalt plants, light-weight, hazardous wastes as fuels in, 22
Assumption of responsibility, state, for TSD facility, 120—121
Assurance, of incinerator performance, 38—39
Atomization
  kinematic viscosity and solids handling limitations of, 553
  in liquid combustors, 14
Atomizing nozzles, applicability of, 253—254
Autoignition temperature, in ranking of compound incinerability, 40
Automobile exhaust, as cancer cause, 46

B

Bankruptcy, of TSD facility, 119
Barium
  emissions of, 32—33
  infrared systems and, 275
BDAT, see Best demonstrated available technology
Belt feeder, in waste feeding to combustion chamber, 12
Benzene
  chlorinated, 518
  destruction and removal efficiencies for, 346
  emissions of, 36

stack effluent concentrations of, 345
VOST blank correction values for, 345
waste feed concentrations of, 344—345
Beryllium, emissions of, 32
Best demonstrated available technology (BDAT)
 in hazardous waste management, 4
 incineration as, 65
Biannual report, for TSD facility, 104
bis(2-ethylhexy)phthalate, emission rates of, 36
Blending, as pretreatment option for wastes, 254
Boilers, industrial
 dioxin emissions from, 37
 emission rates of specific compounds from, 36
 furan emissions from, 37
 hazardous wastes as fuels in, 22
 summary of performance of, 31
Briefings, to community organizations, for public
  involvement in hazardous waste siting, 207,
  209
Brochures, for public involvement in hazardous
  waste siting, 207, 209
Bromine, emissions of, 28
Building codes, local, 83
Bulletins, for public involvement in hazardous waste
  siting, 209
Butane, thermodynamic properties of, 488
Butylbenzylphthalate, emission rates of, 36

C

Cadmium
 emissions of, 32—33
 infrared systems and, 275—276
 waste feed solid matrix properties and, 274
California, ranking of in use of public participation
  in siting hazardous waste facilities, 193
Cancer risk, 45—46
Candidate site identification, 205
Capacity, commercial incineration, shortfalls of,
  64—65
Carbon, waste feed solid matrix properties and, 274
Carbon dioxide
 continuous monitoring data for, 347
 thermodynamic properties of, 486—487
Carbon disulfide, thermodynamic properties of, 490
Carbon monoxide
 continuous monitoring data for, 347
 measurement of emission of, 27—28
 as performance indicator, 40—42, 48
Carbon tetrachloride
 destruction and removal efficiencies for, 346
 emissions of, 36
 plasma arc and, 282
 stack effluent concentrations of, 345
 thermodynamic properties of, 490
 VOST blank correction values for, 345
 waste feed concentrations of, 344—345
 wet air oxidation and, 287
Cement kilns, 22, 29, 32, 37

Centrifugal blowers, calculation of performance of,
  548—552
Centrifugal reactor, testing status of, 257
CERCLA, see Comprehensive Environmental
  Response, Compensation, and Liability Act
  of 1980
CFR, see Code of Federal Regulations
CF Technology, Inc., market trends surveyed by,
  65—66
Charcoal, in volatile organic sampling train, 26
Chemical kinetics, of organic hazardous constituents,
  170
Chemicals, valuable, recovery of, 1
Chemical Waste Management, Inc., market trends
  surveyed by, 61—62
Chemiluminescent monitor, summary data for, 27
Chlorine
 emissions of, 28
 incineration of wastes containing, 12
 waste feed solid matrix properties and, 274
Chlorobenzene, emissions of, 36
Chloroform
 emissions of, 36
 molten salt combustion and, 271
 wet air oxidation and, 287
2-Chlorophenol, wet air oxidation and, 287
Chromium
 emissions of, 32—33
 infrared systems and, 275—276
 waste feed solid matrix properties and, 274
Cigarette smoke, as cancer cause, 46
Circulating bed combuster
 ash quantity produced for, 307
 characteristics of, 257
 combustor mass balance for, 296—298
 fuel oil site remediation and, 233
 heat balance for, 299—302
 schematic representation of, 227—228
 system advantages of, 230—231
 system description for, 226—230
 system disadvantages of, 231
 system operating experience of, 233
 system performance of, 231—233
 testing result summary for, 291
 testing status of, 257
 transportable, 233
Citizen advisory committee, establishing, 208—209
City law, in overall statutory scheme, 74
Clay products kiln, summary of performance of, 32
Clean Air Act
 40 CFR Subchapter C references in, 85—86
 general overview of, 77—78
 municipal waste incineration standards under, 8
 New Source Performance Standards, 6
 PL 88-206, 12/17/63, 77 Stat. 392, 73
 statutory and regulatory overview of, 83—85
Clean Air Act Amendments of 1966, PL 89-675, 73
Clean Air Act Amendments of 1969, PL 9-137, 73
Clean Air Act Amendments of 1970, with technical

amendments in the Comprehensive
Manpower Training Act of 1971, PL 604, 73
Clean Air Act Amendments of 1977, (PL 95-190),
PL 95-95, 8/7/77, 91 Stat. 685, 74
Clean Air Act Extension, PL 93-15, 4/5/78,
87 Stat. 11, 74
Clean Water Act
general overview of, 81—82
Title III, 82
Title IV, 82
Title V, 82
Closure certificate, for TSD facility, 107—109
Closure cost estimate
in RCRA application process, 154—155
for TSD facility, 108
Closure insurance, for TSD facility, 114—115
Closure letter of credit, for TSD facility, 113—114
Closure performance standard, for TSD facility, 105
Closure plan
in RCRA application process, 152—154
for TSD facility, 105—107
Closure schedule, for TSD facility, 107
Closure trust fund
surety bond guaranteeing payment into, 111—112
for TSD facility, 109—112
Coal power plant, thermal destruction process
emissions from, 36
Coastal zone regulations, in states bordered by
ocean, 83
Code of Federal Regulations (CFR)
40 CFR 60, Appendix A, 28
40 CFR Part 124, 79
40 CFR Part 124 Subpart A, 124
40 CFR Part 260, 79
40 CFR Part 261, 79
40 CFR Part 261.24, 38
40 CFR Part 262, 79
40 CFR Part 263, 79
40 CFR Part 264, 79
40 CFR Part 264.343, 7
40 CFR Part 265, 79
40 CFR Part 267, 79
40 CFR Part 270, 79
40 CFR Part 271, 79
40 CFR Subpart O, 79
Colorado, ranking of in use of public participation in
siting hazardous waste facilities, 193
Colorimetric monitor, summary data for, 27
Combustion
by-products of, 49—50
calculations for, 294—307
parameters for, 506—516
Combustion chambers
in general orientation of incinerator subsystems, 13
incorporation of into waste incineration system, 12,
14—19
Combustion efficiency
computation of, 172
in PCB incineration, 8

Combustion gas conditioning, in general orientation
of incinerator subsystems, 13
Combustor feed characteristics, in combustion
calculations, 295—296
Combustor heat balance, 300—301
Combustor mass balance, 296—298
Combustors, vendor provision of, 1
Combustor waste mass flow calculations, for rotary
kiln, 298—299
Commercial alloys, high-temperature, properties of,
497—503
Commercial incineration industry, 60—62, 64
Communications equipment, access to in TSD
facility, 99
Community organizations, for public involvement in
hazardous waste siting, 209
Compound heat of combustion, as ranking of
compound incinerability, 40
Comprehensive Environmental Response, Compensation, and Liability Act of 1980 (CERCLA),
see also Superfund entries
ability to respond to market growth and, 63
general overview of, 80—81
in hazardous waste management, 4
priority cleanup list of, 74
Computerized comment storage and retrieval system,
for public involvement in hazardous waste
siting, 208
Conditional use permits, in siting of new incinerators, 83
Conflict resolution, for public involvement in
hazardous waste siting, 208
Congressional Budget Office, current incineration
practice survey by, 9
Congressional Office of Technology Assessment,
current incineration practice survey by, 9
Connecticut, ranking of in use of public participation
in siting hazardous waste facilities, 193
Consensus building, for public involvement in
hazardous waste siting, 208
Construction procedures, in RCRA application
process, 150
Consultants, incinerator, 1
Container, definition of, 93
Containerization, as pretreatment option for wastes,
254
Contingency plan
in RCRA application process, 148—150
for TSD facility, 100—101
Continuous emission monitors
combustion gas component measurement by, 27
in hazardous waste incinerator sampling, 25
summary of, 27
Continuous process plant-type incinerators,
characterization of, 1
Controlled air incinerators, see Fixed hearth
incinerators
Conversion factors, table of, 574—576
Converted stoker, summary of performance of, 31

Converted watertube, summary of performance of, 31

Copper
  infrared systems and, 276
  waste feed solid matrix properties and, 274

Corporate guarantee of closure, for TSD facility, 115—117

Correction factor, calculation of for corrected particulate matter emissions, 7

Cost estimate, for closure, in RCRA application process, 154—155

County law, in overall statutory scheme, 74

Crating, of new compounds in combustion zone, 35

D

DDT, molten salt combustion and, 271

Delaware, ranking of in use of public participation in siting hazardous waste facilities, 193

Delphi, for public involvement in hazardous waste siting, 208

Demister, in general orientation of incinerator subsystems, 13

Demographic analysis, in staff identification, 188

Density, of various solids, 418—419

Destruction and removal efficiency (DRE)
  calculation of, 122
  for chlorinated dioxins, 8
  formula for, 7
  infrared systems and, 273—275
  in RCRA application process, 157
  trial burn and, 327, 341—348
  waste feed concentrations and, 29—30

Destruction effectiveness, on untested and unique wastes, 48—49

Detoxification, of hazardous wastes, 1

Development agreements, international, 83

Dewatering, as pretreatment option for wastes, 254

Dibenzo-*p*-dioxins, chlorinated, 8

Dibenzofurans, chlorinated, 8

Dibutylphthalate, 36

Dichlorobiphenyl, waste feed solid matrix properties and, 274

1,2-Dichloroethane, wet air oxidation and, 287

Diethylphthalate, 36

Difluorodichloromethane, thermodynamic properties of, 489

2,4-Dimethylphenol, wet air oxidation and, 287

2,4-Dinitrotoluene, wet air oxidation and, 287

Dioxin rule, characterization of, 8

Dioxins
  chlorinated, 5
  emission of, 29, 35—37, 48
  infrared systems and, 276—277
  thermal destruction facility emission of, 37

Diphenylamine hydrochloride, molten salt combustion and, 271

1,2-Diphenylhydrazine, wet air oxidation and, 287

Disposal, definition of, 91—92

Draft permit, issuance and review of, 128—129

DRE, see Destruction and removal efficiency

Droplet formation, in hydrosonic gas scrubber system, 250

Dry chemical scrubber, 249—250

Dye sludges, quantities of generated in United States in 1983, 10

E

ECD, see Electron capture detector

Economic incentives
  for encouraging communities to host facilities, 192—193
  for hazardous waste regulation, 1

Electrocatalytic monitor, summary data for, 27

Electron capture detector (ECD), summary data for, 27

Electrostatic precipitators
  in air pollution control, 20—21
  in general orientation of incinerator subsystems, 13

Emergency coordinator, for TSD facility, 101

Emergency procedures
  in RCRA application process, 148—150
  for TSD facility, 101—102

Energy, recovery of, 1

Energy Supply and Environmental Coordination Act of 1974, PL 93-319, 6/24/74, 98 Stat. 246, 74

Engineering services, for typical hazardous waste incineration project, 217

Enora, Pennsylvania, pollution incident of 1948 in, 71—72

Environmental agreements, international, 83

Environmental design, in RCRA application process, 150

Environmental health implications, of hazardous waste emissions, 42—43

Environmental law, brief history of, 71—77

Environmental permits, general permit application process for, 124—172

Environmental Protection Agency (EPA)
  bubble policy, 85
  combustion research facility, 351—358
  current incineration practices and, 8—10
  D001-D003 waste codes, 9
  discretionary review power of, 130
  establishment of, 73
  F003 waste code, 9
  F020-F028 waste codes, 8
  K011 waste code, 9
  Method 3, 25—26
  Modified Method 5, 25—27, 317, 337, 347
  National Priority List, 1
  New Source Performance Standard, Reference Method 10, 28
  notification of hazardous waste activity, 95
  P063 waste code, 9
  performance measurement standards and, 23
  Permit Writer's Guide to Test Burn, Data-

Hazardous Waste Incineration, 29
Regulatory Impact Analysis program, 38
Science Advisory Board, 46
SW-846, 23
three-phase regulatory path of, 6
waste fuel use data of, 22
waste treatment hierarchy of, 1—2
Environmental regulation, brief history of, 71—77
EPA, see Environmental Protection Agency
Equipment, see also specific equipment
  disposition of contaminated during closure of TSD
    facility, 107
  performance of, 548—577
Ethyl benzene, waste feed solid matrix properties
    and, 274
Ethyl ether, thermodynamic properties of, 490
Executive Order 12291, regulatory impact analyses
    required by, 44
Experimental flame failure mode, in ranking of
    compound incinerability, 40
Explosions, TSD facility reporting of, 105

F

Facilitation, for public involvement in hazardous
    waste siting, 208
Facility
  definition of, 91
  in RCRA application process, 136, 151—152
Fact sheets, 206—207
Fail-safe corrective action, triggering of, 39
Fan and stack, in incineration technology, 216
Federal air pollution legislation, summary of, 73
Federal law, in overall statutory scheme, 74
FID, see Flame ionization detector
Field erected watertube, summary of performance of,
    31
Filtration, as pretreatment option for wastes, 254
Financial assurance for closure, of TSD facility,
    108—117
Financial requirements, in RCRA application
    process, 154—155
Financial test
  for closure of TSD facility, 115—116
  for liability coverage, 117—118
Fire codes, local, 83
Fires, TSD facility reporting of, 104
Fishbowl planning, for public involvement in
    hazardous waste siting, 209
Fixed hearth incinerators
  applicability of to wastes of various physical
    forms, 14
  destruction and removal efficiency of, 30
  dioxin and furan emissions from, 37
  incorporation of into waste incineration systems,
    17—18
  schematic representation of, 244—245
  system description of, 243—246
  typical, 17

Flame ionization detector (FID), 27—28
Flame photometry monitor, summary data for, 27
Florida, ranking of in use of public participation in
    siting hazardous waste facilities, 193
Flue gas streams, removal of particulates and acid
    gases from, 554—555
Fluidized bed combustion
  applicability of to wastes of various physical
    forms, 14
  capacity estimation for, 11
  case study results for, 267
  characteristics of, 257
  current technology status of, 268
  destruction and removal efficiency and, 30
  incorporation of into waste incineration systems,
    18—19
  process description for, 267
  schematic representation of, 235—237, 268
  system advantages of, 239
  system description for, 235—239
  system disadvantages of, 239
  system operating experience of, 240
  system performance of, 240
  technology advantages of, 268—269
  technology disadvantages of, 269—270
  testing results of, 269
  testing result summary for, 291
Forced regulation, characterization of, 75
Formic acid, wet air oxidation and, 287
Forums, public, for public involvement in hazardous
    waste siting, 208
Freons, as surrogates, 40
Fuel oil site remediation, via transportable
    circulating bed combustor, 233
Fuel savings, for oxygen-enriched incineration
    system, 259
Fuels, hazardous waste-derived, use of, 22
Fume incinerators, capacity estimation for, 11
Furans
  chlorinated, 5
  emission of, 29, 35—37, 48
  infrared systems and, 276—277
Furnaces, industrial, hazardous wastes as fuels in, 22

G

Gas bags, in hazardous waste incinerator sampling,
    25—26
Gas chromatography, in continous emission
    monitoring, 27
Gas chromatography/mass spectrometry
  in continuous emission monitoring, 27
  in performance measurement, 23
  trial burn and, 340, 343
Gases
  limits of inflammability of in air and oxygen, 513
  viscosity of, 425—427
Gas phase thermal stability, in ranking of compound
    incinerability, 40

GB chemical warfare agent, molten salt combustion and, 271

General inspection requirements, in RCRA application process, 140—142

Generator, hazardous waste incinerator facility as, 93

Geographic analysis, in staff identification, 188

Georgia, ranking of in use of public participation in siting hazardous waste facilities, 193

Gravity feed, to combustion chamber, 12

Growth process, in hydrosonic gas scrubber system, 251

Guarantee, for liability coverage, 117—118

Guest speaking, for public involvement in hazardous waste siting, 207

H

Halo acids, removal of, 19

Halogenated sludges, quantities of generated in United States in 1983, 10

Halogenated solids, quantities of generated in United States in 1983, 10

Halogenated solvents, quantities of generated in United States in 1983, 10

Hawaii, ranking of in use of public participation in siting hazardous waste facilities, 193

Hazardous and Solid Waste Act of 1984 (HSWA), 4—5, 8

Hazard regulations, local, 83

Health codes, local, 83

Hearth-type incinerators, 11, 244—246

Heat capacity, of organic liquids and vapors at 25°C, 428—429

Heat of combustion, for organic compounds, 158—170, 506—512

Heat of formation
  of gaseous atoms from elements in their standard states, 514
  for organic compounds, 512

Heat recovery systems
  in incineration technology, 216
  vendor provision of, 1
  waste, 252—253

Heavy metals, see also Metals and specific metals
  emissions of, 48—49
  infrared systems and, 275

Heptachlorobiphenyl, waste feed solid matrix properties and, 274

Herbicide orange, ocean incineration of, 21

Herbicides, quantities of generated in United States in 1983, 10

Hexachlorobiphenyl, waste feed solid matrix properties and, 274

HFID, see Hydrogen flame ionization detector

High-performance liquid chromatography (HPLC), in performance measurement, 23

Historical analysis, in staff identification, 188

Historical guidelines, in siting and operation of incinerators, 83

Homogenization, as pretreatment option for wastes, 254

HPLC, see High-performance liquid chromatography

HSWA, see Hazardous and Solid Waste Act of 1984

Hydrocarbons
  continuous monitoring data for, 347
  values of chemical thermodynamic properties of, 475—477

Hydrochloric acid emissions, in RCRA application process, 157

Hydrogen bromide, emissions of, 28

Hydrogen chloride
  emissions of, 28
  removal of, 7, 19, 21

Hydrogen flame ionization detector (HFID), combusted hydrocarbon emissions monitored via, 40

Hydrogen fluoride, emissions of, 28

Hydrosonic gas scrubber system, 250—251

I

ICF Technology, Inc., market trends surveyed by, 60—61, 63—64

Idaho, ranking of in use of public participation in siting hazardous waste facilities, 193

Identification number, for TSD facility, 94

IEM, see Inhalation Exposure Methodology

Ignitable waste
  general requirements for at TSD facility, 98
  in RCRA application process, 150—151

Ignition delay time, as ranking of compound incinerability, 40

Illinois, ranking of in use of public participation in siting hazardous waste facilities, 193

Incinerability, parameters of, 518—546

Incineration, see also specific aspects of incineration
  current practice of, 8—12
  generic functional flow diagram of system for, 215
  historical perspective of, 5—6
  as problem and solution, 75
  technology for, 12—21

Incinerator
  choice of, 214, 216
  definition of, 92—93

Incinerator ship, emissions from, 37

Incompatible waste
  general requirements for at TSD facility, 98
  in RCRA application process, 150—151

Indiana, ranking of in use of public participation in siting hazardous waste facilities, 193

Individuals, lists of, for staff identification, 188

Industrialists, in siting process, 186

Industrial Source Complex Long Term (ISC-LT) Model, for assessment of risks from recurring emissions, 43

Inflammability, limits of, 513

Information contact, for public involvement in hazardous waste siting, 207

Information fairs, for public involvement in hazardous waste siting, 207, 209
Infrared absorption, in continuous emission monitoring, 27
Infrared band centers, of common gases, 568
Infrared systems
  ash leaching and, 275
  case study results for, 272—273
  characteristics of, 257
  current technology status and, 277
  destruction and removal efficiency of
    acid gas emissions and, 274—275
    particulate emissions and, 273—274
  dioxins and, 276—277
  effluent solids and, 275
  furans and, 276—277
  leaching test results for, 275
  metal emissions and, 276—277
  metals analysis for, 276
  metals fixation and, 275
  process description for, 272
  stack emissions data for, 274
  technology advantages and, 277
  technology demonstration summary for, 359—373
  technology disadvantages and, 277
  testing result summary for, 291
  testing status of, 257
  waste feed solid matrix properties and, 274
Inhalation Exposure Methodology (IEM), for assessment of risk from recurring emissions, 43
Innovative technology, role of, 49—51
Input controls, characterization of, 75—76
Inspections
  of hazardous waste incineration facilities, 96—97, 124
  in RCRA application process, 140—142
Instrumentation, performance of, 558—577
Insurance, for TSD facility, 114—115, 117—120
Intuitive information, in staff identification, 188
Involuntary risk, public view of, 189
Ionizing wet scrubbers, 251—252
  in air pollution control, 20
  in general orientation of incinerator subsystems, 13
Iowa, ranking of in use of public participation in siting hazardous waste facilities, 193
Iron-making furnaces, hazardous wastes as fuels in, 22
ISC-LT Model, see Industrial Source Complex Long Term Model
Isobutane, thermodynamic properties of, 489
Issues papers, for public involvement in hazardous waste siting, 207

K

Kansas, ranking of in use of public participation in siting hazardous waste facilities, 193
Kentucky, ranking of in use of public participation

in siting hazardous waste facilities, 193
Kiln puff reduction, for oxygen-enriched incineration system, 259—261
Kilns
  emission rates of specific compounds from, 36
  summary of performance of, 32
Kodak Park incinerator, chemical incineration at, 397—414
Koken's training pyramid, in personnel training, 143

L

Lagoon sediment, contaminated, material properties calculations for, 294—295
Lagoon surface oil, material properties calculations for, 294—295
Lances, water-cooled, 12, 253—254
Land ban variances, expiring, market effects of, 65
Land use, in siting of new incinerators, 83
Latex, quantities of generated in United States in 1983, 10
Law, definition of, 71
Lead
  emissions of, 32—33
  infrared systems and, 275—276
  waste feed solid matrix properties and, 274
Letter of credit
  closure, 113—114
  for liability coverage, 118
Liability coverage, 117—120
Lime kilns, 29, 32, 37
Liquid injection incinerators
  applicability of to wastes of various physical forms, 14
  capacity estimation for, 11
  destruction and removal efficiency of, 30
  dioxin emissions from, 37
  furan emissions from, 37
  incorporation of into waste incineration systems, 14
  schematic representation of, 241—242
  system advantages of, 243
  system description for, 240—243
  system disadvantages of, 243
  system performance of, 243
  typical, 15
Liquids
  devices for measurement of flow of, 567
  heat capacity of organic at 25°C, 428—429
  quantities of incinerable generated in United States in 1983, 10
  viscosity of, 420—424
Local decision making, role of in final siting decision, 192—194
Local laws, impacting incineration, 83
Local regulations, impacting incineration, 83
Location standards, for TSD facility, 98—99
Louisiana, ranking of in use of public participation in siting hazardous waste facilities, 193

Love Canal, New York, discovery of hazardous waste site in, 72

M

Mailing lists, for public involvement in hazardous waste siting, 207

Maine, ranking of in use of public participation in siting hazardous waste facilities, 193

Maintenance schedule, as part of general inspection requirements in RCRA application process, 142

Malathion
molten salt combustion and, 271
wet air oxidation and, 287

Manifest discrepancies, definition of, 103

Manifest system, use of, 102—103

Man-made risk, public perception of, 190

Market, incineration, 62—64

Maryland, ranking of in use of public participation in siting hazardous waste facilities, 193

Massachusetts, ranking of in use of public participation in siting hazardous waste facilities, 193

Mass balance, for rotary kiln, 302—304

Master plans, for siting of new incinerators, 83

Media relations programs, for public involvement in hazardous waste siting, 207, 209

Mediation
between facility developers and proposed host communities, 192—193
for public involvement in hazardous waste siting, 208

Meetings, public, for public involvement in hazardous waste siting, 208—209

Mercury
emissions of, 32
infrared systems and, 275—276

Metals, see also Heavy metals and specific metals
average stack emissions of, 33
cancer risk due to emissions of, 45—46
emissions of, 29, 32—34, 48—49

Methylene chloride
emissions of, 36
waste feed solid matrix properties and, 274

Michigan, ranking of in use of public participation in siting hazardous waste facilities, 193

Minnesota, ranking of in use of public participation in siting hazardous waste facilities, 193

Mississippi, ranking of in use of public participation in siting hazardous waste facilities, 193

Missouri, ranking of in use of public participation in siting hazardous waste facilities, 193

Mist eliminator, in air pollution control, 47

Mixing process
in hydrosonic gas scrubber system, 250
as pretreatment option for wastes, 254

Mobile incineration systems, as thermal destruction systems, 21—22

Modified firetube, summary of performance of, 31

Moisture, waste feed solid matrix properties and, 274

Molten salt combustion, 257, 270—272, 291

Monitoring
of hazardous waste incineration facilities, 124
infrared systems and, 272
trial burn and, 347

Monomers, quantities of generated in United States in 1983, 10

Montana, ranking of in use of public participation in siting hazardous waste facilities, 193

Motor Vehicle Exhaust Study Act, PL 86-493, 6/8/60, 74 Stat. 162, 73

Municipal incinerators, in thermal destruction process emission, 36

Mustard, as chemical warfare agent, molten salt combustion and, 271

Mutagenic emissions rates, comparative, 46

N

NAAQS, see National Ambient Air Quality Standards

Naphthalene
blank correction values for, 346
destruction and removal efficiencies for, 346
emission rates of, 36
waste feed concentrations of, 344—345

National Ambient Air Quality Standards (NAAQS), establishment of, 77

National Emission Standards for Hazardous Air Pollutants (NESHAPS), establishment of, 77

National Environmental Policy Act (NEPA), 71—72, 74

Natural Resources Defense Council (NRDC), EPA agreement with, 8

Natural risk, public perception of, 190

NDIF monitor, see Nondispersion infrared monitor

NDUV monitor, see Nondispersion ultraviolet monitor

Nebraska, ranking of in use of public participation in siting hazardous waste facilities, 193

Negotiation, between facility developers and proposed host communities, 192—193

NEPA, see National Environmental Policy Act

NESHAPS, see National Emission Standards for Hazardous Air Pollutants

Net heating values, approximate, for common auxiliary fuels, 515

Nevada, ranking of in use of public participation in siting hazardous waste facilities, 193

New Hampshire, ranking of in use of public participation in siting hazardous waste facilities, 193

New Jersey, ranking of in use of public participation in siting hazardous waste facilities, 193

New Mexico, ranking of in use of public participation in siting hazardous waste facilities, 193

Newsletters, for public involvement in hazardous

waste siting, 207
Newspaper clipping service, establishing, 206
Newspaper inserts, with response form, for public
    involvement in hazardous waste siting, 209
News release, first, issuance of, 206
New York, ranking of in use of public participation
    in siting hazardous waste facilities, 193
Nickel, emissions of, 32—33
Nitroethane, molten salt combustion and, 271
Nitrogen oxides, emissions of, 27—28
4-Nitrophenol, wet air oxidation and, 287
N-Nitrosodimethylamine, wet air oxidation and, 287
Nominal group technique, for public involvement in
    hazardous waste siting, 208
Nomograph, for checking internal consistency of
    proposed excess air rate and combustion
    temperature in hazardous waste incinerators,
    516
Nonaqueous phase liquids, material properties
    calculations for, 294—295
Nondispersion infrared (NDIF) monitor, summary
    data for, 27
Nondispersion ultraviolet (NDUV) monitor,
    summary data for, 27
Nonflame thermal stability, in ranking of compound
    incinerability, 40
Nonhalogenated sludges, quantities of generated in
    United States in 1983, 10
Nonhalogenated solids, quantities of generated in
    United States in 1983, 10
Nonhalogenated solvents, quantities of generated in
    United States in 1983, 10
North Carolina, ranking of in use of public
    participation in siting hazardous waste
    facilities, 193
North Dakota, ranking of in use of public participa-
    tion in siting hazardous waste facilities, 193
Notices, required, for TSD facility, 94
NRDC, see Natural Resources Defense Council

O

Ocean incineration, as thermal destruction system,
    21
Ohio, ranking of in use of public participation in
    siting hazardous waste facilities, 193
Oily sludges, quantities of generated in United States
    in 1983, 10
Oklahoma, ranking of in use of public participation
    in siting hazardous waste facilities, 193
Open forums, for public involvement in hazardous
    waste siting, 207
Open houses, for public involvement in hazardous
    waste siting, 207, 209
Operating procedures, in RCRA application process,
    150
Operating record, for TSD facility, 103
Operating requirements, for hazardous waste
    incineration facilities, 123—124

Opinion leaders, interviews with, 206—208
Oregon, ranking of in use of public participation in
    siting hazardous waste facilities, 193
Organic chlorides, oxidation of, supercritical fluid
    and, 290
Organic compounds, 506—512
Organic liquids, incinerable, quantities of generated
    in United States in 1983, 10
Organizations, lists of, for staff identification, 188
Organochlorine waste, mixed, ocean incineration of,
    21
Output controls, characterization of, 76—77
Oxygen
    continuous monitoring data for, 347
    limits of inflammability of gases and vapor in, 513
Oxygen-enriched incineration, 257—261, 291

P

Packaged firetube, summary of performance of, 31
Packaged watertube, summary of performance of, 31
Packed bed scrubbers, in air pollution control,
    20—21
Packed tower absorbers, in air pollution control,
    19—20
Packed tower scrubbers, design of, 306—307
Paint sludges, quantities of generated in United
    States in 1983, 10
Panel hearing procedures, in permitting process,
    131—134
Paramagnetic monitor, summary data for, 27
Particulates
    in general orientation of incinerator subsystems, 13
    infrared systems and, 273—274
    removal of from flue gas streams, 554—555
PCBs, see Polychlorinated biphenyls
Pennsylvania, ranking of in use of public participa-
    tion in siting hazardous waste facilities, 193
Pentachlorobiphenyl, waste feed solid matrix
    properties and, 274
Pentachlorophenol, wet air oxidation and, 287
Perchloroethylene, molten salt combustion and, 271
Performance, incinerator
    assurance of, 38—39
    indicators for, 40—41
    measurement of, 23—27
    prediction of, 41—42
Performance of closure, surety bond guaranteeing,
    112—113
Performance standards, in RCRA application
    process, 157
Permits
    administrative appeals and, 130—131
    agency review and, 125—128
    delays of, 63—64
    Environmental Protection Agency discretionary
        review power in, 130
    general application process for, 124—172
    issuance and review of, 128—130

market growth and, 63—64
overview of, 124—125
panel hearing procedures and, 131
public review in, 129—130
RCRA application process and, 155—157
schematic of, 126—128
submitting of application and, 125
Personnel training
  in RCRA application process, 141—148
  for TSD facility, 97—98
Pesticides, quantities of generated in United States in
      1983, 10
Phenols
  blank correction values for, 346
  chlorinated, 8
  destruction and removal efficiencies for, 346
  emissions of, 36
  waste feed concentrations of, 344—345
  wet air oxidation and, 287
Phosphorus pentoxide, emissions of, 28
Photoionization detector (PID), 27—28
Physical properties, of materials, 418—503, see also
      specific properties
PICs, see Products of incomplete combustion
PID, see Photoionization detector
Plant equipment, performance of, 548—557
Plasma arc, 257, 280—285, 291
Plastics, commercially available, properties of, 494
Plate tower scrubbers, in air pollution control, 21
POHCs, see Principal organic hazardous constituents
Point of contact, public, establishing, 206
Political leadership, in siting process, 186—187
Polychlorinated biphenyls (PCBs)
  decline in incineration of, 64
  incineration of, 5, 8, 10
  molten salt combustion and, 271
  ocean incineration of, 21
  plasma arc and, 283—284
  regulation of, 4
  regulatory effects on incineration market and, 62
  thermal decomposition studies for, 34
  in TSCA application process, 172
  waste feed solid matrix properties and, 274
Polyethylene rings, in packed bed scrubbers, 21
Polyethylene saddles, in packed bed scrubbers, 21
Polymeric materials, brand names of, 491
Postclosure care, of hazardous waste disposal
      facilities, 105, 107—108
Postclosure plan, in RCRA application process,
      152—154
Potassium thiocyanate, wet air oxidation and, 287
Power plant emissions, as cancer cause, 46
Precooler design, for rotary kiln, 306
Prediction, of incinerator performance, 41—42
Preliminary notification of hazardous waste activity,
      in RCRA application process, 134
Presidential Advisory Panel Recommendation, in
      1970, 73
Presidential Study on Reorganization of Federal
      Pollution Control Programs, in 1969, 73

Prevention of Significant Deterioration (PSD),
      concept of, 77
Primary combustion chamber, 216, 272
Principal organic hazardous constituents (POHCs)
  cancer risk due to emissions of, 45—46
  combustion by-product emissions and, 34—35
  destruction and removal efficiency for, 7
  performance measurement and, 23, 25
  in RCRA application process, 156—170
  trial burn and, 323—324, 326—328, 344—346
Process failure, detection of, 48
Process management, infrared systems and, 272
Process monitoring, of incinerator operating
      parameters, 27—28
Process performance, 22—28
Product emissions, combustion by, 29, 34—35
Products of incomplete combustion (PICs)
  cancer risk due to emissions of, 45—46
  emissions of, 34—35
  in RCRA application process, 156—157
Progressive cavity pumps, in waste feeding, 12
Propane, thermodynamic properties of, 489
PSD, see Prevention of Significant Deterioration
Public
  acceptance of hazardous waste incinerators by,
      197—204
  characterization of, 185—187
  communicating to about environmental risk,
      198—201
  environmental risk viewed by, 189—190
  identification of, 187
  involvement of in hazardous waste facility siting,
      191—210
  strategy for involvement of in hazardous waste
      facility siting, 204—210
  values of, 197—198, 208
Publically owned treatment works, in general
      orientation of incinerator subsystems, 13
Public health implications, of hazardous waste
      emissions, 42—43
Public review, of draft permit, 129—130
Pulsed fluorescence monitor, summary data for, 27
Pyramid training concept, in personnel training, 141,
      143
Pyrene, wet air oxidation and, 287
Pyretron thermal destruction system, technology
      demonstration summary for, 351—358
Pyrolytic incinerators, see Fixed hearth incinerators

Q

Quality controls, for trial burn, 328
Quench, in air pollution control, 19—20, 47
Questionnaires, mailed, for public involvement in
      hazardous waste siting, 207

R

Radiation, total, emissivity of for various materials,
      430

Ram feeders
    applicability of, 253—254
    in waste feeding to combustion chamber, 12
RCRA, see Resource Conservation and Recovery
    Act of 1976
Reactive waste
    general requirements for at TSD facility, 98
    in RCRA application process, 150—151
Reading room, establishing, 206
Real-time performance, assurance of, 49—50
Recurring emissions, assessment of risks from,
    43—47
Recycling, of waste material, 77
Reduction, in waste volume, 1
Refuse Act, history of, 71
Regional administrator releases, TSD facility
    reporting of, 104
Regulations, 1, 71 see also specific regulations
Regulators, incinerator, 1
Regulatory impact analyses, long-term air pollution
    risks assessed via, 44
Residue handling, 12—13, 20—21
Resins, quantities of generated in United States in
    1983, 10
Resource, Conservation, and Recovery Act of 1976
    (RCRA)
    Appendix VIII, 7
    application of to hazardous waste incineration
        facilities, 91—93
    application process for TSD facility
        Part A contents, 134—136
        Part B contents, 136—155
        Subpart O technical requirements, 155—171
    40 CFR parts concerning, 79
    40 CFR Subpart O and, 79
    current incineration practice and, 8—9
    destruction and removal efficiency and, 29
    enforcement mechanisms and, 90—91
    exemptions from, 93
    general overview of, 78—80
    Hazardous and Solid Waste Amendments to, 72
    in hazardous waste management, 4—6, 8
    incinerator regulations in, 79
    incinerator standards in
        applicability of, 121
        closure and, 124
        hazardous waste incineration permits and, 123
        inspections and, 124
        monitoring and, 124
        operating requirements and, 123—124
        performance standards and, 122—123
        waste analysis and, 121—122
    performance standards in, 29—32, 87—88
    permitting requirements in, 88—90
    regulatory effects on incineration market and, 62
    Section 3004, 6
    state authorization and, 90
    statutory overview of, 85—87
    Subtitle C hazardous waste management, 78
    trial burns and, 328—329

TSD facility standards of general applicability and,
    93—124
Rhode Island, ranking of in use of public participa-
    tion in siting hazardous waste facilities, 193
Rivers and Harbors Act of 1899, Section 13, for
    prohibition of refuse dumping, 71
Rotary kiln incinerators
    afterburner design and, 305
    afterburner heat and, 302—304
    applicability of to wastes of various physical
        forms, 14
    ash quantity produced for, 307
    capacity estimation for, 11
    combuster mass balance for, 296—298
    design of, 305
    destruction and removal efficiency of, 30
    dioxin emissions from, 37
    furan emissions from, 37
    heat available for recovery and, 305
    heat balance for, 299—302
    history of, 6
    incorporation of into waste incineration systems,
        14, 17
    mass balance and, 302—304
    mobile, 21
    packed tower scrubber design and, 306—307
    parameter employed to trigger waste feed shutoff
        for corrective actions in, 225
    performance tests for, 224
    precooler design and, 306
    typical, 16
    schematic representation of, 221—222
    system advantages of, 223
    system description for, 220—223
    system disadvantages of, 224
    system operating experience of, 224—226
    system performance of, 224
    trial burn data from, 225
    Venturi scrubber design and, 306
Rubber, property comparisons of natural and
    synthetic, 492—493

S

S004, in hazardous waste incinerator sampling, 25
S006, in hazardous waste incinerator sampling, 25
S007, in hazardous waste incinerator sampling, 25
Safety design, in RCRA application process, 150
Sand particles, in fluidized bed incinerators, 18
SARA, see Superfund Amendements and Reauthori-
    zation Act of 1986
Scoping meetings, public, for public involvement in
    hazardous waste siting, 209
Screw feeders
    applicability of, 253—254
    in waste feeding to combustion chamber, 12
Scrubbers
    in air pollution control, 20—21, 47
    performance measurement and, 23
Secondary combustion chamber

in incineration technology, 216
in infrared systems, 272
Security, of TSD facility, 96
Security procedures, in RCRA application process,
    139—140
Selenium, infrared systems and, 275
Self-identification, in identification of pertinent
    public, 187
Self-regulation, characterization of, 75
Semivolatile organics, in RCRA application process,
    157
Shale aggregate kiln, summary of performance of,
    32
Shirco infrared incineration system, technology
    demonstration summary for, 359—373
Shredding, of bulk solid wastes, 12
Silent Spring
    environmental laws and problems, 72
    role of in modern environmental movement, 184
Silica, general characteristics of, 495
Silver, infrared systems and, 275
Single-event emissions, risks from, 42—43
Site selection, 201—205
Siting board review, public involvement in, 193
Slide shows, for public involvement in hazardous
    waste siting, 207
Sludges, feeding of, 12
Smoke control laws, history of, 71
Sodium carbonate, in fluidized bed incinerators, 18
Sodium cyanide, wet air oxidation and, 287
Sodium, infrared systems and, 276
Software requirements, for typical hazardous waste
    incineration project, 217
Soils, contaminated, disposition of during closure of
    TSD facility, 107
Solids, density of, 418—419
Solid waste management units, releases from, 105
Solvents, quantities of generated in United States in
    1983, 10
Source assessment sampling system, in hazardous
    waste incinerator sampling, 26
South Carolina, ranking of in use of public
    participation in siting hazardous waste
    facilities, 193
South Dakota, ranking of in use of public participa-
    tion in siting hazardous waste facilities, 193
Stack emissions
    data summary for, 32
    detection limits for in trial burn, 326—328
    infrared systems and, 276
    metal, 33
    performance measurement and, 23
    toxic compounds in, 5
Staff identification, in identification of pertinent
    public, 188
Start-up services, vendor provision of, 1
Starved air incinerators, see Fixed hearth incinerators
State laws

assumption of responsibility for TSD facility and,
    120—121
impacting incineration, 82—83
in overall statutory scheme, 74
State-required mechanisms, use of, for TSD facility
    operation, 120
24 Statute 329, for prohibition of refuse dumping, 71
Steam, tables for, 477—485
Steel pipes
    data for, 570—573
    flow of air in, 557
    flow of water in, 556
Steels, standard types of stainless and heat resisting,
    496—497
Stockton source test, 234
Stoichiometric oxygen requirements, combustion
    product yields and, 514
Storage, definition of, 91
Strontium, 274, 276
Structures, contaminated, disposition of during
    closure of TSD facility, 107
Subpart E evidentiary hearings, in permitting
    process, 132—134
Subpart F hearing procedures, in permitting process,
    131—132
Substitutions, product and raw material, 77
Sulfur, waste feed solid matrix properties and, 274
Sulfur dioxide
    reduction of pollution by, 76
    thermodynamic properties of, 487—488
Sulfur hexafluoride, as surrogate, 40
Sulfur oxides, emissions of, 28
Sulfur trioxide, emissions of, 28
Summary and evaluation reports, for public
    involvement in hazardous waste siting, 208
Supercritical fluid, 257, 288—291
Superfund Amendments and Reauthorization Act of
    1986 (SARA), 4, 64
Superfund sites
    cleanup of, 60, 375—394
    on-site incineration at, 62
Surety bond, 111—113, 118
Surface soil, material properties calculations for,
    294—295
Surrogates, in performance estimation, 39—40
Surveys, mailed, for public involvement in
    hazardous waste siting, 207
Swanson River PCB demonstration test, operating
    conditions for, 232

T

Tangentially fired watertube, summary of perform-
    ance of, 31
Technical assistance grants, to communities,
    192—193
Technocrats, in citing process, 186
Technology, for incineration, 12—21

Tenax resin, in volatile organic sampling train, 26
Tennessee, ranking of in use of public participation
    in siting hazardous waste facilities, 193
Test sieves, standard, 569
Tetrachlorobiphenyl, waste feed solid matrix
    properties and, 274
Tetrachloroethylene, emissions of, 36
Texas, ranking of in use of public participation in
    siting hazardous waste facilities, 193
Theoretical flame mode, as ranking of compound
    incinerability, 40
Thermal decomposition
  data summary of for selected hazardous organic
    compounds, 517
  parameters of for selected chlorinated benzenes in
    air at 2.0-s residence time, 518
Thermal destruction facilities, emissions from, 37
Thermal destruction systems, for hazardous waste,
    21—22, see also specific systems
Thermocouples, 558—566
Thermodynamic properties
  of ammonia, 485—486
  of butane, 488
  of carbon dioxide, 486—487
  of carbon disulfide, 490
  of carbon tetrachloride, 490
  chemical
    hydrocarbon values for, 475—477
    selected values of, 431—474
  of difluorodichloromethane, 489
  of ethyl ether, 490
  of isobutane, 489
  of propane, 489
  of sulfur dioxide, 487—488
Third-party identification, in identification of
    pertinent public, 188—191
Third-party liability coverage requirements, for TSD
    facility, 117—119
Throughput increase, for oxygen-enriched incinera-
    tion, 258—259
TNMO analyzer, see Total gaseous nonmethane
    organics analyzer
Toluene
  destruction and removal efficiencies for, 346
  emissions of, 36
  stack effluent concentrations of, 345
  VOST blank correction values for, 345
  waste feed concentrations of, 344—345
  waste feed solid matrix properties and, 274
  wet air oxidation and, 287
Total gaseous nonmethane organics (TNMO)
    analyzer, TUHC emissions measured via, 28
Total unburned hydrocarbons (TUHCs)
  measurement of emission of, 27—28
  as performance indicators, 40—42, 48
Toxic Substances Control Act (TSCA)
  application process for TSD facility and, 171—172
  general overview of, 80

  in hazardous waste management, 4, 7
  polychlorinated biphenyl rules under, 62
Transportable incineration systems, as thermal
    destruction systems, 21—22
Trash disposal laws, history of, 71
Tray tower absorber, in air pollution control, 20
Treatment, definition of, 91
Treatment, storage, and disposal (TSD) facilities
  closure of, 105—107
  contingency plan for, 100—101
  emergency coordinator for, 101
  emergency procedures for, 101—102
  financial requirements and, 108—121
  general facility standards for, 94—98
  manifest system for, 102—103
  postclosure and, 107, 108
  preparedness and prevention in, 99—100
  RCRA application process for, 134—171
  recordkeeping for, 103—104
  releases from solid waste management units and,
    105
  reporting for, 104—105
  TSCA application process for, 171—172
Trial burns
  actual testing and, 338—339
  aqueous waste analysis and, 347
  conducting, 329—348
  continuous monitoring and, 347
  contractor selection and, 319
  cost factors and, 320—321
  data conversion and, 341—344
  data labels and, 333—335
  data reporting and, 344—348
  data sheets and, 333—334
  destruction and removal efficiency and, 327,
    341—343, 346, 348
  detection limits and, 326—328
  equipment for, 319, 321—322, 330, 333, 337
  facilities and, 319, 330
  instrumentation requirements and, 321—322
  length of runs and, 325
  miniburns and, 322—323
  Modified Method 5 sampling train and, 317, 337,
    347
  number of runs and, 325
  observers of, 336
  operating conditions and, 321—323, 344
  organic waste analysis and, 348
  overview of, 311—321, 340
  permit conditions and, 312
  plan for in RCRA application process, 155—156,
    169—171
  planning for, 321—329
  POHCs and, 7, 323—324, 326—328, 344—346
  potential problems and, 313—314, 339—341, 345
  preliminary testing and, 338
  pretests and, 322—323
  process data and, 333

quality controls and, 328
RCRA requirements and, 328—329
regulatory limits and, 312
result reporting and, 344—348
rotary kiln incinerators and, 225
safety precautions and, 336
sampling and analysis activities and, 312—318,
    339—341
sampling crew for, 330
scheduling, 329—330
staffing for, 319, 325—326
time requirements and, 313
volatile organic sampling train and, 318, 340, 342,
    345—346
waste requirements and, 324—325
Trichlorobiphenyl, waste feed solid matrix properties
    and, 274
Trichloroethane
  emissions of, 36
  molten salt combustion and, 271
Trichloroethylene
  destruction and removal efficiencies for, 346
  emissions of, 36
  stack effluent concentrations of, 345
  VOST blank correction values for, 345
  waste feed concentrations of, 344—345
Trihalomethanes, in potable water used for scrubber
    water make-up, 35
Trust fund
  closure, 109—111
  for liability coverage, 118—119
TSCA, see Toxic Substances Control Act
TSD facilities, see Treatment, storage, and disposal
    facilities
TUHCs, see Total unburned hydrocarbons
Turnkey service, vendor provision of, 1

U

Ultraviolet absorption, in continuous emission
    monitoring, 27
Unmanifested waste report, for TSD facility, 104
Unopened capacitor destruction, plasma arc for, 283
Utah, ranking of in use of public participation in
    siting hazardous waste facilities, 193

V

Vanadium
  infrared systems and, 276
  waste feed solid matrix properties and, 274
Vapors
  heat capacity of at 25°C, 428—429
  limits of inflammability of in air and oxygen, 513
Vehicle Air Pollution Control Act, PL 89-272,
    10/20/65, 79 Stat. 992, 73
Vehicular traffic description, in RCRA application
    process, 151

Vendors, of incinerator products and services, 1
Venturi scrubber, wet, 19—21, 47, 247—248, 306
Vermont, ranking of in use of public participation in
    siting hazardous waste facilities, 193
Vibratory feeders, in waste feeding to combustion
    chamber, 12
Victims, in siting process, 186
Videos, for public involvement in hazardous waste
    siting, 207
Virginia, ranking of in use of public participation in
    siting hazardous waste facilities, 193
Viscosity
  of gases, 425—427
  kinematic, 553
  of liquids, 420—424
Viscosity conversion, table for, 577
Visible vapor plume elimination, see Demister
Volatile organic sampling train (VOST)
  in hazardous waste incinerator sampling, 25—27
  trial burn and, 318, 340, 342, 345—346
Volatile organics, in RCRA application process, 157
Voluntary risk, public perception of, 189
VOST, see Volatile organic sampling train
VX chemical warfare agent, molten salt combustion
    and, 271

W

Washington, ranking of in use of public participation
    in siting hazardous waste facilities, 193
Waste analysis
  RCRA standards for incinerators and, 121—122
  for TSD facility, 94, 96
Waste feeding
  applicability of, 254
  characterization of, 253
  detection limits for in trial burn, 326
  in general orientation of incinerator subsystems, 13
  in incineration technology, 216
  incorporation of into waste incineration system, 12
  POHC concentrations and, 344—345
Waste oils, quantities of generated in United States
    in 1983, 10
Waste preparation, 12—13
Waste quantities and characteristics, in combustion
    calculations, 294—295
Waste survey form, schematic representation of,
    218—219
Waste volume, reduction in, 1
Water, flow of in steel pipe, 556
Watertube stoker, summary of performance of, 31
Westat mail survey, of current incineration practice,
    9
Westinghouse electric pyrolyzer, testing status of,
    257
Westinghouse/O'Connor combustor
  case study results for, 266—267
  characteristics of, 257

current technology status of, 267
emission compliance test results for, 266
flow diagram for, 264
O$_2$ control effects and, 262
process description for, 261—263, 265—266
schematic representation of, 265
technology advantages of, 267
testing result summary for, 291
Westinghouse plasma arc, testing status of, 257
West Virginia, ranking of in use of public participation in siting hazardous waste facilities, 193
Wet air oxidation
  case study results for, 286—287
  characteristics of, 257
  current technology status of, 287
  process description for, 285
  schematic representation of, 286
  technology advantages of, 288
  technology disadvantages of, 288
  test results for, 287, 291
Wet scrubbers, in air pollution control, 20
Wetting process, in hydrosonic gas scrubber system, 250

Wisconsin, ranking of in use of public participation in siting hazardous waste facilities, 193
Wood stoves, mutagenic emission rates of, 46
Workshops, public, for public involvement in hazardous waste siting, 207, 210
Wyoming, ranking of in use of public participation in siting hazardous waste facilities, 193

X

XAD-2 resin, as sorbent module, 26
Xylenes, waste feed solid matrix properties and, 274

Z

Zinc
  emissions of, 32
  infrared systems and, 276
  waste feed solid matrix properties and, 274
Zirconium oxide monitor, summary data for, 27
Zoning, in siting of new incinerators, 83